PRINCIPLES OF SUSTAINABLE SOIL MANAGEMENT IN AGROECOSYSTEMS

Advances in Soil Science

Series Editors: Rattan Lal and B. A. Stewart

Published Titles

Interacting Processes in Soil Science
R. J. Wagenet, P. Baveye, and B. A. Stewart

Soil Management: Experimental Basis for Sustainability and Environmental Quality
R. Lal and B. A. Stewart

Soil Management and Greenhouse Effect
R. Lal, J. M. Kimble, E. Levine, and B. A. Stewart

Soils and Global Change
R. Lal, J. M. Kimble, E. Levine, and B. A. Stewart

Soil Structure: Its Development and Function
B. A. Stewart and K. H. Hartge

Structure and Organic Matter Storage in Agricultural Soils
M. R. Carter and B. A. Stewart

Methods for Assessment of Soil Degradation
R. Lal, W. H. Blum, C. Valentine, and B. A. Stewart

Soil Processes and the Carbon Cycle
R. Lal, J. M. Kimble, R. F. Follett, and B. A. Stewart

Global Climate Change: Cold Regions Ecosystems
R. Lal, J. M. Kimble, and B. A. Stewart

Assessment Methods for Soil Carbon
R. Lal, J. M. Kimble, R. F. Follett, and B. A. Stewart

Soil Erosion and Carbon Dynamics
E.J. Roose, R. Lal, C. Feller, B. Barthès, and B. A. Stewart

Soil Quality and Biofuel Production
R. Lal and B. A. Stewart

Food Security and Soil Quality
R. Lal and B. A. Stewart

World Soil Resources and Food Security
R. Lal and B. A. Stewart

Soil Water and Agronomic Productivity
R. Lal and B. A. Stewart

Principles of Sustainable Soil Management in Agroecosystems
R. Lal and B. A. Stewart

Advances in Soil Science

PRINCIPLES OF SUSTAINABLE SOIL MANAGEMENT IN AGROECOSYSTEMS

Edited by

Rattan Lal and B. A. Stewart

CRC Press
Taylor & Francis Group
Boca Raton London New York

CRC Press is an imprint of the
Taylor & Francis Group, an **informa** business

CRC Press
Taylor & Francis Group
6000 Broken Sound Parkway NW, Suite 300
Boca Raton, FL 33487-2742

© 2013 by Taylor & Francis Group, LLC
CRC Press is an imprint of Taylor & Francis Group, an Informa business

No claim to original U.S. Government works

Printed on acid-free paper
Version Date: 20130422

International Standard Book Number-13: 978-1-4665-1346-4 (Hardback)

This book contains information obtained from authentic and highly regarded sources. Reasonable efforts have been made to publish reliable data and information, but the author and publisher cannot assume responsibility for the validity of all materials or the consequences of their use. The authors and publishers have attempted to trace the copyright holders of all material reproduced in this publication and apologize to copyright holders if permission to publish in this form has not been obtained. If any copyright material has not been acknowledged please write and let us know so we may rectify in any future reprint.

Except as permitted under U.S. Copyright Law, no part of this book may be reprinted, reproduced, transmitted, or utilized in any form by any electronic, mechanical, or other means, now known or hereafter invented, including photocopying, microfilming, and recording, or in any information storage or retrieval system, without written permission from the publishers.

For permission to photocopy or use material electronically from this work, please access www.copyright.com (http://www.copyright.com/) or contact the Copyright Clearance Center, Inc. (CCC), 222 Rosewood Drive, Danvers, MA 01923, 978-750-8400. CCC is a not-for-profit organization that provides licenses and registration for a variety of users. For organizations that have been granted a photocopy license by the CCC, a separate system of payment has been arranged.

Trademark Notice: Product or corporate names may be trademarks or registered trademarks, and are used only for identification and explanation without intent to infringe.

Library of Congress Cataloging-in-Publication Data

Principles of sustainable soil management in agroecosystems / editors: Rattan Lal, B.A. Stewart.
    p. cm. -- (Advances in soil science)
  Includes bibliographical references and index.
  ISBN 978-1-4665-1346-4 (hardcover : alk. paper)
  1. Soil conservation. 2. Soil management--Environmental aspects. 3. Sustainable agriculture. I. Lal, R. II. Stewart, B. A. (Bobby Alton), 1932- III. Series: Advances in soil science (Boca Raton, Fla.)

S623.P75 2013
631.4'5--dc23
                                                   2013014243

Visit the Taylor & Francis Web site at
http://www.taylorandfrancis.com

and the CRC Press Web site at
http://www.crcpress.com

Contents

Preface ix
Editors xi
Contributors xiii

Chapter 1 Principles of Soil Management 1

  Rattan Lal

Chapter 2 Marginality Principle 19

  Jerry L. Hatfield and Lois Wright Morton

Chapter 3 Principles of Soil Management in Neotropical Savannas: The Brazilian Cerrado 57

  Yuri L. Zinn and Rattan Lal

Chapter 4 Facts and Myths of Feeding the World with Organic Farming Methods 87

  Bobby A. Stewart, Xiaobo Hou, and Sanjeev Reddy Yalla

Chapter 5 Building upon Traditional Knowledge to Enhance Resilience of Soils in Sub-Saharan Africa 109

  Moses M. Tenywa, Julius Y. K. Zake, and Rattan Lal

Chapter 6 Soil Fertility as a Contingent Rather than Inherent Characteristic: Considering the Contributions of Crop-Symbiotic Soil Microbiota ... 141

  Norman Uphoff, Feng Chi, Frank B. Dazzo, and Russell J. Rodriguez

Chapter 7 Human Dimensions That Drive Soil Degradation 167

  Tomas M. Koontz, Vicki Garrett, Respikius Martin, Caitlin Marquis, Pranietha Mudliar, Tara Ritter, and Sarah Zwickle

Chapter 8 Managing Soil Organic Carbon Concentration by Cropping Systems and Fertilizers in the North China Plain 189

  Jin Qing, Xiangbin Kong, and Rattan Lal

Chapter 9 Global Extent of Land Degradation and Its Human Dimension 203
Ephraim Nkonya, Joachim von Braun, Jawoo Koo, and Zhe Guo

Chapter 10 Cost–Benefit Analysis of Soil Degradation and Restoration 227
Fred J. Hitzhusen and Sarah E. Kiger

Chapter 11 Spiritual Aspects of Sustainable Soil Management 257
Bruce C. Ball

Chapter 12 Theological and Religious Approaches to Soil Stewardship 285
Gregory E. Hitzhusen, Gary W. Fick, and Richard H. Moore

Chapter 13 Traditional Knowledge for Sustainable Management of Soils 303
B. Venkateswarlu, Ch. Srinivasarao, and J. Venkateswarlu

Chapter 14 Sustainable Soil Management Is More than What and How Crops Are Grown 337
Amir Kassam, Gottlieb Basch, Theodor Friedrich, Francis Shaxson, Tom Goddard, Telmo J. C. Amado, Bill Crabtree, Li Hongwen, Ivo Mello, Michele Pisante, and Saidi Mkomwa

Chapter 15 Mining of Nutrients in African Soils Due to Agricultural Intensification 401
Eric T. Craswell and Paul L. G. Vlek

Chapter 16 Carbon Sink Capacity and Agronomic Productivity of Soils of Semiarid Regions of India 423
Ch. Srinivasarao, B. Venkateswarlu, Rattan Lal, A. K. Singh, Sumanta Kundu, and Vijay Sandeep Jakkula

Chapter 17 Soil Renewal and Sustainability 477
Richard M. Cruse, Scott Lee, Thomas E. Fenton, Enheng Wang, and John Laflen

Contents

Chapter 18 Organic Carbon Sequestration Potential and the Co-Benefits in China's Cropland 501

Genxing Pan, Kun Cheng, Jufeng Zheng, Lianqing Li, Xuhui Zhang, and Jinwei Zheng

Chapter 19 Soil Management for Sustaining Ecosystem Services 521

Rattan Lal and Bobby A. Stewart

Index 537

Preface

This volume describes the basic laws of soil management to enhance ecosystem services while restoring degraded soils and promoting sustainable use. World-class soil scientists, ecologists, and social scientists have contributed chapters to address the following generic themes. (1) The biophysical process of soil degradation is driven by economic, social, and political forces, and people do not tolerate being kept hungry. (2) When people are poverty stricken, desperate, and starving, they pass on their sufferings to the land. (3) Similar to a bank account, it is also not possible to take more out of a soil than what is put in it without degrading its quality. (4) Marginal soils cultivated with marginal inputs produce marginal yields and support marginal living. (5) Plants cannot differentiate nutrients supplied through chemical fertilizers and organic amendments. (6) Mining carbon from soil organic matter affects atmospheric chemistry similar to that by fossil fuel combustion. (7) Soil can be a source or sink of atmospheric CO_2 depending on the land use and management. (8) Even the elite varieties cannot extract water and nutrients from a soil where they do not exist. (9) Sustainable management of soil is the engine of economic development. (10) Today's soil management problems cannot be solved by yesterday's technology. This 19-chapter volume addresses these basic laws for sustainable soil management.

This book provides the basis of an optimism that through adoption of sustainable management, world soils have the capacity to meet food demands of the present and projected future population. The goal is to usher a soil-based Green Revolution, because the seed-based Green Revolution of the early 1970s bypassed sub-Saharan Africa. No matter how powerful the seed technology, the seedling emerging from it can flourish only in a healthy soil.

If modern, high-level soil management technology is used, Africa has the capacity to feed several billion people. Yet, with a few exceptions, food production has stagnated since the 1960s. Presently, more than 12 million people in the Horn of Africa, suffering from drought-induced famine, have joined in 2011 the ranks of about 850 million already prone to hunger and malnutrition worldwide. Yet, crop yields in Africa, South Asia, and the Caribbean can be doubled or tripled through adoption of technologies based on laws of sustainable soil management.

Amid perpetuating hunger and malnutrition, degrading and desertifying soils, and the ever-depleting and dwindling soil resource base, the time to understand basic laws of soil management and promote recommended technology is now. This compendium of chapters, written by the most knowledgeable researchers, is aimed at promoting "The Soil Quality as a Basis of the Green Revolution."

The relevance of this book is enhanced with the increase in frequency of extreme events (e.g., the century drought experienced by the United States in 2012). The frequency and intensity of these events will be exacerbated in the future. Therefore, the book *Principles of Sustainable Soil Management in Agroecosystems* is extremely relevant towards an attempt to change the agriculture from being a cause and victim to a solution to adapt to and mitigate the abrupt climate change while advancing global food security and improving the environment.

The editors thank all the authors for their outstanding contributions to this volume and for sharing their knowledge and experiences with others. Despite their busy schedule and numerous commitments, preparation of the manuscripts in a timely manner by all authors is greatly appreciated. The editors also thank the editorial staff of Taylor & Francis for their help and support in publishing this volume. The office staff of the Carbon Management and Sequestration Center provided support in the flow of the manuscript between authors and editors and made valuable contributions, and their help and support are greatly appreciated. It is a challenging task to thank, by listing names, all those who contributed in one way or another to bringing this volume to fruition. Thus, it is important to build upon the outstanding contributions of numerous soil scientists and ecologists whose research is cited throughout the book.

Rattan Lal
Bobby A. Stewart

Editors

Rattan Lal is a distinguished university professor of soil physics in the School of Environment and Natural Resources and the director of the Carbon Management and Sequestration Center, of the College of Food, Agricultural, and Environmental Sciences/Ohio Agriculture Research and Development Center, at The Ohio State University. Before joining The Ohio State in 1987, he was a soil physicist for 18 years at the International Institute of Tropical Agriculture, Ibadan, Nigeria. In Africa, Professor Lal conducted long-term experiments on land use, watershed management, soil erosion processes as influenced by rainfall characteristics, soil properties, methods of deforestation, soil-tillage and crop-residue management, cropping systems including cover crops and agroforestry, and mixed/relay cropping methods. He also assessed the impact of soil erosion on crop yields and related erosion-induced changes in soil properties to crop growth and yield. Since joining The Ohio State University in 1987, he has continued research on erosion-induced changes in soil quality and developed a new project on soils and climate change. He has demonstrated that accelerated soil erosion is a major factor affecting emission of carbon from the soil to the atmosphere. Soil erosion control and adoption of conservation-effective measures can lead to carbon sequestration and mitigation of the greenhouse effect. Other research interests include soil compaction, conservation tillage, mine soil reclamation, water table management, and sustainable use of soil and water resources of the tropics for enhancing food security.

Professor Lal is a fellow of the Soil Science Society of America, American Society of Agronomy, Third World Academy of Sciences, American Association for the Advancement of Sciences, Soil and Water Conservation Society, and Indian Academy of Agricultural Sciences. He is a recipient of the International Soil Science Award of the Soil Science Society of America, the Hugh Hammond Bennett Award of the Soil and Water Conservation Society, the 2005 Borlaug Award, and the 2009 Swaminathan Award. He also received an honorary degree of Doctor of Science from Punjab Agricultural University, India; the Norwegian University of Life Sciences, Aas, Norway; and the Alecu Russo Balti State University in Moldova. He is a past president of the World Association of the Soil and Water Conservation, the International Soil Tillage Research Organization, and the Soil Science Society of America. He has been a member of the US National Committee on Soil Science of the National Academy of Sciences (1998 to 2002, 2007 to 2011). He has served on the Panel of Sustainable Agriculture and the Environment in the Humid Tropics of the National Academy of Sciences. He has authored and coauthored about 1500 research papers. He has also written 15 books and edited or coedited 48 books.

B.A. Stewart is a distinguished professor of soil science at the West Texas A&M University, Canyon, Texas. He is also the director of the Dryland Agriculture Institute and a former director of the US Department of Agriculture (USDA) Conservation and Production Laboratory at Bushland, Texas; a past president of the

Soil Science Society of America; and a member of the 1990–1993 Committee on Long-Range Soil and Water Policy, National Research Council, National Academy of Sciences. He is a fellow on the Soil Science Society of America, American Society of Agronomy, Soil and Water Conservation Society; a recipient of the USDA Superior Service Award; a recipient of the Hugh Hammond Bennett Award of the Soil and Water Conservation Society; and an honorary member of the International Union of Soil Sciences in 2008.

Dr. Stewart is very supportive of education and research on dryland agriculture. The B.A. and Jane Anne Stewart Dryland Agriculture Scholarship Fund was established in West Texas A&M University in 1994 to provide scholarships for undergraduate and graduate students with a demonstrated interest in dryland agriculture.

Contributors

Telmo J. C. Amado
Federal University of Santa Maria
Santa Maria, Brazil

Bruce C. Ball
SRUC Crop and Soil Systems Research Group
Edinburgh, United Kingdom

Gottlieb Basch
Institute of Mediterranean Agricultural and Environmental Sciences (ICAAM)
University of Évora
Évora, Portugal

Kun Cheng
Center of Agriculture and Climate Change
Nanjing Agricultural University
Nanjing, China

Feng Chi
University of Southern California Keck School of Medicine
Los Angeles, California

Bill Crabtree
Crabtree Agricultural Consulting
Beckenham, Western Australia, Australia

Eric T. Craswell
School of Environment and Society
Australian National University
Canberra, Australia

Richard M. Cruse
Department of Agronomy
Iowa State University
Ames, Iowa

Frank B. Dazzo
Department of Microbiology and Molecular Genetics
and
Department of Crop and Soil Sciences
Michigan State University
East Lansing, Michigan

Thomas E. Fenton
Department of Agronomy
Iowa State University
Ames, Iowa

Gary W. Fick
College of Agriculture and Life Sciences
Cornell University
Ithaca, New York

Theodor Friedrich
Plant Production and Protection Division
Food and Agriculture Organization
Rome, Italy

Vicki Garrett
School of Environment and Natural Resources
The Ohio State University
Columbus, Ohio

Tom Goddard
Alberta Agriculture and Rural Development
Edmonton, Alberta, Canada

Zhe Guo
International Food Policy Research Institute
Washington, District of Columbia

Jerry L. Hatfield
National Laboratory for Agriculture and the Environment
Ames, Iowa

Fred J. Hitzhusen
Department of Agricultural, Environmental and Development Economics
The Ohio State University
Columbus, Ohio

Gregory E. Hitzhusen
School of Environment and Natural Resources
The Ohio State University
Columbus, Ohio

Li Hongwen
Conservation Tillage Research Centre
China Agriculture University
Beijing, China

Xiaobo Hou
Dryland Agriculture Institute
West Texas A&M University
Canyon, Texas

Vijay Sandeep Jakkula
Central Research Institute for Dryland Agriculture
Andhra Pradesh, India

Amir Kassam
School of Agriculture, Policy and Development
University of Reading
Reading, United Kingdom

Sarah E. Kiger
School of Natural Resources and Environment
University of Michigan
Ann Arbor, Michigan

Xiangbin Kong
College of Resources and Environmental Science
China Agricultural University
Beijing, China

Jawoo Koo
International Food Policy Research Institute
Washington, District of Columbia

Tomas M. Koontz
School of Environment and Natural Resources
The Ohio State University
Columbus, Ohio

Sumanta Kundu
Central Research Institute for Dryland Agriculture
Andhra Pradesh, India

John Laflen
Agricultural and Biosystems Engineering
Iowa State University
Ames, Iowa

Rattan Lal
Carbon Management and Sequestration Center
The Ohio State University
Columbus, Ohio

Scott Lee
Department of Agronomy
Iowa State University
Ames, Iowa

Lianqing Li
Center of Agriculture and Climate Change
Nanjing Agricultural University
Nanjing, China

Contributors

Caitlin Marquis
School of Environment and Natural Resources
The Ohio State University
Columbus, Ohio

Respikius Martin
School of Environment and Natural Resources
The Ohio State University
Columbus, Ohio

Ivo Mello
No-Till Federation of Brazil (FEBRAPDP)
Paraná, Brazil

Saidi Mkomwa
African Conservation Tillage (ACT) Network
Nairobi, Kenya

Richard H. Moore
School of Environment and Natural Resources
The Ohio State University
Wooster, Ohio

Lois Wright Morton
College of Agriculture and Life Sciences
Iowa State University
Ames, Iowa

Pranietha Mudliar
School of Environment and Natural Resources
The Ohio State University
Columbus, Ohio

Ephraim Nkonya
International Food Policy Research Institute
Washington, District of Columbia

Genxing Pan
Center of Agriculture and Climate Change
Nanjing Agricultural University
Nanjing, China

Michele Pisante
University of Teramo
Teramo, Italy

Jin Qing
Chinese Academy of Land Resource Economics
Beijing, China

Tara Ritter
School of Environment and Natural Resources
The Ohio State University
Columbus, Ohio

Russell J. Rodriguez
Symbiogenics
Seattle, Washington

Francis Shaxson
Land Husbandry Group
Tropical Agriculture Association (TAA)
United Kingdom

A. K. Singh
Indian Council of Agricultural Research
New Delhi, India

Ch. Srinivasarao
Central Research Institute for Dryland Agriculture
Andhra Pradesh, India

Bobby A. Stewart
Dryland Agriculture Institute
West Texas A&M University
Canyon, Texas

Moses M. Tenywa
College of Agricultural and
 Environmental Sciences
Makerere University
Kampala, Uganda

Norman Uphoff
Cornell International Institute for Food,
 Agriculture and Development
Cornell University
Ithaca, New York

B. Venkateswarlu
Central Research Institute for Dryland
 Agriculture
Andhra Pradesh, India

J. Venkateswarlu
Central Research Institute for Dryland
 Agriculture
Andhra Pradesh, India

Paul L. G. Vlek
Ecology and Natural Resources
 Division
Center for Development Research
 (ZEF)
Bonn, Germany

Joachim von Braun
Center for Development Research
 (ZEF)
University of Bonn
Bonn, Germany

Enheng Wang
Northeast Forest University
Harbin, China

Sanjeev Reddy Yalla
Dryland Agriculture Institute
West Texas A&M University
Canyon, Texas

Julius Y. K. Zake
College of Agricultural and
 Environmental Sciences
Makerere University
Kampala, Uganda

Xuhui Zhang
Center of Agriculture and Climate
 Change
Nanjing Agricultural University
Nanjing, China

Jinwei Zheng
Center of Agriculture and Climate
 Change
Nanjing Agricultural University
Nanjing, China

Jufeng Zheng
Center of Agriculture and Climate
 Change
Nanjing Agricultural University
Nanjing, China

Yuri L. Zinn
Depto. de Ciência do Solo
Universidade Federal de Lavras
Lavras, Brazil

Sarah Zwickle
School of Environment and Natural
 Resources
The Ohio State University
Columbus, Ohio

1 Principles of Soil Management

Rattan Lal

CONTENTS

1.1 Introduction 2
1.2 Agronomic Yield Trends and Resource Use 2
1.3 Yield Gap 6
    1.3.1 Soil Quality Management 7
1.4 Key Soil Parameters Important to Soil Sustainability 8
    1.4.1 Technological Options for Soil Sustainability 9
1.5 Principles of Soil Management 10
1.6 Laws of Sustainable Soil Management 11
    1.6.1 Policy Makers 11
        1.6.1.1 Causes of Soil Degradation and Desertification 11
        1.6.1.2 Human Needs and Stewardship of Natural Resources 11
        1.6.1.3 Poverty and Soil Degradation 13
        1.6.1.4 Soil Degradation Is a Cause of Global Warming 13
        1.6.1.5 Desertification Control and Mitigation of Climate Change ... 13
    1.6.2 Land Managers 13
        1.6.2.1 Nutrient Bank 13
        1.6.2.2 Organic and Inorganic Sources of Plant Nutrients 14
        1.6.2.3 Modern Issues and Ancient Technologies 14
        1.6.2.4 Masters of Their Own Destiny 14
        1.6.2.5 Being Proactive 14
    1.6.3 Researchers 14
        1.6.3.1 Relative Importance of Natural Resources versus Improved Germplasm 14
        1.6.3.2 Indicators of Soil Quality Improvement 15
        1.6.3.3 Training Cadre of Young Researchers 15
        1.6.3.4 Building Bridges across Nations 15
        1.6.3.5 Lack of Technology Adoption 15
1.7 Conclusions 15
Abbreviations 16
References 16

1.1 INTRODUCTION

World soil resources are finite, unequally distributed among diverse regions, fragile and prone to degradation by land misuse and soil mismanagement, and vulnerable to extreme events related to the abrupt climate change (ACC). The land area prone to degradation processes is estimated at 3500 Mha affecting livelihood and wellbeing of a large proportion of underprivileged population living in regions characterized by a harsh climate and agriculturally marginal lands (Bai et al. 2008). Further, the resource-poor farmers are unable to invest in soil and water conservation techniques and in replenishing soil fertility. Thus, a perpetual use of extractive farming practices by small landholders of the tropics and subtropics and cutting corners for quick economic returns by large-scale commercial farmers have exacerbated the problem of a widespread distribution of depleted and degraded soils, often with a truncated topsoil layer because of the accelerated soil erosion (Lal 2001, 2003). Whereas the area equipped for supplementary irrigation has expanded to ~287 Mha (FAO 2012), it has also caused severe problems of groundwater depletion (Lal and Stewart 2012) and secondary salinization (Oldeman 1994). Yet, the food production must be drastically increased to feed a world population of 10 billion (Borlaug and Dowswell 2005), to meet the food demands of the growing population and changing the dietary preferences toward more animal-based than plant-based diet. By 2030, global cereal demand for food and animal feed is expected to be 2.8 billion Mg/year, which is double the production in 2005 (Lobell et al. 2009). With a small, if any, scope for bringing new land under cultivation, productivity must be increased from the existing agricultural land per unit area, time, and use of energy-based input (i.e., fertilizer, irrigation). In accord with the concept of zero net land degradation (Lal et al. 2012), any new land degradation must be negated by restoration of prior degraded land.

Therefore, the objective of this chapter is to describe the basic principles of soil management in the context of the present status of agronomic productivity and the availability of the soil and water resources to meet the growing demands for global food production.

1.2 AGRONOMIC YIELD TRENDS AND RESOURCE USE

The world cereal grain yield increased 2.63 times from 1353 kg/ha in 1961 to 3564 kg/ha in 2010 (Table 1.1). The highest increase of 3.9 times occurred in East Asia, from 1414 kg/ha in 1961 to 5497 kg/ha in 2010. The lowest increase of 1.3 times occurred in Central (middle) Africa from 713 kg/ha in 1961 to 953 kg/ha in 2010 (Table 1.1). The cereal yield in Africa, ranging from 810 kg/ha in 1961 to 1501 kg/ha in 2010, has lagged behind that of all other continents (Figure 1.1). Indeed, the Green Revolution of the 1960s and 1970s bypassed the entire continent of Africa. Despite the adoption of improved varieties, agronomic yields in Sub-Saharan Africa have stagnated probably because of the widespread problem of soil degradation and nutrient depletion. The adverse effects of degraded/depleted soils have been exacerbated by the harsh and uncertain climate. For example, the fertilizer use in Africa is the lowest, at 3.2×10^6 Mg of NPK. In comparison, nutrient (NPK) use in Asia is $\sim 90 \times 10^6$ Mg or 63.5% of the world consumption (Table 1.2). Indeed, the rate of fertilizer (NPK nutrients) use in Africa is merely 13 kg/ha, compared with 156 kg/ha in Asia and 90 kg/ha in the

Principles of Soil Management

TABLE 1.1
Temporal Changes in Cereal Grain Yield on Regional Basis

Continent	Region	\multicolumn{6}{c}{Cereal Grain Yields (kg ha^{-1})}
		1961	1971	1981	1991	2001	2010
Africa	Continent	810	986	1241	1236	1392	1501
	Eastern	885	995	1252	1284	1363	1595
	Middle	713	737	734	754	839	953
	Northern	843	1277	1087	1672	1650	1852
	Southern	1049	1415	2495	1860	2188	3600
	Western	975	690	962	921	1009	1167
Asia	Continent	1212	1673	2146	2823	3176	3634
	East	1414	2308	3182	4285	4823	5497
	South	1012	1166	1448	1930	2430	2667
Americas	Continent	1912	2076	3209	3353	4218	5337
	North	2203	3376	3840	4025	5054	6342
	South	1347	1499	1941	2145	3146	4216
	Caribbean	1176	1472	1886	1878	2185	2022
Oceania/Pacific		1115	1310	1434	1652	2255	1761
Europe		1379	2050	2046	2712	3418	3683
World		1353	1892	2247	2684	3136	3564

Source: FAO. FAOSTAT, 2012. Available at http://faostat.fao.org/Rome.

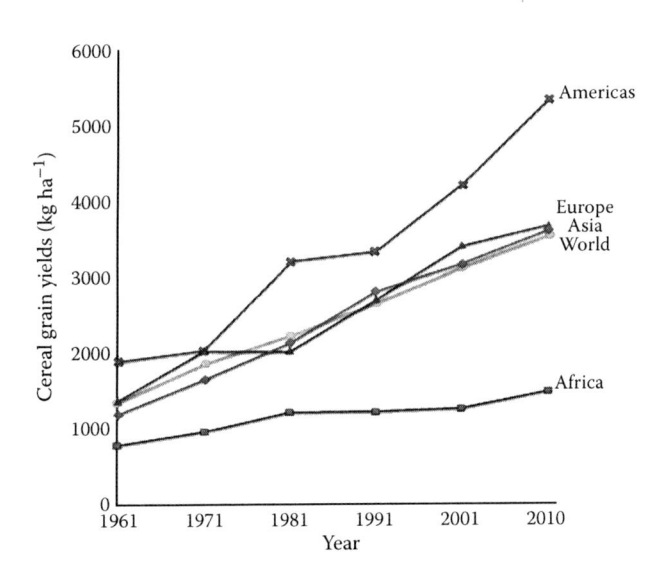

FIGURE 1.1 Temporal changes in cereal grain yields on regional basis between 1961 and 2010. (Redrawn from the data from FAO, FAOSTAT, 2012. Available at http://faostat.fao.org/Rome.)

TABLE 1.2
Fertilizer Use (N, P, K) on Continental Basis in 2009

Region	N	P	K	Total
World	106.0	17.1	17.8	140.9
Africa	2.5	0.4	0.3	3.2
North America	12.5	1.7	2.9	17.1
Central America	1.4	0.1	0.2	1.7
Caribbean	0.05	0.05	0.02	0.1
South America	4.8	1.7	2.7	9.2
Asia	69.4	11.0	9.1	89.5
Europe	13.5	1.4	2.6	17.5
Oceania	0.9	0.4	0.06	?
Australia and New Zealand	0.9	0.4	0.05	1.3

Nutrients (10^6 Mg)

Source: FAO. FAOSTAT, 2012. http://faostat.fao.org/Rome.

world (Table 1.3). Of the global irrigable land area of 287 Mha, only 13.6 Mha (4.7%) is irrigated in Africa, and most of it is in the North Africa/Middle East countries. Indeed, the per capita irrigated land area is the lowest (0.013 ha) in Africa (Table 1.3). Therefore, two important management factors, nutrients and water, are being used at a suboptimal level throughout the continent. It is not surprising, therefore, that the current relative crop yield with references to the highest regional yield anywhere in the world is the lowest in Africa. The data in Table 1.4 show that the relative yield of

TABLE 1.3
Total Population and Per Capita Fertilizer Use and Irrigated Land Area in 2007

Region	Population (10^6)	Arable Land (10^6 ha) Total	Arable Land (10^6 ha) Irrigated	Per Capita Land Area (ha) Total	Per Capita Land Area (ha) Irrigated	Fertilizer Use (kg/ha)
World	6974	1553.7	286.8	0.223	0.041	90.7
Africa	1046	246.5	13.6	0.235	0.013	13.0
North America	346	225.3	23.8	0.647	0.068	76.0
Central America	158	35.2	8.2	0.222	0.052	48.3
Caribbean	42	7.1	1.3	0.169	0.031	14.0
South America	397	125.8	10.4	0.317	0.026	73.1
Asia	4208	573.3	201.3	0.136	0.048	156.1
Europe	739	293.4	26.3	0.397	0.036	59.6
Oceania	9	1.7	0.1	0.189	0.011	–
Australia and New Zealand	27	45.3	3.1	1.678	0.115	29.8

Source: Recalculated from FAOSTAT.2012. http://faostat.fao.org/Rome.

maize (*Zea mays*), with reference to the highest yield of 9.34 Mg/ha in North America, is only 14.6% and 17.3% in eastern and western Africa, respectively. Compared with the highest yield of 10.84 Mg/ha of rice (paddy) in Australia, the relative yield is merely 19.3% in eastern Africa. In comparison with the highest soybean yield of 3.2 Mg/ha in southern Europe, the relative yield is 30.9% in eastern Africa and 40.5% in western Africa (Table 1.4). The low relative yields of major crops (e.g., wheat, maize, rice, soybean) can be mostly attributed to mismanagement of soil and water resources under the prevailing harsh and uncertain climate.

Global energy use (primarily based on coal, oil, and gas) increased by a factor of 19 from 28.6 EJ in 1850 to 542.8 EJ in 2008 (Table 1.5). Whereas the energy-based input (fertilizer, irrigation, machinery) must be increased in Africa and South Asia, the importance of enhancing the use efficiency of inputs (e.g., fertilizer, water, farm operations), by minimizing losses and conserving energy, cannot be overemphasized.

TABLE 1.4
Current Relative Yield Levels of Major Crops in Different Regions

Region	Wheat	Maize	Rice	Soybean
Western Europe	100 (6.85 Mg/ha)	–	45.9	85.3
Northern Europe	94.3	–	–	–
Southern Europe	45.3	–	63.0	100 (3.20 Mg/ha)
Australia and New Zealand	19.3	–	100 (10.84 Mg/ha)	59.3
North America	38.7	100 (9.34 Mg/ha)	69.6	91.3
Eastern Asia	65.4	55.6	60.0	54.6
Southern Asia	36.6	25.4	31.9	33.6
Northern Africa	30.2	–	385.8	85.2
South America	35.6	35.4	42.3	89.6
Eastern Europe	34.5	44.5	47.9	45.3
Central America	–	24.4	35.0	39.0
Southeastern Asia	–	34.6	38.2	44.0
Eastern Africa	–	14.6	19.3	30.9
Western Africa	–	17.3	–	40.5
Caribbean	–	–	32.7	–
Eastern Africa	–	–	24.4	–
World	40.1	65.4	36.3	72.5

Source: Recalculated from Hengsdijk, K.H. and J.W.A. Langeveld. Yield trends and yield gap analysis of major crops in the world. Wageningen Wettelijke Onderzoekstaken Natuur & Milieu, wOt-werkdocument 170 blz. 57; fig. 15; tab. 13; ref. 40, Wageningen, Holland, 2009.

Note: The minimum yield of a specific crop in a region (given in parenthesis as Mg/ha) is considered 100. Yield for all other regions are relative to 100.

TABLE 1.5
Temporal Changes in Primary Energy Source from 1850 to 2008

			Primary Source (EJ)					
Year	Biomass	Coal	Oil	Gas	Hydropower	Nuclear	Other Renewables	Total
1850	27.13	1.43	–	–	–	–	–	28.56
1875	28.56	7.14	–	–	–	–	–	35.70
1900	30.35	19.63	1.07	–	–	–	–	51.05
1925	34.27	29.99	10.71	3.57	1.79	–	–	80.33
1950	39.27	41.06	23.20	7.14	3.57	–	–	114.24
1975	30.35	69.61	117.81	39.27	17.85	3.57	–	278.46
2000	46.41	96.18	158.86	89.25	28.56	28.56	3.15	450.97
2008	49.98	142.37	173.15	116.02	32.13	24.99	4.14	542.78

Source: Johansson, T.B. et al., *Global Energy Assessment: Towards a Sustainable Future.* Cambridge University Press, Cambridge, UK, 2012.

Increasing energy use and enhancing its efficiency in agroecosystems are high priorities in countries with low relative agronomic yields of food crop staples (Table 1.4).

1.3 YIELD GAP

While there is some potential to expand the arable land area, such as in Latin America and Africa (Bruinsma 2003; Rosegrant et al. 2001), the focus must be on increasing productivity from the existing cropland by abridging the yield gap. The latter is defined as the difference between the technical yield potential (at the national or regional level) and the average farmers' yield for a specific crop (Lobell et al. 2009). The technical yield potential refers to the agronomic productivity of the modern variety grown under the best management practices (BMPs). With these criteria, the yield gap of major crops in India is 5%–60% for wheat and 33%–70% for rice. Similarly, the yield gap for maize is 54%–84% in Sub-Saharan Africa and 60%–70% in Latin America. In comparison, the yield gap for rice in China is only 17%–20% (Lobell et al. 2009). In general, the lower the national/regional farmers' yield and the more degraded the soil, the larger the yield gap. High yield gaps have been reported for crop production under semiarid conditions (Fischer et al. 2009; Rockström and Faulkenmark 2012), for wheat in northeastern Spain (Abeledo et al. 2008), rice in Southeast Asia (Laborte et al. 2012), and major food crops in West and Central Africa (Nin-Pratt et al. 2011). Witt et al. (2009) used an agronomic model based on "yield gap" analysis, fertilization for attainable yield, and the area growth in Indonesia. There are several options of closing the yield gap, which are crop specific (Tran 2010). Furthermore, there is a wide range of controls that affect the yield gap (Figure 1.2). Important among these are the following: (1) natural resources including climate, terrain, soil, water, and so forth; (2) technology adoption comprising management of soil, water, nutrients, climate, and human capital; (3) institutional support such as research and

Principles of Soil Management

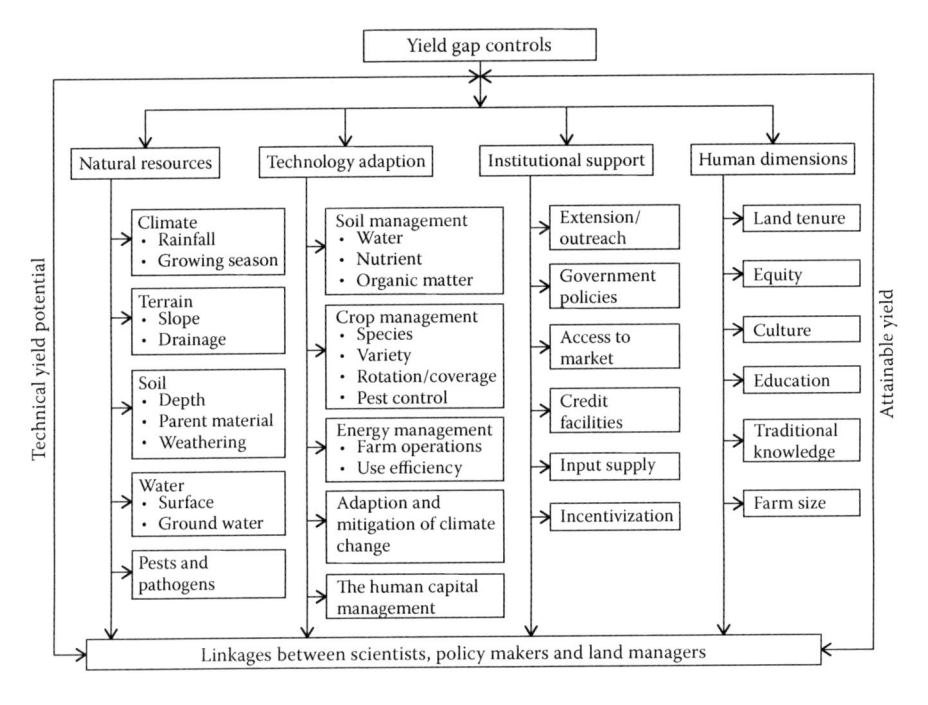

FIGURE 1.2 Determinants of a crop yield gap.

extension, government policies, access to market and credit, and so forth; and (4) the human dimensions such as land tenure, equity (gender and social), education, farm size, and so forth. The strategy is to identify and fine-tune soil-based technology for site/crop-specific situation. The goal is to optimize soil conditions that support favorable crop growth even under harsh climatic conditions. Adoption of this strategy implies enhancing soil/ecosystem resilience through improvement in soil quality. It is because of the lack of focus on soils, by taking soil resources for granted, that the Green Revolution of the 1970s bypassed Sub-Saharan Africa.

1.3.1 Soil Quality Management

Soil degradation is a serious threat especially in the tropics and subtropics of a Sub-Saharan Africa, South/Central Asia, and the Caribbean (Bai et al. 2008; Oldeman 1994). It is caused by the interactive effects of natural and anthropogenic factors (Figure 1.3). The effects of natural factors (e.g., soil quality, climate—especially rainfall, temperature, and erosivity) and land (slope gradient) are strongly moderated by the management (i.e., land use, cropping/farming system). Socioeconomic factors are important determinants of soil degradation, especially of accelerated soil erosion. It is in this context that Halim et al. (2007) developed a methodology that integrated both biophysical and socioeconomic aspects into a framework for soil erosion hazard assessment using principal component analysis at the land unit level. Therefore, any program designed to alleviate the problem of soil erosion must be consistent with farmers' knowledge and their perception of the erosion and its short- and long-term

```
                    Soil quality
                    • Physical
                    • Chemical
        Land        • Biological      Climate
   • Slope (gradient,                 • Precipitation
     aspect, length)                  • Temperature
                    Interactive
   • Drainage        effects          • Growing season
   • Vegetation cover                 • Erosivity
                                      • Extreme events
                    Management
                    • Land use
                    • Land management
                       (i) Soil
                       (ii) Crop/vegetation
                       (iii) Water
```

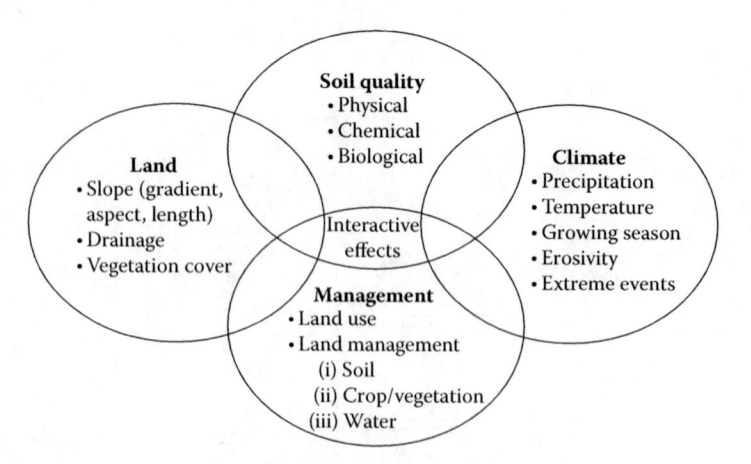

FIGURE 1.3 Natural and anthropogenic factors affecting the extent and severity of soil degradation.

impacts (Moges and Holden 2007). Wind erosion is an important feature of desertification on cultivated fields in regions receiving <500 mm/year of rainfall. In the Sahel, Toure et al. (2011) observed that the absence of crop residue cover is a major factor exacerbating the wind erosion hazard on cultivated millet fields. Stocking rate, grazing intensity, and uncontrolled grazing are important factors affecting erosion on grazing lands (Keay-Bright and Boardman 2007).

The slash-and-burn agriculture, with a short fallow period, is another cause of degradation (Styger et al. 2007) by erosion and nutrient and soil organic C (SOC) depletion. Desertification, soil degradation in arid climates, can alter both the patch index and the Shannon landscape diversity index. The latter is affected by a severe decline in the vegetation cover (Hirche et al. 2011). In a chronosequence study conducted in Western Kenya, Moebius-Clune et al. (2011) reported that physical, chemical, and biological soil quality indicators (e.g., SOC, aggregation, available water capacity, pH, electrical conductivity, cation exchange capacity). Thus, soil management must be done to ensure and sustain soil quality over decadal to millennial scales.

1.4 KEY SOIL PARAMETERS IMPORTANT TO SOIL SUSTAINABILITY

Soil physical, chemical, and biological properties relevant to sustainable management are outlined in Table 1.6. Key soil physical properties are texture, structure, available water capacity, rooting depth, water transmission, soil strength, crusting, and erodibility. Important soil chemical properties include pH, total acidity, electrical conductivity, total soluble salts, effective cation exchange capacity, total and available plant nutrients, and concentration of heavy metals (Pb, Hg, As, etc.). Similarly, relevant soil biological characteristics comprise total SOC and different fractions, microbial biomass C (MBC), activity and species diversity of soil fauna, soil enzymes, and other rhizospheric properties and processes. Threshold/critical levels of these key soil parameters must be maintained to sustain numerous ecosystem services and functions.

TABLE 1.6
Key Parameters Important to Soil Quality and Sustainability

Soil Parameters

Physical	Chemical	Biological
1. *Structure*: aggregation, penetration resistance, bulk density, texture, mineralogy, SOC concentrations, sesquioxides consistency, shrinkage characteristics	1. *Reaction*: pH, electrical conductivity, total acidity and exchangeable Al, soluble salt concentration, sodium absorption ratio	1. *SOC*: total and different fractions (e.g., labile, passive, heavy, light), MBC
2. *Water retention*: field capacity, wilting point, clay content, SOC concentration, pore size distribution	2. *Charge properties*: total charge cation and anion exchange capacity, point of zero charge, effective cation exchange capacity, relative abundance of different cations	2. *Soil macrofauna*: earthworms, termites, centipedes, millipedes, nematodes
3. *Water transmission*: aggregation, total and drainage on macroporosity, bulk density, crusting, horizonation, hydraulic conductivity, infiltration capacity	3. *Plant nutrients*: concentration of total and available macronutrients and micronutrients	3. *Rhizospheric*: disease suppressive properties, soil enzymes
4. *Erodibility*: texture, aggregation, clay minerals, SOC concentration, infiltration rate	4. *Elemental imbalances*: toxicity (Al, Fe) and deficiency (macron and micro nutrients) heavy metal concentration (Pb, Hg, As, Cd)	

1.4.1 Technological Options for Soil Sustainability

The basic strategy is to (1) replace what is harvested; (2) respond prudently and rationally to what is transformed naturally or anthropogenically; (3) predict, anticipate plan for, and adapt to potential perturbations; and (4) use technology wisely. The SOC concentration [including total stock, depth distribution to 1 m, different fractions, and the mean residence time (MRT)] is a key determinant of soil chemical, physical, and biological quality and biogeochemical cycles (Manlay et al. 2007). Thaer's "humus theory" proposed in 1809 is indicative of the recognition of the importance of SOC to agronomic productivity.

Soil degradation by accelerated erosion leads to rapid depletion of the SOC pool and thus to adverse changes in soil properties. Land use, and specifically uncontrolled and intensive grazing, can also affect the SOC pool and its variability across the landscape (du Preez et al. 2011a,b). Land use and soil management influence the

MRT or turnover rate of SOC by altering its accessibility to microbes (Dungait et al. 2012). An objective of soil management is to moderate/control the accessibility of SOC to soil biota.

Conservation agriculture or no-till (NT) farming, adopted in conjunction with cover cropping and integrated nutrient management (INM), is another strategic option of enhancing the SOC pool in the surface layer, reducing risks of soil erosion, and sustaining productivity. Toliver et al. (2012) compared agronomic yields from 442 paired tillage experiments across the United States. They observed that mean yields of sorghum (*Sorghum bicolor*) and wheat (*Triticum aestivum*) were higher with NT than plow tillage (PT). Further, there was a trend of similar or more yield with NT than PT for crops grown on loamy soils in the Southern Seaboard and Mississippi Portal regions. In contrast, NT performed poorly on sandy soils except for cotton (*Gossipium hirsutum*) and corn (*Zea mays*). Toliver and colleagues concluded that soil and climatic factors impact NT yields relative to PT. Another principal advantage of NT farming is the reduction in diesel consumption for plowing. Labreuche et al. (2011) observed from long-term experiments at Biogneville in the Paris basin (France) that diesel consumption was reduced by 28 L/ha with conversion from PT to reduced tillage and by 41 L/ha with conversion to NT, reducing total energy consumption by 6%–11%. Further, the magnitude of savings of diesel can be double this in heavy-textured (clayey) soils with a large draft power. Considering higher N_2O emissions and more SOC storage in NT than PT, Labreuche et al. (2011) estimated a decrease of 200 kg/ha/year of CO_2 equivalent with conversion from PT to NT farming.

Grazing management is another important factor affecting soil quality. Over and above the effects of grazing on the SOC pool and soil quality as discussed above (du Preez et al. 2011a,b), grazing induces alterations in several parameters. In South Africa, Keay-Bright and Boardman (2007) observed that high stock numbers and less benign management practices in the nineteenth century and early twentieth century underlie much of the degradation observed today. In Burkina-Faso, Savadogo et al. (2007) observed that heavy grazing reduced the aboveground biomass and vegetation cover and decreased water infiltrability because of the trampling pressure (static load) exerted by the animals. Further, even the prescribed fire reduced the soil water infiltration rate: 42.2 ± 27.5 mm/h in burnt areas versus 78 ± 70.5 mm/h in unburnt areas. The adverse effects of overgrazing on soil quality are also documented by Mekuria et al. (2007) for the Tigray region of Ethiopia. Mekuria and colleagues observed that exclosures of grazing are effective not only in restoring vegetation but also in improving SOC concentration and nutrient status and reducing the erosion hazard. Conversion of grazing areas to exclosures for an extended period can restore the quality of degraded soils.

Thus, conversion to a restorative land use and adoption of BMPs can restore degraded/desertified soils and advance sustainable management of soil and water resources.

1.5 PRINCIPLES OF SOIL MANAGEMENT

Assessing fertilizer/nutrient requirements for increasing crop yields has been a major challenge ever since the Industrial Revolution of around 1750. Synthesis of a vast amount of literature on sustainable management of soil would indicate several basic

Principles of Soil Management

principles. Notable among these is the "law of the minimum" by Sprengel (1832) and Justus von Liebig (1843). German agronomist Carl Sprengel (1737–1859) disproved Thaer's humus theory and formulated the law of minimum. It states that the agronomic yield of a specific field is determined by the nutrient(s) in least supply. When the most limiting nutrient (or any other factor) is corrected, yields are then determined by the next important limiting factor. Mitscherlich (1909) proposed the "law of the diminishing returns." It states that an incremental increase in crop yield decreases with progressive increase in the input of the limiting factor [$dy/dx = C(A - Y)$, where x is the input, A is the maximum possible yield, Y is the yield corresponding with the amount of X supplied, and C is a proportionality constant]. Wallace (1993) proposed the "law of maximum." It states that the effect of specific input on agronomic yield progressively increases as deficiency of other limiting factors is corrected, and the maximum yield (when factors are corrected) exceeds the sum of the effects of individual inputs. Ramamoorthy and Velayutham (2011) proposed "the law of optimum." It is defined as the concept of soil test based on major plant nutrients' (NPK) application to crops. It is also the basis of "precision farming" for sustainable and enduring agriculture.

In the context of the severe and global problem of soil degradation, an important strategy is to enhance soil and ecosystem resilience, particularly in dryland agroecosystems (Enfors and Gordon 2007). It is also argued that an effective approach to improving soil quality is by using farmers' knowledge for the soil/site-specific situations (Tesfahunegn et al. 2011) and social/cultural factors (Anley et al. 2007). Agricultural intensification based on enhanced diversification (Tappan and McGahuey 2007) can improve agronomic yields while also improving the environment and alleviating poverty (Brown et al. 2011).

1.6 LAWS OF SUSTAINABLE SOIL MANAGEMENT

Lal (2009a,b) proposed 10 tenets of soil management (Table 1.7). These generic laws emphasize the following.

1.6.1 Policy Makers

Establishing a dialogue with policy makers is important to identifying policies that reverse the degradation trends. Five basic laws of soil management, which are related to appropriate policies, are the following.

1.6.1.1 Causes of Soil Degradation and Desertification

The biophysical process of soil degradation and desertification is driven by economic, social, and political forces. The effectiveness of managing biophysical processes in minimizing degradation risks and enhancing restoration mechanisms depend on addressing the human dimensions that affect land misuse, soil management, and prevalence of extractive farming practices.

1.6.1.2 Human Needs and Stewardship of Natural Resources

Because humans are always dependent on agriculture, improving agriculture is essential to the stewardship of natural resources. In this context, it is important to

TABLE 1.7
Basic Principles of Soil Management

Law	Description
1. Causes of soil degradation	The biophysical process of soil degradation is driven by economic, social, and political forces. Vulnerability to degradation depends on "how" rather than "what" is grown.
2. Soil stewardship and human suffering	When people are poverty stricken, desperate, and starving, they pass on their sufferings to the land.
3. Nutrient, carbon, and water bank	It is not possible to take more out of a soil than what is put in without degrading its quality. Only by replacing what is taken can a soil be kept fertile, productive, and responsive to inputs.
4. Marginality principle	Marginal soils cultivated with marginal inputs, produce marginal yields, and support marginal living. Recycling is a good strategy especially when there is something to recycle.
5. Organic versus inorganic source of nutrients	Plants cannot differentiate the nutrients supplied through inorganic fertilizers or organic amendments.
6. Soil carbon and greenhouse effect	Mining C has the same effect on global warming, whether it is through mineralization of soil organic matter and extractive farming, or burning fossil fuels or draining peat soils. Soil can be a source of sink of GHGs depending on land use and management.
7. Soil versus germplasm	The potential of elite varieties can be realized only if grown under optimal soil conditions. Even the elite varieties cannot extract water and nutrients from any soil where they do not exist.
8. Soil and global warming	Soil is integral to any strategy of mitigating global warming and improving the environment.
9. Engine of economic development	Sustainable management of soils is the engine of economic development, political stability, and transformation of rural communities in developing countries.
10. Traditional knowledge and modern innovations	Sustainable management of soil implies the use of modern innovation built upon traditional knowledge. Those who refuse to use modern science to address urgent global issues must be prepared to endure more suffering.

Sources: Lal, R., *Agron Sustain Develop* 29: 7–9, 2009; Lal, R., *J Soil Water Conserv* 64: 20A–21A, 2009.

realize that, when people are poverty stricken, desperate, and starving, they pass on their sufferings to the land. The stewardship concept is important only when the basic needs are adequately met. A sermon about the virtues of saving a tree falls on deaf ears when there is no fuel for cooking the family meal. In other words, starving people do not care about stewardship (Bartlett 2004).

1.6.1.3 Poverty and Soil Degradation

There exists a close relationship between poverty and soil degradation. Poor farmers carve out meager living from marginal and impoverished soils. Marginal soils cultivated with marginal inputs produce marginal yields and support marginal living. The sustainable soil management strategy is to cultivate the best soils by BMPs to produce the best yields so that surplus land can be saved for nature conservancy. Sustainable management of soils is the engine of economic development, political stability, and economic transformation of rural communities in developing countries.

1.6.1.4 Soil Degradation Is a Cause of Global Warming

Degraded soils and ecosystems are sources of CO_2 and other greenhouse gasses (GHGs). Mining C has the same effect on global warming, whether it is by mineralization of soil organic matter (SOM) and releasing nutrients through plowing and extractive farming, or it is through burning fossil fuels (coal, gas, oil), using petrol-based products, or draining and cultivation of peat soils.

1.6.1.5 Desertification Control and Mitigation of Climate Change

Degraded and desertified ecosystems have a large C sink in soils and biota because of historic loss and perpetual misuse of natural resources. Thus, restoration of these ecosystems can be a major sink for atmospheric CO_2 and CH_4 through conversion to a restorative land use adoption of recommended management practices that lead to positive C and nutrient budgets. Filling the C sink capacity of the pedosphere (2–4 Gt C/year) in soils of croplands, grazing lands, and degraded and desertified lands, being a cost-effective and natural process, has numerous ancillary benefits. While advancing food security and improving water quality, C sequestration in the biosphere also mitigates climate change.

1.6.2 LAND MANAGERS

The strategy of establishing dialogue with land managers is to emphasize the importance of adoption of those farming practices that create a positive C and nutrient budget and enhance soil resilience. Basic laws that govern these processes are the following.

1.6.2.1 Nutrient Bank

Soils are analogous to a bank account. Similar to a bank account, it is also not possible to take more out of a soil than what is put into it without degrading its quality. In addition to the amount taken out, soil quality also depends on the rate, timing, method, and form of what is being extracted or replaced. Thus, managed ecosystems are sustainable in the long term if the output of all components produced balances the input into the system. Soils are vulnerable to degradation and desertification when inputs are perpetually less than the output. Soils of Sub-Saharan Africa have had a negative nutrient budget since the 1960s. Therefore, these soils do not respond to other inputs (e.g., improved varieties) because of severe physical and nutrient-related constraints.

1.6.2.2 Organic and Inorganic Sources of Plant Nutrients

The debate on organic versus inorganic sources of plant nutrients is futile and irrational rather than factual. Plants cannot differentiate the nutrients supplied through inorganic fertilizers or organic amendments. Rather than an "either/or" question, it is a matter of logistics and practicality in making nutrients available in sufficient quantity, in appropriate form, and at the critical time needed for optimum crop growth and desired yields. It is logistically difficult to find enough quantity of biofertilizers and transport bulk amounts of manure, especially in arid and semiarid regions.

1.6.2.3 Modern Issues and Ancient Technologies

The problems of the twenty-first century, exacerbated by 7.1 billion people increasing at the rate of 70–80 million/year, cannot be addressed by technologies developed during the middle ages. Thus, it is important to build upon the traditional knowledge and avail of the benefits of modern innovations. Similar to the debate on inorganic versus organic fertilizers, this is also not an "either/or" scenario. Modern science must synthesize the traditional knowledge and build upon it. Those who refuse to use modern science to address urgent global issues of the twenty-first century must be prepared to endure more sufferings.

1.6.2.4 Masters of Their Own Destiny

While numerous hardships and suffering can be attributed to harshness of nature and to poor governance, land managers (farmers, ranchers, foresters) are also masters of their own destiny. Soil restoration and desertification control are more effectively addressed through farmer-driven initiatives in addressing their own problems. The lack of rain has been blamed on nature (e.g., climate change), yet it is the farmer's responsibility to conserve water in the soil, harvest and recycle excess rainfall, and adopt technologies that produce more crop per unit drop (of rain).

1.6.2.5 Being Proactive

Land managers must be proactive in demanding from policy makers implementation of programs that reduce their vulnerability to climate change and from researchers innovative technologies that adapt to harsh environments and reverse degradation. These programs include those that involve expansion in irrigation, making timely availability of seeds and fertilizers, providing institutional support and marketing facilities, and initiating research relevant to the needs of resource-poor farmers. It is the squeaking wheel that gets the grease.

1.6.3 Researchers

Researchers must also undertake some appropriate projects that are relevant to societal needs and those that address issues of soil degradation, desertification, and food security. Some relevant laws of sustainable soil management for researchers are the following.

1.6.3.1 Relative Importance of Natural Resources versus Improved Germplasm

Both are important and must complement one another. However, it is important to realize that even the elite varieties, even those developed through biotechnology and genetic engineering, cannot extract water and nutrients from any soil where they do not exist. The yield potential of improved germplasm can be realized only if grown under recommended management practices of soil, water, and crop husbandry. Being the foundation of agrarian societies, sustainable management of soils is the engine of economic development, political stability, and transformation of rural communities in developing countries.

1.6.3.2 Indicators of Soil Quality Improvement

Even modest improvements in soil quality can have a drastic positive effect on agronomic productivity, food security, farm income, and the environment. Thus, it is important to develop indicators of soil quality improvement (soil health score card system) that farmers can relate to and scientists can quantify.

1.6.3.3 Training Cadre of Young Researchers

Regional problems of soil degradation and desertification (e.g., those in Sub-Saharan Africa and South Asia) will be solved by researchers from the regions who also understand social, cultural, ethnic, and political issues. Therefore, a cadre of young scientists and researchers must be trained who are good world citizens, prepared for life, responsive to societal needs, and useful to humanity. Science without humanity is one of the serious blunders (Mahatma Gandhi). Researchers must be trained in problem-solving skills, creativity, and originality.

1.6.3.4 Building Bridges across Nations

Soil degradation and desertification are global issues; they do not respect political boundaries, equally affect all irrespective of ethnicity, and cannot be addressed in isolation. Research programs must be developed that strengthen linkages among institutions in different countries facing similar problems. Building bridges across nations is essential to mitigating desertification.

1.6.3.5 Lack of Technology Adoption

There has been a serious lack of adoption of improved soil and water management technologies especially by the resource-poor farmers of Sub-Saharan Africa and South Asia (e.g., mulch farming, NT, cover cropping, INM). It is probably because the top–down approach is not effective in technology transfer. Scientists need to identify reasons for the lack of technology adoption and develop programs based on participatory approaches. Scientists must also understand that adoption of "technology without wisdom" is equally problematic (Lal 2007).

Thus, the theme of this volume is elaboration of these and other basic principles of soil management.

1.7 CONCLUSIONS

The trilemma of food insecurity, global warming, and environmental degradation can be effectively addressed by enhancing soil quality. The interest in enhancing agronomic yield of food crops is an important issue from many perspectives including food security (Lal 2006), biofuel production (Gitiaux et al. 2011), and poverty alleviation (Brown et al. 2011). Sustainable economic development, especially in

Sub-Saharan Africa and South Asia, where 60%–70% of the population depends on agriculture, can be realized only through adoption of BMPs based on basic principles of soil management. The per capita agriculture growth rate of 0.36%/year for Sub-Saharan Africa during the 2000s (Pardey 2011) necessitates adoption of soil-based BMPs. The large yield gaps of major crops (Hengsdijk and Langeveld 2009) must be abridged through restoration of soil quality and augmentation of soil and ecosystem resilience. Thus, widespread adoption of BPMs based on sound principles of soil management is more urgent now than ever before in human history.

ABBREVIATIONS

ACC:	abrupt climate change
As:	arsenic
BMPs:	best management practices
Hg:	mercury
INM:	integrated nutrient management
K:	potassium
MBC:	microbial biomass C
Mha:	million hectares
MRT:	mean residence time
N:	nitrogen
NT:	no-till
P:	phosphorus
Pb:	lead
PT:	plow tillage
SOC:	soil organic C

REFERENCES

Abeledo, L.G., R. Savin, and G.A. Slafer. 2008. Wheat productivity in the Mediterranean Ebro Valley: analyzing the gap between attainable and potential yield with a simulation model. *Eur J Agron* 28(4):541–550.

Anley, Y., A. Bogale, and A. Haile-Gabriel. 2007. Adoption decision and use intensity of soil and water conservation measures by smallholder subsistence farmers in Dedo District, Western Ethiopia. *Land Degrad Dev* 18:289–302.

Bai, Z.G., D.L. Dent, L. Olssom, and M.E. Schaepman. 2008. Proxy global assessment of land degradation. *Soil Use Manage* 24:2223–450.

Bartlett, A.A. 2004. *The essential exponential: for the future of our planet*. Univ. of Nebraska, Lincoln, NE.

Borlaug, N.E. and C.R. Dowswell. 2005. Feeding a world of 10 billion people: a 21st century challenge. In *In the wake of the double helix: from green revolution to the gene revolution*, eds. R. Tuberosa, R.L. Phillips, M. Gale, 3–23. Proc Congress, 27–31, May 2003, Bologna, Italy.

Brown, D.R., P. Dettmann, T. Rinaudo, H. Tefera, and A. Tofu. 2011. Poverty alleviation and environmental restoration using the clean development mechanism: a case study from Humbo, Ethiopia. *Environ Manage* 48:322–333.

Bruinsma, J. 2003. *World agriculture: towards 2015/2030: an FAO perspective*. Earthscan, Rome.

du Preez, C.C., C.W. van Huyssteen, and P.N.S. Mnkeni. 2011a. Land use and soil organic matter in South Africa 1: a review on spatial variability and the influence of a rangeland stock production. *S Afr J Sci* 107(5/6):1–8.

du Preez, C.C., C.W. van Huyssteen, and P.N.S. Mnkeni. 2011b. Land use and soil organic matter in South Africa 2: a review on the influence of arable crop production. *S Afr J Sci* 107(5/6):1–8.

Dungait, J.A.J., D.W. Hopkins, A.S. Gregory, and A.P. Whitmore. 2012. Soil organic matter turnover is governed by accessibility not recalcitrance. *Global Clim Bio* doi: 10.1111/j.1365-2486.2012.02665, vol. 18:1781–1796.

Enfors, E.I. and L.J. Gorgon. 2007. Analyzing resilience in dryland agro-ecosystems: a case study of the Makanya catchment in Tanzania over the past 50 years. *Land Degrad Dev* 18:680–696.

FAO. 2012. FAOSTAT. http://faostat.fao.org/Rome.

Fischer, G., E. Teixeira, E. Tothne-Hizsnyik, and H. van Velthuizen. 2009. Land use dynamics and sugarcane production. In: Sugarcane ethanol, contributions to climate change mitigation and the environment. Edited by Peter Zuurbier and Jos van de Vooren, Wageningen Academic Publishers, ISBN 978-90-8686-090-6. Also available as IIASA RP-09-001, IIASA, Laxenburg, Austria.

Gitiaux, X., J. Reilly and S. Paltsev. 2011. *Future yield growth: what evidence from historical data?* University of Colorado, Department of Economics, and MIT Joint Program on the Science and Policy of Global Change, MIT, Boston.

Halim, R., R.S. Clemente, J.K. Routray, and R.P. Shrestha. 2007. Integration of biophysical and socio-economic factors to assess soil erosion hazard in the upper Kaligarang watershed, Indonesia. *Land Degrad Dev* 18:453–469.

Hengsdijk, K.H. and J.W.A. Langeveld. 2009. Yield trends and yield gap analysis of major crops in the world. Wageningen Wettelijke Onderzoekstaken Natuur & Milieu, wOt-werkdocument 170 blz. 57; fig. 15; tab. 13; ref. 40, Wageningen, Holland.

Hirche, A., M. Salamani, A. Abdellaoui, S. Benhouhou, and J.M. Valderrama. 2010. Landscape changes of desertification in arid areas: the case of south-west Algeria. *Environ Monit Assess* 179:403–420.

Johansson, T.B., A. Patwardhan, N. Nakicenovic, and L. Gomez-Echeverri. 2012. *Global energy assessment: towards a sustainable future.* Cambridge University Press, Cambridge, U.K.

Keay-Bright, J. and J. Boardman. 2007. The influence of land management on soil erosion in the Sneeuberg Mountains, Central Karoo, South Africa. *Land Degrad Dev* 18:423–439.

Laborte, A.G., C.A.J.M. de Bie, E.M.A. Smaling et al. 2012. Rice yields and yield gaps in Southeast Asia: past trends and future outlook. *Eur J Agron* 36:9–20.

Labreuche, J., A. Lellahi, C. Malaval, and J. Germon. 2011. Impact des techniques culturales sans labour (TCSL) sur le bilan énergétique at le bilan des gaz à effect de serre des systèmes de culture. *Cah Agric* 20:204–215.

Lal, R. 2001. Soil degradation by erosion. *Land Degrad Dev* 12:519–539.

Lal, R. 2003. Soil erosion and the global carbon budget. *Environ Int* 29:437–450.

Lal, R. 2006. Enhancing crop yields in the developing countries through restoration of soil organic carbon pool in agricultural lands. *Land Degred and Dev* 17:147–209.

Lal, R. 2007. World soils and global issues. *Soil Tillage Res* 97:1–4.

Lal, R. 2009a. Laws of sustainable soil management. *Agron Sustain Dev* 29:7–9.

Lal, R. 2009b. Ten tenets of sustainable soil management. *J Soil Water Conserv* 64:20A–21A.

Lal, R., U. Safriel, and B. Boer. 2012. *Zero net land degradation.* Rio+20 Convention, UNCCD, Bonn, Germany.

Lal, R. and B.A. Stewart (Eds). 2012. Soil water and agronomic productivity. CRC/Taylor and Francis, Boca Raton, FL, 578 pp.

Liebig, V.J. 1843. Chemistry in its application to agriculture and physiology. Report to British Association, Cambridge, U.K.

Lobell, D.B., K.G. Cassman, and C.B. Field. 2009. Crop yield gaps: their importance, magnitudes, and causes. *Annu Rev Environ Resour* 34:179–204.

Manlay, R.J., C. Feller, and M.J. Swift. 2007. Historical evolution of soil organic matter concepts and their relationships with the fertility and sustainability of cropping systems. *Agric Ecosyst Environ* 119:217–233.

Mekuria, W., E. Veldkamp, M. Haile, J. Nyssen, B. Muys, and K. Gebrehiwot. 2007. Effectiveness of exclosures to restore degraded soils as a result of overgrazing in Tigray, Ethiopia. *J Arid Environ* 69:270–284.

Mitscherlich, E.A. 1909. Das gestez abnehmenden. *Land Witsah Jahred* 38:537–552.

Moebius-Clune, B.N., H.M. van Es, O.J. Idowu et al. 2011. Long-term soil quality degradation along a cultivation chronosequence in western Kenya. *Agric Ecosyst Environ* 141:86–99.

Moges, A. and N.M. Holden. 2007. Farmers' perceptions of soil erosion and soil fertility loss in southern Ethiopia. *Land Degrad Dev* 18:543–554.

Nin-Pratt, A., M. Johnson, E. Magalhaes et al. 2011. *Yield gaps and potential agricultural growth in the west and central Africa*. International Food Policy Research Institute, Washington, D.C.

Oldeman, R. 1994. The global extent of soil degradation. In Soil resilience and sustainable land use, eds. D.J. Greenland, I. Szabolas, 99–118. CAB International, Wallingford, U.K.

Pardey, P.G. 2011. African agricultural productivity growth and R&D in a global setting. Stanford symposium series on global food policy and food security in the 21st century. Center on Food Security and Environment (FSE), Stanford University, San Francisco, USA.

Ramamoorthy, B. and M. Velayutham. 2011. The "Law of Optimum" and soil test based fertilizer use for targeted yield of crops and soil fertility management for sustainable agriculture. *Madras Agric J* 98(10–12):295–307.

Rockström, J. and M. Falkenmark. 2000. Semiarid crop production from a hydrological perspective: gap between potential and actual yields. *Plant Sci* 19(4):319–346.

Rosegrant, M., M. Paisner, S. Meijer, and J. Witcover. 2001. *Global food projections to 2020: emerging trends and alternative futures*. International Food Policy Research Institute, Washington, D.C.

Savadogo, P., L. Sawadogo, and D. Tiveau. 2007. Effects of grazing intensity and prescribed fire on soil physical and hydrological properties and pasture yield in the savanna woodlands of Burkina Faso. *Agric Ecosyst Environ* 118:80–92.

Sprengel, C. 1832. Chemie fur landwirthie, forstmcnner and caeralisten, Gottingen.

Styger, E., H.M. Rakotondramasy, M.J. Pfeffer, E.C.M. Fernades, and D.M. Bates. 2007. Influence of slash-and-burn farming practices on fallow succession and land degradation in the rainforest region of Madagascar. *Agric Ecosyst Environ* 119:257–269.

Tappan, G. and M. McGahuey. 2007. Tracking environmental dynamics and agricultural intensification in Southern Mali. *Agric Syst* 94:38–51.

Tesfahunegn, G.B., L. Tamene, and P.L.G. Vlek. 2011. Evaluation of soil quality identified by local farmers in Mai-Negus catchment, northern Ethiopia. *Geoderma* 163:209–218.

Toliver, D.K., J.A. Larson, R.K. Roberts et al. 2012. Effects of no-till on yields as influenced by crop and environmental factors. *Agron J* 104(2):530–541.

Toure, A.A., J.R. Rajot, Z. Garba et al. 2011. Impact of very low crop residues cover on wind erosion in the Sahel. *Catena* 85:205–214.

Tran, D.V. 2010. Closing the rice yield gap for food security. *Cah Options Mediterr* 58:145–162.

Wallace, A. 1993. The law of maximum. *Better Crops Plant Food* 2:20–22.

Witt, C., J.M. Pasuquin, and G. Sulewski. 2009. Predicting agronomic boundaries of future fertilizer needs in AgriStats. *Better Crops* 93(4):16–18.

2 Marginality Principle

Jerry L. Hatfield and Lois Wright Morton

CONTENTS

2.1 Introduction 19
2.2 Dimensions of Marginality..23
2.3 Soil Degradation and Loss of Soil Function...23
    2.3.1 Biological Degradation 24
    2.3.2 Chemical Degradation 26
2.4 Extent of Soil Degradation 27
    2.4.1 Soil Management Practices 28
2.5 Scales of Variation...30
    2.5.1 Water Availability...30
    2.5.2 Crop Productivity...30
    2.5.3 Environmental Quality 33
2.6 Multifunctionality: Optimum Performance of Soil Resource.....................35
    2.6.1 Soil Management to Reduce Marginality..35
    2.6.2 Agroforestry..37
2.7 Social Values of Soil Functions: Definitions and Meanings of Marginality 38
    2.7.1 Social Definitions of Marginality 40
    2.7.2 Willingness and Capacity to Protect Soil Resource 42
    2.7.3 Capacity to Act: Whose Responsibility Is It to Protect the Soil?........45
2.8 Conclusions..47
Abbreviations 48
References..48

2.1 INTRODUCTION

There are 12 different soil orders in the United States representing a wide variation among soils both horizontally and vertically with a range of economic, social, and environmental functions (Hamdar 1999). The distinct traits of these soil orders have been used to create a classification system similar to the way in which we divide animals or plants into genus and species. The names of these soil orders provide us with a general description of the soils: Alfisol (moderately weathered); Andisol (volcanic ash); Aridisol (very dry); Entisol (newly formed); Gelisol (frozen); Histosol (organic, wet); Inceptisol (slightly developed, young); Mollisol (deep, fertile); Oxisol (very weathered); Spodosol (sandy, acidic); Utisol (weathered); and Vertisol (shrinks and swells). Soil classification allows for a comparison of different soil series and their aggregation into soil associations, which describe how different series are positioned adjacent to each other across a landscape. A county soil map provides a visual image of the diversity of soils across a landscape, and if we walk across a field, there are likely to be noticeable differences in color, texture, and vegetative growth. If we were to dig a pit in these soils, there would be a vast array of differences below the soil surface. A county soils map as shown in the following diagram demonstrates the diversity of soils within a relatively small area of the world, and a picture of a soil profile provides a glimpse into the complex nature of the soil system (Figures 2.1 and 2.2).

The characteristics of each soil series affect the functional uses and ecosystem services they have the capacity to provide. A primary function is the soil's ability to grow crops to meet human needs for food, fiber, and fuels. Another critical function is the regulating of the ecosystem directly and indirectly via nutrient filtration, retention, and cycling as well as carbon retention and sequestration and regulation of the water balance—all of which affect vegetative growth and the nitrogen, carbon, and water cycles (Collins et al. 2011). Karlen et al. (1997) elaborate on soil functions as follows: "1) Sustaining biological activity, diversity, and productivity; 2) Regulating and partitioning water and solute flow; 3) Filtering, buffering, degrading, immobilizing, and detoxifying organic and inorganic materials, including industrial and municipal by-products and atmospheric deposition; 4) Storing and cycling nutrients and other elements within the earth's biosphere; and 5) Providing support of socioeconomic structures and protection for archeological treasures associated with human habitation." County soil maps are

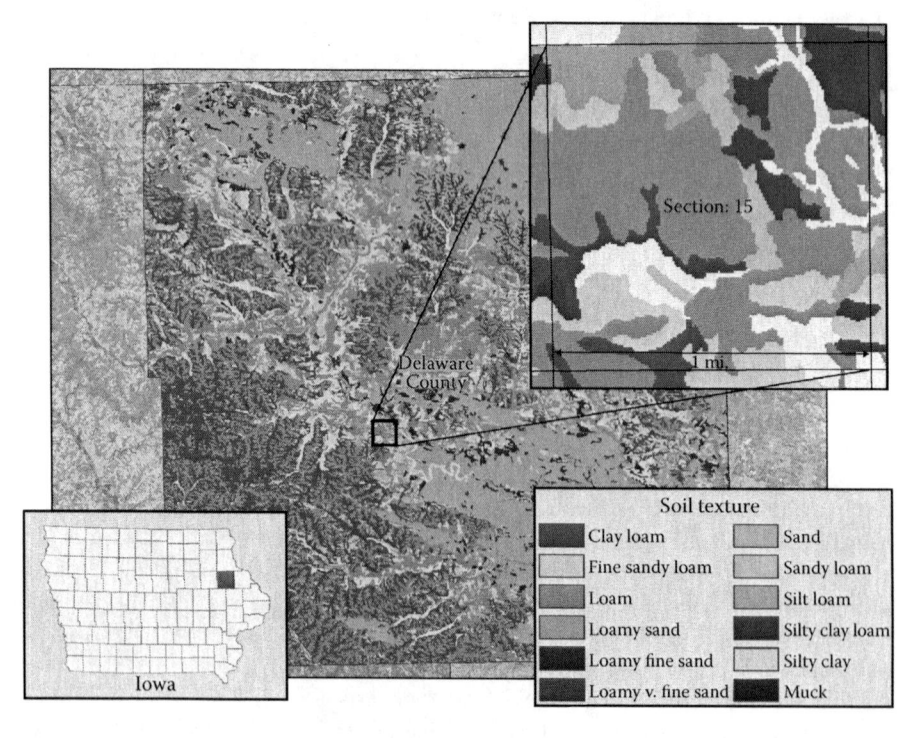

FIGURE 2.1 Map of soils for Delaware County, Iowa, showing the variation at the county level with an inset for an individual landscape (USDA NRCS).

Marginality Principle

FIGURE 2.2 Vertical distribution of soil layers from a Nashville-series soil. (From USDA-NRCS. Available at soils.usda.gov/gallery/photos/profiles, accessed July 13, 2012.)

often accompanied by charts linking soil traits to particular functions to help landowners, engineers and technical specialists, government officials, teachers, and developers in land use investments, planning, and decision making.

The concept of marginality applied to soils has its origins in how humans value the functions of soils and the different characteristics they provide. Although soils have multifunctional capacities, some functions are valued above others. The soil's ability to effectively and efficiently produce food and economic livelihoods underlies common valuations of soil quality and marginality. The US Department of Agriculture (USDA) Natural Resources Conservation Service (NRCS) soil capability classification system (Class 1–Class 8) uses the land's ability to grow crops as a metric to evaluate soil suitability for particular uses with the definitions shown in Table 2.1. These capability classes are given relative to a specific purpose for which the soil is being used, that is, capability classification will depend upon whether the soil is being used for annual crops, perennial crops, forests, and so forth. This method provides a measure of the suitability of the soil to perform its intended use, and most often, these methods are related to agricultural uses. Class 1 is best suited for growing a wide range of crops, and Class 8 is considered unsuitable for growing crops (Hamdar 1999). Classes 2 and 3 have moderate to severe limitations that reduce the choice of plants that can be grown and require special conservation practices to assure crop productivity. Under this classification system, marginal lands are those with severe limitations (Classes 4–8) and considerable reduction in the types of vegetation that will grow and thus require careful management (Hamdar 1999).

TABLE 2.1
Land Capability Classification and Definitions

Classification	Definitions
1	Soils have slight limitations that restrict their use.
2	Soils have moderate limitations that restrict the choice of plants or that require moderate conservation practices.
3	Soils have severe limitations that restrict the choice of plants or that require special conservation practices, or both.
4	Soils have very severe limitations that restrict the choice of plants or that require very careful management, or both.
5	Soils are subject to little or no erosion but have other limitations, impractical to remove, that restrict their use mainly to pasture, rangeland, forestland, or wildlife habitat.
6	Soils have severe limitations that make them generally unsuitable for cultivation and that restrict their use mainly to pasture, rangeland, forestland, or wildlife habitat.
7	Soils have very severe limitations that make them unsuitable for cultivation and that restrict their use mainly to pasture, rangeland, forestland, or wildlife habitat.
8	Soils and miscellaneous areas have limitations that preclude commercial plant production and that restrict their use to recreational purposes, wildlife habitat, watershed, or esthetic purposes.

Source: National Soil Survey Handbook, 2012. Natural Resources Conservation Service, USDA, Washington, D.C., http://soils.usda.gov/technical/handbook.

Note: Subclasses exist to describe the main hazard for each capability classification and are denoted by subscripts to the class, for example, erosion (e), water (w), shallow or droughty (s), and very cold or very dry (c).

They are generally considered unsuitable for cultivation and have pasture, woodland, recreation, wildlife, and water supply uses. This capability classification system is widely applied to agricultural landscapes and often leads to a devaluation of the ecosystem services different soil types may provide.

The characteristics of soils and their functional uses are not static but change over time naturally and through anthropological activities. Soil formation occurs as a result of time, parent material, climate, topography, and organisms all working together to create the vast array of soils we have around the world (Jenny 1941). The time scale of soil formation is hundreds and millions of years compared with the time scale of hours, days, weeks, and years in which soil functions can be lost. It is of particular interest whether soils have maintained their ability to perform their different functions. Soil ability to perform valued functions has been defined by soil scientists as soil quality (Doran et al. 1994; Doran and Parkin 1994). Two questions arise from scientific and nonscientific land manager definitions of marginality and soil quality: (1) Which function of soil does society give high priority value, and how does that affect definitions of marginality? (2) How much of a deviation away from the optimal functioning of a soil is necessary for the soil to be marginalized? For example, if we lose a 10% capability of producing a crop, is that sufficient to consider the soil as marginal? If we suffer a loss of 0.5 cm of topsoil, does that move a quality soil into a marginal category? If the soil water holding capacity decreases by 10% and leaching or runoff increases, is that sufficient to consider this soil as marginal? These are very complex questions that need to be addressed as we consider how to best manage soil, address issues of marginality, and increase soils' functional value to society.

One dilemma we face is that agricultural productivity in many areas of the world continues to increase, and we assume that soil function has continued to improve. However, we suggest that although our crop production has increased, this has come at the expense of a degraded soil resource. We are at the tipping point of being unable to sustain these increases unless we recognize that human activities can create marginal soils and, more importantly, have the capacity to reverse the trend of soil degradation. In this chapter, we first discuss the second question, what constitutes the loss of optimal functioning, and what criteria move a quality soil into marginality. Our findings are then used to address the first question, how society values soil functions, and how it influences land managers' and public policy makers' definition of marginality and willingness and capacities to protect soil quality. Future ability of agriculture to provide multifunctional roles—production of food, fuel, fiber, and livelihoods as well as a suite of ecosystem services—depends upon the quality of the soil resource (Hatfield 2006).

2.2 DIMENSIONS OF MARGINALITY

Marginality is an integrative principle that implies a transition from an optimum state to a less optimum state. However, values of marginality depend upon the potential use of the soil resource. If we consider marginality from a scientific viewpoint, then the capability classification provides a structure for assigning a value to marginality because there would be less productivity from a soil that moved from an optimal to less than optimal state. The classification definitions in Table 2.1 reveal a coupling of different soils with human use and practices reflecting the meaning of marginality as socially constructed based on societal values for the different functions of soil. This is particularly important since it also implies that human intervention to better manage soil to accomplish preferred uses can be successful and is a reminder that the norms of which soil functions are valued can shift over time.

The path a specific soil has in moving between capability classes and soil degradation offers a framework from which to assess changes in the soil and movement toward or away from marginality. Evaluation of soil traits and functions associated with optimum states provides an experimental structure for quantifying soil degradation or soil aggradation.

2.3 SOIL DEGRADATION AND LOSS OF SOIL FUNCTION

Soil degradation represents the changes in soil that reduce its capability to function. Lal et al. (2004) documented the degradation process and how this is set in motion by a combination of human and natural perturbations. Conceptually, soil degradation is the shifting of soil functionality from an optimal state to a less optimal state based on soil water holding capacity as a function of soil organic matter (SOM), as described by Hudson (1994). This can be seen as we look across the landscape and view soils of different colors, indicative of a reduced SOM content in the surface (Figure 2.3).

FIGURE 2.3 Degradation of soil across a landscape showing the variation in surface organic matter. (From USDA-NRCS. Available at photogallery.nrcs.usda.gov, accessed July 13, 2012.)

Degradation of the soil is a complex process linking the physical, chemical, and biological processes within the soil into a set of complex interactions, which are triggered by natural and anthropogenic disturbances. The types of soil degradation are shown in Table 2.2. Scientific evidence suggests that the hierarchy of these factors begins with degradation in the soil biological component followed by physical and chemical changes. If we monitor these types of degradation, it is easy to see that degradation will induce a vertical and horizontal component because the factors that give rise to the spatial variation shown in Figures 2.1 and 2.2 will respond differently as the degradation forces are applied. Changes in the soil are most apparent at the surface more than with depth because the forces causing degradation manifest themselves in this part of the soil profile more quickly than with depth. These changes are not uniform across a landscape because different soils respond differently to degradation forces.

2.3.1 Biological Degradation

Biological degradation is associated with the dynamics of the microbial systems within the soil profile. Microbial activity and soil biodiversity are linked with the soil organic C (SOC) pools in the soil and ultimately are associated with the depletion of SOC and the turnover rates of the SOC pool. Assessments of soil quality must include one or more indicators of microbial activity within the soil profile. Bastida et al. (2008) developed a biological index to assess soil quality and compared different

TABLE 2.2
Types of Soil Degradation

Type	Degradation Process
Biological	Decline in spoil biological activity
	Decline in soil biodiversity
	Depletion of SOC pool
	Increase in soilborne pathogens
Physical	Breakdown in soil structure
	Crusting and surface sealing
	Compaction: surface and subsoil
	Reduction in water infiltration capacity
	Increase in runoff rate and amount
	Decrease in soil water holding capacity as a result of reduced SOC
	Inundation, waterlogging, and anaerobiosis
	Accelerated erosion by water and wind
	Desertification
Chemical	Leaching of bases
	Acidification
	Elemental imbalance with excess of Al, Mn, Fe
	Salinization, alkalinization
	Nutrient depletion
	Increased contamination, pollution

Source: Adapted from Lal, R., *Philos. Trans. R. Soc. Biol. London*, 352, 997–1010, 1997.

soil biological methods to determine differences among techniques. They found that the metabolic quotient (ratio of respiration to microbial biomass) had value as a potential index to evaluate ecosystem disturbance or its maturity. From their analysis, they concluded that any framework for assessing soil quality should incorporate some aspect of soil biological status or function as an indicator of soil quality or soil degradation.

Any factor that modifies water and air exchanges in the soil will lead to changes in the physical conditions linked with soil–atmosphere interactions. Degradation of the soil structure as a result of impaired soil biological activity will no longer allow soil particles to maintain their ability to withstand either mechanical or hydrological forces. Once this breakdown in soil structure begins, a reduction in aggregate strength or stability occurs, resulting in a promotion of slaking, crusting, and bulk density (compaction). Changes in soil structure affect soil porosity and diminish gas exchange, leading to less oxygen and increasing carbon dioxide concentrations in the soil profile. When there is a diminished soil structure at the surface, the potential for erosion increases, because under rainfall events, the stability of the soil surface decreases, and the infiltration rate of water can no longer be maintained. With subsequent rainfall, the instable soil particles begin to move, and if rainfall events continue long enough, there will be erosion. Compaction is a result of the lack of soil structure and poor aggregate stability. Hamza and Anderson (2005) suggest that alleviation of compaction involves soil management practices that lead to an increase in the organic matter content of the soil and incorporation of deep-rooted crops into the rotation to restore structure to the lower soil profile.

2.3.2 Chemical Degradation

Chemical degradation is a result of changes in the chemical processes within the soil, which will, in turn, affect plant growth. A reduction in pH causing soil acidification occurs after leaching of bases through the soil profile or the continual addition of acid-producing fertilizers. Depletion of nutrients by removing plants or leaching without resupplying nutrients is another form of chemical degradation. Conversely, soils can become toxic through the buildup of elements (e.g., high Al, Fe, and Mn concentrations can be toxic to plants). Continual increases in soluble salts in the root zone that increase the electrical conductivity above 4 dS m^{-1} create salinization of the soil. Addition of Na ions through sodic salts can lead to alkalinization of the soil. Erosion can cause nutrient depletion in soils, and the degradation of soil structure can increase denitrification of N because of reduced oxygen content. Vanderpol and Traore (1993) observed these processes to be evident in the soils of Mali with the magnitude of the nutrient losses equivalent to 40% of the annual farm income. In the sandy loam soils of Georgia, Sainju et al. (2002) observed that SOC and nitrogen (N) concentrations in the soil could be maintained by reducing the mineralization and erosion losses through the use of no-tillage coupled with cover crops and N fertilization. In northern China, Su et al. (2004) compared the differences in soil properties in a cultivated and ungrazed area in a degraded grassland ecosystem. They found that cultivation had a significant negative impact on N and phosphorus (P) availability, soil biological activity, and soil enzyme activity. These effects were reversed by maintaining the soil with a permanent grassland cover, and they concluded that tillage has a detrimental impact on soil properties associated with high soil quality.

Lal (1997) proposed that soil degradation could be described as a function of soil properties (*S*), climate (*C*), terrain (*T*), vegetation (*V*), management (*M*), and time (*t*), as shown in Equation 2.1:

$$S_d = f(S, C, T, V, M)_t \tag{2.1}$$

He posited that these factors would interact and affect soil degradation. Soil properties, texture, and the structure of the soil are particularly affected, of which the latter can be greatly affected by SOM and biological activity. Climate factors include precipitation, temperature, evapotranspiration (ET), and seasonal patterns of these parameters. The terrain parameters of slope length and gradient along with slope aspect can determine how factors like soil temperature, vegetative growth, and soil water affect degradation. Slope shape is also likely to affect erosion and, conversely, deposition of eroded soil. The vegetative factors associated with soil degradation include the amount of ground cover and canopy height, species composition, and succession in native systems, which are related to SOC dynamics and nutrient cycling (Lal 1997). The cumulative effect of these different factors is thought to contribute to marginality in soils and the rate of change from a high-quality to a low-quality soil.

2.4 EXTENT OF SOIL DEGRADATION

Soil degradation is extensive throughout the world. It is a major threat to agricultural sustainability and environmental quality and is particularly serious in the tropics and subtropics (Lal 1993). For example, Nyssen et al. (2009) reported that nearly all of the tropical highlands (areas above 1000 m asl covering 4.5 million km^2) are degraded due to medium to severe water erosion. Zhao et al. (2007) evaluated the change from original pasture to cropland in the Horquin sands and found significant changes in crop yield and soil properties after conversion to cropland. In their study, soil parameters exhibited a change away from optimal values after the conversion. Kidron et al. (2010) suggested that the increasing pressure for food alleviating the traditional practice of 10 to 15 years of cultivation followed with 10 to 15 years of fallow with a continuous cropping practice has increased the rate of soil degradation. They found that SOM content showed the strongest relationship to soil degradation and practices which accelerated the removal of SOM increased the rate of degradation.

In the subhumid and semiarid Argentinean Pampas region, Buschiazzo et al. (1998) observed that intensive cultivation for over 50 years resulted in soil degradation leading to moderate to severe erosion. A similar conclusion was reached by dos Santos et al. (1993) for southern Brazil, in which they attributed the severe soil degradation to the widespread use of wheat (*Triticum aestivum* L.)–soybean (*Glycine max* L. Merr.) or barley (*Hordeum vulgare* L.)–soybean double cropping systems coupled with intensive tillage. Krzic et al. (2000) observed in the maritime climate of the Fraser Valley in British Columbia, with over 1200 mm of annual rainfall, that conventional tillage over a number of years contributed to poor infiltration, low organic matter content, and poor soil structure.

In southern Brazil and eastern Paraguay, Riezebos and Loerts (1998) observed that mechanical tillage resulted in a loss of SOM, leading to soil degradation across this region. The conversion of semideciduous forests to cultivated lands has the potential for soil degradation, and proper management will be required to avoid degradation. Degradation of the soil resource occurs in many different forms, and in Nepal, Thapu and Paudel (2002) observed that watersheds are severely degraded from erosion. They found that erosion has impacted nearly half of the land area in the upland crop terraces. This degradation was coupled with depletion of soil nutrients, which in turn is continuing to affect productivity in this area. This is similar to the observation in Ethiopia by Taddese (2001), where severe land degradation caused by the rapid population increase, severe soil erosion, low amounts of vegetative cover, deforestation, and a lack of balance between crop and livestock production will continue to threaten the ability to produce an adequate food supply for the population.

Wang et al. (1985) observed that differences in soil structure and saturated hydraulic conductivity were related to cropping systems and that degradation of soil structure in the profile led to corn yield reductions as large as 50%. Yield declines were related to the shallow root growth and limitations in water availability to the growing plant. Impacts of poor soil structure on plant growth and yield can be large, and continued degradation of the soil resource will have a major impact on the ability of the plant to produce grain, fiber, or forage.

Eickhout et al. (2006) stated that over the next 20 years, in order to meet food demand, there may have to be an additional clearing of forest land for production to offset the declining soil quality in the current land resource base. They advocated the need to consider N dynamics in current and future food production systems, increase our emphasis on N use efficiency, and focus on improvements in agronomic management to offset the impacts of soil degradation.

Soil degradation will have an effect on the marginality of soils and change adequate soils to poor soils and excellent soils to adequate soils. With these changes, there will be negative effects on crop production and crop production efficiency, rendering more risk into our ability to produce sufficient food, feed, and fiber for an ever-increasing world population.

2.4.1 Soil Management Practices

A critical component of soil degradation is the linkage between human soil management practices and the creation of conditions with potential for increased soil degradation. Soils in cultivated row crops are particularly vulnerable to loss of vegetative cover, residue removal, and tillage systems. Implementation of any tillage practice compared to a virgin soil or sod may lead to a decrease in the physical quality attributes of the soil (Reynolds et al. 2007). Reynolds et al. (2007) observed that converting bluegrass (*Panicum dichotomiflorum* Michx.) sod to a corn (*Zea mays* L.)–soybean rotation with moldboard plow (MP) caused the surface soil physical characteristics—for example, bulk density, macroporosity, air capacity, plant available water capacity (AWC), and saturated hydraulic conductivity—to decline within 3 to 4 years to levels similar to long-term corn–soybean with MP systems. Compared to virgin soil and sod systems, even the no-till (NT) system with the corn–soybean rotation showed declines in soil physical quality. Similar results were found by Evrendilek et al. (2004) in the Taurus Mountains of the southern Mediterranean region of Turkey. They evaluated the changes in SOC content and other physical soil properties over a 12-year period from three adjacent ecosystems, cropland (converted from grasslands in 1990), open forest, and grassland, in a Mediterranean plateau. Conversion of grassland into cropland increased the bulk density by 10.5% and soil erodibility by 46.2% and decreased SOM by 48.8%, SOC content by 43%, AWC by 30.5%, and total porosity by 9.1% for the 0- to 20-cm soil depth. They observed that SOC content was positively correlated with AWC, total porosity, mean weight diameter (MWD), forest, and grassland, and negatively correlated with bulk density, pH, soil erodibility, and cropland.

In the upper US Midwest, Olson's (2010) 20-year tillage study quantified the amount and rates of SOC storage and retention as a result of sod conversion to NT, chisel plow (CP), and MP tillage systems. The sloping (6%) and eroding soils in the plot area had been in sod for 15 years prior to the establishment of the tillage treatments. Although the NT and CP plots stored and retained 8.4 and 0.6 Mg C ha^{-1} more SOC in the soil than MP, no SOC sequestration occurred in any of the cultivated systems. The SOC level of the plot area was higher at the start of the experiment (before tillage) than at the end of the study, with NT plots losing a total of 6.8 Mg C ha^{-1}; the CP lost 15.1 Mg C ha^{-1}, and the MP lost 15.2 Mg C ha^{-1}. Olson (2010) concluded that after tillage, soil erosion and transport of SOC-rich sediment off of the sloping plots contributed substantially (29%–39%) to these SOC losses.

Surface sealing and soil crusting from raindrops on the soil consolidate the surface layers when crop residues are removed. After the surface seals, erosive forces increase, leading to sheet and rill erosion. Surface soil properties are sensitive to the maintenance of stable soil aggregates, which are created through the continual addition of organic material. To prevent surface sealing, it is necessary to maintain surface residue (Ruan et al. 2001). Tillage practices, for example, NT, which maintain crop residue on the surface and application of compost and manure as sources of organic materials, reduce surface sealing and crusting (Cassel et al. 1995; Pagliai et al. 2004). Blanco-Canqui et al. (2006a) observed in soils without residue cover that crusts with a thickness of 3 cm and cracks with widths of 0.6 cm were formed during periods with no rainfall. Removal of crop residue affects the stability of aggregates, which depends upon the maintenance of SOM concentration at the surface (Blanco-Canqui et al. 2006b; Rhoton et al. 2002). Changes in the soil structural stability are rapid, and often, degradation of the surface soils occurs within the first year after the residue is removed (Blanco-Canqui and Lal 2009). A regional study by Blanco-Canqui et al. (2009) comparing different tillage systems observed that aggregates under NT systems were more stable under rain but did not show any effect on dry aggregate stability. The formation of stable aggregates and exposure to the forces of raindrops under a rainfall event will lead to increased resistance to the soil erosion. Since erosion is one of the major causes of soil degradation, any change in a soil property to diminish the impact of intense rainfall events on the soil surface will have a positive effect on reducing soil erosion.

Tillage decreases the SOM content in the surface soil and creates conditions fostering a corresponding decrease in soil biological activity (Mahboubi and Lal 1998). Mahboubi and Lal (1998) found a seasonal response to tillage effects on aggregation and soil structure, suggesting that any assessment of tillage on soil properties must account for seasonality. Removal of crop residue from the soil surface decreased the soil microbial biomass C and N concentrations (Salinas-Garcia et al. 2001). Doran et al. (1998) observed a loss of soil C from three tillage systems in a Nebraska study; however, the loss from NT was less than from conventional tillage. In the NT system, they observed an increase in soil microbial activity near the soil surface. Karlen et al. (1994) concluded from a comparison of different systems that removal of crop residue caused soil aggregates to be less stable and decreased soil biological activity. The advantage of the NT system was the maintenance of the protective soil cover and partially decomposed organic material near the soil surface, which reduced the rate of soil degradation. Reeves (1997) concluded that maintenance of SOM is critical for soil quality. SOM is a critical component of the soil. Loveland and Webb (2003) reviewed the literature from around the world and concluded that when organic C declines below 2%, there will be a decline in soil quality. Papiernik et al. (2007) evaluated a glacial till landscape in west central Minnesota as affected by long-term tillage and found that areas of the field with greater than 20 Mg ha^{-1} year^{-1} erosion had shallower soil profiles and reduced inorganic C content. They also observed that SOC and total N in the cultivated areas were less than half of those in adjacent uncultivated areas. The effect of the long-term tillage (over 40 years) was to reduce the depth of the topsoil in the upper hillslope positions by at least 20 cm.

2.5 SCALES OF VARIATION

One of the critical questions scientists must address is how to assign a value to marginality in soils. If we maintain that marginality represents a loss of soil function, then any change from optimal function will have a negative impact. This concept and its implications can be illustrated by examining soil conditions in the context of water availability, crop productivity, and environmental quality, which determine the value of a degraded soil.

2.5.1 WATER AVAILABILITY

Water availability within soils is a function of the texture, SOM content, and depth. As the texture changes from sand to clay, there is an increase in the AWC, and these values can be found in any US county-level soil survey. There can be as much as a threefold difference in soil water holding capacity as the texture decreases from sand to clay soils. As Hudson (1994) pointed out, there is an increase in the available soil water with increasing organic matter content. Finally, the depth of the soil has a large influence since total available water depends upon the cumulative water holding capacity of the different soil layers. Water is central to the proper functioning of all biological activity, for example, plants, microbes, insects, and soil animals, and without water, these life forms would cease to exist. Water availability is complicated by the seasonality of precipitation and ET. If we consider that winter months have low ET rates, there is less water use from the soil compared to the summer, when ET is higher, causing the apparent effect of variable precipitation to be less noticeable on water availability.

Water availability impacts biological activity in the soil, and degraded soils will have a reduced biological activity. A method to evaluate water availability is to examine the soil water balance, as shown in Equation 2.2:

$$SW_i = SW_{i-1} + P_i - E_i - R_i - L_i + I_i \tag{2.2}$$

where SW_i is the soil water content on day i, SW_{i-1} is the soil water content on the previous day, P_i is the precipitation on day i, E_i is the evaporation on day i, R_i is the runoff on day i, L_i is the leaching from the bottom of the soil layer on day i, and I_i is the irrigation water applied on day i. This simple soil water balance model shows the impact of a degraded soil on the ability to maintain soil water content without increasing leaching or runoff during many precipitation events. This aspect of a marginal soil allows for further development of the subsequent discussion of crop productivity and environmental quality.

2.5.2 CROP PRODUCTIVITY

Water availability has a greater source of variability in corn (*Zea mays* L.) yields across different production fields in the Midwest United States than N application

Marginality Principle

rates (Hatfield 2012). An example of this variation is shown in Figure 2.4. Yield variation was related to the seasonal and cumulative patterns of soil water use from different soils within these production fields (Hatfield et al. 2007; Hatfield and Prueger 2011). Differences in cumulative water use were as large as 300 mm during the growing season of a corn crop and as much as 200 mm of transpiration (Hatfield and Prueger 2011). The primary factor causing these differences was soil water holding capacity driven by the SOM content of the different soils within the field. Within-field variations of soil properties have been found to suppress soil water holding capacity and diminish yield. These factors include soils with high sand and low clay content (James and Godwin 2003; Sandras et al. 2003; Jiang and Thelen 2004) and soils with excessive clay content and/or an impervious high argillic horizon (Kitchen et al. 1999, 2003). Differences across the landscape can be seen in the soil water holding capacity on eroded side slopes of an eroded Alfisol and can be 50%–60% less than summit and foot slope landscape positions (Jiang et al. 2008). Over the course of the growing season, separation in the ET among soils within the field was most pronounced during the grain-filling period, and the shortage of available soil water, except in the years with more than adequate rainfall during this period of the growing season, led to a reduction in water use efficiency (WUE) (Hatfield and Prueger 2011). Increasing variation in rainfall associated with climate change will change the seasonal patterns of rainfall and the soil water balance (Equation 2.2). Yield patterns within fields were dependent upon the total rainfall during the growing season, with the upper slope soils producing the largest yields in the years with above-normal rainfall and the lowest yields in years with less than normal rainfall (Jaynes and Colvin 1997; James and Godwin 2003). Sadler et al. (2000a,b) found a relationship between soil map units and grain yield from a field scale study on drought-stressed corn; however, these relationships did not explain the yield variation within the field. They suggested that improved understanding of yield variation would require more attention to within-season observations of crop water stress (soil

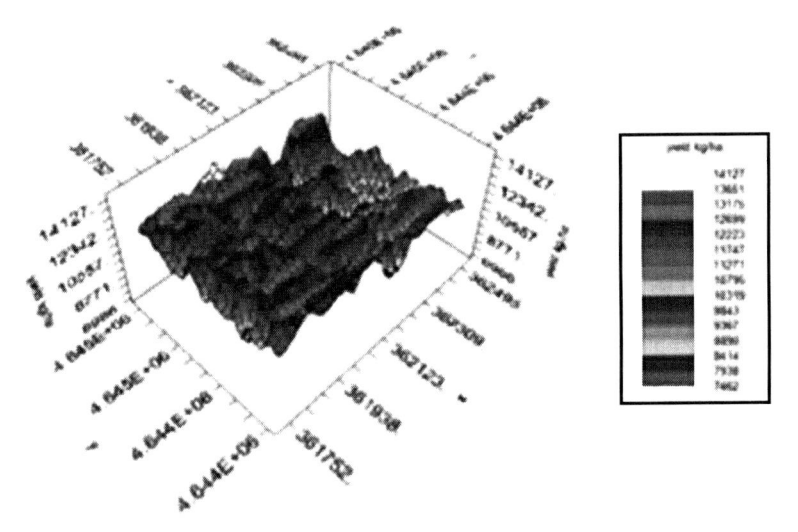

FIGURE 2.4 Yield variation across a corn production field in central Iowa.

water holding capacity, ET, and precipitation) and higher resolution of soil characterization, for example, soil texture, rooting depth, and compaction. The effect of soil topography was found to be a dominant factor by Kaspar et al. (2003) after they evaluated corn yields across a field in central Iowa. They observed in growing seasons with less than normal rainfall that there was a negative relationship relative to elevation, slope, and curvature, while in years with above-normal rainfall, there was a positive relationship with these terrain attributes. Kumhalova et al. (2011) found that yield and crop nutrient concentrations of small grains were spatially related to topography. The relationship between water flow accumulation and grain yield was strongest in the dry years and weak for the wetter years, similar to observations by Kitchen et al. (1999) and Kaspar et al. (2003). Observations by Hatfield and Prueger (2011) on water use differences within fields help explain differences observed in the other studies on yield variation within fields.

Soil water is a critical factor in plant growth, and any limitation in the soil to supplying optimum amounts of water at critical growth stages induces a limitation to plant growth, further reinforcing the concept of marginality in soils. Bouman (2007) addressed this problem in the form of an analysis of crop production systems and increasing crop water productivity with the overall goal of increasing food production and saving water. His approach was based on four principles, which can apply to maximizing the productivity of marginal soils. These principles were as follows: "1) increase transpirational crop water productivity; 2) increase the storage size for water in time and space; 3) increase the proportion of non-irrigation water inflows to the storage pool; and 4) decrease the non-transpirational water outflows of the storage pool" (Bouman 2007). His principle number two is a definition of a marginal soil and is related to yield variation as shown in the previous discussion. Since soil water is related to observed yield variation across fields in central Iowa, management practices oriented toward increasing crop productivity per unit amount of water will have to be based on increasing the size of the water storage pool in the soil profile. In other soils (e.g., claypan soils in central Missouri), where infiltration of water into the soil is a limitation, emphasis will be on increasing the movement of water into the storage pools in order to increase water availability. No single solution to increasing soil water availability in soils is evident; however, understanding the cause of this marginal response provides an opportunity for modification. Machado et al. (2000) showed that corn yield variation across a field was due to a combination of biotic and abiotic factors, and additional soil nitrate–N levels affected yields only when adequate soil water was available. Once soil water availability is quantified in spatial and temporal dimensions, the development of management practices to reduce the impact of these marginal soils can be implemented.

The value of soil water availability on crop yields is reinforced when yield data are transformed into monetary values. In Missouri, Massey et al. (2008) evaluated 10 years of site-specific yield data for corn, soybean, and grain sorghum (*Sorghum bicolor* L.) across a 36.5-ha field with claypan soils to quantify changes in profit and loss response. While some areas within the field were profitable almost every year, other areas of the field showed negative profit most years. Areas of the field with monetary loss were associated with field areas with significant loss of more than half of the original topsoil depth. Brock et al. (2005) observed that high-yielding management zones and profitable zones in a field under a corn–soybean rotation were associated with poorly drained level soils, while low-yielding zones were associated with eroded or more sloping soils. Over the past decades, the impact of landscape-dependent factors causing differences in water availability for growing crops and impacts on yield have been well documented and demonstrate the negative impact of soil degradation on soil water holding capacity (Spomer and Prest 1982; Jones et al. 1989; Wood et al. 1991; McConkey et al. 1997; McGee et al. 1997; Timlin et al. 1998).

Crop productivity is related to the ability of the crop to extract adequate water and nutrients from the soil profile. Limitations of rooting depth or inadequate soil water or nutrients will reduce growth and yield. Marginal soil areas within a field will vary depending upon the growing season because of the availability of soil water. Reducing the water holding capacity will have direct effects on crop water stress, and with the increasing uncertainty in precipitation timing and amounts coupled with increased atmospheric demand for water, the resulting effect will be an amplification of the negative effects of marginal soils on crop productivity.

2.5.3 Environmental Quality

Environmental quality is affected as a result of runoff or leaching of nutrients, sediment, or agricultural chemicals from agricultural fields or landscapes. Rainfall intensity exceeding the infiltration rate of the soil and the soil having limited capacity to store water, causing leaching through the soil carrying soluble nutrients, are key sources of reduction in environmental quality. It is difficult to separate environmental quality impacts from climate impacts because of the linkage between environmental quality and rainfall events. Under future climate scenarios, Gutowski et al. (2007) suggested that high-intensity precipitation events would constitute a larger fraction of total precipitation for all regions and seasons. Across the United States, precipitation patterns have shown regional variability over the past 50 years (Karl et al. 2009). The upper Northwest and Southeast United States have experienced declines in annual precipitation, while throughout the remainder of the United States, there has been variation even within individual states (Karl et al. 2009). Projections for North America for the period from 2040 to 2050 begin to show trends in the seasonal patterns of precipitation, with more established patterns by the end of the twenty-first century of increased spring precipitation and reduced summer precipitation (Karl et al. 2009). The projected decrease in summer precipitation across the United States into Canada has implications for agriculture since marginal soils with low water holding capacity are already exhibiting decreases in productivity. Increases in winter and spring precipitation for the upper portion of the United States and Canada would have potential environmental impacts because of the increased likelihood of surface runoff. Changes in precipitation events have already occurred and are expected to continue to increase throughout the remainder of this century (Kunkel et al. 1999).

The consequences of increases in rainfall intensity show that runoff and sediment movement from agricultural landscapes will increase (Nearing 2001). Increases in surface runoff lead to potential increases in sediment transport carrying herbicides and P from the surface. As one example of the potential impacts of changing climate,

Shipitalo and Owens (2006) observed that extreme events were responsible for a large amount of the herbicide loss from fields. Extreme events will link agricultural systems and the off-site impacts because of the potential impact of increased precipitation amounts. However, extreme events do not have to occur for water quality to be impacted; for example, Hatfield et al. (2009) showed that changes in nitrate concentrations in surface water in the Raccoon River watershed were related to changes in cropping patterns more than changes in precipitation. Bertol et al. (2007) observed that for soybean production in Brazil, there were positive relationships between P and potassium concentrations in runoff water as well as concentrations in sediments. Fullen and Brandsma (1995) observed positive linkages between erosion rates and textural changes, with a continual loss of smaller texture materials along with significant losses of both macronutrients and micronutrients from the topsoil. Their conclusion was that typical erosion rates for Europe are likely to have negative consequences on long-term soil fertility. The change in the textural composition of the soil also has implications for the water holding capacity.

Water quality impacts from farming systems vary. Jaynes et al. (1999) evaluated the surface water quality in a central Iowa watershed and found that herbicide concentrations of atrazine, alachlor, metolachlor, and metribuzin increased in surface water with runoff events, while nitrate decreased. Nitrate concentrations increased with increased subsurface drainage flow from the watershed. Jorden et al. (1997) observed that water impacts from agricultural landscapes increased with the intensity of agricultural practices. Similar results were found earlier by Mason et al. (1990) for different watersheds. Keeney and DeLuca (1993) concluded that intensity of tillage and subsequent N mineralization in the soil profile was the primary source of NO_3–N in the Raccoon River that offset the increase in commercial N fertilizer use in the 1970s. In a companion study across the Raccoon River watershed, Lucey and Goolsby (1993) concluded that climate variation and the subsequent amount of water movement from the watershed and fields were responsible for the temporal variation in nitrate concentrations in the surface water. Mason et al. (1990) observed similar results in watersheds across Wisconsin. Water quality is affected by the water balance in the soil profile and across the landscape. We conclude from these observations that marginal soils with reduced water holding capacity will have a greater likelihood of drainage through the soil profile or runoff from excess precipitation than will soils with an improved water holding capacity.

Soil erosion is related to soil C loss, and Gregorich et al. (1998) evaluated changes in soil C across a landscape and concluded that erosion affected C dynamics in the soil through the redistribution of soil across a landscape. They found that erosion was related to diminished primary productivity of the soil, which would have a long-term impact because of the reduced return of C to the soil. Owens et al. (2002) observed large differences in sediment and C losses from small watersheds in Ohio due to different tillage systems; however, their primary conclusion was that the greatest factor in reducing C movement from the watersheds was to reduce sediment movement. A subsequent study by Larney et al. (2009) on wheat found that over a 16-year period, grain yield declined 2.1% per centimeter of topsoil depth removed on the dryland site and only 1.7% per centimeter on the irrigated site. They concluded that the need to reduce erosion was paramount to avoiding declining productivity in soils.

2.6 MULTIFUNCTIONALITY: OPTIMUM PERFORMANCE OF SOIL RESOURCE

Marginal soils have marginal performance as evidenced by our examination of crop productivity, water availability, and environmental quality. However, there are other dimensions to this problem that need to be considered because of the role soils have in the ecosystem in providing a foundation for optimal primary productivity. Hatfield (2006) has shown that the multifunctionality of agricultural systems is dependent on a high-quality soil resource to optimize ecosystem services. A similar statement can be made about the role of soil in providing resilience to climate change. This aspect has not been fully investigated and needs additional research; however, adequate soil water is critical for optimum plant growth and stress avoidance. This can come only as a result of ensuring that soil water storage is maximized through the addition of SOC content.

2.6.1 Soil Management to Reduce Marginality

Conservation practices encompass a range of different systems employed to reduce the off-site impacts of agricultural systems and to ultimately enhance the natural resources. Delgado et al. (2011) recently reviewed the state of different practices capable of mitigating and adapting to climate change. They listed several examples of potential mitigation strategies including the following: "increasing soil C sequestration to improve soil function; reducing CH_4 emissions from ruminants; using slow release fertilizers, increasing N-use efficiencies for cropping systems; capturing nutrients and energy from manure, crop residues, and cover crop management; and using more efficient power sources and renewable energy." Specifically for soil and water practices, they identified a range of practices dealing with erosion, irrigation infrastructure, more diverse cropping systems, crop varieties more tolerant to drought and heat stress, synchronizing planting and harvesting with shifts in the hydrologic cycle, managing soil and crops to increase WUE, evaluating agricultural commodities for their water footprint and environmental traits, increasing soil C sequestration, increasing N-use efficiency, and implementing precision and targeted conservation practices to increase the effectiveness of practices to handle the increased temporal and spatial variation. These practices represent a range of potential ideas and concepts that, if implemented, would have a positive impact on the resilience of the production system to climate stresses and a positive feedback in terms of contributing to climate change mitigation.

In this process, it is critical to understand the feedback that occurs among all of the components in an agricultural system and their linkages. As an example, ET will increase with warming temperatures and cause crop water use rates to increase, leading to a faster depletion of the soil water reserves. In soils with limited soil water holding capacity, this will create variation in total crop water use and, ultimately, crop yield since it is a function of crop water use. Hatfield and Prueger (2011) found there was variation among soils within the same field, with a corresponding difference in yield between these two soils caused by the difference in soil water use throughout the growing season. The role of soil management on WUE was reviewed by Hatfield et al. (2001); they proposed that practices that increased water availability would lead to improved WUE. In a simple exercise using the WUE relationship developed for corn in the central United States, they observed that the additional amount of required soil water to increase the current level of yields of 10,000 to 18,000 kg ha^{-1} would require an additional 200 mm of soil water to be transpired through the plant during the growing season. If we follow the relationship developed by Hudson (1994) between soil water holding capacity and SOM, achieving sufficient change in the soil profile to result in this amount of extra water in the soil profile will require very aggressive soil management to increase soil organic levels.

This analysis can be extended further with respect to biofuel production because the observations from Hickman et al. (2010) and VanLoocke et al. (2010) suggest that the water use of switchgrass (*Panicum virgatum* L.) and *Miscanthus* (*Miscanthus giganteus*) is greater than that of corn. To achieve the projected yields for these crops to be viable bioenergy crops will require production systems with an increased amount of soil water availability. Increasing available soil water for transpiration will require a combination of techniques; these include increasing the soil water holding capacity, reducing soil water evaporation, and reducing drainage through the soil profile. The combination of strategies that will provide mitigation strategies and enhance adaptation approaches will require a focus on combining soil water and SOM dynamics. The role of conservation practices to enhance marginal soils is not as simple as characterizing their effectiveness by the degree of tillage or management of surface residues in order to improve SOC content. Interactions among soil water balance, soil temperature, soil biological activity, and gas exchange between the soil and the atmosphere are complex and vary by cropping system, soil, and climate (West and Marland 2002). These interactions prevent the development of a standard set of responses on how management of conservation practices will reduce climate change through mitigation. The foundation for soil management changes is related to the change in SOC. Follett (2001) attributed these changes to tillage and soil management systems, management to increase the amount of crop cover, and increased efficiency in the use of inputs (N and water) by the cropping system. Martens et al. (2005) provided a review on the role of soil management practices on the mitigation of greenhouse gas (GHG) emissions, and one of their conclusions was the observation that agricultural impacts on C cycling in soil needed improved understanding before the full potential of soil management practices as C mitigation strategies could be quantified. West and Marland (2002) stated that changing tillage practices could potentially cause a reduction in C emissions from agriculture of 368 kg C ha^{-1} year^{-1}, with variation among cropping systems and climates. Franzluebbers (2005, 2010) concluded that reductions in tillage intensity would increase the SOC content; however, the effects of changing tillage on other GHG emissions were less well defined. These studies direct us toward the conclusion that the role of soil management practices on increasing C storage in the soil profile will be understood only if we can quantify the effect on soil processes by altering soil management practices.

2.6.2 Agroforestry

One family of management practices with great potential and increasing usage as a practice to increase SOC is agroforestry. Agroforestry practices include field windbreaks, silvopastures, alley cropping, forest farming, and riparian buffers. These practices are innately resilient to climatic extremes as they routinely involve multiple perennial species that provide greater plant diversity and less vulnerability to climate stress than is provided by monocropping and annual species. The perennial woody vegetation also modifies the local microclimate by influencing airflow and light interception. Understory and adjacent plant species are thus protected from extremes in temperature and from damaging winds (Stigter 1988; Brenner 1996; Cleugh and Hughes 2002). Deeper root systems of the perennial vegetation afford greater resilience to drought and increased exploitation of soil water and nutrients from soil layers not accessible to more shallow rooted annual crops. Greater nutrient efficiency and WUE of species within agroforestry systems are a key strength and further enhance their potential as a mitigation practice under the uncertainties of climate change (Wallace 1996; Kho 2008).

Agroforestry practices also sequester C in soils. In a review of several Midwestern US sites, Paul et al. (2003) found SOC changes after tree planting of −0.07 to 0.55 Mg C ha^{-1} year^{-1} beneath deciduous trees and −0.85 to 0.58 Mg C ha^{-1} year^{-1} beneath coniferous trees. In another review, Post and Kwon (2000) reported SOC changes following tree planting, from small decreases in cool temperate pine plantings to increases of 3.0 Mg C ha^{-1} year^{-1} in wet subtropical plantings. Sauer et al. (2007) estimated an SOC change of 0.11 Mg C ha^{-1} year^{-1} for the surface 15 cm beneath a 35-year-old eastern red cedar (*Juniperus virginiana*)–Scotch pine (*Pinus sylvestris*) field windbreak in eastern Nebraska. Hernandez-Ramirez et al. (2011) used stable C isotope techniques on soil samples from the Sauer et al. (2007) study and a white pine (*Pinus strobus*) planting in Iowa to determine the source of the SOC found beneath the trees. Their source-partitioning analysis indicated that 53.9% and 47.1% of the SOC in the 7.5- and 10-cm surface layers at the Nebraska and Iowa locations, respectively, was tree derived. When the C sequestration in above- and below-ground tree biomass is also considered (Schroeder 1994; Kort and Turnock 1999; Schoeneberger 2009), agroforestry practices are likely to have C sequestration potential at least equivalent to the oft-cited practice of conversion from tilled to NT crop production (e.g., West and Post 2002).

Although agroforestry practices are inherently more resilient to climate-related stresses, rapidly increasing global food demand is creating intense pressure to produce more food, fiber, and fuel per unit of land area. To avoid losses in food production, agroforestry plantings must either provide a food source themselves (fruit, nut, or other edible produce) or provide ecosystem services (C sequestration, enhanced hydrology, or improved water quality) to compensate for increasing intensification of production practices on adjacent cropland.

Utilization of conservation practices for the mitigation of GHG in association with agricultural systems can reduce CO_2, CH_4, or N_2O emissions. CH_4 emissions from soil systems and conservation practices are relatively small compared to animal production systems including manure storage, while CO_2 and N_2O dynamics offer the potential for mitigation through management of plant and soil systems. The dynamics of these processes and the linkage to soil management practices have been recently reviewed by Hatfield et al. (2012).

2.7 SOCIAL VALUES OF SOIL FUNCTIONS: DEFINITIONS AND MEANINGS OF MARGINALITY

We began this chapter asking how social values and functions of soil influenced the meanings of marginality and society's willingness and capacities to protect soil quality. The processes of soil degradation from natural and anthropological sources have been well documented. Sustainable resource management is one of society's most complex concerns (Popp et al. 2002). Agriculture is the largest human use of land on the planet, with almost 38% of productive soils in climates well suited to farming (Foley 2011). The remainder, deserts, mountains, ice, tundra, cities and suburban places, parks, and other areas, are unsuitable for growing crops, leaving tropical forests and savannas as primary stores of C and biodiversity (Foley 2011). Forest clearing, farming of lands with soils and slopes not well suited to cultivation, and intensified farming of soils in sensitive landscapes present significant threats to long-term soil productivity and sustainability of ecosystem services. Social pressures to feed 7 billion people (projected to be 9 billion by 2050) and reduce food insecurity, predictions of changing and uncertain climate conditions and their impacts on agricultural production, and documented hypoxia zones in gulf and bay waters as the result of nutrient- and sediment-filled rivers are reminders that soil quality and its management underlie much of the human–social–biophysical value chain of systems. To meet world food demand, protecting and enhancing agricultural soils and their functional capacities is a clear priority. However, society is also increasingly realizing that there is a difficult and delicate balance among human uses of land for agriculture and ecosystem regulating functions that are necessary for human and biophysical wellbeing.

If agricultural landscapes are to be sustained into the future, purposeful land management decisions and policies are needed to assure that the multifunctional services of soil are available to meet agricultural productivity and valuable ecosystem services. Public policies and farmer decisions frequently give agricultural land productivity goals the highest social value, with protection of soil and ecosystem concerns ranking considerably lower. "While productivity-enhancement goals have largely been met over the last decades, lack of progress on the environmental side of the agricultural sustainability equation has created increasing levels of conflict over natural resource degradation" (Morton et al. 2013). In some regions, a growing number of farmers are uncertain about the connection between their farm management practices and sustainability of agriculture (Morton et al. 2013). A longitudinal random-sample panel survey of Iowa farmers revealed in 2002 that almost 29% of farmers were uncertain about whether sustainable farming practices helped to maintain their natural resource base, compared to 18.8% in 1989 (Morton et al. 2013). This suggests that there is an increasing disconnect between farmer perceptions of their current farming practices and how they affect long-term sustainability of their natural resource base, of which soil is primary.

Agricultural policy and on-farm management are public concerns since tax dollars are invested and external environmental benefits and liabilities derived from agriculture affect whole populations. There have been increasing levels of conflict and differences of opinions over the extent to which natural resource degradation is a problem among government, environmental groups, a variety of publics, and agricultural sectors (Morton and Brown 2007). In the United States, one trend is a growing expectation that government will pay for the external benefits of protecting the soil resource, and individual farmer confidence that they are doing a good job in conservation but others are not (Lasley 1993; Morton and Brown 2007; Comito and Helmers 2011). A 1993 Iowa survey found that farmers were most likely to believe that soil quality was declining worldwide (58%) but not on their own farm (9%); inversely, 56% reported that soil quality on their own farm was improving and only 8% reported that it was improving worldwide (Table 2.3) (Lasley 1993).

In regions with high portions of land in cultivated row crops, differences of viewpoints and the extent of soil degradation are centered on water quality concerns as the negative impacts of soil loss, sedimentation, and excessive P and N compounds on natural balances are widespread (USEPA 2009; Ribaudo 2011). A US national soil erosion service and system of funding farmers to help pay for soil management practices was established in the 1930s as a result of the Dust Bowl, which was caused by drought and cultivation of sensitive lands. More recently, water quality has emerged as a primary environmental concern with scientists finding that silt, nutrients, and chemicals delivered to water bodies by soil erosion are a principal cause of agricultural nonpoint source pollution (Morton and Brown 2011; USEPA 2009). In the Mississippi River Basin alone, US Geological Survey scientists have estimated that agricultural sources contribute more than 70% of the N and P delivered to the Gulf of Mexico via the Mississippi River and its tributaries (Alexander et al. 2008).

A 2011 environmental NGO report claims that the rich, dark soil of the Midwest corn belt, which made the region the nation's breadbasket, "is being swept away at rates many times higher than official estimates" (Cox et al. 2011). Further, they claim that although the *average* soil erosion in Iowa is officially estimated at 5.2 tons per acre, recent storms have resulted in significantly greater soil loss, estimating up to

TABLE 2.3
Results from the 1993 IFRLP on Views About Soil Quality

In general, would you say soil quality is

	Declining	Remaining the Same	Improving
Worldwide	58	34	8
In United States	31	41	28
In Iowa	21	35	44
In my county	17	35	48
On my farm	9	35	56

Source: Lasley, P., *Iowa Farm and Rural Life Poll 1993 Summary Report*, Department of Sociology, Iowa State University, Ames, IA. 1993.

64 tons of soil per acre lost in single events in some locations (Cox et al. 2011). Under climate change predictions, not only are global temperatures rising but also localized regional precipitation patterns are expected to continue to shift, creating drought in some places (increasing potential of wind erosion to soil when exposed) and more severe flooding in others, increasing the potential for greater soil losses. The soils of the world, with the biota they support, are major sources of C and function as absorbers, depositories, and releasers of organic C (Hillel and Rosenzweig 2011). Human preferences for particular uses of soils will influence the extent to which these soil functions mitigate climate by absorbing and storing C or become additional sources of CO_2, CH_4, and N_2O release.

2.7.1 Social Definitions of Marginality

The social meanings of soil quality and marginality can be broadly grouped into three categories: (1) expert soil scientist classifications based on soil properties; (2) interpretations of scientific classifications as codified and standardized in publically available materials such as soil surveys for use by professionals such as engineers, government agencies and technicians, extension educators, tax assessors, and real estate appraisers as well as landowners evaluating functional and market values for personal uses; and (3) soils made marginal or enhanced by climate and anthropological activities. Human values determine definitions and meanings, which guide decisions and actions (Morton and Padgitt 2005). Expert soil scientist classifications provide objective knowledge or facts, which can be used (or ignored) to make interpretations and judgments about soil quality and marginality. However, human decisions and applications are made outside of science. Science tells us the way the world is and what might happen under certain conditions; it cannot tell us what decision to make (Anderson-Wilk 2008).

The expected use and function of a particular parcel of land and its soil characteristics determines its value from a human point of view. However, there are considerable variations between scientists and nonscientists, between farmers and nonfarmers, and among farmers as to values associated with soils and potential uses. If the parcel does not have the "right" soils for the preferred use, it is likely to be considered marginal from a nonsoil-scientist point of view. Thus, a wetland has marginal soils if you wish to grow grain or vegetable crops, but if the goal is nutrient storage and cycling and wetland habitats, it is likely not to be considered marginal from one who values those services. A rocky gorge or steep mountainside might be considered marginal soils from a soil classification system but considered just right and not marginal at all for a public park, a recreational hiking trail, or a homeowner seeking a perfect vista.

Moisture holding capacity is one characteristic of soil quality highly valued by some farmers, followed closely by soil texture and visible organic matter (Table 2.4; Lasley 1993). There are publics that claim that rural countrysides are becoming agricultural industrial zones and ecological sacrifice areas and advocate the return of cultivated cropland to an original state such as forests, grasslands, and prairies in order to protect soil and ecosystem resources (Jackson and Jackson 2002). There are economic developers that envision vibrant suburban and urban housing or industrial

TABLE 2.4
Results from the 1993 IFRLP and the Value of Different Metrics of Soil Quality

When you judge the quality of soil, how important are the following characteristics?

	Not at All Important (%)	Not Very Important (%)	Moderately Important (%)	Very Important (%)
Moisture holding capacity	0	2	31	67
Compaction	1	4	36	59
Results of a soil test	0	4	37	59
Texture of the soil	1	2	44	53
Presence of earthworms	1	6	43	50
Visible organic matter	0	5	48	47
Amount of weed species	3	14	44	39
Crusting	1	15	51	33
Color of the soil	4	20	51	25
Smell of the soil	6	30	44	20

Source: Lasley, P., *Iowa Farm and Rural Life Poll 1993 Summary Report*, Department of Sociology, Iowa State University, Ames, IA. 1993.

complexes on agricultural lands, and there are engineers skilled in remaking the landscape building levees to protect against flooding and terraces to reduce soil loss. All of these people have their own version of which soils are likely to best match the uses they wish to put them to.

Shifts in precipitation patterns, long-term climate changes, and population growth place significant pressures on agriculture to efficiently and effectively provision society with food, fiber, and energy products, while protecting the environment. Lands with soil capacities to meet these highly valued uses are limited. Agricultural management practices play an important role in whether the soil resource is degraded and loses value or is enhanced in value. Farmer concepts of the risk to their crop productivity and the extent to which their soils are vulnerable to degradation are often different than soil scientists' assessments. Expert judgments of risk are highly correlated with technical estimates; layperson judgments of risk are sensitive to other factors such as catastrophic potential, controllability, and perceived threat to future generations and are often considerably different than expert risk estimates (Slovic 2009).

When a soil is redistributed to a new position in the landscape, it loses its valued functionality in its original location and may or may not provide value in the new location. Erosion of highly productive soils from the hilltop to the toe slope may retain the soil within field but is likely to redistribute the field capacity to produce a high-yielding crop. Similarly, high-quality soil lost to wind erosion or carried in runoff events is lost to one location, becoming sediment and possibly a fertile floodplain downstream, or may become buried in water, under sand, rocks, and lower-quality soils, with its redistribution severely limiting its ability to serve the functions of which it is capable. When soil is no longer able to produce a particular crop or to absorb moisture and support vegetation necessary for high ecosystem functioning, it becomes marginal.

It is marginal in that it no longer has the ability to provide the socially desired service-provisioning society or to facilitate ecosystem functions. This is not a new phenomenon. "The archeological record shows that some ancient people abused their soils and that their civilizations were disrupted by the ecological and environmental consequences" (Olson 1994). Good stewardship of the soil is one factor in continuing prosperity; overcutting forests, overplowing fields, and overgrazing of pastures and rangelands have accelerated erosion, diminishing crop yields and capacities to maintain a reasonable standard of living (Olson 1994). Loss of topsoil from mismatch of soil characteristics, crop choice, and management techniques on highly erodible lands as well as reduced biologically diverse soil-conserving rotations in favor of continuous corn or a corn–soybean rotation have created a quiet crisis in the world economy (Brown and Wolf 1984; Olson 1994). Unlike earthquakes or volcanic eruption, this crisis is largely human-made and unfolds gradually (Brown and Wolf 1984).

It is the gradual nature of soil loss that reduces the sense of risk. People will tolerate higher risks from activities they view as highly beneficial, and voluntariness of exposure is a key mediator of risk acceptance (Slovic 2009). The relationship between perceived risk, perceived benefit, and risk acceptance is influenced by familiarity, sense of control, and level of knowledge (Slovic 2009). Thus, a farmer sense of their soil's vulnerability and being at risk is likely to increase if they expect that an uncontrollable, external force such as extreme precipitation will result in field flooding and runoff accompanied by large quantities of soil loss. However, without the fear of a catastrophic event, agricultural management of soil becomes an unexamined, controllable routine decision in support of a highly beneficial activity, crop production, and earning a good livelihood. And it is exactly this sense of the ordinary that, without purposeful management, puts it at potential risk of losing its value and becoming marginal.

2.7.2 Willingness and Capacity to Protect Soil Resource

The suite of crops, structures, and management practices farmers select to achieve crop productivity and profitability often have unintended but critical consequences for soil resources. Farmer interpretations of the environmental outcomes and the meanings they give to the impacts of their management decisions on soil conditions and water quality can be barriers to addressing difficult and complex soil problems or sources of innovation and solutions (Comito et al. 2012). Farmer management of soil is a social process based on the social construction of the risks and vulnerabilities of weather and under what conditions soil might become increasingly marginal. The farmer's concern and sense of risk are central to their purposeful management of agricultural lands and willingness to respond and adapt to changing long-term weather patterns that make soil losses even more likely (Arbuckle et al. 2013).

Two routine decisions, tillage and management of the water balance, have critical implications for maintaining, enhancing, or reducing soil quality. These decisions are socially constructed, based on personal observations and experiences, the shared knowledge of other farmers, and exposure to scientific information (Coughenour and Chamala 2000). There are both analytical and affect heuristic components to these decisions. The experiential system relies on affect, a subtle form of emotion to make rapid, automatic evaluations to efficiently navigate a complex and uncertain world (Slovic 2009). The analytical component requires deliberative evaluation and purposeful mental processing of facts and information and is highly time consuming, so the farmer must believe that the time invested in reevaluating and changing current practices must have some benefit. Taken together, "…affective and emotional processes interact with reason-based analysis in all normal thinking and indeed are essential to rationality…" (Slovic 2009). Thus, the farmer selects crops and develops a production system based on his/her knowledge and experience with a particular field, its drainage conditions, the soil capacity to hold moisture, his/her existing stock of equipment, the current climate, and economic conditions (Coughenour and Chamala 2000). The extent to which the farmer defines increasing soil nutrients and improving soil structure for a particular crop as a goal will influence whether a conservation approach or conventional tillage system is used. Conservation tillage is defined as at least one-third of the previous crop residue remaining on the soil surface, compared to conventional tillage, where crop residue is removed or turned over. That is, if the land manager believes that the soil conditions are at risk in the short term (and possibly long term) of not being fully capable of producing the desired crop, there is a higher likelihood that some type of minimum or mulch tillage, no-tillage, stubble mulching, or other conservation tillage will be utilized (Coughenour and Chamala 2000). Lasley (1987) reports that there is some evidence that farmers perceive that conservation tillage provides considerable advantage in wind erosion control (64%) and reduction in soil erosion due to water runoff (65%); however, a much lower percentage of these same farmers perceive that there are considerable overall profitability (15%) or yield (3%) advantages (Lasley 1987). This suggests that farmers are likely to view conservation tillage as good for the environment and long-term retention of their soil but without much economic or production value.

A number of researchers have examined farmers' adoption of conservation best management practices (BMPs) and the factors that compel or constrain their capacity and willingness to protect soil (Prokopy 2008). A study by Kurzejeski et al. (1992) of Missouri land use intentions and attitudes about the USDA Conservation Reserve Program (CRP) found that 22.7% of farmers reported that their enrollment decision was based on soil erosion concerns, followed by profitable use of the land (18.4%), low risk associated with payments (14.9%), retirement (14.3%), wildlife habitat (11.9%), reduced labor/more free time (10.8%), and an easy way to meet conservation compliance (7%). About 49% included wildlife concerns as part of the mix of their enrollment decision. Hua et al. (2004) found that conservation tillage decisions by Ohio producers were mainly influenced by age, education, conservation compliance requirements, and attitudes, as well as owner–operators being more likely than renters to adopt practices. This finding is consistent with previous studies by other researchers' (Norris and Batie 1987; Lynne et al. 1988; Featherstone and Goodwin 1993) examination of Kansas farm-level economic and BMP adoption, which show that nutrient BMP (soil testing and split N applications) had positive economic effects on wheat and corn profitability, while soil conservation BMPs showed no significant impact on farm profitability.

Smith et al. (2007) found that a high proportion of Kansas producers were aware of federal programs (97% CRP, 80.5% Environmental Quality Incentives Program [EQIP], and 63.2% Conservation Stewardship Program [CSP]) but had a much lower participation rate (45% CRP, 31% EQIP). Further, they report that 98% of producers use one or more BMPs in farming, but only 36% use four or more on their farm. Smith et al. (2007) offer these observations: (1) Each farmer's adoption decision is a complex mix of attitudes and perceptions, farm operation characteristics, and external policy and economic forces, and (2) innovations to simplify or reduce producer time to enroll in federal programs would be a beneficial investment. A 2006 report to US senate (US GAO 2006) supports their second point, concluding that producers' main reasons for not participating in USDA conservation programs are federal government regulations and paperwork and concern that participation would constrain future production flexibility and decisions.

A meta-analysis by Prokopy et al. (2008) of 55 studies reports that there are no factors that consistently determine farmer BMP adoption of conservation technologies. However, they do find, in support of affect theories, that a number of social variables are significant: access to information, awareness, and social networks are more likely to have a positive rather than a negative relationship to BMP adoption (Prokopy et al. 2008). The synthesis results of Prokopy et al. (2008) are inconclusive about demographic factors that consistently determine BMP adoption but identify the following trends:

1. Capital and income are more positive than negative to BMP adoption, but most studies show insignificant findings for capital; income is never significant (either positive or negative).
2. Age has a negative relationship with adoption more often than positive. There is some evidence that the type of BMP matters. It is never significant for nutrient or water management.
3. Labor has a more positive than negative influence on adoption rates of most BMPs.
4. Diverse farms are more likely to adopt BMPS.
5. Networking (agency, business, and peer) is positive overall for BMPs; there are three negative studies associated with livestock BMP and landscape and soil management.
6. There is a positive attitude toward environment and always positive BMP adoption.
7. There is a predominantly positive relationship among attitude toward risk, profitability of practice, receiving adoption payments, and heritage. Risk variables lose significance at the 0.01 level, so it may not be a consistent driver of adoption.
8. Increases in environment awareness and knowledge are associated with an increase in adoption rates in general.
9. Farm characteristics. Farm location near a river was more often negatively associated with adoption. Animal farms are more likely to have a negative

relationship with adoption; a grain farmer is more likely to have positive relationship. Better soil quality seems to lead farmers to be more likely to practice conservation BMPs.
10. Synthesis results are inconclusive about which factors consistently determine BMP adoption.
11. Social factors appear to determine adoption: education levels, income, acres, capital, diversity, labor, and access to information generally lead to better adoption rates.
12. Social networks are complex to measure but seem to positively influence adoption of BMP rates.

Prokopy concludes that "…determinants of adoption of BMPs may well be different than determinants of adoption for the entire set of agricultural innovations" (Prokopy et al. 2008). This is consistent with the assessment by Coughenour and Chamala (2000) that conservation agriculture is different than adopting one management technology at a time, requiring a systems approach rather than solving one specific problem.

2.7.3 Capacity to Act: Whose Responsibility Is It to Protect the Soil?

Humans have invented the concept of risk to understand and cope with danger and uncertainties they face in life; dangers are real, but there is no such thing as a real risk or objective risk. All risks are subjective; assumptions and inputs of risk assessments depend on judgment (Slovic 2009). Scientific classifications and definitions of marginality aside, in practice, it is the land managers' perceptions and actions that determine the condition of the soil. It is a judgment call if a farmer perceives the soil they farm is at risk of losing its function and becoming marginal. Seasonal production decisions and market variability create a short-term crisis and opportunities that draw farmer and public policy attention. Significant losses of soil that affect production yields are often long-term losses that are less visible and compelling.

Agriculture is at a pivotal point in terms of societal demands for agricultural systems with improved sustainability—systems that address and balance social, economic, and environmental performance (NRC 2010). The challenge is to better match soil characteristics and climate conditions to social and biophysical functions in our working agriculture landscapes. Past uses and the complex matrix of decisions at the individual land manager and policy levels make it difficult to create agricultural systems to meet the many functional goals of society.

There are field-, farm-, and watershed-scale management practices in concert with public policies, which can prevent quality soils from becoming marginal, improve already marginal soils, and better match soil characteristics to ecosystem function needs. As science better understands the ecosystem roles of soils, concepts of marginality are likely to evolve. Lands considered marginal for crop production may not truly be marginal when they offer society other valued services such as regulating water flows; filtering, cycling, and storing nutrients; providing livelihoods from forest lands and grasslands; and detoxifying organic and inorganic materials. A critical question is how to improve the balance between the farm-level productivity and collective-level environmental functional goals. There are examples of working agricultural landscapes that integrate crop production and wildlife diversity areas that serve as diversions to manage floodplain flooding and ecosystem regulation, and reclaim soil redistributed from uplands (Olson et al. 2011). However, integrated practices that create a multifunctional agricultural landscape need many more innovations and widespread adoption, if soils are to fully meet their multifunctional capacities.

Agricultural public policy can no longer simply be about incentivizing greater production alone; if we are to protect the soil resource, policy must simultaneously address social, economic, and environmental sustainability issues. For this to happen, some level of agreement and support from farmers, environmentalists, agribusiness, and the general public will be needed. The Iowa Farm and Rural Life Poll (IFRLP), an annual random-sample, longitudinal survey of farmer beliefs, concerns, and current practices conducted in a state whose land and economy are dominated by row crop agriculture, illustrates what some farmers are thinking regarding agricultural policy and farm bill provisions over almost 30 years. The IFRLP reported in 1984 that Iowa farmers thought the highest priority of the 1985 Farm Bill should be developing new international markets (6.2 on a scale of 1–7) and lowering interest rates (6.1), with protecting agricultural natural resources ranking valued but lower in priority (5.8) (Lasley 1984). This pattern repeats itself in farmer concerns in 1988 and 10 years later in 1998, with concerns about prices for farm products and supplies ranking highest (6.3 in 1988, 6.4 in 1998) and soil erosion ranking sixth (5.7 in 1988, 5.7 in 1998) (Lasley 1998). This does not mean that farmers are not concerned with soil degradation; rather, it means that other more pressing issues convey greater risk and vulnerability to their enterprise. In Iowa, 83% of farmers supported in 1986 the 1985 Farm Bill establishment of a 10- to 15-year CRP with a goal to reduce soil erosion (and reduce production during a time of excess capacity) (Lasley 1986), and in 1987, 72% of Iowa farmers supported tying approved conservation plans and implementation to government payments (Lasley 1987). In 2005, 80% of Iowa farmers agreed or strongly agreed that CRP should be continued, and 56% thought the 2002 Farm Bill overall provided good support to funding conservation efforts (Korsching et al. 2005). To put this in perspective, in the same survey, 84% of farmers wanted more government effort promoting exports; 57% wanted more incentives to address environmental issues; and 87% thought the current government incentives encouraged farmers to reduce soil erosion (Korsching et al. 2005).

The 2009 IFRLP focused on conservation beliefs and attitudes in depth (Arbuckle et al. 2009). Seventy-six percent of Iowa farmers thought conservation programs should give conservation funding preference to lands that were most vulnerable to soil and water problems, and 74% supported targeting as a strategy for funding decisions. There was a high expectation (78%) that environmental groups in the next 10 years would have greater influence on farm policies. It is noteworthy that 66% thought soil erosion would very likely (22%) or somewhat likely (44%) decline because of new government programs. There was a great deal of confidence that "I" am doing enough and "my farm" will be capable of maintaining productivity over the next 25 years (77%). Only 7% strongly agreed that farmers should spend more money to control soil erosion. About 46% of 2011 IFRLP farmers reported currently participating in CRP (Arbuckle et al. 2011). Several patterns emerge over three decades of polling Iowa farmers: (1) Public investments in markets and price supports are higher priorities than soil erosion and conservation interventions. (2) Farmers expect that government should fund conservation and believe that these programs reduce soil erosion. (3) Farmers believe that they are doing a good job on their farm protecting soil resources. A majority of farmers (51%) in the 2011 IFRLP reported that they had incurred no conservation expenditure at all in the 10 years prior to the 2011 survey, with an additional 21% reporting spending less than $5000 (Arbuckle et al. 2011). This strongly suggests that while farmers support conservation initiatives funded by government, they are not willing to invest their own money. More research is needed to understand the reasons. Iowa farmers seem unconvinced that protecting the soil is their responsibility and that the long-term sustainability of their farm enterprise depends on it. Iowa farmers do not represent all farmers, and these results must be interpreted with caution. Beliefs may vary by systems of agriculture, crop choices, state or country locations, economic conditions, and a whole set of other factors. These findings do suggest that whoever controls the definition of risk controls the rational solution to the problem at hand. Soil degradation, which is considered a significant risk by soil scientists, may not be assigned a similar high risk level by agricultural land managers whose management practices affect soil quality and marginality. When farmers and public policy makers define risks to agriculture as short term and economic, then certain options will rise to the top as the most effective, the safest, or the best. If we define risk another way, we will likely get a different ordering of our action solutions (Slovic 2009). The challenge is to help policy makers and farmers to recognize that the quiet crisis at hand will eventually become a social crisis that compromises the soils' capacities to provision society and support a healthy ecosystem.

2.8 CONCLUSIONS

Soil is a fragile resource supplying many goods and services. Given the diversity of soil across the world and within a landscape, there are many different capacities among soils to provide the basic soil functions. Marginality of soils is a difficult process to define because the metrics to define a marginal soil depend upon the use, and if we utilize the framework of soil capability classes, then a marginal soil could be defined as one that moves from a higher to a lower class soil. The processes of soil degradation from a biological, physical, or chemical viewpoint contribute to the marginality of soils, and as a soil degrades, it will become more marginal. However, a view of marginality of soil will depend upon the intended use of the soil. Water availability in the soil is a critical component for crop productivity, and environmental quality and a degraded soil with reduced capacity to absorb, infiltrate, or retain water will provide marginal services for crop production and increase the likelihood of environmental quality concerns. Soil degradation is a major threat to soils around the world and limits the capacity of a given soil to support the different functions, of which providing food for humankind is foremost. Reversal of soil degradation by restoring organic matter and improving soil biological activity offsets the trend toward marginality in soils.

Marginality in soils is not a new concept because we know that degraded soils have contributed to societal problems throughout history. However, the social view of soil and soil management provides a different dimension to our view of marginal soils. The perception is that soils are more degraded on other areas than in areas being managed by an individual and that one's capability to manage and improve soil is better than that of others. The major problem is that the consequence of improper management and degradation is a gradual process rather than an immediate change, and this leads to systems in which we have grown tolerant of reductions in soil quality and trends toward marginality. Soil management practices need to renew their focus on eliminating soil erosion and enhancing soil biological function, with the goal of ultimately increasing soil quality. Through this combination of focusing on soil enhancement, we will be able to achieve the goal of being resilient to climate change and ensuring adequate food for an increasing population.

ABBREVIATIONS

AWC: available water capacity
C: carbon
CH_4: methane
CP: chisel plow
CRP: Conservation Reserve Program
CSP: Conservation Stewardship Program
EQIP: Environmental Quality Incentives Program
ET: evapotranspiration
GHG: greenhouse gas
IFRLP: Iowa Farm and Rural Life Poll
MP: moldboard plow
MWD: mean weight diameter
N: nitrogen
N_2O: nitrous oxide
NT: no till
P: phosphorus
SOC: soil organic carbon
SOM: soil organic matter
WUE: water use efficiency

REFERENCES

Alexander, R.B., R.A. Smith, G.E. Schwarz, E.W. Boyuer, J.V. Nolan, and J.W. Brakebill. 2008. Differences in phosphorus and nitrogen delivery to the Gulf of Mexico from the Mississippi River Basin. *Environ Sci Technol* 42(3):822–830.

Anderson-Wilk, M. 2008. Science and stewardship in a nonmonolithic conservation movement: facilitating positive change. *J Soil Water Conserv* 63:142A–146A.

Arbuckle, J.G., L.W. Morton, and J. Hobbs. 2013. Farmer beliefs and concerns about climate change and attitudes toward adaptation and mitigation. *Climatic Change* DOI 10.1007/s10584-013-0700-0.

Arbuckle, J.G., P. Lasley, P. Korsching, and C. Kast. 2009. *Iowa Farm and Rural Life Poll 2009 Summary Report*. Department of Sociology, Iowa State University, Ames, IA.

Arbuckle, J.G., P. Lasley, and J. Ferrell. 2011. *Iowa Farm and Rural Life Poll 2011 Summary Report*. Department of Sociology, Iowa State University, Ames, IA.

Bastida, F., Z. Zsolany, T. Hernandez, and C. Garcia. 2008. Past, present and future soil quality indices: a biological perspective. *Geoderma* 147:159–171.

Bertol, I., F.L. Engel, A.L. Mafra, O.J. Bertol, and S.R. Ritter. 2007. Phosphorus, potassium and organic carbon concentrations in runoff water and sediments under different tillage systems during soybean growth. *Soil Till Res* 94:142–150.

Blanco-Canqui, H., R. Lal, W.M. Post, R.C. Izaurralde, and L.B. Owens. 2006a. Corn stover impacts on near-surface soil properties of no-till corn in Ohio. *SSSAJ* 70:266–278.

Blanco-Canqui, H., R. Lal, W.M. Post, R.C. Izaurralde, and L.B. Owens. 2006b. Soil structural parameters and organic carbon in no-till corn with variable stover retention rates. *Soil Sci* 171:468–482.

Blanco-Canqui, H., M.M. Mikha, J.G. Benjamin, L.R. Stone, A.J. Schlegel, D.J. Lyon, M.F. Vigil, and P.W. Stahlman. 2009. Regional study of no-till impacts on near-surface aggregate properties that influence soil erodibility. *SSSAJ* 73:1361–1368.

Blanco-Canqui, H. and R. Lal. 2009. Crop residue removal impacts on soil productivity and environmental quality. *Crit Rev Plant Sci* 28:139–163.

Bouman, B.A.M. 2007. A conceptual framework for the improvement of crop water productivity at different spatial scales. *Agric Syst* 93:43–60.

Brenner, A.J. 1996. Microclimate modifications in agroforestry. pp. 159–187. In: Ong, C.K. and P. Huxley (ed.). *Tree-Crop Interactions*. CAB International. Wallingford, UK.

Brock, A., S.M. Brouder, G. Blumhoff, and B.S. Hoffman. 2005. Defining yield-based management zones for corn-soybean rotations. *Agron J* 97:1115–1128.

Brown, L.R. and E.C. Wolf. 1984. *Soil Erosion: Quiet Crisis in the World Economy*. Worldwatch Institute, Washington, D.C.

Buschiazzo, D.E., J.L. Panigatti, and P.W. Unger. 1998. Tillage effects on soil properties and crop production in the subhumid and semiarid Argentinean Pampas. *Soil Till Res* 49:105–116.

Cassel, D.K., C.W. Raczkowski, and H.P. Denton. 1995. Tillage effects on corn production and soil physical conditions. *SSSAJ* 59:1436–1443.

Cleugh, H.A. and D.E. Hughes. 2002. Impact of shelter on crop microclimates; a synthesis of results from wind tunnel and field experiments. *Aust J Exp Agric* 42:679–701.

Collins, S.L., S.R. Carpenter, S.M. Swinton, D.E. Orenstein, D.L. Childers, T.L. Gragson, N.B. Grimm, et al. 2011. An integrated conceptual framework for long-term social-ecological research. *Front Ecol Environ* 9(6):351–357.

Comito, J. and M. Helmers. 2011. The language of conservation. Chapter 6, pp. 67–80. In: Morton, L.W. and S. Brown (eds.). *Pathways for Getting to Better Water Quality: The Citizen Effect*. Springer Science+Business, London.

Comito, J., J. Wolseth, L.W. Morton. 2012. Tillage practices, the language of blame and responsibility for water quality impacts in cultivated row crop agriculture. *Human Ecology Review (Winter)* 19(2).

Coughenour, C.M. and S. Chamala. 2000. *Conservation Tillage and Cropping Innovation: Constructing the New Culture of Agriculture*. Iowa State University Press, Ames, Iowa.

Cox, C., A. Hug, and N. Bruzellius. 2011. *Losing Ground*. Environmental Working Group. www.ewg.org/losingground (accessed June 1, 2012).

Delgado, J.A., P.M. Groffman, M.A. Nearing, T. Goddard, D. Reicosky, R. Lal, N.R. Kitchen, C.W. Rice, D. Towery, and P. Salon. 2011. Conservation practices to mitigate and adapt to climate change. *J Soil Water Conserv* 66:118a–129a.

Doran, J.W. and T.B. Parkin. 1994. Defining and assessing soil quality. pp. 3–21. In: Doran J.W., D.C. Coleman, D.F. Bezdicek, and B.A. Stewart (eds.). *Defining Soil Quality for a Sustainable Environment*. SSSA Spec. Pub. No. 35, SSSA Argon., Madison, WI.

Doran, J.W., E.T. Elliott, and K. Paustian. 1998. Soil microbial activity, nitrogen cycling, and long-term changes in organic carbon pools as related to fallow tillage management. *Soil Till Res* 49:3–18.

Doran, J.W., D.C. Coleman, D.F. Bezdicek, and B.A. Stewart. 1994. *Defining Soil Quality for a Sustainable Environment*. SSSA Spec. Publ. No. 35, Soil Sci. Soc. Am., Inc. and Am. Soc. Agron., Inc., Madison, WI.

dos Santos, H.P., R.P. Zentner, F. Selles, and I. Ambrosi. 1993. Effect of crop rotation on yields, soil chemical characteristics, and economic returns of zero-till barely in southern Brazil. *Soil Till Res* 28:141–158.

Eickhout, B., A.F. Bouwman, and H. van Zeijts. 2006. The role of nitrogen in world food production and environmental sustainability. *Agric Ecosyst Environ* 116:4–14.

Evrendilek, F., I. Celik, and S. Kilic. 2004. Changes in soil organic carbon and other physical soil properties along adjacent Mediterranean forest, grassland, and cropland ecosystems. *J Arid Environ* 59:743–752.

Featherstone, A.M. and B.K. Goodwin. 1993. Factors influencing a farmers' decision to invest in long-term conservation improvements. *Land Econ* 69(1):67–81.

Foley, J.A. 2011. Can we feed the world and sustain the planet? *Sci Am* 305(5):60–65.

Follett, R.F. 2001. Soil management concepts and carbon sequestration in cropland soils. *Soil Till Res* 61:77–92.

Franzluebbers, A.J. 2005. Soil organic carbon sequestration and agricultural greenhouse gas emissions in the southeastern USA. *Soil Till Res* 83:120–147.

Franzluebbers, A.J. 2010. Achieving soil organic carbon sequestration with conservation agricultural systems in the southeastern United States. *SSSAJ* 74:347–357.

Fullen, M.A. and R.T. Brandsma. 1995. Property changes by erosion of loamy soils I east Shropshire, UK. *Soil Technol* 8:1–15.

Gregorich, E.G., K.J. Greer, D.W. Anderson, and B.C. Liang. 1998. Carbon distribution and losses: erosion and deposition effects. *Soil Till Res* 47:291–302.

Gutowski, W.J., Jr., E.S. Takle, K.A. Kozak, J.C. Paton, R.W. Arritt, and J.H. Christensen. 2007. A possible constraint on regional precipitation intensity changes under global warming. *J Hydrometeorol* 8:1382–1396.

Hamdar, B. 1999. An efficiency approach to managing Mississippi's marginal land based on conservation reserve program (CRP). *Resour Conserv Recyc* 26:15–24.

Hamza, M.A. and W.K. Anderson. 2005. Soil compaction in cropping systems: a review of the nature, causes and possible solutions. *Soil Till Resour* 82:121–145.

Hatfield, J.L. 2006. Multifunctionality of agriculture and farming system design: perspectives from the United States. *Bibliotecha Fragmenta Agronomica* 11:43–52.

Hatfield, J.L. 2012. Spatial patterns of water and nitrogen response within corn production fields. In: Aflakpui, G. (ed.). Agricultural Science. Intech Publishers, Rijeka, Croatia. ISBN 978-953-51-0567-1. pp. 73–96.

Hatfield, J.L. and J.H. Prueger. 2011. Spatial and temporal variation in evapotranspiration. pp. 3–16. In: Gerosa, G. (ed.). *Evapotranspiration—From Measurements to Agriculture and Environmental Applications*. Intech Publishers (www.intechopen.com).

Hatfield, J.L., T.J. Sauer, and J.H. Prueger. 2001. Managing soils for greater water use efficiency: a review. *Agron J* 93:271–280.

Hatfield, J.L., J.H. Prueger, and W.P. Kustas. 2007. Spatial and temporal variation of energy and carbon dioxide fluxes in corn and soybean fields in central Iowa. *Agron J* 99:285–296.

Hatfield, J.L. L.D. McMullen, and C.W. Jones. 2009. Nitrate-nitrogen patterns in the Raccoon River Basin as related to agricultural practices. *J Soil Water Conserv* 64:190–199.

Hatfield, J.L., T.B. Parkin, T.J. Sauer, and J.H. Prueger. 2012. Mitigation opportunities from land management practices in a warming world: increasing potential sinks. In: Liebig, M.A. and R.F. Follett (eds.). *Managing Agricultural Greenhouse Gases*. Academic Press, San Diego, CA.

Hernandez-Ramirez, G., T.J. Sauer, C.A. Cambardella, J.R. Brandle, and D.E. James. 2011. Carbon sources and dynamics in afforested and cultivated Corn Belt soils. *SSSAJ* 75:216–225.

Hickman, G.C., A. VanLooke, F.G. Dohleman, and C.J. Bernacchi. 2010. A comparison of canopy evapotranspiration for maize and two perennial grasses identified as potential bioenergy crops. *GCB Bioenergy* 2:157–168.

Hillel, D. and C. Rosenzweig. 2011. *Handbook of Climate Change and Agroecosystems: Impacts, Adaptation and Mitigation.* Hillel, D. and C. Rosenzweig (eds.). Imperial College Press, London.

Hua, W., C. Zulauf, and B. Sohngen. 2004. *Ohio Farmers' Conservation Decisions: 2004 Survey Results.* The Ohio State University AED Economics. AEDE-RP-0045-04.

Hudson, B.D. 1994. Soil organic matter and available water capacity. *J Soil Water Conserv* 49:189–194.

Kurzejeski, E.W., L.W. Burger, Jr., M.J. Monson, and R. Lenkner. 1992. Wildlife conservation attitudes and land use intentions of Conservation Reserve Program participants in Missouri. *Wildl Soc Bull* 20:253–259.

Jackson, D.L. and L.L. Jackson. 2002. *The Farm as Natural Habitat: Reconnecting Food Systems With Ecosystems.* Island Press, Washington, D.C.

James, I.T. and R.J. Godwin. 2003. Soil, water and yield relationships in developing strategies for the precision application of nitrogen fertilizer to winter barley. *Biosyst Eng* 84:467–480.

Jaynes, D.B. and T.S. Colvin. 1997. Spatiotemporal variability of corn and soybean yield. *Agron J* 89:30–37.

Jaynes, D.B., J.L. Hatfield, and D.W. Meek. 1999. Water quality in Walnut Creek Watershed: herbicides and nitrate in surface waters. *J Environ Qual* 28:45–59.

Jenny, H. 1941. *Factors of Soil Formation: A System of Quantitative Pedology.* p. 281. McGraw-Hill, New York.

Jiang, P. and K.D. Thelen. 2004. Effect of soil and topographic properties on crop yield in a North-Central corn-soybean cropping system. *Agron J* 96:252–258.

Jiang, P., N.R. Kitchen, S.H. Anderson, K.A. Sudduth, and E.J. Sadler. 2008. Estimating plant-available water using the simple inverse yield model for claypan landscapes. *Agron J* 100:830–836.

Jones, A.J., L.N. Mielke, C.A. Bartles, and C.A. Miller. 1989. Relationship of landscape position and properties to crop production. *J Soil Water Conserv* 44(4):328–332.

Jorden, T.E., D.L. Correll, and D.E. Weller. 1997. Relating nutrient discharges from watersheds to land use and streamflow variability. *Water Resour Res* 33:2579–2590.

Karl, T.R., J.M. Melillo, and T.C. Peterson. 2009. *Global Climate Change Impacts in the United States.* Cambridge University Press, Washington, D.C.

Karlen, D.L., N.C. Wollenhaupt, D.C. Erbach, E.C. Berry, J.B. Swan, N.S. Eash, and J.L. Jordahl. 1994. Crop residue effects on soil quality following 10 years of no-till corn. *Soil Till Res* 31:149–167.

Karlen, D.L., M.J. Mausbach, J.W. Doran, R.G. Cline, R. F. Harris, and G. E. Schuman. 1997. Soil quality: concept, rationale, and research needs. *Soil Sci Soc Am J* 61:4–10.

Kaspar, T.C., T.S. Colvin, D.B. Jaynes, D.L. Karlen, D.E. James, D.W. Meek, D. Pulido, and H. Butler. 2003. Relationship between six years of corn yields and terrain attributes. *Prec Agric* 4:87–101.

Keeney, D.R. and T.H. DeLuca. 1993. Des Moines river nitrate in relation to watershed agricultural practices: 1945 versus 1980s. *J Environ Qual* 22:267–272.

Kho, R.M. 2008. Approaches to tree-environment-crop interactions. pp. 51–72. In Batish, D.R., R.K. Kohli, S. Jose, and H.P. Singh (eds.). *Ecological Basis of Agroforestry.* CRC Press, Boca Raton, FL.

Kidron, G.J., A. Karnieli, and I. Benenson. 2010. Degradation of soil fertility following cycles of cotton-cereal cultivation in Mali, West Africa: a first approximation to the problem. *Soil Till Res* 106:254–262.

Kitchen, N.R., K.A. Sudduth, and S.T. Drummond. 1999. Soil electrical conductivity as a crop productivity measure for claypan soils. *J Prod Agric* 12(4):607–617.

Kitchen, N.R., S.T. Drummond, E.D. Lund, K.A. Sudduth, and G.W. Buchleiter. 2003. Soil electrical conductivity and topography related to yield for three contrasting soil-crop systems. *Agron J* 95:483–495.

Korsching, P., P. Lasley, and T. Gruber. 2005. *Iowa Farm and Rural Life Poll 2005 Summary Report*. Department of Sociology, Iowa State University, Ames, IA.

Kort, J. and R. Turnock. 1999. Carbon reservoir and biomass in Canadian prairie shelterbelts. *Agroforest Syst* 44:175–186.

Krzic, M., M. Fortin, and A.A. Bomke. 2000. Short-term responses of soil physical properties to corn planting-tillage systems in a humid maritime climate. *Soil Till Res* 54:171–178.

Kumhalova, J., F. Kumhala, M. Kroulik, and S. Matejkova. 2011. The impact of topography on soil properties and yield and the effects of weather conditions. *Prec Agric* 12:813–830.

Kunkel, K.E., K. Andsager, and D.R. Easterling. 1999. Long-term trends in extreme events over the conterminous United States and Canada. *J Clim* 12:2515–2527.

Lal, R. 1993. Tillage effects on soil degradation, soil resilience, soil quality, and sustainability. *Soil Till Res* 27:1–8.

Lal, R. 1997. Degradation and resilience of soils. *Philos Trans R Soc (Biology) London* 352:997–1010.

Lal, R., T.M. Sobecki, T. Iivari, and J.M. Kimble. 2004. *Soil Degradation in the United States: Extent, Severity, and Trends*. Lewis Publishers, Boca Raton, FL.

Larney, F.J., H.H. Janzen, B.M. Olson, and A.F. Olson. 2009. Erosion-productivity-soil amendment relationships for wheat over 16 years. *Soil Till Res* 103:73–83.

Lasley, P. 1984. *Iowa Farm and Rural Life Poll 1984 Summary Report*. Department of Sociology, Iowa State University, Ames, IA.

Lasley, P. 1986. *Iowa Farm and Rural Life Poll 1986 Summary Report*. Department of Sociology, Iowa State University, Ames, IA.

Lasley, P. 1987. *Iowa Farm and Rural Life Poll 1987 Summary Report*. Department of Sociology, Iowa State University, Ames, IA.

Lasley, P. 1993. *Iowa Farm and Rural Life Poll 1993 Summary Report*. Department of Sociology, Iowa State University, Ames, IA.

Lasley, P. 1998. *Iowa Farm and Rural Life Poll 1998 Summary Report*. Department of Sociology, Iowa State University, Ames, IA.

Loveland, P. and J. Webb. 2003. Is there a critical level of organic matter in the agricultural soils of temperate regions? a review. *Soil Till Res* 70:1–18.

Lucey, K.J. and D.A. Goolsby. 1993. Effects of climatic variations over 11 years on nitrate–nitrogen concentrations in the Raccoon River, Iowa. *J Environ Qual* 22:38–46.

Lynne, G.D., J.S. Shonkwiler, and L.R. Rola. 1988. Attitudes and farmer conservation behavior. *Am J Agric Econ* 70(1):12–19.

Machado, S., E.D. Bynum, Jr., T.L. Archer, R.J. Lascano, L.T. Wilson, J. Bordovsky, E. Segarra, K. Bronson, D.M. Nesmith, and W. Xu. 2000. Spatial and temporal variability of corn grain yield: site-specific relationships of biotic and abiotic factors. *Prec Agric* 2:359–376.

Mahboubi, A.A. and R. Lal. 1998. Long-term tillage effects on changes in structural properties of two soils in central Ohio. *Soil Till Res* 45:107–118.

Martens, D.A., W. Emmerich, J.E.T. McLain, and T.N. Johnsen. 2005. Atmospheric carbon mitigation potential of agricultural management in the southwestern USA. *Soil Till Res* 83:95–119.

Mason, J.W., G.D. Wegner, G.I. Quinn, and E.L. Lange. 1990. Nutrient loss via groundwater discharge from small watersheds in western and south-central Wisconsin. *J Soil Water Conserv* 45:327–331.

Massey, R.E., D.B. Myers, N.R. Kitchen, and K.A. Sudduth. 2008. Profitability maps as an input for site-specific management decision making. *Agron J* 100:52–59.

McConkey, B.G., D.J. Ullrich, and F.B. Dyck. 1997. Slope position and subsoiling effects on soil water and spring wheat yield. *Can J Soil Sci* 77:83–90.

McGee, E.A., G.A. Peterson, and D.G. Westfall. 1997. Water storage efficiency in no-till dryland cropping systems. *J Soil Water Conserv* 52:131–136.

Morton, L.W. and S. Padgitt. 2005. Selecting socio-economic metrics for watershed management. *Environ Monit Assess* 103:83–98.

Morton, L.W. and S.S. Brown. 2011. *Pathways for Getting to Better Water Quality: The Citizen Effect*. Springer Science+Business, New York.

Morton, L.W. and S.S. Brown. 2007. *Water Issues in the Four-State Heartland Region: A Survey of Public Perceptions and Attitudes about Water (Iowa, Kansas, Missouri and Nebraska)*. Technical Report SP 289 Iowa State University, Department of Sociology, Ames, IA.

Morton, L.W., J. Hobbs, and J.G. Arbuckle. 2013. Shifts in farmer uncertainty over time about sustainable farming practices and modern farming reliance on commercial fertilizers, insecticides and herbicides. *J Soil Water Conserv* 68(1):1–12.

Nearing, M.A. 2001. Potential changes in rainfall erosivity in the U.S. with climate change during the 21st century. *J Soil Water Conserv* 56:229–232.

Norris, P.E. and S.S. Batie. 1987. Virginia farmers' soil conservation decisions: an application of TOBIT analysis. *SJAE* 19(1):79–90.

NRC (National Research Council). 2010. Toward Sustainable Agricultural Systems in the 21st Century. The National Academies Press. Washington, D.C.

Nyssen, J., J. Poesen, and J. Deckers. 2009. Land degradation and soil and water conservation in tropical highlands. *Soil Till Res* 103:197–202.

Olson, K.R., M. Reed, and L.W. Morton. 2011. Multifunctional Mississippi river leveed bottomlands and settling basins: Sny Island Levee Drainage District. *J Soil Water Conserv* 66(4):90A–96A.

Olson, K.R. 1994. Effects of soil formation, erosion, and management on long-term productivity. In: McIsaac, G. and W.R. Edwards (eds.). *Sustainable Agriculture in the American Midwest: Lessons From the Past, Prospects for the Future*. University of Illinois Press, Urbana, IL.

Olson, K.R. 2010. Impacts of tillage, slope, and erosion on soil organic carbon retention. *Soil Sci* 175:562–567.

Owens, L.B., R.W. Malone, D.L. Hothem, G.C. Starr, and R. Lal. 2002. Sediment carbon concentration and transport from small watersheds under various conservation tillage practices. *Soil Till Res* 67:65–73.

Pagliai, M., N. Vignozzi, and S. Pellegrini. 2004. Soil structure and the effect of management practices. *Soil Till Res* 79:131–143.

Papiernik, S.K., M.J. Lindstrom, T.E. Schumacher, J.A. Schumacher, D.D. Malo, and D.A. Lobb. 2007. Characterization of soil profiles in a landscape affected by long-term tillage *Soil Till Res* 93:335–345.

Paul, E.A., S.J. Morris, J. Six, K. Paustian, and E.G. Gregorich. 2003. Interpretation of soil carbon and nitrogen dynamics in agricultural and afforested soils. *Soil Sci Soc Am J* 67:1620–1628.

Popp, J., D. Hoag, and J. Ascough II. 2002. Targeting soil-conservation policies for sustainability: new empirical evidence. *J Soil Water Conserv* 57(2):66–74.

Post, W.M. and K.C. Kwon. 2000. Soil carbon sequestration and land-use change: processes and potential. *Glob Change Biol* 6:317–328.

Prokopy, L.S. 2008. Understanding farmer adoption of agricultural best management practices. *J Soil Water Conserv* 63(5):169A.

Prokopy, L.S., K. Floress, D. Klotthor-Weinkauf, and A. Baumgart-Getz. 2008. Determinants of agricultural best management practice adoption: evidence from the literature. *J Soil Water Conserv* 63(5):300–311.

Reeves, D.W. 1997. The role of soil organic matter in maintaining soil quality in continuous cropping systems. *Soil Till Res* 43:131–167.

Reynolds, W.D., C.F. Drury, Y.M. Yang, C.A. Fox, C.S. Tan, and Q.T. Zhang. 2007. Land management effects on the near-surface physical quality of a clay loam soil. *Soil Till Res* 96:316–330.

Rhoton, F.E., M.J. Shipitalo, and D.L. Lindbo. 2002. Runoff and soil loss from midwestern and southeastern US silt loam soils as affected by tillage practices and soil organic matter content. *Soil Till Res* 66:1–11.

Ribaudo, M. 2011. Reducing agriculture's nitrogen footprint: are new policy approaches needed? Amber Waves, U.S. Government Printing Office.

Riezebos, H.TH. and A.C. Loerts. 1998. Influence of land use change and tillage practice on soil organic matter in southern Brazil and eastern Paraguay. *Soil Till Res* 49:271–275.

Ruan, H.X., L.R. Ahuja, T.R. Green, and J.G. Benjamin. 2001. Residue cover and surface-sealing effects on infiltration: numerical simulations for field applications. *Soil Sci Soc Am J* 65:853–861.

Sadler, E.J., P.J. Bauer, and W.J. Busscher. 2000a. Site-specific analysis of a droughted corn crop: I. Growth and grain yield. *Agron J* 92:395–402.

Sadler, E.J., P.J. Bauer, W.J. Busscher, and J.A. Miller. 2000b. Site-specific analysis of a droughted corn crop: II. Water use and stress. *Agron J* 92:403–410.

Sainju, U.M., B.P. Singh, and W.F. Whitehead. 2002. Long-term effects of tillage cover crops, and nitrogen fertilization on organic carbon and nitrogen concentrations in sandy loam soils in Georgia, USA. *Soil Till Res* 63:167–179.

Salinas-Garcia, J.R., A.D. Baez-Gonzalez, M. Tiscareno-Lopez, and E. Rosales-Robles. 2001. Residue removal and tillage interactions effects on soil properties under rain-fed corn production in Central Mexico. *Soil Till Res* 59:67–79.

Sandras, V., J. Baldock, D. Roget, and D. Rodriguez. 2003. Measuring and modelling yield and water budget components of wheat crops in coarse-textured soils with chemical constraints. *Field Crops Res* 84:241–260.

Sauer, T.J., C.A. Cambardella, and J.R. Brandle. 2007. Soil carbon and tree litter dynamics in a red cedar-scotch pine shelterbelt. *Agroforest Syst* 71:163–174.

Schoeneberger, M.M. 2009. Agroforestry: working trees for sequestering carbon on agricultural lands. *Agroforest Syst* 75:27–37.

Schroeder, P. 1994. Carbon storage benefits of agroforestry systems. *Agroforest Syst* 27:89–97.

Shipitalo, M.J. and L.B. Owens. 2006. Tillage system, application rate, and extreme event effects on herbicide losses in surface runoff. *J Environ Qual* 35(6):2186–2194.

Slovic, P. 2009. *The Perception of Risk*. Earthscan Publications, Ltd., Sterling, Va.

Smith, C.M., J.M. Peterson, and J.C. Leatherman. 2007. Attitudes of Great Plains producers about best management practices, conservation programs and water quality. *J Soil Water Conserv* 62(5):97A–105A.

Spomer, R.G. and R.F. Prest. 1982. Soil productivity and erosion of Iowa loess soils. *Trans ASAE* 25:1295–1299.

Stigter, C.J. 1988. Microclimate management and manipulation in agroforestry. pp. 145–168. In: K.F. Wiersum (ed.). *Viewpoints on Agroforestry. Second Renewed Edition*. Agricultural University, Wageningen.

Su, Y., H. Zhao, T. Zhang, and X. Zhao. 2004. Soil properties following cultivation and non-grazing of a semi-arid sandy grassland in northern China. *Soil Till Res* 75:27–36.

Taddese, G. 2001. Land degradation: a challenge to Ethiopia. *Environ Manage* 27:815–824.

Thapu, G.B. and G.S. Paudel. 2002. Farmland degradation in the mountains of Nepal: a study of watersheds "with" and "without" external intervention. *Land Degrad Dev* 13:479–493.

Timlin, D.J., Y. Pachepsky, V.A. Snyder, and R.B. Bryant. 1998. Spatial and temporal variability of corn grain yield on a hillslope. *Soil Sci Soc Am J* 62:764–773.

US GAO (US Government Accountability Office). 2006. USDA Conservation Programs: Stakeholder Views on Participation and Coordination to Benefit Threatened and Endangered Species and Their Habitats. Washington, D.C.: USGAO http://www.gao.gov/new.items/.

USDA-NRCS. 2012a. soils.usda.gov/gallery/photos/profiles (accessed July 13, 2012).

USDA-NRCS. 2012b. photogallery.nrcs.usda.gov (accessed July 13, 2012).

USEPA National 2010 Water Quality Inventory Report to Congress, http://www.epa.gov/305b (Retrieved 2011).

Vanderpol, F. and B. Traore. 1993. Soil nutrient depletion by agricultural production in Southern Mali. *Fert Res* 36:79–90.

VanLoocke, A., C.J. Bernacchi, and T.E. Twines. 2010. The impacts of *Miscanthus* x *giganteus* production on the Midwest US hydrologic cycle. *GCB Bioenergy* 2:180–191.

Wallace, J.S. 1996. The water balance of mixed tree-crop systems. pp. 189–233. In: Ong, C.K. and P. Huxley (eds.). *Tree-Crop interactions*. CAB International, Wallingford, UK.

Wang, C., J.A. McKeague, and K.D. Switzer-Howse. 1985. Saturated hydraulic conductivity as an indicator of structural degradation in clayey soils of Ottawa area, Canada. *Soil Till Res* 5:19–31.

West, T.O. and G. Marland. 2002. A synthesis of carbon sequestration, carbon emissions, and net carbon flux in agriculture: comparing tillage practices in the United States. *Agric Ecosyst Environ* 91:217–232.

West, T.O. and W.M. Post. 2002. Soil organic carbon sequestration rates by tillage and crop rotation: a global data analysis. *Soil Sci Soc Am J* 66:1930–1946.

Wood, C.W., G.A. Peterson, D.G. Westfall, C.V Cole, and W.O. Willis. 1991. Nitrogen balance and biomass production of newly established no-till dryland agroecosystems. *Agron J* 83:519–526.

Zhao, H.L., J.Y. Cui, R.L. Zhou, T.H. Zhang, X.Y. Zhao, and S. Drake. 2007. Soil properties, crop productivity and irrigation effects on five croplands of Inner Mongolia. *Soil Till Res* 93:346–355.

3 Principles of Soil Management in Neotropical Savannas
The Brazilian Cerrado

Yuri L. Zinn and Rattan Lal

CONTENTS

3.1 Neotropical Savannas and the Cerrado 58
    3.1.1 Main Soils of the Cerrado and Soil Formation Factors 60
    3.1.2 Oxisols 65
    3.1.3 Quartzipsamments and Sandy Oxisols 66
    3.1.4 Ultisols 67
    3.1.5 Other Relevant Soil Types 67
3.2 Principal Land Use and Soil Management Systems in the Cerrado 69
    3.2.1 Basic Tenets of Soil Management in the Cerrado 70
        3.2.1.1 Clearance of Native Vegetation and Correction of Soil Acidity 70
        3.2.1.2 Fertilization 70
        3.2.1.3 Soil Tillage and Seedbed Preparation 71
        3.2.1.4 Crop Rotations 71
    3.2.2 Pastures 72
    3.2.3 Annual Crops 74
    3.2.4 Perennial Crops 76
    3.2.5 Plantation Forests 77
    3.2.6 Biofuel Crops 78
    3.2.7 Crop/Livestock Integrations 79
3.3 Future Trends: Land Use, Research, and Environment 80
3.4 Conclusions 82
Acknowledgments 82
Abbreviations 82
References 82

3.1 NEOTROPICAL SAVANNAS AND THE CERRADO

Savannas are major land ecosystems comprising a global extent that has been estimated to vary between 15 and 37 million (M) km^2 (House and Hall 2001). This broad range of area reflects the difficulties in defining what exactly is a savanna vegetation, not only due to its wide variability in plant diversity and degree of tree coverage but also because savannic areas shift considerably in decadal time frames due to changes in regional rainfall and land use. In the context of this article, a savanna is defined as a biome in which different combinations of grass, shrub, and tree strata occur, either as a roughly homogeneous mix in the same area or as a dominantly grassy landscape where shrubs and trees are sparse, or concentrated on patches or drainageways. Savannas occur in large areas in Africa and the Americas, but also in Southeast Asia and Australia. The word "neotropical" is traditionally used to refer to the tropics of the "New World," especially South America. Neotropical savannas differ considerably from their African counterparts, more notably by a relatively more humid climate, high plant, and animal biodiversity (Grace et al. 2006) but paucity of large terrestrial animals, and a sparser human occupation (House and Hall 2001; Romero-Ruiz et al. 2012).

Neotropical savannas are estimated to cover 2.69 M km^2, which can be broadly divided into two major areas and a group of much smaller, isolated occurrences (Rippstein et al. 2001). These two main savannas consist of the *Cerrado* of central Brazil and the *Llanos* of Venezuela and Colombia, comprising, respectively, 76% and 17% of the total neotropical savanna area. The remaining 7% consists of smaller formations in the Caribbean, Central America, Amazon basin, and other regions of South America. These isolated savannas evolved under disparate environmental settings, and most do not support intensive agriculture presently and are sparsely studied. For example, the savannas of the inner Amazon basin are so remote and poorly known that earlier aerial images of them were interpreted as clear-felled areas. On the other hand, the Llanos (from the Spanish word for "plains") of the Orinoco basin has for centuries been used as natural pastures, the only land use compatible with the low-fertility, poorly drained soils and the excessive rain in some areas. Pioneering centers for agriculture research, such as Centro Internacional de Agricultura Tropical (CIAT), were established in the region and have done extensive research to study the Llanos environment and stimulate crop production. However, significant infrastructure improvements have only recently been established to support oil production and also intensive annual cropping (Rippstein et al. 2001). As much as 14% of Colombian flooded savannas were converted to croplands, exotic pastures, and oil palm (*Elaeis guineensis* Jacq.) plantations between 1990 and 2010, and 70% of the eastern Llanos is expected to be converted to agriculture in the near future (Romero-Ruiz et al. 2012). Despite recent developments and a solid base of scientific research, the full agricultural potential of the region is yet to be realized, and the predominant land use is still low-density cattle ranching (Rondón et al. 2006).

In contrast, the Brazilian Cerrado (also known as *Cerrados*) covers a contiguous land area of 2.04 M km^2 (Figure 3.1), an area slightly larger than that of Mexico. It is the most important savanna ecosystem for agriculture in the world not only due to its areal extent, favorable climate and soil but also because of recent state-induced

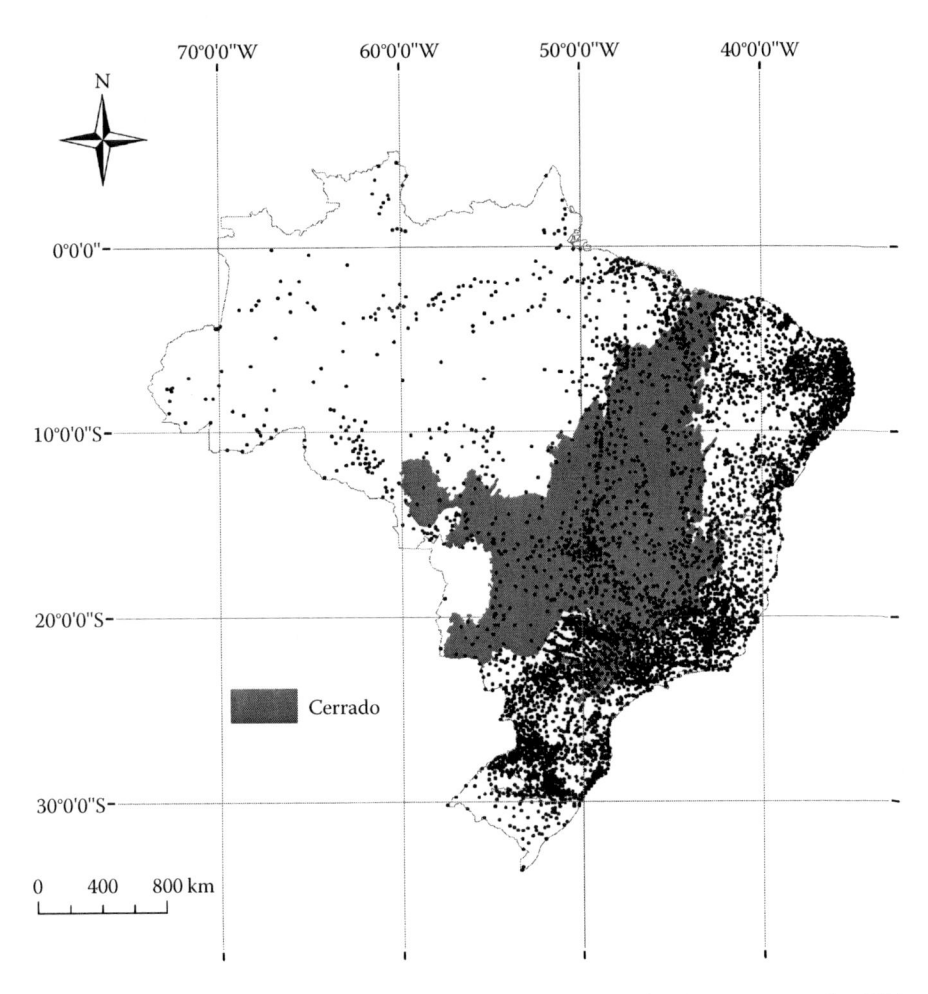

Principles of Soil Management in Neotropical Savannas

FIGURE 3.1 Map showing the Cerrado biome in central Brazil. Dots represent the 5584 cities of Brazil, mostly located in other biomes. Many cities in the Cerrado are concentrated in the core area around the capitals Brasília and Goiânia, but human occupation in the Cerrado is higher only than that in the Amazon basin. (From data available at www.ibge.gov.br.)

development, such as the transfer of Brazil's capital from Rio de Janeiro to Brasília in 1960, and by a range of tax incentives and land settling programs aimed to stimulate agriculture, which had been hitherto restricted to extensive grazing (Sano et al. 2010). These strategic decisions attracted many rural workers from other regions of Brazil where farmland was too expensive. However, the initial productivity was low because of the scarce knowledge on climate and soil properties and the lack of crops adapted to the region. Thus, a major research program was initiated during the mid-1970s through state-sponsored networks of agricultural institutes (e.g., Embrapa, Epamig) and land-grant universities. These initiatives created a portfolio of technologies that dramatically increased productivities of many crops. Currently, the

Cerrado region supplies more than a third of Brazil's grain and cattle production and >90% of cotton (*Gossypium herbaceum* L.) and wood charcoal (Resck et al. 2008). The relatively sparse population of the Cerrado, higher only than that in the Amazon, not only demonstrates its short settling time but also the high productivities achieved, critical to feed Brazil's mostly urban population and also a large export market.

Thus, neotropical savannas are important biomes for which a large, multifaceted research has been conducted in many countries and institutions. With the focus of this volume being on soil management, the objective of this chapter is to discuss the Brazilian Cerrado, its principal soils, land use systems, potential and challenges, and the effects of its ever-increasing importance to global agriculture and food markets.

3.1.1 Main Soils of the Cerrado and Soil Formation Factors

Land use and soil management systems in the Cerrado are primarily determined by soil type and topography, not only due to their obvious influence on mechanization, fertilization, and irrigation practices, but also because soil and relief themselves result from other important environmental controls. It is out of the scope of this chapter to provide a thorough review of soil genetic processes in the Cerrado, but some important aspects are pertinent to be discussed prior to describing the major soil types. The physical environment of the Cerrado and its control on soil genesis and properties can be briefly described by means of the five state factors of soil formation (Jenny 1941).

The most important soil *parent materials* would be deceptively easy to predict because the Cerrado occupies most of the Brazilian Shield, composed of the Archean and Proterozoic (>540 M years B.P., often >1 billion years) granites and gneisses. However, this shield is covered by thick (>1000 m depth in some areas), extensive Proterozoic marine sediments, often metamorphosed as shale and quartzite, and, to a lesser degree, by limestone. Recently, deposited sediments of variable texture are also common, more notably on the São Francisco river basin, and "lateritic" covers are also widespread. There are significant areas covered by Cretaceous (145–65 M years B.P.) basalt flows (especially in the Paraná river basin), mafic intrusions, and, to a lesser degree, ultramafic rocks. The disparate nature of these parent rocks strongly determines soil depth, mineralogy, texture, and fertility (Zinn et al. 2007a), which are discussed in detail for each of the principal soil types. However, it is important to note that most parent rocks are poor in nutrient cations and phosphate, and even basalts and limestones are often highly weathered, thus resulting in a majority of Cerrado soils being inherently nutrient-poor and acidic.

Time of soil formation is seldom accounted for in neotropical savannas, since most of this biome is characterized by soils developed from very old rocks. However, its importance is vividly expressed in special cases, such as in some recent fluvial deposits, or on freshly denuded erosional surfaces. Nonetheless, due to the pre-weathered nature of most parent rocks, the time state factor most often controls only soil depth and horizonation, with little effect on inherent soil fertility except for shallow eutrophic soils developed over recently exposed limestones and basalts.

Topography exerts a major regional control through climate: ~73% of the Cerrado area occurs between altitudes of 300 and 1100 m, with 22% below 300 m (Adámoli et al. 1986). These altitudes, combined with the inland continental location (Figure 3.1), increase solar radiation and decrease the amount of precipitation coming from the ocean. In comparison, the Llanos is much lower (mostly between 50 and 300 m a.s.l.) and closer to the Atlantic Ocean, resulting in higher precipitations. However, topography can be roughly considered as a long-term response of geological structures to erosion, and thus, there are other direct effects of it on soils. As mentioned earlier, most rocks in the Cerrado are extremely old, and although some sparse tectonic uplift has occurred in the Pliocene (2.6–5.3 M years B.P.), most land elevations are much older (Saadi et al. 2005). Therefore, there are few recent mountain ranges, and thus, Cerrado landscapes basically consist of erosional surfaces, which are very large, roughly level areas resembling a plateau from a distance, where streams and drainageways carve typically gentle slopes (<10%) and reliefs (<100 m). An older erosional surface always occupies a higher altitude than a younger one, and their transition is apparent in the field (Figure 3.2). In the Cerrado, the main erosional surfaces are the Gondwana (remnants of rocks denuded during the existence of the supercontinent formed by South America and Africa, >145 M years B.P.), the South American (the dominant surface, with rocks denuded during the late Cretaceous 100–65 M years B.P.), Velhas River (Miocene–Pliocene, 2.6–23 M years B.P.), and Paraguassu River (Quaternary, <2 M years B.P.), according to Thomas (1994) and Marques et al. (2004). Although it is important to understand the formation of

FIGURE 3.2 Smooth landscape (Sul-Americana erosion surface) most commonly associated to Oxisols, between the cities of Brasília and Goiânia. A well-managed pasture is shown in the first place, and a riverine ("gallery") forest is seen as a thin tree strip along the drainageway a few hundred meters away. In the distance is the Pireneus quartzite range (a remnant of the Gondwana erosion surface), where soils are too poor and sandy for cropping.

Cerrado landscapes, these erosion surfaces have little effect on agriculture, since the same erosion surface may encompass a range of altitudes, rock, and soil types. However, the Gondwana surface is generally small in area and often composed of quartzite ranges and plateaus (Figure 3.2), where soils are too steep, unfertile, and rocky for an arable land use. Thus, most agricultural lands in the Cerrado are located on the South American, Velhas, and younger surfaces. This geomorphic setting implies that topography can locally control soil properties in a predictable way: the most extensive uplands are slightly dissected plateaus, where savannic formations and occasional dry forests occur. On these flat or gently rolling surfaces, soils are typically very well drained and very deep, and thus suitable for a large-scale agriculture and urban development. Drainageways along these surfaces comprise soils that are usually higher in soil organic matter (SOM) and nutrients due to convergent flow, and hydromorphic in ca. 20% of the area (Skorupa et al., in press), often poorly drained, and mostly covered by riverine forests (Figure 3.2) or *veredas* (a vegetation consisting of wetland grasses and *Mauritia* spp. palm trees). The plateaus often present sharp borders and steep slopes, where soils are shallow and gravelly, suitable only for extensive grazing. Graphic models of typical topographies and landscapes of the Cerrado are presented by Curi and Franzmeier (1984), Macedo and Bryant (1987), and Marques et al. (2004).

Climate in neotropical savannas and the Cerrado is typically the Aw of the Köppen classification, marked by a stark contrast between the wet and dry seasons (García 2008). The data in Table 3.1 show mean annual temperature (MAT), mean annual precipitation (MAP), and mean annual evaporation (MAE) for selected locations within the Cerrado. The MAT is negatively correlated with altitude and latitude ($r = -0.84$ and -0.82, respectively), whereas MAP and MAE show a more complex distribution. Frosts are very rare and occur only in the southernmost areas, never to the point of freezing soil. MAP is always >1000 mm and exceeds total evaporation in many areas. Since rains are generally concentrated between October and April (Inmet 1992), they can be very intense and highly erosive, but short (~10 days) dry spells are frequent during the growing season and can pose a severe risk to young rainfed crops. The dry season occurs between May and September and is marked by very low precipitation and relative humidity, leading to natural wildfires whose frequency is increased by agricultural land use. The climatic conditions of the Cerrado promote intensive weathering and leaching of bases from all parent rocks, resulting in soils predominantly containing low-activity clays (kaolinite and Fe/Al oxides and oxi-hydroxides). An important consequence of this is that most of the cation exchange capacity (CEC) depends on humified SOM (Silva et al. 1994), which is also the case in most savannas of West Africa (Carsky et al. 2000). However, climate also favors annual cropping by allowing harvest during the dry season. Furthermore, almost all rivers and streams maintain a baseflow even during the driest winters, favoring the cultivation of two or more crops per year with supplemental irrigation of nearby areas.

Organisms were treated by Jenny (1941) separately as microorganisms, vegetation, and man. In this chapter, vegetation and soil fauna are emphasized due to their strong effect on soil properties in the Cerrado region. Although the word "Cerrado" is a Portuguese term for "closed" or "dense," this vegetation is actually a complex of plant formations varying from treeless grasslands (*Campo Limpo*, or

TABLE 3.1
Average (1961–1990) Climatic Data from Selected Sites in the Cerrado Region, Brazil

Location	State	Lat. S	Long. W	Altitude	Mean Annual Temperature	Mean Annual Precipitation	Total Evaporation
Carolina	Maranhão	7.2	47.28	192	26.1	1718	1678
Bom Jesus	Piauí	9.06	44.07	331	26.5	1156	1986
Porto Nacional	Tocantins	10.43	48.25	239	26.1	1668	1740
Barreiras	Bahia	12.09	45	439	24.3	1121	1575
Cuiabá	Mato Grosso	15.33	56.07	151	25.6	1315	1293
Brasília	Distrito Federal	15.47	47.56	1159	21.2	1552	1692
Aragarças	Goiás	15.54	52.14	345	24.9	1575	1565
Posse	Goiás	14.06	46.22	825	23.3	1537	2414
Goiânia	Goiás	16.4	49.15	741	23.2	1576	1577
Rio Verde	Goiás	17.48	50.55	745	22.5	1708	1430
João Pinheiro	Minas Gerais	17.42	46.1	760	22.5	1441	1518
Paracatu	Minas Gerais	17.13	46.52	711	22.6	1439	1314
Patos de Minas	Minas Gerais	18.36	46.31	940	21.1	1474	1222
Sete Lagoas	Minas Gerais	19.28	44.15	732	20.9	1328	1052
Uberaba	Minas Gerais	19.45	47.55	742	21.9	1589	1498
Campo Grande	Mato Grosso do Sul	20.27	54.37	530	22.7	1469	1911
Dourados	Mato Grosso do Sul	22.14	54.59	452	21.9	1406	2151

Source: Extracted from Instituto Nacional de Meteorologia (Inmet). *Normais climatológicas.* Brasília, Inmet, 1992.

if on rock outcrops, *Campo Rupestre* in Portuguese), shrubby grasslands (*Campo Sujo*), savanna woodland (*Cerrado stricto sensu*), and *Cerradão*, which is a forest savanna, mostly tall-tree formation, with few grasses (Lopes and Cox 1977; Grace et al. 2006). Riverine forests and palm-grass wetlands (*veredas*) occur along streams and drainageways, and many dry forests with little or no grass strata occur amidst the savannic formations. Typically, the dominance of grasslands is associated with very low soil fertility and moisture, so treeless grasslands occur on the poorest soils, whereas the Cerradão, the riverine, and dry forests indicate more fertile, less acidic soils. The impact of vegetation biomass on soil properties can be unexpected, since there appears to be no correlation between vegetation biomass and soil organic carbon (SOC) concentrations in surface soils (Lopes and Cox 1977). This effect is also reflected in subsurface layers: Chapuis-Lardy et al. (2002) noted that SOC stocks under treeless grasslands can be higher than under tall-tree savannas with much higher aboveground biomass, because of stronger root biomass in the former vegetation. Since the classic studies on earthworms and soil bioturbation by Charles Darwin (1882), the impact of soil fauna on soil genesis and properties has chiefly been attributed to worms, moles, and other burrowing animals. Although large earthworms and armadillos are abundant in the Cerrado, social arthropods exert major influence on its soils not only by their sheer diversity and ubiquity but also because of their overall biomass (Benito et al. 2004) and amount of soil disturbed annually. Termites and leaf-cutting ants excavate large volumes of soils, disrupting aggregates but also forming new ones, typically smaller and rounded, sometimes spherical and welded (Figure 3.3), which can comprise a significant percentage of total topsoil mass (Zinn et al. 2007b). The activity of termites over considerable geologic time has been ascribed as the cause for the loose granular structure of many Oxisols, especially in subsurface horizons (Schaefer 2001).

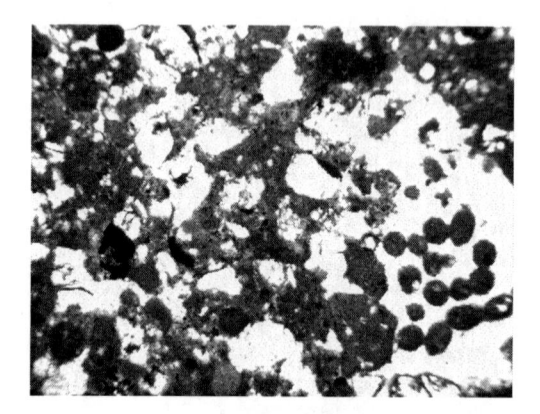

FIGURE 3.3 Typical granular microstructure of Cerrado Oxisols. The high macroporosity is expressed in the form of packing voids, through which saturated flow is very rapid. A cluster of biogenic, spherical aggregates is shown. Image is 2.3 mm wide.

3.1.2 Oxisols

Oxisols are by far the dominant soil order in Brazil and in the Cerrado region, where they cover about 46% of the total land area (Adámoli et al. 1986). The reasons for such dominance are multifold: (1) high MATs and MAPs favoring extreme weathering and intense leaching of bases; (2) Oxisols can develop from most parent rocks, except quartzite and purely siliceous sandstone; (3) Oxisols typically develop on flat or gently rolling uplands, which predominate in the Cerrado; and (4) significant bioturbation by soil fauna, which homogenizes the entire soil profile. In the Cerrado, Oxisols develop under ustic soil moisture regime, and most are classified as Haplustoxes and Kandiustoxes. However, Cerrado Ustoxes are in fact highly variable on the basis of principal soil-forming factors. According to the Brazilian Soil Classification System (Embrapa 2006), Oxisols correspond to the order *Latossolo* (a term often translated as *Latosol*), in which all suborders are defined in terms of soil color (e.g., dusky red, yellow-reddish, and yellow, in decreasing order of hematite content). Thus, Ustoxes vary considerably in color as a consequence of different proportions of the basic minerals (quartz, kaolinite, Fe/Al oxides), but other important differences also occur simultaneously. When kaolinite is the dominant mineral in the clay fraction, a face-to-face arrangement of clay domains is favored and the structure tends to be blocky. Conversely, when goethite, hematite, or gibbsite predominates, a granular structure predominates (Ferreira et al. 1999). Oxisols with blocky structure and high bulk density due to predominantly kaolinitic clay also present much higher soil strength than granular, gibbsitic Oxisols with low bulk density (Ajayi et al. 2009). Texture in Ustoxes is also highly variable, ranging from a minimal clay content of 15% to as high as 85%. Silt content is generally <20% and often <10%, except where siltstones are the parent material.

Although Ustoxes can develop from almost any parent material, the nature of the rock has a clear effect on their properties. Ustoxes developed from granites and gneisses typically present a yellow-reddish hue and sand (mostly quartz) contents of ~40%, whereas those derived from basalts and other mafic rocks are redder and with much less sand, which can be rich in magnetite (strongly attracted to a hand magnet). Ustoxes derived from sandstones and shales differ strongly in texture (respectively, coarse or fine), and their color is highly variable from dusky red (drier climate, higher position on the landscape, or strong internal drainage) to a much lighter yellow-reddish, where climate is wetter or on footslopes (Curi and Franzmeier 1984; Macedo and Bryant 1987). Many Ustoxes developed from sediments or basalt have small contents of ironstones or lateritic fragments. As a general rule, Ustoxes are acidic and low in plant nutrients, but when derived from mafic rocks, they can be somewhat richer in P and other nutrients. Ustoxes derived from limestones and carbonate-laden sediments can contain significant illite (clay mica) and traces of interstratified 2:1 minerals (Zinn et al. 2007a). However, since these carbonatic sediments are very old, and weathering is advanced, most derived Ustoxes are acidic and low in bases, but still more fertile than those developed from other rocks.

Ustoxes are primarily used for the production of grains, cattle, and forest plantation in the Cerrado region. Their high suitability for agricultural land use is not only due to their areal extent and favorable topography but also because of a set of other properties: (1) deep weathering and effective rooting depth (always >1 m, often >4 m); (2) low bulk densities (average 0.9 g cm^{-3}, sometimes as low as 0.7 g cm^{-3}), especially in clayey topsoils with granular structure and large packing voids (Figure 3.3), easing soil preparation (Araujo, M.A. et al. 2011); (3) perfect aeration and internal drainage in most cases, resulting from properties 1 and 2; and (4) general absence of gravels and abrasive materials. On the other hand, the most severe soil-related constraints of Cerrado Ustoxes have been technologically alleviated: (1) high acidity and aluminum saturation, corrected by adequate liming; (2) low basic cation corrected by liming in topsoil and by gypsum in subsoil; (3) strong sorption of phosphates minimized by localized fertilization, especially in no-till systems; (4) low N levels enhanced by symbiotic nitrogen fixation by including legumes in rotation; and (5) low micronutrient levels corrected by judicious liming and fertilization. There are two other consequences of the low bulk density and granular structure that can be mitigated but not fully corrected, which are low volumetric water holding capacity of <10% between −6 and −100 kPa (Resck 1998) and strong susceptibility to gully erosion resulting from low soil strength and weak cohesion of granular peds (Curi, N., personal communication; Figure 3.3).

3.1.3 Quartzipsamments and Sandy Oxisols

These soils are primarily derived from sandstones, either in Cretaceous or recent deposits, and to a lesser extent from Proterozoic quartzites. In the Brazilian System, Quartzipsamments are classified as *Neossolos Quartzarênicos* (formerly *Areias Quartzosas*) and cover ~15% of the Cerrado (Adámoli et al. 1986). Almost all of these soils occur on flat lands, are very deep, and are well drained, except for lowlands with aquic moisture regimes. Clay contents are necessarily <15%, and their color ranges from light gray to light yellow or red, depending on the proportion of goethite and hematite concentrations. Although they are generally acidic and low in basic cations, these soils have moderate fertility when derived from sandstone/basalt strata (common in the Paraná basin). Unlike Ustoxes and Ustults, these soils do not have strong P fixation. Therefore, these soils have been cultivated for annual crops such as soybeans in many Cerrado areas. Nonetheless, arable land use causes severe depletion of CEC and SOC. In western Bahia, 5 years of soybeans cultivation with heavy harrows decreased SOC and CEC in Quartzipsamments by 73% and 61%, respectively, whereas these losses were 45% and 29% in clayey Haplustoxes (Silva et al. 1994). These severe losses are caused by several factors: (1) low antecedent SOC contents in comparison with loamy and clayey soils (Zinn et al. 2005a); (2) occurrence of about half of the total SOC in the sand size fraction as particulate organic matter (POM; Zinn et al. 2007a), which is more readily decomposable than humified SOC; (3) none, low, or weak aggregation because these soils lack cementation by clay, and thus, the few large aggregates are easily disrupted and release occluded POM, which is rapidly oxidized; (4) the low clay fraction in these soils can have very high SOC contents (Zinn et al. 2007a), but SOC sorption capacity is near saturation and SOC can be lost even with low-intensity land use systems such as eucalypt plantations (Zinn et al. 2002, 2011). On the other hand, some Quartzipsamments under native Cerrado vegetation are so low in SOC (<0.7% at the 0–5 cm depth) that soybean cultivation (no-till or conventional) causes only small losses and may even increase SOC and total N concentrations (Frazão et al. 2010).

Some Ustoxes with low clay contents (15%–20%) and a weak granular or blocky structure occur in association with Quartzipsamments and have similar properties and land use restrictions. In the Cerrado, these Ustoxes are called *Latossolo textura média* and probably comprise a significant area whose extent is not exactly known and are especially common in the São Francisco river basin.

3.1.4 Ultisols

Ultisols comprise ~15% of the Cerrado land area (Adámoli et al. 1986), and, as Oxisols, are developed mostly under ustic soil moisture regimes. However, research information on Cerrado Ultisols is rather scanty compared with those in the coastal region of Brazil, where these soils are more common. In the Brazilian System, these soils are classified as *Argissolos* and *Nitossolos* (formerly *Podzólicos*; Embrapa 2006), for their argillic or kandic horizon with low-activity clays. Most of these soils can be classified as Kandiustults, Kanhaplustults, Rodhustults, and Haplustults in Soil Taxonomy (Soil Survey Staff 2010). Although Ustults can develop from the same rocks as Ustoxes, they are more frequently associated with granites and gneisses and rarely with sandstones. Thus, sand contents of these soils are neither too low nor too high. Ustults can occur in association with Ustoxes, but they typically form on midslopes and small summits of hilly terrains. The predominant clay minerals are similar to those in Ustoxes, but in contrast, Fe/Al oxide contents are relatively low and kaolinite almost always predominates in Ustults. Therefore, the granular structure in Ustults is exceedingly rare and restricted to the surface layer, and subangular or angular blocky structure predominates in most horizons. A practical consequence of the blocky structure is that soil bulk density is ~20% higher in uncultivated Ustults than Ustoxes, to 1-m depth (Araujo, M.A. et al. 2011), and root penetration is severely impeded in the subsoil horizons. In general, most Cerrado Ustults do not have a strong textural gradient and are thus distinguished by high bulk density and/or abundance of clay films on ped surfaces. Mechanical strength is also high, often necessitating subsoiling, and land use-induced compaction can severely inhibit root growth. Soil depth and internal drainage are also much less than in Ustoxes, which aside from the hilly terrain is highly conducive to sheet or interrill erosion (Carvalho et al. 2008). On the other hand, leaching of bases can be lower (Curi, N., personal communication) and natural nutrient levels higher than those in Ustoxes. It is because of these factors that these soils are more commonly used for pastures, forest plantations, and perennial crops such as coffee, especially in southwestern Minas Gerais.

3.1.5 Other Relevant Soil Types

Soils with *ironstone* and/or *plinthic* features, as defined in Soil Taxonomy (Soil Survey Staff 2010), can be classified as Ustoxes, Ustults, Plinthic Quartzipsamments, and Petraquepts. Soils with these features cover ~9% of the Cerrado area, and 6% may have aquic soil moisture regimes (Adámoli et al. 1986). The Brazilian System has a very different classification approach to these soils and recognizes the separate order *Plintossolo*, in which suborders are defined based on the presence of clay illuviation and ironstone layers or gravels (Embrapa 2006). The nomenclature used in the study of these soils is variable: ironstone and plinthite are commonly termed "laterite" across disciplines other than Soil Science, and soils rich in these materials are named "Latosols" by Padmanabhan and Eswaran (2002), a term also used by many Brazilian soil scientists as a translation for *Latossolos* (Oxisols). Basically, these soils present significant Fe concentrations and cementations where oxic and anoxic conditions alternate, or where relict materials thus indurated are being exposed or dismantled, often on steep borders of plateaus. Ironstone remnants can thus be distributed throughout the landscape and appear as opaque sands or gravels (Figures 3.3 and 3.4) in almost all types of Cerrado soils. Plinthite-forming conditions exist in many but not all lowlands and drainageways and appear to be rarer where organic soils and peats occur (Skorupa et al., in press). Plinthic soils are commonly derived from Fe-rich parent materials such as clayey sediments, basalts, and especially ultramafic rocks (Trescases et al. 1979), but they can form from low-Fe rocks such as quartzites in small areas where Fe-enriched water flows from uplands for a long time. Thus, the principal limitations to intensive use of these soils are poor drainage, year-round waterlogging, and heavy wear-and-tear of plows and other soil preparation tools. In addition, many of these soils occur on steep slopes, and most are acidic and nutrient-poor. However, about 10% of plinthic soils have high nutrient levels (Adámoli et al. 1985), either due to nutrient flow from uplands or to strong micronutrient sorption by the Fe-oxides. Some of these fertile plinthic soils are unexpectedly young and contain significant amounts of weatherable minerals, as observed in a Plintaquult in the Cerrado of Goiás (Moreira and Oliveira 2008). Some upland clayey Ustoxes of the Cerrado, with ironstone gravel contents too low to have a plinthic diagnostic horizon, present lower internal drainage and thus are able to retain

FIGURE 3.4 Scanned image of soil thin sections of a clayey Haplustox (0- to 5-cm depth) cultivated with annual crops for 20 years under different soil preparation practices. From left to right: native Cerrado vegetation, no-till, chisel plow, disk plow, and moldboard plow. Each image is 2.7 cm wide. (Adapted from Skorupa, A.L.A. et al. Micromorfologia de um latossolo do cerrado sob diferentes sistemas de manejo e preparo por vinte anos. In XXXIII Cong. Bras. Ci. Solo. *Proceedings*. Uberlândia, SBCS, 2011. With permission.)

much more exchangeable bases than other Ustoxes (Zinn et al. 2007a; Zinn and Resck 2008). Thus, since 2000, more and more of these fertile plinthic soils of the Cerrado are being cultivated for rice (*Oryza sativa*) paddy fields, but also for soybean following artificial drainage (Oliveira, G.C., personal communication).

Inceptisols cover ~3% of the Cerrado area (Adámoli et al. 1986), mostly on the steep borders of plateaus or hilly lands, where they are often associated with Ustoxes (summit) and Ustults (less pronounced slopes). Most of these soils are classified as Dystrustepts, derived from preweathered rocks and thus are acidic and low in basic cations. Haplustepts are also common and may have higher fertility when developed from basalt or limestone. Bulk density of these soils is similar to those of Ultisols, ranging from a mean 1.07 Mg m^{-3} in the surface to 1.35 Mg m^{-3} to 1-m depth (Araujo, M.A. et al. 2011). Crop productivity is limited by low or extremely low fertility, shallow rooting depth, steep slopes, and vulnerability to interrill or sheet erosion. Thus, these soils are most commonly used for pastures, forest plantations, and perennial crops. However, high doses of gypsum (>10 Mg ha^{-1}) can greatly enhance depth of root penetration of coffee in Usteptis in southwestern Minas Gerais (Serafim et al. 2011).

Entisols with lithic properties occupy ~7.3% of the Cerrado (Adámoli et al. 1986) and commonly occur on quartzite ranges, steep slopes, or recently exposed rock surfaces. In Brazil, these soils are named *Neossolos Litólicos*, and low or extremely low fertility and water retention are among principal constraints to agronomic productivity. Thus, these soils are commonly used for extensive grazing. However, many low-income farmers depend on these and other unproductive soils to grow annual and perennial crops, especially cassava (*Manihot esculenta* Crantz).

Soils with aquic moisture regimes and *without plinthite* occupy about 2% of Cerrado (Adámoli et al. 1986). While more commonly confined to drainageways and lowlands, they sometimes occur in uplands where perched water tables accumulate over impervious layers. Some of these soils can accumulate deep peat and/or high SOC stock because of slow rates of organic matter decomposition under anoxic conditions. However, this C accumulation can be readily lost upon artificial drainage or by drastic lowering of water tables during the dry season, even causing subsoil fires. In the Brazilian System, these soils are included in the separate orders *Gleissolos* and *Organossolos*, the latter being equivalent to Histosols. Many small landholders cultivate these soils for growing horticultural crops or vegetable gardens, because of abundance of water and relatively high fertility. Rice paddy, although a common land use in Southern Brazil, is not common in aquic soils of the Cerrado.

3.2 PRINCIPAL LAND USE AND SOIL MANAGEMENT SYSTEMS IN THE CERRADO

By the year 2002, approximately 40% of the total Cerrado biome, (or ~80 M ha) had been converted to some kind of anthropogenic use (Sano et al. 2010). This land conversion has been mostly concentrated near the densely populated Brazilian southeast, on the southern reaches of the Cerrado biome (Figure 3.1), where ~85% of the land surface has been reclaimed, in contrast to only 10% in the northernmost regions. Thus, the Cerrado underwent a rapid land settlement leading to an immense economic growth and welfare, which also created some environmental concerns. Most of the ensuing degradation of natural resources, such as biodiversity loss and CO_2 emissions, are direct consequences of removal of the native vegetation. In addition, the attending agricultural management also causes drastic environmental impact, especially on soil properties and water quality. The following sections discuss the most common land use systems in the Cerrado, different soil management options, and impacts on soil properties.

3.2.1 Basic Tenets of Soil Management in the Cerrado

Resck (1998) summarized that successful management of Cerrado soils requires a proper combination of four practice sets. Each practice must be performed in an optimal manner so that the other three are effective. Failure to do so almost always leads to low or declining yields and soil degradation. Although the frequency and intensity of each practice set vary among major land use systems, as discussed below, a brief description of these practices follows.

3.2.1.1 Clearance of Native Vegetation and Correction of Soil Acidity

Removing native trees and shrubs is traditionally done by pulling a heavy metal chain by two tractors, followed by raking of fell biomass in windrows and burning (Folle and Seixas 1985; Klink and Moreira 2002). Decomposition and burning release nutrients from the biomass within a few months, causing nutrient and SOC depletion, especially by the burning of residues that otherwise would have slowly converted into humus. Liming is required to raise pH levels and replace exchangeable Al^{3+}, toxic to most crops, by Ca^{2+} and Mg^{2+}. Liming enhances CEC through the dissociation of carboxyl groups of humic substances, an important process because of low CEC of clay minerals in most Cerrado soils. However, excessive liming or its poor incorporation by harrowing in a <20 cm layer can turn micronutrients insoluble, induce severe deficiency, and produce low crop yields. Also, high Al saturation in the subsurface horizons restricts root growth into the topsoil, exacerbating susceptibility of crops to dry spells during the growing season. Thus, gypsum (calcium sulfate) is often applied on the soil surface at the rate of 0.5–1.5 Mg ha^{-1} (Spehar 1998), because its dissolution and subsequent leaching into subsurface layers can effectively replace Al^{3+} by Ca^{2+}, enhancing root depth and agronomic productivity.

3.2.1.2 Fertilization

Cerrado soils being low or extremely low in all mineral nutrients, require corrective fertilization to "build up" an initial nutrient bank/reserve, followed by replacement of losses through crop harvest and/or erosion. The fertilization must be done only after liming; otherwise, fertilizer nutrients are wasted either by leaching in soils of low CEC or by high P fixation on positively charged sites of Fe/Al oxides at low pH. Nonetheless, high soil pH by excessive liming also leads to micronutrient insolubility, causing deficiencies that can only be corrected by deeper incorporation into the subsoil. Agronomic productivity in most Cerrado soils is mostly P-limited. Corrective fertilization often involves applying rock phosphate on the whole area (typically <1 Mg ha^{-1}), but because of low solubility, its use is recommended only for pastures, forest plantations, and perennial crops, which can absorb P over longer terms than annual crops. Soluble phosphate sources must always be applied in restricted soil volumes (e.g., planting lines or seedling pits) in order to reduce the strong specific sorption to Fe/Al oxides, and there is a growing trend to observe the same practice for all soluble nutrient sources. After a "nutrient bank" is created, soil nutrient levels must be determined by soil testing every other year, or more frequently for intensive cropping, and supplemental fertilizer used as needed.

3.2.1.3 Soil Tillage and Seedbed Preparation

These practices are usually initiated after clearance of native vegetation and residue burning. In many cases, a tractor-pulled rake or tiller gathers residues in windrows for burning, thus causing the initial disruption of soil (excluding soil upheaval when trees and shrubs are pulled off with a chain). Also, every time lime is applied on soil surface, it must be incorporated with disk plows or harrows to a 0.2-m depth or deeper, followed by another harrowing. Alternatively, farmers use a heavy disk plow (*grade pesada*) not only to incorporate lime but also for a routine seedbed preparation. These operations can be repeated in different frequencies and intensities depending on the crop and soil management, varying from several times per year to only once every 14 years or more. Soil tillage for weed control is also common in coffee stands and small farms by tractor or animal/manual driven operations. As a rule of thumb, tillage on an annual or higher frequency strongly disrupts the larger macroaggregates and promotes compaction of most Cerrado soils (Resck 1998; Resck et al. 2001), and also causes significant SOC loss (Zinn et al. 2005b) by oxidation of crop residues and coarse-sized SOC hitherto occluded inside aggregates or within the interaggregate pores (Zinn et al. 2011; Pes et al. 2011). These degradation processes are especially severe when the heavy disk harrow is used rather than a plow or a harrow (Silva et al. 1994; Resck 1998).

3.2.1.4 Crop Rotations

Numerous benefits of crop rotations for diverse regions and soils are widely known by farmers and researchers. In the Cerrado, crop rotations are especially important to improve plant nutrition and to maintain SOC at adequate levels. Upland rice (*Oryza sativa* L.) and corn (*Zea mays* L.), the most common crops grown during the early years of the Cerrado agriculture, generate enough harvest residues to maintain adequate mulching, and whose quality allows relatively slow decomposition and proper humification. However, after P buildup by corrective fertilization, N becomes the limiting nutrient, since N fertilizers are among the most expensive in Brazil. On the other hand, soybean varieties specifically adapted to the Cerrado region supply all their N requirements by symbiotic N fixation. However, the amount of crop residues are low and C/N ratio is narrow. Thus, adequate mulching is not feasible, and SOC can decrease drastically under continuous monoculture. Therefore, alternating cereals and legume crops, such as corn/soybean [*Glycine max* (L.) Merrill] or corn/beans (*Phaseolus* spp.), or even soybeans and pastures (see Section 3.2.7), is a useful strategy that is becoming more and more common. Planting cover crops such as

Pennisetum or *Mucuna* spp. during the dry season after harvest of the main crop provides additional mulch and enhances SOC levels (Nunes et al. 2011).

These tenets indicate that a successful management of Cerrado soils requires a proper management of SOC and must result in its preservation, as stated earlier by van Wambeke (1992) for tropical soils in general. This high importance of SOC is attributed to the fact that (1) fresh organic residues and coarse/light SOC fractions are immediate C sources for soil food webs and nutrient sources to plants; (2) humic SOC is the main source of CEC; and (3) both SOC forms improve soil aggregation and its resistance to compaction and other physical degradation processes, and also enhance water retention at low suctions. In addition, about 90% of N in Cerrado soils is in a NH_4^+ form, mostly bound to SOC (Frazão et al. 2010). Although these beneficial effects of SOC are commonly observed in most soils throughout the world, they are especially critical in highly weathered soils of the humid tropics, where nutrient and CEC levels are remarkably low, and low bulk density due to large macropores causes excessive internal drainage. However, SOC in the Cerrado and elsewhere can be lost by mismanagement, excessive tillage, and erosion. Yet, it can also be sequestered under best management practices (Bayer et al. 2006). Finally, there is growing interest in mitigation of global warming by the sequestration of atmospheric CO_2 as SOC in agricultural soils (Lal 2008). Therefore, choice of best management practices in the Cerrado is inseparable from a discussion of SOC forms, functions, and effects.

The most important combinations of land use systems and soil management techniques for SOC management are discussed below.

3.2.2 Pastures

The oldest settlements by European descendants in the Cerrado region were established in the early eighteenth century by gold miners (Klink and Moreira 2002), most of which used cattle as traction force for their wagons. When gold mining was no longer profitable, these settlements enhanced the use of the "natural" pastures to raise more cattle, which became their main activity. In fact, these were range areas where most or many of the Cerrado trees were removed; thus, the underlying herbaceous native vegetation could be grazed by cattle at low stocking rates, using neither fertilizers nor other inputs. Thus, productivities were always low, which was not a serious problem given the low human population for the entire region. This extensive management could sustain most of the first urban settlements (e.g., Goiás, Pirenópolis, Planaltina, Paracatu, etc.). Following an accelerated agricultural settlement after 1960, many pastures were transformed to monocultures of exotic grass species, more notably of *Brachiaria decumbens* Stapf. and *B. brizantha* Stapf. As in the year 2002, Sano et al. (2010) reported that 54.1 M ha, slightly more than a quarter of the Cerrado biome, was used for pastures, which is the most widespread land use system. Most pastures are seeded with *B. decumbens*, making it the most commonly cultivated species in the region, as well as the main feedstock for bovines, the predominant livestock species in the Cerrado and in Brazil. Pastures are predominant in the Cerrado because they can be established even in an infertile soil (e.g., rocky Entisols) and in climates too dry for annual crops.

Traditionally, pastures were not seeded immediately after clearance of native vegetation, but after growing upland (rainfed) rice for 1 or 2 years (Klink and Moreira 2002). The upland rice was appropriate because of its low nutrient demand, which could be met by the nutrients mineralized from decaying residues of native vegetation, even without liming or fertilization (Lilienfein et al. 2003). Fertilized pastures of the Cerrado typically produce 5–11 Mg ha^{-1} year^{-1} of grass biomass, feeding 2–2.5 animals ha^{-1} with 500–800 kg ha^{-1} year^{-1} of liveweight gain (Lilienfein et al. 2003). Thus, most of these pasturelands are reasonably productive for the low inputs they receive and have a relatively low environmental footprint compared to those of annual crops. Further, pastures improve soil aggregation by their strong root system. Salton et al. (2008) reported that pastures or crop–pasture rotations increased large water-stable aggregates (>4.76 mm) in comparison to native vegetation, in two out of three locations in the Cerrado in the State of Mato Grosso do Sul. Increases in aggregate size were strongly correlated with the higher SOC stocks in 0.2-m depth. For a sandy Haplustox in the southernmost Cerrado, SOC stocks for 0.3-m depth increased from 30 Mg ha^{-1} under native vegetation to 37 Mg ha^{-1} after 80 years under pasture (Maquere et al. 2008). Bustamante et al. (2006) reviewed SOC changes after conversion to pastures and reported an average change of 1.3 Mg C ha^{-1} year^{-1}, with a range of −0.87 to 3.0 Mg C ha^{-1} year^{-1}. Furthermore, some severely compacted soils can be efficiently reclaimed by planting tropical grasses commonly used in pastures, such as *Cynodon* spp. (Severiano et al. 2010).

Although improved management practices are being increasingly adopted, most pastures in the Cerrado show some level of degradation. Therefore, productivity of *Brachiaria* pastures in the Cerrado is much lower than that in Australia (Lilienfein et al. 2003). Vendrame et al. (2010) analyzed >70 Oxisols under pasture in the Cerrado and observed that about 90% had low or very low available P, and most contained other nutrients below critical levels. Little or no nutrient replacement is often the principal cause of pasture degradation, although physical degradation is also common. Some farmers intensify production by adding corrective and replacement fertilization. However, degradation can still occur if the stocking rate exceeds the optimum level, which varies widely according to local soil and climate conditions. Overgrazing aggravates compaction and depletes SOC stocks by as much as 20% in comparison to soils under native vegetation (Costa et al. 2011).

Effects of pasturelands on other environmental indicators have also been studied in the Cerrado. Termite and ant mounds are a common sight in old and degraded pastures. However, pasture restoration by tillage and fertilization may adversely affect the macrofauna diversity and biomass (Benito et al. 2004), also decreasing some ecosystem services such as pest control by natural enemies, and even water infiltration rate because of elimination of biochannels.

Fluxes of greenhouse gases (GHGs) from pastures are also an important environmental issue. Using static chambers in situ on a clayey Haplustox, Siqueira Neto et al. (2011) reported a strong contrast in methane (CH_4) emissions and its uptake under a *Brachiaria* pasture and nearby native Cerrado vegetation and croplands. However, fluxes of CO_2 and N_2O were similar under land uses. Thus, conversion of Cerrado to pastures has complex effects on the ecosystem and its environmental services, which deserve further investigation.

3.2.3 Annual Crops

Croplands are the second most important land use in the Cerrado biome and covered 21.6 M ha by 2002 (Sano et al. 2010). While these statistics do not differentiate between annual and perennial crops, croplands are primarily concentrated in the most developed southern and western areas of the Cerrado. Unlike pastures, productive annual crops are mainly planted on clayey, but also on coarser Ustoxes, as well as on Quartzipsamments and Ustults when slopes are not too steep. Although occupying less than half the area under pastures, croplands pose a higher environmental impact because of their intensive management and often with heavy use of pesticides, fertilizers, and mechanization. However, as mentioned earlier, this was not always the case, since early farmers in the Cerrado only grew rice or corn immediately following the land clearance, with little or no inputs, obtaining low productivities for no more than 2 years. At present, after decades of government incentives and research and development, cultivations of annual crops in the Cerrado have become highly technical and productive, most commonly involving soybeans and corn, but also rice, beans, and more recently, cotton. Liming and fertilizer use have also become more widespread, and in fact, some farmers are applying doses much above the technically recommended rates, aiming to build "nutrient banks," especially of P, which may result in losses but also increase overall soil quality. Even when initial P levels were adequate, annual rates of fertilizer use of 100 kg P ha^{-1} for 11 years to corn/soybean rotations caused small but significant increases in SOC stocks for a 0- to 20-cm depth (Nunes et al. 2011).

Conventional tillage for annual crops in the Cerrado traditionally involves disk plowing and harrowing prior to planting, and in some cases, crop residues are incorporated right after harvesting. Moldboard plowing is less common than disk plowing, whereas a heavy disk harrowing is frequently used instead of plowing and harrowing. When applied to a typical clayey Haplustox of high macroporosity, these implements adversely impact bulk density, saturated conductivity, and compressive behavior in the order of moldboard plow>disk plow>heavy disk harrow (Silva et al. 2003). However, over a long period, these tools cause different effects depending on the crop and soil type. Continuous (>15 years) use of disk plow for a soybean/corn rotation in a clayey Haplustox strongly decreased the proportion of macro (>2 mm) water-stable aggregates from 80% under native Cerrado to only 15%, whereas the heavy harrow was less disruptive, maintaining ~50% of large macroaggregates (Resck et al. 2001).

Whatever the implement, annual tillage operations demanded large amounts of fuel and time of farmers and became expensive in increasingly larger farms. Additionally, widespread soil compaction and erosion resulting from annual tillage encouraged the Cerrado farmers to no-till (NT) systems, which had been used in Paraná, southern Brazil, since the 1970s. The rapid and wide adoption of NT for annual crops is perhaps the most striking shift in soil management systems that ever happened in Brazil and in the Cerrado. Presently, NT systems are almost universally used for large-scale annual crops in the Cerrado, and annual tillage is all but restricted to small areas in family farms. In consequence of the NT boom, the use of herbicides and pesticides has drastically increased.

Principles of Soil Management in Neotropical Savannas

Soil properties are affected differently by the use of conventional tillage or NT in the cultivation of annual crops, and this theme constitutes the scope of a large research program in Brazil and the Cerrado. However, the nature and intensity of these differences vary with soil type and the indicator chosen. Twenty years of continuous cropping of a clayey Haplustox near Brasília increased bulk density from 1.0 to 1.2 g cm^{-3} to a depth of 20 cm, either under NT or disk and moldboard plowing (Jantalia et al. 2007). Although soil compaction implies a considerable reduction of macropore volume, it remains adequate and perhaps at better levels considering the typically low water retention capacity in pristine Oxisols. It is also possible to visualize that the loose granular aggregates under native Cerrado vegetation get closely packed upon cultivation, forming granular clods resembling subangular blocks, separated by planar pores resembling fissures (Figure 3.4). This effect is especially visible under disk plow and NT, where large cavity macropores also occur, and large macropores (either as fissures or cavities) are rarer with moldboard or chisel plowing.

The dynamics and distribution of SOC are particularly sensitive to the choice of tillage systems for annual crops. Cerrado soils under annual tillage typically show low but homogeneous SOC and N concentrations throughout the 20-cm depth, whereas soils under NT have similar values for the 10- to 20-cm layer but much higher values in the top 10 cm, resulting in significantly higher total C and N stocks (Nunes et al. 2011). Such accumulation is a direct consequence of keeping crop residue inputs on the soil surface instead of mixing them into a soil teeming with decomposer biotas. Thus, adoption of NT leads to SOC sequestration when compared to conventional annual tillage, especially in clayey (Jantalia et al. 2007; Nunes et al. 2011) but also in coarse-textured Oxisols (Bayer et al. 2006). The average rate of SOC sequestration by NT in Brazil for 0- to 20-cm depth is +0.37 Mg ha^{-1} year^{-1} (Bayer et al. 2006). While the average rate may change for deeper layers (e.g., 0–40 cm), the data suggest that soil quality can be improved by SOC sequestration. Increase in SOC concentration under NT occurs not only for bulk soil but also for SOC occluded within aggregates, which are often depleted by annual tillage with disk plowing or heavy harrowing (Resck et al. 2001). However, just as most particulate SOC in Cerrado Haplustoxes is not occluded within aggregates but rather is free in bulk soil (Zinn et al. 2007b), annual tillage in Oxisols affects much more organic residues free in soil than SOC occluded within aggregates (Pes et al. 2011). Thus, conventional tillage exacerbates SOC losses when plowing is done before planting and also after harvesting (Jantalia et al. 2007).

The effect of tillage systems on fluxes of GHGs is being more widely studied in the Cerrado. Siqueira Neto et al. (2011) reported no difference in CO_2, N_2O, and CH_4 fluxes and microbial C after 12 years cropping under NT and conventional tillage in southern Goiás, although SOC levels were 25% lower under annual tillage. Such a contradiction suggests that, after a significant amount of SOC pool (most likely POM) is oxidized by annual tillage, soil microorganisms keep a basal respiration by oxidizing different, perhaps more resistant, SOC forms that are similar in pool size to the labile SOC fraction under NT system. Also, hidden C costs of agricultural systems have seldom been reported in the Cerrado, which can have a strong impact especially on NT systems (Batlle-Bayer et al. 2010). Nitrate leaching under annual crops is not a major concern even in sandy soils under soybeans, since nitrate levels are often an order of magnitude lower than ammonium forms (Frazão et al. 2010) and most soils are so deep that nitrate percolation may not reach groundwater. Nevertheless, additional research is needed to enhance scientific understanding on these subjects.

3.2.4 Perennial Crops

By definition, perennial crops have production cycles spanning many years, and thus site preparation and tillage are used more infrequently than in annual crops, sometimes on intervals of 15 years or more. Thus, perennial crops generally cause less impact on properties of Cerrado soil than do annual crops. The principal perennial crops in the Cerrado are coffee (*Coffea* spp.), cassava (*Manihot* spp.), rubber trees (*Hevea* spp.), and a range of fruit trees (sugarcane is discussed separately in Section 3.2.6). Soil management in fruit orchards is highly variable depending on the species and farm size, and the management of rubber tree plantations, except for the annual harvesting, is similar to that of planted forests (see Section 3.2.5). Although cassava plantations are common in other regions of Brazil, cassava in the Cerrado is typically planted in small farms or collective settlements, with little fertilizer or other inputs, and soil preparation is often done with manual- or animal-drawn implements. Therefore, this section focuses on coffee plantations, by far the most prominent perennial crop either in commercial value or in the planted area.

Productive coffee plantations are among the most intensively managed land use systems in Brazil and usually employ large amounts of fertilizers, pesticides, and labor throughout the year. Approximately half of the coffee production in Brazil comes from the Cerrado region (Resck et al. 2008). Further, coffee is only planted in the most fertile soils, although innovative management systems are currently being developed that enhance productivity even in low-fertility Cerrado soils (Serafim et al. 2011). Soil preparation before planting primarily involves liming, plowing, and harrowing, as well as pitting for the seedlings, but these operations typically are done every 15 to 30 years, when the stands are cleared and replanted. Although this long rotation cycle may suggest that soil disturbance is low in coffee stands, in many cases, soil tillage with harrowing is used for weed control between the planting rows. In addition, heavy tanks with fertilizers and pesticides are pulled by tractors several times per year, not to mention harvest machines used on a yearly basis. Therefore, soil compaction and aggregate disruption are common in Haplustoxes and other soils with low bulk densities under coffee plantations. In a clayey Oxisol under coffee for >20 years in southern Minas Gerais, the relative volume of pores >50 μm decreased by 25%–60% at a 0- to 3-cm depth depending on weed control systems and by ~30% at a 10- to 13-cm depth, in comparison to soils under native dry forest (Araujo Jr. et al. 2011). However, in most cases, this compaction does not hinder coffee productivity, and if this occurs, the stands are cut and the soil tilled again for a new plantation. When the coffee plants are spaced densely (2000–3000 plants ha^{-1}), biomass production by the crop and weeds is significant and SOC levels may be preserved depending on weed control systems. At the same site mentioned above, SOC stocks to a 0- to 30-cm depth (after correction for soil compaction) were similar to those under native forest in plots where weeds were allowed to grow or controlled by mowing (Cogo et al., in press). However, SOC stocks decreased by ~20% when weeds were controlled under rototiller, manual hoeing, harrowing, and postemergence herbicide, and 35% under preemergence herbicide. The use of agroforestry systems for coffee production, still mostly experimental, may lead to higher soil quality for coffee production.

3.2.5 PLANTATION FORESTS

Clearing the Cerrado vegetation for agriculture has always resulted in a variable amount of dead tree stems that were mostly short (<15 m long) and so twisted that they could not be used in sawmills. Using these stems and small woody residues for charcoal production in rustic, earthen ovens was an obvious alternative, and this became a common activity by landless workers, becoming widespread with large land use conversion. After 1950, the demand for wood charcoal by large iron-and-steel plants in Minas Gerais spiked, and it soon became obvious that the native Cerrado vegetation could not meet the high demand. Thus, large-scale forest plantations with fast-growing *Eucalyptus* spp. started spreading, supplying wood not only for charcoal but also, more recently, for pulp-and-paper industries. Many small plantations were also established for on-farm use as fences, fuelwood, and building works. Large forest plantations with *Pinus* spp. also became common in some southern Cerrado areas for diverse uses such as timber, resin, and pulp production. All forest plantations in the Cerrado occupied 3.1 million ha in 2002 (Sano et al. 2010) or about half of the total reforested area in Brazil. Although *Eucalyptus* and *Pinus* forests are planted in every type of Cerrado soil, due to high land prices, they are generally established on marginal lands, that is, degraded pastures, eroded Haplustoxes, unfertile Quartzipsamments, and sometimes, hilly or rocky areas. Other plantation forests were initially poorly adapted to Cerrado conditions and thus had low yields, but after decades of genetic improvement, these plantations are highly competitive with other regions of Brazil, where *Eucalyptus* productivities are among the highest in the world.

When not established on degraded pastures or croplands, *Eucalyptus* and *Pinus* stands are planted after clearance of native vegetation. Soil preparation typically involves heavy harrowing of the entire area, but often only in the planting row, followed by digging pits with shovels or hoes for the seedlings. Liming is not traditionally done, and fertilization is mostly restricted to <1 Mg rock phosphate ha^{-1} and ~100 kg ha^{-1} of soluble P and NPK mixed in the pit, before and sometimes after planting (Zinn et al. 2011). Weed control is done chemically or by hand or mechanical hoes, but after the second year, most stands grow so high that shading precludes weed growth. The management of *Eucalyptus* plantations depends on the industrial purpose: plantations for charcoal are typically established with genetically heterogeneous, seed-produced trees, which are cut by age 6 to 7 years and allowed to sprout as coppice stands, which are cut again after additional 6 to 7 years (Zinn et al. 2011), and sometimes allowed to grow a third, low-production cycle. Conversely, plantations for pulp and paper use highly selected, productive clone seedlings, which are cut at ages 5 to 7 years and replanted. Pine stands are typically formed only by seeds and use much less fertilizers and inputs than *Eucalyptus* plantations; also, they are harvested at ages 15 to 20 years and then replanted. For both species, pesticide use in the field is generally low or null, as pests and diseases are prevented by resistance traits selected through genetic breeding and, in the case of insect pests, with biological control by natural enemies introduced or preexistent in nearby areas under native vegetation.

As for pastures, soil management in plantation forests is sparse and of little intensity when compared to croplands, but it still affects soil properties depending on soil type and site preparation. Planting *E. camaldulensis* Dehnh. using heavy harrow on the whole area resulted in lower SOC concentrations at the 0- to 5-cm depth by the age of 7 years, and more intensely on sandy than on loamy Haplustoxes in the central Cerrado region (Zinn et al. 2002). However, after the stands were cut and allowed to sprout as coppice stands without further site preparation for 7 more years, SOC was resequestered and no significant differences in SOC stocks occurred in comparison to the native vegetation (Zinn et al. 2011). On the other hand, when *Eucalyptus* trees were planted without an intensive site preparation, SOC stocks at age 16 years were similar or even larger than those under native Cerrado (Resck et al. 2000). Thus, synthesis and analysis of a large number of published reports show no significant difference between SOC stocks under *Eucalyptus* plantations and native vegetations in Brazil as a whole and in the Cerrado region (Fialho and Zinn 2012). *Pinus* plantations are much less studied, but it is known that the slow decomposition of its needle layers favors wildfires and tends to result in lower surface SOC levels (Zinn et al. 2002).

Soil physical properties are also affected by afforestation, depending on site preparation. Zinn et al. (2011) observed that mean aggregate size under *Eucalyptus* was always smaller than that under native vegetation, suggesting that soil aggregation recovers more slowly than SOC when disturbance is interrupted. Soil compaction following the use of heavy harvesters, feller bunchers, and skidders to extract wood has been reported elsewhere in Brazil, but not yet in the Cerrado region, where these machines are less common and few or no studies have been conducted.

3.2.6 BIOFUEL CROPS

Brazil is well known for having perhaps the most successful industry of first-generation biofuels, based on the production of ethanol from sugarcane (*Saccharum* spp.), used as single fuel or mixed with gasoline for most modern Brazilian cars, which run with both fuels (i.e., "flex" cars). With regards to agricultural and industrial initiatives in Brazil, the success of biofuels is attributed to large public investments on research and technology and intervention on markets (Moraes 2011), mandated after the petroleum crisis of 1973. Recent efforts have also focused on plant-derived oils, which federal law mandates to be mixed with diesel at a 2% rate ("biodiesel") for use by trucks. Although crops such as sunflower (*Helianthus annuus* L.) and others are currently spreading in the Cerrado and other regions to meet this demand, such efforts are at a much lesser scale and profitability than for sugarcane ethanol, which is the focus of this section.

Sugarcane is an important biofuel crop in the world, and Brazil is its major producer, especially along the Atlantic coast and in the State of São Paulo, where it is often planted on former Cerrado land. Currently, sugarcane plantations are spreading through many Cerrado areas where they have never been planted before, such as in Goiás, due to high demand not only for ethanol but also for sugar. Sugarcane plantations are generally large and established on Haplustoxes and Haplustults of low to moderate relief. Typically, sugarcane is planted as stem cuttings with buds, which are inserted by hand or mechanically along planting rows, usually following site preparation with plow and harrow, or only heavy harrow. Harvest takes place annually and is still mostly done by hand and machete after burning, although mechanical harvesting is gradually increasing. Fertilizer use is generally less than 100 kg N ha^{-1} year^{-1}, and it has been estimated that biological N fixation supplies ~60% of crop demands (Lisboa et al. 2011).

Since sugarcane is a perennial crop commonly lasting six annual cycles, soil does not need to be disturbed annually, yet soil degradation and erosion under sugarcane plantations is widespread, especially in coarse-textured soils and where harvest is preceded by burning. Cerri et al. (2011) estimated that conversion from burned to unburned plantations can sequester on average 0.7 and 2.0 Mg SOC ha^{-1} year^{-1} for sandy and clayey Oxisols, respectively. Although sugarcane ethanol is one of the cleanest biofuels, N_2O emissions from N fertilization and C losses from conversion of native vegetation to sugarcane may offset CO_2 mitigation from petrol substitution (Lisboa et al. 2011). The emission of other air pollutants is also significant and probably underestimated, even where burning is being abandoned (Tsao et al. 2011). Despite the availability of technologies for the use of sugarcane residues (bagasse and vinasse), its potential for energy generation is still largely untapped and the disposal of these materials is a major concern.

3.2.7 Crop/Livestock Integrations

Crop/livestock integrations are innovative land use systems based on millennia-old practices, in which crops and pastures are alternated on the same land, aiming at a combination of management practices and effects on soil properties that eventually enhance soil quality, productivity, and biological diversity. There is large global variability on how this system is used, and in the Cerrado, it commonly involves three major options (Vilela et al. 2008): (1) pasture-based farms that plant crops in order to enhance productivity by reducing liming and fertilization costs with the income from grains; (2) crop-based farms that plant forage grasses in the dry season aiming to increase mulching for NT crops (especially maize and *Sorghum*); and (3) farms that systematically do crop/pasture rotations in all or most areas.

Despite the variability of these systems and the relatively recent onset of systematic studies on the theme, there exists research evidence that crop/livestock integrations may enhance soil quality in the Cerrado. In a 13-year-long experiment near Brasília, Marchão et al. (2007) reported increased bulk density and reduced macropore volume in comparison to the native Cerrado vegetation. Although the compaction was statistically significant, the magnitude of changes was too small to raise any concern. In the Cerrado of Mato Grosso do Sul, Salton et al. (2011) reported that SOC sequestration rates in soybean/livestock rotations were about half of those under continuous *Brachiaria* pastures (~1 Mg ha^{-1} year^{-1}), whereas NT and conventional till plots showed no SOC changes even after 10 years. Additional research is needed to address a wider range of combinations of crop and grazing species, and management practices.

3.3 FUTURE TRENDS: LAND USE, RESEARCH, AND ENVIRONMENT

In 1960, no one would predict that the Cerrado region could produce three crops per year, or grow soybean varieties totally independent of N fertilizers yet highly productive, or would adopt NT systems faster than anywhere else in the world. Indeed, the mere vision of the small, twisted trees and the highly variable grass strata on the top of those nutrient-depleted, acidic soils was enough to deem "Cerrado agriculture" an oxymoron. By the same token, predicting how agriculture and soil management will look like in the Cerrado for the next 50 years is perhaps no more than an educated guess (still much clearer, however, than foreseeing the same for the Amazon). Nevertheless, it is apparent that some issues will become more and more important for soil management in the next few decades, which are given as follows.

(1) *Climate change.* Some climatic models simulate that precipitations by 2020 to 2080 in the Cerrado core area will decrease by ~10%, whereas MAT will steadily increase by 2°C, in comparison to the 1961–1990 period (Hamada et al. 2010). These changes may have negative impacts on the productivity and suitability of the area for major crops in Brazil and the Cerrado (Assad et al. 2008). Regardless of the predictions, researchers in land use planning and genetic breeding must consider adaptive or mitigation strategies. Soil scientists must provide answers to questions such as how SOC storage, water retention, GHG emissions, and other critical processes in Cerrado soils will respond to higher atmospheric CO_2 and temperature, both in managed and natural ecosystems.

(2) *Assessment of natural resources.* Sano et al. (2010) stated that, prior to their mapping work, the only wide assessment of land use in the Cerrado had been made in the 1970s (the large-scale Radambrasil project), clearly showing a general deficiency in providing critical information for land use planning. In fact, available estimates of geographic distribution and the estimated areas for the Cerrado biome vary considerably. Thus, the need for precise, digital mapping on land use and changes has often been stressed (Batlle-Bayer et al. 2010). However, soil survey and mapping efforts are mostly based on scales too small to be actually useful for local purposes. For instance, the Brazilian soil map is on the 1:5,000,000 scale, and although more recent efforts include the Minas Gerais State Soil Map on the 1:650,000 scale (Ufv-Cetec-Ufla-Feam 2010), initiatives such as county-level surveys are not generally envisaged in the near future. Innovative approaches such as the construction of thematic maps of critical soil properties (e.g., pH, SOC, and clay content) by geostatistical techniques such as kriging have been made (Silva et al. 2010; Skorupa et al. 2012), but still at a small scale, and only for few soil properties in two Brazilian states.

(3) *Nature conservation.* The Cerrado is the most biodiverse savanna biome of the world, yet conversion of native ecosystems to agriculture is occurring so rapidly that this biome is perhaps even more endangered than the

Amazon forest, with little or no international awareness. Nevertheless, precise statistics on its status are not available. The Brazilian Forest Act, currently under review, mandates that variable but considerable (e.g., >80% in the Amazon) tracts of private lands must be set up for preservation at the owner's expense. In theory, this approach could result in a widespread preservation throughout all biomes. In practice, however, this law was never fully enforced. Wherever enforced, it has resulted in a mosaic of unconnected fragments of native vegetation, which have some ecosystems services (e.g., water infiltration and erosion control), but fall short of providing adequate habitat for biodiversity, especially when surrounded by croplands (Carvalho et al. 2009). In addition, the status of public lands under native vegetation is even less clear: although government-managed nature reserves are common in the Amazon and the Atlantic Rainforest, they comprise merely 2.2% of the Cerrado area (Klink and Machado 2005). Most of these areas are preserved only because either of unsuitable slopes or of too infertile soils for agriculture. With about half of the Cerrado already converted to agriculture (Sano et al. 2010), there is still time for better land use planning that allows for adequate agricultural development and nature conservation, but in fact, the prospects for that are not good since the present laws are mostly superficial and are poorly enforced.

(4) *Physical and economical infrastructure.* These are severe challenges to agricultural development, mostly caused by obsolete public policies. Physical infrastructure limitations (e.g., lack of on-farm grain storage, the sparse and unfunctional network of roads and railways) are obvious hindrances to agricultural development and profitability for Brazil as a whole, especially in the Cerrado (Lopes, A.S., personal communication 2011). Less obvious to international readers and investors are the extremely high interest rates (~10% $year^{-1}$), taxation (~38% of Gross Domestic Product (GDP), more than double of that in the United States), and heavy payroll costs. During the early years of Cerrado settlement, interests on public loans were equally high but got diluted by a steep annual inflation rate (Klink and Moreira 2002), and the net taxation was much lower. With current inflation rates similar to those of industrial countries, interest rates are accountably higher. While this obstacle can be passed by international credit, the same cannot be said for high taxation and payroll costs. Nevertheless, Cerrado agriculture has developed and prevailed despite these limitations, which probably will persist for decades to come. Finally, these hindrances have one unintended bright side, at least from an environmental viewpoint: deforestation and ecosystem degradation in Brazil have historically been related to economic growth and infrastructure development. It is amply clear and evident that unless adequate land use planning is devised and rigorously enforced, the long-due modernization of physical, legal, and fiscal infrastructure in Brazil would undoubtedly develop Cerrado agriculture at the sheer expense of its rich biodiversity.

3.4 CONCLUSIONS

The principal factors of soil formation, soil types, and soil management systems of the Brazilian Cerrado were briefly reviewed, as well as the most likely challenges for the future agricultural development of that biome. Limiting factors for productivity in Cerrado soils involve acidity, low fertility, and low water retention, which can be corrected by relatively inexpensive management practices. In consequence, the Cerrado is a major producer of food and fibers for internal and export markets. The Cerrado, as well as other tropical savannas, represents the last biome in which intensive, highly productive agriculture can expand with lesser environmental degradation, in comparison to the tropical rainforest biome. Thus, the continuous development of proper management practices of Cerrado soils by researchers and their adoption by farmers has global relevance, as other tropical savannas yet to be intensively developed can benefit from the knowledge, experience, and mistakes learned here.

ACKNOWLEDGMENTS

The authors would like to thank Dr. Geraldo C. Oliveira and Dr. Alfredo S. Lopes (Universidade Federal de Lavras) for comments and inputs, and Alba L.A. Skorupa (Graduate Program in Soil Science, Universidade Federal de Lavras) for the geoprocessing work in Figure 3.1.

ABBREVIATIONS

CEC: cation exchange capacity
CH$_4$: methane
M: million
MAE: mean annual evaporation
MAP: mean annual precipitation
MAT: mean annual temperature
NT: no-till
POM: particulate organic matter
SOC: soil organic carbon
SOM: soil organic matter

REFERENCES

Adámoli, J., J. Macedo, L.G. Azevedo, and J.S. Madeira Neto. 1986. Caracterização da Região dos Cerrados. In *Solos dos Cerrados: tecnologias e estratégias de manejo*, ed. W.J. Goedert, 33–74. São Paulo: Nobel.

Ajayi, A.E., M.S. Dias Jr., N. Curi, I. Gontijo, C.F. Araujo Jr., and A.I. Vasconcelos Jr. 2009. Relation of strength and mineralogical attributes in Brazilian Latosols. *Soil Till Res* 102:14–18.

Araujo Jr., C.F., M.S. Dias Jr., P.T.G. Guimarães, and E.N. Alcântara. 2011. Sistema poroso e capacidade de retenção de água em latossolo submetido a diferentes manejos de plantas invasoras em uma lavoura cafeeira. *Planta Daninha* 29:499–513.

Araujo, M.A., A.R. Guerra, A.V. Pedroso, and Y.L. Zinn. 2011. Densidade do solo sob vegetação nativa em Minas Gerais: efeito de profundidade e ordem de solo. Cong. Bras. Ci. Solo, XXXIII. *Proceedings*. Uberlândia, SBCS, 2011 (CD-ROM).

Assad, E.D., F.R. Marin, N.P. Martins, H.S. Pinto, and J. Zullo Jr. 2008. Análise de riscos climáticos para competitividade agrícola e conservação dos recursos naturais. In *Savanas: desafios e estratégias para o equilíbrio entre sociedade, agronegócio e recursos naturais*, eds. F.G. Faleiro and A.L. Farias Neto, 1083–1108. Planaltina: Embrapa Cerrados.

Batlle-Bayer, L., N.H. Batjes, and P.S. Bindraban. 2010. Changes in organic carbon stocks upon land use conversion in the Brazilian Cerrado: a review. *Agric Ecosyst Environ* 137:47–58.

Bayer, C., L. Martin-Neto, J. Mielniczuk, A. Pavinato, and J. Dieckow. 2006. Carbon sequestration in two Brazilian Cerrado soils under no-till. *Soil Till Res* 86:237–245.

Benito, N.P., M. Brossard, A. Pasini, M.F. Guimarães, and B. Bobillier. 2004. Transformations of soil macroinvertebrate populations after native vegetation conversion to pasture cultivation (Brazilian Cerrado). *Eur J Soil Biol* 40:147–154.

Bustamante, M.M.C., M. Corbeels, E. Scopel, and R. Roscoe. 2006. Soil carbon storage and sequestration potential in the Cerrado region of Brazil. In *Carbon sequestration in soils of Latin America*, eds. R. Lal, C.C. Cerri, M. Bernoux, and E. Cerri, 285–304. New York: Haworth Press.

Carsky R.J., S.S. Jagtap, G. Tian, N. Sanginga, and B. Vanlauwe. 2000. Maintenance of soil organic matter and nitrogen supply in the moist savanna zone of West Africa. In *Soil quality and agricultural sustainability*, ed. R. Lal, 223–236. Chelsea: Ann Arbor Press.

Carvalho Jr., W., C.E.G.R. Schaefer, C.S. Chagas, and E.I. Fernandes Filho. 2008. Análise multivariada de argissolos da faixa Atlântica brasileira. *Rev Bras Cienc Solo* 32:2081–2090.

Carvalho, F.M.V., P. Marco Jr., and L.G. Ferreira. 2009. The Cerrado into pieces: habitat fragmentation as a function of landscape use in the savannas of central Brazil. *Biol Conserv* 142:1392–1403.

Cerri, C.C., M.V. Galdosa, S.M.F. Maia, M. Bernoux, B.J. Feigl, D. Powlson, and C.E.P. Cerri. 2011. Effect of sugarcane harvesting systems on soil carbon stocks in Brazil: an examination of existing data. *Eur J Soil Sci* 62:23–28.

Chapuis-Lardy, L.C., M. Brossard, M.L.L. Assad, and J.Y. Laurent. 2002. Carbon and phosphorus stocks of clayey Ferralsols in native Cerrado and agroecosystems, Brazil. *Agric Ecosyst Environ* 92:147–158.

Cogo, F.D., C.F. Araujo Jr., Y.L. Zinn, M.S. Dias Jr., E.N. Alcântara, and P.T.G. Guimarães. 2011. Estoques de carbono orgânico do solo em cafezais sob diferentes controles de plantas invasoras. *Semin-Cienc Agr* (in press).

Costa Jr., C., M.C. Picollo, M. Siqueira Neto, P.B. Camargo, C.C. Cerri, and M. Bernoux. 2011. Carbono total e δ13C em agregados do solo sob vegetação nativa e pastagem no bioma Cerrado. *Rev Bras Cienc Solo* 35:1241–1252.

Curi, N., and D.P. Franzmeier. 1984. Toposequence of Oxisols from the Central Plateau of Brazil. *SSSAJ* 48:341–346.

Darwin, C. 1882. *The formation of vegetable mould, through the action of worms, with observations on their habits*. London: John Murray. 7th ed.

Embrapa. 2006. *Sistema Brasileiro de classificação de solos*. Rio de Janeiro: Embrapa Solos. 306 p.

Ferreira, M.M., B. Fernandes, and N. Curi. 1999. Mineralogia da fração argila e estrutura de Latossolos da região Sudeste do Brasil. *Rev Bras Cienc Solo* 23:515–524.

Fialho, R.C., and Y.L. Zinn. 2012. Changes in soil organic carbon under Eucalyptus plantations in Brazil: a comparative analysis. *Land Degrad Dev* (in press).

Folle, S., and J.M. Seixas. 1985. Mecanização agrícola. In *Solos dos Cerrados: tecnologias e estratégias de manejo*, ed. W.J. Goedert, 385–405. São Paulo: Nobel.

Frazão, L.A., M.C. Piccolo, B.J. Feigl, C.C. Cerri, and C.E.P. Cerri. 2010. Inorganic nitrogen, microbial biomass and microbial activity of a sandy Brazilian Cerrado soil under different land uses. *Agric Ecosyst Environ* 135:161–167.

García, M.R.F. 2008. Agricultural activities, management and conservation of natural resources of Central and South American savannas. In *Savanas: desafios e estratégias para o equilíbrio entre sociedade, agronegócio e recursos naturais*, eds. F.G. Faleiro and A.L. Farias Neto, 263–281. Planaltina: Embrapa Cerrados.

Grace J., J. San José, P. Meir, H.S. Miranda, and R.A. Montes. 2006. Production and carbon fluxes of tropical savannas. *J Biogeogr* 33:387–400.

Hamada, E., R. Ghini, R.R.V. Gonçalves, J.A. Marengo, and M.C. Thomaz. 2010. *Atlas digital dos cenários climáticos futuros projetados para o Brasil com base no Terceiro Relatório do IPCC (2001): variáveis de interesse agrícola*. Jaguariúna, Embrapa Meio Ambiente, Documentos, 82 (CD-ROM).

House, J.I., and D.O. Hall. 2001. Productivity of tropical savannas and grasslands: past, present and future. In *Terrestrial global productivity*, eds. H.A. Mooney, J. Roy, and B. Saugier, 363–400. San Diego: Associated Press.

Inmet (Instituto Nacional de Meteorologia). 1992. *Normais climatológicas*. Brasília: Inmet.

Jantalia, C.P., D.V.S. Resck, B.R.J. Alves, L. Zotarelli, S. Urquiaga, and R.M. Boddey. 2007. Tillage effect on C stocks of a clayey Oxisol under a soybean-based crop rotation in the Brazilian Cerrado. *Soil Till Res* 95:97–109.

Jenny, H. 1941. *Factors of soil formation—a system of quantitative pedology*. New York City, Dover Books. 1994 unabridged edition.

Klink, C.A., and A.G. Moreira. 2002. Past and current human occupation, and land use. In *The Cerrados of Brazil*, eds. P.S. Oliveira and R.J. Marquis, 69–88. New York: Columbia University Press.

Klink, C.A., and R.B.A. Machado. 2005. Conservação do Cerrado Brasileiro. *Megadiversidade* 1:147–155.

Lal, R. 2008. Sequestration of atmospheric CO_2 in global carbon pools. *Energy Environ Sci* 1:86–100.

Lilienfein, J., W. Wilcke, L. Vilela, M. Ayarza, S.C. Lima, and W. Zech. 2003. Soil fertility under native Cerrado and pasture in the Brazilian savanna. *SSSAJ* 67:1195–1205.

Lisboa, C.C., K. Butterbach-Bahl, M. Mauder, and R. Kiese. 2011. Bioethanol production from sugarcane and emissions of greenhouse gases—known and unknowns. *GCB Bioenergy* 3:277–292.

Lopes, A.S., and F.R. Cox. 1977. Cerrado vegetation in Brazil: an edaphic gradient. *Agron J* 69:828–831.

Macedo, J., and R.B. Bryant. 1987. Morphology, mineralogy and genesis of a hydrosequence of Oxisols in Brazil. *SSSAJ* 51:690–698.

Maquere, V., J.P. Laclau, M. Bernoux, L. Saint-Andre, J.L.M. Gonçalves, C.C. Cerri, M.C. Piccolo, and J. Ranger. 2008. Influence of land use (savanna, pasture, Eucalyptus plantations) on soil carbon and nitrogen stocks in Brazil. *Eur J Soil Sci* 59:863–877.

Marchão, R.L., L.C. Balbino, E.M. Silva, J.D.G. Santos Jr., M.A.C. Sá, L. Vilela, and T. Becquer. 2007. Qualidade física de um Latossolo Vermelho sob sistemas de integração lavoura-pecuária no Cerrado. *Pesq Agropec Bras* 42:873–882.

Marques, J.J., D.G. Schulze, N. Curi, and S.A. Mertzman. 2004. Major element geochemistry and geomorphic relationships in Brazilian Cerrado soils. *Geoderma* 119:179–195.

Moraes, M. 2011. Lessons from Brazil. *Nature* 474:s25.

Moreira, H.L., and V.A. Oliveira. 2008. Evolução e gênese de um Plintossolo Pétrico Concrecionário êutrico argissólico no município de Ouro Verde de Goiás. *Rev Bras Cienc Solo* 32:1683–1690.

Nunes, R.S., A.A.C. Lopes, D.M.G. Sousa, and I.C. Mendes. 2011. Sistemas de manejo e os estoques de carbono e nitrogênio em Latossolo de Cerrado com a sucessão soja-milho. *Rev Bras Cienc Solo* 35:1047–1419.

Padmanabhan, E., and H. Eswaran. 2002. Plinthite and petroplintite. In *Encyclopaedia of Soil Science*, ed. Lal, R., 1008–1011. New York: Marcel Dekker Inc.

Pes, L.Z., T.J.C. Amado, N. La Scala Jr., C. Bayer, and J.E. Fiorin. 2011. The primary sources of carbon loss during the crop-establishment period in a subtropical Oxisol under contrasting tillage systems. *Soil Till Res* 117:163–171.

Resck, D.V.S. 1998. Agricultural intensification systems and their impact on soil and water quality in the Cerrados of Brazil. In *Soil quality and agricultural sustainability*, ed. R. Lal, 288–300. Chelsea: Ann Arbor Press.

Resck, D.V.S., C.A. Vasconcelos, L. Vilela, and M.C.M. Macedo. 2001. Impact of conversion of Brazilian Cerrados to cropland and pastureland on soil carbon pool and dynamics. In *Global climate change and tropical ecosystems*, eds. R. Lal, J.M. Kimble, and B.A. Stewart, 169–196. Boca Raton: CRC Press.

Resck, D.V.S., E.A.B. Ferreira, C.C. Figueiredo, and Y.L. Zinn. 2008. Dinâmica da matéria orgânica no Cerrado. In *Fundamentos da matéria orgânica do solo: ecossistemas tropicais e subtropicais*, eds. G.A. Santos, L.S. Silva, L.P. Canellas, and F.A. Camargo, 359–417. Porto Alegre: Metrópole.

Rippstein, G., E. Amézquita, G. Escobar, and C. Grollier. 2001. Condiciones naturals de la sabana. In *Agroecología y biodiversidad de las sabanas en los Llanos Orientales de Colombia*, eds. G. Rippstein, G. Escobar, and F. Motta, 1–21. Cáli: CIAT. Available at http://www.ciat.cgiar.org/downloads/pdf/Agroecologia_y_biodiversidad.pdf (accessed December 23, 2011).

Romero-Ruiz, M.H., S.G.A. Flantua, K. Tansey, and J.C. Berrio. 2012. Landscape transformations in savannas of northern South America: land use/cover changes since 1987 in the Llanos Orientales of Colombia. *Appl Geogr* 32:766–776.

Rondón, M.A., D. Acevedo, R.M. Hernandez et al. 2006. Carbon sequestration potential of the neotropical savannas of Colômbia and Venezuela. In *Carbon sequestration in soils of Latin America*, eds. R. Lal, C.C. Cerri, M. Bernoux, and E. Cerri, 213–243. New York: Haworth Press.

Saadi, A., F.H.R. Bezerra, R.D. Costa, H.L.S. Igreja, and E. Franzinelli. 2005. Neotectônica da plataforma brasileira. In *Quaternário do Brasil*, ed. C.R.G. Souza, K. Suguio, A.M.S. Oliveira, P.E. Oliveira, 211–234. Ribeirão Preto: Holos.

Salton, J.C., J. Mielniczuk, C. Bayer, M. Boeni, P.C. Conceição, A.A. Fabrício, M.C.M. Macedo, and D.L. Broch. 2008. Agregação e estabilidade de agregados do solo em sistemas agropecuários no Mato Grosso do Sul. *Rev Bras Cienc Solo* 32:11–21.

Salton, J.C., J. Mielniczuk, C. Bayer, A.C. Fabrício, M.C.M. Macedo, and D.L. Broch. 2011. Teor e dinâmica do carbono no solo em sistemas de integração lavoura-pecuária. *Pesq Agropec Bras* 46:1349–1356.

Sano, E., R. Rosa, J.L.S. Brito, and L.G. Ferreira. 2010. Land cover mapping of the tropical savanna region in Brazil. *Environ Monit Assess* 166:113–124.

Schaefer, C.E.R. 2001. Brazilian latosols and their B horizon microstructure as long-term biotic constructs. *Aust J Soil Res* 39:909–926.

Serafim, M.E., G.C. Oliveira, A.S. Oliveira, J.M. Lima, P.T.G. Guimarães, and J.C. Costa. 2011. Sistema conservacionista e de manejo intensivo do solo no cultivo de cafeeiros na região do alto São Francisco, MG: um estudo de caso. *Biosci J* 27:964–977.

Severiano, E.C., G.C. Oliveira, M.S. Dias Jr., K.A.P. Costa, M.B. Castro, and E.N. Magalhães. 2010. Potencial de descompactação de um Argissolo promovido pelo capim-tifton 85. *Rev Bras Eng Agric Ambient* 14:39–45.

Silva, J.E., J. Lemainski, and D.V.S. Resck. 1994. Perdas de matéria orgânica e suas relações com a capacidade de troca catiônica em solos da região de Cerrados do oeste baiano. *Rev Bras Cienc Solo* 18:541–547.

Silva, R.B., M.S. Dias Jr., F.A.M. Silva, and S.M. Folle. 2003. O tráfego de máquinas agrícolas e as propriedades físicas, hídricas e mecânicas de um latossolo dos Cerrados. *Rev Bras Cienc Solo* 27:973–983.

Silva, A.R., I.G. Souza Jr., and A.C.S. Costa. 2010. Suscetibilidade magnética do horizonte B de solos do Estado do Paraná. *Rev Bras Cienc Solo* 34:329–338.

Siqueira Neto, M., M.C. Piccolo, C. Costa Jr., C.C. Cerri, and M. Bernoux. 2011. Emissão de gases do efeito estufa em diferentes usos da terra no bioma Cerrado. *Rev Bras Cienc Solo* 35:63–76.

Skorupa, A.L.A., G.C. Oliveira, and Y.L. Zinn. 2011. Micromorfologia de um latossolo do cerrado sob diferentes sistemas de manejo e preparo por vinte anos. In XXXIII Cong. Bras. Ci. Solo. *Proceedings*. Uberlândia: SBCS.

Skorupa, A.L.A., L.R.G. Guilherme, N. Curi, C.P.C. Silva, J.R. Scolforo, and J.J.G.S.M. Marques. 2012. Propriedades de solos sob vegetação nativa em Minas Gerais: distribuição por fitofisionomia, hidrografia e variabilidade espacial. *Rev Bras Cienc Solo* 36:11–22.

Skorupa, A.L.A., M. Fay, Y.L. Zinn, and M. Scheuber. Assessing hydric soils in a gallery forest in the Brazilian Cerrado. *Soil Use Manage* (in press).

Soil Survey Staff. 2010. *Keys to Soil Taxonomy*. Washington: USDA/NRCS, 337 p.

Spehar, C.R. 1998. Production systems in the savannas of Brasil: key factors to sustainability. In *Soil quality and agricultural sustainability*, ed. R. Lal, 301–318. Chelsea: Ann Arbor Press.

Thomas, M.F. 1994. Geomorphology in the tropics: a study of weathering and denudation in low latitudes. Chichester: John Wiley & Sons, 460 p.

Trescases, J.J., A.J. Melfi, and S.M.B. Oliveira. 1979. Nickeliferous laterites of Brazil. In: *Lateritisation processes* (Proceedings of the International Seminar on Lateritisation Processes, Trivandrum, 1979). Rotterdam, A.A. Balkema, pp. 170–183.

Tsao, C.C., J.E. Campbell, M. Mena-Carrasco, S.N. Spak, G.R. Carmichael, and Y. Chen. 2011. Increased estimates of air-pollution emissions from Brazilian sugar-cane ethanol. *Nat Clim Change* 2:53–57.

Ufv-Cetec-Ufla-Feam. 2011. Mapa de solos do Estado de Minas Gerais. Belo Horizonte: Fundação Estadual do Meio Ambiente, 49 p. Available at http://www.feam.br/noticias/1/949-mapas-de-solo-do-estado-de-minas-gerais (accessed February 2, 2011).

Van Wambeke, A. 1992. *Soils of the tropics: properties and appraisal*. New York: McGraw-Hill. 343 p.

Vendrame, P.R.S., O.R. Brito, M.F. Guimarães, E.S. Martins, and T. Becquer. 2010. Fertility and acidity status of latossolos (oxisols) under pasture in the Brazilian Cerrado. *An Acad Bras Cienc* 82:1085–1094.

Vilela, L., G.B. Martha Jr., R.L. Marchão, R. Guimarães Jr., L.G. Barioni, and A.O. Barcellos. 2008. In *Savanas: desafios e estratégias para o equilíbrio entre sociedade, agronegócio e recursos naturais*, eds. F.G. Faleiro and A.L. Farias Neto, 933–962. Planaltina: Embrapa Cerrados.

Zinn, Y.L., and D.V.S. Resck. 2008. O cascalho como indicador de poligênese em Latossolos do Cerrado. In Simpósio sobre o Cerrado, IX, Brasília, 2008. Anais. Available at http://www.cpac.embrapa.br/publicacoes/search_pbl/1?q=Micromorfologia (accessed January 12, 2012).

Zinn, Y.L., D.V.S. Resck, and J.E. Silva. 2002. Soil organic carbon as affected by afforestation with *Eucalyptus* and *Pinus* in the Cerrado region of Brazil. *For Ecol Manage* 166:285–294.

Zinn, Y.L., R. Lal, and D.V.S. Resck. 2005a. Texture and organic carbon relation described by a profile pedotransfer function in Brazilian Cerrado soils. *Geoderma* 127:168–173.

Zinn, Y.L., R. Lal, and D.V.S. Resck. 2005b. Changes in soil organic carbon stocks through agriculture in Brazil. *Soil Till Res* 84:28–40.

Zinn, Y.L., R. Lal, J.M. Bigham, and D.V.S. Resck. 2007a. Edaphic controls on soil organic carbon retention in the Brazilian Cerrado: texture and mineralogy. *SSSAJ* 71:1204–1214.

Zinn, Y.L., R. Lal, J.M. Bigham, and D.V.S. Resck. 2007b. Edaphic controls on soil organic carbon retention in the Brazilian Cerrado: soil structure. *SSSAJ* 71:1215–1224.

Zinn, Y.L., R. Lal, and D.V.S. Resck. 2011. Eucalypt plantation effects on organic carbon and aggregation of three different-textured soils in Brazil. *Soil Res* 49:614–624.

4 Facts and Myths of Feeding the World with Organic Farming Methods

Bobby A. Stewart, Xiaobo Hou, and Sanjeev Reddy Yalla

CONTENTS

4.1 Introduction 87
4.2 Organic Farming..88
    4.2.1 Soil Productivity of Organic Farming 89
    4.2.2 Safety of Organic Crops Compared to Crops Produced with Chemical Fertilizers 92
    4.2.3 Quality of Organic Crops Compared to Crops Produced with Chemical Fertilizers 94
4.3 Limitations of Organic Farming..96
4.4 Need for Better Management of N Fertilizers...99
4.5 Intensification of Animal Agriculture and Translocation of Nutrients 102
4.6 Integrating the Use of Organic Materials, Biological Processes, and Chemical Fertilizers 103
4.7 Conclusions 105
Abbreviations 106
References... 106

4.1 INTRODUCTION

During the past few years, two issues have somewhat dominated the future of world agriculture. The first is that the world population that has passed 7 billion people in 2011 is expected to reach more than 9 billion by 2050. Even more daunting, while the world population is expected to increase by about 30%, food production, particularly cereals, is expected to increase 70% or more because of increasing prosperity that is resulting in changing diets that include more meat, eggs, and dairy products. At the same time, an increasing number of people think our global food system is rapidly approaching, if not already in, a condition of crisis, and a conversion to sustainable agriculture is required. For many of these concerned individuals, synthetic nitrogen fertilizers, irrigation, pesticides, and loss of genetic diversity that have been so important in increasing food production during the past 50 years are not sustainable, and a conversion to more sustainable farming systems is necessary.

For the 50 years between 1961 and 2011, world population increased from 3 billion to 7 billion, but cereal production increased even faster from about 875 million tonnes (Mt) to almost 2500 Mt (FAOSTAT 2012). Therefore, the agricultural community did a remarkable job of increasing food production at a rate faster than population growth, and while there was still a segment of the population that was malnourished, the world food situation improved considerably during this period. Much of the increased cereal production during this 50-year period was due to a doubling of irrigated land from about 140 to 280 million ha and an increase in synthetic N fertilizer from about 9 to 100 Mt (Smil 2011).

Organic agriculture, generally considered as a more sustainable system, has grown rapidly in the past few years. From 1999 to 2010, organic agriculture increased almost fourfold from 11 to 37 Mha. While this is still a tiny portion of the world's cropland, its rapid growth is occurring in more than 100 countries (IFOAM 2012).

While some believe that more sustainable systems like organic farming can produce enough food to feed the world, others believe that the use of synthetic nitrogen fertilizer is absolutely essential. This paper will look at some of the facts and myths of organic farming methods.

4.2 ORGANIC FARMING

As defined by IFOAM (2012), organic agriculture is a production system that sustains the health of soils, ecosystems, and people. It relies on ecological processes, biodiversity, and cycles adapted to local conditions, rather than the use of inputs with adverse effects. Organic agriculture combines tradition, innovation, and science to benefit the shared environment and promote fair relationships and a good quality of life for all involved. IFOAM (2012) further elaborates that organic farming is based on four principles, as follows:

1. *Health:* Organic agriculture should sustain and enhance the health of soil, plant, animal, human, and planet as one and indivisible.
2. *Ecology:* Organic agriculture should be based on living ecological systems and cycles, work with them, emulate them, and help sustain them.
3. *Fairness:* Organic agriculture should build on relationships that ensure fairness with regard to the common environment and life opportunities.
4. *Care:* Organic agriculture should be managed in a precautionary and responsible manner to protect the health and wellbeing of current and future generations and the environment.

Few people, if any, would find fault with these principles. Many, however, would say that, with the present and growing number of people, without inputs, especially irrigation and synthetic N, food production needs simply cannot be met. Even though the land area devoted to organic agriculture has grown almost fourfold in the past

10 years to 37 Mha (IFOAM 2012), it represents only about 2.7% of the 1381 Mha of the cropland estimated by FAO (FAOSTAT 2012).

4.2.1 Soil Productivity of Organic Farming

Numerous literature reviews have been conducted to compare soil productivity and crop yields of organic fields with those of conventional production fields. Not surprisingly, the conclusions drawn by different authors have sometimes varied. Edmeades (2003) reviewed the results of 14 long-term experiments conducted in North America and Europe, all of which had yield data for more than 20 years. The yields from plots treated with balanced chemical fertilizer were similar to those from plots treated with applications of manure when the nutrient rates were equivalent. Edmeades (2003) concluded that only after a very large amount of manure has been applied for a very long time, which may result in significantly higher soil organic carbon, can any additional benefit of manure be observed. Higher soil organic carbon can have beneficial effects on water infiltration and storage, and on soil biological properties that can increase yields of some soils under some conditions. For a specific site, application of manure can result in higher, similar or lower yield compared with chemical fertilizer applications, depending on the previous soil organic C pool, soil management, soil type, rate and frequency of chemical and organic fertilizer application, rainfall and irrigation amounts, and other factors (Edmeades 2003; Lal 2006; Diacono and Montemurro 2010; Miao et al. 2010). Edmeades (2003) stated that it cannot be assumed that the long-term use of manures will enhance soil quality—defined in terms of productivity and potential to adversely affect water quality—in the long term relative to applying the same amounts of nutrients as fertilizer. However, the most important point to be gained from the Edmeades review is perhaps the fact that the input of nutrients of either fertilizers or manures had very large effects (150% to 1000%) on soil productivity as measured by crop yields. Clearly, high yields require high inputs of nutrients.

Badgley et al. (2007) stated that the principal objections to the proposition that organic agriculture can contribute significantly to the global food supply are low yields and insufficient quantities of organically acceptable fertilizers. They carried out an extensive review of the literature specifically to evaluate, and dispel, these claims. For comparing yields, Badgley et al. (2007) compared yields of organic versus conventional or low-intensive food production for a global dataset of 293 examples and estimated the average yield ratio (organic/nonorganic) of different food categories for the developed and developing world. They reported that for most food categories, the average yield ratio was slightly less than 1.0 for studies in the developed world and greater than 1.0 for studies in the developing world. Badgley et al. (2007) also reviewed 77 studies where leguminous cover crops were grown to supply nitrogen to soils and concluded that the amount of nitrogen potentially available from fixation by leguminous crops used as fertilizer could replace the amount of synthetic fertilizer currently in use without adding any additional cropland. In drawing these conclusions, they assumed that the cover crops could be grown between normal cropping periods. This assumption is clearly not valid for much, if not most, of the cropland in the world.

Seufert et al. (2012) reviewed 66 studies representing 62 study sites that reported 316 organic-to-conventional yield comparisons on 34 different crop species. Their analysis was restricted to (1) organic systems defined as those with certified organic management or noncertified organic management, following the standards of organic certification bodies; (2) studies with comparable spatial and temporal scales for both organic and conventional systems; and (3) studies reporting sample size and error or information where these could be estimated. Seufert et al. (2012) stated that, overall, organic yields are typically lower than conventional yields, but the differences were highly contextual, depending on system and site characteristics, and range from 5% lower organic yields (rain-fed legumes and perennials on weak-acidic to weak alkaline soils), 13% lower yields (when best organic practices are used), to 34% lower yields (when the conventional and organic systems are most comparable). Under conditions with good management practices, however, they found that organic systems can nearly match conventional yields for particular crop types. For example, organic fruits and oilseed crops showed slightly lower, but not statistically significant, yields of 3% and 11%, respectively, when compared to conventional crop yields. In contrast, organic cereals and vegetables had significantly lower yields of 26% and 33%, respectively. Seufert et al. (2012) stated that part of the yield response could be explained by differences in the amount of N input received by the systems. Most of the N in organic materials is only slowly available and often over multiple crops, whereas N fertilizer is readily available. Therefore, Seufert et al. (2012) stated that organic systems appear to be N limited, whereas conventional systems are not.

de Ponti et al. (2012) conducted a similar study to that of Seufert et al. (2012) in that they compiled and analyzed 362 published organic–conventional comparative crop yields. They began their review with the hypothesis that the yield gap between organic and conventional agriculture increases as conventional yields increase. The stated rationale behind the hypothesis of de Ponti et al. (2012) was that when conventional yields are high and relatively close to the potential or water-limited level, nutrient stress, as per definition of the potential or water-limited yield levels, must be low, and pests and diseases well controlled, which are conditions more difficult to attain in organic agriculture. They showed that organic yields of individual crops are on average 80% of conventional yields, but variation was substantial having a standard deviation of 21%. They further stated that relative yields differed between crops with soybean, some other pulses, rice, and corn scoring higher than 80% and wheat, barley, and potato scoring lower than 80%. Most regions had relative yields fairly close to the overall average, but Asia and Central Europe had comparatively higher relative yields and Northern Europe had lower relative yields. In Denmark and The Netherlands, countries with very intensive agricultural systems, they found the gap between organic and conventional yields somewhat larger. de Ponti et al. (2012) concluded that the findings gave some support to their hypothesis that the organic–conventional yield gap is higher when conventional yields are high, but that the relationship and hence the evidence underpinning were not strong.

Although studies that compare yields of organic and conventional systems are of interest, it is clear that high yields are dependent on high inputs and the inputs must be available to the crop. From a theoretical viewpoint, there is every reason to believe that a well-managed organic crop can yield as much as a conventional crop as long as sufficient N, P, and other nutrients are supplied and pests are controlled. The reason that producers using conventional practices often have higher yields is generally because they have more options for supplying nutrients and controlling insects and diseases and can often react more timely. To feed a growing world population that is also becoming more prosperous resulting in diets that include more meat, dairy products, and eggs that generally require more grain and other feeds for animals, it is time to accept and promote various types of crop production. The National Research Council (2010) states that there are four goals required for a sustainable agricultural system. These are achieving sufficient productivity, enhancing the natural-resource base and environment, making farming financially viable, and contributing to the wellbeing of farmers and their communities. A combinative organic–conventional system has many advantages. Miao et al. (2010) reviewed a number of long-term experiments. A large number of agricultural experiments were initiated around the world beginning in the 1800s. Lawes and Gilbert established the Broadbalk plot in 1843 in Rothamsted, England (Rothamsted Research 2006). In the United States, the Morrow Plots were established in 1876 on the campus of the University of Illinois, Urbana; the Sanborn Field plots were started in 1888 on the campus at the University of Missouri, Columbia; and the Magruder Plot was initiated at Oklahoma State University in 1892. These plots are still active today, although some of them have been modified substantially. A number of other stations have also been in existence for more than 50 years, and several of these are listed by Miao et al. (2010). Miao et al. (2010) concluded from their review that chemical fertilizer alone is not enough to improve or maintain soil fertility at high levels, and the soil acidification problem caused by overapplication of synthetic N fertilizers can be reduced if more N fertilizer is applied as nitrate relative to ammonium- or urea-N fertilizers. Organic fertilizers can improve soil fertility and soil physical properties and result in high yields. However, long-term applications of organic fertilizers at high rates can lead to more nitrate leaching and accumulation of phosphorus. Stewart et al. (2000) discussed the ratio of N and P in manure compared with that in major crops. They stated that maize (*Zea mays*) and wheat (*Triticum aestivum*) require approximately five times as much N as P, and that the ratio of N to P in manure is only about 2.5 to 1. Thus, if sufficient manure is added to supply N needs, there is often a high accumulation of P. Even when manure supplied only part of the nutrients in a 15-year study in China by Zhang et al. (2009), there was a very high buildup of available P in the soil at the end of the study. Enriched P in the soil may be lost through runoff, or leached in soils with low P retention or in situations of organic P leaching, leading to water pollution (Edmeades 2003).

Organic fertilizers, when properly used, can provide adequate soil fertility for maximum yields. For over 4000 years, Chinese farmers managed to produce modest crop yields and maintain soil fertility using traditional farming practices, emphasizing integrated and efficient utilization of different strategies of crop rotation, intercropping, all possible resources of organic manures aiming at the most complete recycling of nutrients (e.g., animal waste, human excreta, cooking ash, compost, and dredged canal sediments, etc.), and green manures (Wittwer et al. 1987; Ellis and Wang 1997; Gao et al. 2006; Yang 2006). A Chinese farming proverb says "Farming is a joke without manuring." Many ancient Chinese publications mention

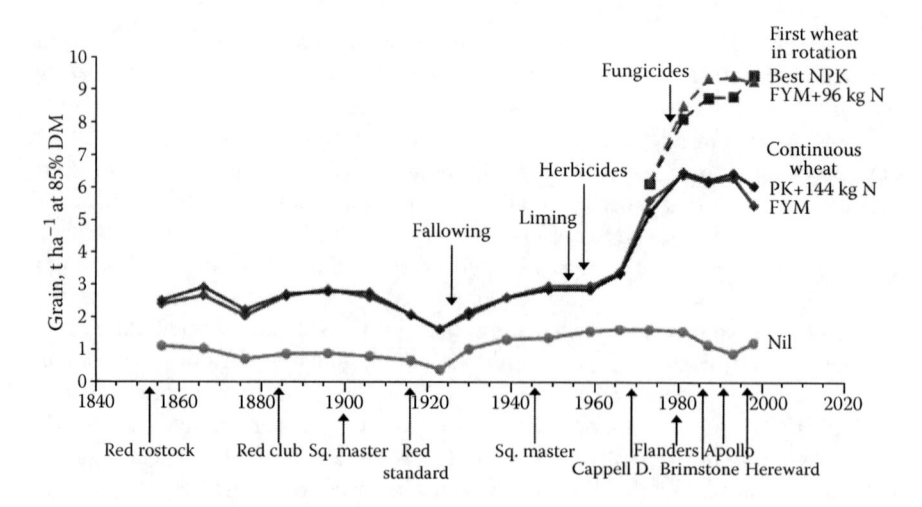

FIGURE 4.1 Mean yields of wheat grain on Broadbalk plots from 1852 to 1999 treated with NPK fertilizers, farmyard manure (FYM), FYM+N, or no fertilizers showing the effects of changing cultivars and management practices. (From Rothamsted Research. Guide to the classical and other long-term experiments, datasets and sample archive. Lawes Agricultural Trust Co. Ltd., 2006. Available at http://www.rothamsted.bbsrc.ac.uk/resources/LongTermExperiments.pdf, accessed March 1, 2012.)

the application of human and animal wastes to the land. For example, according to Yang (2006), Han Feizi (280–233 BCE) stated that human excreta must be applied to restore and improve soil fertility. King (1911) stated that "one of the most remarkable agricultural practices adopted by any civilized people is the centuries-long and well-nigh universal conservation and utilization of all human waste in China, Korea and Japan, turning it to marvelous account in the maintenance of soil fertility and in the production of food." However, these traditional agricultural practices could not produce enough food to meet the demand of a fast-growing population started around 1800. Today, the world population is 7 billion and expected to reach 9.2 billion by 2050. Therefore, as will be discussed later, the production of adequate food and fiber without the use of chemical fertilizers does not appear feasible. Miao et al. (2010) concluded that well-managed combination of chemical and organic fertilizers can overcome the disadvantages of applying a single source of fertilizers and sustainably achieve higher crop yields, improve soil fertility, alleviate soil acidification problems, and increase nutrient-use efficiency compared with only using chemical fertilizers (Figure 4.1). Of course, the combined use of organic wastes and chemical fertilizers is not an approved practice for organic farming.

4.2.2 Safety of Organic Crops Compared to Crops Produced with Chemical Fertilizers

Many people feel that safety is greater with organic than with conventional foods, primarily because of the precautionary principle followed in the formulation of organic regulations and in the assessment of food safety. Organic farming practices and regulations promote high standards of product safety. Organic foods are generally produced using lower nitrogen applications that potentially reduce nitrate concentrations. Pesticides are banned, so no pesticide residues are present. For organic meat and animal products, there are a number of rules directed toward a high status of animal welfare, care for the environment, restricted use of medical drugs, and the production of healthy products without residues of pesticides or drugs. Most developed countries have safeguards for assuring food safety, but a growing number of consumers still feel that there is enough risk for them to prefer organic foods. In developing countries, there are generally less safeguards both in applying pesticides and other chemicals to growing crops and in testing products being sold in the markets. Therefore, if recommendations and regulations for producing organic foods are properly followed and enforced in developing countries, they may very well be safer than conventionally produced food products.

Even though most developed countries have programs to ensure food safety, many people are still concerned. This seems particularly true in Europe where a number of countries show concern about food safety. Public worries relate mainly to specific agri-food-environment problems that have been recently encountered, and the following were noted by Hansen et al. 2002:

- Discovery of animals with bovine spongiform encephalopathy (BSE)
- Increased occurrence of salmonella in meat and eggs
- Increased occurrence of campylobacter in meat
- Finding of listeria in some dairy products
- Increased occurrence of dioxins in food and feed
- Excessive amounts of pesticides, antibiotics, additives, etc., in food
- Presence of toxic fungi in stored foods
- Pollution of drinking water with pesticides and nitrate
- Organic food products contaminated with genetically modified organisms (GMOs)
- Deception in sale of conventional foods as organic products

Although most food is safe, there is always some risk regardless of whether the food is produced according to organic standards or with conventional practices. Organic foods, by the very fact that they are produced without being subjected to many of the practices and chemicals that conventional foods are subjected, have several advantages from the food safety perspective. The growers of organic foods are also exposed to fewer chemicals, and this is probably of greater importance in developing countries than developed countries. Many countries have strict regulations and procedures for applying pesticides and other chemicals to crops that provide adequate protection, but this is not true in all countries. Also, the proper use of pesticides requires following the label as to how and when they are to be applied. Failure to follow the prescribed procedures will certainly result in more risk associated with foods produced under conventional practices. Ultimately, the consumer makes the final decision based on personal choices including availability and cost factors.

4.2.3 Quality of Organic Crops Compared to Crops Produced with Chemical Fertilizers

Organic products are those that are produced under controlled cultivation conditions in accordance with the provisions of the regulations on organic farming and its supplementary statutory provisions or the guidelines of the various recognized farming associations (IFOAM 2012). Organic foods can be divided into plant products and animal products. Organic plant products such as fruits, vegetables, and grains are grown without using conventional methods to fertilize or control pests. Organic animal products such as eggs, dairy products, meat, and poultry are produced without the use of antibiotics or hormones. Organic food can also be divided into fresh and processed products. Fresh organic foods are harvested directly from organic farms, such as fruits, vegetables, grains, eggs, milk, meat, and honey. Processed organic foods are food products like breakfast cereals, snack foods, canned foods, and drinks, which are completely or partially made of organic ingredients.

Both fresh and processed organic foods can be easily found today on market shelves. By reading the label of the products, organic food can be distinguished from others. In countries where organic standards are established, qualified organic foods are clearly labeled as meeting the standards. Based on the IFOAM definition (IFOAM 2012), certified organic products are those that have been produced, stored, processed, handled, and marketed in accordance with precise technical specifications (standards) and certified as "organic" by a certification body. In the United States, for example, the Organic Foods Production Act (OFPA) and the National Organic Program (NOP) assure consumers that organic agricultural products meet the prescribed standards during production, processing, and certification. Though organic farming is practiced on every continent, North America and Europe have comprised most of the global demand for organic food, while most organic production in Africa and Latin America is for export.

The growing awareness of health issues and the increased concern of environmental issues have resulted in a higher percentage of the public to focus on food quality. This in turn has increased attention paid to organic farming because products from organic farming are often perceived as having higher quality. Various rationales that consumers use when purchasing organic foods have been identified by Hughner (2007). Among all motives, including concerns for the environment, food safety, and animal welfare, the concern of health is the primary reason that consumers buy organic foods. In general, health-concerned consumers believe organic foods are more nutritious and of better quality. Several studies also found taste among the most important criteria in organic food purchases (Magnusson et al. 2001). They suggested that people buy organic food because they believe it is somewhat better than conventional food. They believe organic farms grow more nutritious and better tasting food from healthier soil with healthier agronomic methods. Numerous studies confirm that many people believe that organic foods are healthier than conventionally produced foods and that they are produced in a more environmentally compatible manner (Folkers 1983).

Although it is clear that consumers who purchase organic foods think these are better than those produced using conventional practices, there is little or no scientific basis to support their views. Studies about organic food nutrition started soon after organic farming came into being. Woese et al. (1997) reviewed numerous studies that had been published dating back to 1924. More recently, several reviews of comparative studies have been published (Worthington 2001; Williams 2002; Magkos et al. 2003). Since then, more studies on organic food quality had been published, and meta-analysis provided better comparisons. Meta-analysis is a statistical technique for combining the findings from independent studies, the validity of which is highly dependent on the quality systematic review. As more studies on organic food quality had become available in the past decade, meta-analysis was used more often in organic food quality reviews to give complete coverage of available studies.

Numerous research studies have been conducted on organic food quality. However, there is a major challenge on how to define food quality concepts and methods for determination (Kahl et al. 2012). The concepts of food quality and evaluation methods developed as the knowledge of food science expanded and new measuring technology became available. Now, sensitivity of analytical methods to measure nutrients has been increased. Taking plant-based organic food as an example, most of the early studies only evaluated dry matter, total sugars, and mineral contents. Secondary metabolites, with the exception of vitamin C and polyphenolic substances, were seldom included in these studies (Worthington 2001; Williams 2002; Magkos et al. 2003), but have received more attention in recent studies (Brandt et al. 2011).

Whether organic food has higher quality than conventionally produced food is highly dependent on how organic food quality is defined and determined. Food quality can be classified into the following fields: sensory properties, nutrition and health, authenticity and traceability, and specific organic properties (Willer and Kilcher 2012). The last two fields, however, are specific to organic food quality, so only sensory properties and nutrition and health can be used to compare the quality of organic foods and conventional foods. For both fresh and processed food, studies carried out on the sensory difference in terms of taste, smell, texture, and appearance showed a wide variety of results. In some studies, significant differences were found while other studies did not detect any differences (Willer and Kilcher 2012). Even though information on sensory differences is limited and inconclusive, more research is needed because taste and other sensory factors play an important role in purchasing decisions of consumers (Hughner et al. 2007).

Most of the data reported has been from plant products because information on organic animal products is vague and less sufficient. Early research was limited to dry matter; crude protein content; vitamins A, B_1, B_2, and C; minerals; and trace elements. Results indicated that organic plant products including leafy, root, and tuber vegetables contain more dry matter, but there was no significant difference between organically and conventionally produced fruits and fruit vegetables (Magkos et al. 2003). Due to lower nitrogen availability, organic fruits and vegetables generally have lower crude protein content but higher quality protein in some vegetables. Although the majority of comparisons on vitamins and mineral elements revealed no significant difference, Worthington (2001) reported that organic fruits and vegetables contained significantly more iron, magnesium, phosphorus, and vitamin C than conventionally produced ones. However, no general statement could be drawn because there was insufficient data for carbohydrate, protein, and other vitamins.

More recently, Brandt et al. (2011) concluded that organic crops contain significantly higher levels of secondary metabolites (compounds that are believed to protect people against a range of diseases including obesity) than conventionally grown crops.

High nitrogen availability generally results in increased protein synthesis. Thus, when N fertilizers are added to fields where grain crops are produced, these grain crops often, but not always, have higher crude protein than organically grown crops. Magkos et al. (2003) reported that organically produced grains had higher concentrations of some of the essential amino acids, but there was no clear picture. Data for comparisons of vitamins, minerals, and trace elements were too limited to draw any conclusions. More recently, Dangour et al. (2009) surveyed 52,471 articles, identified 162 studies (137 crops and 25 livestock products), and selected 55 as having data of satisfactory quality for further comparisons. From an analysis of the data from these selected studies, they concluded that conventionally produced crops had a significantly higher content of nitrogen, and organically produced crops had a significantly higher content of phosphorus and higher titratable acidity. No evidence of a difference was detected for the remaining 8 of 11 crop nutrient categories analyzed. Analysis of the more limited database on livestock products found no evidence of a difference in nutrient content between organically and conventionally produced livestock products.

4.3 LIMITATIONS OF ORGANIC FARMING

The previous sections have shown that organically grown crops can be similar to those produced using chemical fertilizers, and at least equal or superior in terms of quality and safety. The real question with regards to organic agriculture is to what extent the food and fiber needs of the world population can be produced without the use of chemical fertilizers. The world population is presently 7 billion and expected to reach 9.2 billion by 2050. Protein consumption is essential for humans, and the concentration of N in protein is approximately 16%. Smil (2002) states that published recommendations for ideal protein requirements were 3–4 g/kg of body weight for infants, 1.5–3 g for teenagers, and 0.3–1 g for adults. Smil (2011) estimated that global consumption of N contained in food was 30 Mt and essentially all of this is excreted. Furthermore, since more than 50% of people on all continents, except Africa, now live in cities, most of this waste is released directly into sewers. Most of the sewage water is either released to streams or coastal water and little is ever recycled for crop production. Therefore, a tremendous quantity of N is lost from human consumption, and there are also large losses of N each year from animals. These losses must be balanced by inputs to sustain the system. There are natural ways for providing N for crops that have been long known. Von Liebig (1840) listed three ways: (1) recycling of organic wastes (mainly crop residues and animal wastes); (2) crop rotations including N-fixing leguminous species; and (3) growing leguminous cover crops and plowing them under as green manures. While all of these can provide N to growing crops, only the N fixed from N_2 in the atmosphere by symbiotic bacteria associated with leguminous crops is added. Other sources of added N for crop production are atmospheric deposition and irrigation waters from aquifers. Ladha et al. (2005) estimated that additions to crop production from atmospheric deposition and biological N-fixation were 24 and 33 Mt, respectively. They also estimated leaching, runoff, and erosion losses, and they estimated ammonia volatilization losses from soil and vegetation as 37 and 21 Mt, respectively. These estimates indicate that losses are slightly greater than inputs from natural sources, and the differences would likely be even greater if only cropland was considered. Much of the N in crops, particularly grain crops, is removed from the crop land and is not recycled. Therefore, at the present time, additional inputs of N are essential for producing adequate food and fiber for the growing population.

In 1909, Fritz Haber, a German scientist, developed a process that takes N_2 from the air and synthesizes it to ammonia that can be used by plants or transformed into other N compounds that they can use. Within 4 years, Carl Bosch, an engineer for Germany's largest chemical company, developed the process commercially and was named the Haber–Bosch process. Haber was awarded the Nobel Prize in Chemistry in 1918, and Bosch shared the Nobel Prize in 1931 for industrializing the Haber invention. Without this synthesized N, many scientists do not think the population could have grown to today's level. Ladha et al. (2005) stated that 50% of the human population relied on N fertilizer for food production. Smil (2002) stated, "the Haber–Bosch synthesis provides the very means of survival for about 40% of humanity; that only half of today's population could be supplied by prefertilizer farming with overwhelmingly vegetarian diets; and that traditional cropping could provide today's average diets to only about 40% of the existing population." Since Smil made that statement, the world population has increased almost a billion people, calorie intake has increased, and the percent of protein coming from animal products has increased. During the first decade of the new century, world population increased by about 16%, while cereal production increased by about 20%. Maize, of which about 65% is fed to animals, increased by more than 35% (FAOSTAT 2012). Cereal production is extremely important because cereals provide a large part of people's food. The Worldwide Institute (2011) reported that, on average, people consume 48% of the world's grain supply directly, roughly 35% becomes livestock feed, and 17% is used to make ethanol and other fuels. The Worldwatch Institute (2011) further stated that humans get about 48% of their calories from eating grains directly. This has declined only slightly from 50% over the past 40 years. However, as incomes rise, the percent drops dramatically as people change diets to consume more meat, milk, eggs, and other foods that require more grain to produce. Because of projected increases in population and rising incomes, it is projected that cereal production will need to increase by 50% to 70% by 2050.

Ladha et al. (2005) estimated that during 2001 and 2002, about 60% of global N fertilizer consumed was used for cereal production. Specifically, three cereals (rice, wheat, and maize) accounted for about 56% of the N fertilizer used (IFA 2002). Norman Borlaug, the Nobel Laureate known as the father of the Green Revolution, developed short-straw wheat varieties that were credited for saving countless lives in Asia. However, he credited N fertilizers with much of the success because without added N, production would have been significantly less. Figure 4.2 shows how cereal production increased between 1961 and 2010, and also how the amount of N fixed by the Haber–Bosch process has increased. Although the Haber–Bosch process was first commercialized in 1913, it was not widely used for fertilizer until following

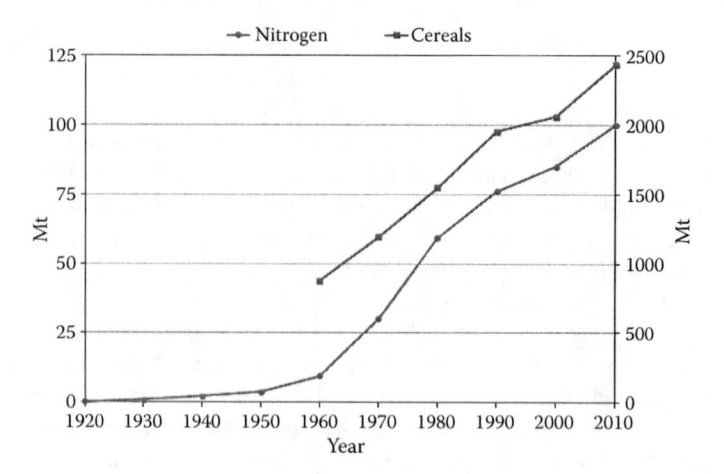

FIGURE 4.2 Nitrogen fixed by Haber–Bosch process (Mt, left axis) and world production of cereals (Mt, right axis). (N data from Smil, V., *World Agric* 2:9–1, 2011. Cereals data from FAOSTAT. Food and Agriculture Organization of United Nations, Rome, 2012. Available at http://faostat.fao.org, accessed April 3, 2012.)

World War II. During both World War I and World War II, the process was used to make N for bombs. Following World War II, many of the facilities that made bombs during the war were converted to N fertilizer plants, and this was the beginning of a major expansion of the use of fertilizers. The data shown in Figure 4.2 clearly show how closely cereal production has been linked with ammonia synthesis. Assuming that the N concentration of cereals is 1.8%, there was about 45 Mt of N contained in the cereals produced in 2010. Much, if not most, of this N will not be recycled through the crop production system. Therefore, while synthetic N fertilizer appears essential, some believe that the world's population can be fed without using N-based fertilizers. Nielsen (2005) makes this argument and also believes that the problems created by the use of synthetic fertilizers have outnumbered their benefits. He challenges two claims: (1) that the human-invented process of fixing nitrogen caused a global population explosion, and (2) without nitrogen-based fertilizers, we would not be able to feed the world. He points out that the population explosion was already advanced and well on the way when nitrogen-based fertilizers were introduced, and while the use of fertilizers may have supported the explosion, even this is doubtful. Nielsen (2005) stated that the world could be fed without using N fertilizers. He argued that this could be done by prudent organic farming, improving irrigation efficiency, improving the environment for nitrogen-fixing organisms, healing the soil that has been destroyed by using N-based fertilizers and other agricultural chemicals, and changing our food consumption habits. Even if this is possible, it is certainly not feasible because the reality is that the trends are in the opposite direction. In 1961, the number of calories per capita for the world population was 2201 compared to 2798 in 2007, and calories from meat were 110 and 218, respectively (FAOSTAT 2012). A more striking example is China. In 1961, China was in a period of famine, and the number of calories per person was only 1469, of which only 29 were from meat. Today, China is a rapidly developing economic power, and in 2007, the number of calories per capita was 2981, and 420 were from meat (FAOSTAT 2012). Therefore, it is most likely that if the world population becomes more prosperous, the demand for grain and other feeds to produce more meat and animal products will increase that will require more nitrogen. Unless synthesized ammonia is used, the only other feasible method to add additional N to the crop production system is by growing leguminous crops and incorporating them into the soil so the N can be used for growing nonleguminous crops such as cereals. This would, however, require enormous amounts of added cropland. Most leguminous crops, even when grown in favorable areas, will generally not supply more than 100 to 150 kg N ha^{-1}, so this would almost double the land requirement for cereal production, which is not a feasible alternative.

4.4 NEED FOR BETTER MANAGEMENT OF N FERTILIZERS

It is apparent to most that crop production in the future will continue to depend largely on the application of N fertilizer that is produced by the Haber–Bosch process. At the same time, there is increasing awareness that along with the benefits of this amazing and important process that has played such an important role in food production, there is a real potential that the environment can be seriously affected. A few, including Nielsen (2005), have expressed great concern over the large use of N fertilizer and went so far as to state that the Haber–Bosch process was a "fantastic invention—or was it?" Townsend et al. (2003) also expressed great concern about human health effects of a changing global nitrogen cycle. They stated that changes to the global nitrogen cycle affect human health well beyond the associated benefits of increased food production, and that many intensively fertilized crops become animal feed, helping to create disparities in world food distribution and leading to unbalanced diets, even in wealthy nations. Townsend et al. (2003) suggested that the net public health consequences of a changing N cycle are largely positive at lower levels, but they eventually peak and then become increasingly negative as our creation and use of fixed N continues to climb (Figure 4.3). Based on the conceptual model shown in Figure 4.3, Townsend et al. (2003) postulate that low to moderate increases in fertilizer use in developing countries will improve food availability and overall nutrition with only minor elevated losses of reactive N to the environment. Ladha et al. (2005) stated that about 60% of the global N fertilizer is used for producing the world's three major cereals: maize, wheat, and rice. These three cereals account for almost 90% of all cereals (FAOSTAT 2012). Figure 4.2 shows the world production of cereals increased almost three times between 1961 and 2009 (FAOSTAT 2012), but the amount of atmospheric N_2 fixed as ammonia that was almost entirely used for the production and use of N fertilizer increased about 10 times during the same time period (Smil 2011). Assuming the average N content of cereals to be 1.6%, the amount of N removed with all cereals in 1961 was approximately 14 Mt compared to about 40 Mt in 2009. In contrast, about 10 Mt of N was fixed in 1961 compared to about 100 Mt in 2009. This clearly indicates that much of the N added as fertilizer is not utilized directly by the plants. Although this fact has been recognized for many years, it is still not entirely clear where the N that is not used by the plants ends

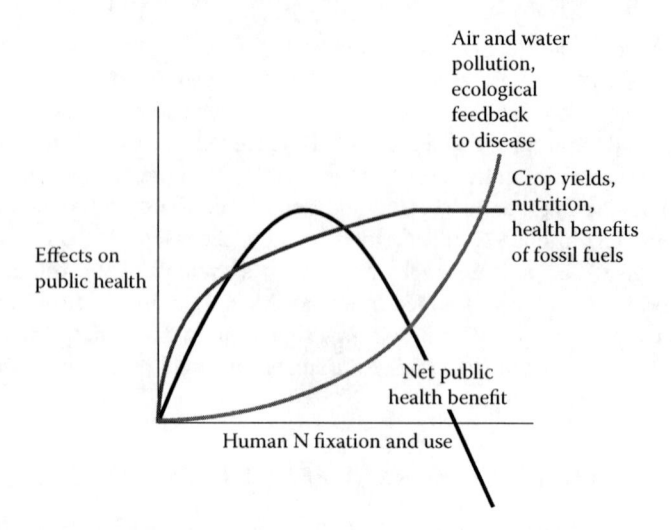

FIGURE 4.3 Conceptual model of the overall net public health effects of increasing human fixation and use of atmospheric N_2. (From Townsend, A.R. et al., *Ecol Environ* 1:240–248, 2003.)

up and what affect it has on the environment and public wellbeing. Galloway and Cowling (2002) estimated that for every 100 kg of N fertilizer used for producing crops, only 14 kg N was actually consumed by humans when they ate the products directly, which would represent a vegetarian diet (Figure 4.4a). They further estimated that if the cereals produced were fed to animals and the meat produced was consumed by humans, only 4 of the 100 kg N added as fertilizer was actually consumed by humans (Figure 4.4b).

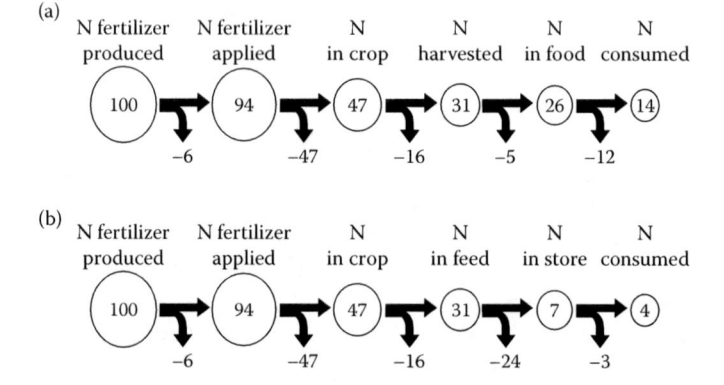

FIGURE 4.4 Fate of N fertilizer produced by the Haber–Bosch process from the factory to the mouth for (a) vegetarian diet and (b) carnivorous diet. (From Galloway, J.N., and E.B. Cowling, *Ambio* 31:64–71, 2002.)

Ladha et al. (2005) estimated that synthetic fertilizers supplied approximately 84 Mt of N, or about 45% of the total N input for global food production in 2005. They estimated other inputs to crop production as 33 Mt from biological nitrogen fixation, 16 Mt from recycling of N from crop residues, 20 Mt from atmospheric deposition, and 24 Mt from irrigation water. However, it is important to remember that the N from crop residues and much of that from irrigation water was not newly introduced to the system, but recycled within the system. Ladha et al. (2005) estimated that the amount of N removed by the harvested crop was almost equal to the amount of N added as fertilizer, but that 37 Mt was lost to the environment by leaching, runoff, and erosion; 21 Mt by ammonia volatilization from animal wastes, soil, and vegetation; 14 Mt from denitrification; and 8 Mt from nitrous oxide and nitric oxide emissions.

Although the losses discussed above are large and may be causing serious disruptions in ecosystem functions, much research has been conducted in recent decades to improve nitrogen use efficiency. There is also increasing awareness among producers, as well as the public, that practices must be developed and used that better synchronize the supply and requirement of N fertilizer for crop production. While more improvement is needed, nitrogen use efficiency has improved significantly in recent years. The use of synthetic N as fertilizer began in earnest following World War II, and the results were astonishing. The positive results, coupled with the low cost of the product, led farmers to apply more and more N. Although studies showed that as the amount of N added increased there was less increase in yield for each added increment, the guideline was to continue to add N until it was evident that additional N would not lead to a further increase in yield. There was no thought at the time among most scientists and producers that there were negative environmental effects associated with excess N additions. However, things began to change when scientists such as Rachel Carson (1962) in *Silent Spring* and Barry Commoner (1971) in *The Closing Circle* began raising concerns about chemicals used in crop production. Commoner was particularly concerned about nitrates in groundwater and their potential for causing methemoglobinemia (commonly known as blue-baby syndrome) in infants. Also, the US Environmental Protection Agency (EPA) was created in 1970. These concerns became great enough that scientists began seriously considering the effects of agricultural chemicals and practices on the environment, and particularly how they affected air and water quality and human health. This change is clearly reflected with the data shown in Figure 4.3. Between 1964 and 1976, the rate of N fertilizer applied per unit area of corn in the United States increased almost threefold while yield increased by about 50%. Since 1976, the average rate of N fertilizer has actually declined slightly, while the yield has increased almost every year and is almost two times higher. These data show that nitrogen use efficiency has greatly increased, and assuming the N content of corn grain is 1.6%, the amount of N removed with the grain is essentially equal to the amount of N applied as fertilizer. Therefore, the percentage of N fertilizer used for crop production that is lost to the environment has been significantly reduced in recent years. Europe has also made significant progress, but other areas such as China (Miao et al. 2010) have increased N fertilizer use so rapidly and to such high levels that serious environmental concerns are being raised.

4.5 INTENSIFICATION OF ANIMAL AGRICULTURE AND TRANSLOCATION OF NUTRIENTS

During the past 50 years and continuing today, there has been a dramatic increase in the concentration of animals into Concentrated Animal Feeding Operations (CAFOs). Beef animal feedlots with less than 1000 head of capacity comprise the vast majority of US feedlots but market a relatively small share of fed cattle. In contrast, lots with 1000 head or more of capacity comprise less than 5% of total feedlots but market 80% to 90% of fed cattle (USDA 2012). Feedlots with 32,000 head or more of capacity market around 40% of fed cattle, and some feedlots have a one-time capacity of 100,000 or more. Beef animals are usually fed only about 140 days, so 60,000 to 75,000 beef animals are commonly fattened annually in a 30,000-capacity feedlot. Similarly, dairy cows are becoming increasingly concentrated. In 2000, 10.5% of milk production was from dairy farms with more than 2000 cows, and this grew to 23.4% by 2006 (MacDonald et al. 2007). Poultry animals have long been highly concentrated, and swine are also increasingly being concentrated. An even more important factor is that more and more of these large operations are located in areas far removed from where most of the grain and other feed stuffs fed to these animals are produced, so massive amounts of feeds are transported into the area by trucks and trains. This has led to the translocation of millions of tons of N and P from areas where crops are produced to areas where animals are fed. Most of the N and P fed to animals are excreted and remain in the area as manure, and there is often not enough cropland to efficiently utilize the manure. On average, about 12 Mt of N and 2 Mt of P have been used as fertilizer every year in the United States for the past few decades, and approximately 40% of these nutrients are applied to corn and 40% to 50% of the corn is fed to animals. Because more and more of the corn fed to animals is being fed in areas far removed from where the corn was grown, and often in areas where there are limited amounts of cropland, these nutrients are not being recycled. The 12 Mt of N comes from atmospheric N_2 and the 2 Mt of P comes from mines. Thus, huge amounts of N and P are added to the environment each year, some of which damages surface and groundwater supplies and increases greenhouse gases. The disconnect between the location of nutrients used for production of grain crops, and where the nutrients ultimately reside in the environment after excretion by both humans and animals, is a growing and serious concern. In retrospect, it is clear that there was not enough thought given to the total system as needed, and this lack of a systems approach is still evident today. Recycling of nutrients should be a high priority, and this is the foundation of organic farming. Therefore, while much more attention should be given to the recycling of nutrients, organic farming as defined by IFOAM (2012) cannot meet the food and fiber needs of the world according to many scientists as discussed above. Norman Borlaug, the Nobel Peace Prize Laureate, credited for saving more human lives than any man in history by being the father of the "Green Revolution," saving countless millions of people from hunger in India and other countries, responded to a reporter that stated that a lot of people claim organic food is better for human health and the environment than conventionally grown food by stating, "That's ridiculous. This shouldn't even be a debate. Even if you could use all the organic material that you have—the animal manures, the human waste, the plant residues—and get them back on the soil, you couldn't feed more than 4 billion people. In addition, if all agriculture were organic, you would have to increase cropland area dramatically, spreading out into marginal areas and cutting down millions of acres of forests" (Borlaug 2009).

4.6 INTEGRATING THE USE OF ORGANIC MATERIALS, BIOLOGICAL PROCESSES, AND CHEMICAL FERTILIZERS

Crop yields, particularly cereals, must continue to increase to support a growing and more prosperous world population that is demanding more meat, dairy products, eggs, and other products requiring more animal feeds. At the same time, serious concerns are increasing about the effect that the heavy use of synthetic nitrogen, phosphorus fertilizers, and pesticides (Galloway and Cowling 2002; Townsend et al. 2003; Pimentel et al. 2005; Ladha et al. 2005) is having on the environment and perceived health issues. While organic production is growing and seen by some as a solution, most scientists believe that food and fiber needs can only be met by continued high use of fertilizer and pesticide inputs. The primary question is not whether high yields can be produced from organic systems. Pimentel et al. (2005) stated that various organic agricultural technologies have been used for about 6000 years to make agriculture sustainable while conserving soil, water, energy, and biological resources. They also reviewed results from 1981 to 2002 of the Rodale Institute farming systems trial conducted in Kutztown, PA. They concluded that there were many benefits from organic technologies. Soil organic matter and nitrogen were higher in the organic farming systems and helped conserve soil and water resources during drought years. Fossil energy inputs were about 30% lower than for conventionally produced corn. Labor inputs averaged about 15% higher but were more evenly distributed over the year. Regarding yields, they concluded that depending on the crop, soil, and weather conditions, organically managed crop yields on a per-hectare basis can equal those from conventional agriculture, although it is likely that organic cash crops cannot be grown as frequently over time because of the dependence on cultural practices to supply nutrients and control pests. In an earlier study conducted in California, Sean et al. (1999) found that nitrogen deficiency and weed competition were the two primary problems associated with organic farming. Pimentel et al. (2005) acknowledged that the favorable geographical soil characteristics present at the Rodale Institute farm may not be universally applicable.

Unfortunately, there has been too much conflict between people supporting use of synthetic fertilizers and those that use only organic materials. There are several reasons this has happened, but it is important that both sides become more fully informed. Many users of organic products believe that synthetic fertilizers are anthropogenic substances. Although synthetically fixed ammonia is produced by human activities, there is absolutely no difference in an NH_4^- ion formed in the soil from synthetically fixed ammonia and one that is formed from ammonia added with manure or from biologically fixed nitrogen. Likewise, phosphorus ions taken up by plants are the same regardless of their source. In contrast, many of the pesticides used in conventional agriculture are anthropogenic substances, and supporters of organic foods can make a more valid case for concern. From a nutrient standpoint, however, plants cannot distinguish between sources when taking up ions such as N and P. This is highly important because there are ample studies showing that, in many cases, current nutrient practices are not sustainable and more efficient management systems are needed. A review of long-term experiments conducted around the world indicated that chemical fertilizer alone is not enough to improve or maintain soil fertility, and the soil acidification problem caused by overapplication of synthetic N fertilizer can be reduced if more N fertilizer is applied as NO_3^- relative to ammonium- or urea-based N fertilizers (Miao et al. 2010). Organic fertilizers can improve soil fertility and quality, but long-term application of high rates can also lead to nitrate leaching, and accumulation of P, if not managed well. The review by Miao et al. (2010) clearly showed that well-managed combinations of chemical and organic fertilizers can overcome the disadvantages of applying single sources of fertilizers and sustainably achieve higher crop yields, improve soil fertility, alleviate soil acidification problems, and increase nutrient-use efficiency compared with only using chemical fertilizers.

A major disadvantage of relying on organic materials as a source of N and P for crop production is that enough material is added to supply adequate N, and P accumulations are common that can result in excessive amounts of P in runoff water causing eutrophication. The primary reason for this is that the N/P ratios of freshly excreted animal manures are about 3 to 3.5, and the ratio decreases to about 2 for aged manure (Stewart et al. 2000). Furthermore, the N in manures is not as readily available to plants as the P, so relatively large amounts of P can accumulate in soils when manures are added to supply N. Also, most crops require a higher ratio of N to P than is present in manures. For example, maize and wheat plants require approximately five times as much N as P. Therefore, the use of N fertilizer in combination with manure can result in more balanced soil fertility and reduce the threat of P pollution (Figure 4.5).

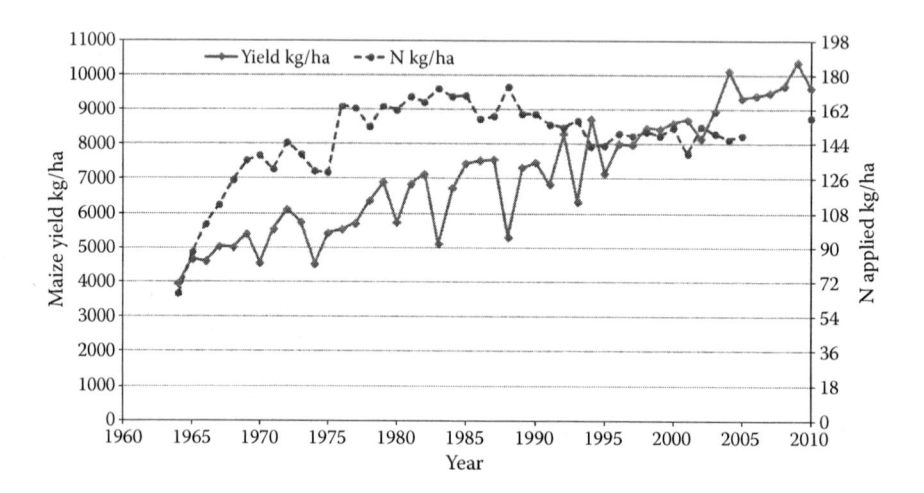

FIGURE 4.5 Amounts of nitrogen applied to maize in the United States to grain yield. (Data from National Agricultural Statistics Service, 2012.)

The use of manure and other forms of organic wastes should be highly encouraged and practiced. As already discussed, there is growing concern about the changes to the global nitrogen cycle being caused by synthetic N fertilizers and the effect this might have on human health and the environment (Townsend et al. 2003). The recycling of as much N as feasible could reduce the need for synthetically produced N substantially. Perhaps as important, or even more so, is the recycling of P. Unlike N, which has no limit to the amount that can be synthesized as long as an energy source is available, commercial P fertilizer is manufactured from phosphate rock obtained from underground mines and is a finite resource. As recent as 1950, about 80% of the P used for fertilizer globally was from organic sources, whereas about 90% today comes from phosphate rock (Cordell et al. 2009). Although the amount of phosphate rock present globally is not known, the supply is finite. Cordell et al. (2009) estimated that peak production of phosphate rock known to be economically available for mining and processing would occur between 2030 and 2040; more recent estimates have been revised upward (Jasinski 2011). Compared to the estimate of about 16 billion Mt by Jasinski (2011), the International Fertilizer Development Center estimated reserves as 60 billion Mt and would be sufficient for 300 to 400 years (Syers et al. 2011). Regardless of the amount, the fact that phosphate rock is a finite resource and an element that is essential for food production requires recycling for long-term sustainability. Cordell et al. (2009) estimated that only one-fifth of the phosphorus mined in the world is consumed by humans as food.

4.7 CONCLUSIONS

While the production and sales of organically grown foods continue to increase at significant rates, they still account for only about 2% to 3% of the total. There is ample evidence showing that the yields of organically grown crops can be as high, or even higher, than conventionally grown crops. Although consumers of organic products generally consider the quality to be higher and also safer, the scientific evidence to support these views is limited at best, particularly in developed countries. While there is every reason to think the production of organic foods will continue to increase, there is little likelihood that it will ever be a major source of the world's total food supply. With a few exceptions, agricultural scientists suggest that world food needs can only be met with the use of commercial fertilizers, and that even with their use, the task will be challenging in view of increasing world population and changes in diets that increase the demand for grain. At the same time, an increasing number of scientists believe that changes in the global nitrogen cycle are becoming so serious that major changes must be made in the fixation and use of atmospheric nitrogen. Many long-term experiments have shown that the combined use of organic materials and commercial fertilizers results in the highest yields and the best soil quality. Therefore, every effort should be made to recycle nutrients to the fullest extent feasible. This will not be easy because commercial fertilizers have historically been relatively inexpensive, easy to use, highly concentrated, free of weed seed, and of known composition of single or multiple nutrients. In contrast, organic materials such as manure are highly variable in terms of moisture, concentration and availability of nutrients, kind and number of weed seed, ease of handling, and uniformity of spreading. Many producers chose, and were often encouraged by scientists, industry representatives, extension workers, and other change agents, not to use organic materials even when they were available because they felt the short-term benefits of using commercial fertilizers were the best.

ABBREVIATIONS

BSE: bovine spongiform encephalopathy
CAFOs: concentrated animal feeding operations
GMOs: genetically modified organisms
Mt: million tonnes
NOP: National Organic Program
OFPA: The Organic Foods Production Act
EPA: US Environmental Protection Agency

REFERENCES

Badgley, C., J. Moghtader, E. Quintero, E. Zakem, M.J. Chappell, K. Avilés-Vázquez, A. Samulon, and I. Perfecto. 2007. Organic agriculture and the global food supply. *Renew Agr Food Syst* 22:86–108.

Borlaug, N. 2009. Norman Borlaug on organic farming. http://www.coyoteblog.com/coyote_blog/2009/03/norman-borlaug-on-organic-farming.html (verified June 6, 2012).

Brandt, K., C. Leifert, R. Sanderson, and C.J. Seal. 2011. Agroecosystem management and nutritional quality of plant foods; the case of organic fruits and vegetables. *Crit Rev Plant Sci* 30:177–197.

Carson, R. 1962. *Silent Spring.* Houghton Miffin Publishers, Boston, MA.

Commoner, B. 1971. *The Closing Circle: Nature, Man, and Technology.* Knopf Publishers, New York.

Cordell, D., J.-O. Drangert, and S. White. 2009. The story of phosphorus: global food security and food for thought. *Global Environ Change* 19:292–305.

Dangour, A.D., S.K. Dodhia, A. Hayter, E. Allen, K. Lock, and R. Uauy. 2009. Nutritional quality of organic foods: a systematic review. *Am J Clin Nutr* 90:680–685.

De Ponti, T., B. Rijk, and M.K. van Ittersum. 2012. The crop yield gap between organic and conventional agriculture. *Agr Syst* 108:1–9.

Diacono, M. and F. Montemurro. 2010. Long-term effects of organic amendments on soil fertility: a review. *Agron Sustain Dev* 30:401–422.

Edmeades, D.C. 2003. The long-term effects of manures and fertilizers on soil productivity and quality: a review. *Nutr Cycl Agroecosyst* 66:165–180.

Ellis, E.C. and S.M. Wang. 1997. Sustainable agriculture in the Tai Lake Region of China. *Agric Ecosyst Environ* 61:177–193.

FAOSTAT. 2012. Food and Agriculture Organization of United Nations. Rome. http://faostat.fao.org (verified April 3, 2012).

Folkers, D. 1983. Biologischer anbau von obst und gemüse—die einstellung der Bundesbürger. *Ernährungsumschau* 30:B36.

Galloway, J.N. and E.B. Cowling. 2002. Reactive nitrogen and the world: 200 years of change. *Ambio* 31:64–71.

Gao, C., B. Sun, and T.L. Zhang. 2006. Sustainable nutrient management in Chinese agriculture: challenges and perspective. *Pedosphere* 16:253–263.

Hansen, B., H.F. Alrøe, E.S. Kristensen, and M. Wier. 2002. Assessment of food safety in organic farming. Danish Research Centre for Organic Farming (DARCOF) Working Paper no. 52, Post Box 50, DK-8830 Tjele, Denmark. http://orgprints.org/00000206 (verified April 10, 2012).

Hughner, R.S., P. McDonagh, A. Prothero, C.J. Shultz, and J. Stanton. 2007. Who are organic food consumers? A compilation and review of why people purchase organic food. *J Consum Behav* 6:94–110.

IFA. 2002. Statistics, 2nd edn. International Fertilizer Industry Association. http://www.fertilizer.org/ifa (verified April 3, 2012).

IFOAM. 2012. International Federation of Organic Agriculture Movements. www.ifoam.org. (verified April 3, 2012).

Jasinski, S.M. Phosphate rock. 2011. In: *Mineral Commodity Summaries*. United States Geological Survey, United States Government Printing Office, Washington, DC.

Kahl, J., T. Baars, S. Bügel, N. Busscher, M. Huber, D. Kusche, E. Rembiałkowska, O. Schmid, K. Seidel, B. Taupier-Letage, A. Velimirov, and A. Załęcka. 2012. Organic food quality: a framework for concept, definition and evaluation from the European perspective. *J Sci Food Agric* 92:2760–2765.

King, F.H. 1911. *Farmers of Forty Centuries: Organic Farming in China, Korea, and Japan*. Dover Publications, Inc. Mineola, NY, USA.

Ladha, J.K., H. Pathak, T.J. Krupnik, J. Six, and C. van Kessel. 2005. Efficiency of fertilizer nitrogen in cereal production: retrospects and prospects. *Adv Agron* 87:85–156.

Lal, R. 2006. Enhancing crop yields in the developing countries through restoration of the soil organic pool in agricultural lands. *Land Degrad Dev* 17:197–209.

MacDonald, J.M., E.J. O'Donoghue, W.D. McBride, R.F. Nehring, C.L. Sandretto, and R. Mosheim. 2007. Profits, Costs, and the Changing Structure of Dairy Farming. Economic Research Report No. (Err-47), USDA Economic Research Service, Washington, D.C.

Magkos, F., F. Arvaniti, and A. Zampelas. 2003. Organic food: nutritious food or food for thought? A review of the evidence. *Int J Food Sci Nutr* 54:357–371.

Magnusson, M.K., A. Arvola, U. Hursti, L. Aberg, and P. Sjoden. 2001. Attitudes towards organic foods among Swedish consumers. *Br Food J* 103:209–227.

Miao, Y., B.A. Stewart, and F. Zhang. 2010. Long-term experiments for sustainable nutrient management in China. A review. *Agron Sustain Dev* 31:397–414.

National Agricultural Statistics Service. 2012. Quick Stats. http://www.nass.usda.gov/QuickStats/Create_Federal_All.jsp (verified June 6, 2012).

National Research Council. 2010. *Toward Sustainable Agricultural Systems in the 21st Century*. The National Academies Press, Washington, D.C.

Nielsen, R. 2005. Can we feed the world? Is there a nitrogen limit of food production? http://home.iprimus.com.au/nielsens/nitrogen.html (verified April 3, 2012).

Pimentel, D., P. Hepperly, J. Hanson, D. Douds, and R. Seidel. 2005. Environmental, energetic, and economic comparisons of organic and conventional farming systems. *BioScience* 55:573–582.

Rothamsted Research. 2006. Guide to the classical and other long-term experiments, datasets and sample archive. Lawes Agricultural Trust Co. Ltd., available at http://www.rothamsted.bbsrc.ac.uk/resources/LongTermExperiments.pdf (verified March 1, 2012).

Sean, C., K. Klonsky, P. Livingston, and S.T. Temple. 1999. Crop-yield and economic comparisons of organic, low-input, and conventional farming systems in California's Sacramento Valley. *Am J Altern Agric* 16:25–35.

Seufert, V., N. Ramankutty, and J.A. Foley. 2012. Comparing the yields of organic and conventional agriculture. *Nature* 485:229–232.

Smil, V. 2002. Nitrogen and food production: proteins for human diets. *Ambio* 31:126–131.

Smil, V. 2011. Nitrogen cycle and world food production. *World Agric* 2:9–1.

Stewart, B.A., C.A. Robinson, and D.B. Parker. 2000. Examples and case studies of beneficial reuse of beef cattle by-products. In *Land Application of Agricultural, Industrial, and Municipal By-Products*. J.F. Power and W.A. Dick, eds., 387–407. SSSA Book Series 6, Soil Science Society America, Madison, WI, USA.

Syers, K., M. Bekunda, D. Cordell, J. Corman, J. Johnston, A. Rosemarin, and I. Salcedo. 2011. *Phosphorus and Food Production*. UNEP Year Book. United Nations Environmental Program, Nairobi, Kenya.

Townsend, A.R., R.W. Howarth, F.A. Bazzaz, M.S. Booth, C.C. Cleveland, S.K. Collinge, A.P. Dobson, P.R. Epstein, E.A. Holland, D.R. Keeney, M.A. Mallin, C.A. Rogers, P. Wayne, and A.H. Wolfe. 2003. Human health effects of a changing global nitrogen cycle. *Ecol Environ* 1:240–248.

USDA. 2012. Cattle: Background. Economic Research Service, Washington, DC. http://www.ers.usda.gov/briefing/cattle/background.htm (verified April 14, 2012).

Von Liebig, J. 1840. *Chemistry in Its Application to Agriculture and Physiology*. Taylor & Walton, London.

Willer, H. and L. Kilcher (Eds.). 2012. *The World of Organic Agriculture—Statistics and Emerging Trends*. Research Institute of Organic Agriculture (FiBL), Frick, and International Federation of Organic Agriculture Movements (IFOAM), Bonn.

Williams, C.M. 2002. Nutritional quality of organic food: shades of grey or shades of green. *Proc Nutr Soc* 61:19–24.

Wittwer, S., Y. Youtai, H. Sun, and L. Wang. 1987. *Feeding a Billion: Frontiers of Chinese Agriculture*. Michigan State University Press, Michigan, USA.

Woese, K., D. Lange, C. Boess, and K.W. Bögl. 1997. A comparison of organically and conventionally grown foods—results of a review of the relevant literature. *J Sci Food Agric* 74:281–293.

Worldwatch Institute. 2011. Grain harvest sets record, but supplies still tight. August 12, 2011. http://worldwatch.org/node/5539 (verified April 14, 2012).

Worthington, V. 2001. The nutritional quality of organic versus conventional fruits, vegetables, and grains. *J Altern Complement Med* 7:161–173.

Yang, H.S. 2006. Resource management, soil fertility and sustainable crop production: experiences in China. *Agric Ecosyst Environ* 116:27–33.

Zhang, H., B. Wang, M. Xu, and T. Fan. 2009. Crop yield and soil responses to long-term fertilization on a red soil in southern China. *Pedosphere* 19:199–207.

5 Building upon Traditional Knowledge to Enhance Resilience of Soils in Sub-Saharan Africa

Moses M. Tenywa, Julius Y. K. Zake, and Rattan Lal

CONTENTS

5.1 Introduction 110
5.2 Nature of Problem of Lack of Integration of TK and SK 110
    5.2.1 Institutional Failures—Policy Aggressively Promoting Western S&T 110
    5.2.2 Idealization of TK in Past Studies 112
    5.2.3 Lack of Framework/Mechanism for Integration of TK and SK 117
    5.2.4 Justification of Promotion of TK—Failure of Western Solutions to Usher in Sustainable Soil Management 119
5.3 Resilience Approach to Soil Management 120
    5.3.1 Soil Management Dimensions; Socioecological Resilience, Ecosystem Resilience, and Soil Resilience 121
    5.3.2 At the Socioecological Scale 122
    5.3.3 At the Ecological Scale 122
    5.3.4 At the Soil Scale 123
5.4 Resilience Model 123
    5.4.1 Managing for Resilience 123
    5.4.2 Conventional View of Resilience 125
    5.4.3 Alternative View to Managing for Resilience 127
5.5 Renaissance of TK 129
5.6 Union of SK and TK 132
    5.6.1 Conceptualization Framework for Management of Explicit TK 133
5.7 Conclusion 133
Acknowledgments 133
Abbreviations 133
References 134

5.1 INTRODUCTION

More than ever before, the debate of the role of traditional knowledge (TK) in accentuating the processes of sustainable soil management in Sub-Saharan Africa (SSA) has been rekindled and fuelled by the continuously increasing global pressures, namely, demographic, urbanization, climate change, cyclic drought–floods, and associated economic growth and development theories and policies (Jones and Thornton 2008). Previously, the pursuit of sustainable soil management in terms of planning, utilization, protection, and monitoring has largely been based on science and technologies from outside, with little attention paid to TK (Reij et al. 1996). However, land degradation has persisted, becoming widespread and believed to be increasing (Scherr and Yadav 1997). The discourse of scholars of TK has not helped the situation. Rather than addressing the fundamental constraints for the TK and scientific knowledge (SK) systems to symbiotically come together to pave the way for informing policy action on integration of TK in curricula, and integration with SK for dissemination and development, they have largely been defensive, focusing on justification of TK. One benefit that can accrue from understanding TK is to make it explicit in forms that can be recorded for easier distribution and reuse. Therefore, this chapter is aimed at improving the understanding of TK for building synergies with SK for sustainable soil management. The chapter specifically covers the nature of the problem of lack of integration of TK and SK, justification for strengthening TK, renaissance of TK resource and soil management for enhancing resilience, and promoting union of SK and TK.

5.2 NATURE OF PROBLEM OF LACK OF INTEGRATION OF TK AND SK

Agrawal (1995) argues that pristine indigenous knowledge is more likely to have been interacting with external knowledge at least to some extent for the last 500 years. Ever since the colonization of SSA, Western science and technology have been the predominant external knowledge interacting with TK for aiding decisions for managing the capitals (natural, human, social, cultural, physical, and financial) for economic growth and development. However, the widespread and increasing land degradation provides evidence that the modernization through science and technology alone has not yielded the desired impacts to agriculture. TK has not been effectively used for promoting the soil management aspects of increasing resilience and recovery, nor used to build the confidence of communities in making their management decisions, following their own procedures and objectives and for development workers to add value. Several reasons are cited for this discrepancy, including institutional failures (policy thrust aggressively promoting Science and Technology (S&T) even without relevance), idealization of TK, and lack of a framework/mechanism for integration.

5.2.1 Institutional Failures—Policy Aggressively Promoting Western S&T

Modern sciences can be traced back to the scientific revolution of the seventeenth century and are linked to the ancient Greeks, who used empirical systems to make observations and quantitative measurements, logical thinking, rational reasoning, and experimentation to accumulate knowledge and improve the understanding of the physical, biological, and social reality (Evenson and Gollin 2003). Because of their robust methods of empirical research combining induction and deduction based on theories, systematic collection and processing of data to test formulated researchable hypotheses, value neutrality, objectivity, and capacity to transcend time and space, modern sciences are generally considered to be universally valid (Shiva 1991). It is against this strength that modern science derives its strength. Its introduction to indigenous cultures has variably resulted in different permutations of interactions, sometimes competing, replacing, or confronting each other and rarely forming an intercultural integration (Akabogu 2000). The introduction of modern science in SSA dates back to the nineteenth century and is inextricably linked to the great technological advances that drove the market-oriented Industrial Revolution in Europe (Evenson and Gollin 2003). However, indiscriminate use of scientific ideas and methods causes disruption not only to existing social and economic relationships but also to TK (Larson 1998). According to Haverkort and Reijntjes (2007), the SK intolerant of indigenous epistemologies has been promoted as the dominant knowledge system to substitute for TK, perceived to be superstitious, subjective, and only intuitive. In its aggressive budding stages, SK dominated, prevailed, and marginalized TK that was characterized by low rigor, vitality, and comprehensiveness in ways of knowing, learning, and explaining the physical, biological, and social phenomena.

In most traditional societies, TK has been marginalized, whereas conventional SK is recognized and taught in schools from kindergarten, through primary education to university, forming the basis for formal research and decision making at development and government levels. Adas (1990) argues that in SSA, there were no opportunities for advanced training to develop TK into a robust science for lack of investments in methods and knowledge development, and scientific research was not promoted among the indigenous population. In the past, as reflected in literature, the tendency and policy thrust has been to focus on how to promote the modern science among indigenous cultures (Orvik and Barnhardt 1974). Thus, TK has not been developed and is disappearing fast (Galaty 1999). Modernization in general has been about government policies, research, training, and extension, promoting scientific solutions often at the expense of traditional farming. Many indigenous practices that were based on experiences and insights of preservation of soils over generations were either abandoned or underutilized in preference of technologies that promise greater short-term benefits and are unsustainable (Gayton and Michel 2002). According to Patton (1978), time has also been a factor in the substitution of TK for SK as the notion that modern science is superior to TK has gone on for a considerably long time. Consequently, most of the indigenous cultures were left ill prepared to address the emerging issues, where the extraneous concepts of science, law, nation building, and administration were still dominant and determined the economic competitions, development planning, and intellectual discourse. The institutional failures including lack of access to incentives, information, and institutions for promoting traditional ways of knowing and learning, have not created space for integration of SK and TK, and should ultimately be addressed by policy makers.

5.2.2 Idealization of TK in Past Studies

Although indigenous practices have evolved and survived the changing biophysical and socioeconomic conditions, through continuous responsive changes and adaptation, they are not perfect (Briggs 2005). If TK and indigenous soil and water conservation (ISWC) were truly effective, there would not be a problem of food shortages and widespread land degradation (Blaikie and Brookfield 1987; Reijntjes et al. 1992). Perhaps a major disservice to TK has been its romanticization, a view that idealizes that TK is basically "good" and should be preserved and used as it is without pointing out its major drawbacks that can be used as the entry/opportunity points to leveraging the SK. The fundamental difficulty in integrating TK stems from the limited understanding of its nature; the limited literature on TK is often used by scholars largely of explicit SK background that use the scientific method to falsely judge the TK that is largely tacit as being shallow. Trying to analyze and validate TK systems by using external (scientific) criteria carries the risk of distorting such systems in the process (Briggs 2005). This is a typical case of the accuser being the judge whereby scientists use the scientific method and that glaringly lack the appropriate tools for extracting tacit TK into literature, causing contradiction, and perhaps explains why TK is often despised (Berkes 1993).

A value judgment assigned to TK based solely on literature can be misleading. According to Altieri et al. (1983), Berkes (1993), and Gata (1994), there is a long tradition of defining TK systems in opposition to SK. The frequent neglect of local social and cultural priorities further reduces the relevance of scientific surveys and reports pertaining to land use and management (de la Cruz 2005). Very few studies make an effort to make distinction between traditional, indigenous, and local knowledge. Berkes (1993) note that no universal definition is available for TK, and many terms are used to describe what indigenous people know. The TK in literature is often used interchangeably with traditional ecological knowledge, indigenous knowledge, indigenous science, local knowledge, folk knowledge, farmers' knowledge, fishers' knowledge, community knowledge, and rural people's knowledge on the premise that they have common epistemic ground and there is no difference among them (Warren 1989; Tick 1993; Ellen et al. 2000; Cools et al. 2003; Adepide et al. 2004). By this analysis, scholars have instead grossed over the problem; the fundamental problem lies in failure to recognize that TK is largely tacit and, rather than explicitly written down, is embodied in the beholder.

"The TK interweaves empirical, spiritual, social and other components. In general, by isolating elements from such a holistic worldview one runs the risk of misrepresenting both the elements and the whole" (ICSU 2002). The word "traditional," for example, places the emphasis on the transmission of knowledge along a cultural continuity but might ignore the ability of traditional societies to adapt to changing circumstances (Briggs 2005). At present, traditional ecological knowledge is interpreted as a cumulative body of knowledge, practices, and representations that describes the relationships of living beings with one another and with their physical environment, which evolved by adaptive processes and has been handed down through generations by cultural transmission (Berkes et al. 2000). Another widely used word, "indigenous," is meant to highlight the autochthonous nature of this knowledge, but it might overlook knowledge from populations who are not officially recognized as indigenous (Nakashima and Roué 2002). There is a general understanding as to what constitutes indigenous knowledge. Accordingly, it refers to a set of truths, beliefs, values, perspectives, concepts, judgments, expectations, methodologies, and know-how that people in a given community possess based on experience and adaptation to a local culture and environment and has developed over time and continues to develop (Berkes 1999). Warren et al. (1995) defines it as the intimate everyday knowledge of the local environment held by indigenous people or a local knowledge unique to a given culture or society. "Indigenous knowledge represents the accumulated experience, wisdom and know-how unique to cultures, societies, and for communities of people, living even before colonization in an intimate relationship of balance and harmony with their local environments" (Emery 2000). It serves as the basis for community-level decision making in areas pertaining to governance, food security, human and animal health, childhood development and education, natural resource management, and other vital socioeconomic activities (Aoki 2003). The word "local" can be applied to different geographic contexts, but it lacks specificity. Critchley and Mutunga (2003) define local knowledge as a dynamic and continuously evolving combination of pristine indigenous knowledge and external knowledge including SK.

A common weakness in TK management is the overemphasis on introduced technology at the expense of TK (Briggs 2005). Typically, research & development (R&D) does not address either knowledge management or the cross-functional, cross-organizational process by which knowledge is created, shared, and applied (acquisition through use). Understanding these relationships is fundamental to understanding land use and management decisions and thereby influencing them toward improved resilience. In many cases, past R&D approaches actually interfered with sustainable resource use, by taking the access and decision-making power away from those who understand best both the resources and what they need from those resources. Climate change is likely to be the dominant driver of ecological change in the twenty-first century, and removing local stressors may not be enough to maintain soil quality (Graham et al. 2008); focus should be placed on finding ways and means for protecting soil ecosystems in a changing climate. Moreover, TK systems interpret reality not on the basis of a linear conception of cause and effect but, rather, as a world made up of constantly forming multidimensional cycles in which all elements are parts of an entangled and complex web of interactions (Freeman 1992). Of course, there is always the risk of oversimplifying by reducing the things of interest to essentials and/or dichotomies. However, from this brief overview of the dissimilarities, we can gain an understanding of how hard it is to compare two systems of knowledge that are so profoundly different. At the same time, we cannot extract just those parts of TK that seem to measure up to scientific criteria and ignore the rest. This process of cognitive mining would atomize the overall system and threaten TK with dispossession (Nakashima and Roué 2002).

What is lacking is the capacity to effectively utilize the TK system and ways of knowing that are embedded in TK communities to enrich school curriculum and enhance teaching and learning of students, that is, making TK "policy relevant." The traditional ecological knowledge has been compromised by the erosion of the intimate relationship between farmer- and farm-caused modern science and education, rural–urban migration, slavery, civil conflicts, natural hazards, land grabbing, and so forth (Rutatora 1997). Capacity building is vital for the local people to develop their own knowledge base and methodologies to promote activities that leverage both TK and SK in such a way that the positive sum produces beneficial outputs (Warren et al. 1995; Teklu and Gezahegn 2003). Attributes of TK of soil can be summarized below:

1. *Spatial–temporal attributes*: Soils exhibit wide spatial–temporal variability that is also reflected in the TK for their management. For example, during a given cropping season, the land use and management practices for a hillslope may differ across the landscape positions.
2. *Gender connectedness*: The embodiment of TK for soil management can be gender disaggregated depending on the situations. For example, in the cattle corridor district of Teso, Uganda, the women and girls, who are more of crop cultivators, had different knowledge of soils from that of the men and boys, who are more of cattle keepers. The women and girls were richer in knowledge of upland soils, which they cultivate, while the men and boys were more resourceful in describing valley soils, which support the pastures where they graze the livestock. Using semistructured interviews regarding constraints for different soil types in the district of Teso, Uganda, the differences and similarities were captured for men and women, illustrated in Table 5.1.

In Table 5.1, pairwise ranking of problems associated with soil types by men and women of Teso District, Uganda, is shown. Women perceived water retention as a major problem in lacustrine/light-textured soils that dry up a few days after the rains have stopped. The men looked at the constraints more in terms of loss incurred as a result of a particular constraint. For the same reason as women, poor water and erosion were highly ranked.

TABLE 5.1
Pairwise Ranking of Problems Associated with Soil Types by Men and Women of Teso District, Uganda

Row	Soil Constraint	Ranking Women	Ranking Men
1	Poor water retention	1	3
2	Erosion	2	2
3	Pests and diseases	4	1
4	Infertility	3	6
5	Weeds	5	5
6	Sickness	7	4
7	Floods	6	7

3. *Systemic nature*: The TK for soil management is not monolithic but diverse and dynamic, varying from community to community and in time in response to the external forces. Its evolution is inextricably linked to the changing social and economic needs of the communities and, implicitly, the reigning economic growth and development theories. As such, if the utility of TK is to be increased to contribute to sustainable soil management, its study must take into account the various constructs of development theory and conceptualization of management–social–ecological resilience, community resilience, and soil resilience.
4. *Livelihood connectedness*: Humans and communities interact with ecosystems in such a way to maintain long-term sustainable resource yields (Ostrom 1990). For the survival of communities whose livelihoods are dependent on soil, communities must efficiently and effectively create, locate, capture, and share their TK and expertise and have the ability to bring that knowledge to bear on problems and opportunities. Ostrom's work has considered how societies have developed diverse institutional arrangements for managing natural resources and avoiding ecosystem collapse in many cases, even though some arrangements have failed to prevent resource exhaustion.

    The polycentric approach to livelihoods postulates that the farmer's practices at any time are inexplicably interwoven with the technological, social, economic development. It is a reflection of the pooling of all knowledge (traditional and nontraditional) deemed relevant to withstand the challenges. The external pressures on soil to meet the increasing demands for food, feed, fiber, and fuel cause changes under some circumstances exceeding its ability to recover using available knowledge in the community. The ability of soil to absorb the pressure at any given time depends on the aggregate utilization of knowledge.
5. *Dynamism*: Innovation is commonly triggered by a desire to escape from poverty and meet increasing demands for food, feed, and fiber. Thus, rural innovation is usually linked to production and economic (market) opportunity. When land is put to use, it gradually experiences external pressures (e.g., erosion, salinization, nutrient mining, burning), causing progressive diminution in soil fertility. In order to maintain the demands, farmers have to continuously adapt. As such, TK is dynamic, responding to the external forces. The strategy is to take such innovative thinking and practices, understanding innovation, stimulating it, and adding value through combining with SK and then using a form of farmer-to-farmer horizontal transfer.
6. *Sensitivity to context*: Since TK is generated through observations, communication, interpretation, and actions of cause and effect in specific environments, it may not necessarily be directly transferred for application to another area. A very clear example is the use of names like Eitela (Toroma, Uganda) and Itongo (Iteja, Tanzania), which in principle are used as land use descriptions but initially caused confusion with soil types at the beginning of the study (Payton et al. 2003). Both mean an area in the uplands in which several soil types can occur. For example,

in humid hillslopes overlain by fragile volcanic Andosols that are highly permeable to and experience mass wasting, the farmers practice soil conservation rather than water conservation. Instead of laying water and soil erosion control measures across the slope as Miiro et al. (1995) reported for the neighboring district of Kabale or as is usually done in arid and semiarid lands (ASALs), they lay them along the slope. A case in point is the stone lines that are laid up and down the highlands of Kisoro District of Uganda to tortuously and safely rid the soil of excess water. Similarly, the crops are planted in rows running up the hillsides. The practice works as evidenced by the absence of a soil fertility gradient. Highland areas characterized by runoff and soil erosion usually exhibit differences in plant vigor down the slope resulting from nutrient mining and deposition gradient.

7. *Practicability*: The TK is not monolithic; it has different facets and varies from place to place and time, aimed at addressing real-life problems. The different facets include nonuniformity, dynamism, codified assimilation of outside knowledge, adaptiveness, and innovativeness. The TK may be applicable more to a specific soil attribute (e.g., soil organic carbon) than to several characteristics combined together (Lal 1999).

The TK is dynamic (each knowledge tradition has its own ways of learning, experimenting, teaching, and consensus seeking). The data used to generate TK are obtained right in the field in real time, processed, and put into action for survival. The above attributes of TK of soil are summarized in Figure 5.1.

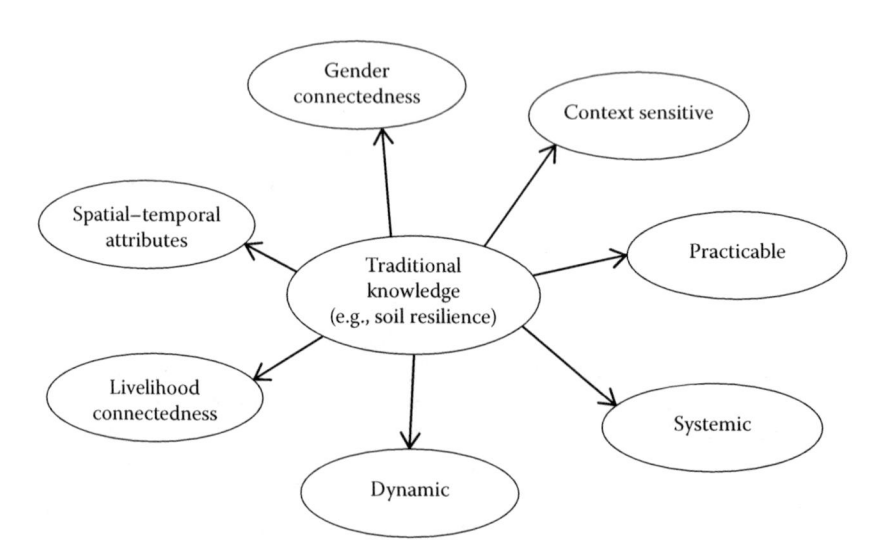

FIGURE 5.1 Attributes of TK of soil.

5.2.3 Lack of Framework/Mechanism for Integration of TK and SK

Postmodern science's school of thought transcends the modern notions of the material reality and subscribes to application of a wider variety of methods of knowing and learning in pursuit of the ultimate truth (Kamil 2011). Under this school of thought, both TK and SK for soil management are invariably in concordance with the principles of conservation, namely, planning, protection, and monitoring, but variably differ in terms of practice quality. Understanding these differences is fundamental to the development of a comprehensive framework for integration of the two divergent knowledge systems, assessment of geographic scale (local to global) of application, standardization of methods for measuring soil quality (e.g., soil fertility loss or gain), and a variety of other elements that increase the benefits to soil management. Very few countries have applied TK to inform decision-making processes (e.g., land evaluation in Swaziland) (Osunade 1992). Yet the integration of both TK and SK sources is a promising new stream of research and application for land evaluation and land use planning using Geographical Information System (GIS) (Barrera-Bassols and Zinck 1986), managing for increasing ecosystem resilience to global climate change by minimizing the impacts of local stressors or disturbances. Scientific research associated with resilience is useful in influencing policy making and subsequent environmental decision making in several ways:

> *Protection*: Observed resilience within specific ecosystems drives management practice. When resilience is observed to be low, or impact seems to be reaching the threshold, management response can be to alter human behavior to result in less adverse impact to the ecosystem (Brand 2009).
> *Planning*: Ecosystem resilience influences environmental decision making, similar to the way that existing ecosystem health impacts upon what development pathway is adopted (Brand 2009). For example, the management of fragile mountainous soils on steep slopes is influenced by their stability to external forces, and the ecosystem conditions influence the land use (Tamang 1993). For threatened ecosystems, environmental impact assessment of the development pathway to be taken must be informed by their health and resilience.
> *Monitoring*: Improvement of the *socioecological resilience* of ecosystems is now a worldwide concern for scientists and other experts as demonstrated by the Millennium Ecosystem Assessment (MEA) initiative. This initiative's stated objective is "to assess the consequences of ecosystem change for human wellbeing and the scientific basis for action needed to enhance the conservation and sustainable use of those systems and their contribution to human wellbeing."

The lack of a framework for integration of TK and SK *inter alia* emanates from the differences in management attributes captured (Table 5.2). Farmers' management practices are based on subjective rationality, while scientific technologies are based on objective measurements (Beckford 2002). Farmers' knowledge is well developed

TABLE 5.2
Comparison of Western Science and TK

Aspect	Western Science	TK
Place of knowing and learning	Formal classrooms, laboratories, and workshops	Informal visual observations relating practices and impacts captured within the real-world context and specific local conditions from early childhood working in the field
Source and nature of data and information processed into knowledge	Samples/objects of study often detached from their vital context into simplified and controllable experimental environments	Nondualistic (does not distinguish between research subject and object, between mind and matter or man and nature; sacred persons, places, and books may be important sources of knowledge
Approach to acquisition	Objective and quantitative empiricism	Subjective and qualitative spiritualism
Methods of collection	Uses rational theoretical concepts and robust reductionist methods, quantification, randomizing, replications, statistical analyses, logical reasoning, and controlled experiments	Intuition and meditative methods with holistic view and learning from inside in real contexts sharpened by experience, participation, insight of seers, communities of practice, councils of elders, dreams, initiation rites, use of alcohol or stimulants, reflection, and communication with ancestral spirits
Methods of dissemination	Academic and literate transmission (e.g., classroom, laboratory)	Folklore knowledge orally passed on from one generation to the next by elders in real-life situations
Basis for acquisition and assessment of level of knowledge	Competence and objective tests	Survival under diverse conditions or extinction and yield benefits
Nature of knowledge	Reductionist/ compartmentalized/ fragmented into disciplines (physical sciences, biology, psychology, and social sciences); also science versus technology	Holistic, often very sophisticated, embracing all aspects of the ecosystem as a whole and its knowledge, including socioeconomic and physical sciences, biology, and psychology
Form of knowledge	Explicitly recorded in forms that can be disseminated	Tacit, embodied in beholder, and not easily written down in a form that can be accessed by outsiders
Notion of time	Linear (going from the past through the present to the future)	Time is often considered to be cyclic or spiral; times do not always have the same quality; auspicious for certain activities

Source: Nakashima, D. and M. Roué: Indigenous Knowledge, Peoples and Sustainable Practice, *Encyclopedia of Global Environmental Change*, ed. T. Munn. 2002. Copyright Wiley-VCH Verlag GmbH & Co. KGaA. Reproduced with permission.

and has answers to the relatively straightforward cause–effect interrelationships (Warren 1991) but is inept to address the outgoing questions/subtle issues pertinent to sustainable development (Nitesh and Shefali 2004). Farmers usually focus on topsoil and are interested in soil productivity, appropriate management practices, and indicators they can experience, yet scientific surveys can be very deep in covering national and international scales (King et al. 1983). Examining the literature on the relationship between TK and soil resilience shows that there is lack of conclusive supporting data. Generally, TK lacks the tools to measure resilience and associated processes of resistance and recovery to inform the management approaches (Adger 2000). For sustainable soil management, a framework for configuring communities and technical resources and capabilities to leverage its TK is required.

The integration of SK and TK is variably envisaged to take different forms, including parallelism (coexistence without interaction); utilitarianism and selective inclusion of elements of local knowledge amenable to scientific validation (e.g., adoption of aspirin used by the ancient Egyptians and Greeks and of Chinese gunpowder by Western scientists); paternalism (TK as a foundation on which scientific contributions are built, e.g., extension programs); syncretism (SK and TK systems merge mutually); complementarity (SK and TK maintaining diversity and mechanisms for exchange and mutual learning aimed at complementing each other); and coevolution (SK and TK to evolve simultaneously on the basis of their own and in response to their interaction with other forms of knowledge).

5.2.4 Justification of Promotion of TK—Failure of Western Solutions to Usher in Sustainable Soil Management

The general expectation has been one of ongoing science and technology to bring about the desired changes in sustainable soil management, improved productivity, and reduced poverty, yet the practices applied have been unsustainable and have led to such soil-related problems as salinity, acidity, nutrient decline, erosion, and structure decline. Currently, save for the few bright spots, the management of soils is worse than before the introduction of the modern technologies for half-hearted adoption. For example, many farmers are increasingly registering negative nutrient balances because they do not apply fertilizers for fear of the presumed damage (Kirchmann et al. 2007). However, TK persists in local communities and until today plays a very important role in the communal and individual decision making about, for example, farming, health care, land use, local governance, family affairs, and interpersonal relations and exchanges. Traditional or indigenous knowledge is now increasingly being used not only with the aim of finding new drugs but also to derive new concepts that may help us to reconcile empiricism and science (Iaccarino 2003). Often in the past, farmers used the fertilizers in many ways, ranging from indiscriminate application of any available fertilizers without soil testing to nonapplication of any fertilizers in the subsequent seasons, yet expected higher yields (Lu et al. 2005). This is a typical case of efforts by the farmers to switch from TK to adopt modern knowledge and technologies, but ending up in a worse situation with local knowledge that cannot ensure sustainable soil management. Knowledge is that which we come to believe and value based on the meaningfully organized accumulation of information (messages) through experience, communication, or inference.

Technological solutions often offer short-time benefits and undermine sustainability. Technological solutions (e.g., terraces, mulching, and fertilizers) to land degradation at the field scale are well understood. Barrow (1991) claims that despite decades of humans talking of an impending environmental crisis, including the breaking point of soils' fertility, threats to the environment have continued to grow faster than the willingness to control them. Even with good intentions and best farming practices, soils are still vulnerable to cyclic drought or floods, which exacerbate soil degradation. Arguably, development interventions have failed to induce people to meaningfully participate and realize impacts from science-driven development because of a lack of tools and mechanisms to facilitate them to use their own knowledge accumulated over generations of using the land. Western science and education is compartmentalized, contextualized, and taught in detached settings (e.g., classroom, laboratory). Indigenous people acquire knowledge through direct exposure to the real world. Particulars are understood in relation to whole; laws are tested continually in the context of everyday survival. Scientists gain knowledge through the scientific method. In this method, scientists start by finding a problem, which generates questions. A scientist then picks a question of interest and, based on previous knowledge, develops a hypothesis. The scientist then designs a controlled experiment, which will allow for testing the hypothesis against the real world. Then one makes predictions about the outcome of the test, based on the hypotheses.

5.3 RESILIENCE APPROACH TO SOIL MANAGEMENT

Soil degradation in the tropics is most often caused by neglect, misuse, and mismanagement, and once the degradative processes are set in motion, they are exacerbated by unfavorable socioeconomic and political factors (Lal 2000). It is in the interest of humans to sustain soils as this is the essence of human existence. Resilience has become a central concept in the management of natural ecosystems and is closely linked to vulnerability (Schoon 2005). In the context of natural disaster studies, Blaikie et al. (1994) define vulnerability as a limited capacity to "anticipate, cope with, resist, and recover from the impact of natural hazard." The concept of soil resilience is useful in understanding the stability, recovery, and transformation of when and why people would innovate to prevent land degradation. There is general consensus about a strong link between resilience, vulnerability, and sustainability; therefore, assessing changes in resilience as a result of management action is critical because successfully increasing the resilience of natural systems may, therefore, have important implications for human welfare in the face of global climate change (Hughes 2003). Vulnerability to a hazard can be defined as an act of resistance or lack of resistance to external stresses, as a state of absence of resistance, and as a nature of limited inherent capacity. Blaikie et al. (1994) defines vulnerability as a nature of limited capacity to "anticipate, cope with, resist, and recover from the impact of a natural hazard." Others have described vulnerability as the inverse of development model or system sustainability. The root causes of vulnerability and risk are fundamentally tied to preexisting social, economic, and environmental challenges. Sustainable development is geared toward minimizing risks (Pasteur 2011). Vulnerability, resources, adaptive capacity, and certain aspects of TK may, hypothetically, promote or hinder certain virtues that increase soil resilience (Holling 1973): stability, recovery, and transformation (Smit and Wandel 2006; Adger 2000; Folke 2006; Maguire and Hagan 2007; Pimm 1984; Schoon 2005; Fenton et al. 2007). Natural hazard risks are greatly increased by diverse socioeconomic factors such as poverty, uncontrolled and unsound land use and urbanization, weak governmental capacity for disaster mitigation and planning, and environmental degradation. Measuring resilience is quite challenging and only assessed qualitatively using TK by determining changes in productivity. According to Bridges and Catizzone (1996), soil is a grossly undervalued component of the natural ecosystem. Soil scientists must learn how to communicate with society about the role of its proper management to food security, environmental sustainability, climate change mitigation, and livelihood improvements. One possible strategy is to adopt an ecosystem approach to the studying of soils and linking to TK.

5.3.1 Soil Management Dimensions; Socioecological Resilience, Ecosystem Resilience, and Soil Resilience

An important step toward attaining sustainable soil management is to understand the TK and integrated soil and water management practices. The TK is useful for learning and behavior change to reduce risk and enhance resilience. It consists of dimensions of the natural, social, and spiritual inseparably integrated. The natural domain captures knowledge of soils, plants, climates, and animals that translates into land use, agricultural, and health practices. The social domain covers knowledge of local institutions, social capital, natural resource management, conflict resolution, gender relations, art, and language, while the spiritual domain concerns beliefs about the invisible world, divine beings, spiritual forces, and ancestors that regulate cultural values, attitudes, and related practices (e.g., rituals, festivals). Soil management dimensions are summarized in Figure 5.2.

Social resilience	• Of communities
Ecological resilience	• Of ecosystem
Soil resilience	• Of soil system

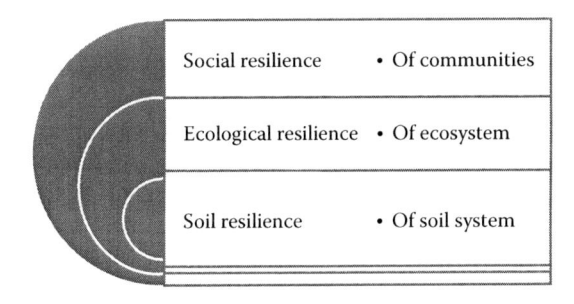

FIGURE 5.2 Conceptual framework depicting the interrelated effects of resilience of ecosystem components.

5.3.2 AT THE SOCIOECOLOGICAL SCALE

The resilience approach recognizes that for the social–ecological system (SES), consisting of both human (social system) and nature (ecological), there is no single stable state, but the system is exposed to different external force "shocks" that challenge its fundamental identity and make it dynamic. According to Grenville and Riverstone (2009), one of the major challenges of operationalizing the resilience approach is defining what constitutes the fundamental structure and function of the system. A resilient system is one that is able to absorb shocks and adapt without changing its fundamental structure and function (Gunderson and Holling 2002). Shocks may be random (e.g., tsunami, land policy reform, major macroeconomic changes) or cyclical (monsoon), or they may occur at different temporal scales—decadal (e.g., drought), annual (e.g., hurricane, labor migration), or finer time scales.

5.3.3 AT THE ECOLOGICAL SCALE

The ecological concept of resilience recognizes the capacity of an ecosystem to respond to a disturbance by resisting damage and recovering quickly (Holling 1996). The disturbances include fires, flooding, erosion, windstorms, insect population explosions, and human activities such as deforestation and the introduction of exotic plant or animal species. Disturbances of sufficient magnitude or duration can profoundly affect an ecosystem and may force an ecosystem to reach a threshold beyond which a different regime of processes and structures predominates (Folke 2006). Human activities that adversely affect ecosystem resilience such as reduction of biodiversity, exploitation of natural resources, pollution, land use, and anthropogenic climate change are increasingly causing regime shifts in ecosystems, often to less desirable and degraded conditions (Adger et al. 2005). Ecological resilience and the thresholds by which resilience is defined are closely interrelated in the way that they influence environmental policy making, legislation, and subsequently, environmental management (Holling 1996). The ability of ecosystems to recover from certain levels of environmental impact is not explicitly noted in legislation; however, because of ecosystem resilience, some levels of environmental impact associated with development are made permissible by environmental policy making and ensuing legislation. An ecosystem is a biological environment consisting of all the organisms living in a particular area, as well as all the nonliving (abiotic), physical components of the environment with which the organisms interact, such as air, soil, water, and sunlight (Ives and Carpenter 2007). The entire array of organisms inhabiting a particular ecosystem is called a community. Central to the ecosystem concept is the idea that living organisms interact with every other element in their local environment.

Are degraded communities more susceptible to environmental stressors? There is increasing recognition by ecologists that, in a variety of ecosystems, species loss following disturbance is not random. Selective deaths following disturbance have a direct impact on community structure, by changing the absolute and relative abundances of flora and fauna species (Vinebrooke et al. 2004). Shifts in community assemblages have been observed following diverse natural and anthropogenic disturbances (e.g., pollution, erosion, sedimentation) (Scheffer et al. 2001). Interdisciplinary discourse on resilience now includes consideration of the interactions of humans and ecosystems via socioecological systems and the need for a shift from the maximum sustainable yield paradigm to environmental management, which aims to build ecological resilience through "resilience analysis, adaptive resource management, and adaptive governance" (Hughes et al. 2005).

5.3.4 AT THE SOIL SCALE

Soil resilience refers to a process by which a soil subjected to an external force resists the change imposed in order to maintain its healthy state and functions (Seybold et al. 1999). Soil resilience should first be looked at as a subset of a notion of environmental resilience along with social and ecosystem resilience. In doing so, the holistic perspective of TK becomes relevant in informing sustainable soil management. External forces constitute the pressure imposed on the soil in terms of biosphere (direct human impacts included), and climate constitutes environmental change—the rate of this change compared to other changes over pedological time constitutes a shock.

5.4 RESILIENCE MODEL

Risk factors are related to poor or negative outcomes. Resilience is a two-dimensional construct concerning the exposure of adversity and the positive adjustment outcomes of that hazard (e.g., degradative forces, extreme climate events); this two-dimensional construct implies two dimensions: one about an "adaptive capacity" and the other about the significance of risk (or vulnerability):

$$\text{Vulnerability}(V) = \frac{\text{Risk} * \text{Capacity}(C)}{\text{Hazard}(H)}$$

This synthesis has drawn on the growing literature on vulnerability and resilience to develop a conceptual framework for understanding the interrelated effects of natural disasters and how capacity can be improved to mitigate these effects (Figure 5.3).

5.4.1 MANAGING FOR RESILIENCE

Resilience is usually defined as the capacity of a system to absorb disturbance without shifting to an alternative state and losing function and services (Folke 2006). Resilience describes systems that are expected to adapt successfully even though they experience risk factors that "place burdens" against them experiencing good development. The resilience of a soil is limited by the rate and extent of change imposed compared with the time that soil requires to recover (Seybold et al. 1999). The concept of managing for resilience aims to alleviate local stressors in an effort to increase system resilience to degradative forces and climate change (Carilli 2009). Such a management philosophy is premised on the belief that eliminating local drivers of ecological change will increase the ability of the system to resist future

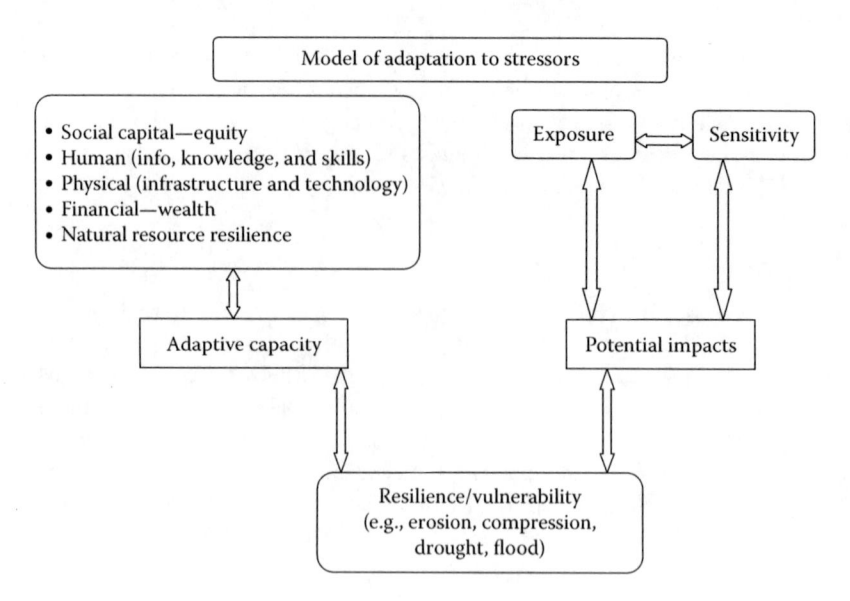

```
                    ┌─────────────────────────────────┐
                    │ Model of adaptation to stressors │
                    └─────────────────────────────────┘
┌─────────────────────────────────────┐    ┌──────────┐      ┌─────────────┐
│ • Social capital—equity             │    │ Exposure │ ⇔    │ Sensitivity │
│ • Human (info, knowledge, and skills)│    └──────────┘      └─────────────┘
│ • Physical (infrastructure and technology)│     ⇕                 ⇕
│ • Financial—wealth                  │
│ • Natural resource resilience       │
└─────────────────────────────────────┘
              ⇕                              ┌──────────────────┐
   ┌───────────────────┐                     │ Potential impacts │
   │ Adaptive capacity │                     └──────────────────┘
   └───────────────────┘                              ⇕
              ⇕                                       ⇕
                    ┌─────────────────────────────────┐
                    │     Resilience/vulnerability     │
                    │   (e.g., erosion, compression,   │
                    │        drought, flood)           │
                    └─────────────────────────────────┘
```

FIGURE 5.3 Response dimensions of soil to external stressors.

stressors and climate disturbances, its ability to recover from such disturbances, or both (Precht and Miller 2006).

According to Lal (1999), soil quality (depth of solum; physical, chemical, and biological properties; and productivity) and soil resilience are a function of the dynamic interaction among the five environmental components (pedosphere, atmosphere, biosphere, hydrosphere, and lithosphere). For example, under low socioeconomic status (poverty) and for pedologically inferior soils, it should be possible to answer the question: How long can soils "hold out" or adjust, and how long may it realistically take in terms of pedological time? Will soil quality remain constant—to continue to provide the same environmental and commercial services as at present—or will it establish a new equilibrium? Extreme climatic events are expected to increase under most climate change scenarios; however, direct empirical research specifically linking TK and soil resilience is lacking to answer the question of whether it may be prudent to focus on managing for resistance or recovery to stress. Farmers' knowledge is well developed and has answers to the relatively straightforward cause–effect interrelationships (Warren 1991) but is inept to address the outgoing questions/subtle issues pertinent to sustainable development (Nitesh and Shefali 2004). According to Baskin (1997), the maintenance of a fertile soil is one of the most vital ecological services the living world performs; the "mineral and organic contents of soil must be replenished constantly as plants deplete soil elements and pass them up the food chain." It is the change in active factors in soil formation that constitutes an environmental change or shock. Critical to sustainable soil management is the ability to determine if the changes in soil properties over time have been significant, constant, or evolving.

The question remains as to whether we should choose a strategy of bolstering soil resistance or recovery. We would argue that the focus should be on resistance rather than recovery on two grounds: (1) Management action, such as the implementation of protection, may promote recovery but not necessarily bolster resistance. (2) The frequency of extreme climatic events (e.g., increasing temperatures, rainfall, floods) is expected to increase under most climate change scenarios; thus, the window available between climate disturbances may be less than the time needed for soils to recover. Besides, not all climate disturbances will be acute (Hoegh-Guldberg et al. 2007). Thus, enhancing soil resistance to climatic stress may be a better long-term goal.

5.4.2 Conventional View of Resilience

Under the conventional approach to managing for resilience are two management strategies: (1) protection of fragile and/or undegraded lands [e.g., conservation agriculture (CA)] and (2) exposed conventional system (Table 5.3).

According to Pasteur (2011), the conventional school of thought postulates that conserved or protected undegraded soils are highly resilient to human-induced degradation and that climate change effects undermine the ability of soils to resist the impacts of degradative forces, tipping degraded soils into alternative, less desirable states sooner than protected ones (conceptual model shown in Figures 5.4a and 5.4b). CA is perhaps the most popular form of land conservation widely thought to have the potential to increase soil resilience. But does it really?

First, the frequency of resilience is the process by which the soil system tends to cope with stress and adversity (Seybold et al. 1999). This coping may result in the system "bouncing back" to a previous state of normal functioning or using the experience of exposure to adversity to produce a "steeling effect" and function better than expected. Resilience refers to the capacity to resist a sharp decline in functioning even though a soil system temporarily appears to get worse. Resilience is the result of a soil system being able to interact with its environment and the processes that are either degradative or constructive that protect it against the overwhelming influence

TABLE 5.3
Management Strategy as Is Informed by Sensitivity to External Forces

Management Strategy	Increasing Level of Stress Plus Climate Change	
	Low Level of Stressors	High Level of Stressors
Protection (e.g., CA)	Bouncing back—good. Not all disturbances will be acute. When impact seems to be reaching the threshold, management response can be to alter human behavior to result in less adverse impact to the ecosystem	Stability—constant competence under stress
Exposed (e.g., conventional agriculture)	Alleviation of local stressor's original state following disturbance vis-à-vis an unconserved or more degraded state. Protected or less degraded land returns more quickly to original state	On severely degraded land, resistance is lost. Recovery is limited under persistent stressors Beyond a certain threshold, recovery is no longer possible

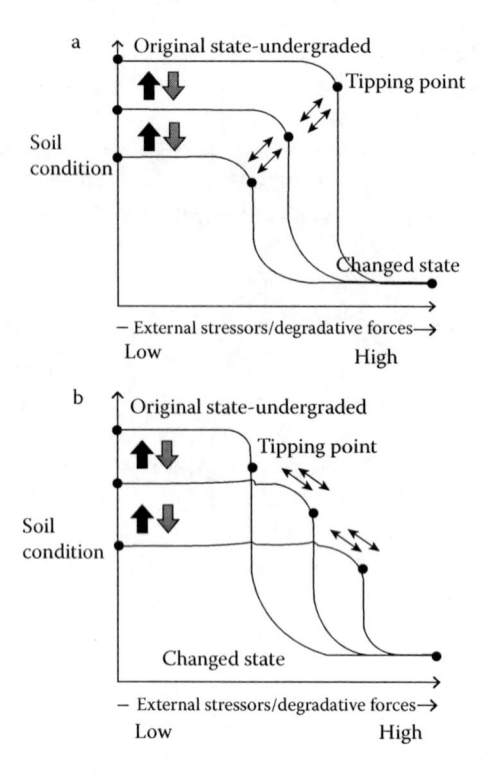

FIGURE 5.4 Managing degraded lands for resilience to degradative forces and climate change.

of risk factors (Seybold et al. 1999). These processes can be enhanced by internal coping strategies or may be helped along by community adaptive capacity, institutions, and social policies that make resilience more likely to occur (Adger 2000). In this sense, "resilience" occurs when there are cumulative "protective factors." These factors are likely to play a more and more important role the greater the soil system exposure to cumulative "risk factors." Resilience is a dynamic process whereby soil systems exhibit a positive response when they encounter significant adversity, hazard, threats, shock, or even significant sources of stress (Enfors and Gordon 2007). It is different from strengths or developmental assets, which are a characteristic of an entire ecosystem, regardless of the level of adversity it faces.

We illustrate our argument against the conventional view of resilience using *protected systems (CA)* vis-à-vis the *exposed systems* (conventional system). CA aims at mitigating local stressors by ensuring continuous minimal mechanical soil disturbance, permanent organic soil cover, and diversified crop rotations (Selig and Bruno 2010). External stressors can cause proportionally greater degradation of a protected than an unprotected soil system. This effect is probably due to the difference in soil genesis between protected and unprotected sites (Cleary et al. 2008). The dominant impact of climate change can override any advantage provided by protection from degradative forces. The conventional agriculture system is exposed to stressors. The general trend of such a system shifts from vulnerability to resilience that survives the disturbance and adaptation following a disturbance. The conventional view of resilience predicts that these shifted or "degraded" systems should be more vulnerable to external stressors. A conventional approach depicts the potential relationships between soil ecosystem state and the strength of disturbances; natural communities are highly resilient to degradative forces, that is, the tipping point (Figure 5.4, block circle), leading to an alternative ecosystem state that is far to the right and attained only at high levels of stress (Donner 2009). As anthropogenic disturbances persist, gradually degrading the original soil system (light block arrows), the tipping point in response to degradative forces gradually shifts to the left, making the soil system less resilient to degradative forces. Thus, management that controls local stressors to reverse degradation and recover the original condition of soil may actually increase the vulnerability of the soil system and may effectively decrease soil system resilience to external disturbance and climate change (Jupiter et al. 2008). A counterintuitive model implies that protected or more conserved soils will cross a tipping point and subsequently shift into an alternative adaptive state, but other alternative states are possible (Coelho and Manfrino 2007)—only at high levels of climate disturbance (Figure 5.4). As nonclimatic, local disturbances degrade the original ecosystem (Figure 5.4, dark block arrows), the tipping point, in response to external stressors and climate change, shifts to the left (Figure 5.4), making the ecosystem less resistant. Management that seeks to control local stressors and reverse degradation (Figure 5.4, black block arrows) is therefore expected to increase resilience by shifting the tipping point back to the right and keeping soil climate disturbance. Using TK, farmers may protect fragile soils and leave less sensitive soils unprotected or under conventional systems of production.

5.4.3 Alternative View to Managing for Resilience

The postulation that protected systems should be more resilient to disturbances from external forces is poorly supported by empirical evidence pertaining to resistance and recovery (Rachello-Dolmen and Cleary 2007). With this perspective, we argue that the expectation of increased resilience of soil to external stressors and climate change through the reduction of local stressors may be fundamentally incorrect and that a conventional approach to resilience may, in fact, result in greater vulnerability to degradative forces and climate impacts. The alternative view of resilience encompasses two separate processes: resistance (the magnitude of disturbance that causes a change in structure) and recovery (the speed of return to the original structure), which are fundamentally different but rarely distinguished (Schoon 2005). If the tolerance of a soils system to a nonclimatic disturbance is correlated with its tolerance to external and climatic impacts, that is, positive cotolerance (Hughes et al. 2003), then degradation can actually increase the resilience of a soil within an ecosystem (Tenywa et al. 2001; Fenton et al. 2007) and thus the ability of the system to resist the impacts of anthropogenic and climate disturbance. Anthropogenic disturbances gradually degrade the soil; the tipping point in response to degradative forces gradually shifts to the left, making the system less resilient (West and Salm

2003). This alternative view, which is more consistent with the majority of empirical observations, is depicted in Figure 5.4b. Thus, with continued degradation caused by local stressors, an altered soil system becomes resistant, and the tipping point in response to external stressors and climate change will shift to the right (Figure 5.4b), making the ecosystem more resilient to external disturbance (Norström et al. 2009). Management that seeks to control local anthropogenic disturbances and reverse degradation (Figure 5.4b, dark block arrows) moves the tipping point back to the left, toward lower resilience to anthropogenic and climate disturbance.

Noteworthy, is the fact that the alternative states depicted in Figures 5.4a and 5.4b are not assumed to be stable. Moreover, the conceptual model works with or without thresholds. If the ecosystem's state declines linearly with external disturbance, it is expected that the slope of this relationship will decrease as degradation increases (i.e., as the intercept decreases). The general view held that reducing local stressors will mitigate the impacts of external stressors, such as degradative forces and climate change, may be fundamentally flawed, at least in terms of one facet of resilience, namely, the ability of soil system to resist externally induced stress (Hughes et al. 2003). The other facet of resilience is recovery. There is growing evidence that protected or less degraded systems return more quickly to their original state following a range of disturbances (including external stress) than unprotected or more degraded ecosystems. Thus, the alleviation of local stressors can potentially enhance soil recovery from degradative forces and climate change impacts (Knowlton and Jackson 2008). If resilience to degradative forces varies in relation to soil system state as depicted in Figure 5.4a, then two general postulations arise:

1. Soil systems exposed to local or persistent degradative forces are more susceptible to external forces than less degraded communities.
2. Soil systems that are conserved/protected from degradative forces are less susceptible to anthropogenic perturbations than those in areas without similar management.

These impacts stem from a multiplicity of local stressors, such as soil erosion, continuous cultivation, and sedimentation. It is therefore not surprising that the concept of resilience—to climate change in particular—is perhaps more strongly advocated as an underpinning of management for tropical soils than for any other ecosystem.

This conceptual model implies that soil resistance (i.e., the extent to which the tipping point is shifted to the right; Figure 5.4b) should covary with increasing degradation. This is true only up to a point. Beyond a threshold level of degradation, changes in species composition and interactions may become irreversible, impairing ecosystem function and (both aspects of) resilience (Srivastava and Vellend 2005). Here, soil with a high probability of experiencing heavy degradation appears to degrade at lower threshold values of stressors (e.g., temperatures) than soil ecosystems in more permanently conserved locations (Graham et al. 2008), leading to the suggestion that management to improve soil quality could increase soil resistance (Seybold 1999). On severely degraded soils such as these, managing for resistance may be unsuccessful, and removing local stressors could offer the only hope for recovery in between disturbances. The challenge for managers will be to identify the levels of local stress under TK on which to build SK for maximizing soil system resistance. Salinization is expected to become a major threat, and nutrient depletion is likely to be a serious problem in large areas of Africa. This analysis of soil resilience in the context needs to be deepened in terms of soil formation and development over pedological time and what resilience means in terms of human expectations and practices. Soil formation and development (pedogenesis), a continuous process taking thousands of years, puts into context the short time that humans have so extensively utilized, changed, and depended directly on soil (Pidwirny 2006). Pedogenesis is the result of five factors: the first two are parent material and topography, which are passive and contribute to soil mass and position; the next two are climate and the biosphere, which are active and supply the energy in soil formation and time (Olson 2005).

5.5 RENAISSANCE OF TK

Since the beginning of the twentieth century, modern sciences have been challenged by new scientific insights [e.g., relativity (Einstein 1920), quantum physics (Heisenberg 1927), and chaos theory (Prigogine and Stengers 1984)]; the changing interpretations of ways of knowing from other cultures have influenced/diversified the Western approach to knowledge. Indigenous practices are based on TK, experience, and insights on how to counteract external forces. The practices are aimed at maximizing benefits and minimizing the adverse impacts of external forces (Garibaldi and Musimwa 2000). With regard to the tragedy of the commons, Hardin's commons theory supports the notion of an inextricable relationship (ties) between sustainable development, and meshing economic growth and environmental protection, and has had an effect on numerous current issues, including the debate over global warming. The approaches are interlinked with reigning economic theory and cannot be divorced. However, due to intensification of external factors, the practices can be overstretched and are unable to cope beyond the limits. However, over time, the pressures to meet the increased demands for food, feed, fiber, and fuel have negatively impacted the TK (Gata 1994).

Initially, research on TK increased as part of the efforts to capture TK in participatory approaches to sustainable land management (Winkleprins 1999). More recently, the awakening to the depth and value of TK has come from climate change adaptation studies that have revealed the tremendous power of TK and also the failure of current development models (Fenton et al. 2007). An increasingly emerging view is that TK and crop varieties may prove even more important for adaptation. In fact, modern agriculture has made many rural communities more vulnerable to climate change by increasing their reliance on external resources (Gata 1994). In the recent past, there has been increasing recognition of TK based on the persistent failure of agricultural growth and development approaches (Senyonjo 2009); there is growing interest in understanding how TK can be combined with SK to strengthen the capacity of soil to resist changes. There is increasing recognition of TK, giving credence to the very practices that were previously ignored or derided. ISWC has been mostly developed under harsh conditions (e.g., dry and marginal lands) and/or on steep hillslopes with high demographic pressure (Critchley et al. 1994). Literature on TK and ISWC in SSA is scanty, but Critchley et al. (1994) provide a comprehensive list.

TK is a strategic asset that is used by communities as an investment to strengthen resilience against external forces (Adepide et al. 2004). Critchley and Mutunga (2003) argue that indigenous environmental knowledge is essential to improved management of resources (e.g., "green water" in drought-prone areas of the tropics). The traditional and innovative technologies described comprise eight technology groups: mulching, no-till farming, home garden systems, terraces, live barriers, gully gardens, forms of riverbank protection, and waterborne manuring (Kirchmann et al. 2007). The efforts to strengthen TK and integrated soil and water conservation practices must be well informed. Intellectual Property Rights (IPRs) also raise the price of seeds and restrict access to seeds by farmers and scientists. Yet the challenges of climate change may require the widest possible circulation of germplasm to enable effective and timely adaptation. TK for soil resilience embodies experiences, practices, and insights utilized by local communities to minimize negative effects resulting from external pressures to meet the increasing demands for food, feed, fiber, and fuel. This knowledge is accumulated and refined purposively to sustain a community and its culture for its continued survival. Depth of knowledge is rooted in long inhabitation and use of places, lessons, and a more satisfying and sustainable complex knowledge system with adaptive integrity (Barnhardt and Kwagley 2005). Indigenous people are known to exhibit adaptive integrity, sustaining unique perspectives, core values, beliefs, practices, and associated knowledge systems even while going through changes beyond their control (Aoki 2003). However, with the intervention of modern teaching and learning, the knowledge associated with the symbiotic relationship between man and nature is slowly being eroded (Briggs 2005).

Regardless of the degree to which local people have embraced modernity, they continue to prefer their own knowledge that covers spatial and temporal dimensions deemed necessary for specific purposes. According to Warren (1991), people who survive by tradition and folklore perceive change as risky and trust what has been tried and tested by long-established custom. Many studies (Talawar 1996; Talawar and Rhodes 1998; Teklu and Gezahegn 2003) provide evidence that farmers have detailed knowledge of their soils in terms of physical, economic, and social environment to meet their goals and exhibit the ability to translate this knowledge into agronomic management options. Habarurema and Steiner (1997), in their studies in southern Rwanda, reported farmers to have profound knowledge of their soils to the extent that they classify soils according to their agricultural potential and tillage properties into nine major soil types based on the following criteria: crop productivity, soil depth, soil structure, and soil color corresponding to soil suitability classes. In the Siaya District of Kenya, farmers base their classification on the topsoil characteristics (color, depth, texture, and ease of workability) (Werner 2001). In northern Ethiopia, three different soil types are distinguished by farmers on basis of yields, topography, soil depth, color, texture, water holding capacity, and stoniness (Corbeels et al. 2000). Yoruba farmers in Nigeria, in assessing the physical properties of soil (texture, color, drainage condition, organic matter, presence or absence of earthworm casts, bulk density), focus on topsoil using their senses of sight, touch, and smell (Osunade 1988). The Wasukuma people of northwest Tanzania, basing on soil color, texture, and structure, recognize nine major soil classes and specific management practices associated with each soil type (Rounce 1949; Malcolm 1953).

Tamang (1992) observed that in all studies, soil color was the most widely used indicator by farmers to classify their soils, but its importance in soil classification along with texture can vary depending on the area and ethnic group considered. TK can be inconsistent at a regional scale, with members of the same ethnic group but from different villages naming and characterizing differently soil types and land classes. SK, on the other hand, perceives soil in three dimensions and emphasizes presence or absence of diagnostic horizons.

Under circumstances where land use systems are not practiced in accordance to potential or suitability of land, these practices can often be traced back to socioeconomic challenges to development faced by society (FAO 1976; Ryder 1994; Bouma 1999). According to Mafalacusser (1995), farmers base their choice of enterprises on four factors: (1) interest/goal; (2) available resources (land, water, labor); (3) external factors (access to credit, timely and relevant information); and (4) risk associated with production and market prices. The extension workers who facilitate learning and modernization strategies for farmers should be open and willing to modify the cold rationalism to accommodate the subjective rationality of farmers (Beckford 2002).

Field studies in China, Kenya, and Bolivia, summarized in a new International Institute for Environment and Development (IIED) briefing paper (2011), show that communities severely impacted by changes in climate have survived thanks to traditional crops (Climate Development Knowledge Network [CDKN]). Traditional crops are hardier and more resilient to impacts such as drought and new pests, because their genetic makeup is more varied and better suited to local conditions. In southwest China, for example, most local Landraces survived the big spring drought in 2010, while most modern hybrids were lost. Villages growing only hybrids lost all their production due to a shortage of hybrid seed in the market. In all three cases, farmers understand the value of sustaining a diversity of crops to reduce the risk of crop failure.

Traditional varieties are also more accessible because they come from farmers' own saved seeds and so do not have to be bought in markets and are cheap. The research also shows that traditional conservation practices—such as kaya forests in coastal Kenya—may need to be reestablished to enable adaptation where government systems have not been effective (Darling et al. 2010).

In southwest China, where remaining areas of traditional agriculture are located, maize and rice staple crops have been highly commercialized, and the area cultivated with traditional varieties has rapidly decreased in the last 10 years (by 44% and 21%, respectively).

Incentives are urgently needed to encourage soil conservation by governments, public breeding institutes, and farmers. Policy and institutional reforms are needed to support both modern and traditional agriculture. The capacity of the world's poorest and most affected communities to adapt depends on the interlinked TK, culture, and ecosystems—or biocultural systems—from which new innovations can develop and spread. Policy makers at Durban need to recognize the value of traditional farming systems and identify ways to support them, including through reform of IPRs.

Understanding local cultural beliefs, changes, local technologies, their efficiencies and how they can be improved, and the institutional capacity for data collection, dissemination, and use can identify economic opportunities for adaptation and integration of resilience into the local level.

5.6 UNION OF SK AND TK

The approach used to support development in SSA consisting of transfer of technologies, knowledge, and values from the modern world to the underdeveloped world needs to be revised. Gadamer (2001) argues that there is a need to prepare pathways for science to accept the truth from other worldviews, which could result in a fusion of the horizons of human wisdom. In many communities, indigenous people are often critical of universities, yet at the same time, they prefer their children to gain Western educations and high-level qualifications but not at the expense of their identities, culture, languages, values, and practices (Smith Tuhiwai 1999; Haverkort and Reijntjes 2007). As such, communities see intracultural learning and endogenous development as an important first step toward intercultural dialogue and sustainable development (Haverkort and Rist 2007). Endogenous development refers to the process that takes the indigenous perspective as a starting point in understanding the traditional ways of knowing and learning aimed at revitalizing and enhancing the TK and value systems (Smith Tuhiwai 1999; Rist 2002; Coetzee and Roux 1998; Emagalit 2004; Wiredu 2005). A study on sites in Uganda (Wera and Toroma) and Tanzania (Mahiga and Iteja) noted the importance of combining elements of a broad-scale scientific survey with a localized assessment of indigenous knowledge. Contrary to the first impression that farmers' knowledge is inconsistent, farmers' soil classifications are more often comparative than hierarchical, with qualitative distinctions made in terms of darker/lighter color or greater/lesser fertility. According to Gowing et al. (2004), evidence of hierarchical subgroups was collected from both sites where farmers divided their shallow soils into different categories on the basis of soil surface texture. *Ingaroi, Aputon naingaroikitos,* and *Apokor naingaroikit* are all shallow soils that have ironstone gravel and indurated ironstone near the surface. The term *ingaroi* is common to all and means gravel in the Ateso language. Soils with sandy surface texture are mapped as *Aputon naingaroikitos,* while those clayey surface textures are mapped as *Apokor naingaroikit.* Comparative analysis of LK and SK using GIS generally showed poor correspondence, but in some cases, there was a direct relationship (Table 5.2).

The strength of SK lies in observation and experimentation, but this is undermined by its acquisition from nontypical environments that are either strictly controlled (e.g., laboratory) or highly simplified from the real-life complexity (van Regenmortel 2004). This approach is criticized for being reductionist but no longer sufficient to analyze and understand higher levels of complexity (Kellenberger 2004; van Regenmortel 2004). Thus, the integration of methods and results from different approaches and levels of analysis becomes essential. Western science and TK constitute different paths to knowledge, but they are rooted in the same reality. We can only gain from integrating them.

TK and Western science are divergent systems that coexist (Gayton and Michel 2002). Understanding the processes of learning in TK can help increase efficiency of knowledge utilization (Pant and Odame 2008).

Tacit knowledge is possessed and subconsciously understood only by an individual from direct experience and action, but is difficult to transfer to another person by means of writing it down or verbalizing it via words and symbols, and is usually shared through highly interactive conversation, storytelling, and shared experience (Ellen et al. 2000). Its transfer is very sensitive to social context.

5.6.1 Conceptualization Framework for Management of Explicit TK

1. Repositories for accumulation and storage of explicit TK
2. Refinery for organization, creation of new knowledge, and sharing knowledge
3. Knowledge management function
4. Information technologies to support storage and archival of data

Any choice and decision with regard to the practices/technology solely rest with the community itself and cannot be made by outsiders.

5.7 CONCLUSION

Knowledge structures and institutions provide the context for interpreting accumulated content, but these have been weakened by ushering in ST. Soils face many environmental changes and shocks from which they will recover or to which they will adjust if sufficient "pedological" time is allowed. Because of interest in short-term benefits to meet needs for improved livelihoods, farmers adopt soil management practices that promise a lot of benefits in a short time frame but are unsustainable in the long term leading to a declining soil health and negatively impacting human livelihoods. The resilience of the soil in terms of human expectations and time frames will depend on its ability to recover to an equilibrium state once improved practices have been extensively applied. Appropriate policies, methods for theory and knowledge development, learning, training, and research that enhance diversity, revitalization, and coevolution of different ways of knowing, TK-based endogenous development processes for conventional scientists, intergenerational transmission of knowledge, and joint learning for sustainable soil management are essential for enhancing resilience of soils in SubSaharan Africa. TK targeted toward policy goals will be the key to ensuring that societies realize the true value of ecosystem services, sooner rather than later. It should present to policy makers consequences of decisions (e.g., ignoring TK).

ACKNOWLEDGMENTS

We thank the farmers, extension workers, local government, and students Geofrey Gabiri and Robert Kaliisa, who helped with organizing the references.

ABBREVIATIONS

ASALs: arid and semiarid lands
CA: conservation agriculture

ISWC: indigenous soil and water conservation
R&D: research and development
SES: social–ecological system
SK: scientific knowledge
TK: traditional knowledge

REFERENCES

Adas, M. 1990. *Machines as the Measure of Man: Science, Technology, and Ideologies of Western Dominance*. New York: Cornell University Press.

Adepide, N.O., P.A. Kuneye, and L.A. Ayinde. 2004. Relevance of local and indigenous knowledge for Nigeria agriculture. Presented at the international conference on Bridging scales and Epistemologies; linking local knowledge with Global science in multi-scale Assessments, March 16–19, Alexandria, Egypt.

Adger, W.N. 2000. Social and ecological resilience: are they related? *Prog Hum Geogr* 24(3):347–364.

Adger, W.N., T.P. Hughes, C. Folke, S.R. Carpenter, and J. Rockström. 2005. Social ecological resilience to coastal disasters. *Science* 309(5737):1036–1039.

Agrawal, A. 1995. Dismantling the divide between indigenous and scientific knowledge. *Dev Change* 26:413–439.

Akabogu, E. 2000. Indigenous knowledge systems, integrity of the commons and emerging regimes of intellectual property rights in a globalising world. Presented at "The Commons in an Age of Globalisation," the Ninth Conference of the International Association for the Study of Common Property, Victoria Falls, Zimbabwe, June 17–21.

Altieri, M.A., D.K. Letourneau, and J.R. Davis. 1983. Developing sustainable agroecosystem. *Bioscience* 33(1):45–49.

Aoki, K. 2003. "(Intellectual) Property and Sovereignty: Notes Toward a Cultural Geography of Authorship," *Stanford Law Review 48 (May 1996)*, 1293–1355. Cited in Butt, D., J. Bywater, N. Paul (eds.). 2008. *"Place: Local Knowledge and New Media Practice,"* 214 pp. ISBN 9781847184849. Cambridge Scholars Publishing. http://www.c-s-p.org.

Barnhardt, R. and A.O. Kawagley. 2005. Indigenous knowledge systems and alaska native ways of knowing. *Anthropology and Education Quarterly* 36(1):8–23, ISSN 0161-7761, electronic ISSN 1548-1492 by the American Anthropological Association.

Barrera-Bassols, N. and J.A. Zinck. 1986. The other pedology: empirical wisdom of local people. ITC, The Nelletands Symposium 45:16–36.

Barrow, E.G. 1991. Evaluating the effectiveness of participatory agroforestry extension programmes in a pastoral systems, based on existing traditional values: a case study of the Turkana in Kenya. *Agroforest Syst* 14(1):1–38.

Baskin, Y. 1997. The Work of Nature, the Scientific Community on Problems of the Environment SCOPE. Washington to DC: Island Press.

Beckford, C.L. 2002. Decision-making and innovation among small-scale yam farmers in Central Jamaica: a dynamic, pragmatic and adaptive process. *Geogr J* 168:248–259.

Berkes, F. 1993. Traditional ecological knowledge in perspective. In Inglis, J.T. (ed.). *Traditional Ecological Knowledge: Concepts and Cases*, 1–9. Ottawa: Canadian Museum of Nature/International Development Research Centre.

Berkes, F. 1999. Role and significance of "tradition" in indigenous knowledge. *IKDM* 7:19.

Berkes, F., J. Colding, and C. Folke. 2000. Rediscovery of traditional ecological knowledge as adaptive management. *Ecol Appl* 10:1251–1262.

Blaikie, P. 1994. *At Risk*. London: Routledge.

Blaikie, P.M. and H. Brookfield. 1987. *Land Degradation and Society*. London: Methuen.

Bouma, J. 1999. Land evaluation for landscape units. In M.E. Summer (ed.) *Handbook of Soil Science*. pp. E393–E412. Boca Raton, Florida: CRC Press.

Brand, F. 2009. Critical natural capital revisited: ecological resilience and sustainable development. *Ecol Econ* 68:605–612.

Bridges, E.M. and M. Catizzone. 1996. Soil science in a holistic framework. *Geoderma* 71:275–287.

Briggs, J. 2005. The use of indigenous knowledge in development: problems and challenges. *Prog Dev Studies* 5:99–114.

Carilli, J.E., R.D. Norris, B.A. Black, S.M. Walsh, and M. McField. 2009. Local stressors reduce coral resilience to bleaching. *PLoS One* 4:6324.

Cleary, D.F.R., L. De Vantier, V.L. Giyanto, and P. Manto. 2008. Relating variation in species composition to environmental variables: a multitaxon study in an Indonesian coral reef complex. *Aquat Sci* 70:419–431.

Coelho, V.R. and C. Manfrino. 2007. Coral community decline at a remote Caribbean island: marine no-take reserves are not enough. *Aqua Conserv Mar Freshwat Ecosyst* 17:666–685.

Coetzee, P.H. and A.P. Roux. 1998. *The African Philosophy Reader*. London: Routledge.

Cools, N., E. De Pauw, and J. Deckers. 2003. Towards an integration of conventional land evaluation methods and farmers' soil suitability assessment; a case study in Northwestern Syria. *Agric Ecosyst Environ* 95(1):327–342.

Corbeels, M., A. Shiferaw, and M. Haile. 2000. Farmers' knowledge of soil fertility and management strategies in Tigray, Ethiopia. Managing Africa's Soils No. 10. Nottingham, Russell Press. http://www.iied.Org/docs/drylands/soils_10.pdf.

Critchley, W. and K. Mutunga. 2003. Local innovation in a global context: documenting farmer initiatives in land husbandry through WOCAT. *Land Degrad Dev* 14:143–16.

Critchley, W.R.S., C. Reij, and T.J. Willcocks. 1994. Indigenous soil and water conservation: a review of the state of knowledge and prospects for building on traditions. *Land Degrad Rehab* 5:293–314.

Darling, E.S., T.R. McClanahan, and I.M. Côté. 2010. Combined effects of two stressors on Kenyan coral reefs are additive or antagonistic, not synergistic. *Cons Lett* 3:122–130.

de La Cruz. 2006. Regional Study in the Andean Countries: Customary Law in the Protection of Traditional Knowledge. Unpublished Final Report. World Intellectual Property Organization (WIPO).

Donner, S.D. 2009. Coping with commitment: projected thermal stress on coral reefs under different future scenarios. *PLoS One* 4:5712.

Einstein, A. 1920. *Relativity. The Special and General Theory*. Third edition, translated by R.W. Lawson. New York: Henry Holt and Company. p. 168.

Ellen, R., P. Parkes, and A. Bicker. 2000. *Indigenous environmental knowledge and its transformation*. Amsterdam, the Netherlands: Harwood Academic publishers.

Emagalit, Z. 2004. Contemporary African philosophy, web publication: http://faculty.msmc.edu/lindeman/af.html (accessed March 8, 2013).

Emery, A.R. 2000. *Guidelines: integrating indigenous knowledge in project planning and implementation*. Canada: KIVU Nature Inc.

Enfors, E. and L.J. Gordon. 2007. Analyzing resilience in dryland agro-ecosystems: a case study of the Makanya catchment in Tanzania over the past 50 years. *Land Degrad Dev* 18:680–696.

Evenson, R.E. and D. Gollin. 2003. Assessing the impact of the green revolution, 1960 to 2000. *Science* 300(5620):758–762.

FAO. 1976. *A Framework for Land Evaluation*. Soil Bulletin 32. Rome: Food and Agriculture Organization of the United Nations.

Fenton, M., G. Kelly, K. Vella, and J. Innes. 2007. Climate change and the Great Barrier Reef: industries and communities. In: Johnson, J.E. and P.A. Marshall (eds.). *Climate Change and the Great Barrier Reef: A Vulnerability Assessment*. Australia: Great Barrier Reef Marine Park Authority and Australian Greenhouse Office.

Folke, C. 2006. Resilience: the emergence of a perspective for social-ecological systems analyses. *Global Environ Change* 16:253–267.

Freeman, N. 1992. "International Economic Responses to Reform in Vietnam: An Overview of Obstacles and Progress," Studies in Comparative Communism, September. www.brad.ac.uk/acad/management/external/pdf/people/publicationsnfreeman.pdf (accessed March 8, 2013).

Gadamer, H.G. 2001. Gadamer in Conversation. Translated by R.E. Palmer (ed.). New Haven and London: Yale University Press. pp. 115–132.

Galaty, J. 1999. Losing Ground: Indigenous Rights and Resources Across Africa. *Cultural Survival Quarterly* 22(4):28–31.

Garibaldi, V. and E. Musimwa. 2000. *Indigenous knowledge systems and National Resource Management in Southern Africa.* Harare, Zimbabwe: IUCN-ROSA Distributor.

Gata, N.R. 1994. Indigenous science and technologies for sustainable agriculture, food systems and natural resource management with special reference to Zimbabwe. In J.Z.Z. Matowanyia, V. Garibaldi, and E. Musimwa (eds.). Harare, Zimbabwe: IUCN-ROSA Distributor.

Gayton, D. and H. Michel. 2002. *Linking Indigenous Peoples' Knowledge and Western Science in Natural Resource Management: Conference Proceedings.* Kamloops, Canada: Southern Interior Forest Extension and Research Partnership (SIFERP Series, no. 4).

Gowing, J., R. Payton, and M. Tenywa. 2004. Integrating indigenous and scientific knowledge on soils: recent experiences in Uganda and Tanzania and their relevance to participatory land use planning. *Ug J Agric Sci* 9:184–191.

Graham, N.A.J., T.R. McClanahan, M.A. Mac-Neil, S.K. Wilson, and N.V.C. Polunin. 2008. Climate warming, marine protected areas and the ocean-scale integrity of coral reef ecosystems. *PLoS One* 3:3039.

Grenville, B. and G. Riverstone. 2009. Exploring vulnerability and resilience in land tenure systems after Hurricanes mitch and Ivan, University of Florida. Available at: http://siteresources.worldbank.org/INTIE/Resources/Paper_Barnes_Riverstone.pdf.

Gunderson, L.H., C.S. Holling, and (Eds.). 2002. *Panarchy: Understanding Transformations in Human and Natural Systems.* p. 508. Washington, D.C: Island Press.

Habarurema, E. and K.G. Steiner. 1997. Soil suitability classification by farmers in southern Rwanda. *Geoderma* 75:75–87.

Haverkort, B. and C. Reijntjes (eds). 2007. *Moving Worldviews: Reshaping Sciences, Policies and Practices for Endogenous Sustainable Development.* ETC/Compas Series on Worldviews and Sciences, The Netherlands: Leusden. ISBN-10:90-77347-10-0; ISBN-13: 978-90-77347-10-2, 433 pp.

Haverkort, B. and C. Reijntjes. 2007. Diversities of worldviews, knowledge communities and sciences and the challenges of its co-evolution.

Haverkort, B. and S. Rist. 2007. *Endogenous Development and Bio-cultural Diversity: The Interplay Between Worldviews, Globalization and Locality.* Compas Series on Worldviews and Sciences Nr. 6. The Netherlands: Leusden. pp. 67–81.

Heisenberg, W. 1949. The Physical Principles of the Quantum Theory. Courier Dover Publications, 1949. *Science*, 183 pp. Translated by C. Eckart and F.C. Hoyt, ISBN 0486601137.

Hoegh-Guldberg, O., P.J. Mumby, A.J. Hooten, R.S. Steneck, and P. Greenfield. 2007. Coral reefs under rapid climate change and ocean acidification. *Science* 318:1737–1742.

Holling, C.S. 1996. Engineering resilience versus ecological resilience. In Schulze, P.C. (ed.). *Engineering Within Ecological Constraints*, 31–44. Washington: National Academy Press.

Holling, C.S.1973. Resilience and stability of ecological systems. *Annu Rev Ecol Syst* 4:1–23.

Hughes, T.P., A.H. Baird, D.R. Bellwood, M. Card, and S.R. Connolly. 2003. Climate change, human impacts, and the resilience of coral reefs. *Science* 301:929–933.

Hughes, T.P., D.R. Bellwood, C. Folke, R.S. Steneck, and J. Wilson. 2005. New paradigms for supporting the resilience of marine ecosystems. *Trends Ecol Evol* 20:380–386.

Iaccarino, M. 2003. Science and culture. *EMBO Rep* 4:220–223.
ICSU (International Council for Science). 2002. *Annual Report*. Paris, France.
IIED Briefing Paper. 2011. *Adapting Agriculture with Local Knowledge*.
Ives, A.R. and S.R. Carpenter. 2007. Stability and diversity of ecosystems. *Science* 317:58–62.
Jones, P. and P.K. Thornton. 2008. Croppers to livestock keepers: livelihood transitions to 2050 in Africa due to climate change. *Environ Sci Policy* 12:427–437.
Jupiter, S., G. Roff, G. Marion, M. Henderson, and V. Schrameyer. 2008. Linkages between coral assemblages and coral proxies of terrestrial exposure along a cross-shelf gradient on the southern Great Barrier Reef. *Coral Reefs* 27:887–903.
Kamil, N.M. 2011. The quagmire of philosophical standpoints (paradigms) in management research. *Postmod Openings* 5(5):13–18.
Kellenberger, E. 2004. The evolution of molecular biology. *EMBO Rep* 5:546–549.
King, G.J., D.F. Acton, and R.J. St. Arnaud. 1983. Soil-Landscape Analysis in Relation to Soil Distribution and Mapping at a Site Within the Weyburn Association. Canadian Journal of Soil Science 63(4):657–670, 10.4141/cjss83-067.
Kirchmann, H., L. Bergströma, T. Kätterera, L. Mattssona, and S. Gessleinb. 2007. Comparison of long-term organic and conventional crop–livestock systems on a previously nutrient-depleted soil in Sweden. *Agro J* 99:960–972.
Knowlton, N. and J.B.C. Jackson. 2008. Shifting baselines, local impacts, and global change on coral reefs. *PLoS Biol* 6:0215–0220.
Lal, R. 1999. Soil management and restoration for C-sequestration to mitigate the accelerated greenhouse effect. *Prog Environ Sci* 1(4):307–326.
Lal, R. 2000. Physical management of soils of the tropics: priorities for the 21st century. *Soil Sci* 165(3):191–207.
Larson, J. 1998. Perspectives on indigenous knowledge systems in Southern Africa. Environment Group, Africa Group World Bank Discussion Paper No. 3. April.
Lu, Y., J. McDonagh, O. Semalulu, M. Stocking, and C. Owuor. 2005. Enhancing the impacts of research in soil management—development of practical tools in the hillsides of Eastern Uganda. In Stocking, M., H. Helleman, and R. White, 209–220. Renewable Natural Resources Management for Mountain Communities. Kathmandu: International Centre for Integrated Mountain Development. 314pp.
Mafalacusser, J.M. 1995. The use of indigenous knowledge for land use planning in a part of Xai-Xai District, Gaza Province, Mozambique. Unpublished M.Sc. Thesis, ITC, Enschede.
Maguire, B. and P. Hagan. 2007. Disasters and communities: understanding social resilience. *AJEM* 22:16–20.
Malcolm, D.W. 1953. *Sukumaland. An African People and Their Country. A Study of Land Use in Tanganyika*. Oxford: Oxford University Press.
Miiro, H.D., J. Tumuhairwe, K.J. Kabananukye, A. Lwakuba, W.R.S. Critchley, J. Ellis-Jones, and T.J. Willcocks. 1995. *A Participatory Rural Appraisal (PRA) in Kamwezi SubCounty, Kabale District, Uganda*. Unpublished report no OD/95/13. Silsoe: Silsoe Research Institute.
Nakashima, D. and M. Roué. 2002. Indigenous knowledge, peoples and sustainable practice. In T. Munn (ed.). *Encyclopedia of Global Environmental Change*. Chichester, UK: John Wiley & Sons, Ltd.
Nitesh, T. and B. Shefali. 2004. Integrating indigenous knowledge and GIS for participatory natural resource management: state of the practice. *EJISDC* 17(3):1–13.
Norström, A.V., M. Nyström, J. Lokrantz, and C. Folke. 2009. Alternative states on coral reefs: beyond coral-macroalgal phase shifts. *Mar Ecol Prog Ser* 376:295–306.
Olson, K.R. 2005. Factors of soil formation/parent material. In Hillel, D. (ed.), *Encyclopedia of Soils in the Environment*, 532–535.
Orvik, J. and R. Barnhardt (eds). 1974. *Cultural Influences in Alaska Native Education*. Fairbanks: Center for Northern Education Research, University of Alaska Fairbanks.

Ostrom, E. 1990. *Governing the Commons: The Evolution of Institutions for Collective Action.* Cambridge, UK: Cambridge University Press.

Osunade, M.A.A. 1988. Soil suitability classification by small farmers. *Prof Geogr* 40(2):194–201.

Osunande, M.A.A. 1992. Land resource appraisal by small farmers in Swaziland. *UNISWA Res J* 6:93–104.

Pant, L.P. and H.H. Odame. 2008. Innovation systems in agriculture: civic engagement in participatory research in Nepal. In *World University Services of Canada (WUSC)'s National Research Forum, November 7–9, 2008.* Ottawa, Canada: WUSC.

Pastuer, K. 2011. *From Vulnerability to Resilience: a framework for analysis and action to build community resilience.* Bourton on Dunsmore, Rugby, Warwickshire CV23 9QZ, UK: Practical Action Publishing Ltd, Schumacher Centre for Technology and Development. www.practicalactionpublishing.org.

Patton, M.Q. 1978. *Utilization Focused Evaluation.* Beverly Hills, CA: Sage Publications.

Payton, R.W., J. Barr, A. Martin, P. Sillitoe, J. Deckers, J. Gowing, N. Hatibu, S. Naseem, M. Tenywa, and M. Zuberi. 2003. Contrasting approaches to integrating indigenous knowledge about soils and scientific soil survey in East Africa and Bangladesh. *Geoderma* 111:355–386.

Pidwirny, M. 2006. *Soil Pedogenesis. Fundamentals of Physical Geography*, 2nd Ed. http://www.physicalgeography.net/fundamentals/10u.htm (accessed March 8, 2013).

Pimm, S. 1984. The complexity and stability of ecosystems. *Nature* 307:321–326.

Precht, W.F. and S.L. Miller. 2006. Ecological shifts along the Florida reef tract: the past as key to the future. In: Aronson, R.B (ed.). *Geological Approaches to Coral Reef Ecology.* New York: Springer. pp. 237–312.

Prigogine, I. and I. Stengers. 1984. *Order Out of Chaos: Man's New Dialogue With Nature.* Bantam.

Rachello-Dolmen, P.G. and D.F.R. Cleary. 2007. Relating coral species traits to environmental conditions in the Jakarta Bay/Pulau Seribu reef system. *Estuar Coast Shelf Sci* 73:816–826.

Reij, C., I. Scoones, and C. Toulmin (eds.). 1996. *Sustaining the Soil; Indigenous Soil and Water Conservation in Africa.* London: Earthscan. ISBN 1-85383-372 X, 260 pp.

Reijntjes, C., B. Haverkort, and A. Waters-Bayer. 1992. *Farming for the Future. An Introduction to Low-External Input and Sustainable Agriculture.* London: Macmillan. www.leisa.info.

Rist, S. 2002. *Si Estamos de Buen Corazon, Siempre Hay Produccion.* La Paz: Ed. Plural.

Rounce, N.V. 1949. *The Agriculture of the Cultivation Steppe; Department of Agriculture, Tanganyika Territory.* Cape Town: Longmans, Green and Co.

Rutatora, D.F. 1997. Strength and weaknesses of the indigenous farming system of the Matengo people of Tanzania. *IKDM* 5:6–9.

Ryder R. 1994. Land evaluation for steepland agriculture in the Dominican Republic. *The Geographical Journal* 160(1):74–86.

Scheffer, M., S. Carpenter, J.A. Foley, C. Folke, and B. Walker. 2001. Catastrophic shifts in ecosystems. *Nature* 413:591–596.

Scherr, S. and S. Yadav. 1997. *Land Degradation in the Developing World: Issues and Policy Options for 2020, 2020 Vision Policy Brief No. 44.* Washington, D.C.: IFPRI.

Scherr, S.J., 1999. *Soil degradation; a threat to developing country food security by 2020?*, IFPRI Discussion paper 27, Washington DC, USA.

Schoon, M. 2005. *A Short Historical Overview of the Concepts of Resilience, Vulnerability and Adaptation (Working Paper W05-4); Workshop in Political Theory and Policy Analysis, Indiana University.* Cited in: Obrist, B., C. Pfeiffer, and R. Henley. 2010. Multi-layered social resilience: a new approach in mitigation research. *Progress in Development* 10(4):283–293. Online version: http://pdj.sagepub.com/content/10/4/283 (accessed March 8, 2013).

Selig, E. and J. Bruno. 2010. A global analysis of the effectiveness of marine protected areas in preventing coral loss. *PLoS One* 5:9278.
Senyonjo, D. 2009. Land evaluation for rainfed agriculture based on farmer's indigenous knowledge: a case study of Mityana District. Unpublished M.Sc. thesis, Makerere University.
Seybold, C.A., J.E. Herrick, and J.J. Brejda. 1999. Soil resilience; a fundamental component of soil quality. *Soil Sci* 166:224–234.
Shiva, V. 1991. *The Violence of the Green Revolution: Third World Agriculture, Ecology and Politics*. London: Zed books. 257 pp.
Smit, B. and J. Wandel. 2006. Adaptation, adaptive capacity, and vulnerability. *Global Environ Change* 16:282–292.
Smith Tuhiwai, L. 1999. Decolonising methodologies: research and indigenous peoples, London: Zedbooks Ltd.
Srivastava, D.S. and M. Vellend. 2005. Biodiversity-ecosystem function research: is it relevant to conservation? *Annu Rev Ecol Evol Syst* 36:267–294.
Talawar, S. 1996. *Local Soil Classification and Management Practices: Bibliographic Review. Laboratory of Agriculture and National Resources of Anthropology*. Athens, GA, USA: University of Georgia.
Talawar, S. and R.E. Rhoades. 1998. Scientific and local classification and management of soils. *Agric Hum Values* 15(1):3–14.
Tamang, D. 1992. Indigenous soil fertility management in the hills of Nepal: lessons from an east–west transect. Report No. 19. HMG Ministry of Agriculture/Winrock International Research.
Tamang, D. 1993. *Living in a Fragile Ecosystem: Indigenous Soil Management in the Hill of Nepal*; IIIED Gatekeeper Series, 41, 23, London: International Institute for Environment and Development (IIED).
Teklu, E. and A. Gezahegn. 2003. Indigenous Knowledge and Practices for Soil and Water Management in East Wollega, Ethiopia; Göttingen, October 8–10, 2003. Conference on International ARD.
Tenywa, M.M., R. Lal, and M.J.G. Majaliwa. 2001. Characterization of the stages of soil resilience to degradative stresses: erosion. In Stott, E., R.H. Mohtar, and G.C. Steinhardt (eds.). *Sustaining the Global Farm-Selected Papers from the 10th ISCO Meeting Held in West Lafayette, Indiana-USA*, 606–610. http://topsoil.nserl.purdue.edu/nserlweb-old/isco99/pdf/ISCOdisc/SustainingTheGlobalFarm/P201-Tenywa.pdf (accessed march 8, 2013).
Tick, A. 1993. Aspects of indigenous knowledge. *IKDM* 1(2):6–10.
Van Regenmortel, M.H. 2004. Reductionism and complexity in molecular biology. *EMBO Rep* 5:1016–1020.
Vinebrooke, R.D., K.L. Cottingham, J. Norberg, M. Scheffer, and S.I. Dodson. 2004. Impacts of multiple stressors on biodiversity and ecosystem functioning: the role of species co-tolerance. *Oikos* 104:451–457.
Warren, D.M. 1989. Linking scientific and indigenous agricultural systems. In Compton, J.L. (ed.). *The Transformation and International Agricultural Research and Development*, 153–170. Boulder, Colorado, USA: Lynne Rienner Publishers.
Warren, D.M. 1991. *Using indigenous knowledge in agricultural development*. World Bank Discussion paper Washington DC: The World Bank.
Warren, D.M., L.J. Slikkerveer, and D. Brokensha. 1995. *The Cultural Dimension of Development: Indigenous Knowledge Systems*. London, UK: Intermediate Technology Publications.
Werner, S. 2001. Environmental knowledge and resource management: Sumatra's Kerinci-Seblat National Park. PhD Dissertation. Series: Berliner Beiträge zu Umwelt und Entwicklung. Berlin: Berlin Technische Universität.

West, J.M. and R.V. Salm. 2003. Resistance and resilience to coral bleaching: implications for coral reef conservation and management. *Conserv Biol* 17:956–967.

Winklerprins, A.M.G.A. 1999. Local soil knowledge: a tool for sustainable land management. In *Society and Natural Resources* 12(2):151–161(11). Routledge, part of the Taylor and Francis Group, UK.

Wiredu, K. 2005. Towards decolonizing African philosophy and religion. *Afr Stud Q* 1(4):3.

Worldviews and Sciences, Compas/CDE.

6 Soil Fertility as a Contingent Rather than Inherent Characteristic: *Considering the Contributions of Crop-Symbiotic Soil Microbiota*

Norman Uphoff, Feng Chi, Frank B. Dazzo, and Russell J. Rodriguez

CONTENTS

6.1	Introduction	141
6.2	Biological Actors within Soil Systems	142
6.3	High Crop Productivity from "Poor Soils" in Madagascar	144
6.4	Increases of Beneficial Microorganisms in Rhizospheres in Response to Management Practices	148
6.5	Evidence on the Beneficial Effects of Root Endophytes	151
6.6	Effects of Soil Microbial Endophytes in Plant Canopies	152
6.7	Effects of Seed Endophytes for Root Emergence and Growth	154
6.8	Influence of Soil Microbes on Gene Expression in the Plant Canopy	157
6.9	Implications for Advancement of Soil Science	160
6.10	Conclusion	161
Abbreviations		161
References		162

6.1 INTRODUCTION

Usually soil fertility is assessed in terms of certain inherent, fixed properties at the time of sampling, rather than over a period of time and reflecting the dynamic interaction of multiple factors. Also, soil fertility assessments are made as a rule in terms of specific chemical properties of the mineral portion of the soil, mostly the concentrations of essential plant nutrients in plant-available or unavailable forms, plus various physical properties including soil texture, porosity, and water retention, among others.

Biological factors are a third classification to be reckoned with; but the many difficulties in measuring as well as interpreting the soil's biological components have made this aspect of soil systems a less-considered category. A basic principle of soil science is that all three aspects of soil systems warrant concurrent attention, because they function as three dimensions of the same phenomenon. However, work on the biological dimension has lagged behind the chemical and physical focuses in soil science. Research findings reviewed in this chapter should encourage a narrowing of this disparity as they show profound and intimate relationships between the soil microbiota and crop production, the latter heretofore attributed primarily to the soil's chemical and physical properties. The chapter throws new light on the contributions made by soil organisms to crop performance, not just in the soil but in plants as well.

6.2 BIOLOGICAL ACTORS WITHIN SOIL SYSTEMS

It is well understood that many of the physical and chemical properties of soil such as its aggregation and available forms of nitrogen are reflective of "the life in the soil," influenced by the presence and activity of earthworms and many other fauna, as well by its abundance and diversity of microflora. Soil structure and functioning are directly affected by fungi, for example, which produce the compound glomalin that "glues" soil particles together. Bacteria and algae likewise produce extracellular, adhesive hygroscopic polymers within the soil that contribute to its aggregation and water-retaining properties. Both the stocks and availability of plant nutrients in the soil are affected by uncountable numbers and species of microbes, through well-known processes such as biological nitrogen fixation (BNF); phosphorus (P) solubilization; nitrogen (N) and sulfur (S) cycling; and the conservation and concentration of nutrients in soil, referred to as immobilization. This latter process coexists with the complementary processes of nutrient mobilization and mineralization, critical for both micronutrients and macronutrients, which are mediated by microbial activity (Coleman et al. 2004).

At a larger scale, soil nutrients are cycled through immensely complex food chains or food webs with myriad organisms operating at multiple trophic levels (Thies and Grossman 2006; Wolfe 2002). Their biodiversity parallels and even surpasses that observed above-ground (Wardle 2002). Much is being learned about how plants and microbes interact to mutual benefit (Barea et al. 2005; Berg 2009; Badri and Vivanco 2009; Hartmann et al. 2009). Still, the roles that soil organisms play have been mostly considered in terms of how they *modify* the chemical and physical parameters of soil systems.

Less recognized—except where soil organisms have observable negative effects on crop yield as pathogens or parasites—is how biological agents in the soil themselves contribute to crop production, directly affecting the measures of yield that are used as indicators of soil fertility. What appear to be the effects of soil physics or soil chemistry may, on closer examination, be biological effects. Put another way, they may be the outcome of inextricably meshed effects of all three dimensions of soil system capabilities and performance—in which biological actors are pivotal and animating.

Studies in recent years reviewed here shed some different light on the present paradigm for crop production. They give evidence that certain soil organisms can directly enhance crops' productivity by affecting crops' expression of their genetic potential, rather than by the indirect route of just improving the chemical and physical characteristics of the soil systems that the plants inhabit. This microbiological knowledge may modify our current understanding of what constitutes, and contributes to, soil fertility.

The current paradigm regards the soil itself as a vast, differentiated substrate for the nurture of crop plants, which are viewed as beneficiaries receiving nutrients from the soil under favorable chemical and physical conditions. Biological activity enhances these parameters to varying extents. Soil fertility in this view is a quality that resides in the soil itself. Such a stylization accepts the premise that soil and crop sciences can be pursued as separable, albeit complementary, endeavors. The respective domains of soil and crops are linked by transactions of nutrients and water (and root exudates) between soil systems and plants, with some amount of plant material and residues subsequently returned to the soil. Much attention is focused on anthropogenic soil nutrient amendments. This view of separate domains, however, misses that connective role between the domains of soil organisms that directly influence the vegetative growth and reproductive success of crop plants.

Both crop and soil scientists should be taking into account the contributions that are made by mutually symbiotic microbial endophytes. These organisms live within plant tissues and not just in their many and varied soil habitats. That such organisms do more than facilitate crops' access to nutrients presents a more complicated connection between the crop and soil sciences than presently considered. Soil organisms inhabit not only the rhizosphere, being many times more populous in the thin zone of soil adhering to and immediately surrounding the roots than in the bulk soil (for a recent review and references therein, see Dazzo and Gantner 2009); they also live within root interiors, as discussed below. Moreover, soil organisms can benefit host plants while residing within their sheaths, leaves, and even seeds, as also discussed here. When present in plant tissues, soil microbial symbionts, through processes not yet well understood, affect the expression of tissue cells' genetic potentials, thereby contributing to plants' vegetative growth and reproductive success, which are the principal markers of soil fertility.

Advances in soil science are likely to come from greater appreciation that soil systems' support for crop production is more than a consequence of certain measurable, inanimate qualities of the soil itself, considering the easily assessed mineral components of soil and its organic matter content made inert for more precise and replicable analysis. Soil fertility will be understood differently and better if it is regarded as an *emergent and contingent property* that can only be known and validated outside (or above) the soil system, measured in terms of desired kinds and extents of plant growth and productivity. Seeing how crop performance is more intimately affected by the unseen realm of soil microbes than previously appreciated will expand our understanding of what constitutes and causes soil fertility. It should also contribute to more prominence for the biological principles in soil science.

6.3 HIGH CROP PRODUCTIVITY FROM "POOR SOILS" IN MADAGASCAR

In the mid-1990s, some anomalous results at variance with established principles of soil science were encountered in Madagascar by the first author, and these eventually led him into communication with the other authors. While being the director of the Cornell International Institute for Food, Agriculture and Development (CIIFAD), Uphoff became involved with the implementation of a conservation-and-development project funded by the US Agency for International Development. Its objective was to help save the rain forest ecosystems and their biodiversity within Ranomafana National Park.

Food-insecure Malagasy households living and farming in the peripheral zone around the park needed alternatives to their use of slash-and-burn methods to grow rice (*Oryza sativa*) within the nearby rain forest. Their conventional practices used on irrigated patches of land outside the Park gave them low paddy rice yields of only 2 Mg ha^{-1}. This made shifting cultivation within the rain forest necessary for households to feed themselves—unless their lowland rice yields could be substantially increased.

An evaluation of the soils in the peripheral zone around the park, done by the Agronomy Department of North Carolina State University, had concluded that these were some of the least fertile soils that had been analyzed at NC State (Johnson 1994). Systematic sampling of soils to a depth of 1.5 m had shown them to be very acidic (pH 3.8–5.0), with low or very low cation exchange capacity (CEC) in all horizons. Iron (Fe) and aluminum (Al) toxicity were serious and common in acid soils throughout the peripheral zone. Most constraining, available P was very low, usually in the range of 3–4 mg kg^{-1} (ppm), less than half the 10 mg kg^{-1} usually considered as a minimum for productive cropping.

It was thus unexpected that, on the same land and with the same rice varieties that had been giving them very low yields, farmers using different agronomic practices taught to them by a Malagasy nongovernmental organization were able to raise their average yields fourfold, to 8 Mg ha^{-1}. The alternative methods, known as the System of Rice Intensification (SRI), had been synthesized by a French priest who had lived and worked for 34 years in Madagascar (Laulanié 1993; see also Stoop et al. 2002; Uphoff 2007). After such results were achieved for three consecutive years, it was unlikely that they occurred by chance. Also, the same methods were used for measuring both SRI and conventional yields, so the *relative* increase was hard to dismiss, even if the absolute yield levels might be disputed.

This quadrupling of yields did not require the use of new, high-yielding varieties, or the purchase and application of chemical fertilizers, or the use of agrochemical crop protection. Moreover, farmers were *reducing* their applications of irrigation water, keeping their fields just moist or alternatively wetted and dried, rather than continuously flooded. Even labor requirements, greater at first while new techniques were being learned, could be reduced once farmers had mastered them (Barrett et al. 2004).

Increased production was achieved, counterintuitively, with many fewer plants. SRI methods reduced plant population densities per square meter by 80% to 90%: young seedlings (<15 days) were carefully and quickly transplanted just one per hill in a widely spaced square pattern, 25 × 25 cm apart (16 plants m^{-2}). Weeds were controlled by using a simple push weeder that actively aerated the soil as it churned them back into the soil as a kind of green manure.

One self-evidently beneficial practice was the application to the soil of as much compost as possible, made from rice straw and any available biomass. While it was not expected that compost application could promote higher yields than chemical fertilizer, nevertheless, this result was seen in large-scale, replicated factorial trials (Uphoff and Randriamiharisoa 2002).* More output was being achieved with fewer inputs, which seemed illogical.

There was usually a wide range of yield results with SRI management, and sometimes there was no improvement in yield, usually when water control was not maintained and the paddy soils were mostly or always anaerobic. With the same methods used to measure both SRI and conventional paddy yields, the fourfold increase in average yield could not be attributed to measurement error or to grain moisture content (Surridge 2004). In this situation, crop yields appeared to have been "decoupled" from inherent soil fertility as assessed by standard agronomic criteria.

The extreme phosphorus constraint noted above might have been counteracted by the SRI practice of alternately wetting and drying paddy soils. Such water management could mobilize N from atmospheric sources through biological processes and also available P from the soil's usually large reserves of unavailable P. Alternating aerobic and anaerobic soil conditions expands and contracts the in-field populations of soil microbes, including P-solubilizing bacteria (Birch 1958; Magdoff and Bouldin 1970; Turner and Haygarth 2001; Turner 2005). However, research on factors contributing to SRI effects on paddy yield suggested that possibly more was involved here than microbes enhancing nutrient availability in the soil.

In 2000 and 2001, large-scale factorial trials were conducted in two contrasting agroecosystems, comparing the effects of SRI management practices, respectively and collectively, with standard practice. The first set of trials ($N = 288$) at Morondava on the west coast, at sea level, with tropical climate, and on poor (sandy) soils, found that conventional methods (older seedlings, 3 per hill, with continuous flooding and NPK fertilization) gave average yields of 2.11 and 2.84 Mg ha^{-1} from local and improved varieties (*riz rouge* and 2798), respectively. In contrast, when SRI methods were used (young seedlings, 1 per hill, with intermittent irrigation and compost), the average yields were 5.96 and 6.83 Mg ha^{-1} from traditional and new varieties, respectively (Uphoff and Randriamiharisoa 2002).

---

* Two evaluations done in the latter 1990s for the French aid agency in Madagascar reported similar average yields of more than 8 Mg ha^{-1} when SRI management was used on farmers' fields, where standard methods gave only 2 to 3 Mg ha^{-1} (Bilger 1996; Hirsch 2000). The 2000 report showed SRI methods producing average yields of 8.55 Mg ha^{-1} from an area that had expanded from 34.5 ha in 1994/1995 to 542.8 ha in 1998/1999. Use of "modern methods with inorganic fertilizer produced an average yield of 3.77 Mg ha^{-1}, which was 60% more than the 2.36 Mg ha^{-1} that farmers obtained during this same period in the same irrigation schemes with their usual methods. Some farmers who used SRI methods very rigorously obtained yields in the 10–15 Mg ha^{-1} range, and a few had even higher yields, considered unattainable according to prevailing agronomic thinking" (Dobermann 2004; Sheehy et al. 2004). Very high yield with SRI methods has been recently demonstrated in India, with a world-record paddy rice yield of 22.4 Mg ha^{-1} measured by technicians and accepted by the Indian Council for Agricultural Research (Diwakar et al. 2012). The same hybrid varieties cultivated on these same soils gave only one-third as much yield as did SRI methods.

The second set of factorial trials at Anjomakely ($N = 240$) held the rice variety constant (all used a traditional cultivar, *riz rouge*) and compared the effects of management practices on soils of contrasting quality; half were done on better (clay) soil versus half on poorer (loam) soil. The climate was temperate, and the elevation was 1200 masl, so growing conditions were quite different from the first trials.

On these better and poorer soils, the respective yields with conventional practices were 3.00 and 2.04 Mg ha^{-1}, while with SRI methods, paddy yields were more than tripled on both soil types, 10.35 and 6.39 Mg ha^{-1}. Details on both sets of factorial trials are given in Uphoff and Randriamiharisoa (2002) and Randriamiharisoa and Uphoff (2002). In both sets of trials, the patterns of yield response as well as other parameters measured (numbers of tillers and panicles, panicle length, root length, and root density) were very similar, even though climatic and soil parameters at the sites were quite different.

The advisor for these studies, Prof. Robert Randriamiharisoa, director of research for the University of Antananarivo's Faculty of Agriculture, suspected from the first trials that there could be some soil biology influences involved. Accordingly, the research in the second trials examined the populations of the N_2-fixing bacterium *Azospirillum* associated with the roots of the rice plants grown under the different treatments, as part of the evaluation of interaction among six factors using random block design (Andriankaja 2001).

The data in Table 6.1 are averages for six replications of each of four treatments on the same clay soil: (1) conventional practice: 20-day-old seedlings, 3 per hill, continuous flooding without any soil amendments; and three versions of SRI: (2) SRI practices with 8-day-old seedlings, one per hill, with intermittent irrigation, and without any soil amendments; (3) SRI practices as in (2) but with fertilizer amendments (NPK 11-11-16 at 300 kg ha^{-1}); and (4) SRI practices as in (2) with compost additions (5 Mg ha^{-1}) rather than NPK. The compost was made from cow dung, rice straw, and the legumes tephrosia and crotalaria, and had become stabilized before use. These treatments resulted in very different numbers of tillers and in different yield levels as seen in Table 6.1.

That grain yields correlated strongly with the number of grain-bearing tillers per plant was expected. The almost sixfold production increase in yield associated with SRI practices when these were combined with compost applications, compared to conventional practice without soil amendments, was unanticipated. Such results could have been dismissed as implausible if similar increases had not been seen on some farmers' fields in the Ranomafana area.

The most interesting results from the analysis were the numbers of *Azospirillum* associated with root tissue taken from plants in the respective plots and analyzed at Institut Pasteur in Antananarivo. Because the roots were not surface-sterilized before analysis, the organisms counted could have been endophytes or epiphytes or residing in neighboring rhizosphere soil. The research was exploratory to see whether SRI practices had an effect on microbial populations in, on, and around the roots, not designed for hypothesis testing. Thus, the results were regarded as indicative of SRI practice effects on the population dynamics of beneficial microorganisms associated with rice, recognizing that more conclusive research on this subject would have to be done.

TABLE 6.1
Populations of *Azospirillum* Associated with Rice Yield and Tillering Using Alternative Cultivation Practices with Different Nutrient Amendments, on Clay Soil, Anjomakely, Madagascar, 2001

Methods	Yield (t ha^{-1})	Tillers Plant^{-1}	*Azospirillum* CFUs in Roots[a] (10^3 mg^{-1})
Conventional cultivation methods: without nutrient amendments	1.8[b]	17	65
SRI cultivation methods: without nutrient amendments	6.1	45	1100
with NPK amendments	9.0	68	450
with compost amendments	10.5	78	1400

Source: Andriankaja, A. H., *Mise en evidence des opportunites de developpement de la riziculture par adoption du SRI et evaluation de la fixation biologique de l'azote: Cas des rizieres des hautes terres.* Memoire de fin d'etudes. ESSA-Agriculture, University of Antananarivo, Antananarivo, 2001.

Note: The data reported here are means from six replications of the respective treatments on randomized plots, 2.5 × 2.5 m.

[a] CFUs: colony-forming units.

[b] Note that average yield of conventional cultivation methods with NPK amendments was 3 t ha^{-1}; *Azospirillum* counts were not made for this treatment as the researcher focused on without-amendment comparisons.

- With *conventional practices and no soil amendments*, the average paddy yield was 1.8 Mg ha^{-1}, approximating the national average yield in Madagascar. The associated population density of *Azospirillum* was 65,000 colony-forming units (CFU) mg^{-1} of sampled root biomass. When conventional practices were used with NPK fertilizer, the grain yield was 3.0 Mg ha^{-1}, two-thirds more than with no fertilization. Unfortunately, the researcher did not analyze root-sample bacteria from these plots.
- SRI practices *without soil amendments* gave much greater grain yield, 6.2 Mg ha^{-1}, and were accompanied by a 20-fold increase in the density of associated *Azospirillum*, to 1.1 million CFU mg^{-1}. This was twice as much yield without soil amendments as resulted from conventional practices supported by inorganic fertilizer applications. No causal connection was examined between microbial populations and yield, but some causal positive association could be hypothesized.
- When *NPK fertilizer* was used together with SRI practices, rice grain yield increased by another 50%, to 9.1 Mg ha^{-1}. However, the population density of *Azospirillum* in the root samples declined by 60%, an effect not surprising when soil is supplied with inorganic nutrients (Pariona-Llanos et al. 2010). The substantial yield increase observed would probably have been mostly attributable to the inorganic N supplied.

- In the fourth set of treatments, it was seen that when *compost* was used with SRI methods instead of NPK, there was a further yield increase with SRI methods, to 10.2 Mg ha^{-1}, and the *Azospirillum* population density increased to 1.4 million CFU mg^{-1} root tissue. In this case, the increase in grain yield increase could be attributed to increased availability of N through biological processes or to other effects from microbial-plant symbiosis. Unfortunately, these processes and effects were not studied in this research as it was exploratory, and a few laboratory facilities were available in the country.

It should not be inferred that these effects were due just to variations in the population density of *Azospirillum* in, on, and around the roots. This bacterial genus was chosen for analysis because it was relatively easier and more reliable to count than other microbes within the endophytic community. In the research, *Azospirillum* was regarded as a proxy or indicator for overall levels of bacteria associated with plant roots. As a diazotroph, *Azospirillum* could have been contributing fixed N to the soil and to the plants because it has the ability to fix atmospheric N. However, as seen in Section 6.4, why or how this and other endophytes enhance crop plant performance appears more complex than just increasing the availability of N. BNF appears to be just one of several mutually beneficial functions that symbiotic endophytes perform.

These results from the high plateau in central Madagascar fortunately directed some other researchers' attention to the effects that soil microorganisms living not just around but also within the roots could have on rice plants when managed with SRI practices that create conditions favoring greater root growth. Young seedlings transplanted singly and carefully, with wide spacing, in soil well-endowed with organic matter and actively aerated by mechanical weeding, and having mostly aerobic soil conditions with no continuous flooding, produced larger and longer-lived root systems (Barison and Uphoff 2011). These practices also favor larger populations of aerobic soil organisms, as seen in Section 6.4.

6.4 INCREASES OF BENEFICIAL MICROORGANISMS IN RHIZOSPHERES IN RESPONSE TO MANAGEMENT PRACTICES

The findings reported from Madagascar prompted researchers in India and Indonesia to undertake their own evaluations of the effects of SRI management on the soil biota.* Their findings are summarized in Table 6.2. The increases in microbial populations seen there—both from direct counts and from the biochemical "footprints

---

* In 2002, two theses were done at Tamil Nadu Agricultural University (TNAU) in India, assessing differences in soil microbial populations and activity in the rhizospheres of rice plants under SRI compared with conventional management. Then, the soil biota in over 200 farmers' fields across 11 districts of Andhra Pradesh state of India were evaluated over four seasons in 2005 and 2006 under a joint project of the Worldwide Fund for Nature (WWF) and the International Crop Research Institute for the Semi-Arid Tropics (ICRISAT) in Hyderabad, India. Similar research was conducted in 2009 by the Soil Biotechnology Laboratory in the Department of Soil Sciences and Land Resources at the Institut Pertanian Bogor (IPB) in Indonesia. Each study used standard methods for making soil biological measurements, as reported in Uphoff et al. (2009) and Anas et al. (2011). Some similar research was undertaken in China during this period by agronomists evaluating the effects of SRI management, reported below.

TABLE 6.2
Increases in Population Densities of Soil Microorganisms in the Rhizospheres of SRI Rice Plants Compared to Conventionally Grown Plants, from Evaluations in India and Indonesia (in percent)

Microorganisms	*India:* Tamil Nadu Agric. University study (2002)	*India:* ICRISAT-WWF study (2005–2006)[a]	*Indonesia:* IPB study (2009)
Total bacteria	312%	ND	65%
Total diazotrophs[b]	61%	6.4%**	NM
Azospirillum[b]	32%	NM	211%
Azotobacter[b]	36%	NM	94%
P-solubilizing microbes	53%	3.6%ns	78%
Dehydrogenase (μg TPF g^{-1} 24 h^{-1})	140%	22.5%**	125%
Microbial biomass N (mg kg^{-1} soil)	NM	20%**	NM

Source: Adapted from Anas, I. et al. *Paddy Water Environ* 9:53–64, 2011.
Notes: ND: no difference; NM: not measured.
[a] These trials, over four seasons, included wet-season results when water control was incomplete and when aerobic soil conditions were therefore difficult to maintain.
[b] N-fixing bacteria.
** Significant at 0.05 level of confidence; ns: not significant.

[dehydrogenase and microbial biomass N (MBN)] that are indicative of the total amount of life in the soil—are consistent in direction although variable in amount. The differences in percentage increases across the studies reflect the variability that characterizes the soil biota, year to year, season to season, even month to month and week to week. The abundance and activity of soil organisms are affected by differences in soil structure, aeration, humidity, pH, and nutrients, especially carbon supply, which can be quite large. The absolute sizes of microbial populations can thus be expected to vary considerably.

Of interest here is that the numbers of microbes known to have positive impacts on soil fertility status, for example, through N_2 fixation, P solubilization, or nutrient cycling, were greater around the roots of rice plants that have been grown under SRI management than around the roots of the same variety of rice when grown conventionally. These populations were of microorganisms in the rhizosphere soil rather than of endophytes within the root interior; but as indicated from other studies, many of the microbes that live on and around the roots also commonly inhabit the root interior and other plant parts as well.

How does SRI management contribute to such effects? It starts with very young seedlings, planted individually (not in clumps) in soil that is both well aerated and well supplied with soil organic matter (SOM). Such management is associated with rice plants having root systems that are larger and healthier, with less or delayed senescence, and also larger, more photosynthetically active canopies (Thakur et al. 2010; Barison and Uphoff 2011). Larger root systems together with more leaf area and light interception (Thakur et al. 2010) are associated with increased root exudation that delivers more carbohydrates, amino acids, and other compounds into the rhizosphere, supporting larger populations of microbes and the web of fauna that feed upon them.

It is not self-evident why SRI practices should enhance the populations of *beneficial* soil organisms, rather than of pathogens, in the root zone. The empirical results so far simply show increases particularly for known beneficial organisms. We do not know of any studies that compare population densities of pathogens in response to SRI management practices; but the evidently better health and measurably higher yields of SRI plants suggest that numbers of adverse organisms and the diseases they cause are not increased nor intensified.

Aerobic organisms tend to be more beneficial than anaerobic ones for plants. A study done at the China National Rice Research Institute examined the impact of two key SRI management practices—intermittent versus continuous irrigation, and the degree of organic fertilization, keeping constant across all trials the total amount of N added to the soils—on the populations of actinomycetes in the rhizospheres of rice plants (Lin et al. 2011). Actinomycetes are a Gram-positive order of soil bacteria known to enhance soil fertility when there are sufficient organic sources of nutrients in the soil to sustain their active populations. They also commonly have antagonistic antimicrobial activity against a wide range of soil and plant pathogens.

Table 6.3 shows that when 100% organic fertilization was used with intermittent irrigation so that aerobic soil conditions were maintained, the population density of actinomycetes was almost five times greater than with 25% organic fertilization (same total N) and continuous flooding giving anaerobic soil. Similar research done at Zhejiang University in China comparing the rhizosphere soils of rice plants grown with SRI methods (more organic matter applications and intermittent irrigation)

TABLE 6.3
Effect of Organic Matter Treatments and Water Regime on Actinomycete Population Densities in Rhizosphere Soil (g^{-1} dry soil)

Fertilization	Actinomycetes (×10^6) Intermittent Irrigation	Continuous Irrigation
25% organic	66.3[cC]	52.3[Bb]
50% organic	119.7[bB]	84.4[aAB]
100% organic	259.6[aA]	93.3[aA]

Source: Lin, X.Q. et al. *Paddy Water Environ* 9:33–39, 2011.
Notes: Lower-case letters indicate significance at 5% level; capital letters show significance at 1% level.

versus conventional practices (inorganic fertilization and continuous flooding) reported significant increases in both microbial biomass C (MBC) and MBN under SRI management, regardless of the sampling dates during the growing season (Zhao et al. 2010).

Crop management practices, particularly those affecting soil C availability and aerobic soil status, thus are associated with enhanced soil microbial populations. The greater abundance of microbes such as diazotrophs and P-solubilizing bacteria seen in Table 6.2 would be consistent with the increases observed in the populations of *Azospirillum* in and on the roots of SRI-grown rice plants reported in Table 6.1. This invites more research on the effects of endophytic microbes in crop plant roots.

6.5 EVIDENCE ON THE BENEFICIAL EFFECTS OF ROOT ENDOPHYTES

Independently of the research being conducted on the effects on rice plants of SRI management in Madagascar and several rice-producing countries in Asia, research as been undertaken on rice root endophytes in the Nile delta of Egypt. Rice has been grown there for centuries in alternation with a leguminous fodder crop (berseem clover), so that the soils there are well stocked with Rhizobia. It is well known that Rhizobia inhabit root nodules in legume plants in a classic mutually symbiotic relationship, fixing N_2 for the benefit of plant growth and receiving organic acid photosynthate supplies from the plant in return. However, similar symbiotic relationships of soil microbes with nonleguminous plants that had no root nodulation were not well established.

That a variety of free-living diazotrophic bacteria found in the rhizosphere of nonleguminous plants, particularly of gramineae (grass-family) species, could fix N_2 and provide other benefits to plants was documented (for an early summary on this, see Döbereiner and Pedrosa 1987). However, their beneficial effects, which include plant growth promotion stimulated by hormones and induced systemic resistance that protects plants against pathogens, are still not well or fully understood.

Much of the early work on these beneficial microbes that live in the root zone and are found also in plant roots as endophytes was done on rice and sugarcane (*Saccharum officinarum*) in Brazil. Considerable research was carried out on this subject in the 1990s (Peoples et al. 1995; Ladha et al. 1997), but interest waned thereafter. Perhaps this was because the attempts to induce and increase BNF, treating bacterial inoculation of soil and plants as a kind of alternative to fertilizer, did not produce the anticipated gains in crop productivity.

That status changed considerably when studies by Dazzo and Yanni with a large number of colleagues (Yanni et al. 1997, 2001; Dazzo and Yanni 2006) showed that the root interiors of field-grown rice plants were naturally inhabited by plant growth-promoting rhizobial microbes as endophytes. After providing experimental proof of the endophytic infecting ability of rhizobia in rice using the criteria of molecular Koch's postulates, they found that certain endophytic strains of rhizobia significantly benefited rice growth through the vegetative and reproductive stages of its life cycle, and also enhanced the protein content of rice grains (Biswas et al. 2000b; Chi et al. 2005; Yanni et al. 2001).

The physiological benefits of the endophytic rhizobia–rice association are manifested in many ways. These include increased phytohormone production (auxins and gibberellins) accompanied by an enhanced root architecture that enables the plants to be better miners of essential nutrients within the soil; increased nutrient uptake efficiency and photosynthetic capacity; solubilization of both inorganic and organic P reserves; and modulation of plant phenolic metabolism involved in innate plant defenses (Biswas et al. 2000a; Chi et al. 2005; Mishra et al. 2006; Yanni et al. 2001).

More recently, extensive field inoculation tests (Yanni and Dazzo 2010) have demonstrated the efficacy of using rhizobial endophytes as biofertilizer inoculants for rice based on findings that certain selected strains boost fertilizer-N use efficiency and grain production of rice. This enables farmers to significantly reduce their requirements for chemical fertilizer inputs to achieve maximal grain yield potential under real-world agronomic field conditions.

6.6 EFFECTS OF SOIL MICROBIAL ENDOPHYTES IN PLANT CANOPIES

The research in Egypt was consistent with the effects seen in the research reported in Section 6.2 from Madagascar, where populations of beneficial symbiotic endophytes, *Azospirillum* rather than rhizobia, were apparently enhanced by the crop, soil, water, and nutrient management practices introduced by SRI. That soil microbes would enter into plant roots and live within them to mutual benefit was somewhat surprising, but could be easily understood given that roots were intimately associated with the rhizosphere soil surrounding them, and this soil supported high concentrations of beneficial soil organisms (Vessey 2003 and references therein).

These insights were taken a step further by research carried out in China by Chi and associates, together with Dazzo and Yanni (Chi et al. 2005). Their data showed that certain microorganisms migrate from the soil into the roots, as discussed below, and then ascend upward through the roots and stems into the sheaths and lower leaves of rice plants. As seen below, the presence of soil rhizobia in the canopy of plants is associated with both morphological and physiological changes that are advantageous for plant productivity.

Chi and his colleagues used five different strains of soil rhizobia that are known to inhabit the root nodules of leguminous plants as endosymbionts. The roots of five sets of rice plants (Japonica varieties Zhonghua 8 and Nipponbare) were inoculated, each set with a different strain of rhizobia, with an additional set of uninoculated plants grown as a control. The soil had been sterilized so that the trials were gnotobiotic. The experimenters had genetically tagged the rhizobia with a reporter gene that constitutively expresses the green fluorescence protein so that movements of these microbes could be definitively tracked and their local population densities in the plant could be evaluated by computer-assisted fluorescence microscopy, so as to assess their influence on the growth physiology of rice plants.

The migration of rhizobia within the plants after inoculation was documented by laser scanning confocal microscope examination, as was their surface and endophytic colonization of healthy rice plant tissues. The study definitely confirmed earlier findings (Prayitno et al. 1999; Reddy et al. 1997; Yanni et al. 1997, 2001), together indicating that rhizobia use "crack entry" at lateral root emergence sites and between epidermal cells to gain access into the interior of rice plants.

The invasive bacteria can also utilize a dynamic infection process within the plant that permits them to migrate endophytically from within the roots upward into the stem base, leaf base, leaf sheaths, and lower leaves. There they grow transiently into large local populations that are metabolically active, and some of the rhizobia persist throughout the vegetative phase and into the reproductive phases of development. Geostatistical analysis by CMEIAS Image Analysis software on images acquired by computer-assisted microscopy indicated that endophytic populations were spatially autocorrelated in aggregated patterns that could reach local densities as high as 9 × 10^{10} rhizobia cm^{-3} within the infected host tissues (Chi et al. 2005).

The researchers found that the rice plants inoculated with different rhizobia strains produced significantly greater root and shoot biomass, accelerated photosynthetic rate, more water utilization efficiency (WUE), and higher grain yield (Table 6.4). Other morphological and physiological parameters that were significantly greater for inoculated rice plants compared with those that were uninoculated included stomatal conductance, transpiration velocity, CO_2 concentration within intracellular leaf spaces, area of the flag leaf (considered to possess the highest photosynthetic activity), and length of flag leaves relative to their width (Chi et al. 2005). There were also significant differences in the levels of phytohormones such as indole-3-acetic acid (IAA) and gibberellic acid 3 (GA_3) within rice plant tissues, which would have a positive impact on the growth physiology of the rice plant.

TABLE 6.4
Growth Response, Photosynthetic Rate, WUE,[a] and Grain Yield of Potted Rice Plants 180 days after Inoculation with Various *gfp*-Tagged Strains of Wild-Type Rhizobia[b]

Rhizobium Test Strain	Total Plant Root vol/pot (cm³) ± SE	Shoot Dry wt/pot (g) ± SE	Net Photosynthetic Rate (μmol of CO_2 m^{-2} s^{-1}) ± SE	Water Utilization Efficiency ± SE	Grain Yield/pot (g) ± SE
Ac-ORS571	210 ± 36[A]	63 ± 2[A]	16.42 ± 1.39[A]	3.63 ± 0.17[BC]	86 ± 5[A]
Sm-1021	180 ± 26[A]	67 ± 5[A]	14.98 ± 1.64[B]	4.02 ± 0.19[AB]	86 ± 4[A]
Sm-1002	168 ± 8[AB]	52 ± 4[BC]	13.70 ± 0.73[B]	4.15 ± 0.32[A]	61 ± 4[B]
R1-2370	175 ± 23[A]	61 ± 8[AB]	13.85 ± 0.38[B]	3.36 ± 0.41[C]	64 ± 9[B]
Mh-93	193 ± 16[A]	67 ± 4[A]	13.86 ± 0.76[B]	3.18 ± 0.25[CD]	77 ± 5[A]
Control	130 ± 10[B]	47 ± 6[C]	10.23 ± 1.03[C]	2.77 ± 0.69[D]	51 ± 4[C]

Source: Adapted from Chi, F. et al. *Appl Environ Microbiol* 71:7271–7278.

[a] Water utilization efficiency = net photosynthetic rate/transpiration velocity.

[b] Data presented are means per pot from three pot replicates, each containing 15 transplanted plants per strain. Values followed by a different superscript capital letter are significantly different at the 95% confidence level according to Duncan's multiple-range test.

Considered collectively, these findings indicated that the natural endophytic *Rhizobium*-rice association is more complex, inclusive, dynamic, and invasive than previously thought, with soil microbes colonizing both above-ground and below-ground tissues. Physiological and morphological parameters were demonstrably enhanced in response to inoculation with several rhizobial species. In addition, significant agronomic benefits to inoculation of rice with selected strains of endophytic rhizobia have been repeatedly established under field conditions. But the specific mechanisms for these effects, beyond some positive effect on phytohormone production, remained to be identified and demonstrated. Results from further investigation into mechanisms are reported in Section 6.7 after considering some results of contemporaneous research on fungal endophytes in rice seeds.

6.7 EFFECTS OF SEED ENDOPHYTES FOR ROOT EMERGENCE AND GROWTH

That microbial endophytes by inhabiting rice plant tissues can enhance crop growth and production outcomes attributed to soil system fertility are documented (Biswas et al. 2000a,b). Less known is that microorganisms can also inhabit seeds to mutual benefit; and while much is known about rhizobacteria as endophytes, fungi can also have symbiotic relationships with plants (Rodriguez et al. 2009a,b). Note that we are not considering here the well-known contributions of mycorrhizal fungi that also live symbiotically in the roots of most terrestrial plants and confer many functional advantages to their plant hosts, particularly water and nutrient uptake.

We focus here on other fungal endophytes that have been shown to protect plants from both biotic and abiotic stresses and that can have profound impacts on plant physiology. Plants colonized with symbiotic fungal endophytes can be more robust, e.g., with greener leaves, more tillers in rice, and larger root systems, and can grow faster and use less water than nonsymbiotic plants without endophytes (Rodriguez et al. 2009b; Aly et al. 2011; Redman et al. 2011).

Fungal endophyte-induced changes in plant physiology occur very early in seedling development (Rodriguez et al. 2009a). Upon germination, symbiotic seedlings develop root systems more rapidly, allocate carbon differently, and produce more biomass (shoots and roots) compared to nonsymbiotic seedlings. Trials showed that rice seedlings emerging from seeds inoculated with the fungus *Fusarium culmorum*, generally considered a plant pathogen, had more rapid root growth after germination, as much as five times faster in the first 5 days, with root hairs emerging 2 days sooner (Figure 6.1). Such endophyte-induced plant growth responses occur under greenhouse or field conditions, with symbiotic plants accumulating 25% to 50% more biomass than nonsymbiotic plants (Redman et al. 2002, 2011) (Figure 6.2).

Why more of the inoculated plant's growth resources should be allocated in early growth to the root relative to the shoot is not clear; but the effect of inoculation with fungal endophytes on seedlings' performance is very visible and statistically significant. Plants with more vigorous root growth in their early stages of development are better positioned for subsequent nutrient and water uptake and for supporting the growth of the canopy. It has also been determined that plants inoculated with

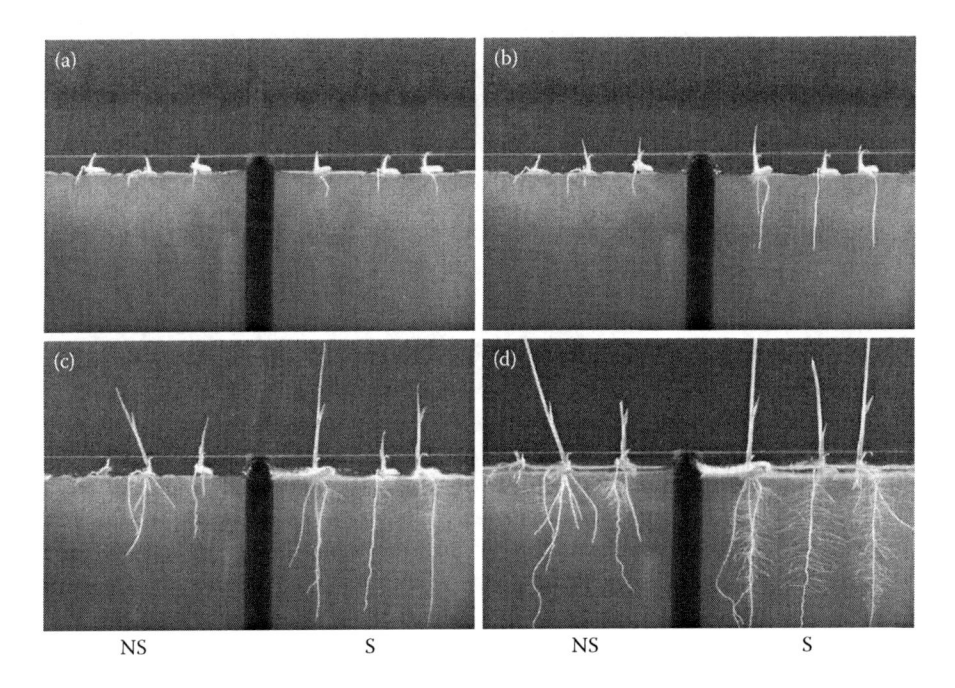

Soil Fertility as a Contingent Rather than Inherent Characteristic 155

FIGURE 6.1 Growth of rice seedlings: nonsymbiotic (NS) on left and symbiotic (S) on right, shown over an 8-day period. a = day 0; b = day 2; c = day 4; d = day 8. Germinated seeds were placed on top of an agarose medium that was poured between two glass plates and were then grown under constant light for 8 days at 22°C (daylight balanced fluorescent bulbs, 22 µmol m^{-2} s^{-1}). S-seedling roots grew more rapidly and developed root hairs earlier than NS seedlings. Preparation of the endophyte (*F. culmorum*) and plant inoculations are described in Rodriguez et al. (2009a).

selected fungi have more tolerance of drought and salinity stress (Rodriguez et al. 2008; Redman et al. 2011).

Current research is examining how endophytes can increase plant growth rate and biomass while decreasing water consumption. Although the genetic and biochemical mechanisms for these symbiotic benefits are not yet established, there are several possible explanations, including fungal-generated or fungal-induced plant hormones, increased metabolic efficiency, decreased metabolic rates, and/or enhanced photosynthesis.

Many fungi are known to produce plant hormones that regulate plant growth, development, and water relations (Chung et al. 2003; Bomke et al. 2009; Tsavkelova et al. 2012). Our own studies have demonstrated that fungal endophytes produce substantial amounts of indole acetic acid (IAA) in culture. However, documentation of fungal-produced plant hormones *in planta* is limited, so it is not yet possible to determine the significance of fungal-derived hormones with regard to symbiotic benefits. For some initial work regarding phytohormone-induced tolerance of salinity, see Khan et al. (2012).

FIGURE 6.2 Endophyte-induced growth responses in eudicot and monocot plants. Nonsymbiotic (NS) and symbiotic (S) plants are the same age in each panel. Difference in biomass between NS and S plants varied from 2× (tomato) to 10× (rice). Tomato plants were grown under field conditions, and the others in a growth chamber.

Growth and development of inoculated plants could also be accelerated by increased metabolic efficiency so that more biomass is produced per photosynthetically derived carbon building block. This could occur as a result of endophytes influencing plants' gene expression so that some genes are up-regulated and more biochemical building blocks become available for tissue generation. To address this possibility, the overall gene expression patterns and metabolic activity were analyzed in symbiotic and nonsymbiotic plants. This research indicated increases in the number of up-regulated plant genes in symbiotic versus nonsymbiotic plants when grown in the absence of abiotic stress (Woodward et al. 2012). However, metabolic activity as measured by differential scanning calorimetry revealed little to no difference in either metabolic rates or efficiencies between symbiotic and nonsymbiotic plants (rice or tomato). Thus, there was no apparent relationship between the number of up-regulated genes and metabolic activity or efficiency and the growth metrics in symbiotic plants.

Comparative analysis of photosynthetic activity in symbiotic and nonsymbiotic plants yielded similar results to the metabolic studies (Woodward et al. 2012). Although there were no measureable differences in photosynthetic rates or stomatal activity, symbiotic plants had significantly greater levels of photosynthetic efficiency than nonsymbiotic plants. This suggests that symbiotic plants more efficiently transfer energy from photons into C building blocks, and therefore have greater amounts of photosynthate to use for generating shoot and root biomass. More research on these relationships is needed, however.

Considered together, these data suggest that there are active forms of communication between fungal endophytes and plant hosts affecting the expression of genetic potentials and metabolism of the latter. Individual fungal endophytes have broad host ranges and confer similar benefits to both monocot and eudicot plants, suggesting that the symbiotic communication involved in these fitness benefits is highly conserved, predating the divergence of these plant lineages approximately 200 mya (Chaw et al. 2004; Wolfe et al. 1989; Yang et al. 1999). Since endophytes alter plant gene expression and phenotype without causing changes in the underlying plant DNA sequences, as discussed in the next section, and can be vertically transmitted between generations via seed coats, these symbioses can be defined as intergenomic epigenetic phenomena.

The ability of fungal endophytes to confer fitness benefits to plants makes them ideal potential tools to improve agricultural sustainability. These tools represent a symbiogenic technology (symbio = symbiosis and genic = gene influence), and methods for large-scale endophyte production, seed treatments, and field applications to implement this technology are currently under development.

6.8 INFLUENCE OF SOIL MICROBES ON GENE EXPRESSION IN THE PLANT CANOPY

Establishing the underlying molecular mechanisms that explain how these various plant growth and health responses are achieved by microbial endophytes living in plant roots, shoots, and seeds has been challenging. Accepting empirical results, even if well documented, is often resisted until the mechanisms involved are demonstrated. Since the growth and functioning of plant tissues is regulated ultimately at the level of gene expression, a proper understanding of endophytic effects needs to be linked to what is occurring in terms of the plants' DNA and in their protein expression. Proteomic analysis, which investigates accumulative changes and modifications of proteins, can provide a more precise and comprehensive understanding of the physiological responses that occur in host plants when in association with soil microbes.

These relationships have been explored by Chi and associates (2010), following up their observations reported in Section 6.5. Proteomic analysis of the expression of specific genes in rice plant cells, causing certain proteins to be produced in plant tissues above or below ground, has shown that the presence of a certain rhizobia *Sinorhizobium meliloti* 1021—versus its absence—induces the production of identifiable proteins that contribute to better plant performance and health. These effects of rhizobial inoculation are expressed in the leaves, sheaths, and roots of rice plants.

Two-dimensional gel-based comparative proteomic approaches coupled with a mass spectrometric strategy were used to investigate endophyte-induced changes in protein expression that could improve the growth of rice and resistance to disease. The protein profiles in tissues sampled from the roots, the leaf sheaths, and the leaves of inoculated rice seedlings were compared to the profiles for similar tissues taken from uninoculated plants, all grown under gnotobiotic conditions.

Using the criterion of at least a twofold change in the abundance of particular proteins in the analyzed tissues, 21, 19, and 12 protein-expression spots were found to have such large differentials displayed in the respective gel electrophoresis profiles of samples from roots, leaf sheaths, and leaves.

These divergences are indicative of either up-regulation or down-regulation of gene expression. According to their gene annotation in the database, the identified proteins analyzed in the three sets of rice tissues could be classified into nine functional groups. These groups are primary metabolism proteins, energy-associated proteins, defense-related proteins, protein destination and storage agents, transporters, signal transduction proteins, cell growth/division proteins, cell-structure proteins, and unknown-function proteins (Figure 6.3).

All of the seven energy-associated proteins identified in the roots were down-regulated by rhizobium inoculation, while most of the energy-associated proteins identified in the leaf sheaths and leaves were up-regulated, as were the proteins associated with the photosynthesis subgroup. In roots, all the up-regulated proteins were classified as ones involved in signaling or as defense-related, whereas only two defense-related proteins were identified in leaf sheaths.

Promotion of plant growth in response to rhizobial infection: Several proteins produced in cell chloroplasts that contribute to photosynthesis enhancement in the plant canopy were up-regulated in association with or in response to rhizobial colonization. These photosynthesis-related proteins include

- Rubisco activase and pyruvate orthophosphate dikinase, which could increase CO_2-fixation efficiency.

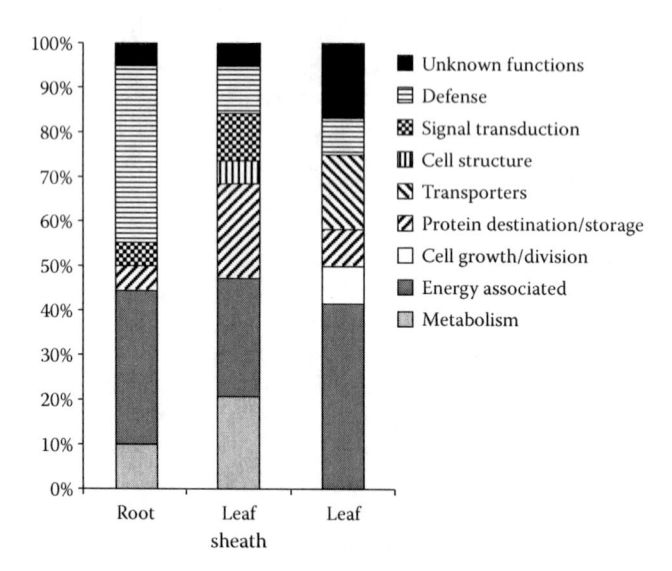

FIGURE 6.3 Functional categorization of the differentially expressed proteins in the root, leaf sheath, and leaf tissues in rice plants resulting from inoculation with *S. meliloti* 1021.

- Some major proteins involved in both light and dark reactions of photosynthesis, which suggests that there would be an overall increase of photosynthetic efficiency.
- Some chloroplast membrane and envelope proteins, which increase nutrient transportation and protein folding, protein translocation through biological membranes, and signal transduction.

Some changes were also detected in the levels of proteins involved in plant hormone metabolism. One auxin-induced protein was detected in the up-regulated set of proteins, which is consistent with increased auxin content in inoculated plants (Chi et al. 2005). At the same time, IAA-amino acid hydrolase, which hydrolyzes IAA-l-amino acid conjugates into IAA, was detected as being down-regulated in roots. This suggests that the increase in plant-synthesized IAA biosynthesis that is induced by rhizobia might be through a tryptophan-independent route, even though tryptophan and rice root exudates significantly enhance IAA synthesis in endophytic rhizobia cultured *ex planta* (Yanni et al. 2001).

Defensive reactions in roots and shoots: Most of the up-regulated defense-related proteins were found to be changed in root tissues, which are the primary loci for rhizobial colonization and infection. These defense-related proteins could be grouped into two categories based on their functional mechanisms. The first set of up-regulated protein expression relating to bacterial killing includes the following:

- Exoglucanase, which affects important pathogenesis-related (PR) proteins; these can antagonize external pathogens by lysing their cell walls.
- Subtilisin-like proteinase, which is a calcium-activated endopeptidase that acts in the posttranslational modification of proteins that participate in the defense response.
- Aminopeptidase N, which is an enzyme releasing the amino-terminal amino acid residues from proteins and peptides, thereby inactivating their function.
- A protein with sequence similarity to 30-N-debenzoyltaxol N-benzoyl-transferase, which is involved in the synthesis of a phytoalexin that is an important plant antipathogenic antibiotic.
- Peroxidase, which can detoxify reactive oxygen species (ROS), a common strategy used by hosts to kill infective bacteria, and a cross-linking enzyme that strengthens plant wall polymers that have been broken during bacterial ingress (Salzwedel and Dazzo 1993).

The second group of up-regulated protein expression that relates to plant cell wall synthesis to inhibit bacterial colonization includes

- Arabinoxylan arabinofuranohydrolase isoenzyme (AXAH-II), which participates in the accumulation of arabinoxylan that thickens the secondary plant host cell wall.
- Germin protein 4, which could improve plant resistance to pathogens by strengthening the host cell wall structure.

Certain defense-related proteins were also up-regulated in the shoot tissues, including S-adenosylmethionine synthetase, which is involved in methyl group transfers and polymerization of lignin monomers to reinforce the plant cell walls, and catalase, a well-known ROS [H_2O_2] detoxification enzyme. These results indicated that the rhizobial-mediated activation of the plant's defensive reactions occurs not only in the roots but also in above-ground tissues, which might improve the stress resistance of inoculated rice plants.

Taken together, these results predict that plant growth-promotion mechanisms affected by endophytic rhizobia may operate in these modes: (1) activation of defense mechanisms in different plant tissues below and above ground to minimize the negative effects of environmental and pathogenic factors; (2) enhancement of anabolism, for example, photosynthesis, to increase the biomass of the plant; and (3) regulation of auxin levels or status that promotes vegetative growth.

Exactly what signals the rhizobia send to trigger these genetic-expression responses remain to be determined. But the molecular details involved in the intimate and mutually beneficial association between these soil microorganisms and crop plants have been experimentally validated. This relationship is one in which soil scientists, crop scientists (especially physiologists), and microbiologists (and microbial ecologists) should all have mutual interest.

6.9 IMPLICATIONS FOR ADVANCEMENT OF SOIL SCIENCE

It is noteworthy that many of the phenotypical advantages that have been reported for rice plants grown under SRI management based on many observations and validated by scientific evaluations match the physical characteristics that Chi, Dazzo, Rodriguez, and their colleagues have reported as associated with symbiotic endophytes: more vigorous growth of roots and canopy; increased plant nutrient uptake efficiency; more resistance to water and other abiotic stresses; more resistance to insects and pathogens; slower or delayed senescence; higher levels of chlorophyll and higher rates of photosynthesis; higher agronomic N-fertilizer use efficiency; and higher grain productivity (see particularly the findings in Thakur et al. 2010).

Correlations of course do not establish causation, so there is much scientific research to be done in this promising area. But the recently reported findings of Chi and associates identifying differences in gene expression—with up-regulation or down-regulation of specific genes fitting patterns of observed phenotypical variation—are consistent with observed relationships between certain management practices for plants, soil, water, and nutrients and improved crop outcome. While the latter could be taken as evidence of soil fertility, probably we should be constructing a different language and analytical framework to encompass the more complex biological relationships seen here in relation to the results now associated with soil fertility.

We should get beyond the tacit assumptions that plants operate like organic, C-based machines that extract inorganic nutrients and water from the soil, transforming them into carbohydrates and other compounds by utilizing solar energy in photosynthetic processes that lead to biomass outputs from a mechanistic process.

Both plants and the microbes that inhabit them acquire more implications of agency in this more biologically shaped view of crop production.

One implication of a less sequential/more interactive understanding of crop productivity is that the separation of crop science from soil science becomes less tenable. The effects of soil biota on plant gene expression need to be taken into account since assessments of soil fertility can only be validated in terms of crop productivity. This makes genetics—and particularly epigenetics—along with soil biology more directly relevant to soil science than in the present disciplinary division of labor. Currently, the evaluation of soils under axenic (sterile) conditions is considered not only acceptable but good science. While there are some issues for which such analysis is appropriate, in general, it places the same kind of limitations on the understanding of soil systems as doing all of medical research on cadavers rather than studying living bodies.

6.10 CONCLUSION

Even more than soil scientists, plant scientists will probably have to make adjustments in their mental models to integrate the roles of microbial symbiotic endophytes into their paradigm. The research findings reported here on the contributions to crop productivity attributable to soil microorganisms are likely to draw both disciplinary domains more closely together. It makes sense for them to become more mutually informative rather than pursue even further the long-standing strategy of disciplinary specialization. This makes a virtue of knowing more about smaller and smaller subject areas, in effect to the exclusion of knowledge from other fields. Yet other fields such as microbiology, soil ecology, and genomics and epigenetics are becoming more relevant to soil science than before.

Soil fertility could become a bellwether concept for the integration of knowledge across the agricultural sciences. The emerging understanding of plant–soil–microbial interactions could lead to a correction of the imbalance among soil chemistry, soil physics, and soil biology, kindling greater interest in and research on the latter dimension of soil systems. These are indeed dimensions rather than aspects or domains. Soil systems should be understood as phenomena that function concurrently in three dimensions, each affecting and interacting with the other. While the soil biota constitutes only a small percentage of soil volume, it is like "the tail that wags the dog." We now are learning that they are also "wagging" the plants that grow in soil systems. Soil fertility is thus a concept that is due for some revision and expansion.

ABBREVIATIONS

Al: aluminum
BNF: biological nitrogen fixation
CEC: cation exchange capacity
CFU: colony-forming units
CIIFAD: Cornell International Institute for Food, Agriculture and Development
CMEIAS: Center for Microbial Ecology Image Analysis Software
CO_2: carbon dioxide

DNA:	deoxyribonucleic acid
Fe:	iron
GA$_3$:	gibberellic acid
IAA:	indole-3-acetic acid
ICRISAT:	International Crop Research Centre for the Semi-Arid Tropics (Hyderabad, India)
IPB:	Institut Pertanian Bogor (Bogor, Indonesia)
masl:	meters above sea level
MBC:	microbial biomass carbon
MBN:	microbial biomass nitrogen
N:	nitrogen
NPK:	nitrogen/phosphorus/potassium fertilizer
NUE:	nitrogen use efficiency
P:	phosphorus
PR:	pathogenesis-related
ROS:	reactive oxygen species
S:	sulfur
SOM:	soil organic matter
SRI:	System of Rice Intensification
TNAU:	Tamil Nadu Agricultural University (Coimbatore, India)
WUE:	water use efficiency
WWF:	Worldwide Fund for Nature

REFERENCES

Aly, A.H., A. Debbab, and P. Proksch. 2011. Fungal endophytes: unique plant inhabitants with great promises. *App Microbiol Biotechnol* 90:1829–1845.

Anas, I., O.P. Rupela, T.M. Thiyagarajan, and N. Uphoff. 2011. A review of studies on SRI effects on beneficial organisms in rice soil rhizospheres. *Paddy Water Environ* 9:53–64.

Andriankaja, A.H. 2001. *Mise en evidence des opportunites de developpement de la riziculture par adoption du SRI et evaluation de la fixation biologique de l'azote: Cas des rizieres des hautes terres.* Memoire de fin d'etudes. ESSA-Agriculture, University of Antananarivo: Antananarivo.

Badri, D.V. and J.M. Vivanco. 2009. Regulation and function of root exudates. *Plant Cell Environ* 32:666–681.

Barea, J.-M., M.J. Pozo, R. Azcon, and C. Azcon-Aguilar. 2005. Microbial cooperation in the rhizosphere. *J Exp Botany* 56:1761–1778.

Barison, J. and N. Uphoff. 2011. Rice yield and its relation to root growth and nutrient-use efficiency under SRI and conventional cultivation: An evaluation in Madagascar. *Paddy Water Environ* 9:65–78.

Barrett, C.B., C.M. Moser, J. Barison, and O.V. McHugh. 2004. Better technologies, better plots or better farmers? Identifying changes in productivity and risk among Malagasy rice farmers. *Am J Agric Econ* 86:869–888.

Berg, G. 2009. Plant-microbial interactions promoting plant growth and health: Perspectives for controlled use of microorganisms in agriculture. *App Microbiol Biotechnol* 84:11–18.

Bilger, E. 1996. *Avantages et contraintes du SRI, enquêtes auprès de 108 exploitants des régions d'Antananarivo et d'Antsirabe.* Rapport du Ministère de l'Agriculture et Association Tefy Saina: Antananarivo.

Birch, H.F. 1958. The effect of soil drying on humus decomposition and nitrogen. *Plant Soil* 10:9–31.

Biswas, J.C., J.K. Ladha, and F.B. Dazzo. 2000a. Rhizobia inoculation improves nutrient uptake and growth in lowland rice. *Soil Sci Soc Am J* 64:1644–1650.

Biswas, J.C., J.K. Ladha, F.B. Dazzo, Y.G. Yanni, and B.G. Rolfe. 2000b. Rhizobial inoculation influences seedling vigor and yield of rice. *Agron J* 92:880–886.

Bomke, C. and B. Tudzynski. 2009. Diversity, regulation, and evolution of the gibberellin biosynthetic pathway in fungi compared to plants and bacteria. *Phytochemistry* 70:1876–1893.

Chaw, S., C. Chang, H. Chen, and W. Li. 2004. Dating the monocot–dicot divergence and the origin of core eudicots using whole chloroplast genomes. *J Mol Evol* 58:424–441.

Chi, F., S.H. Shen, H.P. Chang, Y.X. Jing, Y.G. Yanni, and F.B. Dazzo. 2005. Ascending migration of endophytic rhizobia, from roots to leaves, inside rice plants and assessment of benefits to rice growth physiology. *App Environ Microbiol* 71:7271–7278.

Chi, F., P.F. Yang, F. Han, Y.X. Jing, and S.H. Shen. 2010. Proteomic analysis of rice seedlings infected by *Sinorhizobium meliloti* 1021. *Proteomics* 10:1861–1874.

Chung, K.R., K.-R. Chung, T. Shilts, U. Erturk, and L.W. Timmer. 2003. Indole derivatives produced by the fungus *Colletotrichum acutatum* causing lime anthracnose and postbloom fruit drop of citrus. *FEMS Microbiol Lett* 226:23–30.

Coleman, D.C., D.A. Crossley, and P.F. Hendrix. 2004. *Fundamentals of soil ecology*, 2nd edition. Elsevier: Amsterdam.

Dazzo, F.B. and S. Gantner. 2009. Rhizosphere. In *Encyclopedia of microbiology*, ed. M. Schaechter, 3rd edition, vol. 3, 335–349, Elsevier: Oxford, UK.

Dazzo, F.B. and Y.G. Yanni. 2006. The natural *Rhizobium*-cereal crop association as an example of plant-bacterial interaction. In *Biological approaches to sustainable soil systems*, eds. N. Uphoff, A. Ball, E.C. Fernandes, H. Herren, O. Husson, M. Laing, C.A. Palm, J. Pretty, P.A. Sanchez, N. Sanginga, and J.E. Thies, 109–127. CRC Press: Boca Raton.

Diwakar, M.C., A. Kumar, A. Verma, and N. Uphoff. 2012. Report on the world-record SRI yield in kharif season 2011 in Nalanda district, Bihar state, India. *Agric Today*, 54–56, June.

Dobereiner, J. and F.O. Pedrosa. 1987. *Nitrogen-fixing bacteria in nonleguminous crop plants*. Springer: Berlin.

Dobermann, A. 2004. A critical assessment of the system of rice intensification (SRI). *Agric Syst* 79:261–281.

Hartmann, A., M. Schmid, D. van Tuinen, and G. Berg. 2009. Plant-driven selection of microbes. *Plant Soil* 321:235–257.

Hirsch, R. 2000. *La riziculture malgache revisitée: Diagnostic et perspectives, 1993–1999*. Agence Française de Développement, Départment des Politiques et des Etudes: Antananarivo.

Johnson, B.K. 1994. Soil survey. In: *Final report for the agricultural development component of the Ranomafana National Park Project in Madagascar*, 5–12. Soil Science Department, North Carolina State University: Raleigh, NC.

Khan, A.L., M. Humayun, S.M. Kang, Y.H. Kim, Y.H. Jung, J.H. Lee, and I.J. Lee. 2012. Endophytic fungal association via gibberellins and indole acetic acid can improve plant growth under abiotic stress: An example of *Paecilomyces formosus* LHL10. *BMC Microbiol*, 12:3 (doi:10.1186/1471-2180-12-3).

Ladha, J.K., F.J. de Bruijn, and K.A. Malik, eds. 1997. *Opportunities for biological nitrogen fixation in rice and other non-legumes*. Kluwer: Dordrecht, Netherlands.

Laulanié, H. 1993. Le système de riziculture intensive malgache. *Tropicultura* (Brussels) 11:110–114.

Lin, X.Q., D.F. Zhu, and X.J. Lin. 2011. Effects of water management and organic fertilization with SRI crop practices on hybrid rice performance and rhizosphere dynamics. *Paddy Water Environ* 9:33–39.

Magdoff, F.R. and D.R. Bouldin. 1970. Nitrogen fixation in submerged soil–sand–energy material media and the aerobic-anaerobic interface. *Plant and Soil* 33:49–61.

Mishra, R.P.N., R.K. Singh, H.K. Jaiswal, V. Kumar, and S. Maurya. 2006. Rhizobium-mediated induction of phenolics and plant growth promotion in rice (*Oryza sativa* L.). *Curr Microbiol* 52:383–389.

Pariona-Llanos, R., F. Ibañez de Santi Ferrera, H.H. Soto Gonzales, and H.R. Barbosa. 2010. Influence of organic fertilization on the number of culturable diazotrophic endophytic bacteria isolated from sugarcane. *Eur J Soil Biol* 46:387–393.

Peoples, M.D., D.F. Herridge, and J.K. Ladha. 1995. Biological nitrogen fixation: An efficient source of nitrogen for sustainable agriculture? *Plant Soil* 174:3–28.

Prasanna, R., E. Sharma, P. Sharma, A. Kumar, R. Kumar, V. Gupta, R.K. Pal, Y.S. Shivay, and L. Nain. 2011. Soil fertility and establishment potential of inoculated cyanobacteria in rice crop grown under non-flooded conditions. *Paddy Water Environ* 11:175–183 (DOI 10.1007/s10333-011-0302-2).

Prasanna, R., A. Rana, V. Chaudhury, M. Joshi, and L. Nain. 2012. Cyanobacteria-PGPR interactions for effective nutrient and pest management strategies in agriculture. In: *Microorganisms in sustainable agriculture and biotechnology*, eds. T. Satyanarayana, N. Johri, and A. Prakash. Springer: Berlin, 173–196 (DOI 10.1007/9798-94-007-2214-9_10).

Prayitno, J., J. Stefaniak, J. McIver, J.J. Weinman, F.B. Dazzo, J.K. Ladha, W. Barraquio, Y.G. Yanni, and B.G. Rolfe. 1999. Interactions of rice seedlings with bacteria isolated from rice roots. *Aust J Plant Physiol* 26:521–535.

Randriamiharisoa, R. and N. Uphoff. 2002. Factorial trials evaluating the separate and combined effects of SRI practices. In: *Assessments of the System of Rice Intensification: Proceedings of an international conference, Sanya, China, April 1–4, 2002*, 40–46. Cornell International Institute for Food, Agriculture and Development: Ithaca, NY (http://sri.ciifad.cornell.edu/proc1/sri_10.pdf).

Reddy, P.M., J.K. Ladha, R.B. So, R.J. Hernandez, M.C. Ramos, O.R. Angeles, F.B. Dazzo, and F.J. de Bruijn. 1997. Rhizobial communication with rice roots: Induction of phenotypic changes, mode of invasion, and extent of colonization. *Plant Soil* 194:81–98.

Redman, R.S., Y.O. Kim, C.J. Woodward, C. Greer, L. Espino, S.L. Doty, and R.J. Rodriguez. 2011. Increased fitness of rice plants to abiotic stress via habitat-adapted symbiosis: A strategy for mitigating impacts of climate change. *PLoS One* 6(7):e14823. doi:10.1371/journal.pone.0014823.

Redman, R.S., M.R. Rossinck, S. Maher, Q.C. Andrews, W.L. Schneider, and R.J. Rodriguez. 2002. Field performance of cucurbit and tomato plants colonized with a nonpathogenic mutant of *Colletotrichum magna* (teleomorph: *Glomerella magna*; Jenkins and Winstead). *Symbiosis* 32:55–70.

Rodriguez, R.J., J. Henson, E. Van Volkenburgh, M. Hoy, L. Wright, F. Beckwith, Y.O. Kim, and R.J. Rodriguez. 2008. Stress tolerance in plants via habitat-adapted symbiosis. *ISMEJ* 2:404–16.

Rodriguez, R.J., D.C. Freeman, E.D. McArthur, Y.O. Kim, and R.S. Redman. 2009a. Symbiotic regulation of plant growth, development and reproduction. *Commun Integr Biol* 2:1–3.

Rodriguez, R.J., J.F. White, Jr., A.E. Arnod, and R.S. Redmanl. 2009b. Fungal endophytes: Diversity and functional roles. *New Phytol* 182:314–330.

Rupela, O.P., S.P. Wani, M. Kranthi, P. Humayun, A. Satyanarayana, V. Goud, B. Gujja, V. Shashibhusan, D.J. Raju, and P.L. Reddy. 2006. Comparing soil properties of farmers' fields growing rice by SRI and conventional methods. Paper for WWF-ICRISAT Dialogue Project, Hyderabad, India (http://sri.ciifad.cornell.edu/countries/india/Ap/InAP_Rupela_soil_bio_Hyderabad06.pdf).

Salzwedel, J.L. and F.B. Dazzo. 1993. pSym *nod* gene influence on elicitation of peroxidase activity from white clover and pea roots by rhizobia and their cell-free supernatants. *Mol Plant Microbe Interact* 6:127–134.

Sheehy, J.L., S. Peng, A. Dobermann, P.L. Mitchell, A. Ferrer, J. Yang, Y. Zou, X. Zhong, and J. Huang. 2004. Fantastic yields in the system of rice intensification: Fact or fallacy? *Field Crops Res* 88:1–8.

Stoop, W., N. Uphoff, and A.H. Kassam. 2002. A review of agricultural research issues raised by the System of Rice Intensification (SRI) from Madagascar: Opportunities for improving farming systems for resource-poor farmers. *Agric Syst* 71:249–274.

Surridge, C. 2004. Rice cultivation: Feast or famine? *Nature* 428:360–361.

Thakur, A.K., N. Uphoff, and E. Antony. 2010. An assessment of physiological effects of System of Rice Intensification (SRI) practices compared with recommended rice cultivation practices in India. *Exp Agric* 46:77–98.

Thies, J.E. and J.M. Grossman. 2006. The soil habitat and soil ecology. In: *Biological approaches to sustainable soil system*, eds. N. Uphoff, A. Ball, E.C. Fernandes, H. Herren, O. Husson, M. Laing, C.A. Palm, J. Pretty, P.A. Sanchez, N. Sanginga, and J.E. Thies, 59–78. CRC Press: Boca Raton, FL.

Tsavkelova, E., B. Tudzynski, M. Israeli, B. Oeser, L. Oren-Young, Y. Sasson, and A. Sharon. 2012. Identification and functional characterization of indole-3-acetamide-mediated IAA biosynthesis in plant-associated *Fusarium* species. *Fungal Genet Biol* 49:48–57.

Turner, B.L. 2005. Organic phosphorus transfer from terrestrial to aquatic environments. In: *Organic phosphorus in the environment*, eds. B.L. Turner, E. Rossard, and D.S. Baldwin, 269–294. CAB International: Wallingford, UK.

Turner, B.L. and P.M. Haygarth. 2001. Phosphorus solubilization in rewetted soils. *Nature*, 411:258.

Uphoff, N. 2007. Increasing water saving while raising rice yields with the System of Rice Intensification. In: *Science, technology and trade for peace and prosperity: Proceedings of the 26th International Rice Congress, 9–12 October, 2006, New Delhi*, eds. P.K. Aggrawal, J.K. Ladha, R.K. Singh, C. Devakumar, and B. Hardy, 353–365. International Rice Research Institute: Los Baños, and Indian Council for Agricultural Research: New Delhi.

Uphoff, N. and R. Randriamiharisoa. 2002. Reducing water use in irrigated rice production with the Madagascar System of Rice Intensification. In: *Water-wise rice production: Proceedings of the international workshop on Water-wise Rice Production, 8–11 April 2002*, eds. B.A. Bouman, H. Hengsdijk, B. Hardy, P.S. Bindraban, T.P. Tuong, and J.K. Ladha, 71–88. International Rice Research Institute: Los Baños.

Uphoff, N., I. Anas, O.P. Rupela, A.K. Thakur, and T.M. Thiyagarajan. 2009. Learning about positive plant-microbial interactions from the System of Rice Intensification (SRI). *Aspects Appl Biol* 98:29–52.

Vessey, J.K. 2003. Plant growth-promoting rhizobacteria as biofertilizers. *Plant Soil* 255:571–586.

Wardle, D.A. 2002. *Communities and ecosystems: Linking above ground and below ground components*. Princeton Univ. Press: Princeton, NJ.

Wolfe, D. 2002. *Tales from the underground: A natural history of the subterranean world*. Perseus Press: Cambridge, MA.

Wolfe, K.H., M. Guoy, Y.W. Yang, P.M. Sharp, and W.H. Li. 1989. Date of the monocot-dicot divergence estimated from chloroplast DNA sequence data. *Proc Natl Acad Sci* 86:6201–6205.

Woodward, C., L. Hansen, F. Beckwith, R.S. Redman, and R.J. Rodriguez. 2012. Symbiogenics: An epigenetic approach to mitigating impacts of climate change on plants. *HortScience* 47:699–703, in press.

Yang, Y.W., K.N. Lai, P.Y. Tai, and W.H. Li. 1999. Rates of nucleotide substitution in angiosperm mitochondrial DNA sequences and dates of divergence between *Brassica* and other angiosperm lineages. *J Mol Evol* 48:597–604.

Yanni, Y.G., R.Y. Rizk, V. Corich, A. Squartini, K. Ninke, S. Philip-Hollingsworth, G. Orgambide, et al. 1997. Natural endophytic association between *Rhizobium leguminosarum* bv. *trifolii* and rice roots and assessment of its potential to promote rice growth. *Plant Soil* 194:99–114.

Yanni, Y.G., R.Y. Rizk, F.K. Abd El-Fattah, A. Squartini, V. Corich, A. Giacomini, F. de Bruijn, et al. 2001. The beneficial plant growth-promoting association of *Rhizobium leguminosarum* bv. trifolii with rice roots. *Aust J Plant Physiol* 28:845–870.

Yanni, Y.G. and F.B. Dazzo. 2010. Enhancement of rice production using endophytic strains of *Rhizobium leguminosarum* bv. trifolii in extensive field inoculation trials within the Egypt Nile delta. *Plant Soil* 336:129–142.

Zhao, L.M., L.H. Wu, Y.S. Li, S. Animesh, D.F. Zhu, and N. Uphoff. 2010. Comparisons of yield, water use efficiency, and soil microbial biomass as affected by the System of Rice Intensification. *Comm Soil Sci Plant Anal* 41:1–12.

7 Human Dimensions That Drive Soil Degradation

*Tomas M. Koontz, Vicki Garrett,
Respikius Martin, Caitlin Marquis,
Pranietha Mudliar, Tara Ritter, and Sarah Zwickle*

CONTENTS

7.1 Human Dimensions 167
7.2 Systems Models of Human Dimensions That Drive Soil Degradation 168
7.3 Forms of Soil Degradation and Land Management as Proximate Cause..... 171
    7.3.1 Lack of Investment in Conservation 171
    7.3.2 Intensification of Agricultural Management 172
    7.3.3 Conversion of Land Cover 173
7.4 Factors Affecting Land Management Choices 173
    7.4.1 Population Growth and Urbanization 173
    7.4.2 Poverty and Access to Land and Credit... 174
    7.4.3 Globalization and Other Economic Forces.. 175
    7.4.4 Government Policies 177
    7.4.5 Technology.. 178
    7.4.6 Culture and Religion 179
    7.4.7 Psychological Factors in Land Manager Decision Making.............. 179
        7.4.7.1 Social Risk Tolerances... 180
        7.4.7.2 Worldviews... 180
        7.4.7.3 Rationality in Decision Making... 180
        7.4.7.4 Heuristics in Decision Making 181
        7.4.7.5 Dual Processing 182
7.5 Conclusion 182
Abbreviations 183
References... 184

7.1 HUMAN DIMENSIONS

The interconnection between human and natural systems is widely recognized. This is no less true of agroecosystems and sustainable soil management. Soil is the foundation for agricultural production and other ecological services that sustain human life, so we can draw the causal arrow from soil conditions to human wellbeing (Stocking 2006). At the same time, human behaviors can greatly affect soil resources.

Scientific research about human–soil interactions cuts across many disciplines and fields of study. Of course, it is impossible to fully catalogue this diverse scholarship within the confines of one chapter (or one book, for that matter). However, we can provide a broad sketch of key knowledge pieces and how they fit together. In particular, we emphasize how human dimensions affect soil degradation.

We begin with a high-level view of broad categories of variables to represent human dimensions. These categories have been illustrated visually in several different systems models, which we describe below. While each model differs in its particular configuration and demarcation of different variables, common themes emerge. We explore these themes as different variable sets, for which we synthesize research findings from the relevant literature.

7.2 SYSTEMS MODELS OF HUMAN DIMENSIONS THAT DRIVE SOIL DEGRADATION

Numerous models have been developed to portray visually the interconnections among various components of human–ecological systems. We start with a broad model relating to environmental change from the Millennium Ecosystem Assessment (2003). Figure 7.1 portrays a causal connection from indirect to direct drivers of change. Here the indirect drivers are human dimensions such as demographic, economic (globalization, trade, market), sociopolitical (governance, institutional, and legal framework), science/technology, and cultural/religious (beliefs, consumption choices). These human dimensions lead to a wide range of direct drivers of environmental change, including land use and land cover, technology use and adaptation, external inputs to agriculture (e.g., fertilizer, pest control, irrigation), species introduction, and resource consumption.

Another model that focuses on direct and indirect drivers of environmental change was developed for the Intergovernmental Panel on Climate Change Synthesis Report (2007). In Figure 7.2, human dimensions are included as components of the "Human Systems," while natural system components are labeled "Earth Systems." Here particular "climate process drivers" such as greenhouse gas emissions are the proximate cause of climate change, but these drivers result from human dimensions such as governance, technology, literacy, production and consumption patterns, trade, health, equity, population, and sociocultural preferences.

Another systems model that integrates human and ecological systems comes from Nobel prize laureate Elinor Ostrom, whose work on common-pool resources and institutions led to the "multitier framework for analyzing a social–ecological system" shown in Figure 7.3. This model portrays resources, resource users, and governance systems interacting to yield outcomes embedded in broader ecosystems and social, political, and economic contexts. Human dimensions in the broader context are enumerated in accompanying text to include economic development, demographic trends, political stability, government settlement policies, market incentives, and media organization (Ostrom 2007). Additional human dimensions to describe resource users and governance systems include property rights systems, rules, monitoring and sanctioning processes, social capital, dependence on the resource, technology used, and socioeconomic attributes of users, among others.

Human Dimensions That Drive Soil Degradation

```
Global                    ← Short-term →
                          ← Long-term →
  Regional
    Local
```

Human wellbeing and poverty reduction
- Basic material for a good life
- Health
- Good social relations
- Security
- Freedom of choice and action

Indirect drivers of change
- Demographic
- Economic (e.g., globalization, trade, market, and policy framework)
- Sociopolitical (e.g., governance, institutional, and legal framework)
- Science and technology
- Cultural and religious (e.g., beliefs, consumption choices)

Ecosystem services
- Provisioning (e.g., food, water, fiber, and fuel)
- Regulating (e.g., climate regulation, water, and disease)
- Cultural (e.g., spiritual, aesthetic, recreation, and education)
- Supporting (e.g., primary production, and soil formation)

Life on earth—biodiversity

Direct drivers of change
- Changes in local land use and cover
- Species introduction or removal
- Technology adaptation and use
- External inputs (e.g., fertilizer use, pest control, and irrigation)
- Harvest and resource consumption
- Climate change
- Natural, physical, and biological drivers (e.g., evolution, volcanoes)

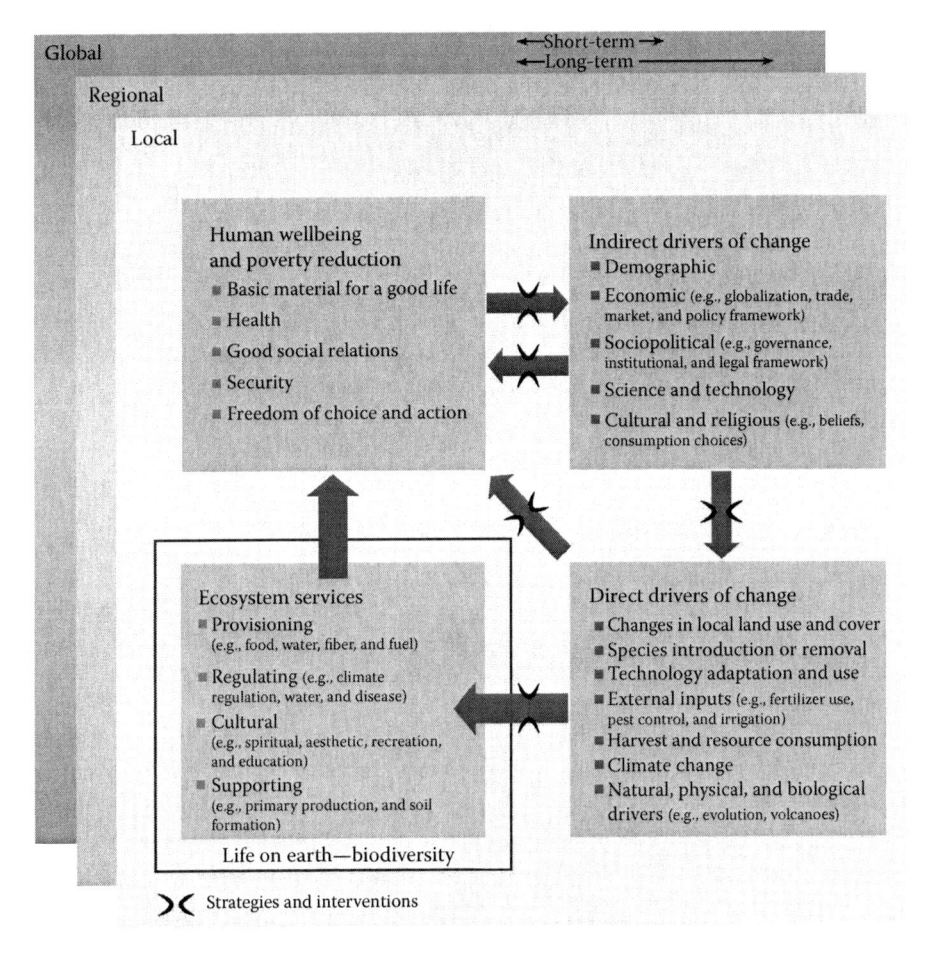

)(Strategies and interventions

FIGURE 7.1 Indirect and direct drivers of environmental change. (From Millennium Ecosystem Assessment, *Ecosystems and Human Well-Being Synthesis*, Washington, DC: Island Press, 2005, vii.)

Comparing across these three models, two things become clear. First, unpacking a concept such as "human dimensions" leads to many diverse variables of interest. Many of these variables have been the subject of considerable study and further unpacking. For example, the concept of "social capital" has spawned a rich field of inquiry in sociology and political science. Similarly, "governance" has been examined not only by political scientists but also by legal, policy, and public administration scholars. Second, there is considerable overlap among these models in the identified variables of importance for understanding social–ecological systems. Our aim in this chapter is to focus on these common themes as they apply to understanding the human dimensions that cause soil degradation. Our key variables include the proximate causes (or direct drivers) of soil degradation as well as the precursor factors (or indirect drivers) that affect the proximate causes. Figure 7.4 shows these variables. To better understand the variables and their interactions, we start with the end result—soil degradation—and work backward.

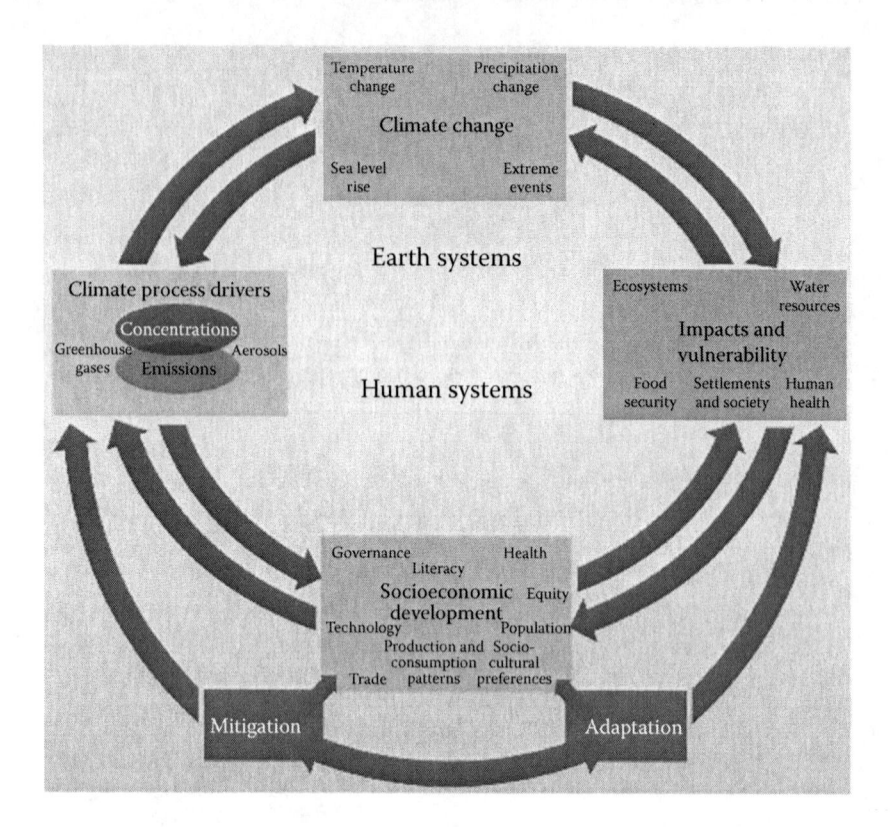

FIGURE 7.2 Earth and human systems. (From Climate Change 2007: Synthesis Report. Contribution of Working Groups I, II and III to the Fourth Assessment Report of the Intergovernmental Panel on Climate Change, Figure I.1. IPCC, Geneva, Switzerland.)

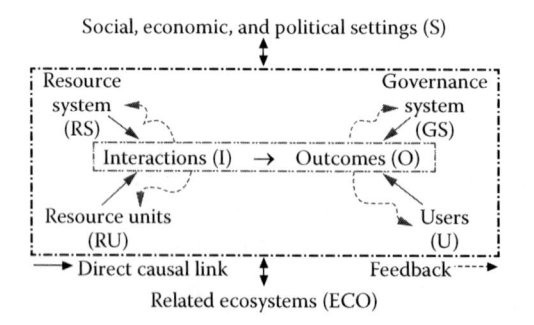

FIGURE 7.3 Multitier framework for analyzing social–ecological systems. (From Ostrom, E., *Proceedings of the National Academy of Sciences* 104(39), 15181–15187, Copyright (2007) National Academy of Sciences, USA.)

Human Dimensions That Drive Soil Degradation

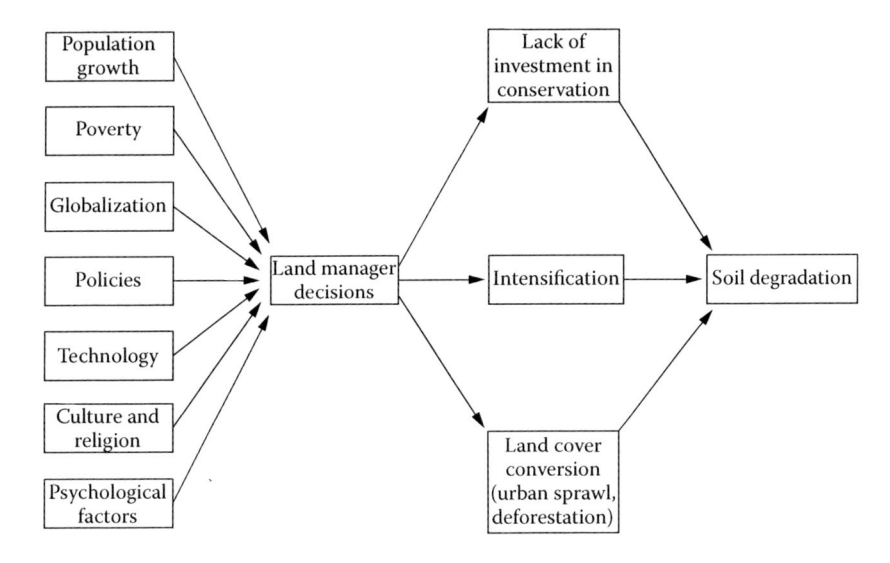

FIGURE 7.4 Human dimensions affecting soil degradation.

7.3 FORMS OF SOIL DEGRADATION AND LAND MANAGEMENT AS PROXIMATE CAUSE

In agroecosystems, soil degradation is indicated by a loss of soil fertility (Majule 2003; Reed 2008). Soil fertility includes the ability of the soil to supply essential nutrients in the right amounts and at the right time to produce desired results (Rowell 1993). It also includes the quality of soil physical properties (soil structure) such as the arrangements of soil granules and the pore spaces between them (Marshall and Holmes 1979; Boix-Fayos et al. 2001; Ayres et al. 2009).

Since our interest in this chapter is the human dimension that drives soil degradation, we focus on forms of degradation that result from human activities. These include forms relating to soil chemical properties such as nutrient depletion, chemical contamination, and salinization, and structural properties such as compaction, waterlogging, and erosion. As will be described below, each of these forms of degradation has proximate causes related to land management (direct drivers). Land management choices of particular relevance for soil degradation have been described in three broad categories: lack of investment in conservation, intensification of agricultural management, and conversion of land cover.

7.3.1 Lack of Investment in Conservation

Lack of investment in conservation contributes to soil degradation. Although a variety of best management practices (BMPs) and projects to protect soil have been developed for agricultural practices in a wide range of contexts, underinvestment remains a problem worldwide. In fact, entire fields of study have developed around the question, "What factors promote the adoption and diffusion of innovations throughout a community?" Soil conservation practices and projects have long been studied in this field (see Prokopy et al. 2008 for a comprehensive review). While BMPs differ by local context, some common examples include grass waterways, control drainage, reduced-till and no-till practices, vegetated buffers along streams, phosphate-free fertilizers, manure management and nutrient management planning, and fencing animals out of streams.

7.3.2 Intensification of Agricultural Management

Intensification of agricultural management refers to increased inputs (e.g., labor, pesticides, herbicides, fertilizer, water, mechanized equipment) in an effort to increase agricultural yield per unit area. This often involves greater use of technology. From 1961 to 1996, intensification permitted a doubling of the world's food production with only a 10% increase in arable land (Tilman 1999).

While intensification may increase soil fertility in the short run, many researchers have identified numerous ways that intensification causes soil degradation. The introduction of machinery compacts soil, reducing its capacity to absorb rainwater, thus speeding up runoff and resulting in erosion. This is also known as "mining" of the soil's organic matter and nutrients because they are removed rather than renewed with use (Gudger and Barker 1993). The negative effects of intensive agricultural practices on the sequestration of soil carbon are well documented (Burke et al. 1989; Paul et al. 1997).

Furthermore, the "green revolution" model of intensive agriculture is highly water intensive and requires increased use of chemical fertilizers and pesticides. The increased irrigation necessary for high-yield plant varieties causes soil salinization and waterlogging (Rosegrant et al. 2009; Urama 2005; Howes 2008). Waterlogging is caused by irrigation water seeping through the soil and raising the water table. Salinization occurs when the highly mineralized groundwater rises. If it reaches the zone where plant roots are growing, it dehydrates the plants. The dehydrated plants then wilt and finally die. The FAO (1996) declared that there was 20 to 30 million ha of land severely degraded by salinization and an additional 60 to 80 million ha that was affected to some extent by saturation and salinization.

Additionally, the application of chemical fertilizers can cause excesses of soil nitrogen, which can ultimately harm the cultivation potential of the soil (Matson et al. 1997; Turner and Ali 1996). Application of inorganic nitrogen, which has greater environmental effects than most other inputs (Vogt et al. 2010), has grown to about seven times what it was in the 1960s and is still growing. Besides nitrogen's well-known effects on surface and groundwater, it also has a detrimental effect on soil bacteria. Although beneficial at optimal application levels, this is often exceeded and overapplication increases the above-mentioned harmful effects (Spiertz 2010; Guo et al. 2011), decreasing yield and the effectiveness of manufactured fertilizers (Marenya and Barrett 2009).

Although pesticides have greatly increased agricultural production, their indiscriminate use can be harmful. The greatest concerns over pesticides revolve around their adverse effects to humans and the environment (Gaby 2004; Weiss et al. 2004), but they also can decrease soil quality by reducing beneficial insect populations (Gudger and Barker 1993). Pesticides have been known to accumulate in soil and be transferred to rivers via irrigation, and then deposited into other soils when downstream farmers irrigate (Stocking 2006; Tse-Yan Lee 2005). Unfortunately, inputs of nitrogen increase pesticide use. Nitrogen encourages growth of weeds better adapted to high levels of nitrogen. In addition, nitrogen can lead to greater damage from animal pests, because plants become more nutritious for herbivores as nitrogen level increases (Vogt et al. 2010).

7.3.3 CONVERSION OF LAND COVER

Conversion of land cover often degrades soil resources. Two especially problematic types of land cover conversion are deforestation and urban sprawl (conversion of land to impervious surfaces such as roads and buildings).

Deforestation has been well documented as a leading cause of land degradation, especially in developing countries and tropical regions (Barbier 1997, 2000; Lambin et al. 2001). Deforestation is estimated to account for 40% of soil erosion in Asia and South America, 22% of soil erosion in Mexico and Central America, and 14% of soil erosion in Africa (Barbier 1997, 892). The conversion of forest to farmland or exotic tree plantations has been found to cause a depletion of major nutrients such as organic matter, total nitrogen, and available phosphorus, while at the same time increasing soil erosion (Fu 2000; Majule 2003). Deforestation also reduces land cover, which reduces rainwater infiltration into the soil and affects soil aeration (Majule 2003).

Urban sprawl reduces surface permeability, thus inhibiting infiltration of water into the soil, and the associated compaction degrades soil structure. Where urban uses completely cover the soil, as in concrete foundations and asphalt roadways, this severely impairs soil ecological functions and processes. Moreover, studies in the United States and China indicate that urban development follows soil resources and that, in general, the best soils are converted to nonagricultural uses by urban sprawl (Imhoff et al. 2004; Chen 2002). Urban sprawl and its associated uses also increase soil contamination by industrial and consumer by-products such as manufacturing pollutants and residential chemicals (Chen 2002).

7.4 FACTORS AFFECTING LAND MANAGEMENT CHOICES

Land managers make decisions and take actions on the landscape. Their decisions about investing in conservation, intensifying agricultural production, and converting land cover are a function of many precursor factors (indirect drivers). Our focus here is on land managers involved in agricultural production. Our review of the literature revealed extensive scholarship seeking to explain land management decision-making, drawing on several key variables: population growth and urbanization, poverty and access to land and credit, globalization and other economic forces, government policies, technology, culture and religion, and individual psychological factors.

7.4.1 POPULATION GROWTH AND URBANIZATION

Population growth and distribution both impact activities that degrade soil. As human populations grow, there is corresponding pressure for shorter fallow periods and migration to even less productive and more fragile soils, for example, drier land or steeper slopes (Warren et al. 2001; El-Swaify 1997). Moreover, as populations grow, people are increasingly concentrating in urban areas and sprawling ex-urban settlements (Alig et al. 2004). These clusters of people fragment the landscapes of large areas, and in doing so, they threaten ecosystem processes. In fact, human demographic shifts are said to be the leading source of soil degradation (Tolba and El-Kholy 1992). This is due to the expansion of populations into what was previously prime agricultural land, thereby triggering both real and perceived needs for intensified agricultural production (Meyer and Turner 1992). This chain of processes driven by human demographic shifts affects the ability of biological systems to support human needs (Vitousek et al. 1997).

Between 1980 and 2000, the population of the United States increased by more than 50 million people, about a 24% increase (Mackun and Wilson 2011). During the same time frame, the amount of land used for urban and built uses increased by roughly 34% (USDA Natural Resources Conservation Service 2009). This growth is not projected to slow in the future; the United Nations estimates that the world's population will surpass 9 billion people by 2050 and exceed 10 billion in 2100 (United Nations 2011). All of these extra people will require expanded infrastructure and development, and in the past, increases in developed land have predominantly come from the conversion of croplands and forests (Alig et al. 2004).

In the United States, the quantity of land in agricultural production peaked around 1950 at 1161 million acres. Agricultural land has been declining since then; there were 922 million acres in 2007, and the current rate of decline is estimated to be 5.39 million acres per year (US Department of Agriculture 2007). The conversion of agricultural land to urban uses is caused by increasing urban demand for land, especially low-density residential and commercial development at the urban fringe (Heimlich and Anderson 2001). In addition to conversion, this leads to adaptation of the agricultural system to increase yields on smaller areas of land (Dasgupta et al. 2000). More generally, agricultural intensification is often triggered by growth in population and density, whether caused by natural increase or migration (Ostrom et al. 1999). The harmful effects of such intensification were described above (see Section 7.3).

Human demographic shifts are at the core of land use and land cover changes. As the population grows, so does the amount of land needed for infrastructure and other development. This increased amount of land oftentimes comes at the expense of land previously allocated to agricultural uses. Furthermore, larger populations require more food, which forces agricultural producers to increase yields on smaller tracts of land. This chain of events leads to soil degradation.

7.4.2 Poverty and Access to Land and Credit

Causal links have been established between differential access to land and credit among rural populations in developing regions, and the abandonment of degraded agricultural lands for new agricultural and forest land frontiers (Barbier 1997). This process of exploitation and abandonment has been found to perpetuate a cycle of land degradation among the rural poor (Munasinghe and Cruz 1995).

Lack of access to credit and capital for purchasing technological inputs among the rural poor in developing countries fosters increased dependence on labor to sustain farming operations. Rather than invest in labor to increase yield, poor rural farmers often adopt practices such as cropping more densely and weeding and harvesting more rigorously. "Over time," Barbier (1997, 303) says, "this type of intensification is 'unsustainable' as it depletes soil nutrients and cannot be sustained without shifting toward 'capital-led' investments such as inorganic fertilizers". The unsustainability of this type of agricultural intensification drives farmers off the depleted lands in pursuit of nutrient-rich frontiers to convert to agriculture. These frontiers are often forests.

The development of sustainable methods for large-scale agricultural operations, such as low or no-till cultivation, use of cover crops, application of organic fertilizers, and low-input production, may help to reduce the loss of soil nutrients and organic matter (Follett 2001). Although incentives for adopting these practices may help to mitigate rates of deforestation and land degradation among larger farmers, a variety of economic factors influence the ability of the rural poor to adopt conservation practices. Such factors include insecure tenure or ownership of land, distorted market prices for inputs and outputs, imperfect competition, and incomplete markets. Also, lack of access to credit is a major factor preventing the adoption of soil conservation practices among the rural poor, since conservation practices can be costly to implement in terms of lost production (Lipper and Osgood 2004) and lack of capital necessitates borrowing toward such practices. However, credit lenders tend to reserve credit for those they perceive to be less likely to default on their loans (Thampapillai and Anderson 1990).

One aspect of poverty is the size of landholdings. As parcel size decreases, so does the ability of farmers to economically benefit from agriculture alone. Small landholdings often drive farm families to seek off-farm income, which creates labor constraints to implementing soil conservation practices. The distance farmers have to travel to their fields is associated with erosion, because the less time farmers have for conservation practices, the less attention the fields receive. This is especially damaging in windy regions such as the West African Sahel (Warren et al. 2001). For near-landless households, the lack of rural employment opportunities also factors into the decision to migrate to frontier lands in pursuit of greater economic returns (Barbier 1997).

There is a feedback look from soil conditions to poverty at work here, interacting with fertilizer inputs. Typically, poorer farmers are on the most degraded lands. These lands have low levels of organic matter. Because application of nitrogen fertilizer increases production in the presence of soil organic matter, but shows poor returns when organic matter is low, fertilizer inputs on these soils become unprofitable. Poor farmers investing in nitrogen fertilizers are not getting a good return on their investments, which creates a "poverty trap" (Marenya and Barrett 2009). This interaction is a reminder that although our description in this chapter focuses on the link from human dimensions to soil degradation, feedback mechanisms further complicate our understanding of the system.

7.4.3 Globalization and Other Economic Forces

Over the last 30 years, the economic forces influencing environmental decision making have shifted from state-based economic policies to global free market structures.

Francesco Di Castri (2000, 324) includes the following changes in the reach and scope of this economic shift:

> the shift from protectionism, centralized planning, and state-dominated economies to freer trade, freer markets, and privatization; from local to national scale economic activities to globalization characterized by increased international flows of capital, information, and goods; from industrial concentration in a few sites and countries to diffusion over space throughout the world of productive activities; and from rather strict dependence of development on local natural resources to much larger dependence on human culture (technology, infrastructure, and institutions).

Although some countries are still in the process of making these transitions, while others are fully integrated into a globalized, liberalized, and free market economy, it has been argued that the current global economic milieu is a major underlying cause of environmental degradation worldwide, especially in developing countries (Pearce 1998).

Globalization of markets creates a greater disconnect between the sources of demand and the location of production that occurs when a region becomes incorporated into the world economy. The loss of this connection is associated with rapid land-use changes through new market cultivation leading to the intensification and specialization of plant and livestock operations (Lambin et al. 2001). This shift in agricultural practices leads to increased exploitation of water resources and, in turn, soil salinization and desertification (Duraiappah 1998).

The formation of international free trade agreements such as the NAFTA, GATT, and the WTO reduces trade tariffs and creates market competition from agricultural imports (McMichael 2003). This competition forces farmers to "scale up" their agricultural operations to produce commodities that can compete in the global market, or else miss out on the economic benefits of industrialization (Reardon and Barrett 2008). Additionally, interventionist economic development agendas led by states, donors, or nongovernmental organizations on the world stage pursue agricultural intensification as a means of poverty alleviation through the promotion of agricultural specialization and transitions to export-oriented crops (Lambin et al. 2000).

Globalization exhibits strong interaction effects with poverty. Agricultural smallholders in developing countries are particularly sensitive to the price shocks driven by economic globalization, including the emergence or international reorganization of processing industries, loss of traditional export markets, competition from imported substitutes, and exposure to global market fluctuations. Thus, poor farmers with limited access to land tenure, capital, credit, and technologically advanced agricultural inputs may counter these economic impacts by abandoning exhausted soils for untouched, nutrient-replete sites (Barbier 2000).

Globalization can also interact with government policies to increase soil degradation. Economic liberalization policies can increase the productivity and export orientation of agriculture by influencing both agricultural input and output prices. State policies reduce the price of export commodities through subsidies that help those commodities compete on the global market, while simultaneously decreasing subsidies to inputs, such as fertilizers (McMichael 2003). Thus, a drop in the ratio of economic return on production to cost of inputs, combined with limited access to inputs among the rural poor and weak regulations in the forest industry, can drive increased clearing and cultivation of nutrient-rich forested land in pursuit of increased economic returns (Barbier 2000).

Agricultural corporations can take advantage of competitive markets and lower their production costs, so that consumers also pay less. However, to keep up, small farmers carry out agricultural intensification and extensification contributing to additional land degradation and deforestation (Hazell and Wood 2007).

On the other hand, globalization may lead to improved agricultural practices for niche markets. The rise of certified products, such as organic and shade-grown coffee for export, takes advantage of the possibility that consumers overseas may place a high value, and willingness to pay, for products produced more sustainably (Perfecto et al. 1996).

7.4.4 Government Policies

Macroeconomic policies affecting soil productivity include expansive fiscal and monetary policies, high interest rates, high inflation, overvalued exchange rates, high debt service ratios, and protectionism (Knowler 2004). The first three of these reduce economic stability and encourage maximizing present output rather than investing for the future. Overvalued exchange rates make inputs cheaper and thus more widely used, while high debt service ratios encourage production of exportable crops rather than more sustainable crops. Protectionism can induce greater cultivation of irrigated crops, which can cause salinization and waterlogging of soils.

Government policies are often technology- and physical development–centered, without consideration of political and economic forces (Grossman 1997; Urama 2005) or local institutions (Ostrom 1990). For example, many governments fund interventions to persuade farmers to adopt chemical fertilizers by subsidizing their prices (Andersen and Lorch 1994). While fertilizer is a profitable investment on average, as soil becomes degraded, fertilizer inputs lose their effectiveness on crop yield, which makes the marginal costs too high for many farmers to afford (Lipper and Osgood 2004; Marenya and Barrett 2009).

Government policies may directly promote activities that degrade soil. One such example is the European Union's subsidizing of the cost of soil movement and land leveling up to 50% (Marti'nez-Casasnovas and Ramos 2009), when the detrimental effects of land leveling and deep plowing are well documented.

Policies that promote irrigation are common, despite evidence of the environmental consequences (Urama 2005; Howes 2008). From the late nineteenth to the late twentieth century, Australian farmers in state of Victoria noticed, and were alarmed by, the salinization of their soil through irrigation. Policy-makers paid little attention to landowners' and scientists' concerns. Their resistance made sense in terms of the substantial investments made in the irrigation system, but their response was to blame individual farmers for the degradation of the soil (Howes 2008).

At a broader scale, policy and government structures have influenced land management practices. Traditional and indigenous arrangements of common property regimes and usufruct arrangements have been replaced by Northern institutions and organizations. These institutions in developing countries are often modeled to resemble the Northern institutions, and therefore encourage energy intensive, mechanized agricultural models in the hope of promoting agricultural modernization. Private property regimes are given preference, in spite of literature showing that common property regimes may have more durable and economically sustainable land use regime (Ostrom 1990).

Finally, it should be noted that government policies exhibit strong interaction effects with poverty. Insufficient social support policies for disadvantaged rural populations exacerbate poverty (Munasinghe and Cruz 1995), and poverty can have negative effects on soil as described above (see Section 7.4.2).

7.4.5 Technology

Technology, or the application of scientific knowledge for practical purposes, allows people to advance economically and, in the case of agriculture, increases the capacity to alter landscapes (Scherr and Yadav 1995). Technologies such as agricultural intensification, irrigation methods, fertilizers and pesticides, and biotechnology hold the promise of increasing food productivity to feed a rapidly growing population. At the same time, technologies can increase soil degradation. The theory of induced innovation posits that farmers are driven to intensify by using labor and/or capital such as chemical inputs, organic matter, equipment, and land conservation infrastructure to increase yields per hectare of land (Reardon et al. 1998). These same technologies are capable of degrading soil and natural resources due to improper management practices such as excessive use of water, overgrazing, and insufficient or untimely application of fertilizers (Andersen and Lorch 1994; Hazell and Wood 2007).

Many agricultural technologies are developed with a narrow focus on securing short-term profits for farmers, without considering the sustainability of these technologies. For example, pesticides and herbicides can improve productivity, but their negative effects on the environment, human health, and long-term yields have been well documented (Hazell and Wood 2007). Also, modern irrigation systems can create waterlogging and salinization problems (Hazell and Wood 2007).

Technologies developed for one agroclimatic zone may spread to less suitable areas, where they may further degrade resources. For example, barley cultivation techniques became widespread in the Middle East, after they became popular in high rainfall regions, which led to land degradation (Hazell and Wood 2007). Here lies the danger of technology transfer between countries without considering the climate and the suitability of the technology. In a critical examination of the drivers of technology spread, Biggs (1990) argues that unsuitable technology is often forced on farmers by scientists who design innovations based on political allocation of funds for research. These elite researchers, he argues, fail to consider the importance of knowledge of nonresearchers or suitability to local conditions.

Technology has been used to dramatically alter physical landscapes. In the United States, for example, terraces, shelterbelts, and hedgerows have been flattened to give easy access to heavy machinery. This has modified the existing contours of land, which has intensified the process of soil erosion (Office of Technology Assessment 1982).

Technology interacts with other factors. Market forces such as the prices of agricultural commodities and wages directly influence farmers, with respect to deciding which technology will give them maximum returns (Angelsen and Kaimowitz 1999). Labor-saving technology can result in specific kinds of land cover conversion. Technological changes that increase agricultural yields without altering labor or capital have led to increased deforestation as more resources were freed up for working on additional land (Angelsen and Kaimowitz; Reardon et al. 2001; Southgate 1999).

7.4.6 Culture and Religion

Culture and religion are among the indirect drivers of changes in ecosystem (Millennium Ecosystem Assessment [MEA] 2005). In order to understand the link between culture and soil degradation, one can view culture as a set of values, beliefs, and norms shared by a group of people. Culture shapes perceptions of individuals and how individuals perceive the world. In turn, the perceptions influence what individuals consider important and suggest which courses of action are appropriate and inappropriate (MEA 2005).

Lambin et al. (2003) argue that the attitudes, values, beliefs, and individual perceptions of land managers influence their land use decisions, which have intended and unintended consequences on ecosystems. This argument fits into a long line of studies explaining human behavior using the theory of planned behavior (Ajzen 1991). This psychological theory posits that individual behavior, such as undertaking particular land management practices, is a function of attitudes, norms, perceived behavioral control, and intentions.

One important aspect of culture is religion. Studies of the influence of religion on environmental concern and behavior have reached divergent conclusions (Sherkat and Ellison 2007). While some studies have found negative effect of religious factors (Guth et al. 1993, 1995), others have found that religious beliefs have a positive effect or no influence (Hayes and Marangudakis 2000, 2001; Sherkat and Ellison 2007). Invoking Genesis, Chapter 1, verses 27 and 28, in his essay, Lynn White Jr. (1967) argued that poor environmental management was due to religious beliefs that humans were superior to nature and could deal with it as they wished with no obligation to care for it. White's argument was supported by Eckberg and Blocker (1996), but others have concluded that religion contributes to individuals being more concerned with the protection of the environment rather than less (Sherkat and Ellison 2007; Schultz et al. 2000; Sauer and Nelson 2011). For instance, Sherkat and Ellison (2007) found a significant positive effect of church attendance on stewardship orientation. They also found that beliefs in the inerrancy of scripture have a strong positive impact on support for environmental stewardship. While such studies typically do not examine soil degradation, they do point to environmental protection more generally. Writing specifically about soil, Sauer and Nelson (2011) argue that sustaining soil productivity requires incorporation of ethical principles into management decisions.

7.4.7 Psychological Factors in Land Manager Decision Making

Most of the indirect drivers discussed above are linked to factors outside of the individual. That is, the context of a society (population growth, urbanization, poverty, government policies, culture, and religion) or global forces (globalization, economics, technology) affect land manager choices. Next we turn to key individual psychological factors affecting land manager choices: risk perception and risk tolerance. In particular, we discuss the substantial literature focusing on one aspect of agricultural land management, weed control.

The health of soil is, in part, the result of how each individual farmer perceives the risks of weeds and decides to act on it. Key drivers of farmer weed management decisions that either improve or degrade the health of the soil are social risk tolerances, worldviews, and decision making based on rational, heuristic, or dual processes.

7.4.7.1 Social Risk Tolerances

Many farmers are risk-averse and would rather stick with what has been working than try something new, even if a new weed management practice may reduce soil erosion (Finnoff et al. 2006). In addition, perceived risks from the presence of any weeds may lead farmers to seek eradication. Conventional farmers have very low or zero tolerance for weeds (Johnson and Gibson 2006). This may stem in part from established norms that encourage farmers to make their fields look "clean" for the neighbors (Wilson et al. 2008). A clean field, an easy harvest, and preventing the future spread of weeds are all cited as reasons to stay with the status quo.

7.4.7.2 Worldviews

Since the middle of the twentieth century, farmers in the United States have largely shared a dominant social paradigm (DSP) that includes trust in progress, growth, and private property rights; faith in science and technology; and the belief that nature must be subdued and made useful (Beus and Dunlap 1994). But an emerging group of alternative farmers exhibits a new environmental paradigm (NEP) (Beus and Dunlap 1991). As a whole, farmers operating under the NEP reject economic growth as inherently good and resist the notion that nature is primarily for human use. These two worldviews can influence a farmer's decision making regarding land use in significant ways. Farmers who fall more in line with the DSP may find it difficult to try alternative farming methods that emphasize environmental health rather than economic health. Their implicit trust in science and technology may make it difficult to consider alternative strategies (Bell 2004). Many farmers believe that "science" will consistently produce new technology to compensate for the lost in soil nutrients and quality (Lyson and Welsh 1995; Lyson 2004).

Eckert and Bell (2005) found that conventional farmers split their "domains of influence" when making decisions. They put the environment, family, and community health and sustainability, all characteristics of the NEP, in a different domain than economic profit and efficiency, part of the DSP. The decision to use chemicals is facilitated by avoiding trade-offs between these domains of influence. Such decisions are consistent with theories of judgment and decision making that highlight how difficult it is for individuals to make trade-offs with values that are difficult to quantify (Tetlock et al. 2000).

7.4.7.3 Rationality in Decision Making

While social risk tolerances and worldviews may play a key role in a farmer's likelihood to be open or averse to weed management practices that decrease soil degradation, current theories of decision making might also explain farmer behavior. Early studies of human decision making emphasized that normative models that assumed rational decisions about a risk were devoid of emotion and entirely based on maximizing utility. This perspective of decision making calculates risk with economic utility functions that identify all possible options and their probable consequences (Plous 1993). Normative models are based on the idea that humans are capable of using decision rules to make consistent, rational choices. Risk, in this sense, is a simple calculation of the probability of an outcome multiplied by the consequence.

Modern, chemical, and capital intensive agriculture is an excellent example of the normative model of decision making within an agricultural setting. The production function, used by the USDA and other private agribusinesses, measures the risks and benefits of agriculture according to efficiency, replicability, and standardization (Lyson and Welsh 1995) in order to increase crop yield and maximize profit. Soil health is not included in this equation. Weeds, in this domain, are a threat to yield and therefore high risk (Wilson et al. 2008). Management decisions that follow this normative line of reasoning will be based on the most cost efficient way to control weeds and improve yields through technology, chemical applications, and mechanical tillage that may damage soil structure.

There are two drawbacks to understanding farmer management decisions using normative models. Research on organic weed management has shown that ecological complexity blurs causation between management choices and weed populations (Zwickle et al. 2011). A farmer cannot know all the possible outcomes and consequences of weeds and weed management choices for the health of the soil or any aspect of the agroecosystem. Rational calculations of maximum utility are impossible in this case. Physical attributes of the farm, the farm operation itself, and the farmer all introduce further limitations to maximizing utility. The farm's limits are resource based (e.g., available equipment, time, labor, and money) and physical (e.g., soil type, geography, and perhaps the most limiting of all: climate and weather patterns), while farmer limitations are cognitively based (e.g., memory, cognitive ability, and attention span).

The second reason to abandon a normative model for farmer decision making is that even normative models of risk cannot be free from the influence of value judgments (Slovic 1987). Judging the consequences of a risk, no matter how scientific or analytically based the assessment may be, involves value-laden judgment (Slovic 1999). In contrast to farms that follow industrial methods and scale up production to meet market demand in spite of environmental costs, some organic farmers may view the risks of managing weeds to extend beyond economic values to include the health of soil, people, animals, and other components of the agroecosystem (Berry 1977; Lyson 2004). Previous mental model studies found that farmers do indeed combine financial and lifestyle domains when making decisions (Eckert 2006). Other, noneconomic sources of information in the decision making process are often considered to be as important as the economic viability of the farm (Eckert and Bell 2005).

7.4.7.4 Heuristics in Decision Making

The theory of bounded rationality explains that humans may intend to be rational and consistent, but their judgments and decisions are bound by time, resources, and limited attention and memory (Simon 1959; March 1994; Kahneman 2003). As a consequence, humans often use heuristics, or rules of thumb, to help simplify the decision making process (Kleindorfer 1999). In the face of both complexity and constraints, farmers are often forced to simplify the decision making process. On the one hand, farmers are participating in the complex and unpredictable rhythms of the agroecosystem, while on the other hand, available resources, regulations, and the physical properties of the farm restrict their options within this system.

Because conditions set by the farm parameters and farm attributes are less than ideal, farmers exhibit clear use of "satisficing" in their decision making process. Satisficing describes choice behavior as one that only satisfies an individual's most important concerns (Simon 1959), whether those concerns are economic, ecological, social, or agricultural, rather than maximizing expected utility across all preferences.

7.4.7.5 Dual Processing

Descriptive models such as dual processing recognize the vital role of risk perceptions, affect, and emotion in the decision making process without ignoring the role of analytical processing. Affect is defined as a feeling of "goodness" or "badness" about weeds and weed management built up over time through each farmer's experience. Dual process models describe human decision making as a combination of experiential/emotionally based, or affective, thought processes (system 1) and deliberative, analytical thinking (system 2) (Damasio 1994; Epstein 1994; Kahneman 2003). System 1 is the first response to a risk and often includes heuristics, or rules of thumb, based on experience and affect in order to simplify and speed up decisions.

According to dual process theory, a balance between systems 1 and 2 is necessary for rational decisions. Both systems provide types of information that guide and motivate the decision making process (Plous 1993; Kahneman 2003; Peters 2008). In the absence of experience and/or emotional system (system 1), a decision maker would be unable to choose efficiently between different options in simple decisions (Damasio 1994; Hsee 1996), while the absence of the analytical system (system 2) would cause an individual to become narrow-minded and fail to consider more beneficial options. These two systems influence decision processes concerning a risk that an individual cares about such as soil health or economic viability.

7.5 CONCLUSION

Human dimensions encompass many aspects of human behavior. Scientific studies encompassing human–ecological systems have yielded a variety of models to describe the interrelations among many variables. It is evident that a wide range of social science disciplines can inform our understanding of human dimensions that lead to soil degradation. Contributions to knowledge described in this chapter come from political science, economics, geography, planning, rural sociology, religious studies, sociology, decision science, forestry, demography, history, environmental management, and psychology, among other fields.

Knowing that proximate causes are themselves caused by precursor variables can help us understand some otherwise puzzling choices made by land managers. In particular, it can help us see that proposed solutions that fail to address the precursor variables may not be effective. For example, subsidies to reduce the cost of conservation practices may not induce different land manager choices if the managers perceive that those practices will increase risk of crop loss, or if the necessities of off-farm income leave them with few labor hours, or if the practice goes against their worldview or community norms.

This review and synthesis of literature also highlight the fact that the impact of some factors is not unidirectional. For example, technology can both encourage and discourage activities that increase soil degradation. As one study concluded, poverty often reduces access to technology and thus leads farmers to labor more intensely, which degrades soil (Barbier 1997). On the other hand, studies in different contexts describe cases where greater use of particular technologies degrades soil (Andersen and Lorch 1994; Hazell and Wood 2007). As another example, globalization often encourages land manager choices that degrade soils, but there is at least some evidence that opening up to global consumers may spur niche markets for sustainably grown products such as coffee.

Mung'ong'o (2009) recommends a political ecology approach that recognizes the complexity of the sources of soil degradation. Many agree. Because farmers naturally tend to maximize their profits, Botterweg (1998) showed better results by integrating economic and ecological models. Voluntary programs with incentives and education have been found to work better than compulsory measures in affecting land manager choices (Grossman 1997).

More effective policies are needed at all levels—farm, community, national, and international—and they need to be developed with appropriate attention to feedback mechanisms (Knowler 2004). Giving more attention to feedback mechanisms would lead to an integrated soil fertility management approach, as Marenya and Barrett (2009) suggest, not just promoting fertilizer use in all cases. Maybe the most important aspect of the design of management measures is farmers' input. This both respects local knowledge and will be much more acceptable to potential adopters (Sulieman and Buchroithner 2009; Posthumus et al. 2011; El-Swaify 1997).

In the end, starting with soil degradation and tracing back to the activities that degrade soil, and then back further to the factors that affect these activities, reveals a complex system connecting humans to soil. There is no simple solution to the problem of soil degradation. But better understanding of the variety of factors at play, how they interact, and which factors are most critical in which contexts may provide a foundation for success.

ABBREVIATIONS

BMPs: best management practices
DSP: dominant social paradigm
GATT: General Agreement on Tariffs and Trade
MEA: Millennium Ecosystem Assessment
NAFTA: North American Free Trade Agreement
NEP: new environmental paradigm
OTA: Office of Technology Assessment
USDA: United States Department of Agriculture
WTO: World Trade Organization

REFERENCES

Ajzen, I. 1991. The Theory of Planned Behavior. *Organ Behav Hum Decis Proc* 50(2):179–211.
Alig, R., J. Kline, and M. Lichtenstein. 2004. Urbanization on the US Landscape: Looking Ahead in the 21st Century. *Landsc Urban Plan* 69(2–3):219–234.
Angelsen, A. and D. Kaimowitz. 1999. Rethinking the Causes of Deforestation: Lessons from Economic Models. *World Bank Res Obs* 14(1):73–98.
Ayres, E., H. Steltzer, S. Berg, M.D. Wallenstein, B.L. Simmons, and D.H. Wall. 2009. Tree Species Traits Influence Soil Physical, Chemical, and Biological Properties in High Elevation Forests. *PLoS ONE* 4(6):e5964. doi:10.1371/journal.pone.0005964
Barbier, E.B. 1997. The Economic Determinants of Land Degradation in Developing Countries. *Phil Trans R Soc Land* 352:891–899.
Barbier, E.B. 2000. Links Between Economic Liberalizaton and Rural Resource Degradation in the Developing Regions. *Agricultural Economics* 23:299–310.
Bell, M.M. 2004. *Farming for Us All: Practical Agriculture and the Cultivation of Sustainability*. University Park, PA: Pennsylvania State University Press.
Berry, W. 1977. *The Unsettling of America: Culture and Agriculture*. San Francisco: Sierra Club Books.
Beus, C.E. and R.E. Dunlap. 1991. Measuring Adherence to Alternative vs. Conventional Agricultural Paradigms: A Proposed Scale. *Rural Sociol* 56(3):432–460.
Beus, C.E. and R.E. Dunlap. 1994. Agricultural Paradigms and the Practice of Agriculture. *Rural Sociol* 59(4):620–635.
Biggs, S. 1990. A Multiple Source of Innovation Model of Agricultural Research and Technology Promotion. *World Dev* 18(11):1481–1999.
Boix-Fayos, C., A. Calvo-Cases, A.C. Imeson, and M.D. Soriano-Soto. 2001. Influence of soil properties on the aggregation of some Mediterranean soils and the use of aggregate size and stability as land degradation indicators. *Catena* 44:47–67.
Botterweg, P. 1998. Erosion Control under Different Political and Economic Conditions. *Soil Til Res* 46(1–2):31–40.
Burke, I.C., C.M. Yonker, W.J. Parton, C.V. Cole, D.S. Schimel, and K. Flach. 1989. Texture, Climate, and Cultivation Effects on Soil Organic Matter Content in U.S. Grassland Soils. *Soil Sci Soc of America J.* 53(3):800–805.
Damasio, A. 1994. *Descartes' Error: Emotion, Reason, and the Human Brain*. New York: Cambridge University Press.
Dasgupta, P., S. Levin, and J. Lubchenco. 2000. Economic Pathways to Ecological Sustainability. *BioScience* 50(4):339–345.
Di Castri, F. 2000. Ecology in a Context of Economic Globalization. *BioScience* 50(4):321–332.
Duraiappah, A.K. 1998. Poverty and Environmental Degradation: A Review and Analysis of the Nexus. *World Dev* 26(12):2169–2179.
Eckberg, D.L. and T.J. Blocker. 1996. Christianity, Environmentalism, and the Theoretical Problem of Fundamentalism. *J Sci Study Relig* 35(4):343–355.
Eckert, E. 2006. Continuity and Change: Themes of Mental Model Development Among Small Scale Farmers. *Journal of Extension* 44(1).
Eckert, E. and A. Bell. 2005. Invisible Force: Farmers' Mental Models and How They Influence Learning and Actions. *J Ext* 43(3) (June), http://www.joe.org/joe/2005june/a2.php.
Eckert, E. and A. Bell. 2006. Continuity and Change: Themes of Mental Model Development Among Small-Scale Farmers. *J Ext* 44(1) (February), http://www.joe.org/joe/2006february/a2.php.
El-Swaify, S.A. 1997. Factors Affecting Soil Erosion Hazards and Conservation Needs for Tropical Steeplands. *Soil Technol* 11(1):3–16.
Epstein, S. 1994. Integration of the Cognitive and Psychodynamic Unconscious. *Am Psychol* 49(8):709–724.

FAO. 1996. *Food Production: The Critical Role of Water.* Rome: World Food Summit.

Finnoff, D., J. Shogren, B. Leung, and D. Lodge. 2006. Take a Risk: Preferring Prevention over Control of Biological Invaders. *Ecol Econ* 62(2):216–222.

Fu, B., L. Chen, K. Ma, H. Zhou, and J. Wang. 2000. The Relationships Between Land Use and Soil Conditions in the Hilly Area of the Loess Plateau in Northern Shaanxi, China. *CATENA* 39(1):69–78.

Gaby, A.R. 2004. Pesticides: Less is More. *Townsend Letter for Doctors and Patients* 256:25.

Grossman, L.S. 1997. Soil Conservation, Political Ecology, and Technological Change on St. Vincent. *Geogr Rev* 87(3):353–374.

Gudger, W.M. and D.C. Barker. 1993. Banking for the Environment. FAO Agricultural Services Bulletin 103, http://www.fao.org/docrep/T0719E/T0719E00.htm.

Guo, P., C. Wang, Y. Jia, Q. Wang, G. Han, and X. Tian. 2011. Responses of Soil Microbial Biomass and Enzymatic Activities to Fertilizations of Mixed Inorganic and Organic Nitrogen at a Subtropical Forest in East China. *Plant Soil* 338:(1/2):355–366.

Guth, J.L., J.C. Green, L.A. Kellstedt, and C.E. Smidt. 1995. Faith and the Environment: Religious Beliefs and Attitudes on Environmental Policy. *Am J Pol Sci* 39: 364–382.

Guth, J.L., L.A. Kellstedt, C.E. Smidt, and J.C. Green. 1993. Theological Perspectives and Environmentalism among Religious Activists. *J Sci Study Relig* 32:373–382.

Hayes, B.C. and M. Marangudakis. 2000. Religion and Environmental Issues within Anglo-American Democracies. *Rev Relig Res* 42(2):159–174.

Hayes, B.C. and M. Marangudakis. 2001. Religion and Attitudes towards Nature in Britain. *Br J Sociol* 52: 139–155.

Hazell, P. and S. Wood. 2007. Drivers of Change in Global Agriculture. *Biol Sci* 363(1491):495–515.

Heimlich, R. and W. Anderson. 2001. Development at the Urban Fringe and Beyond: Impacts on Agriculture and Rural Land. Agricultural Economic Report No. (AER-803). U.S. Department of Agriculture Economic Research Service. http://www.ers.usda.gov/publications/aer-agricultural-economic-report/aer803.aspx.

Howes, H.S. 2008. The Spectre at the Feast: The Emergence of Salt in Victoria's Irrigated Districts. *Environ Hist* 14:217–239.

Hsee, C.K. 1996. The Evaluability Hypothesis: An Explanation for Preference Reversals between Joint and Separate Evaluations of Alternatives. *Organ Behav Hum Decis Process* 67(3):247–257.

Imhoff, M.L., D. Stutzer, W.T. Lawrence, and C. Elvidge. 2003. Assessing the Impact of Urban Sprawl on Soil Resources in the United States Using Nighttime "City Lights" Satellite Images and Digital Soils Maps. Land Use History of North America. US Geological Survey. http://biology.usgs.gov/luhna/chap3.html (accessed February 22, 2012).

Jie, C., C. Jing-zhang L. Man-zh, and G. Zi-tong. 2002. Soil degradation: A global problem endangering sustainable development. *Geographical Sciences* 12(2):243–252.

Johnson, W.G. and K.D. Gibson. 2006. Glyphosate-Resistant Weeds and Resistance Management Strategies: An Indiana Grower Perspective. *Weed Technol* 20(3):768–772.

Kahneman, D. 2003. A Perspective on Judgment and Choice: Mapping Bounded Rationality. *Am Psychol* 58(9):697–720.

Kleindorfer, P. 1999. Understanding Individual's Environmental Decisions: A Decision Sciences Approach. In *Better Environmental Decisions: Strategies for Government, Businesses, and Communities*, Sexton, K., A.A. Marcus, K. William Easter, and T.D. Burkhardt (eds). pp. 37–56. Washington, DC: Island Press.

Knowler, D.J. 2004. The Economics of Soil Productivity: Local, National and Global Perspectives. *Land Degrad Dev* 15(6):543–561.

Lambin, E.F., B.L. Turner, H.J. Geist, S.B. Agbola, A. Angelsen, J.W. Bruce, O.T. Coomes, et al. 2001. The Causes of Land-Use and Land-Cover Change: Moving Beyond the Myths. *Glob Environ Change* 11:261–269.

Lambin, E.F., H.J. Geist, and E. Lepers. 2003. Dynamics of Land-Use and Land-Cover Change in Tropical Regions. *Annu Rev Environ Resour* 28:205–241.
Liebman, M. and C.L. Mohler. 2001. *Ecological Management of Agricultural Weeds*. Cambridge: Cambridge University Press.
Lipper, L. and D. Osgood. 2004. Two Essays on Socio-economic Aspects of Soil Degradation, FAO Economic and Social Development Paper 149. http://www.fao.org/DOCREP/004/Y1796E/Y1796E00.HTM.
Lyson, T.A. 2004. *Civic Agriculture: Reconnecting Farm, Food, and Community*. Lebanon, NH: Tufts University Press.
Lyson, T. and R. Welsh. 1995. The Production Function, Crop Diversity, and the Debate between Conventional and Sustainable Agriculture. *Rural Sociol* 58(3):424–439.
Mackun, P. and S. Wilson. 2011. Population Distribution and Change: 2000 to 2010. 2010 Census Briefs. U.S. Census Bureau. http://www.census.gov/prod/cen2010/briefs/c2010br-01.pdf.
Majule, A.E. 2003. Impacts of Land Use/Land Cover Changes on Soil Degradation and Biodiversity on the Slopes of Mt. Kilimanjaro, Tanzania. LUCID Project Working Paper Series No. 26. [online HYPERLINK "http://www.lucideastafrica.org/publications/newworkingpapers/Majule%20LUCID%20WP26.pdf" www.lucideastafrica.org/publications/newworkingpapers/Majule%20LUCID%20WP26.pdf]
March, J.G. 1994. *A Primer on Decision Making: How Decisions Happen*. New York: The Free Press.
Marenya, P.P. and C.B. Barrett. 2009. State-conditional Fertilizer Yield Response on Western Kenyan Farms. *Am J Agric Econ* 91(4):991–1006.
Marshall, T.J. and J.W. Holmes. 1979. *Soil Physics*. New York: Cambridge University Press.
Martı'nez-Casasnovas, J.A. and M. Concepción Ramos. 2009. Soil Alteration Due to Erosion, Ploughing and Leveling of Vineyards in North East Spain. *Soil Use Manag* 25:183–192.
Matson, P.A., W.J. Parton, A.G. Power, and M.J. Swift. 1997. Agricultural Intensification and Ecosystem Properties. *Science* 277:504–509.
Maxwell, B. 1992. Weed Thresholds: The Space Component and Considerations for Herbicide Resistance. *Weed Technol* 6(1):205–212.
McMichael, P. 2003. The Impact of Global Economic Practices on American Farming. *Challenges for Rural America in the Twenty-First Century*. University Park: Pennsylvania State University Press.
Meyer, W. and B.L. Turner. 1992. Human Population Growth and Global Land-Use/Cover Change. *Annu Rev Ecol Syst* 23:39–61.
Millennium Ecosystem Assessment. 2005. *Ecosystems and Human Well-Being: Synthesis*. Washington, DC: Island Press.
Mungasinghe, M. and W. Cruz. 1995. Economywide Policies and the Environment: Lessons from Experience. *World Bank Environment Paper Number 10*. The World Bank: Washington, DC.
Mung'ong'o, C.G. 2009. Political Ecology: A Synthesis and Search for Relevance to Today's Ecosystems Conservation and Development. *Afr J Ecol* 47 (Issue Supplement s1):192–197.
Office of Technology Assessment (OTA). 1982. *Impacts of Technology on US Cropland and Rangeland Productivity*. Washington, DC: Government Printing Office.
Ostrom, E. 1990. *Governing the Commons: The Evolution of Institutions for Collective Action*. New York: Cambridge University Press.
Ostrom, E. 2007. A Diagnostic Approach for Going Beyond Panaceas. *Proc Natl Acad Sci* 104(39):15181–15187.
Ostrom, E., J. Burger, C.B. Field, R.B. Norgaard, and D. Policansky. 1999. Revisiting the Commons: Local Lessons, Global Challenges. *Science* 284(5412):278–282.
Paul, E.A., K. Paustian, E.T. Elliott, C.V. Cole (Eds.). 1997. Soil Organic Matter in Temperate Agroecosystems: Long-term Experiments in North America. Boca Raton: CRC Press.
Pearce, D. 1998. Auditing the Earth. *Environment* 40(2):23–28.

Perfecto, I., R.A. Rice, R. Greenburg, and M. Van der Voort. 1996. Shade Coffee: A Disappearing Refuge for Biodiversity. *BioScience* 46(8):598–608.

Peters, E. 2008. The Functions of Affect in the Construction of Preferences. In *The Construction of Preference*, eds. S. Lichtenstein and P. Slovic, pp. 454–463. Cambridge: Cambridge University Press.

Pinstrup-Andersen, P. and R. Pandya-Lorch. 1994. *Alleviating Poverty, Intensifying Agriculture and Effectively Managing Natural Resources*. Washington, DC: International Food Policy Research Institute.

Plous, S. 1993. *The Psychology of Judgment and Decision Making*. Philadelphia: Temple University Press.

Posthumus, H., L.K. Deeks, I. Fenn, and R.J. Rickson. 2011. Soil Conservation in Two English Catchments: Linking Soil Management with Policies. *Land Degradation Dev* 22(1):97–110.

Prokopy, L.S., K. Floress, D. Klotthor-Weinkauf, and A. Bamgart-Getz. 2008. Determinants of Agricultural Best Management Practice Adoption: Evidence from the Literature. *J Soil Water Conserv* 63(5):300–311.

Reardon, T. and C.B. Barrett. 2000. Agroindustrialization, Globalization, and International Development: An Overview of Issues, Patterns, and Determinants. *Agricultural Economics* 23(3):195–205.

Reardon, T., C.B. Barrett, V. Kelly, and K. Savadogo. 2001. Sustainable Versus Unsustainable Agricultural Intensification in Africa: Focus on Policy Reforms and Market Conditions. In *Tradeoffs or Synergies? Agricultural Intensification, Economic Development and the Environment*, eds. David R. Lee and Christopher B. Barrett, pp. 365–382. New York: CABI Publishing.

Rearon, T., K. Stamoulis, A. Balisacan, M.E. Cruz, J. Berdegue, and B. Banks. 1998. "Rural Nonfarm Income in Developing Countries," Special Chapter in *The State of Food and Agriculture 1998, Rome*: Food and Agricultural Organization of the United Nations.

Reed, M.M., A.J. Dougill, and T.R. Baker. 2008. Participatory Indicator Development: What can ecologists and local communities learn from each other? *Ecological Applications* 18(5):1253–1269.

Rosegrant, M.W., C. Ringler, and T. Zhu. 2009. Water for Agriculture: Maintaining Food Security under Growing Scarcity. *Annu Rev Environ Resour* 34:205–222.

Roswell, D.L. 1994. *Soil Sciences: Methods and Applications*. Longman, London. England. pp. 153–173.

Sauer, T. and M. Nelson. 2011. Science, Ethics, and the Historical Roots of Our Ecological Crisis—Was White Right? In *Sustaining Soil Productivity in Response to Global Climate Change-Science, Policy, and Ethics*, eds. Thomas J. Sauer, John M. Norman, and Mannava V. K. Sivakumar, pp. 3–16. West Sussex, UK: Wiley-Blackwell.

Scherr, S.J. and S. Yadav. 1995. *Land Degradation in the Developing World: Implications for Food, Agriculture and the Environment to 2020*. Food Agriculture and the Environment Discussion Paper 14. Washington, DC: International Food Policy Research Institute. http://pdf.usaid.gov/pdf_docs/PNABY622.pdf.

Schultz, P.W., L. Zelezny, and N.J. Dalrymple. 2000. A Multinational Perspective on the Relation between Judeo-Christian Religious Beliefs and Attitudes of Environmental Concern. *Environ Behav* 32(4):576–591.

Sherkat, D.E. and C.G. Ellison. 2007. Structuring the Religion-Environment Connection: Identifying Religious Influences on Environmental Concern and Activism. *J Sci Study Relig* 46(1):71–85.

Simon, H.A. 1959. Theories of Decision Making in Economics and Behavioral Science. *Am Econ Rev* 49(3):253–283.

Slovic, P. 1987. Perception of Risk. *Science* 236(4799):280–285.

Slovic, P. 1999. Trust, Emotion, Sex, Politics, and Science: Surveying the Risk-Assessment Battlefield. *Risk Anal* 19(4):689–701.

Southgate, D. 1990. The Causes of Land Degradation along "Spontaneously" Expanding Agricultural Frontiers in the Third World. *Land Econ* 66(1):93–101.

Spiertz, J.H.J. 2010. Nitrogen, sustainable agriculture and food security. A review. *Agron Sustain Dev* 30(1):43–55.

Stocking, M. 2006. Land Degradation as a Global Environmental Issue. Washington, DC: United Nations Environment Programme, Scientific and Technical Advisory Panel, Global Environmental Facility. http://www.thegef.org/gef/sites/thegef.org/files/documents/C.30.Inf_.8%20STAP_Land%20Degradation%20as%20a%20Global%20Environmental%20Issue.pdf.

Sulieman, H.M. and M.F. Buchroithner. 2009. Degradation and Abandonment of Mechanized Rain-Fed Agricultural Land in the Southern Gadarif Region, Sudan: The Local Farmers' Perception. *Land Degradation Devel* 20(2):199–209.

Swanton, C.J. and S.D. Murphy. 1996. Weed Science Beyond the Weeds: The Role of Integrated Weed Management (IWM) in Agroecosystem Health. *Weed Sci* 44(2):437–445.

Tetlock, P.E., O.V. Kristel, S.B. Elson, M.C. Green, and J.S. Lerner. 2000. The Psychology of the Unthinkable: Taboo Trade-Offs, Forbidden Base Rates, and Heretical Counterfactuals. *J Pers Soc Psychol* 78(5):853–870.

Thampapillai, D.J. and R. Anderson. 1994. A Review of the Socio-Economic Analysis of Soil Degradation Problems for Developed and Developing Countries. *Review of Marketing and Agricultural Economics* 62(3):291–315.

Tilman, D. 1999. Global Environmental Impacts of Agricultural Expansion: The Need for Sustainable and Efficient Practices. *Proc Natl Acad Sci* 96(11):5995–6000.

Tolba, M.K. and O.A. El-Kholy (eds.). 1992. *The World Environment 1972–1992: Two Decades of Challenge*. London: Chapman & Hall on behalf of the United Nations Environment Programme.

Tse-Yan Lee, A. 2005. Vegetable-borne poisoning." *Internet J Tox.* (Jan.)

Turner, B.L. and A.M. Ali. 1996. Induced Intensification: Agricultural Change in Bangladesh with Implications for Malthus and Boserup. *Proc Natl Acad Sci* 93(25):14984–14991.

United States Department of Agriculture Economic Research Service. 2007. 2007 Census of Agriculture. Washington, DC: USDA. http://www.ers.usda.gov/statefacts/us.htm.

Urama, K.C. 2005. Land-Use Intensification and Environmental Degradation: Empirical Evidence from Irrigated and Rain-Fed Farms in South Eastern Nigeria. *J Environ Manage* 75(3):199–217.

US Department of Agriculture. 2009. Summary Report: 2007 National Resources Inventory. Washington, D.C.: Natural Resources Conservation Service and Iowa State University, Ames, Iowa: Center for Survey Statistics and Methodology. http://www.nrcs.usda.gov/technical/NRI/2007/2007_NRI_Summary.pdf.

Vitousek, P.M., H.A. Mooney, J. Lubchenco, and J.M. Melillo. 1997. Human Domination of Earth's Ecosystems. *Science* 277(5325):494–499.

Vogt, K.A., T. Patel-Weynand, M. Shelton, D.J. Vogt, J.C. Gordon, C.T. Mukumoto, A.S. Suntana, and P.A. Roads. 2010. *Sustainability Unpacked: Food, Energy and Water for Resilient Environments and Societies*. London, Washington, DC: Earthscan.

Warren, A., S. Batterbury, and H. Osbahr. 2001. Soil Erosion in the West African Sahel: A Review and an Application of a Local Political Ecology Approach in South West Niger. *Glob Environ Change* 11(1):79–95.

Weiss, B., S. Amler, and R.W. Amler. 2004. Pesticides. *Pediatrics* 113(4):1030–1036.

White, L., Jr. 1967. The Historical Roots of Our Ecological Crisis. *Science* 155(3767):1203–1207.

Wilson, R.S., M.A. Tucker, N.H. Hooker, J.T. LeJeune, and D. Doohan. 2008. Perceptions and Beliefs about Weed Management: Perspectives of Ohio Grain and Produce Farmers. *Weed Technol* 22(2):339–350.

Zwickle, S., R.S. Wilson et al. 2011. Weeds and Organic Weed Management: Investigating Farmer Decision Making with a Mental Models Approach. Master's Thesis, The Ohio State University.

8 Managing Soil Organic Carbon Concentration by Cropping Systems and Fertilizers in the North China Plain

Jin Qing, Xiangbin Kong, and Rattan Lal

CONTENTS

8.1 Introduction 189
8.2 Data and Methods 190
    8.2.1 Study Area 190
    8.2.2 Data Collection 191
    8.2.3 Data Analysis 193
8.3 Results and Discussions 193
    8.3.1 Change in SOC Concentration from 1980 to 1999 193
    8.3.2 Change in SOC Concentration from 1999 to 2006 195
    8.3.3 Change in SOC Concentration as a Result of Changes in the Cropping System 195
    8.3.4 Results of Changes in the Cropping System and the Increased Use of Fertilizer over Time 197
8.4 Conclusions 198
Acknowledgments 199
Abbreviations 199
References 199

8.1 INTRODUCTION

Concentration of soil organic carbon (SOC) under different cropping systems depends mainly on the types of tillage and residue management practices. Haas et al. (1957) reported an average decrease of 42% in the SOC concentration in the 15-cm depth after 30 to 40 years of dry land management at experiment stations in the Great Plains of the United States. Jenny (1933) proposed that SOC concentration would reach a new equilibrium level after 50 years of cultivation. Other studies have reported similar levels of decline in the concentration of SOC in the surface layer as a result of cultivation (Reeder et al. 1998). Some studies have reported that increase in SOC concentration is correlated with the amount of crop residues that have been returned to the topsoil (Hendrix et al. 1998; Peterson et al. 1998; Janzen et al. 2004).

Many studies have been conducted on the effect of different tillage practices on SOC concentration in different regions of China, such as hilly regions (Guo et al. 2001; Li et al. 2006b), the Loess Plateau (Fu et al. 1994, 1999a,b, 2000; Li et al. 2006a; Pang and Huang 2006), and the Karst area (Long et al. 2002). The North China Plain (NCP) is the region with the most intensive agriculture in China, and the effects of change in the cropping system and its impact on the SOC concentration must be studied in relation to quality of soil and the environment. There have also been studies on the effects of intensive farming (Qi et al. 2001) in this region and its impact on the SOC concentration (Kong et al. 2006). However, the change in the SOC concentration with respect to the effects of change in cropping systems in relation to transition of family farming to profit farming are not yet well understood.

China, as the most populous and rapidly developing country, has experienced great economic development, which has strongly shifted the family farm objectives. This transition has led to the increased use of fertilizer and changes in the cropping system. During the 1980s, the goal of the family farm was to meet its own food needs. However, the farmers had no choice but to increase crop yields by converting native land to farmland. The major cropping system in the region is maize (*Zea mays*), wheat (*Triticum aestivum*), and cotton (*Gossypium hirsutum*).

Therefore, the purpose of this study was to analyze the change in SOC concentration in different cropping systems upon transition from family farm to profit maximization with different economic development. The expansion of farms in relation to choice of cropping systems and increased use of fertilizer according to different economic development phases were the major variables (Kong et al. 2004a; Angelsen 1999). Change of SOC concentration in two crop rotation systems, including continuous cultivation of winter wheat–maize from 1980 to 2006, and the conversion from winter wheat–maize (1980–1999) to cotton (1999–2006) were also assessed.

8.2 DATA AND METHODS

8.2.1 STUDY AREA

The experiment was conducted in Quzhou County (36°34′–57′N, 114°13′–50′E), northeast of Handan, Hebei Province, in the center of the NCP. This county is approximately 67,500 ha with 50,900 ha (75%) of cultivated land. The annual rainfall is 535 mm, with 67% received during the summer. The annual average temperature is 13.2°C, and the frost-free period is 201 days. The principal soil types are light, medium, and heavy saline fluvo-aquic inceptisols. The region has saline phreatic water reserves (2–10 g L^{-1}) (Shi 2003), which may exacerbate the risks of salinization.

The region is characterized by intensive use of irrigation and chemical fertilizers, and the predominant cropping system is double cropping of winter wheat followed by summer maize. This county has undergone rapid economic development in recent years, which caused the gross domestic product (GDP) to increase by approximately 15% per year. However, the proportion of agricultural production to total production decreased from 72.3% in 1980 to 34.6% in 2006. At the same time, the per capita income increased from 55 Yuan in 1980 to 2447 Yuan in 1999 and 3160 Yuan in 2006. During this time, fertilizer use in farmlands increased from 26,331 Mg in 1980 to 180,483 Mg in 1999 and to 211,966 Mg in 2006 (SBQC 1980, 1999, 2006).

8.2.2 Data Collection

The land use objectives for farmers in this region have transitioned from maximization of crop production to maximization of profit with economic development (Kong et al. 2003a,b). This led to the conversion of the cropping system and the increased use of fertilizers. Thus, these changes have impacted the SOC concentration in the NCP region.

Questionnaires were distributed and field investigations were conducted among the different farming households because these farmers are the most familiar with the cropping system changes and the increased use of fertilizers between May 1999 and 2006 (Figure 8.1). The major purpose was to obtain a general understanding of the following points: (1) the conversion of the cropping systems; (2) the history of the application rate of fertilizer to soil, including chemical fertilizer and organic matter; (3) and the criteria and technique the farmers use to assess soil quality. The data were collected on the impact of the cropping system change, land policy, population, and income, etc. (Table 8.1).

FIGURE 8.1 Field investigations and soil sampling in Quzhou county.

TABLE 8.1
Results of the 2006 Survey of Farming Households in Quzhou County

Item	Contents
Characteristic of household	Gender; labor structure; income and outcome; land area
	Cropping system
Soil site	Crop pattern, application rate of fertilizer, crop yield
Mindset of household	Reasons for choice of crop pattern; restrictive conditions within the cropping system
	Criteria used to assess soil quality
	Measures adopted to increase SOC concentration

Soil samples were collected at 79 sites in the Quzhou County in May 1980 and again at the same sites in May 1999 before the wheat harvest. At each site, 3–5 kg of sample of 0- to 20-cm depth was collected for laboratory analysis. Sampling sites were identified via Global Positioning System (GPS) at the same location and time of year in 2006. The distribution of sample sites for SOC concentration analysis was based on the Second National Soil Survey of China (SNSS) map (1:50,000), shown in Figure 8.2.

Soil samples obtained in 2006 were then transferred into 20 × 15 cm plastic bags and labeled with the soil sample number, the farmers' names and addresses, and the date of collection. After cleaning off the plant residues, removing stones and other debris, samples were air dried. Then each soil sample was divided into four equal parts, from which two nonadjacent samples were retained and the remaining two were stored. This process was repeated until the two nonadjacent samples were reduced to approximately 0.5 kg. The final soil samples were gently ground and sieved through 0.15-mm sieves prior to C analysis.

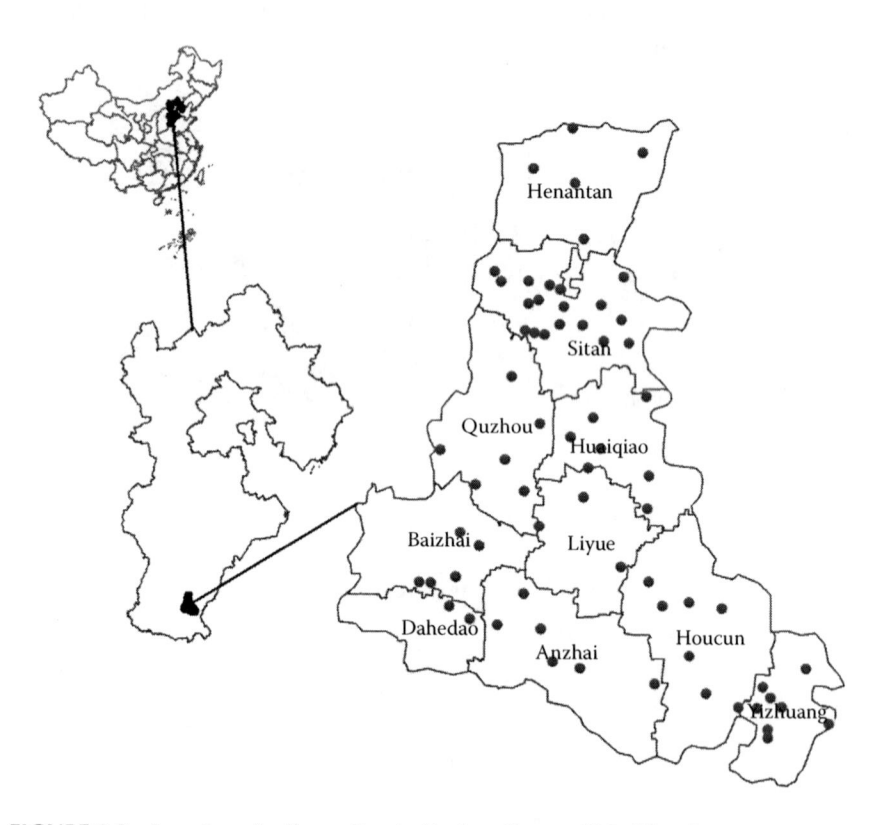

FIGURE 8.2 Location of soil sampling in Quzhou County, Hebei Province.

8.2.3 DATA ANALYSIS

Soil organic matter concentration was analyzed using the rapid dichromate oxidation method (Tiessen and Mioir 1993) and later converted to SOC using a factor of 0.58 (Post et al. 1982). Of these, 67 samples were selected for further analysis after checking all the data.

Changes in SOC concentrations were analyzed using the paired samples t-test using SPSS 11.5 for Windows. First, the results from the entire sampling area obtained in 1980 and 1999 were analyzed via contrast analysis using the year as a variable. Then the results obtained between 1999 and 2006 were similarly tested. Second, the data were analyzed based on the different cropping systems used during 1999 and 2006. Third, the effects of the SOC concentration were evaluated by comparing the data for changes in the cropping system from 1999 to 2006.

8.3 RESULTS AND DISCUSSIONS

8.3.1 CHANGE IN SOC CONCENTRATION FROM 1980 TO 1999

The results show that the SOC concentration was significantly different between samples obtained in 1980 and 1999 (Tables 8.2 and 8.3). The minimum and maximum

TABLE 8.2
Variation of SOC Concentrations in 67 Samples Obtained from Quzhou in 1980, 1999, and 2006

| Year | SOC Concentration (g kg^{-1}) | | | SD | C.V. (%) |
	Minimum	Maximum	Average		
1980	2.84	7.59	5.01	1.82	36
1999	3.39	11.48	6.93	2.7	39
2006	2.73	11.12	6.79	3.06	45

TABLE 8.3
SOC Concentration in Quzhou between 1980–1999 and 1999–2006

| Period of Assessment of SOC | Paired Differences Mean (g kg^{-1}) | Standard Deviation | Standard Error Mean | 95% Confidence Interval of the Difference | | t | df | Signification (Two-Tailed) |
				Low	Upper			
1980–1999	−1.92	1.42	0.17	−2.26	−1.57	−11.01	6	0.00*
1999–2006	0.15	2.29	0.28	−0.41	0.70	0.52	6	0.60**

*$P < 0.01$; **$P > 0.05$.

SOC concentrations were 2.84 and 7.59 g kg^{-1}, respectively, in 1980. The data in Table 8.2 illustrates that the SOC concentration gradually increased by approximately 38% from 1980 to 1999, which is in contrast to other studies (Zeidler et al. 2002), which argue that the SOC concentration is reduced to below the critical levels because of high land use intensity. The coefficient of variation (CV) of the SOC concentration was 36 and 39 in 1980 and 1999, respectively (Tables 8.2 and 8.3). The CV was similar between these two years, although SOC concentration increased from 5.01 to 6.93 g kg^{-1}.

In the 1980s, the application rate of N was 66 kg ha^{-1}, while P and K fertilizers were not in use at the time. Farmers produced crops only for personal consumption under low economic development, so crops were grown by extensive rather than intensive cultivation. Thus, increase in production was achieved by expanding farmland areas and use of natural fertilizers because they had no money to buy chemical fertilizers. Furthermore, the standard of living on a family farm in the 1980s was low due to fuel shortages in rural areas. For this reason, crop residues were the main source of fuel. And only some straw and roots were returned to the soil. Therefore, the SOC concentration in the 1980s was low.

However, family farms improved production by increasing the rate of fertilizer application during the periods of rapid economic development. The application rate of N increased from 66 kg ha^{-1} in 1980 to 241 kg ha^{-1} in 1999, an increase of 260% over <20 years (Table 8.4). At the same time, the crop yield increased from 1538 kg ha^{-1} in 1980 to 5070 kg ha^{-1} in 1999, an increase of 229%. Consequently, the total crop production increased from 87,517 to 324,943 Mg over <20 years. In contrast, crop residues were no longer considered as an important fuel source because of economic development and availability of electricity and natural gas. As a result, the survey data showed that almost 85% of family farms returned crop residues to the soil directly or indirectly.

The increase in crop production and the widespread practice of returning the crop residues to the soil led to buildup of the SOC concentration, which attained equilibrium at a higher level. Therefore, increase in the SOC concentration during the first period can be explained by the input of biomass-C through retention of large amounts of crop residues returned to the soil upon transition of farm household's

TABLE 8.4
Change in Fertilizer Use and Crop Yield in 1980, 1999, and 2006

Year	GDP per Capita (Yuan)	Yield (kg ha^{-1})	N	P	K
1980	167	1538	66	No data	No data
1999	4425	5070	241	101	No data
2006	6842	5295	277	125	40

Application Rate of Fertilizer (kg ha^{-1})

land use objectives. Similar results were reported by Rasmussen et al. (1998) and Nyakatawa et al. (2001).

8.3.2 Change in SOC Concentration from 1999 to 2006

Unlike the variable and increasing SOC concentration levels observed in 1980 and 1999, there was apparently no drastic change in the SOC concentration in soil sampled in 1999 and 2006. The average SOC concentration was 6.93 g kg^{-1} in 1999 and 6.79 g kg^{-1} in 2006. Therefore, SOC concentration slightly decreased by 0.14 g kg^{-1} over this period (Table 8.2). The CV of SOC concentration increased from 39 to 45 during the second period. The SOC concentration was similar to those observed between 1999 and 2006. The results presented here are similar to those reported by Niu et al. (2005).

The results suggest that change in application rate of fertilizers and the cropping system led to a slight decrease in the SOC concentration within the county between 1999 and 2006. However, the total grain yield in 2006 for the wheat–maize cropping system increased by 225 kg ha^{-1} as compared to 1999 (Table 8.4).

The cropping systems changed drastically between 1999 and 2006. The major crops were wheat and maize in 1999 but cotton in 2006 (Table 8.5). However, the percentage of area under wheat and maize decreased by 73.1%, and that of cotton increased by 80.6% between 1999 and 2006. This change may account for the gradual decrease in SOC concentration on the whole.

8.3.3 Change in SOC Concentration as a Result of Changes in the Cropping System

The change in pattern of SOC concentration in the same cropping system from 1999 to 2006 (Tables 8.6 and 8.7) was as follows: (1) SOC concentration in the cropping system of wheat–maize increased by 0.88 g kg^{-1}; (2) SOC concentration increased by 7% in the cropping systems of wheat and maize; (3) SOC concentration decreased by 0.21 g kg^{-1} when the cropping system of wheat–maize was converted to cotton. The

TABLE 8.5
Changes in the Cropping System from 1999 to 2006

Cropping System	Change (%) 1999	Change (%) 2006	Variability (%)
Wheat–maize	88.06	14.93	−83
Cotton	2.99	83.58	2700
Vegetable	7.46	1.49	80
Native grassland	1.49	0	100

Source: Data from 2006 household survey.

TABLE 8.6
Descriptive Statistics of SOC Concentrations in Different Cropping System in 1999 and 2006

		SOC Concentration (g kg^{-1})	
Cropping System	Statistics	1999	2006
Wheat–maize	N	59	10
	Mean	6.84	7.34
	Minimum	3.39	4.84
	Maximum	9.69	10.34
	S.D.	1.37	1.57
	CV (%)	20.0	21.4
Cotton	N	2	56
	Mean	4.07	6.72
	Minimum	3.42	2.73
	Maximum	4.72	11.12
	S.D.	0.91	1.80
	CV (%)	22.0	26.8

TABLE 8.7
Change in SOC Concentrations in Same Cropping System in 1999 and 2006

Cropping System	Year	Sample	SOC Concentration (g kg^{-1})
Wheat–maize	1999	59	6.84
	2006	10	7.34
Cotton	1999	2	4.07
	2006	56	6.72

CV of SOC concentration ranged from 20.0% to 26.8% in different cropping systems (Table 8.6). This change was probably related to conversion of cropping system and fertilizer application rates, since the soils were relatively uniform.

An increase in biomass accumulation was the main reason for the observed increase in the SOC concentration when similar crops were cultivated. Compared to the cotton-based system, more crop residues were returned to soil during the wheat–maize cultivation. The decrease in SOC concentration under the cotton-based system, converted from wheat–maize system, was attributed to low quantity of crop residues. For example, 85% of the family farms returned residues of wheat and maize to the soil, while only 15% of cotton residues were returned to the soil. Similar results were reported by Peterson et al. (1998).

Many studies have reported that stable SOC concentration is generally achieved after several years depending on the crop management system, soil type, climate, and other environmental factors (Hendrix et al. 1998). Lal (2004) hypothesized that land use change and adoption of recommended management practices (RMPs) lead to improvement in SOC concentration, which follows a sigmoid curve and attains the maximum value in 20 to 25 years after adoption of RMPs; moreover, the increase may continue until the SOC attains equilibrium once again. The common RMPs include mulch farming, conservation tillage, diverse cropping systems, cover crops, agroforestry, and integrated nutrient management (Lal 2004). Thus, the incorporation of biomass-C increases the SOC concentration and improves soil quality under intensive agricultural systems.

8.3.4 Results of Changes in the Cropping System and the Increased Use of Fertilizer over Time

The regional cropping systems have been transformed from diverse crops in 1980 to monoculture in 1999, and back to diversity in 2006. In addition, there has been an increase in use of chemical fertilizers over time. There are several factors that influenced these changes.

The first factor is the transition from family farms to maximum profit farms, which triggered the change from subsistence farming to focus on increased crop production from 1980 to 1999 in response to economic development. Thus, these farms increased the rate of fertilizer application. The application rate (kg ha^{-1}) of N was 66, without use of P and K fertilizer in 1980, 241 in 1999, and 277 in 2006 (Table 8.4) for the cropping system of winter wheat and maize, respectively.

When the family's food needs were met, these farmers were motivated to transition to the maximum profit farming. Cotton cultivation yielded more profit than wheat–maize rotation during the second period, which led to the conversion of the cropping system from wheat–maize to cotton. The surveys from farming households show that the labor input for cotton was five times that of wheat and 2.6 times that of maize, respectively. However, the application rate of N was 183 kg ha^{-1} during the cotton cultivation in 2006 (Table 8.8). These data show that the farming households preferred to dedicate much more labor and less materials to meet their goals of profit

TABLE 8.8
Different Application Rates of Fertilizers among Different Cropping Systems in 2006

	Fertilizer Rate (kg ha^{-1})			Number of Irrigation	Human Labor MJ ha^{-1}
	N	P	K		
Wheat	277	206	19	3.6	35
Maize	203	0	0	3	70
Cotton	183	168	100	2.5	182

Source: Data from the 2006 household survey.
Note: Approximate value of 0.8 MJ/working hour or 6–7 MJ/man day (Leach 1976).

maximization. As a result, the cropping systems changed dramatically from a cereal cropping for sustaining the family to the economically profitable cotton cultivation.

Compared to 1999, the per capita GDP increased to 6842 Yuan, an increase of 54% (Table 8.4). From the survey in 2006, the average area of farmland per household was 0.96 ha, with seven persons per household and 6.33 plots per household. This means that each household was faced with a choice of which cropping system would be most profitable given the limited farm area.

Unlike the first period, the survey showed that family farms made decisions on the basis of economics—98% believed that cotton was economically advantageous compared with other crops. Yet, the demand for wheat was still high. For example, 98% of the farms planned to grow wheat for family consumption. These trends indicate that maximizing crop production was no longer the main goal, but rather profit maximization was the principal driver (Kong et al. 2004a). With the change in household's land use objective, the way forward was changing the cropping system and increasing the rate of fertilizer application.

The second factor is the availability of biomass-C in the form of crop residues, animal dung, poultry litter, and manure used in an attempt to increase the SOC concentration and improve soil quality. The survey showed that it is more difficult for family farms to return cotton stalks to soil than to retain the straw of wheat or maize. Thus, only 15% of the cotton residues were returned to the land (Table 8.5). Because of the high cost, only 43% of the farm households applied poultry litter to the land.

The third issue is that the availability of water resources affected the change in the cropping system and application rate of fertilizers. The limited availability of freshwater resources and the high salt content led to an increase in the cultivation of cotton. Soils of the region are highly prone to salinization (Li et al. 2005), which increased demand for irrigation water (Kong et al. 2004a,b, 2006). In particular, cultivation of the monoculture of wheat and maize depleted the groundwater reserves and reduced the amount of water supply. The data of survey showed that 65% of farm households believed that lack of adequate water supply was the major challenge to agricultural intensification, followed by soil quality (13%), and the capital availability (10%).

8.4 CONCLUSIONS

The data presented support the following conclusions:

Economic development led to the availability of biomass-C (crop residues, animal dung, poultry manure), which could be used as soil amendments, and increased SOC concentration. Increasing yields of grain crops was necessitated by the growing demand for food and was achieved by high input of nitrogenous fertilizers.

Conversion of wheat–maize to the cotton-based system was driven by the high profit margin. However, low input of biomass residues in a cotton-based system decreased the SOC concentration.

Factors such as the transition of family farms, economic development, limited availability of water resources, and high salt contents affected the change in rate of fertilizer application and conversion of the cropping system. More crop residues had been returned to the soil since the transition from family farming to commercial farming.

Thus, an increase in the SOC concentration over time is related not only to the adoption of advanced agricultural techniques, but also to the transition of land use objective and the economic development. However, depletion of groundwater and low availability of water are major constraints to increasing in SOC concentration and agricultural intensification in the NCP.

ACKNOWLEDGMENTS

Funding for this research was supported by a Special Fund for National Land Resource-scientific Research from the National Ministry of Land and Resources, China (201011006-3), Natural Science Foundation of China Nos. 70573111 and 70673104.

ABBREVIATIONS

CV: coefficient of variation
GDP: gross domestic product
GPS: global positioning system
RMPs: recommended management practices
SNSS: Second National Soil Survey of China
SOC: soil organic carbon

REFERENCES

Angelsen, A. 1999. Agricultural expansion and deforestation: modeling the impact of population, market forces and property rights. *Journal of Development Economics* 58(1): 185–218.

Fu, B. J., Chen, L. D., Ma, K. M., Zhou, H. F., Wang, J. 2000. The relationship between land use and soil conditions in the hilly area of Loess Plateau in northern Shanxi, China. *CATENA* 39: 69–78.

Fu, B. J., Chen, L. D., Ma, K. M. 1999a. The effect of land use change on the regional environmental in the Yangjuangou catchment in the Loess Plateau of China. *Acta Geographica Sinica* 54(3): 241–246.

Fu, B. J., Gulinck, H., Masum, M. Z. 1994. Loess erosion in relation to land-use changes in Ganspoel catchment, central Belgium. *Land Degradation and Rehabilitation* 5(4): 261–270.

Fu, B. J., Ma, K. M., Zhou, H. F. 1999b. The effect of land use structure on the distribution of soil nutrients in the hilly area of the Loess Plateau, China. *Chinese Science Bulletin* 44(8): 732–736.

Guo, X. D., Fu, B. J., Chen, L. D., Ma, K. M., Li, J. R. 2001. Effects of land use on soil quality in a hilly area—A case study in Zunhua County of Hebei Province. *Acta Geographica Sinica* 56(4): 447–455.

Haas, H. J., Evans, C. E., Miles, E. F. 1957. Nitrogen and Carbon Changes in Great Plains Soils as Influenced by Cropping and Soil Treatments. USDA Tech. US Gov. Print. Office, Washington D.C. pp. 1164.

Hendrix, P. F., Franzluebbers, A. J., McCracken, D. V. 1998. Management effects on C accumulation and loss in soils of the southern Appalachian Piedmont of Georgia. *Soil and Tillage Research* 47: 245–251.

Janzen, H. H. 2004. Carbon cycling in earth systems—A soil science perspective. *Agriculture, Ecosystems and Environment* 104: 399–417.

Jenny, H. 1933. Soil fertility losses under Missouri conditions. *Missouri Agricultural Experiment Station Bulletin* 324: 285.

Kong, X. B., Zhang, F. R., Qi, W. 2004a. The driving forces of land use in intensive agricultural region based on the behavior of household—A case study in Quzhou County in Hebei Province (Chinese). *Progress in Geography* 23(3): 50–57.

Kong, X. B., Zhang, F. R., Qi, W., Xu, Y. 2003a. The influence of land use change on soil fertility intensive agricultural region: A case study of Quzhou County, Hebei. *Acta Geographica Sinica* 58(3): 333–341.

Kong, X. B., Zhang, F. R., Qi, W. 2004b. The effects of land use change on water resource in intensive agricultural region—A case study of Quzhou county in Hebei Province (Chinese). *Journal of Natural Resources* 19(8): 747–753.

Kong, X. B., Zhang, F. R., Qi, W. 2006. Influence of land use change on soil nutrients in an intensive agricultural region of North China. *Soil and Tillage Research* 88: 85–94.

Kong, X. B., Zhang, F. R., Xu, Y., Qi, W. 2003b. Arable-land change and driving force in Intensive Agricultural Region (Chinese). *Natural Science* 25(3): 57–63.

Lal, R. 2004. Soil carbon sequestration impacts on global climate change and food security. *Science* 304: 1623–1627.

Leach, G. 1976. *Energy and Food Production*. London: IPC Business Press Ltd.

Li, H. Y., Jia, G. M., Fang, X. W., Wang, G. 2006a. Dynamics of soil microbial biomass and organic matter under bare fallow and wheat growth conditions (Chinese). *Journal of Lanzhou University (Natural Sciences)* 42(4): 34–36.

Li, X. B., Hao, J. M., Ding, Z. Y., Li, Y., Li, P. 2005. Effect of land use change on ground water resource in salinity transforming region—A case study of Quzhou County in Hebei Province (Chinese). *Journal of Soil and Water Conservation* 19(5): 152–200.

Li, Z. W., Zhang, G. M., Zh, Q., Yuan, X. Y., Fang, Y., Yang, X. C. 2006b. Analysis of the spatial heterogeneity of the soil organic matter in hilly red soil region—A case study in Changsha City (Chinese). *Journal of Huan University (Natural Sciences)* 33(4): 102–105.

Long, J., Huang, C. Y., Li, J. 2002. Effects of land use on soil quality in Karst hilly area (Chinese). *Journal of Soil and Water Conservation* 16(1): 76–79.

Niu, L. A., Hao, J. M., Tan, L., Li, S. H., Shi, P. F. 2005. Spatio-temporal variability of soil nutrients in salt-affected soil under amelioration. *Acta Pedologica Sinica* 42(1): 84–90.

Nyakatawa, E. Z., Reddya, K. C., Sistani, K. R. 2001. Tillage, cover cropping, and poultry litter effects on selected soil chemical properties. *Soil and Tillage Research* 58: 69–79.

Pang, L., Huang, G. B. 2006. Impact of different tillage method on changing of soil carbon in semi-arid area (Chinese). *Journal of Soil and Water Conservation* 20(3): 110–113.

Peterson, G. A., Halvorson, A. D., Havlin, J. L., Jones, O. R., Lyon, D. J., Tanaka, D. L. 1998. Reduced tillage and increasing cropping intensity in the Great Plains conserves soil C. *Soil and Tillage Research* 47: 207–218.

Post, W. M., Emanuel, W. R., Zinke, P. J., Stangenberger, A. G. 1982. Soil carbon pools and world life zones. *Nature* 298: 156–159.

Qi, W., Zhang, F. R., DongYe, G. L. 2001. Study on land quality indicators for assessing stainable land management (Chinese). *Journal of Shandong Agricultural University* 16(3): 205–210.

Rasmussen, P. E., Albrecht, S. L., Smiley, R. W. 1998. Soil C and N changes under tillage and cropping systems in semi-arid Pacific Northwest agriculture. *Soil and Tillage Research* 47: 197–205.

Reeder, J. D., Schuman, G. E., Bowman, R. A. 1998. Soil C and N changes on conservation reserve program lands in the Central Great Plains. *Soil and Tillage Research* 47: 207–218.

Shi, Y. C. 2003. Comprehensive reclamation of salt-affected soils in China's Huang-Huai-Hai Plain. *Journal of Crop Production* 7(1–2): 163–179.

Statistics Bureau of Quzhou County (SBQC), 1980. 1999. 2006. Annual social and economical statistics for Quzhou County. Quzhou County, Hebei Province, People's Republic of China.

Tiessen, H., Mioir, J. O. 1993. Total and organic carbon. In: *Soil Sampling and Methods of Analysis*, Ed, M. R. Carter, 187–199. Boca Raton: Lewis Publishers.

Zeidler, J., Hanrahan, S., Scholes, M. 2002. Land-use intensity affects range condition in arid to semi-arid Namibia. *Journal of Arid Environments* 52: 389–403.

9 Global Extent of Land Degradation and Its Human Dimension

Ephraim Nkonya, Joachim von Braun, Jawoo Koo, and Zhe Guo

CONTENTS

9.1 Introduction 203
9.2 Analytical Methods and Data 205
    9.2.1 Impact of Land Degradation on Food Security 205
    9.2.2 Relationship between Land Degradation and Its Major Drivers 207
9.3 Extent of Land Degradation and Its Relationship with Poverty 209
    9.3.1 Extent of Land Degradation 209
    9.3.2 How Is Land Degradation/Improvement Related to Poverty at Global Level? 209
    9.3.3 Relationship of Poverty and Land Degradation at Household Level—Case Studies 213
9.4 Drivers of Land Degradation 214
    9.4.1 Agricultural Intensification 215
    9.4.2 Population Density 215
    9.4.3 Economic Growth (GDP) 215
    9.4.4 Government Effectiveness 216
9.5 Impact of Land Degradation on Food Security 218
    9.5.1 Relationship of Land Degradation and Crop Yield Gap 218
    9.5.2 Impact of Land Degradation on Maize and Rice Yield: The Case of Mali and Nigeria 219
9.6 Conclusions and Policy Implications 221
Abbreviations 222
References 222

9.1 INTRODUCTION

Since the poor depend heavily on natural resources, land degradation has a direct impact on their livelihoods. For example, 29% of national wealth in low income countries is based on natural resources compared to only 2% in high income countries (World Bank 2008).

Land degradation—defined as the reduction of the capacity of land to provide ecosystem goods and services over time—reduces both the agricultural land area and land productivity. About 24% of the global land area was affected by land degradation between 1981 and 2003 (Bai et al. 2008). This is equivalent to a degradation of about 1% of global land area each year or about 130 Mha. The annual degradation on approximately 1–2.9 Mha is so severe that it is unsuitable for crop production (Lambin and Meifrodt 2011). The annual land area lost due to severe degradation is equivalent to 0.1% of the 2005 to 2009 average global cropland of 1.527 billion ha (FAOSTAT 2009). Globally, 1.5 billion people live on degraded lands (UNCCD 2011). It is also estimated that 42% of the very poor and 32% of the moderately poor live on degraded lands (Nachtergaele et al. 2010). Recent estimates also show that land degradation accounted for about 25% increase in the number of people with hunger in sub-Saharan Africa (SSA) between 2000 and 2010 (Global Mechanism 2011).

Attempts have been made to estimate the global cost of land degradation. Even though such estimates remain contested, they reveal the large cost of land degradation. Land degradation caused a total loss of 9.56×10^8 tons of carbon between 1981 and 2003, which amounts to $48 billion in terms of lost carbon fixation (Nkonya et al. 2011a). Other studies have mainly focused their estimation on on-farm cost of soil erosion (Requier-Desjarding et al. 2011). Such studies have reported the cost of land degradation to be less than 1% up to about 10% of the agricultural gross domestic product (GDP) for various countries worldwide (Nkonya et al. 2011a). Off-site costs of land degradation are large. For example Basson (2010) estimated that the annual global cost of siltation of water reservoirs is about $18.5 billion. Land degradation and land use change also contribute about 20% of greenhouse gas emissions (IPPC 2007).

Global investment to address land degradation has been limited. Allocation of funds to the three key Rio summit conventions reflects the limited investment in addressing land degradation. Only US$5.9 million was allocated in 2010 to the United Nations Convention to Combat Desertification (UNCCD)—the Rio convention charged with addressing land degradation (Ivanova and Delina 2012). This is only 5% of the US$107.9 million allocated to the UN Framework Convention on Climate Change (UNFCCC) and 50% of the US$12.36 million allocated to the UN Convention on Biological Diversity (CBD) during the same period (Ivanova 2012). Such investment patterns across the three conventions is comparable to the Rio summit convention funding pattern since 1998 (OECD 2011). In SSA countries, where land degradation is the most severe (Bai et al. 2008), the average budget allocation to agriculture—which in most cases includes other land-based sectors like forest and the environment—has remained at about 5% of the total government budget since 1990 (Fan and Breisinger 2011). This allocation is only 50% of the 10% of Africa's target expenditure allocation set through the Maputo Declaration—which has been ratified by all countries in SSA (Benin 2010). However, only six countries have reached or surpassed the Maputo Declaration target of allocating 10% of government budget to agriculture, which include Burkina Faso, Guinea, Mali, Niger, and Senegal, Ethiopia (Benin et al. 2010). The limited investment in land-based sectors in SSA implies that investment in prevention of land degradation or rehabilitation of degraded lands is even more limited.

At the international community, there have been increased efforts recently to address land degradation. The recent Rio +20 summit called for achieving zero land degradation (UNCCD 2012). As part of efforts contribute to providing empirical evidence for achieving zero net land degradation, this study analyzes land degradation and its impact on human welfare. The study also discusses actions, which should be taken to prevent land degradation and/or rehabilitate degraded lands. Given that land degradation has a large impact on food security, the paper examines agricultural productivity across regions and actions that could be taken to enhance food security in developing countries experiencing hunger.

9.2 ANALYTICAL METHODS AND DATA

Consistent with other global studies on land degradation and improvement, we use land cover data captured by satellite (Vitousek 1994; Morawitz et al. 2006). We note its weaknesses as a degradation proxy (e.g., see Vlek et al. 2010). For example, river basin scale studies in West Africa have shown that the biomass productivity [normalized difference vegetation index (NDVI)]-based land degradation assessment may underestimate the extent to which soil is being lost, unless a correction is included to account for atmospheric fertilization (Le et al. 2012). Use of land cover change as an indicator of land degradation or improvement is increasing since it is easy to capture and that its cost continues to fall. We use change in the NDVI to represent land degradation or improvement. We overlay the global land cover change with poverty in order to understand the relationship between land degradation and poverty and draw conclusions on their patterns across major regions. We then analyze the impact of land degradation on food security. Finally, we examine the drivers of land degradation in order to draw policy and strategies for taking action against land degradation. Details of the analysis of the impact of land degradation on crop productivity and drivers of land degradation are given below.

9.2.1 IMPACT OF LAND DEGRADATION ON FOOD SECURITY

We examine the impact of land degradation on yield of maize and rice, which are major crops in the world. We first examine the yield gap—that is, the difference between potential yield and actual yield across major regions of the world— and examine its relationship with land degradation. We then examine the impact of unsustainable soil nutrient management on the yield of maize and rice in Mali and Nigeria as case study countries. Selection of the two countries was based on data availability. Previous work by Nkonya et al. (2011c,d) used a well-calibrated crop simulation model to provide 30-year data showing the impact of crop residue, manure, and nitrogenous fertilizer applications on maize and rice yield. Nkonya et al. (2011c,d) used the Decision Support System for Agrotechnology Transfer (DSSAT)-CENTURY (Gijsman et al. 2002) crop simulation model, which is more suitable for simulating low-input cropping systems in SSA. The DSSAT-CENTURY model was calibrated using soil, water, and weather data collected from the countries included in the case study.* For brevity, we use only two land management treatments. The first represents a land degrading practice commonly used in the countries included in the case study, and the second represents fairly sustainable land management practices deemed feasible in both countries. Details of the two land management practices are given below:

(1) *Land degrading practice.* No chemical fertilizer or manure application and 50% harvesting of crop residues. This is a practice that reflects land management practices of most farmers in SSA—where it is estimated that only 10 kg/ha of nutrients [nitrogen, phosphorus, and potassium (NPK)] is used (FAOSTAT 2009). Crop residue harvesting and in situ feeding by livestock are also common practice in SSA.

(2) *Integrated soil fertility management (ISFM) practices.* The use of organic inputs, improved crop varieties, and judicious amounts of chemical fertilizers (Vanlauwe and Giller 2006; Tittonel et al. 2008). We use 40 kg of nitrogen per hectare of chemical fertilizer, 1.7 tons of manure/ha, and 50% of crop residues as the treatment representing ISFM. The recommended fertilizer rate for maize is about 80 kg N/ha for subhumid areas in West Africa (Aduayi et al. 2002). We use the lower rate of fertilizer application to reflect the benefit of ISFM, which typically reduces the chemical fertilizer use requirement (Sharma and Biswas 2004; Vanlauwe and Giller 2006; Roy et al. 2006), and to reflect the low rates used by farmers in the SSA region. Akinnifesi et al. (2010) also observed that agroforestry that uses leguminous trees planted on maize plots in Eastern and Southern Africa reduced the requirement of nitrogen from chemical fertilizers by up to 75%.

Controlling for other drivers of crop yields and varying only the two soil fertility management practices listed above, it is possible to estimate the impact of land degradation on crop yield. We show the trend of yields using a linear trend model to give a rough estimate of the yield loss per year. Linear trend models are commonly used in estimating crop yield trends (e.g., see Bruinsma 2009; Moss and Shonkwiler 1993). However, the model uses normality assumption. The Shapiro–Wilk normality test (Shapiro and Wilk 1968, 1972) showed that maize and rice yields in each country were normally distributed—something that was expected given that the data used are simulated:

$$Y_t = \beta_0 + \beta_1 t + \varepsilon_t \qquad (9.1)$$

where Y_t = crop yield in year t, and β_1 = slope of yield trend depicting the annual change of crop yield. A negative sign will show declining yield and an indicator of impact of land degradation. It should be noted that this model only presents land

---

* Details of the calibration and other simulation procedures are available from the authors upon request.

degradation in terms of the provisioning services (crop yield) and does not include impact of land degradation on other ecological services.

9.2.2 RELATIONSHIP BETWEEN LAND DEGRADATION AND ITS MAJOR DRIVERS

To better understand the relationship between land degradation and its drivers, we use the first difference econometric method.* The drivers of land degradation are divided into two major groups: the proximate causes, that is, the primary or direct causes of land degradation, and the underlying causes, which affect land degradation indirectly. Using change of the NDVI as an indicator of land degradation or improvement (Bai et al. 2008), the following theoretical first difference model represents the key proximate and underlying causes of land degradation, which can change over time:

$$\Delta \text{NDVI} = \beta_0 + \beta_1 \Delta \text{precip} + \beta_2 \Delta \text{AgI} + \beta_3 \Delta \text{pop} + \beta_4 \Delta \text{Pov} + \beta_5 \Delta \text{govt}$$
$$+ \beta_6 \Delta \text{road} + \beta_7 \Delta \text{GDP} + \beta_8 \Delta \text{GDP}^2 + \beta_9 \Delta \text{rurs} + \beta_{10} \Delta \text{tenure} + \varepsilon_i \quad (9.2)$$

where Δ = first difference, $\Delta = X_{t=2} - X_{t=1}$, where t_1 = baseline period and t_2 = end line period; pop = population density; precip = precipitation; pov = poverty; AgI = agricultural intensification; road = access to roads; GDP = gross domestic product; govt = government effectiveness; rurs = rural services (e.g., agricultural extension services); tenure = land tenure; and ε_i = normally distributed random error. Of the causes of land degradation discussed in Nkonya et al. (2011a), topography is not included because it does not change over time and therefore cannot be included in the difference model.

However, due to data availability, not all causes of land degradation that change over time will be included. Global level poverty, access to road networks, and other socioeconomic underlying causes of land degradation are either not available as time series or not available at all. Poverty and road density are available in only one period and will not be included in the empirical model. The variables included in the theoretical model are given in Table 9.1. Two proximate causes of land degradation, namely, precipitation and agricultural intensification, are included in the model. Precipitation represents the biophysical characteristics, while agricultural intensification represents land management on agricultural land. Three underlying causes, namely, government effectiveness, GDP, and population density, are used in the model. Government effectiveness—government's capacity to implement policies with independence from political pressures and with respect to the rule of law (Kaufmann et al. 2010)—will represent the institutions, which play a key role in land management. Government effectiveness indicator ranges from –2.5 for weak government effectiveness to 2.5 for strong government effectiveness (Kaufmann et al. 2010). The relationship between change in GDP and the NDVI is presented using a quadratic relationship between economic growth and the environment. The environmental Kuznet curve (EKC) (Kuznet 1955; Grossman and Krueger 1991; Dinda

---

* For a detailed discussion of the drivers of land degradation, see Nkonya et al. (2011a).

TABLE 9.1
NDVI and Its Key Drivers Change of Agricultural Area (FAOSTAT)

	Resolution	Baseline	End Line	Source
NDVI	8 × 8 km	1982–1986	2002–2006	Global Land Cover Facility (www.landcover.org), Tucker et al. (2004); NOAA AVHRR NDVI data
Proximate Causes of Land Degradation				
Precipitation	0.54° × 054°	1981–1984	2003–2006	Climate Research Unit (CRU), University of East Anglia; www.cru.uea.ac.uk/cru/data/precip/
Agricultural intensification	Country	1990–1992	2007–2009	FAOSTAT: http://faostat.fao.org/site/
Underlying Causes of Land Degradation				
Population density	0.5° × 0.5°	1990	2005	CIESIN: http://sedac.ciesin.columbia.edu/gpw/
GDP	Country	1982–1986	2002–2006	IMF: www.imf.org/external/pubs/ft/weo/2010/02/
Government effectiveness	Country	1996–1998	2007–2009	Worldwide Governance Indicators: http://info.worldbank.org/governance/wgi/index.asp
Poverty–Infant Mortality Rate (per 1000 live births)	Subnational administrative divisions	2005	2005	CIESIN: http://sedac.ciesin.columbia.edu/povmap/

Sources: Benin, S. et al. *Monitoring African Agricultural Development Processes and Performance: A Comparative Analysis.* International Food Policy Research Institute (IFPRI), Washington, DC, 2010. World Bank, Country data on agriculture, 2011. Available at http://data.worldbank.org/indicator/NV.AGR.TOTL.ZS, accessed July 7, 2012.

Notes: Agricultural expenditure as percent of allocated government budget and agriculture contribution to GDP.

2004) hypothesizes that environmental degradation first increases as the economy grows but later reaches a plateau and then decreases. The EKC pattern is driven by a number of factors including industrialization and consequent reduced dependence on natural resources, improved agricultural technologies that reduced conversion of forest land to agriculture, rural–urban migration, improved environmental perception, and stronger regulatory institutions (Kuznet 1955; Grossman and Krueger 1991; Dinda 2004). The GDP also represents national level policies, which has an indirect impact on land degradation or improvement. For example, Graham and Wada (2001) show that China's policy—which allowed foreign direct investment—was one of the major drivers of the country's rapid economic growth. This and other policy reforms led to industrialization, which in turn contributed to rapid urbanization and reduced dependence on the agricultural sector (Hazell and Wood 2008).*

To address outliers in the first difference model, we use an average of four consecutive years for the baseline and end line periods. However, not all data were available for the 4 years of the baseline and end line periods. In such cases, we used time periods closest to the two NDVI time periods.

9.3 EXTENT OF LAND DEGRADATION AND ITS RELATIONSHIP WITH POVERTY

9.3.1 Extent of Land Degradation

In accordance with the focus of this paper, we first examine the extent and severity of land degradation or improvement. We then discuss the relationship between poverty and the NDVI. A recent study that used change of the NDVI in 1981 to 2003 as an indicator of land degradation or improvement showed that the humid areas accounted for 78% of the global degraded land area while arid and semiarid areas accounted for only 13% (Figure 9.1). This pattern is contrary to the current global focus on combating desertification—a form of land degradation in the arid and semiarid areas—upon which the United Nations to Combat Desertification (UNCCD) derives its name. Accordingly, a study by Bai (2008) found a negative correlation between aridity and land degradation.

9.3.2 How Is Land Degradation/Improvement Related to Poverty at Global Level?

Geographically, the regions that experienced the most severe land degradation are Africa south of the equator, which accounted for 13% of the global degraded land area, some parts of China, Indonesia, Myanmar, Australia, and South American plains (the Pampas) (Figure 9.2). Excluding Australia, this pattern shows the strong correlation between land degradation and poverty. Using infant mortality rate (IMR)

---

* China's rural population as share of total population fell from 82% between 1961 and 1970 to 58% between 2000 and 2009, but during the same period, agricultural area per capita of rural population increased by 14% from about 0.6 to 0.7 ha per capita (FAOSTAT 2011).

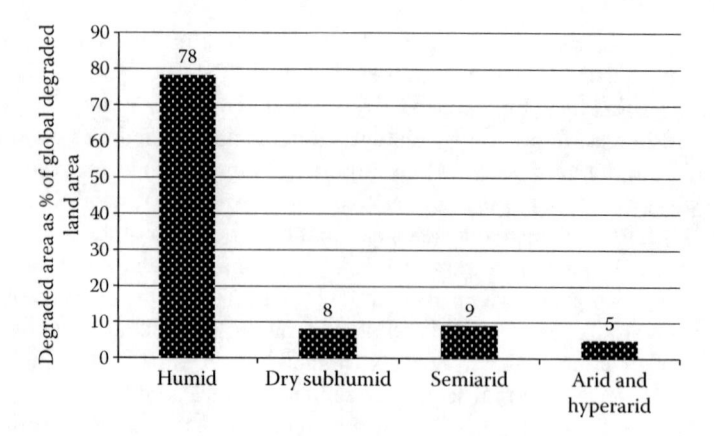

FIGURE 9.1 Degraded area across agroclimatic zones as share of total global degraded area, 1981–2003. (Calculated from Bai, Z.G. et al., *Global assessment of land degradation and improvement 1: identification by remote sensing. Report 2008/01.* Rome/Wageningen: FAO/ISRIC, 2008.)

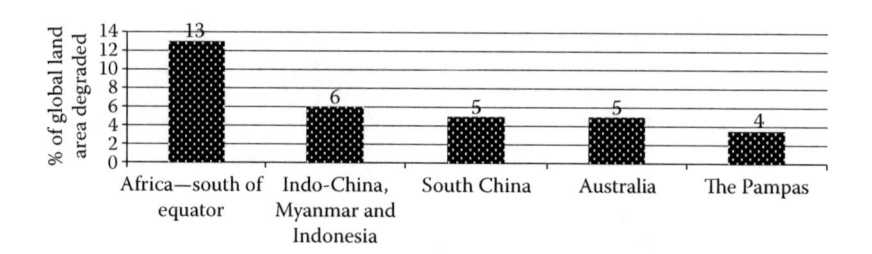

FIGURE 9.2 Degraded land area, 1981–2003. The severely degraded regions account for 27% of the total area degraded. The rest of the degraded area is located in areas with less severe land degradation. (Calculated from Bai, Z.G. et al., *Global assessment of land degradation and improvement 1: identification by remote sensing. Report 2008/01.* Rome/Wageningen: FAO/ISRIC, 2008; UNICEF, http://www.childinfo.org/mortality_ufmrcountrydata.php.)

as an indicator of poverty, Bai et al. (2008) observed a strong correlation between poverty and land degradation.

However, disaggregated analysis shows an interesting picture. Consistent with the work of Bai et al. (2008), North America, Europe, and Australia show low poverty and increase in the NDVI, while Africa south of the equator shows high poverty and decrease in the NDVI (Figure 9.3). Contrary to the work of Bai et al. (2008), the NDVI increased in most western and Central African countries north of the equator and south of the Sahelian region. Likewise, the NDVI increased in much of northern and western India, Pakistan, and Papua New Guinea while poverty levels were high. This shows that other factors mediate the impact of poverty on the NDVI. We briefly

Global Extent of Land Degradation and Its Human Dimension 211

FIGURE 9.3 Relationship between poverty and NDVI change.

discuss this interesting pattern below and focus on western Africa, where there is high poverty and the NDVI has increased.

Of interest is the change in government effectiveness in West and Central Africa. Table 9.2 shows that in 10 of the 16 countries where the NDVI increased, government effectiveness increased. In two countries (Equatorial Guinea and Guinea), where government effectiveness decreased, agricultural area as share of total land area decreased, and accordingly its contribution to GDP decreased. This suggests reduced dependency of the population on agriculture—especially in Equatorial Guinea where the GDP almost doubled and the contribution of agriculture to GDP fell by 94%. The dramatic growth was due to the oil boom in the country (Frynas 2004). The reduced agricultural area and its contribution to GDP could have contributed to

TABLE 9.2
Change of Government Effectiveness, GDP, and Agricultural Area in West and Central African Countries Where NDVI Increased

	Government Effectiveness Change (Δgovt)[a]	GDP	Change of Agricultural Area	Change of Contribution of ag to GDP	Agric. Expenditure as % of Total Allocated Government Budget[b]
		Percent Change (%)			
Benin	0.3	117.8	65	−2.4	6.0
Burkina Faso	0.0	111.0	22	16.6	19.2
Cameroon	0.3	61.0	1	−23.1	4.0
Central African Republic	0.001	28.4	5	37.1	3.1
Congo	−0.3	54.3	0	−32.9	1.0
Equatorial Guinea	−0.4	825.2	−4	−94.4	NA
Gabon	0.001	47.1	0	−13.3	NA
Gambia	−0.2	88.1	8	−28.1	5.0
Ghana	0.2	121.1	23	−31.0	8.7
Guinea	−0.2	83.8	−1	−0.8	13.7
Guinea Bissau	0.4	19.3	14	21.8	1.4
Liberia	−0.3	59.3	0	NA	5.1
Nigeria	0.1	95.5	7	NA	3.6
Senegal	0.4	93.7	−1	−22.7	12.1
Sierra Leone	−1.0	21.9	29	32.7	2.8
Togo	0.6	49.7	13	19.7	4.7

Note: NA = data not available in the baseline or end line period.
[a] Government effectiveness ranges from −2.5 (very weak) to 2.5 (very strong).
[b] Average of 2003–2009.

a higher NDVI since agriculture remains the primary driver of deforestation in the SSA region. In Congo and Liberia, government effectiveness fell, but the agricultural area remained unchanged. Additionally, agricultural contribution to GDP in Congo decreased suggesting diversification to nonagricultural sectors.

Changes in Sierra Leone are puzzling and require additional information for them to be understood. While government effectiveness fell—largely as a result of the civil war in the early 2000s—both the area and contribution of agriculture to GDP increased by almost one-third. Additional data are required to fully understand this puzzling pattern. Of interest also are, the trends and patterns in Burkina Faso. Government effectiveness, agricultural area, and its contribution to GDP increased. Burkina Faso's agricultural total factor productivity (TFP) index (1961 = 100) was 125 in 2001 and 109 in 2008 suggesting increased productivity (Fuglie and Rada 2011). The increased TFP could have been due to the higher budgetary allocation to agriculture, which was the highest in the countries considered and almost twice the Maputo declaration target of allocating 10% of government budget to agriculture. This suggests that agricultural investment and higher TFP could have contributed to a higher NDVI—in addition to the role played by improved government effectiveness and other factors.

Land improvement in Southern Niger (Figure 9.3) illustrates the potential role played by institutions in land improvement. Government effectiveness increased and one of the factors that could have directly contributed to land improvement is the statue (*rural code*) that gave landowners tenure security of any tree that they plant or protect (Larwanou et al. 2006). It is estimated that at least 3 million ha of land has been rehabilitated through tree protection, which allowed for natural regeneration (Adam et al. 2006). However, the rural code was not the only deciding factor that led to this remarkable success. The prolonged drought that spanned the 1970s and 1980s led to loss of trees, hiking the price of tree products. This provided strong incentive to farmers to plant and protect trees. Planted forest area as a share of total forest area in Niger was 12% in 2010 and was among the highest in SSA (FAO 2010). This achievement was a result of a combination of efforts by local communities, change in government policies and statutes, support from NGOs and religious organizations, and environmental stress, which prompted communities for a solution.

Overall, the results show the potential contribution of government effectiveness, agricultural intensification, and diversification to other sectors to land improvement.

9.3.3 Relationship of Poverty and Land Degradation at Household Level—Case Studies

Some studies have been done at the household level to determine the relationship between poverty and land degradation. A study in Malawi showed that 10% deforestation reduced rural household consumption by 0.1% (Bandyopadhyay 2006), and a similar study in India found comparable results (Baland et al. 2006). The impact of deforestation on consumption appears small, but the overall impact is greater since the studies only considered consumption and ignored other forest ecological services such as loss of medicinal plants and timber, regulating services, and supporting services.

9.4 DRIVERS OF LAND DEGRADATION

The discussion below is based on the model that simultaneously controls all major drivers. Hence, the results derive stronger correlation between land degradation/improvement and its drivers than the simple correlations discussed above using maps and descriptive statistics. As emphasized above, however, these results should not be interpreted as causality but rather as only correlations between the land degradation/improvement and its drivers. The focus of the discussion is on the direction of relationship and their statistical significance rather than the magnitude of the coefficients. Hence, Table 9.3 reports only the direction of the relationship (positive or negative) and the statistical significance of the coefficients. We focus on the underlying drivers of land degradation from which we can draw policy implications for actions against land degradation. We also discuss agricul-

TABLE 9.3
Regression Analysis on Association of Change in NDVI and Major Drivers of Land Degradation, 1981–2006

Variable	Global A	Global S	East Asia A	East Asia S	European Union A	European Union S	LAC A	LAC S	MENA A	MENA S	South Asia A	South Asia S	SSA A	SSA S
Δ Population density	+	NS	+	*	−	*	−	NS	+	*	−	**	−	***
Δ Precipitation	+	*	+	NS	+	**	+	*	+	*	+	*	+	***
Δ Agricultural intensification	+	*	+	NS	+	***	+	*	−	*	+	*	+	NS
Δ GDP	−	*	+	***	−	**	−	*	−	*	+	*	+	***
Δ GDP^2	+	*	−	**	+	NS	+	*	+	*	−	NS	−	***
Δ Government effectiveness	+	*	−	NS	+	*	+	*	+	*	−		+	***

Source: Modified from Nkonya, E. et al. The Economics of Land Degradation. Toward An Integrated Global Assessment, Peter Lang, Berlin, 2011.

Notes: + and − respectively mean positive and negative relationship of change of the NDVI and driver of land degradation. A = association of change of NDVI with associated variable; S = statistical significance of association: NS = not significant (at $p = 0.10$) and *,**,***, respectively, mean significant at 0.10, 0.05, and 0.01. LAC = Latin American countries; MENA = Middle East and North Africa; SSA = sub-Saharan Africa; Δ = change end line period − baseline period. Since GDP and agricultural intensification (fertilizer application per hectare) were at a country level, it was not possible to run regional level regression in the Oceania and North America regions—both of which have few countries.

tural intensification since its actions could directly lead to land degradation or improvement.

9.4.1 AGRICULTURAL INTENSIFICATION

Agricultural intensification—quantity of fertilizer applied per hectare—is positively associated with the NDVI at a global level. This is not surprising since agricultural intensification increases yield, thereby reducing the need for agricultural expansion. Accordingly, Bai et al. (2008) also observed that cropland and rangelands, respectively, accounted for 18% and 43% of the 16% of global land area where the NDVI increased. In SSA where fertilizer application is the lowest, agricultural intensification did not have a significant effect on the NDVI. In the Middle East and North Africa (MENA), agricultural intensification was negatively correlated with the NDVI. This result could be due to the fact that fertilizer is not a big driver of the NDVI in the hyperarid region—where water is the driver of the NDVI.

9.4.2 POPULATION DENSITY

At a global level, there was positive but nonsignificant relationship between change in population density and the NDVI (Table 9.3). This is consistent with Bai (2008) who observed a positive correlation between global level population density and the NDVI. The same pattern is observed in East Asia and MENA regions. These results demonstrate that it is possible to achieve sustainable land management even in areas with high population density. This has been achieved in China and India—the first and second most populous countries in the world. As Tiffen et al. (1994) demonstrated in their work in Kenya—"more people less erosion"—this is achievable even in low income countries, but certain conditions are required. Boyd and Slaymaker (2000) show that access to markets contributed to the "more people less erosion" case in Kenya. In South Asia and SSA, where intensity of poverty is high, change in population density is negatively related with change in the NDVI. This is consistent with past studies, which have shown land degradation to be positively correlated with population density (e.g., Greperrud 1995; WCED 1987). This further demonstrates that the impact of population density on land degradation is dependent on other conditioning factors (Bosio et al. 2008)—including those discussed below.

9.4.3 ECONOMIC GROWTH (GDP)

Consistent with the EKC, Table 9.3 shows the increase in national level GDP at a global level; EU, Latin American countries (LAC), and MENA are first associated with a decrease in the NDVI, but later both increase after reaching a threshold. However, given that forest extent in EU and North America* decreased in the nineteenth century and recovered in the twentieth century (Meyfroidt and Lambin 2011), the forest transition observed in EU in the reference period (1981–2006) is a

---

* As noted earlier, North America was not included in the regression because the region has only a few countries.

reflection of the stricter environmental regulations in the region, which has led to greater forest extent.

Results in East Asia, South Asia, and SSA show a pattern that is contrary to the EKC. In these regions, GDP has an inverted U relationship with the NDVI, that is, as GDP increases, the NDVI increases to a certain threshold, beyond which it decreases. This unexpected result could be due to other factors that are not included in the model. For example, China, India, Bangladesh, Indonesia, and other countries in South Asia have implemented reforestation programs, which in turn have contributed to an increasing NDVI as their economies grew (Foster and Rosenzweig 2003; Fang et al. 2001; Murdiyarso et al. 2011; Pender et al. 2008). In SSA, both NDVI and GDP increased simultaneously in many countries in West and Central Africa. As argued above, agricultural intensification and diversification in West and Central Africa could have contributed to this pattern. Government tree-planting programs, which were implemented to head off flooding, wind storms, and other disasters attributable to deforestation, also helped to increase forest area (Foster and Rosenzweig 2003; Fang et al. 2001).

It should also be noted that the predictive power of the forest transition model has been weakened by globalization and the increased role of international trade. Hence, we may see increased deforestation or any form of land degradation in one country, which is due to demand of products in the global market. For example, Rudel et al. (2009) observed that Brazil deforestation of the Amazon increased with the GDP well beyond the expected threshold, and this was largely due to the increasing demand for biofuel and meat products in Brazil and for the international market. In the past 5 years, however, Brazil implemented payment for ecosystem services, which rewarded soybean producers and ranchers who do not clear the Amazon. The country also enhanced enforcement of forest regulations at the municipal level.

9.4.4 Government Effectiveness

Improvement of government effectiveness is associated with significant increase in the NDVI across all regions—except East Asia, where the association is not significant. This shows the great role played by government effectiveness in addressing land degradation. The relationship is most significant in SSA, where governance is the weakest in the world (Kaufmann et al. 2010). These results are consistent with a priori expectations that a country with effective government is likely to effectively enforce regulations preventing land degradation—such as deforestation and other forms of land degradation. For example, weak governance and the attendant high prevalence of corruption could lead to issuance of illegal forest harvesting concessions, which in turn leads to deforestation (Karsenty and Ongolo 2012).

These results are consistent with the work of Bhatarai and Hammig (2001) who observed that strength of political institutions and governance significantly reduced deforestation in Latin America, Asia, and Africa.

However, strength of local institutions is key to achieving sustainable land management. Long-term studies by the International Forest Research Institute (IFRI) demonstrated that local communities are better managers of forests than central governments (Ostrom and Nagendra 2006; FAO 2011). A comprehensive review by

Blaikie (2006) also showed that strong local governments and other local organizations are a necessary but not sufficient condition for sustainable natural resource management. As noted above, other factors (such as access to markets, remunerative prices, etc.) that provide incentive for sustainable natural resource management also play a key role. Additionally, strong local institutions vertically linked with national institutions and policies can provide mandates and empower local communities to manage natural resources and related services (Birner and von Braun 2009). A study done in four African countries to determine drivers of response to climate change at a community level using sustainable land and water management (SLWM) showed an interesting pattern of country level decentralization policies with local council propensity to enact SLWM bylaws (Nkonya et al. 2011b). The study sought to understand the SLWM bylaws enacted in the past 30 years at a community level to adapt to climate change. SLWM bylaws in each community were counted and compared against performance of decentralization in each country. Figure 9.4 shows a strong correlation between the number of SLWM bylaws per community and performance of decentralization. This underscores the importance of the linkage between the local communities and the national government institutions—that is, the more decentralized a country is, the more likely local communities will take collective action to address natural resource challenges. Additionally, a study in Uganda comparing compliance with land management regulations observed greater compliance with locally enacted regulations than with regulations enacted by higher authorities (Nkonya et al. 2008). This further demonstrates the potential of local institutions in achieving sustainable land management.

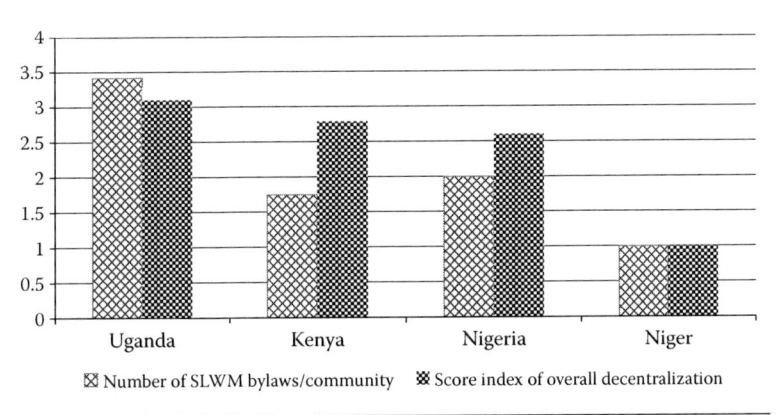

Notes: Overall decentralization includes 12 performance and structural indicators of decentralization. The larger the index, the greater the performance of decentralization.

FIGURE 9.4 Relationship between SLWM bylaws with performance of decentralization in SSA. (From Ndegwa, S. and B. Levy 2004, "The Politics of Decentralization in Africa: A Comparative Analysis." In *Building State Capacity in Africa: New Approaches, Emerging Lessons.* eds. Brian Levy B. and S.J. Kpundeh, pp. 283–322. Washington, DC: World Bank, 2004; Nkonya, E. et al., Climate Risk Management through Sustainable Land Management in Sub-Saharan Africa. IFPRI Discussion Paper 01126, Washington, DC: IFPRI, 2011.)

9.5 IMPACT OF LAND DEGRADATION ON FOOD SECURITY

9.5.1 Relationship of Land Degradation and Crop Yield Gap

As noted earlier, land degradation has contributed to increasing the number of hungry people. Rough estimates by the World Bank show that between 2000 and 2010, land degradation contributed about 20% of the hungry people in SSA (Word Bank Development Indicators 2010). Land degradation has been one of the factors that have led to lower agricultural productivity in SSA and other regions with severe land degradation. This has contributed to wider yield gaps in SSA, South Asia, and Central Asia—regions that experienced land degradation between 1981 and 2003 (Figure 9.5). Many studies have shown that sustainably increasing agricultural productivity on the existing agricultural land—rather than converting forests and other nonagricultural lands to agriculture—is the most feasible solution to addressing the increasing population and its need for additional crop resources, as well as losses of biodiversity and other such challenges (Rockstrom et al. 2009). With a slowing rate of increase in crop yields arising from narrowing gaps between potential and actual yields in much of the developed regions and East Asian countries (Bruinsma 2009), meeting the increase in food demand through yield increases means directing global and national efforts in regions where there is a wide yield gap. Recent simulation scenarios have shown that increases in land productivity will have to account for 77% of the expected growth in global food demand by 2050—demand that is stemming from a growing population and changing dietary preferences (Bruinsma 2009).

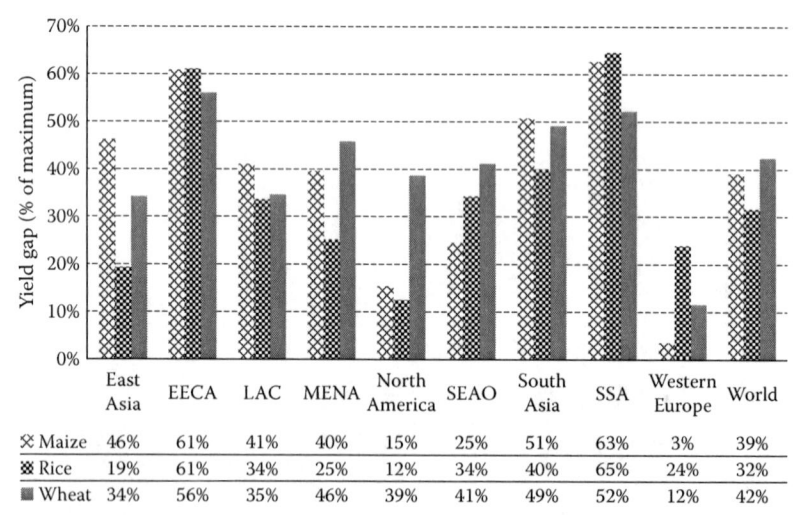

	East Asia	EECA	LAC	MENA	North America	SEAO	South Asia	SSA	Western Europe	World
Maize	46%	61%	41%	40%	15%	25%	51%	63%	3%	39%
Rice	19%	61%	34%	25%	12%	34%	40%	65%	24%	32%
Wheat	34%	56%	35%	46%	39%	41%	49%	52%	12%	42%

Key: EECA = East Europe and Central Asia; LAC = Latin America and Caribbean Countries; MENA = Middle East and North Africa; SEAO = Southeast Asia and Oceania; SSA = Sub-Saharan Africa

FIGURE 9.5 Yield gaps of maize, rice, and wheat across regions. (From Licker, R. et al., *Global Ecol Biogeogr* 119(6):769–782, 2010.)

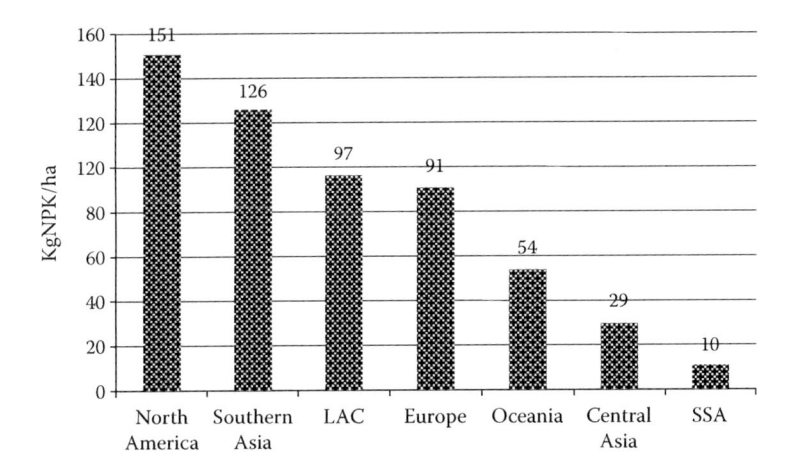

FIGURE 9.6 Average fertilizer nutrient application rate, 2005–2009. (Courtesy of FAOSTAT.)

Soil nutrient depletion is a major form of land degradation and is perceived to be the main driver of low agricultural productivity in SSA (Van Reuler and Prins 1993; Sanchez et al. 1997). SSA applies to only about 10 kg/ha of elemental NPK, a level that is the lowest in the world and lower than 10% of the rate applied in southern Asia (Figure 9.6). As a result of this, soil nutrient depletion is severe. Estimates show that the annual value of depleted nutrients in SSA is about $4.5 billion (Gregersen et al. 2007), which is about 0.5% of the region's US$895 billion GDP in the same year (IMF 2012). The low adoption of chemical fertilizer in SSA is largely due to its high cost, which is precipitated by poor transportation infrastructure in the region.

Depletion of soil organic matter can also lead to significant yield reduction. Studies have shown that increasing soil carbon in the root zone by 1 ton ha^{-1} could increase yields by 20–70 kg ha^{-1} for wheat, 10–50 kg ha^{-1} for rice, and 30–300 kg ha^{-1} for maize (Lal 2006). It is estimated that increasing the global soil carbon pool in the root zone could increase cereal production by 32 million tons per year (Lal 2006), a level that is about 3% of the 1 billion ton cereal production in food-deficit poor countries (FAOSTAT 2009). Organic inputs could also improve soil ecology, as well as a score of other ecosystem services (Powlson et al. 2011).

Below, we demonstrate the impact of soil fertility management on maize and rice yields using Mali and Nigeria as case studies. The analysis is based on only soil nutrient and soil carbon depletion. We also examine the profitability of ISFM and land degrading practices, as well as their adoption rates in the countries included in the case study.

9.5.2 Impact of Land Degradation on Maize and Rice Yield: The Case of Mali and Nigeria

Table 9.4 shows that the 30-year average yield of ISFM practice is more than twice the corresponding yield for the land degrading management practice for both maize

TABLE 9.4
Impact of Land Degradation on Maize and Rice Yields in Mali and Nigeria

	Mali		Nigeria	
	No Inputs	ISFM (40 kg N ha^{-1}, 1.7 manure ha^{-1})	No Inputs	ISFM (40 kg N ha^{-1}, 1.7 manure ha^{-1})
	\multicolumn{4}{c}{Average Yield (tons/ha)}
Maize	0.44	1.99	1.0	3.4
Rice (paddy)	1.17	3.24	1.7	4.3
	\multicolumn{4}{c}{Trend (Annual Change in Yield, kg/ha)[a]}
Maize	−4.8	17.1	−8.6	21.8
Rice (paddy)	−0.5	39.3	−0.4	66.1

[a] Based on trend analysis of 30-year crop simulation results.

and rice in Mali and Nigeria. Additionally, the 30-year yield trend shows a positive slope for plots receiving ISFM treatment. This demonstrates the potential for doubling food production in SSA if farmers could adopt ISFM. A recent study in SSA by Nin Pratt et al. (2011) showed that adoption of ISFM has the potential to more than double the yield.

More profitable land management practices are more likely to be used. For example, studies have shown higher fertilizer use and other inputs for high value crops such as horticultural crops (Gockowski and Ndoumbé 2003) and export crops (Crawford et al. 2003). Table 9.5 shows that the profit on plots receiving ISFM was 28% to 46% higher than the corresponding profit for the land degrading practice. A similar study covering five countries further showed that ISFM was more sustainable and more profitable than practices using chemical fertilizer or organic inputs alone (Nkonya et al. 2012).

These results support other studies that have shown that ISFM in SSA is more profitable than the use of organic inputs or chemical fertilizers alone (e.g., Doraiswamy et al. 2007; Sauer and Tchale et al. 2007; Mekuria and Waddington 2001).

TABLE 9.5
Net Present Value (US$000) Per Hectare for Maize in Mali and Nigeria for 30-Year Planning Horizon

Treatment	Mali	Nigeria
All zero input, 50% crop residues	17.64	16.13
40 kg N ha^{-1}, manure 1.7 tons ha^{-1}, 50% crop residues	22.55	23.47
Difference in returns (%)	28	46

TABLE 9.6
Adoption Rate of ISFM in Selected SSA Countries

Land Management Practice	Nigeria	Mali
	% Adoption	
Fertilizer and animal manure	7.5	17.7
Animal manure	12.1	38.7
Fertilizer	45.3	16.3

Source: Nkonya, E. et al., Land under Pressure. In *2011 Global Food Policy Report*. IFPRI, Washington DC, 2012.

However, marketing and other constraints may limit adoption of practices with higher returns. For example, despite its win-win-win advantage of greater yield, profit, and environmental services, adoption of ISFM is lowest in the selected case study countries (Table 9.6).

The drivers of land degradation discussed, contribute to farmers not adopting sustainable land management practices, some of which could be more profitable than the land degrading practices.

9.6 CONCLUSIONS AND POLICY IMPLICATIONS

Since the poor and women heavily depend on natural resources (FAO 2009), investments to prevent land degradation and rehabilitate degraded lands will have the largest impact on them (World Bank 2012). Unfortunately, land investment at the global level and SSA in particular is limited. Our study shows that, for such investment to be effective, it has to be accompanied with strengthening government effectiveness, improving rural services, and designing policies, which will provide land users incentives to invest in land improvement. For example, our study shows that the ISFM practice—a combination of improved seeds, organic inputs, and judicious amount of chemical fertilizer—gives the highest yield, is most profitable, and is likely to have sustainable yield, but its adoption rate is lower than that of chemical fertilizer and organic inputs. This means that achieving sustainable land management will require multipronged strategies to address multiple land management constraints in developing countries. Such strategies will also help to address constraints leading to the wide yield gaps in regions with severe land degradation.

While some strategies require financial resources, which poor countries may not afford, other strategies only require changes in policies and institutions to provide incentives for land users to use sustainable land management practices. For example, the Niger's rural code, which gave tree tenure to land owners, provided incentives that contributed to the regreening of the Sahel (Larwanou et al. 2009; Herrman et al. 2005). Tree planting programs in Bangladesh, India, and other countries also contributed to forest recovery even in areas with severe poverty.

As demonstrated in the rapid reduction of deforestation in Brazil and other countries, the international community has a role to play to enhance sustainable land management. But evidence shows that international efforts are more effective when the national and local level policies and local institutions are conducive for sustainable land management. This remains a challenge in low income countries, but the few success stories of progress in reducing land degradation in such countries provide hope for addressing land degradation in poor countries, where degradation is most severe.

ABBREVIATIONS

CBD: UN Convention on Biological Diversity
DSSAT: Decision Support System for Agrotechnology Transfer
EKC: environmental Kuznet curve
GDP: gross domestic product
IFRI: International Forest Research Institute
IMR: infant mortality rate
ISFM: integrated soil fertility management
MENA: Middle East and North Africa
NDVI: normalized difference vegetation index
SLWM: sustainable land and water management
SSA: sub-Saharan Africa
TFP: total factor productivity
UNCCD: United Nations to Combat Desertification

REFERENCES

Adam, T., C. Reij, T. Abdoulaye, M. Larwanou, and G. Tappan. 2006. *Impacts des Investissements dans la Gestion des Resources Naturalles (GRN) au Niger: Rapport de Synthese*. Niamey, Niger: Centre Régional d'Enseignement Specialise en Agriculture.

Aduayi, E., V. Chude, B. Adebusuyi, and S. Olayiwola. 2012. *Fertilizer Use and Management Practices for Crops in Nigeria*. Third Edition. Abuja, Nigeria: Federal Fertilizer Department, Federal Ministry of Agriculture and Rural Development.

Akinnifesi, F., O.C. Ajayi, G. Sileshi, P.W. Chirwa, and J. Chianu. 2010. Fertiliser trees for sustainable food security in the maize-based production systems of East and Southern Africa. A review. *Agron Sustain Dev*, 30:615–629.

Bai, Z.G., D.L. Dent, L. Olsson, and M.E. Schaepman. 2008. *Global Assessment of Land Degradation and Improvement 1: Identification by Remote Sensing. Report 2008/01*. Rome/Wageningen: FAO/ISRIC.

Baland, J.M., P. Bardhan, S. Das, D. Mookherjee, and R. Sarkar. 2006. *Managing the Environmental Consequences of Growth: Forest Degradation in the Indian Mid-Himalayas*. Berkeley: University of California, Berkeley.

Bandyopadhyay, S., P. Shyamsundar, and A. Baccini. 2006. *Forests, Biomass Use, and Poverty in Malawi*. Policy Research Working Paper 4068. Washington, DC: World Bank.

Basson, G. 2010. *Sedimentation and Sustainable Use of Reservoirs and River Systems*. International Commission on Large Dams (ICOLD) Bulletin. http://www.waterpowermagazine.com/.

Benin S., A. Kennedy, M. Lambert, and L. McBride. 2010. *Monitoring African Agricultural Development Processes and Performance: A Comparative Analysis.* Washington DC: International Food Policy Research Institute (IFPRI).

Bhattarai M. and M. Hammig. 2001. Institutions and the environmental Kuznets Curve for deforestation: a cross-country analysis for Latin America, Africa and Asia. *World Dev* 29(6):995–1010.

Birner R. and J. von Braun. 2009. Decentralization and public service provision—a framework for pro-poor institutional design. In *Does Decentralization Enhance Poverty Reduction and Service Delivery?* eds. Ahmad E. and G. Brosio, 287–315. Cheltenham, UK: Edward Elgar.

Blaikie P. 2006. Is small really beautiful? Community-based natural resource management in Malawi and Botswana. *World Dev* 34(11):1942–1957.

Bossio D., K. Geheb, and W. Critchley. 2008. Managing water by managing land: addressing land degradation to improve water productivity and rural livelihoods. *Agr Water Manage* 97:536.

Boyd C. and T. Slaymaker. 2000. Re-examining the "more people less erosion" hypothesis: special case or wider trend? *Nat Resour Perspect* 63 (November):1–6.

Bruinsma J. 2009. *By How Much Do Land, Water and Crop Yields Need to Increase by 2050? Expert Meeting on How to Feed the World in 2050.* Rome: Food and Agriculture Organization of the United Nations.

Crawford E., V. Kelly, T.S. Jayne, and J. Howard. 2003. Input use and market development in Sub-Saharan Africa: an overview. *Food Policy* 28(4):277–292.

Dinda S. 2004. Environmental Kuznets curve hypothesis: a survey. *Ecol Econ* 49(4):431–455.

Doraiswamy P.C., G.W. McCarty, E.R. Hunt Jr., R.S. Yost, M. Doumbia, and A.J. Franzluebbers. 2007. Modeling soil carbon sequestration in agricultural lands of Mali. *Agric Syst* 94:63–74.

FAOSTAT. 2011. http://faostat.fao.org/site/.

FAOSTAT. 2009. http://faostat.fao.org/site/.

Fan S. and C. Breisinger. 2011. Development assistance and investment in agriculture: promises and facts. *TATuP* 20(2):20–28.

FAO (Food and Agriculture Organization). 2010. State of World's Forests.

FAO (Food and Agriculture Organization). 2011. The State of World's Land and Water Resources for Food and Agriculture. Managing Systems at Risk. FAO, Rome.

Foa R. 2009. *Social and Governance Dimensions of Climate Change Implications for Policy.* Policy Research Working Paper 4939. Washington, DC: World Bank.

Fuglie K. and N. Rada 2011. *Policies and Productivity Growth in African Agriculture.* ASTI/FARA Conference December 5–7, 2011, Accra, Ghana.

Fynas J.G. 2004. The oil boom in Equatorial Guinea. *Afr Affairs*, 103(413):527–546.

Gijsman A.J., G. Hoogenboom, W.J. Parton, and P.C. Kerridge. 2002. Modifying DSSAT crop models for low-input agricultural systems using a soil organic matter-residue module from CENTURY. *Agron J* 94:462–474.

Global Mechanism. 2011. Integrated economic assessment of land resources and ecosystem services. Bringing policy makers the evidence they need. Paper presented at the United Nations to Combat Dessertification (UNCCD) Conference of Parties (COP) 10, Changwon, South Korea.

Gockowski J. and M. Ndoumbé. 2003. The adoption of intensive monocrop horticulture in southern Cameroon. *Agr Econ* 30(3):195–205.

Graham E. and E. Wada. 2001. *Foreign Direct Investment in China: Effects on Growth and Economic Performance.* Working Paper No. 01-03. http://ssrn.com/abstract=300884 or http://dx.doi.org/10.2139/ssrn.300884 (Accessed July 7, 2012).

Gregersen H.M., P.F. Folliott, and K.N. Brooks. 2007. *Integrating Watershed Management—Connecting People to Their Land and Water.* Wallingford, United Kingdom: CABI.

Grossman G.M. and A.B. Krueger. 1991. Environmental impacts of North America Free Trade. National Bureau of Economic Research (NBER) Working Paper No. 3914.

Hazell P. and S. Wood. 2008. Drivers of change of global agriculture. *Philos T Roy Soc B* 363:495–515.

Herrmann S., A. Anyamba, and C.J. Tucker. 2005. Resent trends in vegetation dynamics in the African Sahel and their relationship to climate. *Glob Environ Change* 15(4):394–404.

IMF (International Monetary Fund). 2012. World Economic Outlook Data. available at http://www.imf.org/external/pubs/ft/weo/2012 (Accessed July 7, 2012).

IPCC Climate Change. 2007. *Impacts, Adaptation and Vulnerability*. Washington, DC: Cambridge University Press.

Ivanova M. 2012. Institutional design and UNEP reform: historical insights on form, function and financing. *International Affairs* 88(3):565–584.

Karsenty A. and S. Ongolo. 2012. Can "fragile states" decide to reduce their deforestation? The inappropriate use of the theory of incentives with respect to the REDD mechanism. *Forest Policy Econ* 18:38–45.

Kaufmann D., A. Kraay, and M. Mastruzzi. 2010. The Worldwide Governance Indicators: Methodology and Analytical Issues. World Bank Policy Research Working Paper No. 5430. SSRN. http://ssrn.com/abstract=1682130.

Kuznet S. 1955. "Economic Growth and Inequality." *American Economic Review* 49:1–28.

Lal R. 2006. Enhancing crop yields in the developing countries through restoration of the soil organic carbon pool in agricultural lands. *Land Degrad Dev* 17(2):197–209.

Lambin E.F. and P. Meyfroidt. 2011. Global land use change, economic globalization, and the looming land scarcity. *Proceedings of the National Academy of Sciences* 108(9):3465–3472.

Larwanou M., M. Abdoulaye, and C. Reij. 2006. *Etude de la régénération naturelle assistée dans la Région de Zinder (Niger): Une première exploration d'un phénomène spectaculaire*. Washington, DC: International Resources Group for the U.S. Agency for International Development.

Le Q.B., L. Tamene, and P.L.G. Vlek. 2012. Multipronged assessment of land degradation in West Africa to assess the importance of atmospheric fertilization in masking the processes involved, *Global Planet Change* 92–93:71–81.

Licker R., M. Johnston, J.A. Foley, C. Barford, C.J. Kucharik, C. Monfreda, and N. Ramankutty. 2010. Mind the gap: how do climate and agricultural management explain the "yield gap" of croplands around the world? *Global Ecol Biogeogr* 119(6):769–782.4.

Mekuria M. and S. Waddington. 2002. Initiatives to encourage farmer adoption of soil fertility technologies for maize-based cropping systems in Southern Africa. In: Barrett C., F. Place, and A. Abdillahi (eds.). *Natural Resource Management in African Agriculture: Understanding and Improving Current Practices*. Wallingford: CAB International.

Meyfroidt P. and E. Lambin. 2011. Global forest transition: prospects for an end to deforestation. *Annu Rev Environ Resour* 36:343–371.

Morawitz D.F., T.M. Blewett, A. Cohen, and M. Alberti. 2006. Using NDVI to assess vegetative land cover change in Central Puget Sound. *Environmental Monitoring and Assessment* 114(1):85–106.

Moss C.B. and J.S. Shonkwiler. 1993. Estimating yield distributions with a stochastic trend and nonnormal errors. *Am J Agric Econ* 75(4):1056–1062.

Murdiyarso D., S. Dewi, D. Lawrence, and F. Seymour. 2011. *Indonesia's Forest Moratorium: A Stepping Stone to Better Forest Governance?* CIFOR Working Paper 76. Bogor, Indonesia: Center for International Forestry Research.

Nachtergaele F., M. Petri, R. Biancalani, G. Van Lynden, and H. Van Velthuizen. 2010. *Global Land Degradation Information System (GLADIS). Beta Version*. An Information database for Land Degradation Assessment at Global Level. Rome: FAO.

Ndegwa S. and B. Levy. 2004. The politics of decentralization in Africa: a comparative analysis. In *Building State Capacity in Africa: New Approaches, Emerging Lessons*, eds. Levy B. and S.J. Kpundeh, 283–322. Washington, DC: World Bank.

Nin-Pratt A., M. Johnson, M. Magalhaes, L. You, X. Diao, and J. Chamberlin. 2011. *Yield Gaps and Potential Agricultural Growth in West and Central Africa.* Washington, DC: IFPRI Research Monograph.

Nkonya E., J. Pender, and E. Kato. 2008. Who knows, who cares? Determinants of enactment, awareness and compliance with community natural resource management regulations in Uganda. *Environ Dev Econ* 13(1):79–109.

Nkonya E., N. Gerber, P. Baumgartner, J. von Braun, A. De Pinto, V. Graw, E. Kato, J. Kloos, and T. Walter. 2011a. *The Economics of Land Degradation. Toward an Integrated Global Assessment.* Berlin, Peter Lang.

Nkonya E., F. Place, J. Pender, M. Mwanjololo, A. Okhimamhe, E. Kato, S. Crespo, J. Ndjeunga, and S. Traore. 2011b. *Climate Risk Management Through Sustainable Land Management in Sub-Saharan Africa.* IFPRI Discussion Paper 01126. Washington, DC: IFPRI.

Nkonya E., F. Bisong, J. Koo, H. Xie, R.C. Izaurralde, E. Kato, B. Wielgosz, S. Salau, and S. Traore. 2011c. *Review of Costs, Benefits, and Public Expenditure on Sustainable Land and Water Management in Nigeria.* Washington, DC: IFPRI.

Nkonya E., B. Coulibaly, J. Koo, K. N'diaye, H. Xie, R.C. Izaurralde, S. Traore, and E. Kato. 2011d. *Review of Costs, Benefits, and Public Expenditure on Sustainable Land and Water Management in Mali.* Washington, DC: IFPRI.

Nkonya E., J. Koo, P. Marenya, and R. Licker. 2012. Land under pressure. In *2011 Global Food Policy Report.* Washington, DC: IFPRI.

OECD. 2011. Global aid raw data. http://stats.oecd.org/Index.aspx?DatasetCode=CRSNEW.

Ostrom E. and H. Nagendra. 2006. Insights on linking forests, trees, and people from the air, on the ground, and in the laboratory. *Proc Natl Acad Sci USA* 103(51):19224–19231.

Powlson D.S., P.J. Gregory, W.R. Whalley, J.N. Quinton, D.W. Hopkins, A.P. Whitmore, P.R. Hirsch, and K.W.T. Goulding. 2011. Soil management in relation to sustainable agriculture and ecosystem services. *Food Policy* 36:S72–S87.

Requier-Desjardins M., B. Adhikari, and S. Sperling. 2011. Some notes on the economic assessment of land degradation. *Land Degrad Dev* 22:285–298.

Rockström J., W. Steffen, K. Noone et al. 2009. A safe operating space for humanity. *Nature* 461:472–475.

Roy R.N., A. Finck, G.J. Blair, and H.L. Tandon. 2006. *Plant Nutrition for Food Security. A Guide for Integrated Nutrient Management.* Fertilizer and Plant Nutrition Bulletin 16. Rome: FAO.

Rudel T.K. 2009. Tree farms: driving forces and regional patterns in global expansion of forest plantation. *Land Use Policy* 26(3):545–550.

Sanchez P.A., K.D. Shepherd, M.J. Soule et al. 1997. Soil fertility replenishment in Africa: an investment in natural resource capital. In *Replenishing Soil Fertility in Africa*, eds. Buresh R.J., P. Sanchez, and F. Calhoun, 1–46. Madison, WI: Soil Science Society of America.

Sauer J., H. Tchale, and P. Wobst. 2007. Alternative soil fertility management options in Malawi: an economic analysis. *Journal of Sustainable Agriculture* 29(3):29–53.

Shapiro S. and M. Wilk. 1972. An analysis of variance test for the exponential distribution (complete sample). *Technometrics* 14(2):353–370.

Shapiro S. and M. Wilk. 1968. Approximations for the null distribution of the W statistic. *Technometrics* 10(4):861–866.

Sharma P. and P. Biswas. 2004. IPNS packages for dominant cropping systems in different agro climatic regions of the country. *Fert News*, 49(10):43–47.

Tiffen M., M. Mortimore, and F. Gichuki. 1994. *More People, Less Erosion: Environmental Recovery in Kenya*. London: Wiley and Sons.

Tittonel P. 2008. Msimu wa Kupanda: targeting resources within diverse, heterogenous and dynamic farming systems of East Africa.

Tittonell P., M. Corbeels, M. van Wijk, B. Vanlauwe, and K. Giller. 2008. Combining organic and mineral fertilizers for integrated soil fertility management in smallholder farming systems of Kenya: explorations using the crop-soil model FIELD. *Agron J* 100(5):1511–1526.

Tucker C.J., J.E. Pinzon, and M.E. Brown. 2004. *Global Inventory Modeling and Mapping Studies*. College Park, MD: University of Maryland.

UNCCD (United Nations Convention to Combat Desertification). 2011. *Land and Soil in the Context of a Green Economy for Sustainable Development, Food Security and Poverty Eradication*. The Submission of the UNCCD Secretariat to the Preparatory Process for the Rio+ 20 Conference. Online at http://www.unccd.int/knowledge/menu.php (Accessed July 7, 2012).

UNCCD. 2012. *Zero Net Land Degradation. A Sustainable Development Goal for Rio+20*. http://www.unccd.int/Lists/SiteDocumentLibrary/Publications/UNCCD_PolicyBrief_ZeroNetLandDegradation.pdf (Accessed July 7, 2012).

Vanlauwe B. and K.E. Giller. 2006. Popular myths around soil fertility management in sub-Saharan Africa. *Agr Ecosyst Environ* 116:34–46.

Van Reuler H. and W.H. Prins 1993. The role of plant nutrients for sustainable food production in sub-Saharan Africa. The Dutch Association of Fertiliser Producers, Netherlands, p. 232.

Vitousek P.M. 1994. Beyond global warming: ecology and global change. *Ecology* 75:1861–1876.

Vlek P.L., Q.B. Le, and L. Tamene. 2010. Assessment of land degradation, its possible causes, and threat to food security in Sub-Saharan Africa. In *Food Security and Soil Quality*, eds. Lal R. and B.A. Stewart, 57–86. Boca Raton, FL: CRC Press.

WCED (World Commission on Environment Development). 1987. *Our Common Future*, 124 pp. Oxford, UK: Oxford University Press.

World Bank. 2008. *Poverty and the Environment. Understanding Linkages at the Household Level*. Washington, DC: World Bank.

World Bank. 2011. Country data on agriculture. http://data.worldbank.org/indicator/NV.AGR.TOTL.ZS (Accessed July 7, 2012).

World Bank. 2012. *Inclusive Green Growth. The Pathway to Sustainable Development*, Washington, DC: World Bank.

World Bank Development Indicators. 2011. Online data, http://data.worldbank.org/indicator/ (Accessed July 7, 2012).

10 Cost–Benefit Analysis of Soil Degradation and Restoration*

Fred J. Hitzhusen and Sarah E. Kiger

CONTENTS

10.1 Magnitude and Economic Impacts of Soil Erosion 227
10.2 Measuring Costs and Benefits 230
10.3 Types and Measures of Environmental Economic Value 233
10.4 Case Studies 235
    10.4.1 Soil and Land Degradation and Ag Growth Model (Case 1) 235
        10.4.1.1 Analysis and General Results 237
        10.4.1.2 Summary and Conclusions 238
        10.4.1.3 Follow-Up Analysis 239
    10.4.2 DR Valdesia Hydro Reservoir Sedimentation (Case 2) 239
        10.4.2.1 Study Methodology 240
        10.4.2.2 Results and Implications of Analysis 241
    10.4.3 Some Ohio Evidence of Soil Erosion Off-Site Economic
        Impacts (Case 3) 243
        10.4.3.1 Methods and Results of Studies 243
    10.4.4 OSU Studies on Ohio Downstream Benefits of CRP (Case 4) 245
        10.4.4.1 Methods and Results of Studies 245
10.5 Summary and Conclusions 250
Abbreviations 252
References 252

10.1 MAGNITUDE AND ECONOMIC IMPACTS OF SOIL EROSION

Soil erosion and resulting sedimentation problems continue to be major concerns throughout the world. Dregne (1972) evaluated the general state of soil and land degradation worldwide in the 1970s. Much of Southwest Asia, large areas of China, India and Southeast Asia, Northern Africa, Central America, and Mexico suffer

---

* Helpful review comments have been provided by OSU colleague and former Chief of Soil and Water, Ohio DNR, David Hanselmann. This paper will appear as an invited chapter in the forthcoming book *Advances in Soil Science*, with a focus on principles of sustainable soil management in agroecosystems.

from severe land degradation. Dregne concludes that desertification "will blunt crop production increases in the subhumid and semiarid regions of the United States and Mexico, as well as in the Central American mountains." In South America, "land degradation is most severe on the cultivated lands of the Andes Mountains" and "water and wind erosion have damaged some Argentine farmland." Dregne also emphasizes the fragility of soils in the Amazon Basin, where deforestation and encouragement of development have been intensifying in more recent years.

Among the many environmental problems now facing less developed countries (LDCs), soil erosion has the most serious economic consequences. If allowed to continue at current rates, soil loss will reduce many countries' capacity for agricultural production (Wolman 1993). Also, erosion is causing sizable external costs in the third world; soil washed from agricultural and other land is being deposited in reservoirs, irrigation canals, and navigable waterways at high rates.

Rapp (1977) and Robinson (1981) suggest that this experience is by no means restricted to Latin America, and Linsley and Franzini (1979) argue that it is not limited to developing countries. National aggregate off-site cost estimates for soil erosion established the magnitude of the problem in the United States over 25 years ago. Clark et al. (1985) estimated total annual off-farm costs for all agricultural erosion sources to range between $3 and $13 billion with a point estimate of $6.1 billion, of which $2.2 billion was attributed to cropland erosion (in 1980 dollars). Damage to recreational uses accounted for the largest share of costs—comprising nearly 33% of total costs—and boating was the largest recreation subgroup. Other high impact receptors or users and their percent of total costs included municipal and industrial (14.8%), water storage facilities (11.3%), dredging (8.5%), and preservation values (8.2%). Cropland erosion was the largest source at 38% of total erosion. A reanalysis of these results suggested a $7.1 billion point estimate (Ribaudo 1986a).

Later analysis of the Conservation Reserve Program (CRP) by Ribaudo et al. (1989) found that water quality benefits from the first 23 million acres enrolled in the CRP totaled $2.05 billion or an average of $219.83 per hectare. This suggests that Ribaudo's 1986 point estimate should be adjusted downward by about $2 billion. However, both the Clark and Ribaudo analyses omit several categories of downstream impacts (e.g., dredge spoil disposal, delays to commercial shipping, biological impacts, etc.), which may make their estimates very conservative. More recent concerns in the United States have shifted to loss of CRP acres due to tight federal budgets and a reduced cap on CRP enrollment, perceived to be needed to get these acres back into crop production with high corn and soybean prices driven in part by corn ethanol tariffs and subsidies (Kiger 2009; Wright 2011).

Clearly, any threat to the longevity of existing and future water resource developments could have serious repercussions and warrants careful evaluation and planning. The primary and most immediate threat to hydroelectric and other water storage projects is sedimentation. A major cause is excessive erosion from watershed overuse or misuse. In many LDCs, the general pattern of heavy land use leading to high erosion rates results from a combination of traditional shifting agriculture and population growth. In reservoir watersheds, this pattern may be preexisting, or the reservoir project may, as with the Anchicaya reservoir in Columbia reported by Allen (1972), provide the access route for transient farmers into a previously relatively undisturbed area. Another contributor to watershed erosion and sedimentation is deforestation. This may initially be by large-scale legal (or illegal) logging, but a deforested state is often maintained by subsequent agricultural use including extensive livestock grazing in many areas. Other important contributors to sedimentation include stream bank cutting and what Sanches (1979) calls "civil engineering erosion" from improper location, construction, and maintenance of roads, culverts, bridges, and channels.

The exploitation of natural resources, perhaps even more so in developing countries, continues in large part because these resources are not priced at their marginal social values. This underpricing in turn occurs because many centrally planned as well as private market economies, with imperfectly defined and enforced property rights, fail to fully internalize the external costs of environmental service benefits related to the use of these natural resources. Examples of environmental services include raw material supply, assimilative capacity of natural systems for the residuals of economics production and consumption activities, esthetic values, human and wildlife habitat, and genetic biodiversity. This suggests the need for more comprehensive measures of the social benefits and costs of resources or their related environmental services (Hitzhusen 1993). Environmental economists usually talk about four categories of marginal social costs, including (1) direct costs to current users, (2) external costs borne by others now and in the future, (3) foregone benefits of future users from a depleted resource, and (4) existence values for the sustainable maintenance of a given resource.

The planned economy or private market imperfections at the microeconomic or watershed level, in the case of soil erosion, also manifest themselves as imperfections in national income accounting at the macroeconomic level. Repetto and Faeth (1990) argue that by ignoring natural resources (or the broader notion of environmental services), statistics such as the gross national product (GNP) can record illusory gains in income and mask permanent losses in wealth. As a result, a nation could exhaust its minerals, erode its soils, pollute its aquifers, and hunt its wildlife to extinction—all without affecting the measured national income. For example, Indonesia's high 7.1% economic growth rate as measured by gross domestic product (GDP) from 1971 to 1984 is only 4.0% when GDP is adjusted for unsustainable soil erosion, forest harvest in excess of annual growth, and oil reserve depletion.

The economic costs of soil erosion and degradation have been generally but not comprehensively established. Reduced agricultural food production and recreation and habitat losses as well as dredging, water treatment, and other cleanup costs from erosion and downstream sedimentation are often cited in the literature. Less evidence exists for the economic benefits, particularly those not traded (bought and sold) in well-functioning markets from restoration of soils. Economists often refer to these as nonmarket or extra market benefits and prescribe a range of estimation methods under the rubric of benefit–cost analysis (BCA). It is understandable that noneconomists in particular might be a bit skeptical about the validity of these methods, even though they have a relatively long history and are being discussed more frequently in popular trade publications (see *Corn and Soybean Digest*, January 2012). Thus, the following section provides a detailed development of their conceptual basis and estimation protocols.

10.2 MEASURING COSTS AND BENEFITS

What passes for "economic" analysis of various soil degradation and restoration scenarios/management strategies varies widely and can be placed within an "accounting stance" continuum regarding both space and time. One end includes private individual or firm-oriented, engineering-type financial analysis utilizing current market or administered prices of inputs and outputs. At the other end are societal and intergenerational efficiency concerns and income distribution analysis, including consideration of both weighted and unweighted income distribution impacts. In between lie a series of adjustments or shadow pricing methods to account for full opportunity cost, willingness to pay, elasticity of supply and demand, unemployed factors, externalities, economic surplus, and overvalued currency considerations.

Margolis (1969) suggests why private market prices may not reflect full social benefits or costs with the following:

> ... there are many cases where exchange occurs without money passing hands; where exchanges occur but they are not freely entered into; where exchanges are so constrained by institutional rules that it would be dubious to infer that the terms were satisfactory; and where imperfections in the conditions of exchange would lead us to conclude that the price ratios do not reflect appropriate social judgments about values. Each of these cases gives rise to deficiencies in the use of existing price data as the basis for evaluation of inputs or outputs.

Costs generated from engineering data and future revenues based on current market prices can be misleading, particularly if one is concerned with societal costs and benefits. These "costs" generally do not represent full opportunity costs or highest use value of all factors of production such as the value of the farmer's time in implementing soil erosion and related sediment control during the busy spring planting season. Alternatively, the opportunity cost of labor is less than the wage rate in those situations involving underemployed or unemployed labor. Costs based only on engineering data may also omit major technological externalities from soil erosion such as flood damage, water pollution and ditch, harbor, and reservoir sedimentation.

"Revenues" may also be overstated, particularly in those cases where local currency is overvalued (which is the case in many developing countries) or where a relatively inelastic demand exists for the end product(s). The inelastic demand situation refers to those cases where an increase in supply results in a disproportionate decrease in price. Ward (1976a,b) discusses the use of shadow exchange rates to adjust for overvaluation of local currencies and the Bruno Criterion to evaluate the foreign exchange saved or earned by alternative erosion control strategies. This is particularly relevant in evaluating erosion control strategies in developing countries. Frequently, these countries are dependent on hydroelectric power as a major domestic energy source. Sedimentation of reservoirs reduces power output and may lead to increased importation of oil.

Gittinger (1982) argues for distinguishing between financial and economic analysis, where financial analysis refers to net returns to private equity capital based on market or administered prices. Financial analysis also treats taxes as a cost and subsidies as a return; interest paid to outside suppliers of money or capital is a cost, while any imputed interest on equity capital is a part of the return to equity capital. By contrast, Gittinger sees economic analysis as concerned with net economic returns to the whole society, frequently based on shadow prices to adjust for market or administered price imperfections. In economic analysis, taxes and subsidies are treated as transfer payments, that is, taxes are part of the total benefit of a project to society and subsidies are a societal cost. For purposes of this chapter, financial and economic analysis will refer to private and social concepts of economic efficiency analysis, respectively.

This financial versus economic distinction is important, but the complementarity of these analytical approaches is equally relevant. Financial analysis provides information on the profitability of a given soil erosion reduction practice (e.g., cover crops or terraces) to individual entrepreneurs or investors and thus gives an indication of the incentive structure and potential adoption rate. Economic or social cost–benefit analysis attempts to determine profitability from a societal standpoint, taking into consideration externalities or environmental costs, pricing of underemployed or unemployed factors, currency evaluation, etc. The appropriateness of these analytical alternatives depends on the question one is asking. Generally speaking, it is relatively straightforward to assign values to the cost and benefit streams in financial analysis; market prices suffice. However, this is substantially more difficult in full social analysis.

Social costs and benefits or gains and losses from an economic perspective refer to the aggregation of individual producer and consumer measures of full willingness to accept or pay compensation. Individual preferences count in the determination of social benefits and cost and are weighted by income or more narrowly by market power. Since most policy changes involve economic gainers and losers, economists have developed the concept of potential Pareto improvement (PPI) to add up gains and losses to get net benefits. Simply stated, the concept holds that any policy change is a PPI or an increase in economic efficiency if at least one individual is better off after all losers are compensated to their original or before the policy change income positions. The compensation need not actually occur but must be possible (Dasgupta and Pearce 1979).

These measures of social costs and benefits are often not fully reflected in current market prices (or in government-regulated prices) as in the case of crop production in an area with high rates of soil erosion. This divergence results from several factors. First, government subsidies of inputs and/or outputs can lead to levels of input use and outputs in agriculture that are not economically efficient or environmentally sustainable, particularly in the case of fertilizer, irrigation, and agricultural sediments. Second, because there are consumers willing to pay more and producers willing to sell for less than prevailing market or regulated prices, they receive what economists call consumer and producer surpluses. Third, technological externalities in agricultural production exist to the extent that, external to the production and consumption of the resulting output, individuals, households, or firms experience uncompensated real economic losses (or gains) from soil erosion, agricultural chemicals, or other residuals. Finally, there may be willingness to pay to keep future economic options such as hydroelectric generation open (see Veloz et al. 1984 and the related case study later in this chapter) or willingness to pay for existence value of plant or animal species threatened by water pollution, which are not reflected in the market or government-regulated prices of agricultural inputs and/or outputs.

Figure 10.1 illustrates both the concepts of economic surplus and technological externality. For example, at market price *P*, consumer surplus is equal to area PEC and producer surplus is equal to area PEA. One might think of *P* as the market price at equilibrium point *E* where marginal private cost (MPC) of the farmer is equal to demand or marginal willingness to pay. Farm output at *Q* also includes the joint production of soil erosion and/or surplus chemicals that impose social costs on downstream watershed residents represented by *P** and *E**. One might also think of this as producing residuals (sediment and chemicals) that exceed environmental assimilative capacity. These external social costs are fully internalized at *P'E'Q'*. The shaded area *BE'EA* represents the net loss in producer and consumer surpluses from the presence of the soil and chemical externalities or alternatively the net gains from internalizing these external effects.

Once all quantifiable cost and benefit streams have been given prices or shadow values, one must decide on an appropriate rate of discount or time value and a criterion for evaluating and ranking the economic efficiency of alternative soil-conserving strategies or programs. A long-standing controversy on the appropriate discount rate centers primarily on those who support various private opportunity cost versus social time preference measures. Baumol (1968) has written a classic article dealing with the discount rate controversy.

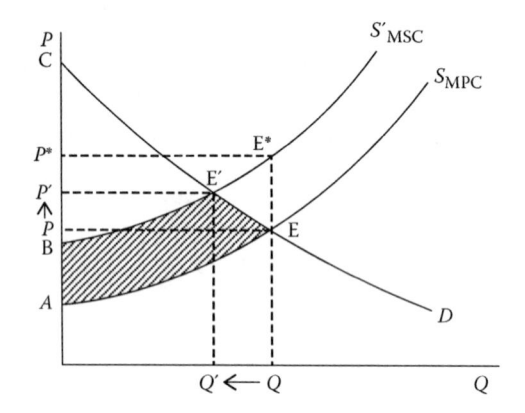

Technological externality defined (Dasgupta and Pearce 1978)
1. Necessary condition
   Phycical interdependence of production
   and/or utility functions
2. Sufficient condition
   Not fully priced or compensated

S = marginal private (e.g., upstream farmer) cost function
S' = marginal social (e.g., watershed) cost function
D = demand or marginal benefit function
Q = output quantity
P = price unity of output

FIGURE 10.1 Soil erosion as an externality.

The alternative efficiency criteria include (1) the ratio of benefits to costs, (2) the net present value, (3) the internal rate of return, (4) the payout period, and several other lesser known criteria related to optimal time phasing of projects, and the optimal utilization of scarce foreign exchange. Several authors, including Dasgupta and Pearce (1979), Gittinger (1982), and Ward (1976a,b), have explored the decision criteria issue in depth. These criteria and the resulting ranking of alternative sediment management strategies are heavily influenced by the nature of future benefit and cost streams, the ratio of future operating costs to initial capital outlay, and the nature of the capital or budget constraint.

Analysts should also be concerned with the equity or income distribution impacts of alternative soil management strategies. For example, Korsching and Nowak (1982) argue that low income farmers may be disproportionately impacted by the costs of many soil erosion or sediment control strategies. This can result from their farming of a higher proportion of erodible soils, possessing fewer savings to cover initial investment demands, and having farms too small to capture any major scale economies in erosion control practices.

Economists use several alternative methods for handling income distribution impacts including (1) explicit weighting of net benefits by income class, group, or region, (2) provision of alternative weighting functions and their distributional consequences to decision makers, (3) estimation of nonweighted net benefits by income class, group, or region, and (4) a constrained maximum or minimum target approach that maximizes economic efficiency subject to an income constraint or vice versa (see Ahmed and Hitzhusen 1988).

10.3 TYPES AND MEASURES OF ENVIRONMENTAL ECONOMIC VALUE

It is possible to develop specific values, measures, instruments, and options for economic assessment of environmental service flows related to soil and other natural resources. The service flows include raw material supply, assimilative capacity, amenities/aesthetics, human habitat, and plant and animal biodiversity. Figure 10.2 summarizes this process. *Direct current use value* refers to use of environmental service flows such as hydropower and water. *External values* are those uncompensated costs or benefits (externalities) from upstream production or consumption processes that are borne or received now or in the future but not reflected in current prices to producers or consumers. *Option value* refers to the willingness to pay to delay the use of something until some future time, while *bequest value* refers to a willingness to preserve something for the use of future generations. This is related to the notion of foregone benefits to future users from current exhaustion of a finite resource without any close substitutes. *Existence value* is the willingness to pay for preservation of plant and/or animal species without regard for their use by humans (see Dixon et al. 1994; Hoehn and Walker 1993).

Economists use a variety of measures or methods to infer or discover these foregoing values. Sometimes it is possible to directly observe values in existing prices, and in other cases, it is necessary to infer values from prices of closely related complementary goods. In the first case, reduction in commercial fish catch from

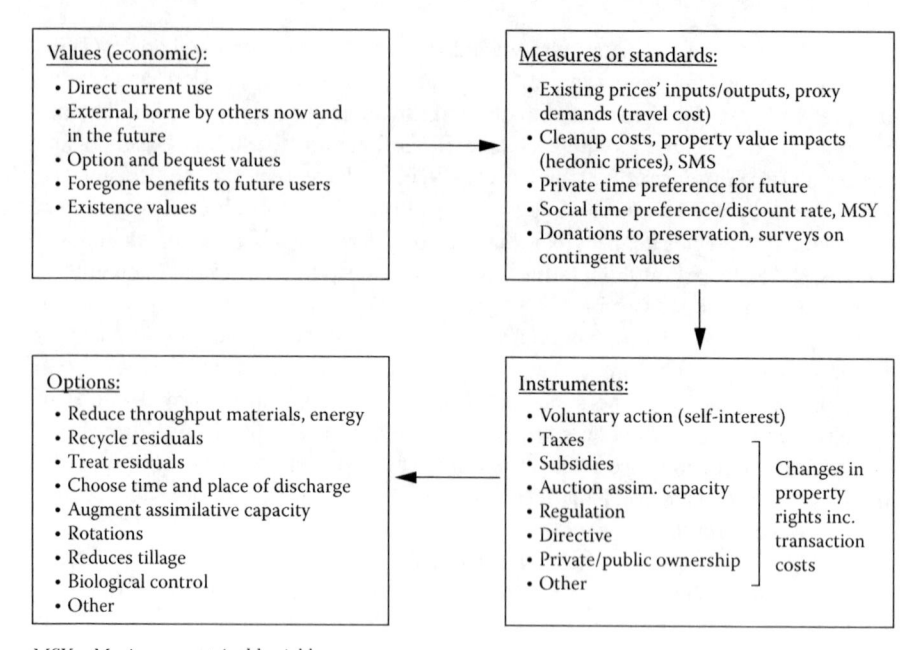

```
┌─────────────────────────────────────┐      ┌──────────────────────────────────────────┐
│ Values (economic):                  │      │ Measures or standards:                   │
│ • Direct current use                │      │ • Existing prices' inputs/outputs, proxy │
│ • External, borne by others now and │─────▶│   demands (travel cost)                  │
│   in the future                     │      │ • Cleanup costs, property value impacts  │
│ • Option and bequest values         │      │   (hedonic prices), SMS                  │
│ • Foregone benefits to future users │      │ • Private time preference for future     │
│ • Existence values                  │      │ • Social time preference/discount rate, MSY│
│                                     │      │ • Donations to preservation, surveys on  │
│                                     │      │   contingent values                      │
└─────────────────────────────────────┘      └──────────────────────────────────────────┘
                                                               │
                                                               ▼
┌─────────────────────────────────────┐      ┌──────────────────────────────────────────┐
│ Options:                            │      │ Instruments:                             │
│ • Reduce throughput materials, energy│     │ • Voluntary action (self-interest)       │
│ • Recycle residuals                 │      │ • Taxes                         ┐        │
│ • Treat residuals                   │      │ • Subsidies                     │Changes in│
│ • Choose time and place of discharge│◀─────│ • Auction assim. capacity       │property │
│ • Augment assimilative capacity     │      │ • Regulation                    │rights inc.│
│ • Rotations                         │      │ • Directive                     │transaction│
│ • Reduces tillage                   │      │ • Private/public ownership      │costs    │
│ • Biological control                │      │ • Other                         ┘        │
│ • Other                             │      │                                          │
└─────────────────────────────────────┘      └──────────────────────────────────────────┘
```

MSY = Maximum sustainable yield

FIGURE 10.2 Monetizing environmental service flows and implementing change/reform.

agricultural pollution (externality) of a reservoir can be measured in lost fishing revenues. However, any reduction in sport fishing in the same lake would require assessment of any decrease in expenditures on goods and services related to boating and sport fishing activity, that is, the development of a travel cost or proxy demand function method (see Macgregor et al. 1991).

The value of an externality can also be conservatively estimated in some cases by replacement, cleanup, or avoidance costs such as reservoir dredging and water treatment related to soil sediments and agricultural chemicals. In addition, the impact of an externality on private property values can frequently be estimated by hedonic pricing, which is a method for statistically decomposing the sources of value or demand in a property market to allow independent estimation of an environmental amenity or disamenity (see Hitzhusen et al. 1997). Contingent valuation refers to a survey method that estimates willingness to pay values directly from respondents for some change in an environmental service flow (see Randall et al. 1996). This method is the most comprehensive for simultaneously estimating all of the types of economic value outlined in Figure 10.2, but it requires careful development to avoid strategic behavior of respondents.

In cases of environmental service flows with a critical zone or threshold, it may be necessary to establish a safe minimum standard (SMS) or maximum sustainable yield (MSY) (Barbier et al. 1990). The objective is to avoid irreversible effects to human health or ecosystems such as in the case of nitrate–nitrogen contamination of groundwater in parts of the United States. Contamination is considered critical to human health at 10 mg/L, which was established by the US Environmental Protection Agency. The T-value in the universal soil loss equation (USLE) is an example of a somewhat more flexible or reversible SMS relative to long-run productivity of the soil. One could envision a similar safe standard of sediment inflows into a water reservoir to maintain minimum storage capacity. The MSY of a water aquifer may be a withdrawal rate equal to or less than the annual recharge rate, which becomes critical in cases where the aquifer is covered by a heavy rock overburden. In these cases, excess withdrawal can result in costly irreversible loss in aquifer capacity.

Once some basic economic estimates have been established for environmental service flows relative to water reservoirs, etc., it is possible to select instruments to accomplish more efficient and/or equitable outcomes. Economists prefer instruments that provide incentives and allow a range of choices as opposed to command and control instruments. Examples include taxes, subsidies, and auctioning of assimilative capacity up to some resource constraint of SMS. Well-defined property rights are a recurring theme of economists, and this is equally true of any changes in property or use rights related to environmental service flows.

10.4 CASE STUDIES

The following four case studies of economic analysis of soil degradation and restoration represent variation in scale, location, method, and analysis. The first case study utilizes a metaproduction function approach to identify the factors (including soil and land degradation) related to growth in agricultural and food output in a sample of 23 developing countries from 1971 to 1980. Follow-up analysis by Enver and Hitzhusen in 2006 is also discussed. The second case is focused on estimating the benefits of extended hydropower production from reduced soil erosion and sedimentation of the Valdesia Reservoir and hydroplant in the Dominican Republic (DR). The third case presents the results of several benefit–cost studies in Ohio to determine the economic off-site impacts of soil agriculture and coal strip mining soil erosion on lake, harbor, and drainage ditch dredging costs, lake-based recreation and residential housing values, and water treatment costs. The fourth case presents the results of two studies of the downstream benefits of the CRP in Ohio including reduced off-site erosion impacts, wildlife habitat, and potential production of cellulosic-based bio (ethanol) energy.

10.4.1 SOIL AND LAND DEGRADATION AND AG GROWTH MODEL (CASE 1)

A study by Zhao et al. (1991) focused on identifying the factors that determine the agricultural production growth rate and on testing the effects these factors have on agricultural growth in developing countries. Specifically, this study involved statistical estimation of an aggregate agricultural growth function based on cross-country data for 23 developing countries. Special attention was devoted to environmental degradation, agricultural pricing policy, and the policy implications resulting from the effects these variables have on agricultural and food production growth.

The methodology used is based on the concept of a metaproduction function hypothesized by Hayami and Ruttan (1985). Following Lau and Yotopoulos (1987), this metaproduction function can be written as

$$Y_t = f(X_{it}, \ldots, X_{mt}, t) \tag{10.1}$$

where Y_t is the quantity of output, X_{it} is the quantity of the ith input, $i = 1, \ldots, m$, and t is time. This production function can be used to represent the input–output relationship of agriculture or food production. As defined by Lau and Yotopoulos (1987), the embedded hypothesis for this metaproduction function is that all producers (or countries) have potential access to the same set of technology options, but each may choose a particular one, depending upon its natural endowment and relative prices of inputs. In the Zhao (1988) research, a metaproduction function similar to Equation 10.1 was estimated. However, since his focus was not on the estimation of productivity changes, the time variable t was replaced by two variables, soil and land degradation and price distortion. These two variables were hypothesized to have impacts on technology choices among different countries during the study period.

This study was concerned with estimating the relative changes in output rather than the absolute levels of output. Thus, the dependent variables are expressed in relative terms as a percent change or average level of output during the study period. Following Hayami and Ruttan (1971), the inputs may be categorized as follows: (1) internal resource accumulation, including an expansion of the arable and permanent cropland, and the growth of labor force, and (2) technical inputs supplied by the nonagricultural sector. Two industrial inputs, fertilizer and machinery, represent proxies for the whole range of inputs that include modern, biological, and mechanical technologies.

It is widely recognized that government policies may greatly affect agricultural production growth in developing countries (IIED and World Resource Institute 1986; Argawala 1983; Krishna 1982). Most of these studies conclude that many of the economic policies pursued in developing countries have limited the growth of agricultural production and hampered efforts to reduce rural poverty.

Although price is not a complete measure of incentives in agricultural production, it is one of the most important policy devices. Agricultural price policy has been actively used by virtually all governments to pursue a wide variety of resource allocation, government revenue generation, and income distribution objectives. The fact that most developing countries discriminate against agriculture is reflected in price distortions.

The agricultural or food production growth function can be established by estimating the coefficients between agricultural/food production growth and the changes in the relevant independent variables. The growth of total agricultural production (TAP) or total food production (TFP) is affected by changes in the same factors. Based upon discussions above, the aggregate agricultural or food production growth function can be expressed in the following form:

$$Y_g = f(A_g, L_g, Q, F_g, M_g, G) \tag{10.2}$$

where Y_g = growth rate of agricultural or food production; A_g = rate of change in labor input; L_g = rate of change in land cropped; F_g = rate of change in fertilizer consumption; M_g = rate of change in machinery power utilization; Q = quality of arable land or soil; and G = government policies, for example, price and land use.

Those variables with subscript "g" are flow variables, and they should be measured in growth rate. Q and G can be seen as stock variables, and the average values are to be used in estimation.

The approach used in this study involves estimating a cross-country agricultural growth function based on a sample of 23 developing countries. Variations in agricultural growth rate are accounted for by differences in the growth rate of agricultural inputs and related factors. All the data used in this study are from the period of 1971 to 1980. The data for flow variables are the average of 1971 to 1980, and the data for stock variables are the 1975 actual figure. More detailed development of the study methodology is presented in Zhao et al. (1991).

10.4.1.1 Analysis and General Results

Zhao et al. (1991) present a detailed discussion of the main results of the estimation of the aggregate agricultural production and the food production growth function. Considering the aggregate nature of the secondary data, the levels of statistical significance of several of the estimated coefficients are quite good. The six independent variables in the model can explain as high as 82% of the variance in total agricultural production and about 78% of the variance of food production when growth is measured by the average index. However, the models based on percentage growth measure are less satisfactory, and the R^2's for the TAP and TFP models are 0.66 and 0.68, respectively. The F tests show that one can be 95% to 99.99% confident of rejecting the hypothesis that all the estimated coefficients are zero for the four models. These models, as a whole, are quite well defined, and the multicollinearity problem in the statistical analysis does not appear to be serious based on the SAS collinearity diagnostics procedure.

Even though the data are categorical (three classifications) and thus gross measures, the statistical significance is relatively strong. Based on the percentage measure, the estimate of the moderate level of land degradation is significant at the 5% level in the total agricultural production growth model, and the estimate for food production growth is significant at the 2% level. If the growth is measured by the average index, the estimate for moderate land degradation is not significant in the total agricultural production growth model. It is, however, still significant at the 15% level in the good production growth model.

When comparing the absolute values of the estimates and their corresponding significance levels, the soil and land degradation variable has higher coefficient values, and the estimates are more significant in the TFP than in the TAP models. This difference indicates that soil and land degradation tends to affect food production more significantly than nonfood agricultural production. The result seems to confirm the belief that soil and land degradation does threaten food production growth. It also impedes income increases in rural areas because of a direct relationship between farmers' income and food production growth. Notice that the estimated coefficients of the high degradation level have wrong signs and are insignificant at the 15% level.

This may be due to the fact that there are only three countries with a high degradation level, which results in large standard errors of the coefficient estimates.

The positive relationship between the land-area-change rate and agricultural production growth or food production growth is quite strong with both significant at the 1% level based on the index measure of growth. However, the estimates are insignificant in percentage measured growth models. In contrast to the negative influence of the soil and land degradation factor (LD) discussed above, the amount of arable and permanent cropland is strongly related to agricultural and food production growth. Reduction in severe soil and land degradation should increase the future availability of arable and permanent land, which may increase agricultural output.

10.4.1.2 Summary and Conclusions

The result of the Zhao et al. (1991) analysis shows that price distortions in the economy and land degradation had statistically significant negative impacts, while the change in arable and permanent land was positively related to the growth of agricultural production and food production in 23 developing countries from 1971 to 1980. These results emphasize the importance of "getting prices right" and implementation of sustainable land and water management practices if future growth in food and agricultural output is to be sustained in developing countries.

After price distortion, the variable of greatest significance is the degree of soil and land degradation. The reduction in overall agricultural (as opposed to food) production growth caused by soil and land degradation is smaller in magnitude and less significant statistically. The regression results strongly suggest that soil and land degradation in developing countries does constitute an immediate as well as long-term threat to these countries' capacity to produce food. The estimation in this study failed to capture the off-site damage from soil degradation, for example, water pollution and siltation of hydropower reservoirs and harbors. Thus, the actual negative effects are likely to be much larger and more significant than estimated, since off-site impacts of soil erosion are generally much greater than on-site impacts.

Most developing countries are more dependent on their natural resources, notably land and water, and soil and land degradation significantly threatens agricultural growth. Soil and water conservation is of great importance to sustainable economic development. Past development efforts have been based on the exploitation of natural resources in many developing countries. In the long run, soil and land protection and agricultural growth are complementary rather than competitive, even though there might be some trade-off between the two in the short run. Not only conservation projects but also policy reforms are needed to protect the soil base. Policy reforms require that soil conservation, proper drainage in irrigation projects, etc., be incorporated as an integral part of a development program.

The policy reforms should focus on increasing economic incentives for conservation. Since in the majority of developing countries, most agricultural activities are done by small operational units, such as households and small farms, appropriate economic incentives for millions of farmers are vital to channel development activities into sustainable development patterns. Studies show that the serious degradation is primarily due to the cumulative effects of many small agricultural operations that are not affected by environmental regulations (Repetto 1987; IIED and World

Resource Institute 1986). The appropriate economic incentives may include increasing agricultural prices to the competitive level, reducing taxation on agricultural production, establishing effective property rights, providing subsidies and assistance for conservation practices, and eliminating input subsidies.

Correcting farm level disincentive problems is inadequate because soil erosion causes major off-site impacts that are not borne by farmers. In addition, farmers' time horizons are much shorter and their discount rates much higher than the society at large. Therefore, public interventions and national actions are usually required to ameliorate the effects of soil erosion and degradation including better defined property rights, government regulations on land use, and the traditional approach to environmental problems—public authorities investing in reforestation and pollution control projects.

Price distortions result in great losses of agricultural production in many developing countries. Thus, to have a high economic growth and at the same time have a high agricultural growth, "getting prices right" is a necessary condition. In some sense, "getting prices right" also applies to the argument for a sustainable pattern of growth. One of the most important causes of environmental degradation is that environmental services are undervalued. Activities that exploit soil and land resources and cause degradation are not fully priced or taxed, or, at least, not valued at their marginal social valuation. This in turn leads to faulty measures of national income and economic wellbeing. The problem of inadequate soil and water management or exploitation of soil and water resources is similar to the undervaluation of commodities or credit—the natural resources are overused and they are undervalued. Thus, the arguments for well-defined property rights, pollution taxes, revised national income accounts, and implementation of conservation measures is "getting prices right on environmental services."

10.4.1.3 Follow-Up Analysis

In 2006, Enver and Hitzhusen completed a follow-up analysis to the earlier Zhao et al. (1991) effort. Data from the Global Assessment of Soil Degradation (GLASOD) in 1991 attempt to assess the diverse causes, processes, and consequences of soil degradation on a globally consistent basis (Wood et al. 2003). These data were utilized by Enver and Hitzhusen (2006) to estimate a production function based on the earlier Zhao formulation combined with a spatial mixed autoregressive model and a spatial error term. The authors found that a reduction of soil quality from the GLASOD estimates of percent of agricultural land degraded in each country due to poor agricultural practices resulted in an average of 9% reduction in productivity on these lands. However, spatial autocorrelation in the error term of the model suggests that unobservable factors are spatially correlated across neighboring countries. More analysis is needed to address this concern.

10.4.2 DR Valdesia Hydro Reservoir Sedimentation (Case 2)

To illustrate some of the concepts introduced in the BCA concepts section of this chapter, this case summarizes an earlier evaluation of soil conservation project for a hydroelectric watershed located in the DR. After some background information about the study area has been presented, the method used to estimate reservoir sedimentation under different assumptions regarding land resource management upstream as well as the approach for estimating the benefits and costs of soil conservation are outlined. A more complete description of this research can be found in Veloz et al. (1985).

Like many other LDCs, the DR has looked to hydroelectric development to reduce fuel oil imports. Many promising dam sites are found in the Cordillera Central, where most of the country's major rivers originate. Elevations vary in that region from 100 to 3000 m above sea level, and precipitation in many areas exceeds 1500 mm/year.

As late as the 1950s, most of the Cordillera Central was forested. But after the death in 1961 of the country's longtime dictator, Rafael Trujillo, who owned most of the country's standing timer, many forested areas were clear cut. Peasants then settled on the newly cleared land. The CRIES Study (1980) shows that by the late 1970s, most of the region had been converted to rangeland and cropland. High rates of soil loss are now a problem throughout the Cordillera Central, according to Hartshorn (1981).

The Valdesia Dam is one of several projects initiated in order to decrease the DR's dependence on imported oil. By developing Valdesia and one other site during the 1970s, de La Fuente (1976) states that the DR reduced the share of electricity generated at oil-burning facilities from over 90% to less than 80% by 1981. However, the Direccion General de Foresta (1976) suggests that sedimentation has been a continuous concern at the reservoir. Even before a major hurricane swept over the island in 1979, annual sediment yield in the reservoir watershed was estimated to be 1.4 million metric tons.

10.4.2.1 Study Methodology

Although the Dominican government had determined that the Valdesia reservoir catchment area would have a watershed management program in the near future, a specific plan for reducing soil erosion there had not yet been developed. Without a more detailed plan, it was possible to characterize in a general way the land uses and land management techniques to be encouraged under the project. Consultation of the literature wand with professionals who have worked in the DR suggested the following guidelines for four slope classes:

Slope class A (3%–20% slope): Continue crop farming on existing cropland. Encourage mulching and contour farming. Renovate all rangeland with a slope of 13% to 20%.
Slope class B (21%–35% slope): Encourage mixed cropping/agroforestry enterprises on existing cropland. Renovate all rangeland.
Slope class C (36%–50% slope): Convert all existing cropland to agroforestry. Renovate all rangeland.
Slope class D (over 50% slope): Reforest all existing cropland and rangeland.

To estimate the impacts of project implementation on resource management, a 1:50,000 scale topographic map of the 85,090-ha watershed was consulted. The watershed was divided into 10-ha cells, each of which was placed into one of the four slope classes. By contrasting existing land use based on the CRIES (1980) results with recommended land use, it was found that 57% of the catchment area would not be affected by project implementation, while 11% would have to undergo a change in land use. In the remaining 32%, soil conservation goals would be met by mulching, contour farming, and range renovation.

Reductions in soil loss resulting from compliance with project guidelines were estimated by applying the USLE (see Veloz 1982) using soil survey data and the estimates of slope that were used to categorize land into the four slope classes. In general, this approach yields conservative estimates of erosion inasmuch as the USLE explains only gross sheet and rill erosion and not gullying, stream bank scour, and other types of soil loss.

Average sedimentation rates were calculated by multiplying an estimated sediment delivery ratio from Onstad et al. (1977) by estimated erosion rates. Given current land use and land management practices and evidence from Espinal (1981), it was estimated that average yearly accumulation of sediment in the reservoir is 921×10^3 metric tons, which is 59% of the rate obtained by dividing total sediment in the reservoir by the reservoir's age. However, the latter rate includes the silt deposit by Hurricane David in 1979 as well as the bed load of sediment. Thus, the observed rate probably overstates the long-term sedimentation rate. Implementation of the project throughout the watershed would reduce sedimentation by an estimated 86%, to 130×10^3 metric tons a year.

10.4.2.2 Results and Implications of Analysis

Sedimentation was assumed to have a negligible impact on electricity production as long as it remains possible to operate the hydroplant. However, without any change in land resource management, it was estimated that the generator intake would be clogged with silt after 19 years. As little as a 25% reduction in erosion, though, would extend the hydroplant's lifetime by about 6 years. The group affected most by project implementation would be hillside farmers. Researchers familiar with the study area supplied information about farm income without the project. Budgets for soil conserving activities were obtained from United States Agency for International Development (USAID) and other sources (see Veloz 1982).

Economic evaluation of the watershed management project outlined above strongly suggests that the net benefits associated with soil conservation depend on the accounting stance. The results of private-level analysis reported in Table 10.1

TABLE 10.1
Results of Private Level Analysis

Slope Class	Hectares	PNPVi	PNPVi/Hectare
A	7640	DR$12,500,000	DR$1635
B	17,800	−550,000	−30
C	8600	−1,925,000	−225
D	2350	−1,100,000	−470
Total	36,390	DR$8,925,000	DR$245

indicate that some farmers would benefit from project implementation, while others would lose, assuming that all use a real discount rate of 5% and a 25-year time horizon. Those with land in slope class A would, on average, benefit; the net returns per hectare they would realize by adopting improved tillage practices would exceed status quo net returns per hectare. However, full project implementation would force farmers who work on steeper land (slope classes B, C, and D) to make a change in land use, which is, on average, not profitable for them.

Social-level analysis of soil conservation's on-site economic effects (Table 10.2) yields results that are similar to the results of private analysis of those same effects. In the former analysis, labor is shadow-priced at less than the minimum wage because of high underemployment and unemployment in the DR. Also, the project's exported outputs and imported inputs are evaluated using the parallel market exchange rate rather than the official exchange rate, which overvalues the local currency. Even after these adjustments have been made, social level analysis yields the same two conclusions as private-level analysis, assuming the same discount rate, time horizon, and accounting stance. First, as an entire group, farmers would benefit from project implementation. Second, farmers with land in the higher slope classes, who would be required to switch from traditional agriculture to agroforestry or reforestation, would lose because of project implementation.

Broadening the accounting stance so as to incorporate the off-site effects of erosion yields substantially increased estimates of the net benefits of soil conservation. Shown in Table 10.2 are the net present values of extensions in the lifetime of the Valdesia Dam obtained by conforming to project guidelines in selected slope classes. For example, given project implementation in slope class A, the additional external benefits of reducing erosion in slope class B (see Table 10.2) would greatly exceed the private costs of doing so (see Table 10.1).

Proper conceptualization and analysis of reservoir sedimentation from an economic perspective is extremely important if optimal corrective measures are to be implemented. A core concept is the private versus social perspective or accounting stance with respect to both space and time. A social or economic accounting stance is primarily concerned with both on-site and off-site costs and returns or returns to the total society over the long run. Several forms of corrective or shadow pricing may be necessary in doing the social cost–benefit estimates; this complicates the analysis and results in greater data needs.

TABLE 10.2
Results of Public Level Analysis

Slope Classes Where Project Guidelines are Followed	Dam Lifetime	Off-Site Benefits of Compliance with Guidelines
None	19 years	0
A	20	DR$1,755,000
A + B	25+	9,350,000+
D	25	9,350,000

Both the private and social perspectives or accounting stances are important as illustrated in the DR case application. The private level analysis demonstrates the on-site profitability of alternative soil conserving or sediment reducing practices to hillside farmers and shows those on steeper slopes as net losers. The social level analysis includes both on-site and hydropower-related off-site impacts and includes corrective or shadow prices for both labor and overvalued local currency. This analysis shows social benefits of soil conserving practices to be substantially higher than the private costs in the Valdesia Watershed and reservoir in the DR, which has important implications for transfer payments to farmers for soil conserving practices.

The tools generally used to determine soil loss rates and sediment transport in LDCs are, by and large, rudimentary. Thus, it is hardly surprising that available estimates of the external costs of erosion (e.g., the costs of lost hydroelectricity, of reductions in irrigation water, of flood control, etc.) are imperfect. Similarly, in order to understand better how the hillside farmer makes the decision of whether or not to utilize land in a soil conserving fashion, better information about his institutional environment and about production functions for both traditional and less erosive agricultural activities is needed. Finally, better demographic information about the populations that inhabit environmentally sensitive areas would improve our understanding of how renewable resource management problems evolve over time.

With more research on the on-site and off-site impacts of reductions in soil loss, it will be possible to perform economic analysis that is more comprehensive than the analysis described in this case. The case application estimates of the external benefits of soil conservation are probably conservative (only one type of benefit—the extension in dam lifetime—was quantified), and the per hectare returns associated with traditional agriculture have probably been overstated. Thus, the results of the Valdesia case argue for more research that allows for the design of better strategies for dealing with pressing soil erosion and sedimentation problems in developing countries throughout the world.

10.4.3 SOME OHIO EVIDENCE OF SOIL EROSION OFF-SITE ECONOMIC IMPACTS (CASE 3)

10.4.3.1 Methods and Results of Studies

Resource economists at The Ohio State University have evaluated the downstream economic impacts of soil erosion in Ohio on several receptors including natural lakes and man-made water reservoirs. For example, Macgregor (1988) used the Ohio Department of Natural Resources State Park Lakes data on lake characteristics, visitations, and dredging to estimate the boater value losses and dredging costs due to sedimentation in 46 state park lakes. Macgregor et al. (1991) report that sediment deposition was significantly and negatively related to boater visitor days, and this coefficient was multiplied by a mean value ($23.92) for a motorized boating recreational day to get boater value loss. These findings indicate an average boater value loss in the 46 lakes of $0.54 per metric ton of sediment, but the values ranged from $0.009 to $13.18 per metric ton of sediment. This emphasizes the need for targeting of soil conservation funds based on off-site economic impacts. The average cost was $1.42 to dredge 1 metric ton of sediment in 11 state park lakes where dredging was being done in 1987, and costs ranged from $1.31 to $1.61 per metric ton. This dredging is funded by boater license fees and a portion of the gasoline tax. Ironically, farmers are exempt from this tax on fuels used in the farm operation.

Forster et al. (1987) estimated the relationship between water treatment costs and soil erosion in 12 public water treatment plants in western Ohio, depending on surface water storage reservoirs. Independent variables other than soil erosion used in the regression analysis included treatment plant size, storage time of untreated water, and turbidity improvement due to water treatment. Results indicate that a 10% reduction in annual gross soil erosion results in a 4% reduction in annual water treatment costs. The average increase in water treatment costs per metric ton of sediment delivered was $0.35 at the 12 treatment plants. For all Ohio communities, it is estimated that annual water treatment costs would decline by $2.7 million with a 25% reduction in soil erosion.

Hitzhusen et al. (1995) have done an economic evaluation of state park lake dredging and sedimentation impacts on lake recreators (particularly boaters) and lakeside property values. A majority of the lakes studied are man-made water reservoirs. The earlier cross-sectional travel cost results from Macgregor (1988) were updated and combined with a two-stage hedonic pricing model (developed by Bejranonda et al. 1999) used to estimate the impacts of rates of sediment inflow and accumulation as well as dredging on lakeside property values. The sedimentation economic impacts on property values were statistically significant and generally larger than the impacts on boaters. Some simulation analyses also suggested that lake residents and recreational users were willing to pay more for upstream erosion control than for dredging.

The farm level implications of the foregoing downstream research results are summarized for a hypothetical 202-ha farm in Table 10.3 utilizing observed low (L), mean (M), and high (H) downstream impacts and comparing 33.6 and 11.2 metric ton on-site soil losses/ha/year. The 33.6 metric ton soil loss represents 3T, a threshold for the current USDA CRP. The 11.2 metric ton soil loss represents a sustainable (T) rate that is slightly larger than the average T value for Ohio conditions. With the exception of lost boater value, these downstream economic impacts are representative of cleanup cost rather than willingness to pay measures. Nevertheless, the results in Table 10.3 suggest that a given level of soil loss may show significant variations in its downstream economic impacts depending on where (in which watershed and upstream from which receptor) the hypothetical farm is located. For example, the variation in boater value loss from the hypothetical farm's 3T versus T level of erosion is from $4 to $5975, while the mean value of annual costs by receptor ranges from $160 for water treatment to $2805 for drainage ditch dredging.

These Ohio downstream research results must be combined with related on-site or upstream farming practices and their associated costs and returns if efficient and fair soil conservation policies are to be established. Previous Ohio studies indicate that many soils [Forster et al. (1987) estimates 60% of the state's soils] are responsive to crop rotations and reduced tillage. Thus, net incomes of farmers using these systems on these soils are potentially equal to or greater than net incomes with conventional systems. However, reduced tillage, particularly no-till, usually results in the use of more chemicals, which may adversely impact surface and groundwater quality. On the remaining 40% of Ohio's soils, if the cost of controlling erosion is less than the

TABLE 10.3
Farmland Soil Erosion Downstream Impacts: Some Ohio Research Results

Type and Number of Receptor/	$/ton Sediment[a]			$ Off-Site Impacts/Year[b]		
Environmental "Sinks" Analyzed	I_L	I_M	I_H	I_L	I_M	I_H
1. State park lakes (*n*-46)						
a. Lost boater value (*n*-46)	0.009	0.54	13.18	$4	$245	$5975
b. Dredging costs (*n*-11)	1.32	1.42	1.61	600	645	730
2. Lake Erie harbors (*n*-9)						
a. Dredging costs	2.49	3.20	5.67	1130	1450	2570
3. Ohio River						
a. Dredging costs	1.96	2.78	3.93	890	1260	1780
4. Drainage ditches (*n*-6 counties)						
a. Dredging costs		2.06			2805[c]	
5. Water treatment costs (*n*-12)		0.35			160	

[a] Multiply by 0.1 to get $/metric ton of gross erosion.
[b] Assumes 202 acres of row crops with average gross erosion of 3T under conventional tillage. The assumed value for *T* is 11.2 metric tons/ha/year, which by definition is sustainable. Sediment delivery ratio of 10% (average). Multiply by 0.1 to get the downstream cost/year for each additional ton of gross erosion from this 202-ha hypothetical farm.
[c] Sediment delivery ratio –30%.

cost of lost recreation, dredging, etc., then control rather than cleanup strategies generate more net benefits to society. The reverse may be true in some situations.

In addition to the foregoing analyses of agricultural soil erosion and related downstream impacts in Ohio, another study looked at the impact of coal surface mining erosion and acid spoils on downstream lakeside property and lake-based recreation utilizing hedonic pricing and travel cost models, respectively. This study by Hitzhusen et al. (1997) also estimated the potential lake-based property and recreation benefits from increased reclamation of existing coal surface mines utilizing fluidized gas desulfurization (FGD) waste or by-product from scrubbers on coal burning power installations.

The study did a paired comparison of Piedmont and Leesville Lakes in Eastern, Ohio, which were surface mine impacted and nonimpacted, respectively. The main finding is that reclamation of coal surface mines upstream from Piedmont Lake would result in increased downstream residential property values of $73,159 and an increase in annual lake-based recreation of $256,345. Although these benefits alone will not cover the full costs of upstream surface mine reclamation, they are a good start.

10.4.4 OSU STUDIES ON OHIO DOWNSTREAM BENEFITS OF CRP (CASE 4)

10.4.4.1 Methods and Results of Studies

The US CRP was established by the Food Security Act of 1985 to establish 10- to 15-year contracts with agricultural producers and landowners to retire highly erodible and environmentally sensitive cropland and pasture from production (USDA-FSA 2009). However, millions of acres of erodible land may be removed from CRP in the following years due to farmers adopting different soil conservation practices on lands leasing CRP or reverting to potentially erosive row crop production (RCP). Shakya and Hitzhusen (1997) carried out a BCA concerned with the sustainable future use of the most erosive and downstream damaging croplands. The question addressed was whether planting trees on these lands could be one economically viable choice. BCA of CRP, white pine plantation (WPP), and RCP suggests that WPP is a viable option for some more erosive CRP lands in the Midwest with proximity to a paper mill. Furthermore, federal expenditure could be minimized by reducing corn subsidies to a level where WPP would maintain its economic feasibility when compared to both CRP and RCP. Besides meeting all the goals of CRP, WPP also ensures longer-term environmental protection and provides a marketable product.

A more recent BCA of CRP in an M.S. thesis by Sarah Kiger has been concerned about the reduced CRP enrollment from higher corn prices resulting from increased demand for food, feed, and ethanol fuel feedstock (Kiger 2009). The latter has been compounded by corn ethanol subsidies and protective tariffs (Wright 2011). The focus of the Kiger thesis research is twofold: to extend Shakya's (1992) research on the economic estimates of the downstream environmental benefits of and costs of CRP and to do a preliminary assessment of the technical and economic potential of ethanol production from CRP land cellulosic feedstock.

M.S. thesis research by Kiger conducted both financial and economic analysis of CRP land in Ohio. These analyses are different but complementary since financial analysis takes the accounting stance of the individual CRP-enrolled farmer, while economic analysis takes a societal accounting stance. The following variables are considered as benefits (+) and costs (−) for CRP depending on whether one is doing financial or economic analysis:

CRP:

$$\text{financial: } F_s + B_{on} + 1/2C_{ec} - 1/2C_{mc} = R_1 \qquad (10.3)$$

$$\text{economic: } B_{of} + B_{on} + B_w + B_{rh} - C_{op} - C_{cc} - C_{ec} - C_{mc} = R_s \qquad (10.4)$$

where B_{of} = off-site benefits from reduced soil erosion; B_{on} = on-site benefits from reduced soil erosion; B_{rh} = benefits of recreation, including freshwater recreation, hunting; B_w = benefits of wildlife viewing; C_{ec} = establishment costs for CRP cover crops; C_{mc} = maintenance costs for CRP land, including mid-harvest management; C_{op} = social opportunity cost of land, includes rental value of land without considering federal crop subsidies; F_s = government rental payment for CRP; R_1 = return to land (to the farmer); and R_s = return to society.

For the financial analysis of CRP, the annual rental rate paid by the government (F_s) is considered a benefit as it is a source of farmer income for enrolling in the CRP and not producing a crop. The other benefits of CRP included in this analysis are the on-site benefits from reduced soil erosion (B_{on}) calculated for the average of Ohio in the economic analysis. The costs of CRP to farmers are 50% of the yearly maintenance costs (C_{mc}) and 50% of the cost for establishing a cover crop (C_{ec}). The summary of costs and benefits considered for CRP are presented in Table 10.4.

For this analysis, two different government rental payments (F_s) are considered, a high payment based on the average of all acres enrolled in Ohio CRP [regular CRP plus land enrolled in the Conservation Reserve Enhancement Program (CREP)] and a low payment based on just acres enrolled in the regular CRP. The cost for cover crop establishment (C_{ec}) is calculated based on the combined cost of seed, fertilizer, herbicides, and seeding. The maintenance cost (C_{mc}) is based on the cost of two mows in the first year to aid in establishment, fertilizer, and additional seeding in year 4, and mid-harvest management, which typically will involve disking of the land. Both of these costs are shared, with the program paying 50% of the total cost. Since some of the costs occur yearly throughout the length of the contract and others occur only once or intermittently throughout the contract, all inputs for this analysis are considered on a yearly basis.

The results of the financial analysis show annual financial returns to the farmer ranging from $276.87 per hectare when rental payments are higher, to $193.97 per hectare when they are lower. When farmers are considering enrollment, reenrollment, or extension in CRP, it is expected that they will weigh the potential financial benefits of putting their land in CRP versus RCP. For example, the large exit of farmers from the program in 2008 was likely due to a combination of factors including high crop prices, which increased the financial returns of RCP relative to CRP (Streitfeld 2008; USDA-FSA 2008).

In order to understand the economic value of CRP, one must calculate the benefits received from reduced soil erosion. This looks at the amount of soil erosion with CRP compared to the amount of soil erosion that would be expected if the land was in RCP. There is a wide array of private and social costs resulting from soil erosion. On-site impacts include loss of fine soil particles, organic matter, and other nutrient

TABLE 10.4
Values for Financial Analysis of CRP

Description	Variable	Total Price ($/ha)	Frequency	Yearly Price ($/ha/year)
Government	F_s – High	$299.98	Yearly	$299.98
Rental payment	F_s – Low	$217.09		$217.09
On-site benefits	B_{on}	$6.97	Yearly	$6.97
Cover crop establishment	1/2 C_{ec}	$177.30[a]	1st year	$17.73
Maintenance	1/2 C_{mc}	$123.50[b]	Total for 10 years	$12.35
Return to land	R_1 – High			$276.87
	R_1 – Low			$193.97

[a] Harvey 2000; University of Minnesota 2008; Ward and Freytag 2008; Ferrell and Sellers 2011; Prairie Seed Farms 2011; DTN 2012.
[b] USDA FSA 2006.

elements, which can result in lower productivity (Pimental et al. 1995; Lal et al. 1998; Olson et al. 1998). Off-site impacts include sedimentation of local water bodies, impairment of water quality, disruption of wildlife habitats and ecosystems, and reduction of recreational and aesthetic water resources (Ribaudo 1986b; Hitzhusen 1991, 2007; Bejranonda et al. 1999). From a societal point of view, the reduction in these costs needs to be accounted for in order to determine the true economic benefits of CRP land. Through different methods such as contingent valuation method, hedonic pricing, and travel cost methodology, economists can take the physical environmental cost and benefits of CRP land and equate them to economic returns.

Both off-site and on-site costs of soil erosion are associated with the amount of soil loss per acre (Ribaudo 1986b). In order to calculate the benefits of reduced soil erosion, we must first calculate the cost of soil erosion from RCP and from CRP. If we subtract the cost of soil erosion from CRP from the cost of soil erosion from row crops, we can get the benefits of reduced soil erosion from CRP (B_{on} and B_{of}).

The annual damage from soil erosion on a per ton basis (studies by Ribaudo 1986b; Shakya 1992) is used for estimation of the cost off-farm. Such estimates were established for different regions of the United States by Ribaudo (1986b). Shakya (1992) recommends using the estimates for three of the regions calculated in this study—Corn Belt, Appalachian, and Northeast—since Ohio lies on the borders of these three regions and displays characteristics of all three. Ribaudo's estimates for these three regions are used for the calculation of off-site costs (Table 10.5) and have been adjusted to 2011 values. In this analysis, the average values from the three

TABLE 10.5
Off-Site and On-Site Costs and Benefits of Soil Erosion in Ohio

Region	Downstream Costs $/t	Off-Site Cost $/ha ($C_{of}$) CRP	RC	On-Site Costs $/t	On-Site Cost $/ha ($C_{on}$) CRP	RC	Total ($C_{of} + C_{on}$) CRP	RC
Appalachia	2.49	1.11	13.96	0.86	0.40	4.82	1.27	18.77
Corn Belt	2.17	0.96	12.18	1.54	0.69	8.65	1.24	20.82
Northeast	12.60	5.66	70.59	1.65	0.74	9.26	5.96	79.86
Average	5.76	2.58	32.24	1.35	0.61	7.57	2.82	29.66

For CRP (average): $/hectare/yr
Off-site Benefit: $B_{of} = C_{of}$ (RC) $- C_{of}$ (CRP) $= 29.66$
On-site Benefit: $B_{on} = C_{on}$ (RC) $- C_{on}$ (CRP) $= 6.97$
Total = 36.63

For CRP (Corn Belt): $/hectare/yr
Off-site Benefit: $B_{of} = C_{of}$ (RC) $- C_{of}$ (CRP) $= 11.21$
On-site Benefit: $B_{on} = C_{on}$ (RC) $- C_{on}$ (CRP) $= 7.95$
Total = 19.17

Notes: Average erosion under row crop (RC) = 5.6 metric tons/hectare/year; average erosion under CRP = 0.45 metric tons/hectare/year.

regions are used as the medium-level soil erosion off-site costs, and the Corn Belt values alone are used for the low-level soil erosion off-site costs. It is important to note that this value does not include the costs of biological impacts of soil erosion and the damage from wind erosion, thus making it a conservative estimate of soil erosion economic impacts (Shakya 1992).

Estimations of the on-farm costs are based on the amount of soil lost per acre per year (Colacicco et al. 1989). These values have also been established for the three regions. They are similarly averaged across the three regions and adjusted to 2011 values (Table 10.5). On-farm benefits not only affect the economic analysis but farmers directly incur these benefits in the financial analysis as well (B_{on}).

The average soil loss for Ohio cropland was estimated to be 5.6 metric tons per hectare per year (USDA-NRCS 2009). The amount of soil saved under CRP can be considered as a societal benefit. CRP land in Ohio has been shown to reduce erosion to about 0.45 metric ton per hectare per year (USDA-NRCS 2009). So while there is still some soil loss on CRP land, it is far less than that of cropland. This makes CRP more sustainable and results in lower off-site costs. Using the same methods and formula created by Shakya (1992), off-site and on-site costs and benefits of soil erosion were determined (Table 10.5).

Since environmental benefits from CRP land extend beyond soil loss, additional benefits were also considered. Erosion from agricultural sites can have a large impact on Ohio lakes and streams and in turn can influence visitation at water-recreation sites (Hitzhusen 2007). CRP in Ohio has been shown to provide habitat for a large number of grassland bird species, including pheasants, which have been marginalized by agricultural production (Swanson et al. 1999). Feather et al. (1999) used data on recreation and land use characteristics from multiple surveys collected by the USDA's Economic Research Service. These data were then used to generate recreation models that estimate the benefits of CRP land.

This study included Ohio in the Northeastern region. Since many other Corn Belt states (Indiana, Illinois, and Wisconsin) were included in this region, averaging across regions was not necessary. This study reports environmental benefits for three aspects of CRP land—freshwater-based recreation, pheasant hunting, and wildlife viewing—and uses trip and travel cost data to determine the consumer surplus due to CRP. Consumer surplus measures the benefits consumers derive from the consumption of goods and services; it is the difference between what consumers are willing to pay for a good or service and what they actually pay. The values reported in this study, totaling $182.95 per hectare, have been adjusted for inflation and are added to the total economic benefits of CRP reported in this study (Table 10.6).

In total, the economic values calculated indicate that the total economic benefit of CRP lands in Ohio is $88.90 per acre per year (Table 10.6). This represents a low-bound value for the economic benefits of Ohio CRP land, as it does not include all economic benefits such as nutrient load reductions to streams and lakes, carbon sequestration, biological improvements, and potential use of energy crops grown on CRP lands for biofuel production. Additional analysis is underway on potential on-site benefits of biofuel production on CRP lands. These on-site and off-site benefits will then be compared to farmers' land opportunity and other costs minus CRP government payments to get at net benefits to both farmers and society.

TABLE 10.6
Review of Economic Benefits Provided by Ohio CRP Land

Explanation of Benefits	Economic Value of Benefit (2011$)	Source
Off-site benefits of reduced soil erosion in Ohio	$29.66/ha/year	Shakya 1992; Ribaudo 1986b
On-site benefits of reduced soil erosion in Ohio	$6.97/ha/year	
Consumer surplus for freshwater-based recreation	$10.18/ha/year	Feather et al. 1999
Consumer surplus for pheasant hunting	$25.86/ha/year	
Consumer surplus for wildlife viewing	$146.92/ha/year	
Total	$219.583/ha/year	

10.5 SUMMARY AND CONCLUSIONS

This chapter starts with a brief review of the extent and economic consequences of soil degradation in a global context with particular emphasis on the United States and developing countries. Next, a rationale is developed for more BCA of soil degradation and restoration to improve private and public decision making. This is followed by a detailed discussion of the range of ways in which resource economists measure or quantify major costs and benefits when market prices either do not exist or "miss the mark." This discussion includes a clarification of the specific sources of environmental economic value and how they are empirically estimated.

The chapter next moves to presentation of a series of examples that apply the foregoing benefit–cost methods in a wide range of global developing country and US domestic case studies. The intent is to demonstrate how to go from theory to application of BCA regarding soil degradation and restoration and develop some recommendations for private decision making and public policy.

Several general conclusions flow from the discussions in this chapter. First, BCA has had a relatively long (at least 60 years) history of theoretical and practical development. Second, BCA has been applied to a broad range of problems including soil degradation where market price signals are either distorted from subsidies, tariffs, etc. or incomplete due to technological externalities that are not priced or fully compensated. Careful and rigorous application of BCA in these circumstances can lead to more inclusive measures of economic wellbeing and more enlightened and sustainable public policy recommendations.

The first case from Zhao et al. (1991) and Enver and Hitzhusen (2006) on soil and land degradation and agricultural growth in a sample of developing countries demonstrates that this degradation is second only to price distortions as the most important limiting factor to increased per capita food and agricultural production in these countries. The second case on sedimentation of a hydroelectric reservoir in the DR demonstrates that soil conservation in the watershed draining into the Valdesia

Reservoir makes economic sense when the more inclusive accounting stance of BCA is applied. The series of studies in case 3 on the extra market downstream economic impacts of soil erosion from agriculture and surface mining in Ohio consistently demonstrate nontrivial economic impacts with important implications for soil conservation policy. Finally, the BCA by Shakya and Hitzhusen (1997) and Kiger and Hitzhusen (2012) on the downstream benefits and potential on-site energy benefits of the USDA CRP in Case 4 has important implications for building an economic rationale for continuing to protect the highly erodible and environmentally sensitive cropland and pasture land enrolled in CRP.

Although a strong case can be made for the importance of BCA in reducing soil degradation and enhancing soil restoration, it is best viewed as a necessary but not sufficient condition for more efficient and enlightened public policy on soil conservation and restoration. The physical and biological technologies for improved soil conservation and restoration are obviously important but not a detailed part of the scope of this chapter. A very large literature exists on the application of linear programming and other constrained optimization models to soil erosion, but these are not part of the benefit–cost paradigm since they only look at net returns to farmers from various soil loss constraints. Changing definition and enforcement of property rights or entitlements to land use and the application of insights from institutional economics and public choice theory are also critical (see Satish et al. 1992; Southgate et al. 1984). Many implications for improved soil conservation policy and more sustainable agriculture in the US context were discussed in case 4 and flow from this broader perspective.

Clearly, more empirical evidence is needed regarding on-site and downstream costs (particularly groundwater contamination) and returns of alternative tillage and rotation systems if socially optimal systems are to be identified. However, the evidence to date suggests that, on average, downstream costs of soil erosion are not trivial and that they exceed the average on-site costs of soil erosion. This implies that some form of tax, subsidy, technical assistance, or regulatory intervention may be appropriate and necessary. The evidence also suggests that downstream costs per unit of soil loss can vary dramatically from site to site. This points to the extreme importance of targeting control measures (even if chemical input taxes or penalties are based on average downstream impacts) and to the need for revision of property rights or institutions to assure both efficiency and fairness in remediation.

More comprehensive economic assessment, particularly of the downstream costs and benefits of alternative farming systems, is likely to favor those systems that are less erosive and chemically intensive. This in turn leads to the need to reassess the entitlements and property rights related to alternative farming systems and their downstream impacts. Evidence to date suggests shifts in favor of the impacted downstream users, and these trends will probably continue. Thus, sustainable agriculture is an idea that is currently ecologically, and in many cases, economically attractive. In addition, its future economic attractiveness is likely to increase.

Finally, for a more holistic and global context regarding applying the foregoing concepts in sustainable agricultural development in developing countries, see the debate series by that title published in the journal *Choices* (First Quarter 1997). An ecologist colleague at OSU, Craig Davis, joined me (at Dr. Harry Ayers request) in responding to an invited article by Dennis T. Avery from the Hudson Institute. The feedback on this debate series continued for almost another year and hopefully brought some clarity to the concept of agricultural sustainability (Hitzhusen and Davis 1997).

ABBREVIATIONS

BCA: benefit–cost analysis
CREP: Conservation Reserve Enhancement Program
CRP: Conservation Reserve Program
DR: Dominican Republic
FGD: fluidized gas desulfurization
GDP: gross domestic product
GNP: gross national product
H: high
LD: land degradation factor
LDCs: less developed countries
L: low
M: mean
MPC: marginal private cost
MSY: maximum sustainable yield
PPI: potential Pareto improvement
RCP: row crop production
SMS: safe minimum standard
T: sustainable
USLE: universal soil loss equation
WPP: white pine plantation

REFERENCES

Ahmed, H. and F. Hitzhusen. 1988. Income Distribution and Project Evaluation in LDCs: An Egyptian Case. Paper presented at the International Association of Agricultural Economics Meeting, Argentina.

Allen, R.N. 1972. The Anchicaya Hydroelectric Project in Columbia: Design and Sedimentation Problems, *The Careless Technology: Ecology and International Development*, eds. M. Taghi Farvar and J.P. Milton, Garden City, NY: Natural History Press.

Argawala, R. 1983. Price distortion and growth in developing countries. World Bank Staff Work. Pap. 575, World Bank, Washington, DC.

Barbier, E., B.A. Markandya, and D.W. Pearce. 1990. Sustainable Agricultural Development and Project Appraisal. *Eur Rev Agric Econ* 17:181–196.

Baumol, W.J. 1968. On the Social Rate of Discount. *Am Econ Rev* 58:788–802.

Bejranonda, S., F. Hitzhusen, and D. Hite. 1999. Agricultural Sedimentation Impacts on Lakeside Property Values. *Agric Resour Econ Rev* 28(2):208–218.

Clark, II, E.H., J. Havercamp, and W. Capman. 1985. *Eroding Soils: The Off-Farm Impacts*, Washington, DC: The Conservation Foundation.

Colacicco, D., T. Osborn, and K. Alt. 1989. Economic Damage from Soil Erosion. *J Soil Water Conserv* 44:35–39.

CRIES. 1980. Land Cover/Use Inventory for The Dominican Republic Through Visual Interpretation of Land Sat Imagery. CRIES/USDA/AID/Michigan State University, Santo Domingo, Dominican Republic.

Dasgupta, A.K. and D.W. Pearce. 1979. *Cost–Benefit Analysis: Theory and Practice*. London, England: The Macmillan Press Ltd.

De La Fuente, S. 1976. *Geografia Dominicana*, Ed. Colegial Quisqueyana, S. A., Santo Domingo, Republica Dominicana.

Direccion General de Foresta (DGF). 1976. Estudio de Conservaciond e Suelos de la Cuenca Hidrograficad el Rio Nizao: Parte I. Subprograma Forestal Pidagro. Santo Domingo, Republica Dominicana.

Dixon, J.A., L. Scura, R. Carpenter, and P. Sherman. 1994. *Economic Analysis of Environmental Impacts*. London, England: Earthscan Publications, Ltd.

Dregne, H.E. 1972. Impact of Land Degradation on Future World Food Production. ERS-677. U.S. Department of Agriculture, Washington, DC.

DTN/The Progressive Farmer. 2012. DTN Retail Fertilizer Trends. Available at http://www.dtnprogressivefarmer.com/dtnag/view/ag/printablePage.do?ID=NEWS_PRINTABLE_PAGE&bypassCache=true&pageLayout=v4&vendorReference=38a3dbe2-40eb-4e4d-90c1-ad5793f098cf_1326906655094&articleTitle=DTN+Retail+Fertilizer+Trends&editionName=DTNAgFreeSiteOnline, accessed January 2012.

Enver, A. and F. Hitzhusen. 2006. The Impact of Distorted Agricultural Policy on Global Food Security: A Spatial Econometrics Approach. Research report to The Mershon Center, The Ohio State University.

Espinal, M.H. 1981. Informe Final Sobre los Levantamientos Batimetricosd el Embalsed e Valdesia, CDE, Santo Domingo, Republica Dominicana.

Feather, P., D. Hellerstein, and L. Hansen. 1999. Economic Valuation of Environmental Benefits and the Targeting of Conservation Programs: The Case of the CRP. Resource Economics Division, Economic Research Service, US Department of Agriculture. Agricultural Economic Report No. 778.

Ferrell, J.A. and B.A. Sellers. 2011. Approximate Herbicide Pricing. University of Florida. Available at http://edis.ifas.ufl.edu/wg056, accessed January 2012.

Forster, D.L., C.P. Bardoes, and D. Southgate. 1987. Soil Erosion and Water Treatment Costs. *J Soil Water Conserv* 42:349–353.

Gittinger, J.P. 1982. *Economic Analysis of Agricultural Projects*, 2nd edition. Baltimore: The Johns Hopkins University Press.

Hartshorn, G. 1981. *The Dominican Republic, Country Environmental Profile, A Field Study*. Washington, DC: USAID.

Harvey, R.G. 2000. Establishing Prairie Plants for CRP or Wildlife Habitat with Herbicides. University of Wisconsin, Weed Science. Available at http://www.soils.wisc.edu/extension/wcmc/proceedings/4C.harvey.pdf, accessed March 2009.

Hayami, Y. and V.W. Ruttan. 1971. Agricultural Productivity Differences among Countries. *Am Econ Rev* 60:895–911.

Hayami, Y. and V.W. Ruttan. 1985. *Agricultural Development: An International Perspective*. Baltimore: Johns Hopkins University Press.

Hitzhusen, F. 1991. The Economics of Sustainable Agriculture: Adding a Downstream Perspective. *J Sustainable Agric* 2:75–89.

Hitzhusen, F.J. 1993. Land Degradation and Sustainability of Agricultural Growth: Some Economic Concepts and Evidence from Selected Developing Countries. *Agric Ecosyst Environ* 46:69–79.

Hitzhusen, F. 2007. Codification, case studies, and methods for economic analysis of river systems. *Economic Valuation of River Systems*, ed. Frederick Hitzhusen, pp. 19–34. Northampton, MA: Edward Elgar.

Hitzhusen, F.J. and C. Davis. 1997. Environmentally Sustaining Agriculture: A Response. *Choices Debate Series*, First Quarter:143–153.

Hitzhusen, F., S. Bejranonda, T. Lehman, and R. Macgregor. 1995. Economics and Political Analysis of Dredging Ohio's State Park lakes. In *Water Quantity/Quality Management and Conflict Resolution: Institutions, Processes and Economic Analyses*, eds. A. Divar and E. Tusak Lochman, pp. 485–499. Westport, CT: Praeger.

Hitzhusen, F., L. Friedman, K. Silva, and D. Hite. 1997. Hedonic Price and Travel Cost Estimation of Stripmine Impacts on Lake-Based Property and Recreation Values. ESO 2376, Dept. of Agric. Econ., Ohio State University.

Hoehn, J.P. and D.R. Walker. 1993. When Prices Miss the Mark: Methods of Evaluating Environmental Change. *EPAT/MUCIA Policy Brief* No. 3. University of Wisconsin.

IIED and World Resource Institute. 1986. *World Resource 1986*. New York: Basic Books.

Kiger, S. 2009. Environmental and Energy Benefits from Conservation Reserve Program Lands vs. Returns from Row Crops. M.S. Thesis, Ohio State Univ.

Kiger, S. and F. Hitzhusen. 2012. Are Environmental Benefits of CRP at Risk? AEDEcon Working Paper, Ohio State Univ.

Korsching, P.F. and P. Nowak. 1982. Environmental Criteria and Farm Structure: Flexibility in Conservation Policy. In: *Farms in Transition: Interdisciplinary Perspectives on Farm Structure*. Iowa: Iowa State University Press.

Krishna, R. 1982. Some Aspects of Agricultural Growth, Price Policy and Equity in Developing Countries. *Food Res Inst Stud* 18:219–254.

Lal, R., D. Mokma, and B. Lowery. 1998. Relation between Soil Quality and Erosion. *Soil Quality and Soil Erosion,* ed. Rattan Lal, pp. 237–258. Boca Raton, FL: CRC Press.

Lau, L.J. and P.M. Yotopoulos. 1987. The metaproduction function approach to technological change in world agriculture. Memo. 270, Stanford University.

Linsley, R.K. and J.B. Franzini. 1979. *Water-Resource Engineering*, 3rd edition. New York: McGraw-Hill Book Company.

Macgregor, R. 1988. The Value of Lost Boater Value Use and the Cost of Dredging: Evaluation of Two Aspects of Sedimentation in Ohio's State Park Lakes. Ph.D. Dissertation, The Ohio State University.

Macgregor, R., J. Maxwell, and F. Hitzhusen. 1991. Targeting Erosion Control With Off-Site Damage Estimates: The Case of Recreational Boating. *J Soil Water Conserv* 46:301–304.

Margolis, J. 1969. Shadow Prices for Incorrect or Nonexistent Market Prices. In: *The Analysis and Evaluation of Public Expenditures: The PPB System*, Vol. 1, U.S. Congress, Joint Economic Committee, pp. 533–546. Washington, DC: U.S. Government Printing Office.

Olson, K.R., D. Mokma, R. Lal, T. Schumacher, and M.J. Lindstorm. 1998. Erosion Impacts on Crop Yield for Selected Soils of the North Central United States. *Soil Quality and Soil Erosion,* ed. Rattan Lal, pp. 259–284. Boca Raton, FL: CRC Press.

Onstad, C.A., C.K. Mutchler, and A.J. Bowie. 1977. Predicting Sediment Yields. Soil Erosion and Sedimentation. *Proc Natl Symp Am Soc Agric Eng* 4:3–58.

Pimental, D., C. Harvey, P. Resosudarmo, K. Sinclair, D. Kurz, M. McNair, S. Crist et al. 1995. Environmental and Economic Costs of Soil Erosion and Conservation. *Science* 267:1117–1123.

Prairie Seed Farms. 2011. Iowa CRP Mixes. Available at http://www.prairieseedfarms.com/prairie-seed-farms-crp-mixes/prairie-seed-farms-iowa-crp-mixes.html, accessed January 2012.

Randall, A., D. de Zoyza, and F. Hitzhusen. 1996. Groundwater, Surface Water, and Wetlands Valuation for Benefits Transfer. In: *An Economic Analysis of Sustainable Agriculture Adoption in the Midwest: Implications for Farm Firms and the Environment*. USDA/EPA Research and Education Grants Program.

Rapp, A. 1977. Soil Erosion and Reservoir Sedimentation—Cash Studies in Tanzania. In *Soil Conservation and Management in Developing Countries*. FAO Soils Bulletin 33, Food and Agriculture Organization, pp. 123–132. Rome, Italy.
Repetto, R. 1987. Economic Incentives for Sustainable Production. *Ann Reg Sci* 21(3):44–59.
Repetto, R. and P. Faeth. 1990. The Economics of Sustainable Agriculture. Draft Paper, World Resources Institute, Washington, DC.
Ribaudo, M.O. 1986a. Targeting Soil Conservation Programs. *Land Econ* 62:402–411.
Ribaudo, M.O. 1986b. Reducing soil erosion: Off-site benefits. Report No. 561, U.S. Department of Agriculture, Economic Research Service.
Ribaudo, M.O., S. Piper, G.D. Schaible, L.L. Langner, and D. Colacicco. 1989. CRP: What Economic Benefits. *J Soil Water Conserv* 44:421–424.
Robinson, A.R. 1981. Erosion and Sediment Control in China's Yellow River Basin. *J Soil Water Conserv* 36:125–127.
Sanches, P.A. 1979. Soil Fertility and Conservation Considerations for Agroforestry in the Humid Tropics of Latin America. In *Soil Research in Agroforestry*, eds. H.O. Mongi and R.A. Hurclay, pp. 79–124. Nairobi, Kenya: International Council for Research in Agroforestry.
Satish, S., F. Hitzhusen, and K.V. Bhat. 1992. Watershed Management in India: In Search of Alternative Institutional Arrangements. Paper presented at the International Symposium on Soil and Water Conservation: Social, Economic and Institutional Considerations, Honolulu, HI.
Shakya, B. 1992. Finance and Economic Analysis of White Pine vs. CRP and Row Crop Production on Erodible Lands of Southern Ohio. Master's Thesis, The Ohio State University.
Shakya, B. and F. Hitzhusen. 1997. A Benefit Cost Analysis of the Conservation Reserve Programs: Are Trees a Part of a Sustainable Future in the Midwest? *J Reg Anal Policy* 27:13–29.
Southgate, D., F. Hitzhusen, and R. Macgregor. 1984. Remedying Third World Soil Erosion Problems. *Am J Agric Econ* 66:879–884.
Streitfeld, D. 2008. As Prices Rise, Farmers Spurn Conservation Program. *The New York Times*, April 9, 2008.
Swanson, D.A., D.P. Scott, and D.L. Risley. 1999. Wildlife Benefits of the Conservation Reserve Program in Ohio. *J Soil Water Conserv* 54:390–394.
US Department of Agriculture, Farm Service Agency (USDA-FSA). 2006. Preparing for CRP General Signup 33 in Ohio—Ohio Notice CRP-06-09. Available at ftp://ftp-fc.sc.egov.usda.gov/OH/pub/Admin/Bulletins/FY-06/FSA_Notice_CRP-06-09.pdf.
US Department of Agriculture. Farm Service Agency (USDA-FSA). 2008. FY 2008 CRP Enrollment Activity, and Change from Last Year (as of March 2008). Available at http://www.fsa.usda.gov/Internet/FSA_File/ytychange.pdf.
US Department of Agriculture. Farm Service Agency (USDA-FSA). 2009. Conservation Reserve Program: Annual Summary and Enrollment Statistics-FY 2009. Available at http://www.fsa.usda.gov/Internet/FSA_File/fyannual2009.pdf.
US Department of Agriculture, Natural Resource Conservation Service (USDA-NRCS). 2009. Summary Report: 2007 Natural Resources Inventory. Available at http://www.nrcs.usda.gov/Internet/FSE_DOCUMENTS//stelprdb1041379.pdf.
University of Minnesota Extension. 2008. Machinery Cost Estimates. Available at http://www.extension.umn.edu/distribution/businessmanagement/DF6696.pdf.
Veloz, J.A. 1982. The Economics of Watershed Protection and Erosion Control: A Case Study of The Dominican Republic. M.S. Thesis, The Ohio State University.
Veloz, J.A., D. Southgate, F. Hitzhusen, and R. Macgregor. 1985. The Economics of Erosion Control in a Subtropical Watershed: A Dominican Case. *Land Econ* 61(2):145–155.
Ward, B., B. Freytag. 2008. Ohio Farm Custom Rates—2008. The Ohio State University Extension: Fact Sheet. Available at http://ohioline.osu.edu/ae-fact/pdf/Custom_Rates_08.pdf.

Ward, W.A. 1976a. The Bruno Criterion in Two Numeraires. Course Notes, CN-22 The International Bank for Reconstruction and Development, Washington, DC.

Ward, W.A. 1976b. Adjusting for Over-Valued Local Currency: Shadow Exchange Rates and Conversion Factors. Course Notes, CN-28, The International Bank for Reconstruction and Development, Washington, DC.

Wolman, M.G. 1993. Soil Erosion and Crop Productivity: A World Wide Perspective. Paper presented at Soil Erosion and Crop Productivity Symposium, Denver, CO.

Wood, S., K. Sebastian, and J. Chamberlin. 2003. Land Quality Agricultural Productivity and Food Security, A Spatial Perspective. In *Land Quality, Agricultural Productivity and Food Security*, ed. Keith Wiebe, pp. 47–110. Northampton, MA: Edward Elgar Publishing, Ltd.

Wright, B. 2011. Biofuels and Food Security: Time to Consider Safety Values? *IPC Policy Focus*. Available at http://www.agritrade.org/Publications/documents/biofueldiversion optionspolicyfocusfinal.pdf.

Zhao, F. 1988. Some new evidence on factors related to agricultural growth in the Third World. M.S. thesis, Department of Agricultural Economics and Rural Sociology, Ohio State University.

Zhao, F., F. Hitzhusen, and W. Chern. 1991. Impact and Implications of Price Policy and Land Degradation on Agricultural Growth in Developing Countries. *Agric Econ* 5:311–324.

11 Spiritual Aspects of Sustainable Soil Management

Bruce C. Ball

CONTENTS

11.1 Introduction 257
11.2 Spirituality and Deep Ecology 259
11.3 Soil as a Metaphor of Human Behavior..261
11.4 Connection to the Soil 263
11.5 Spirituality and Agriculture...266
11.6 Engaging the Spirit 268
      11.6.1 Soil Restoration...269
      11.6.2 Soil Conservation..270
      11.6.3 Ecological Intensification 272
11.7 Knowledge and Wisdom Exchange 275
11.8 Social Change and Sustainability 278
References..281

11.1 INTRODUCTION

The term "industrial agriculture" for intensive production with high inputs of fertilizer, pesticides, energy, and water is used widely for conventional agriculture. There has been an increasing controversy between those who favor industrial agriculture and those who prefer alternative and more holistic methods such as organic farming (Beus and Dunlap 1990). However, as inputs run scarce, industrial agriculture is looking to adopt more holistic practices. Both types of agriculture are informed mainly by soil and agronomic science, which can be seen as being based on "logical positivism." This is the idea that only "positive" or evidence-based conclusions about reality are valid. Inclusion of notions of underlying "being" or "soul" or "vision" or "spirit" that imply human capacities that transcend mere logic within science is metaphysical; however, the metaphysics is generally treated as "nonsense" within science (McIntosh 2008a). Yet the concept of systems working together such that the overall effect differs from and is greater than the sum of the parts is gaining ground in ideas such as Gaia (Lovelock 1979) and theories of complexity and self-organization (Capra 1997). Capra realized the importance to science and to humanity of systems that operate as networks. Topsoil is a good example of a well-organized network of remarkable scale and complexity (Haygarth and Ritz 2009).

Farming and farmers, over the past 50 years, have become increasingly dependent on machines and technology to ensure speed, efficiency, and productivity. The principal disadvantages of this approach are that the emphasis on productivity leads to expansion, which ignores the overall wellbeing of the countryside and its inhabitants, and that it is wasteful (Berry 2009). The consequences to the environment of the expansion and intensification of agriculture are well known. These effects along with the adverse effects of the Green Revolution were attributed by Lal (2009a) to the use of "technology without wisdom." He stressed that the management of soil and natural resources depends on how we choose to use the technology. Similarly, Lichtfouse (2009) described the use of individual solutions for individual problems in food production as "painkiller solutions" and recommended that "whole-system" solutions are required.

Agricultural scientists are realizing more and more that the exchange of knowledge with land users and advisors is crucial to successful research and to its uptake (Uekotter 2006; Le Gal et al. 2011). Similarly, there is a recognized need for a change in the behavior of people to achieve goals of moving to a low-carbon economy and reduced consumption of resources (McIntosh 2008a).

The approaches of scientists, farmers, and land users to agricultural management in the West are governed by a worldview that is dominated by Christianity. White (1967) believed that this led to our deeply held views that we are masters of nature and are relatively free to exploit it and that we expect perpetual progress. He considered that we live, as we have for about 1700 years, mainly in the context of Christian axioms. Likewise, Hillel (1992) believes that we have followed too closely the command in Genesis 1, "Be fruitful and multiply and fill the earth and conquer it," and that we tend to forget the charge in Genesis 2, "God took the man and put him in the Garden of Eden to protect it." Such protection is clearly a call to sustainability. Berry (2009) defined sustainable agriculture as "a way of farming that can be carried on indefinitely."

Such farming needs to be compatible with the development of food security in the face of difficult problems. Proposed solutions to food security have been produced with, for example, emphasis on crop genetics (Gregory and George 2011), soil degradation (Lal 2009b), and soil management (Powlson et al. 2011). These all stress that time for action is short. Powlson et al. (2011) also recognize that a key challenge is to develop effective ways of facilitation of cross-disciplinary communication, at different scales from global to local in a range of fora, and also discussion with local communities.

Lal (2009b) recognized that despite our best efforts with science, such as developing biotechnology or genetically modified crops, hunger will continue as "civil strife and political instability plague the world." Carter and Dale (1974) stated that "human traits and frailties" constitute the basic problem to developing "soil-saving programs."

White (1967) believes that since the roots of our trouble are largely religious then, the solution is also largely religious. Science, agriculture, and humanity need to become more and more connected to approach sustainability, and this can be made possible by engaging the spirit. Here I explore how this can be achieved from two approaches. The first is to consider what spirituality and deep ecology can teach us about soil management. The second is to discover what we can learn from using the networking function of the soil as a metaphor of human activity and how these combine to improve our approaches to sustainability through soil management.

11.2 SPIRITUALITY AND DEEP ECOLOGY

My first spiritual experience occurred at home, somewhat bored in the summer holidays, at the age of 10. I was sitting on the path alongside the cabbage patch at the furthest end of our extended garden in the brilliant sunshine of early summer. Everything was still and quiet, and the fields around the garden were silent. The walls surrounding the garden were low, broken down, and made up of the rounded granite boulders typical of the North East of Scotland. These boulders attracted lichens, which could give them a crusty, golden, dappled appearance. I looked at one of these and then started to stare at it. I felt a growing sense of affinity with this stone, as if it and I were as one. The feeling grew intense and then slowly faded. It was a pleasant, warm feeling that I remember to this day.

Later, as a student, I had a similar experience over a longer time scale. During the summer vacation, I would clear the weeds between ridges of turnip seedlings using a tractor-operated mechanical weeder, for several days at a time. A pair of discs guided by cutaway rollers straddled the rows of turnips and cut the soil and weeds away from the sides of the ridges. Nobody seemed to like this job, as you had to drive slowly and the rollers needed adjustment from the tractor seat to follow any crop sown off center. But I loved it. The dark, moist soil carved away from the ridge, leaving a sharp edge between it and the dry soil around the crop. I could never tire of it, the smell of fresh earth, cut vegetation, and well-combusted diesel. And the sound of the gritty swoosh of the cleaved earth and the purring of the untaxed engine.

I later realized that my feeling of being as one with stone in the garden and with the soil during tillage resulted from my being fully aware of the present. I was living in the now and experiencing "presence." This is the basis of spiritual awareness and is what is aimed for during meditation. It also shows the fundamental difference between spirituality and religion. Spirituality is discovery by oneself, whereas religion is learning from the experience of others (Walsch 1999).

The memory of the "presence" never went away, though. I had established a connection with the "spirit" in the landscape. The property of spirit being not just within you and me but within everything and yet transcendent is called panentheism (Macintosh 2008b). Communities and places also have this spirit. The spirit of a place comes from the location, the presence of curves, the topography, the presence of water, the remoteness, and the size. Moreover, if we accept Lovelock's description of the Earth as a living superorganism, Gaia, it too has its own spirit. We are part of this spirit and are held within the biological system of the Earth rather than being on it. Boff (1997) believes that this results in everything being interconnected inextricably to every other particle of matter in the cosmos in a common consciousness, which includes the relationships between persons, continents, and cultures. Thus, the entire cosmos can be seen as a superorganism. The astrophysicist Haisch (2006) extends this further to describe a being that is limited to neither space nor time. He considers this unlimited conscious being or Spirit to be God. Earlier cultures also saw the Earth as the embodiment of a great spirit, the creative power of the Universe, present in all things—rivers, trees, mountains, springs, and caves. McIntosh (2008b) describes this as the "essential Ground of Being." It also brings home to us that we are three-part beings composed of Body, Mind, and Spirit.

Spirituality was defined by McLaren (2010) using four basic characteristics. First is acceptance that life has a sacred dimension that cannot be reduced to formulas, rules, and numbers. Second, spiritual people have an inner sensitivity to aliveness, meaning, and sacredness in and throughout the universe—a universal sense of integration of everything and everyone. Third, this feeling of aliveness needs to be maintained by practices such as meditation or worship. Fourth is that organized religion does not have all the answers. This all boils down to "that which gives life" (McIntosh 2008b) or "seeking vital connection" or, in a word, *love* (McLaren 2010).

The idea of spirit for us in the West has mainly anthropomorphic or human associations, whereas indigenous cultures consider spirits primarily as modes of intelligence or awareness that do not possess human form (Abram 1995). Ancient indigenous peoples have developed an awareness and respect for the spirit, which they express in their everyday life and in their approach to living with their environment. For example, in North East Scotland, an old tradition was to leave an area of land called the "Guidman's Ground," uncultivated or ungrazed in respect to the forces of nature. It was widely believed that cultivation of this ground would bring misfortune, notably in the form of cattle diseases. The Church actively discouraged this practice as they thought that the land was dedicated to the Devil. However, the belief was very strong, and the Church found it hard to impose cultivation of the land despite heavy fines. Another spiritual practice adopted in biodynamic farming is to spray the Three Kings Preparation, a mixture of gold, frankincense, myrrh, and glycerine, on the perimeters of a farm or village on Three Kings Day (January 6) to ensure growth later in the year.

The first writing soil scientist was well acquainted with spiritual forces. She was the Abbess Hildegard von Bingen, who described the creative energy of the soil in providing conditions for germination of seed as "greening power"; she attributed it to the Divine, calling the soil "alive with the fire of God." This "greening power," along with other emotional images of the power or spirit of the soil such as "Mother Earth" or vital force, was described by Patzel (2010) as "inner soil." He believes that such feelings live within the unconscious of soil scientists and drive and guide their actions along with outer observations and inner images. If this is the case, developing the inner life is clearly important for innovation in soil science. He also believes that our loss of respect, wonder, and reverence for the soil has contributed to the accelerated and ongoing soil deterioration and destruction in many parts of the world.

Our connection with the environment is broken by our constant exploitation of it. We have lost a sense of respect for that which is sacred, holy, and mysterious in our surroundings. Boff (1997) describes the sacred as that quality of things that fascinates us, speaks to us of the depths of our being, and gives us the experience of respect, fear, and reverence. In the teachings of Buddha, Earth itself and all of its life forms, right to the very lowest, are spiritually sacred. Lewandowski (1998) believes that restoring the sacred to the land involves applying the principles of sustainable agriculture, some of which are ancient. All of these principles are those of deep ecology.

The relevance of ecology to sustainable management of environments was identified by Hillel (1992). Naess (1973), credited with coining the term *deep ecology*, differentiates shallow and deep ecology. He sees shallow ecology as a fight against pollution and resource depletion with the main objective of preserving the health and affluence of people in developed countries. Deep ecology—in common with spirituality—sees the world as a network of phenomena that are fundamentally interconnected and interdependent. Deep ecology is integrative (Capra 1997). Thinking is intuitive, holistic, and nonlinear. Values are of conservation, cooperation, partnership, and quality. I believe that adopting more of this type of thinking and values will improve our soil management. This involves a paradigm shift from hierarchies to networks, and a good metaphor of this is the soil itself, where all creatures work together so that the soil teems with wildlife (Lewandowski 1998). Root hairs thread through the spaces in the soil, bacteria and fungi decompose vegetable matter, bacteria hunt protozoa, slime molds consume bacteria and fungi, soil fungi intertwine and live with roots, and nematodes and earthworms work down vegetable and mineral material. To develop sustainability, we need to reconnect with ourselves, with the soil, and with the environment. Indeed, to regain our humanity, Capra (1997) believes that we need to reconnect with the whole web of life to reach an understanding and awareness from within. This reconnecting is the spiritual grounding of deep ecology. Deep ecology, in common with spirituality, holds the sacred.

11.3 SOIL AS A METAPHOR OF HUMAN BEHAVIOR

As well as a model of the networks of deep ecology, I see more straightforward parallels between the structure and life of soils and those of ourselves, our communities, and our families. Just as soil has three basic layers, it seems to me that ourselves can be seen as containing three layers. The top layer, at the surface, is the visible character of the person and is self-absorbed. This character, the one we like to project, is strongly determined by many absorbed external stimuli in the same way that organic matter affects the nature of the topsoil. The bottom or third layer is the bedrock or ground of being, contains traits inherited from our parents and ancestors, and is where our inner life connects with a greater reality, the basic creative Power, Spirit, or God. It is the equivalent of the parent material of the soil. The layer between these two, equivalent to the subsoil, links them and is our personal unconscious. This unconscious is a whole world supporting and surrounding our conscious lives and contains our emotional inner images and ideas that influence our actions unconsciously, our "inner soil" (Patzel 2010). These three layers of character are similar to the three types of self—identified by McIntosh (2008b) as the conscious self (top layer), the shadow self (subsoil), and the deep Self (parent material). The shadow self is so-called because it is what we cast on others. He states that "for full awareness we need to bring together the conscious self, through the shadow self, into the Godspace of the deep Self."

We speak of having roots, which we take as being where we come from and the characters we inherit from our parents and grandparents. Place is clearly important for establishing roots. Indeed, when we buy a family home, we talk about "putting down our roots." As in soil, our roots help to stabilize and ground us within an increasingly reckless society. Ideally, our roots grow through our three levels of self, from the conscious self down to the deep Self. Such growth brings stability and is stimulated by thinking at deeper levels and improving awareness of ourselves and our surroundings.

Soil properties can have remarkably similar functions to those in society. We are all well aware that a decline in soil structural quality leads to a loss of porosity that decreases the ability of the soil to grow crops and to provide ecosystem services. Thinking of parallels with human existence, soil with good porosity and a wide range of pore sizes all joined up is like a person who is both broad minded and open minded, who reaches out to understand where others are coming from, and who fosters life and creativity. Within this structure, I consider organic matter to resemble our own energy and vitality—if they work in a healthy body, then we can be resilient to all kinds of stresses. Extending this further, just as organic matter is the rich glue that holds soils together, perhaps there is an invisible "glue" that holds people together in families and communities such as the shared activities of work, eating and drinking, rituals, and religion. If this glue weakens by diminishing any one of these activities, such as drinking at home rather than in the pub or the decline in clubs or community meetings like annual fairs, then our links to each other weaken.

The microbes work in the soil pores. Many of the small pores, less than about 0.1 mm, are wedge shaped rather than being spherical. Wedges of moisture are trapped in these pores during drying (Figure 11.1), and these form the microbial habitats. The wedges are mostly separated from each other so that the different types of microbes are isolated, with the result that many different types develop and exist together. This helps the soil to have high microbial biodiversity. This illustrates how

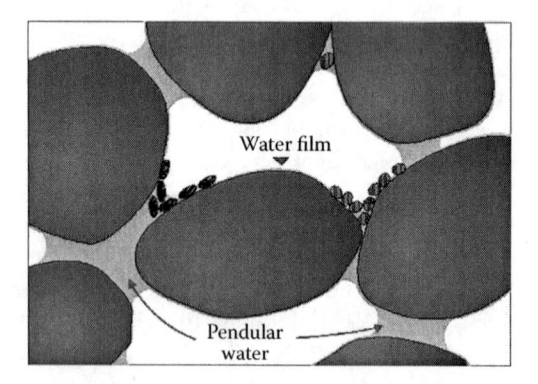

FIGURE 11.1 Soil pore spaces formed by aggregation of particles are typically angular and not cylindrical, allowing dual occupancy of water and air aquatic habitats in "corners" behind air–water interfaces and, thus, biodiversity. (Courtesy of Dr. Dani Or, Swiss Federal Institute of Technology, Zurich.)

we can work together, each to his or her talents, within our communities even though we are of different races, religions, and cultures.

Degradation can seriously reduce soil quality. I think that the process of raindrops causing soil erosion has a parallel with the constant bickering and arguments in some families, which cause them to fall out and fragment, with one or more ending up on the street. Also, we are constantly bombarded with such a relentless overload of information, advertisements, and technology that we begin to lose our sense of what is important in life. Soil erosion is controlled by covering the soil surface with organic crop residues and manure or by growing plants. This protects the soil aggregates and increases soil porosity, which helps the soil to absorb heavy rain and to maintain its vitality. In comparing the restoration of eroded soils with healing, Shapiro (1995) states that "our (inner) soils cry out for a rich inner life and a grounded, diverse community to slow down our lives, to create a holding environment into which we can turn our trials into sources of strength and integration."

Similarly, the anaerobism that the soil endures during waterlogging reminds me of ourselves when we get depressed. Life becomes too much for us, and we start to suffocate under the pressure of work or family commitments, leading us to feeling blue and gray and not motivated to do anything much.

11.4 CONNECTION TO THE SOIL

We need to be reconnected spiritually with our humble origins (Hillel 1992) through recognition of the spirit within components of the environment. Most farmers feel a strong connection to the land, drawing them toward the farming life, perhaps explaining why farms pass through succeeding generations. Working on the land provides food, water, and recreation and creates an emotional attachment, especially if it has been in the family for generations. The land is part of the farmer, supplying a living link back to his or her ancestors. People find it hard to survive when separated from their land, and the land also cannot survive without those who are naturally part of it. Workers of the soil and the land are aware of coming from the soil, of being sustained by it, of the need to care for it, and of eventually returning to it.

Understanding the connection between soil and food is vital, and many children are unaware of the source of their food. We increasingly live in cities where we are remote from contact with the land or cultivation of crops or rearing of animals. City and town dwellers also tend to get cremated and thus break the cycle of returning to the earth. All these serve to lessen our link with the land. Schumacher (1973) saw that one of the main tasks of agriculture is to keep people in touch with living nature, of which they are a vulnerable part.

I think that the love of the land is something that grows within us. I love to handle the soil and find a peace of mind in doing so, yet I have known other soil scientists who do not feel this affinity. They tend to treat the soil as a mixture of separate components of chemistry, biology, and physics, which can each be measured separately. While the methods of these scientists are effective, I prefer the holistic approach, treating the soil as a living organism. In this way, the love of the soil can be understood, a love borne of contact, particularly during crop growth.

Although the soil is most exposed when it is freshly plowed, it is most vulnerable after it is cultivated to produce a seedbed. The soil aggregates are small, and the surface is usually level and fairly firm. The seed and fertilizer are then sown a few centimeters below the soil surface. Everything connects together in hope when seeds are dropped into the soil, and everything is at risk at this stage. The seed can be eaten by birds, the soil can be eroded, and the fertilizer washed away by the rain or blown away by the wind. At this time, the connection between farmer and soil is at its greatest, and the call of nature is most keenly felt.

Hildegard von Bingen was highly creative, composing music, writing books and poetry, healing, learning philosophy, and seeing visions. She first became aware of having visions at the age of 5 and explained that she saw all things in the light of God through the five senses of sight, touch, taste, smell, and hearing. All the senses are involved in engagement with the spirit. She was a remarkable woman, describing the characters of soils from their colors. A red soil was best as it had the right mixture of moisture and dryness and so produced a lot of fruit. A black soil also had a good mixture of moisture and dryness but was cold and less productive. A green soil was the worst because it was both cold and dry.

I believe that by studying the Earth and getting to know it better, we develop a personal relationship with it. When we really observe the soil, we begin to love it and then to treat it with respect and reverence. The theologian Boff (1997) summarized this well: "when you are studying the soil, the cosmos is studying itself."

The act of digging up a spade full of soil and gently pulling it apart as in using a spade test (Ball et al. 2007) is a positive, peaceful, and even a healing experience. This feeling depends on time and place but leads to a sense of the sacred, the "holy ground." I like to handle the soil gently, to treat it with reverence, so that it will reveal more of its true character.

The other way to get up close and personal with the soil is to dig a soil pit. Soil creation takes a very long time. It takes about 10,000 years or longer for ice, rain, wind, sun, vegetation, and bugs to make 30 cm of soil. So when you dig a pit, you dig through thousands of years in a couple of meters of soil (Shaxson 2006). As you look down the sides of the pit, you can imagine how the soil was made, the movement of the ice, the scraping of the rock surface, the particles floating down through the water, the roots spreading down. You are looking through time as well as through space. The other advantage of being in a pit is that you become aware of being held within the biosphere rather than on a planet (Sewall 1995). Sewall believes that this produces a feeling of vulnerability and liberation, as well as a feeling of perception that is like a communion with the soil and that can be experienced as a spiritual practice.

Connection is established to the soil not only by sight but also by the other senses, which need to be reawakened to renew our bond with the Earth. Many people learn about the soil by walking on it, especially when it is soaking wet and you are ruining your good shoes! I reckon that by walking across a harvested field, I can judge what type of tillage is needed to establish the next crop. Hard, smooth areas or areas with wheel ruts will need tillage, whereas softer areas, which sink a little beneath the feet and crumble around the stubble of the previous crop, may need little or no tillage.

In spade tests of soil structure, a lot can be learned from touching the soil. The ease with which the soil breaks up is critical to assessing the quality of the structure.

Handling aggregates is important. Those that feel rounded and light in weight are usually good, whereas sharper, flatter, heavier aggregates are less good. The ultimate way to make contact with the soil is to roll in the mud and rub it over yourself, a primal activity where you are absorbing the land through your skin.

Lines-Kelly (2004) believes that touching the soil literally earths us, connecting our human spirit to our core. She believes that we have a subconscious link to the soil that goes deep into the core of our being and connects us with the spirit of what it means to be alive. She also quotes Wendell Berry, who considers that anyone who contemplates the life of the soil for some time will see it as analogous to the life of the spirit. Many have handled and worked with soil for a long time, yet the soil is such a complex organism that we still have a very poor knowledge of it. Lewandowski (1998) thinks that soils are wild in that they are unknown territory, self-regulating, and beyond our control—just like Spirit.

After handling the soil, I try to avoid saying that my hands are dirty or soiled. In the Maori language, there are no words to say this anyway. Nevertheless, the hands are no longer clean, particles stick to the skin, and the organic matter leaves dark markings. I feel uncomfortable, a bit contaminated, so I wash them or rub them in wet grass. Perhaps it is not really necessary. Lines-Kelly (2004) believes that if we can rid ourselves of the connection between soil and dirt or excrement, we may improve our connection with the soil.

Smell is also important. I sometimes teach the spade test in lecture rooms or village halls, where I will have up to 20 slices of soil. If the room is kept closed, the soil produces a strong smell as it breathes, making all aware of its life. When doing the test, I encourage people to smell the soil. They soon learn the difference between the mellow, clean, sweet smell of a soil rich in organic matter; the fainter, earthy smell of most agricultural soils; and the yucky, sulfurous smell of soil that has been waterlogged and is in poor condition (Shepherd 2009).

Another connection to the soil that is perhaps less obvious is through hearing. I have already mentioned the sound of soil being cultivated. Other direct sounds can occur after heavy rainfall, especially in cereal stubble. As the water drains out of the very large pores and they fill with air, bubbles make their way upward through the pore system. As they burst at the soil surface, they can produce a series of plops. Peasants in ancient Slavic tribes used to listen to what "Moist Mother Earth" was telling them by digging a hole in the ground with their fingers and placing an ear to it. Swan (1993) considers that listening to this language was part of their worship of the Earth. He also thinks that there is a voice deep within us that understands nature and our own nature and that we should listen to its promptings.

These connections to the soil or any other being or thing are enabled by "spirit" and increase our awareness of the environment, motivating us to care. We hear regularly at funerals the words "ashes to ashes" and "dust to dust." This reminds us that we are from the Earth and will return to it in spirit as well as body. Boff (1997) believes that when we become conscious of ourselves as Earth, we can start to feel at one with Creation and recognize that some of the mountain, of the sea, of the air, of the tree, of the animal, of the other, and of God is in us. This is also an important definition of the human spirit in deep ecology, the means of consciousness where one feels connected to the whole and which I call Wholeness.

11.5 SPIRITUALITY AND AGRICULTURE

Environmentally connected agriculture and other land uses need to include spiritual aspects, and these are often associated with organic farming. The early pioneers of organic farming spoke of Mother Earth, implying a link with inner soil and feminine wisdom. The Earth is thus also perceived and treated as a Goddess (Patzel 2010). They also emphasized the "living" soil, usually the organic matter or humus and the cycle of growth and decay. They also spoke of the soil as an "organized whole." Organic farming was originally seen as a holistic system, which assumes that a farm is a single whole, complete organism, and if any part changes, then the whole farm changes. This unity or holism was seen as important for health by Howard (1947), who recognized that "the birthright of all living things is health; this law is true for soil, plant, animal and man: the health of these four is one connected chain." Berry (2009) perceived this health as quality and connected to spiritual, economic, and political health. One of the early pioneers of organic farming in the United Kingdom, Lady Eve Balfour, considered that this holism means that "we cannot escape from the ethical and spiritual values of life for they are part of wholeness."

Organic farming is based on four principles of health, ecology, fairness, and care. In Germany, organic farming originated as part of anthroposophy as described by Rudolf Steiner. Anthroposophy is a spiritual philosophy based on the belief in the existence of an objective, understandable spiritual world or forces and beings accessible by direct experience through inner development. One of Steiner's legacies is biodynamic farming, in which spiritual aspects are included in day-to-day management. Steiner had a model of psychic and spiritual forces and "beings," which would form and inhabit plants, animals, and humans. These interacted and caused transformations at the "farm organism" level (Patzel 2010). Biodynamic farming includes the influence of planetary rhythms on the growth of plants and animals. Soil renewal is important and is achieved through special composted preparations used to treat the soil and plants. Materials are fermented in animal body parts such as cow horn, which is claimed to help concentrate the life forces from the surroundings into the material within. I visited the research center for biodynamic farming at Darmstadt in Germany in 2001. The spirit there was good and bright.

Organic farming is seen by the general public as better than conventional farming, being on a smaller scale and with locally based farmers. It should improve the ecological stability and long-term environmental sustainability of farming because less fossil fuels and chemicals are consumed and more of the beneficial natural processes and renewable resources available on the farm are used. Organic farming is thought to have a good connection to the soil and to improve soil conditions. In my experience in visiting and teaching about soil management in a variety of farms in Scotland, soil structure was consistently better in the organic farms because the farms contained both animals and crops, so that organic matter was being added regularly, and the farms were not managed intensively. The diverse rotations with grass breaks were good for the soil. As in our own existence, a change is as good as a rest for the soil, and doing a variety of things (i.e., providing ecosystem services) and receiving a varied diet is rejuvenating and promotes health and vigor. Organic farming improves soil quality by increasing microbial activity and porosity (Schjønning et al. 2002). However, organic farming depends heavily on tillage to bury organic manures and to spread the nitrogen stored in the roots of legumes through the soil. The compaction produced during tillage can cancel out any improvements in quality, particularly where carrots are grown (Ball and Crawford 2009).

Globally, in 2006, organic farming occupied only 0.65% of agricultural land or 30.4 million ha (de Castro 2009). Organic farming can increase food security in Africa by producing increased yields for small-scale farmers without the need for inputs of fertilizers and pesticides. Biological approaches to increasing fertility are exploited, such as inoculating plants with nitrogen-storing bacteria or adding mycorrhizal fungi to soil to help nutrient exchange between the soil and the roots. In the long term, this should help to reduce the future dependence on food imports in fluctuating world markets.

To me, the best thing about organic farming is not the principles or the defined markets but the mindset of the organic farmers and producers, which is driven by their connection to the land and to the people. They are highly motivated because organic farming demands commitment and a high standard of management. They see their farm as an organic whole. This holism extends beyond the farm to the workers making the food products. In some cases, this extends beyond the farm. Weckmann (2009) described an organic bakery in Germany that is based on a system called corporate social responsibility, where the purpose of the company is not to achieve the maximum return for its shareholders but to produce good products and to contribute a positive benefit to society. The company has a fixed target of providing donations of 10% of net profit to support development initiatives and social engagement of employees in the areas of mentoring, information, and training. It also claims to process 100% ecologically certified raw materials into 100% ecological products. This is an exemplary fair trade model that appears to achieve the triple goals of organic farming of ecological, social, and economic sustainability. This is a good example of holism extending beyond the farm gate, interconnecting the farmer, processor, and consumer.

This and other forms of alternative or low-input agriculture termed "ecologically sustainable" clearly accord with the principles of deep ecology since they emphasize independence, community, harmony with nature, diversity, and restraint (Beus and Dunlap 1990).

Conventional farms also have "spirit." As a student, I worked on several farms all producing similar crops, but the management of the resources and the workers was quite different. Where the farmer still lived in the farmhouse, the spirit of the place felt good as this was close to the traditional farming role of supporting the family. Working the land where no one lived felt more like exploiting a resource for profit. I believe that the nature and the location of the people influenced the spirit of each place and the quality of the food that was produced. Berry (2009) considers that technology allows farming to increase to a scale that is undemocratic because there are not enough people owning and working the land and insufficient attention to the overall quality of the farming system.

The lessons we draw from this for soil management are, first, that the soil needs to be considered as an organized whole. As soil specialists, we tend to label ourselves as soil physicists, soil chemists, soil biologists, and modelers. We need to see our specialism as part of a whole. In particular, I conclude that we need to pay more attention to soil biology.

Second, we need to perceive problems and derive solutions at the whole-farm level. When I visit a farm to teach about improvements in soil management, I need first to discuss with the farmer his overall approach to farming and, preferably, walk over several fields to get the "feel"—or spirit—of the place. Only then do I get him to open up soil pits so that I can make meaningful suggestions for improvements in front of other farmers. I believe that we also need to include the principles of organic farming in thinking up research field trials—especially the need to include rotational cropping—and to put them in context with the ultimate usage of the product. This was neglected in much early work on no-tillage, often leading to negative conclusions on its suitability, particularly in Northwestern Europe (Soane et al. 2012).

We can also learn that perhaps the ultimate standard of agricultural performance is not speed, efficiency, and productivity, but the health of the ecosystem, the farm, the soil, and the human community (Berry 2009).

11.6 ENGAGING THE SPIRIT

The definition of sustainable agriculture by Berry (2009) is a way of farming that can be continued indefinitely. He states that this occurs only "because it conforms to the terms imposed on it by the nature of places and the nature of people." People and place define spirituality. Spiritual aspects include ideas of truth, wisdom, and love. When the spiritual connection is good, then we learn to recognize the "gut" instinct and when to go with it. If our response is to "feel good," then Walsh (1999) believes that this is how you tell yourself that your last thought was truth, your last word was wisdom, and your last action was love.

There are many ways of personally engaging the spirit, such as simple meditation, to increase self-awareness and give a sense of peace and purpose, and prayer, which increases ideas of compassion, expressing gratitude, and increasing hope. There are plenty of other methods of allowing the soul (the spirit within you) to express itself, for example, writing poetry, making music, talking till dawn, singing in the rain, and so forth.

For soil scientists, engaging the spirit is possible just by spending time in field, moor, and forest, among the soils, absorbing their spirit and making the connections. Farmers meditate on the soil and the land by default because they are in the landscape for such long periods that they absorb its spirit. I suggest you try persuading a farmer to let you work the soil or harvest root crops for a day—anything where you have direct soil contact—and your attitude to the soil will change. Engaging the spirit is also possible when down a soil pit, even in awful weather, when the focus is fully on the essence of what the soil has to reveal through us. Here, words that sound good in a warm lecture theater accompanied by nice, carefully selected PowerPoint images sound less convincing as the rainwater runs round your mouth or off your hand full of soil and onto boots heavy with subsoil.

Full-focus knowledge exchange is facilitated by the spirit of the soil. This is not easy, though. It demands letting go of your preconceptions, your protected interests, and your anxieties, but the rewards are tangible. They include the sudden spark of reinvigorated enthusiasm from a struggling farmer or the sudden realization of the variability and value of what is under the feet.

In complete contrast, researchers often generate new ideas in beautiful or unusual places. For example, in 2009, the British Society of Soil Science met in the glorious surroundings of Johnstown Castle in southeast Ireland. Good creativity is fostered in the wonders of creation. I am still reaping the benefits of the ideas formed during the time shared there with other scientists.

11.6.1 Soil Restoration

Much of what was discussed above for engaging the spirit involves the prefix *re-*. Re-generation, re-alization, re-vision, re-clamation, and even re-ligion. This implies a going back, re-vitalization and re-newal. Like those of us who like to restore old cars, motorbikes, or tractors, restoring an ecosystem and the soil below also demands effort and is a "labor of love." A successfully restored ecosystem also gains beauty from the new vegetation or the crop that protects and hides the soil. Indigenous peoples have good contact with the land and believe that they have an obligation to keep the planet alive, entering a reciprocal relationship with the environment. Gray (1995) believes that this is why restoration and conservation are important to them, because tending the world empowers them and gives energy and health. Working with the regenerative powers of nature gives the land back its productivity.

Restoration of compacted or eroded soils should preserve porosity, allowing easy movement of rainwater into the soil and water storage and maintaining the activity of soil organisms to break down organic matter and release the nutrients stored in the soil. Shaxson (2006) emphasizes the need for a positive view in protecting and improving what remains after soil is lost by erosion.

Improving or conserving the soil is best done either by living organisms such as earthworms or by plant roots, which are great natural creators of macropores, rather than by tillage. Soil porosity then increases naturally, and the pores are stable and interconnected, enabling them to maintain the soil's life support functions. Where there is a risk of erosion, maintaining plant cover at the surface and organic matter is important to provide protection and, in drier climates, to increase green water storage and crop yields. In these ways, the soil is rejuvenated, especially under a mixed crop rotation, when supplemented with additions of slow-releasing minerals like lime and phosphates. These restoration techniques also work well in urban agriculture on land damaged during the Industrial Revolution. Many will remember Detroit or Motown, the home of so much well-loved soul music. The motor industry has now largely deserted Detroit. There are large areas of compacted, vacant land left behind by the car industry that are being restored for cropping by soil loosening, removing debris and contaminants, and using plants to absorb toxic chemicals.

Soil compaction damage is the result of stress, which, in extreme cases, can be considered as violence to the soil. In restoring compacted soils, it is helpful to bear in mind that violence usually does not overcome the results of violence. The soil can be cultivated using heavy power-driven tine cultivators, which effectively exert more violence to try to bash the soil back into shape. These can cause more compaction and destroy the natural soil structure. Instead, it is preferable to tease apart the soil by carefully targeted loosening. I think that there is a parallel between this type of tillage and counselling which can free up hard, inflexible areas in our minds to allow creative, positive thinking and awareness of our self-integrity.

In North West Scotland and in the Hebrides, crofters are starting to restore peatland abandoned by their predecessors over the last 50 years to agriculture. The landscape begins to look alive once more. Restoration by drainage, tillage, and liming needs to be guided by the depth of the peat, by the vegetation cover, and by the drainage status, all of which can vary widely over short distances. As with organic farming, a mixed farming system is used, and animals are involved. The overzealous enforcement of environmental pollution regulations, usually designed for large-scale enterprises, to crofts can cause despair and abandonment of the effort. Restoration should aim for the best for the soil, the landscape, and the people dependent on local conditions.

The people involved in land restoration often include volunteers who become aware of the abuse that the soil has suffered and feel the healing of the land and bond with it. Shapiro (1995) considers that this brings connection at a deep level and feelings of pleasure and release, opening up channels to allow the loss of guilt, shame, grief, loneliness, or despair. She also believes that working to restore the land shows our forgiveness and enables us to feel forgiven. The healing of the people and the planet belong together.

11.6.2 Soil Conservation

Carter and Dale (1974) described the basic principle of conservation as "using soil, water, plant and animal resources efficiently in a way that will last and become more productive." They also realized that man needs to cooperate with Nature to achieve this and not assume that he is the master. They astutely observed that "no ancient civilization (that survived) had an effective soil conservation policy." Conservation is considered as being economical or sparing with a resource such as fuel or water. Our parents or grandparents naturally conserved things, turning off lights and heaters without thinking, perhaps prompted by unconscious memories of scarcity, expense, or value. This type of conservation is relatively easy, though many of us choose not to practice it. Indeed, Hartmann (1999) believes that living frugally leads to a feeling of accomplishment and independence.

The farmer is central to the conservation process. Berry (2009) considers that farmers not only produce food but also, by default, conserve soil, conserve water, conserve wildlife, conserve open space, and conserve scenery. Conservation also means asking of soils, agriculture, and people not what we can get out of them, but what we can give back. As soil scientists, we might ask, "What can I learn from this agricultural system that will allow me to manage it for maximum yield?" Instead, we need to ask, "What can I learn from this agricultural system that will enable me to serve it, or conserve it, better?" (Berry 2009). The primary aim of management should be toward sustainability by conserving the soil, which leads to a "win–win" because improving soil quality can increase carbon storage, reduce runoff, and, as a result, improve soil productivity. Berry (2009) considers that conservation involves more than just changes in management; it also involves the heart of the man managing the land. If he loves the soil, he will save it.

Land prone to erosion is often managed by conservation agriculture, the main principles of which are to prolong permanent plant or residue cover, to use diverse crop rotations, to minimize soil disturbance, and to leave crop residues at the surface. An important component of this is the use of conservation tillage (either no-tillage or minimum tillage) instead of plowing. In no-tillage, the seeds are drilled directly into the soil under the residues of the previous crop, which protect the seedlings (Figure 11.2). Minimum tillage uses some shallow fixed blade or disc cultivation before sowing the seeds.

Conservation tillage uses less machinery and fuel than normal plowing but depends heavily on weed control by chemicals and demands high standards of management. It is extensively used in Brazil and Australia, but its continued use is threatened by a buildup of soil compaction.

Perennial crops perhaps offer the ultimate potential for large-scale conservation agriculture. Unlike most crops, the land needs no regular tillage because they do not need to be resown every year. Cox (2008) claims that a field of perennial grain crops would provide food while conserving soils and water and reducing inputs of fertilizers, chemicals, and fuel.

Smaller-scale conservation agriculture methods that may help develop sustainability include the reuse of old farming systems where land was shared and farmed in small parcels such as crofts or allotments and that help in growing more local food crops. More novel systems include agroforestry. For example, in India, wheat is grown for food among poplar trees, which capture carbon in the wood, the roots, and the soil.

The ultimate aim of agricultural sustainability would be to develop less intensive small-scale farming systems that are very similar to natural ecosystems and that match local conditions or resemble gardens. Multiple cropping in the same field or garden is common in some smallholder farming systems. This encourages

FIGURE 11.2 Greening power. Maize seedling emerges from under the protection of the residues of the previous crop in Brazil. (Photo courtesy of Dr. Neyde Fabiola, Universidade Estadual de Ponta Grossa, Brazil.)

many different organisms to flourish, is highly productive, and gives good insurance against the risks of erosion, weather, pests, and disease, but it demands skills and knowledge, which are hard to find. A system that concentrates on producing soil of high quality leading to high productivity is biointensive farming, also known as mini farming. This is based on biodynamic organic farming and intensive raised beds. It builds on the wisdom of our ancestors, having been in use for 4000 years in Asia. It is labor intensive and uses little machinery or fossil fuel inputs. Soil is double dug to twice the normal spade depth, and organic matter is added as compost. Crops are planted close to each other, often in combinations of different types that grow well together. A relatively small area can support someone on a vegan diet and provides a small income. It is particularly relevant to small farmers, often women, in countries where land and resources are scarce, and it could help transform the lives of the rural poor. Good training is required so that all aspects of the system are followed; otherwise, the system can fail and cause soil degradation. It has to be adopted with a spirit of full commitment.

11.6.3 Ecological Intensification

The use of conservation agriculture and organic or low-input systems will not satisfy the current increasing demand for food because they often produce lower yields than intensive agriculture. We still need safe conventional agricultural methods and safe biotechnological methods to feed the hungry bellies (Azadi et al. 2011). We need to increase per-area productivity substantially to avoid taking over yet more land for agriculture. The loss of ecosystem services from this land would be greater than any benefits in reductions of emission of nutrients and pesticides (Glendining et al. 2009). We also need to reduce the number of inputs per product to minimize adverse environmental impacts. Eco-efficient farming systems will need to be highly productive, relying on clean energy sources and using environmentally favorable industrial processes and sustainable agricultural practices. Lal (2009b) showed that these practices should conserve soil quality, increase carbon storage in the root zone, and improve natural water quality. Mineral nitrogen use will need to be reduced by more efficient use, recycling, and exploitation of natural sources.

However, technologies of nitrogen fertilizers, irrigation water, and pesticides continue to be abused by excessive use; all of these can decrease soil quality. This is not confined to developing countries. For example, in New Zealand, the export of dairy products is vital to the economy. The dairy systems are intensive and expanding. I recently spent some time measuring soil quality and nitrous oxide emission in pastures under dairy production in New Zealand. As I worked in the field, my wife waited in the car and watched the cows coming out from milking. So many walked past that she eventually thought that they were going round in circles! Dairy productivity depends on irrigation of the pastures in the spring and summer in certain parts of New Zealand. Dairy production is becoming increasingly unpopular with the local population because the increased use of irrigation water is drying out the rivers, and the removal of tree shelterbelts to accommodate the very wide boom irrigators is increasing the risk of soil erosion. Also, the increased discharge of effluent high in nitrogen is threatening water quality, and there are concerns over the welfare of the animals. Poor soil structure results from compaction by animal hooves and wheel traffic, which restricts the growth of the pasture shoots and roots due to low macroporosity, reduced biological diversity and activity, and restricted nutrient uptake. One such affected soil (Figure 11.3) showed how the aeration status and structural condition of the topsoil were reverting to those of the subsoil such that the actual subsoil appeared to be in better structural condition than the topsoil. Extra mineral nitrogen fertilizer is applied to compensate, and only part is used by the grass, with the rest lost to the environment.

An important approach to improving the sustainability of these systems is to maximize use of the soil biology. This is achieved by improved connection with and response to soil conditions. Visual methods of crop and soil observation can be used to adjust management to maintain high soil and pasture quality. Shepherd (2009) attempts to improve the soil management by ensuring good soil structure and aeration; choosing appropriate mixtures of grass, herbs, and clover in the pasture; using simple pasture reseeding techniques; and using biologically friendly fertilizers that promote carbon sequestration, nutrient turnover and cycling, and the release of nitrogen slowly to the pasture. These combine to encourage earthworm and microbial activity in the soil and allow good pasture production with low nitrogen fertilizer use. Healthy pasture results in healthy animals, and greenhouse gas emissions are reduced relatively cheaply. Conservation shifts the emphasis from agribusiness to husbandry when the farmer aims for a reasonable return rather than productivity.

A system of biological agriculture labeled by some as "beyond organic" is Polyface farming, which was developed by the Salatin family (Salatin 2010). Their

FIGURE 11.3 Gleyed recent alluvial soil in North Island, New Zealand, poached sufficiently by grazing for the topsoil structure and aeration conditions to start reverting to those of the subsoil.

original farm in Virginia, USA, was heavily eroded, with many areas of eroded rock, making the surface appear like pockmarked skin. By applying compost, rotating cattle, and running poultry, these sores developed scabs of soil and healed as the soil reformed over a 50-year period. Enough soil was created to refill many gullies. Polyface farming stimulates soil biology and carbon sequestration by using composting and the rearing of animals and poultry together on perennial pastures of mixed herbage. Plants and animals are allowed habitats that suit their physiology in systems with emphasis on biomimicry. Salatin aims to develop economically viable agricultural enterprises with emphasis on healing of the land, the food, the economy, and the culture. I see these systems as engaging and continually renewing the spirit within the farmer, though Salatin cheerfully calls himself a lunatic farmer! Nevertheless, he describes his spirit soaring as he steps out into the "fresh-scented morning air" of his farm. Aiming for beauty and quality allows both sustainability and productivity.

Application of the biological approaches of organic farming to conventional agriculture is likely to become more important as fossil fuel supplies dwindle. A recent approach proving popular with farmers in their attempts to make the best use of the amendments applied to soil is precision agriculture. Inputs of fertilizer and lime are adjusted at the scale of the harvester using crop yield information gathered the previous year. This is a method of trying to overcome the limitations of the soil's ability to provide uniform levels of fertility for crop production. Perhaps this should be done with care in case soil quality is impaired, and, as Patzel (2010) suggests, "negate soil dignity and autonomy as a living natural body."

Irrespective of the agricultural type, soil management is central to improving sustainability and ecosystem services, with emphasis on soil biology and ecology. Farming needs to conform to the natural laws that govern the local ecosystem (Berry 2009). Key factors to manage are soil carbon, roots, ecology of soil organisms, nutrients, nitrous oxide emissions, and erosion (Powlson et al. 2011). These clearly need to be integrated with other novel approaches in agroscience such as transgenic crops and beneficial microbes (Lichtfouse et al. 2010). Lichtfouse et al. (2010) warns that we need to include sociology in developing these sustainably to improve the balance between the triple-P values of people, planet, and prosperity.

Rains et al. (2011) suggest a redirection of current and emerging technologies such as in pest management practices, genetic engineering, and precision agriculture to provide a more ecologically based and sustainable farming approach. This also needs to include the needs of the world's 900 million small farmers, who are often hampered by the lack of good-quality seed and by severe soil degradation (Kiers et al. 2008). They also emphasize that agricultural innovation is not just technological performance in isolation but is how technology can build knowledge, networks, and capacity. Lamine (2011) elaborates on this. He states that biological farming can increase profitability without necessarily reducing yields but that farmers perceive risks to efficiency. He stresses the importance of farmers belonging to integrated networks. These are not only professional networks set up by advisors and companies but also informal networks where farmers advise each other and provide moral support. Organic farming is good as an alternative as it has a distinct identity that other alternative approaches do not have.

11.7 KNOWLEDGE AND WISDOM EXCHANGE

The combination of spirit from within and complete interconnection with everyone and everything, though wonderful to consider, lays an obligation on us to care for each other and for our environment. Effectively, everything we do to each other and to the environment, we do to ourselves. As soils experts, our primary contact is with farmers and, to a lesser extent, stakeholders and policy makers. I was brought up in a strong rural community in North East Scotland where the soils are highly fertile and productive, but where there is a strong tradition of farming methods handed down by succeeding generations. These traditions are now facing substantial change as farmers age and have difficulty with succession. Although mixed, the farming is intensive with good crop yields from the excellent, fertile brown forest soils derived from basic igneous rocks. Even so, continuous arable cropping without farmyard manure application can impair structural quality (Figure 11.4). Agriculture there, as elsewhere in the West, depends heavily on the use of machinery. I recently attended a vintage tractor rally where we spent a day and a half poring over old tractors and reminiscing. I was the only person to mention the soil over which the tractors passed. Machinery is novel, attractive, labor saving, and a source of pride, but it may distract farmers from learning about and connecting with their soils and crops (Uekotter 2006).

FIGURE 11.4 Brown earth from North East Scotland, which was cloddy and gray in the arable field (a) but brown and with good crumb structure (b) in the grassy area where the vegetation was natural and the soil had not been tilled for at least 10 years.

When advisors or scientists speak to farmers about new crop production methods, the farmers are wary, and their appetite for change and development varies greatly. With the Internet, they have access to a great amount of knowledge but can find it difficult to identify what is appropriate and to develop wisdom for its correct application. The successful application of novel agricultural practices demands a lot of knowledge, with intensive training and continuous access to information and advisory services. Many smallholder farmers do not have education or training and cannot easily get access to information and knowledge.

Solutions for better farming are mostly farmer centered. Developing knowledge and convincing farmers of the benefits of the use of sustainable practices are not easy. The older farmers in my community have developed an instinctive way of conserving the soil over the years. They use rotations with break crops and grass, regular farmyard manure applications, and periods of sheep grazing to improve soil quality. These techniques contain the wisdom of our ancestors and work well, perhaps explaining why farmers are unwilling to change quickly to adopt new management techniques. My organization, SRUC, Scotland's Rural College, has a strong reputation for the provision of advice based on the research of its scientists. Yet there is still disagreement on some management actions and insufficient input by advisors on the details of relevant applied research. If scientists and advisors often do not agree among themselves on the most appropriate courses of action, how does the farmer know where to turn?

Advice is often provided free to farmers by suppliers of fertilizers, pesticides, or machinery and thus can be biased. Knowledge transfer can be hampered by a lack of consistent, unbiased advice. The use of technology without wisdom is considered by Lal (2009a) to be one of the great blunders of humanity. Yet there are examples of wise use of technology. In my community in North East Scotland, a contractor told me that he determined the optimum depth of subsoiling by first crossing the field once with the subsoiler at full depth. He then feels down the subsoiler legs to identify the zone with the warmest metal. This corresponds to the average depth of the compact layer, and he then adjusts the subsoiler to penetrate to just below this depth. This information appeared after one or two whiskies were consumed, a case of one spirit enabling the other!

Farmers and advisors need to get together to develop new production methods from innovative ideas produced by scientists. Agricultural innovation can be seen as having the three main components of biotechnical processes, farm management, and advisory services (Le Gal et al. 2011). Yet very few studies attempt to address these three main components in a single research framework. Le Gal et al. (2011) propose such an integrated framework for conducting integrated research both at the farm and advisory levels. In this, the farmer, researcher, and advisor work together in a triangular relationship.

New agricultural methods need to integrate easily with the old and be developed or adapted to local circumstances, especially where workers are poor, partially skilled, or partially educated. This may require a revitalized connection between the elders and a younger generation (McIntosh 2008b) to carry over wisdom. Development of wisdom was perceived by Schumacher (1973) as the orientation of science and technology toward the holistic, the gentle, the nonviolent, the elegant, and the beautiful and that will lead to indivisible peace.

Much also depends on learning by example. Farmers look over the hedge and, if they see something working, they often adopt it. They like to learn from each other and build on what they know already. Often they stick with what they know, even though it may be suboptimal at the time—particularly in the face of market demands, which can change quickly.

Fostering the desire to increase sustainability and change may need a more direct engagement of the spirit within to help farmers and stakeholders to share and develop further wisdom from their love of the land and its use. I discovered this when I demonstrated the spade test to farmer groups. The careful, respectful handling of the soil by members of the group made me realize the need for its stewardship. As scientists, we tend to forget the obligations of stewardship (Lal 2009c) and of soil husbandry to maintain and improve the health of the soil, plant, animal, and human system (Berry 2009; Batey 1988; Shaxson 2006).

Handling the soil can release a flow of ideas and experiences, which may build on those of others. This is growth from within and is similar to the growth and spread of plants that have rhizomes or corms. Rhizomes spread out sideways in the soil and send up shoots at random. Cutting up the rhizomes causes them to spread rather than to be destroyed. This is a better model of how information can spread than a tree model where information filters from above to the grass roots. La Chapelle and Puzey (1995) suggest that growth from below spreads methods for healing the destruction of the Earth. Each community or village can develop within its own area, protesting and growing, bringing an infection of restoration and conservation. All this helps in the expanding and remolding of knowledge that farmers already have (Shaxson 2006).

The development of more sustainable practices may be helped by changes in our diet. The farmers in my community grow many cereals, mostly barley, which are either fed to cattle or pigs or used to make malt in whisky production. These are great for that most Scottish of meals, the Burns supper, but not so useful for everyday consumption. The only crops grown directly for food are small areas of wheat, oats, and potatoes. World meat production is predicted to double between 2000 and 2050. Grain is fed to animals to make much of the world's meat. This is inefficient as it takes, on average, 6.5 kg of grain, 36 kg roughage (hay, grass, and silage), and 15,000 L of water to make 1 kg of beef. It is more efficient for us to eat the grain rather than feeding it to animals to make meat. Such a transition may help organic farming and conservation agriculture to feed more and more people (Azadi et al. 2011). To achieve this, whole cultures may need to adjust to eating different types of food. This needs a change from within, brought about by the development of a spirit of conservation.

It is important to remember the wisdom of our ancestors. Much is made of the abuse of the Amazonian rain forest as it is cut down for food production, yet the old indigenous peoples had a system of sustainable agroforestry, described by Hillel (1992) and Hartmann (1999), that could still be used. Several trees were cut down in the shape of a circle, like the spokes of a wheel. Legumes and tubers were then planted to stabilize the soil and to capture nitrogen among them. At the end of the growing season, the trees were burned, and the ash was used as fertilizer for next year's food crops. A range of crops was grown, with their locations determined by their tolerance to shade. This cycle continued for about 7 years, when the forest was allowed to reseed with crops still growing around the young trees. We need to build on such traditional, indigenous knowledge with the wise use of modern and innovative techniques (Lal 2009a; Shaxson 2006). A knowledge society differs from the societies that scientists inhabit in that its structure is to a greater extent the result of social action. As we reform it, we need to promote a plurality of opinions and approaches and develop an agricultural learning society (Uekotter 2006).

11.8 SOCIAL CHANGE AND SUSTAINABILITY

The soil scientists Vernon Gill Carter and Tom Dale recognized back in the 1970s the need to "adjust the population and the standard of living to usable resources." They realized that our crisis is one of overconsumption and an unjust distribution of consumption. The growing demand for food by our swelling population and its changing diet is very unlikely to be met in the not-too-distant future because we simply do not have enough fossil fuel, water, minerals, and soil to sustain us. In the Northern Hemisphere, we all need to consume less, waste less, and recycle more of everything. We also need to consider how to reduce the number of mouths to feed by reducing family sizes—particularly in the Southern Hemisphere—so that we can live within the capacity of the land and diminished resources, particularly of fossil fuels.

Carter and Dale (1974) also predicted that a failure to balance people and resources would require us to adapt to a lower standard of living. This is indeed proving to be the case as we see austerity measures being adopted in Europe since the recession of 2008. To achieve this without pain, we need a fundamental change in mindset from all of us. For this, we all need to engage the spirit within. Lichtfouse (2009) realized how agriculture was an important science in tackling more general global society issues because agriculturalists are trained to manage a network of inputs of many disciplines including transport, health, and economics. We need to change our spirit by accepting the direction of the spirit within nature and by developing our connections with the springs of love deep inside us in order to work toward a sustainable future.

In my community in North East Scotland, like many other places in the Western world, twenty-first century living brings problems of obesity, alcoholism, anxiety and depression, dependence on cars and on big supermarkets, materialism, dependence on fossil fuels, isolation from neighbors, and lack of spiritual awareness. We are living in a social recession and try to cope with the threats of violence, environmental destruction, and inequality with the short-term fixes of retail therapy, overeating, and alcohol. The population struggles between the pull of the slower pace of the past and the urgent push of the future crowding them into greater productivity and consumption. Some spiritual input is required to sustain the community, the soil, and agriculture and to give a sense of lasting satisfaction. One approach may be to relearn ideas from our ancestors in their tribal lives, aiming for mutual support, security, and safety.

A wholesale social shift toward low-input simple living was recommended by McIntosh (2008a) to combat climate change, and I am sure that this would help other problems. As Gandhi put it, "we must learn to live simply so that others may simply live." In the United Kingdom, we now throw away about one third of the food we buy, consume double the protein we need, and eat too much processed food. If we could reduce food wastage or overconsumption, we would have less need to increase our crop production.

My vision of a renewed world is where we care for our environment by consuming less, becoming more self-sufficient, and living more simply, but with inward richness, like soils dark with organic matter. We consume within sustainable limits and realize that enough is beautiful. We develop intentional community like groups of porous soils aligning together and coexisting through good interconnection of the pores. Like high-quality soil, we grow deeper and more together, promoting "us" rather than "me." In such ways, we can prosper without growth, move to a low-carbon economy, and help rid the world of poverty (Jackson 2009). "Economic growth," the acquisition and use of things and people, is replaced with a growth of love and respect from within. This type of growth deepens our "topsoil" by using more of what is stored deep within us, enabling our roots to spread down and outward to intertwine with those of others in terms of mutual understanding and love. Mutual understanding resembles how the roots of different types of plants combine to give an increased release of nutrients from the soil around them (mycorrhizal association), thereby feeding and encouraging each other. However, like good soil management, achieving all of these will be difficult.

The soil can provide a clue. Humus is described as the "stuff" of the soil, and the Latin name for humans is derived from it. The words "humble" and "humility" are also derived from it. Perhaps the way to a more sustainable lifestyle may be to live humbly on the Earth by cultivating our inner life or "inner soil" so that we become resistant to degrading processes such as being stifled by domination (like compaction), being addicted to consumer goods (like contamination), and losing our sense of what is important in life (like erosion). Then maybe "the meek shall inherit the earth."

Applying such principles to soil management might involve developing and sustaining soil quality as our first priority by supplying organic matter, taking care of soil structure and the mineral balance so that organisms can live well and make the topsoil rich and deep. Our imperative is to concentrate on the care and the conservation of resources in ecosystems, not on productivity for productivity's sake. This might guide us to start living toward others, as suggested by Bishop Desmond Tutu. To achieve this, we make deeds of love, no matter how small, such as acts of kindness, kind thoughts, or fleeting prayers—acts of mercy. There is evidence that if enough of us do this, then our interconnection will enable a spiritual transformation that can spread across the globe.

Earlier I reported that I was disappointed in the quality of the agricultural soil in my community as I dug in a field under continuous cereals. I left that field and crossed the fence to a nearby natural grassy area that had not been cultivated for years. I put in my spade and there it was (Figure 11.4), just as I remembered it as a child in the garden, the beauty of the broken-up soil revealed in the rounded, nutty crumbs, warm brown in the sunlight of spring. Beauty that can feed us again and again, if we care for it. My vision for the future of soil management and agriculture in my community would have many of the houses in the village with plots of land similar to the former crofts, where a cow is kept for milk, cheese, and butter; hens reared; and food grown. Land for vegetable production would be shared. This is already happening throughout the United Kingdom in land share schemes. The farms would become smaller, more labor intensive, with an emphasis on biological cycling in complex rotations. Farms would begin to resemble gardens. More food crops would be grown like oats, potatoes, vegetables, and fruit. Advanced technology and research would be geared to achieve all these things. Permaculture might flourish (Whitefield 2009). Permaculture is a system of permanent agriculture and permanent culture that goes beyond biodynamics or biointensive farming. It relies on agroforestry and perennial food crops to give self-sufficiency with minimal amounts of energy and recycles most wastes, including human wastes. It has its roots in the highly efficient "permanent farming" practiced for many centuries in China, Korea, and Japan. Soils professor King (1911) described how every square meter of land and liter of freshwater were used and how everything from human waste to canal dredging was recycled in the late nineteenth century. This was combined with an intense economy of practices by the people in their efforts and lifestyle. Permaculture has an ethical component in that it embraces care of the Earth, care of the people, and fair sharing.

Although this vision may never occur, we can make a start now by employing the same deep ecological principles to start to transform our environment and our everyday existence. Harvesting the fruit of the spirit allows us to start to identify the priorities and to realize that everyone and everything interlinks. Above all, we need to understand that the only way to tackle our basic crisis of overconsumption is to consume less, to recycle more, and to avoid maintaining production by use of substitutes. Applying deep ecology is much more than choosing lifestyle options like the purchase of an electric car or going vegetarian. It demands that we give up things and decrease our selfishness and extravagance. Going green or being environmental is not an option; it is an imperative. Research has shown that happiness and fulfillment increase as our possessions decrease and as we renew our interdependence, partnership, and flexibility and accept diversity. We replace domination with a return to our basic instincts of compassion and love. All are the characteristics of cooperation, which maintained our ancestors for thousands of years.

At the center of all this will be the wise, informed farmer who, through the fruits of our research, investment, knowledge, and, above all, the spiritual, will start to move from agribusiness to husbandry and conservation, with increased linkage to consumers. Farming will need to become smaller scaled. Although ecological intensification will be required, this will be much less dependent on oil by-products and capital injections and more dependent on good soil. Political and financial trading systems would favor smaller farmers, especially women in developing countries, leading to better rural development and greater energy and food security. A major priority is to use our oil reserves wisely, principally for producing chemicals and plastics. We cannot recycle oil.

From Oil to Soil

Free oil
Drilling down
and gushing forth
Stored sunlight from ancient earth
Who needs soil when you've got all that oil
Flying free, driving on, population exploding in oil
Artificial fertilizers, plastics, synthetics, tar, cement, and spirits abound
Peak oil, who cares when there's still half left below, deep oil
Gas fracking,* open cast, tarry sand, rip it out as it lasts
It's OK to plunder the ancient landscapes below
Who needs soil when you've got tough oil[†]
Not our problem, business as usual
Technology always advances
Can we make it
digging down
with soil?
Oil free

Although I am not sure what the future would look like, I am sure that when we reach the renewed world, we will be standing on soil that is resilient, dark and porous, and rich and deep with organic matter, wisdom, and love. There will be smooth transitions between topsoil, subsoil, and parent material helped by deep, wide-spreading roots. Beauty that is soil deep. We can start toward this now by walking with respect to the Earth and to all its creatures with joy from within. When we respect the Earth, we tread lightly on it. This is my great hope for the renewal of the Earth, of you and me, and of everyone. Look to the small and the simple, look to the beautiful, look below. From there springs peace.

REFERENCES

Abram, D. 1995. The ecology of magic. In *Ecopsychology: Restoring the Earth, Healing the Mind*, eds. T. Roszak, M.E. Gomes, and A.D. Kanner, 301–315. California: Sierra Club.

Azadi, H., S. Schoonbeek, H. Mahmoudi et al. 2011. Organic agriculture and sustainable food production system: main potentials. *Agric Ecosyst Environ* 144:92–94.

Ball, B.C., T. Batey, and L. Munkholm. 2007. Field assessment of soil structural quality—a development of the Peerlkamp test. *Soil Use Manag* 23:329–337.

Ball, B.C. and C.E. Crawford. 2009. Mechanical weeding effects on soil structure under organic carrots and beans. *Soil Use Manag* 25:303–310.

Batey, T. 1988. *Soil Husbandry: A Practical Guide to the Use and Management of Soils*. UK: Soil and Land Use Consultants Ltd.

Berry, W. 2009. *Bringing it to the Table: On Farming and Food*. Berkeley, USA: Counterpoint.

---

* Gas fracking is a controversial process of extracting gas and oil from shale by hydraulic fracturing of rocks. The chemicals used can pollute water supplies underground, and gas may be released into drinking water supplies.

[†] Tough oil describes much of our remaining oil, which is difficult, risky, and damaging to extract.

Beus, C.E. and R.E. Dunlap. 1990. Conventional versus alternative agriculture—the paradigmatic roots of the debate. *Rural Sociol* 55:590–616.
Boff, L. 1997. *Cry of the Earth, Cry of the Poor*. USA: Orbis.
Capra, F. 1997. *The Web of Life: A New Synthesis of Mind and Matter*. London, UK: HarperCollins.
Carter, V.G. and T. Dale. 1974. *Topsoil and Civilisation*. USA: University of Oklahoma Press.
Cox, S. 2008. *Ending 10 000 years of Conflict between Agriculture and Nature*. Institute of Science in Society Report, London, UK. 7 pp. Available at http://www.i-sis.org.uk/Ending10000YearsOfConflict.php, accessed February 2012.
de Castro, P. 2009. Welcome notes. In *Global Challenges—Organic Approaches. Proceedings of the 2nd European Organic Congress*, 1 Dec. 2009, Brussels, Belgium, 4–5. Brussels: European Union.
Glendining, M.J., A.G. Dailey, A.G. Williams, F.K. van Evert, K.W.T. Goulding, and A.P. Whitmore. 2009. Is it possible to increase the sustainability of arable and ruminant agriculture by reducing inputs? *Agric Syst* 99:117–125.
Gray, L. 1995. Shamanic counseling and ecopsychology. In *Ecopsychology: Restoring the Earth, Healing the Mind*, eds. T. Roszak, M.E. Gomes, and A.D. Kanner, 172–182. San Francisco: Sierra Club.
Gregory, P.J. and T.S George. 2011. Feeding nine billion: the challenge to sustainable crop production. *J Exp Bot* 62:5233–5239.
Haisch, B. 2006. *The God Theory: Universes, Zero-Point Fields and What's Behind It All*. San Francisco, USA: Weiser Books.
Hartmann, T. 1999. *The Last Hours of Ancient Sunlight. Waking Up to Personal and Global Transformation*. London: Hodder and Stoughton.
Haygarth, P. and K. Ritz. 2009. The future of soils and land use in the UK: soil systems for the provision of land-based ecosystem services. *Land Use Policy* 26S:S187–S197.
Hillel, D. 1992. *Out of the Earth: Civilization and the Life of the Soil*. USA: University of California Press.
Howard, A. 1947. *The Soil and Health: A Study of Organic Agriculture*. New York: the Devin-Adair Company.
Jackson, T. 2009. *Prosperity without Growth? The Transition to a Sustainable Economy*. UK: Sustainable Development Commission Report.
Kiers, E.T., R.R.B. Leakey, A.M. Izac, J.A. Heinemann, E. Rosenthal, D. Nathan, and J. Jiggins. 2008. Agriculture at a crossroads. *Science* 320:320–321.
King, F.H. 1911. *Farmers of Forty Centuries: Permanent Agriculture in China, Korea and Japan*. Pennsylvania: Rodale Press.
La Chapelle, D. and J. Puzey. 1995. Interview on deep ecology with Derrick Jensen. In *Listening to the Land: Conversations about Nature, Culture and Eros*, ed. D. Jensen, 232–247. San Francisco: Sierra Club Books.
Lal, R. 2009a. Technology without wisdom. In *Organic Farming, Pest Control and Remediation of Soil Pollutants*. Sustainable Agriculture Reviews, ed. E. Lichtfouse, 1–3:11–14. Netherlands: Springer.
Lal, R. 2009b. Soils and world food security. *Soil Tillage Res* 102:1–4.
Lal, R. 2009c. Tragedy of the global commons: soil, water and air. In *Climate Change, Intercropping, Pest Control and Beneficial Microorganisms*. Sustainable Agriculture Reviews, ed. E. Lichtfouse, 2:9–11. Netherlands: Springer.
Lamine, C. 2011. Transition pathways towards a robust ecologization of agriculture and the need for system redesign. Cases from organic farming and IPM. *J Rural Stud* 27:209–219.
Le Gal, P.Y., P. Dugue, G. Faure, and S. Novak, 2011. How does research address the design of innovative agricultural production systems at the farm level? A review. *Agric Syst* 104:714–728.

Lewandowski, S. 1998. Wild soils. The Crooked Lake Review, Summer issue. Available at http://www.crookedlakereview.com/articles/101_135/108summer1998/108lewandowski.html, accessed February 2012.

Lichtfouse, E. 2009. Sustainable agriculture as a central science to solve global society issues. In *Organic Farming, Pest Control and Remediation of Soil Pollutants*. Sustainable Agriculture Reviews, ed. E. Lichtfouse, 1–3, Netherlands: Springer.

Lichtfouse, E., M. Hamelin, M. Navarrette, P. Debaeke, and A. Henri. 2010. Emerging agroscience. *Agro Sustain Develop* 30:1–10.

Lines-Kelly, R. 2004. *Soil: Our Common Ground—A Humanities Perspective*. 3rd Australian New Zealand Soils Conference, Sydney, Australia. Published on CDROM. Available on http://www.regional.org.au/pdf/asssi/supersoil2004/lineskelly.pdf, accessed February 2013.

Lovelock, J. 1979. *Gaia: A New Look at Life on Earth*. Oxford: Oxford University Press.

McIntosh, A. 2008a. *Hell and High Water: Climate Change, Hope and the Human Condition*. Edinburgh: Birlinn.

McIntosh, A. 2008b. *Rekindling Community: Connecting People, Environment and Spirituality*. Schumacher Briefing 15. Bristol: Green Books for the Schumacher Society.

McLaren, B.D. 2010. *Naked Spirituality*. London: Hodder and Stoughton.

Naess, A. 1973. The shallow and the deep, long-range ecology movement. A summary. *Inquiry* 16:95–100.

Patzel, N. 2010. The soil scientist's hidden beloved: archetypal images and emotions in the scientist's relationship with soil. In *Soil and Culture*, eds. E.R. Lander and C. Feller, 205–226. UK: Springer.

Powlson, D.S., P.J. Gregory, W.R. Whalley et al. 2011. Soil management in relation to sustainable agriculture and ecosystem services. *Food Policy* 36:S72–S87.

Rains, G.C., D.M. Olson, and W.J. Lewis. 2011. Redirecting technology to support sustainable farm management practices. *Agric Syst* 104:365–370.

Salatin, J. 2010. *The Sheer Ecstasy of Being a Lunatic Farmer*. USA, Virginia: Polyface.

Schjønning, P., S. Elmholt, L.J. Munkholm, and K. Debosz. 2002. Soil quality aspects of humid sandy loams as influenced by organic and conventional long-term management. *Agric Ecosyst Environ* 88:195–214.

Schumacher, E.F. 1973. *Small Is Beautiful. A Study of Economics as if People Mattered*. London: Vintage.

Sewall, L. 1995. The skill of ecological perception. In *Ecopsychology: Restoring the Earth, Healing the Mind*, eds. T. Roszak, M.E. Gomes, and A.D. Kanner, 201–215. San Francisco: Sierra Club.

Shapiro, E. 1995. Restoring, habitats, communities and souls. In *Ecopsychology: Restoring the Earth, Healing the Mind*, eds. T. Roszak, M.E. Gomes, and A.D. Kanner, 224–239. San Francisco: Sierra Club.

Shaxson, T.F. 2006. Re-thinking the conservation of carbon, water and soil: a different perspective. *Agron Sustain Dev* 26:9–19.

Shepherd, T.G. 2009. *Visual Soil Assessment. Volume 1. Field Guide for Pastoral Grazing and Cropping on Flat to Rolling Country*, 2nd edition. Palmerston North, New Zealand: Horizons Regional Council.

Soane, B.D., B.C. Ball, J. Arvidsson, G. Basch, F. Moreno, and J. Roger-Estrade. 2012. No-till in Northern and Southern Europe: opportunities and problems for crop production and the environment. *Soil Tillage Res* 118:66–87.

Swan, J.A. 1993. *The Power of Place. Sacred Ground in Natural and Human Environments*. UK: Gateway Books.

Uekotter, F. 2006. Know your soil: transitions in farmers' and scientists' knowledge in Germany. In *Soils and Societies: Perspectives from Environmental History*, eds. J.R. McNeill and V. Winiwarter, 322–339. UK: The White Horse Press.

Walsch, N.D. 1999. *Conversations with God. Book 2. An Uncommon Dialogue.* London: Hodder and Stoughton.
Weckmann, J. 2009. Organic, local and fair—one company's attempt to tackle environmental and climate challenges. Abstract of speech for workshop *Global challenges—organic approaches: organic food and farming in times of climate change, biodiversity loss and global food crisis.* 2nd European Organic Congress, Brussels, pp. 39–41.
White, L. 1967. The historical roots of our ecological crisis. *Science* 155:1203–1207.
Whitefield, P. 2009. *Permaculture in a Nutshell.* 5th edition. UK: Permanent Publications.

12 Theological and Religious Approaches to Soil Stewardship

Gregory E. Hitzhusen, Gary W. Fick, and Richard H. Moore

CONTENTS

12.1 Introduction 285
12.2 Religious Principles and Soil Stewardship 286
    12.2.1 Humans and Soil in Biblical Creation Narratives 287
    12.2.2 Shinto and the Ecological Integrity of the Tohoku Region, Japan...288
    12.2.3 Humans and Soil in World Religious Texts......................................289
12.3 Connecting Best Soil Management Principles with Religious Soil Themes...289
    12.3.1 Recognizing Agricultural Roots within Religions 292
12.4 Religious–Cultural Perspective in Need of Soil Science 293
    12.4.1 Irrigation and Salinization of Soil 293
    12.4.2 Soil Nutrient Management..294
12.5 Examples of Religious–Cultural Integration with Soil Stewardship...........294
    12.5.1 Kayapo Sustainable Anthropogenic Landscape in Brazil 295
    12.5.2 Amish Land Stewardship in Collaboration with Agroecologists.....296
    12.5.3 Soil Stewardship Sundays and the Dust Bowl..................................296
    12.5.4 Emerging Religious Approaches to Land Stewardship in Africa 297
12.6 Conclusion 299
References..300

> Upon this handful of soil our survival depends. Husband it and it will grow our food, our fuel, and our shelter and surround us with beauty. Abuse it and the soil will collapse and die taking man with it.
>
> **Vedas (Hindu scripture in Sanskrit, ca. 1500 BCE)**

12.1 INTRODUCTION

The terms "theology" and "religion" have rarely surfaced in the soil science literature, and this scarcity may be of little surprise. Yet as soil scientists grapple with contemporary challenges of intensified development and pressure on soil resources and ecosystem services, the roles of theology and religion merit closer attention (cf. Lal

2010). Religions impart a multilayered influence on culture and human behavior and are increasingly recognized as playing a critical and complementary role with science in sustainable development (Gardner 2006; Tucker 2007; Palmer and Finlay 2003; Hitzhusen 2007). More than 85% of the world's people identify themselves as members of a religious or spiritual community (Central Intelligence Agency 2012), and as such, religions serve as a lens through which most humans see and understand ethics worldwide. If soil stewardship practices adequate to address contemporary challenges are ever to germinate, set root, and bear fruit in significant measure, resonance of those practices with religious systems of understanding and behavior will be indispensable; soil science alone cannot inspire global soil stewardship. At the same time, religion and theology have potential to play both a positive and a negative role, so it is important to better understand these potentials if we wish religious influences to be positive. Just as religion has historically played a role in mediating soil and land issues, a proper synergy between religious influences, science, and policy can empower soil stewardship as societies strive to meet the needs of growing populations.

12.2 RELIGIOUS PRINCIPLES AND SOIL STEWARDSHIP

Issues of soil stewardship, of course, are as old as religion. Neolithic religions that attended the origins of human agriculture focused on fertility; dharmic religions originating on the Indian subcontinent developed concepts to revere and respect the Earth; the Hebrew prophet Jeremiah bemoaned the desolation of land in ancient Israel; Plato's *Critias* (ca. 360 BCE) lamented the erosion of soil in the Attican countryside surrounding Athens; Jesus likened soil quality to the substance of human souls.* The import of religious influences varies across a range of possible roles, from shaping individual land users' values and behaviors to affecting community and cultural perspectives that lead to local practices or policies and to serving as a direct management force over soil resources. The various examples described below illustrate religious influence at multiple levels and represent different religious and cultural contexts, but each case is informed by particular principles found in the belief systems of the region. In whatever religious or cultural context soil scientists find themselves, understanding the basic principles that might underlie and support local soil stewardship is essential.

One strategy for exploring environmentally relevant religious values is to examine specific soil stewardship instructions within religious traditions, such as Sabbath restrictions as applied to farming practices in Judaism and Christianity,† or the soil-sensitive Balinese Hindu and Japanese Shinto examples described below. But in addition to such soil-specific prescriptions of varying contemporary salience, religious traditions posit fundamental principles that shape perspectives on the relations between humans, soil, and the divine or spiritual realm. Whether through direct

---

* Hughes 1975, 26–27; Lal 2010, 304; Jeremiah 12:10–11, 13; Hughes 1994, 69; Mt 13:3. In this chapter, references to the Hebrew Bible and Christian New Testament will be formatted in footnotes and in text by listing abbreviated book name (http://hbl.gcc.edu/abbreviationsCHICAGO.htm), chapter number, and verse number, as in the following example: Mt 13:3–8 = book of Matthew, chapter 13, verses 3–8.
† Observance varies among Jewish farmers, and regulations governing observance are complex (Kuber 2007); Christian Sabbath practice has probably declined in the face of modern agricultural economics and agribusiness (Wirzba 2006, 37).

religious influence or indirect cultural shaping, these principles affect perspectives on land stewardship and supply values upon which stewardship practices are often based, especially in religiously influenced communities (Kanagy and Nelsen 1995).

12.2.1 Humans and Soil in Biblical Creation Narratives

An example of basic religious principles that connect humans to soil is found in the creation narratives of the Hebrew Bible, setting a frame for Jewish and Christian understanding of the role of humans on Earth. In the creation story in the first chapter of the book of Genesis, light and dark, night and day, heavens and earth, and sea and land are created and ordered as the fecund matrix from which plants and all creatures are brought forth, including humans ('*adam*, humanity, both male and female). The earth brings forth plants and all manner of creatures, the seas bring forth fish, and God blesses creatures to be fruitful and multiply (Gn 1:22,28). Dominion over Earth and its creatures is granted to humans as God's image-bearers; humans are to be stewards of God's very good creation, while at the same time being kin with the other creatures God has blessed.

The second creation story in chapter 2 of Genesis develops further on this story, with even more particular attention to the imprint of soil on humanity. Here, the human ('*adam*) is formed from the soil ('*adamah*), the dust of the ground; the wordplay of the Hebrew parallels *human* from *humus*, or *earthling* from *Earth*. "The '*adamah*, the material for '*adam*, is not just any old ground, it is the soil of arable cropland, good farmland" (Fick 2008). In addition to identifying human substance with the soil, the second chapter of Genesis also declares the role of humans as caretakers of the creation. God places 'adam in the garden of Eden to cultivate (*abad*) and protect (*shamar*) it, and thus in these sacred stories, humans are given a fundamental vocation as gardeners (Gn 2:15; Pollan 1991). Humans will prove themselves poor listeners to God's instructions, however, and the third chapter of Genesis tells the story of human estrangement from God. As a result, the ground ('*adamah*) is cursed (Gn 3:17), and '*adam* must obtain their food from it by toil and sweat (Gn 3:19). In estrangement, humans have become mortal, and in physical death, they return to '*adamah*, the ground from which they and other living things were made (Gn 3:19). The biblical analogy is that we are made from the same kind of soil that produces our food, and we return to this source in our death. With this interpretation, the soil–food–life connection of the Bible runs deep, right down to the rudiments of the nutrient cycle (Fick 2008).

In the following biblical chapters, humanity fails to fulfill its potential, filling the Earth with violence rather than protecting the Earth's fruitfulness (Gn 6:5) and causing God to decide to wipe out wicked humanity with the flood. Only Noah, one who obeyed God's instructions and who is described as "a man of the soil" (Gn 9:20), is saved along with his family. Noah faithfully tends to the call to protect all the creatures by building an ark, and following the flood, Noah and his family are reminded alongside the other creatures to be fruitful and multiply (Gn 9:1,7). In this resetting of human origins, Noah's family is not recharged with having dominion, and God promises never again to curse the ground. Through Noah, future generations are given assurance that the Earth's cycles of seasons and seedtime and harvest will not fail—humans can rely on the provision of the Earth without fearing another deluge from God (Gn 8:21–22).

Generations later, divine guidance in the form of the Jewish law includes distinct Sabbath instructions regarding agricultural practice—fields are to be left fallow once in every 7 years, both to give the land rest and to provide sustenance for wild animals and the poor (Ex 23:10–12; Lv 25:5–12). As stewards of an Earth that is the Lord's, who has blessed all creatures to be fruitful and multiply, such Sabbath practices follow naturally. Hebrew prophets will later warn of the desolation of land that will result from unfaithful humans (Jer 12:4; Hos 4:1–3), to the point even of the land vomiting out faithless inhabitants who do not give the land its Sabbaths (Lv 18:28; 2 Chr 36:21), and the prophet Ezekiel voices encouragement that for faithful people, the earth will yield its increase, and they shall be secure on their soil (Ez 34:27). These sacred stories paint a picture of humans as caretakers, as stewards of land, with soil stewardship and agriculture used as direct reference points to illuminate this human vocation.

12.2.2 Shinto and the Ecological Integrity of the Tohoku Region, Japan

Another example of evocative religious principles comes from the Tohoku Region of Japan, where the indigenous Shinto religion provides an ideological framework for bridging the ecological integrity of the archipelago. First, in the creation myth presented in the *Nihon Shoki* and *Kojiki*, two eighth-century texts detailing the creation mythology and lives of the early emperors of Japan, the mountain goddess *yama no kami* would descend from her winter residence in the mountains to become the *ta no kami* or paddy field god in the spring, residing in the rice fields during the subsequent agricultural season. Following the fall harvest, the deity would return to its winter

FIGURE 12.1 Rural Japanese wedding in Miyagi Prefecture in the Tohoku Region of Japan in 1983. The rice paddy (lower left corner of photo) with water plants coming out of it is included in the wedding because the community observes single-heir succession, so this symbolizes the transfer of land from father to son. The bride and groom also drink sake, which is made from rice and consummates the ceremony. (Photo courtesy of Richard H. Moore.)

home in the mountains. In fact, in the myth, the emperor of Japan is Niniginomikoto, god of the rice fields and grandson of the sun goddess Amaterasu; today, the emperor still symbolically transplants rice by hand in the spring in the imperial grounds, and Shinto shrines dot the mountaintops throughout Japan.

According to the myth, Japan is the land of abundant rice (*mizuho no kuni*). The *yama no kami*, or goddess of the mountain, is identified also as Konohanasakuyajime, granddaughter of Izanami and Izanagi, the divine couple said to have created Japan and the Japanese from *tenchikaibyaku*, the "pillar of heaven." As such, she represents the more mysterious and serene "nature" of the mountains, in contrast to *ta no kami*, "the god of the rice paddies" in the plains, where cultivation occurs (Moore 2009b). *Nihon Shoki's* portrayal of Japan is as the *mizuho no kuni* (瑞穂の国), the "land of abundant rice." Perhaps this is why farmers in the plains of Japan traditionally had such a close relationship with the mountains. Soil is transported from the mountains to enrich the fertility of the rice paddies and is known as "guest soil" (*kyakudo*) (Moore 1990), and a fertile soil and plentiful rice harvests are symbolized during the wedding ceremony of the farm heir (Moore 1990; Figure 12.1). Soil in Japan can also be termed as having a living quality—"fat" (*tsuchi ga futoru*), meaning too much organic matter, or "thin" (*tsuchi ga yaseru*). Here again, religious principles give shape to an active set of relations, and some of the management implications of such principles will be described below.

12.2.3 Humans and Soil in World Religious Texts

References to the soil occur in many of the sacred texts and basic teachings of the world's religions (Table 12.1; Lal 2010). As with biblical texts, since these writings are primarily about the nonmaterial values of various human cultures, sacred scriptures generally do not provide direct instructions on soil management. Instead they use the soil as a metaphor or familiar material example for some lesson deemed to have spiritual significance or some recommended practice thought to build human virtue (Fick 2008). However, using the concept of soil in this way bestows a religious reverence for soil that is almost universal across the religions of the human race. Though religious reverence does not always translate into daily stewardship, for persons who claim to value the virtues of religion, education about their religion should help them become better stewards of the soil.

12.3 CONNECTING BEST SOIL MANAGEMENT PRINCIPLES WITH RELIGIOUS SOIL THEMES

The principles of soil stewardship, or good soil management, are commonly enumerated in modern soil textbooks (Brady and Weil 2008; Gardiner and Miller 2008; Hatfield and Sauer 2011). Those principles can be regarded as the foundations for practices that maintain or improve soil quality or soil health. Because the ancient authors of religious scriptures recognized and described attributes of agricultural soils, they can often be related to modern approaches to soil management that foster good soil quality. As an example, Table 12.2 places six principles of good soil management beside biblical passages related to those practices. It is clear that the Bible is not a text

TABLE 12.1
References to Soil in Some of the Major Religions of the World

American Indian—The old people came literally to love the soil, and they sat or reclined on the ground with a feeling of being close to a mothering power. (Chief Luther Standing Bear, American Indian Quotes, http://www.impurplehawk.com/quotes.html; retrieved 11 January 2012.)

Baha'i—When thou dost plant a tree, its height increaseth day by day. It putteth forth blossoms and leaves and luscious fruits. But after a long time, it doth grow old, yielding no fruitage any more. Then doth the Husbandman of Truth take up the seed from that same tree, and plant it in a pure soil; and lo, there standeth the first tree, even as it was before. (Writings of 'Abdu'l-Bahá 23:52.)

Buddhism—The first truth is that nothing is lost in the universe. Matter turns into energy, energy turns into matter. A dead leaf turns into soil. A seed sprouts and becomes a new plant. (Basic Teachings of Buddha, http://online.sfsu.edu/~rone/Buddhism/footsteps.htm; retrieved January 10, 2012.)

Christianity—A sower went out to sow. And as he sowed, some seeds fell along the path, and the birds came and devoured them. Other seeds fell on rocky ground, where they did not have much soil, and immediately they sprang up, since they had no depth of soil, but when the sun rose they were scorched. And since they had no root, they withered away. Other seeds fell among thorns, and the thorns grew up and choked them. Other seeds fell on good soil and produced grain, some a hundredfold, some sixty, some thirty. (Mt 13:3–8, English Standard Version.)

Daoism—Like plants that flourish, some return to the soil and root they grew from. (Tao Te Ching 16:16.)

Hinduism—[M]other [earth], with your oceans, rivers, and other bodies of water, you give us land [or soil] to grow grains, on which our survival depends…milk, fruits, water, and cereals as we need to eat and drink…our motherland on whom we grow wheat, rice, and barley… (Atharva Veda 12.1.)

Islam—Let the human consider his food! We pour the water generously. Then we split the soil open. We grow in it grains. Grapes and pasture. Olives and palms. A variety of orchards. To provide life support for you and your animals. (80.24–32, Khalifa translation.)

Jainism—Even a small Däna (gift) given to a proper or suitable donee bears much desirable fruit for souls in the fullness of time, just as the tiny seed of a fig tree, sown in good soil, produces a tree, casting magnificent shade. (Ratna-karandaShrävakächär; JAINA Education Series 401—Level 4.)

Judaism—Then God formed the man [*adam*] of dust from the ground [*adamah* = good agricultural soil]… (Gn 2:7, English Standard Version.)

Shintoism—Hence it is said that when the world began to be created, the soil of which lands were formed when Izanami and Izanagi stood on the bridge between heaven and earth and dipped the jewel spear into the ocean and the islands were formed from the brine that dripped from the spear. (The Kojiki and Nihongi, Book 1.)

Sikhism—Root out the choking weeds of lust and anger; loosening the soil, the more thou hoest and weedest, the more lovely grows the soul. (The AdiGranth, Rag Basant, page 1171.)

Zoroastrianism—"Purity is the best thing for men after birth." …Next in order comes the command to cultivate the soil, to produce corn, provender, and fruit bearing trees, to irrigate dry land and to drain marshes … (The teachings of Zoroaster, John Murray, London, 1905: 33.)

Note: The word "soil" is underlined so that it can be more easily found.

on soil management, but the soil specialists working with farmers and other managers of the soil may be able to communicate more effectively in a Christian culture when they realize the appreciation for soil management inherent in the culture. Such study of religious perspectives can therefore foster soil stewardship (Fick 2008).

While the connections in Table 12.2 are suggestive, a wide range of additional connections can be imagined, and the range of religious perspectives that might relate positively to soil stewardship is vast. Individual land users might discover particular tenets that inspire them to better soil stewardship. For example, one author's father

TABLE 12.2
Examples of the Principles of Soil Management for Crop Production and References to Those Principles in the Hebrew Bible and Christian New Testament

Accelerated soil erosion should be controlled.	Dust storms—"heavens like bronze…rain like powder" (Dt 28:23–24). Beating rain that leaves no food (Prv 28:3). Water erosion that leaves no hope (Job 14:18–19).
Soil nutrients removed by crops should be replaced, and soil acidity and alkalinity optimized.	Fertile soils are a blessing; poor soils are a curse (Dt 28:4,18). Ashes applied to the soil [a kind of liming] (Ps 147:16). Composting livestock manure (Isa 25:10). Spreading manure and crop residues on the soil (Ps 83:10; Jer 9:22). Infertile soil produces weeds and profits nothing (Jer 12:13). The parable of fertilizing the soil for the fig tree (Lk 13:8).
Soil water should be managed so there is not too much or too little, and it is available when it is needed.	Fertile soil is well watered (Gn 13:10; Ez 17:8). Irrigation is mentioned (Dt 11:10). Changes in the water regime can destroy or restore the land (Ps 107:33–35). Rain and snow cause the crops to grow (Is 55:10).
Soil organic matter should be maintained by additions of organic matter and by controlling tillage so that soil structure and soil aeration do not limit crop growth.	Sabbath rest of the land gives a fallow period once every 7 years [benefiting soil organic matter, soil structure, and soil aeration] (Ex 28:10–11; Lv 25:2–5). Composting livestock manure [to return organic matter to the soil] (Is 25:10). Soil tillage is for hard ground (Jer 4:3), and it is to be limited (Is 28:24). The parable of the seed sower and the four kinds of soil [good soils give good yields] (Mt 13:4–8; Mk 4:4–8; Lk 8:5–8).
Soil biodiversity should be maintained for ecosystem health and pest management.	Sabbath rest for the land [also helps restore soil biodiversity] (Ex 28:10–11; Lev 25:2–5). Planting a variety of crops (Eccl 11:6). Seed diseases are found in the soil (Joel 1:17).
Soil pollution, including salinization from mismanaged irrigation, should be avoided or corrected.	The land is polluted by bad human behavior such as murder (Nm 35:33–34; Dt 21:23, 24:4). Spreading salt destroys the soil (Jgs 9:45). Fruitful land can become a salty waste (Ps 107:34).

based his stewardship ethic as a farmer on biblical principles; another's grandfather likewise based his farming ethic and soil conservation work on a notion from Leviticus that humans are just tenants on the land, such that we should leave it a little better than we found it in hope that it will continue to provide for the next generation. In addition to specific principles that can inspire individuals, the fact that some respected community leaders hold such views can also help promote stewardship in a given community.

12.3.1 Recognizing Agricultural Roots within Religions

In many cases, religious and cultural connections with agriculture are deep. Some religious systems have evolved from a more agriculturally dominant past, where the rhythms of nature and life on the land were more fully integrated with religious ritual. For instance, in ancient Israel, the harvest festival and agricultural character of Sukkot, the Feast of Booths, was more apparent than in more recent celebration; even so, these deep roots can be drawn on for their contemporary resonance. As detailed below, the Soil Stewardship Sunday tradition in American Catholic rural communities (with Protestant parallels) drew upon traditional Catholic observations of Rogation Days and Ember Days, which had diminished as common Catholic life became less agricultural. The rediscovery and promotion of these themes was simultaneously developed as an expression of soil stewardship and an effort to renew liturgical life by more closely linking the rhythms of farming practice with the worship life of Christian communities. In this way, the meaning of spiritual teachings became more clearly framed within the common sense of cultural relationships with the land, while land stewardship practices were reinforced with spiritual and ethical meaning.

Even in a society like the United States, where agricultural livelihoods now make up less than 2% of the population and relationships with land cannot be assumed as common sense, it has been instructive to witness the recent interest of faith communities in local agriculture, organic farming, food justice, and learning about natural cycles.* Several denominations have generated policy statements and study documents to bring attention to land stewardship issues.[†] Religiously affiliated organic farms have sprung up across the United States,[‡] and in Africa, where rural livelihoods are much more prevalent, faith-based conservation organizations have also

---

* The Unitarian Universalist 2008–2012 Congregational Study/Action Issue is "Ethical Eating: Food and Environmental Justice" (http://www.uua.org/environment/eating/55648.shtml); the Presbyterian Food and Faith program provides related resources (www.pcusa.org/blogs/foodfaith/); "Rooted in God's Word and Lands" is a land stewardship study guide produced by the National Council of Churches (http://www.nccecojustice.org/network/downloads/landresource.pdf).
† The National Religious Partnership for the Environment provides links to food and agriculture (http://www.nrpe.org/index.php?option=com_k2&view=item&layout=item&id=305&Itemid=1132) and other issue statements, as does the National Council of Churches Eco-Justice Programs: Faithful Harvest program (http://nccecojustice.org/food/), and Interfaith Power and Light provides "Cool Harvest" resources (http://action.interfaithpowerandlight.org/site/c.dmJUKgOZJiI8G/b.6605225/k.97F1/Cool_Harvest.htm) that draw upon faith-based principles as a basis for engaging food and agriculture issues.
‡ A few examples include Shepherd's Corner Farm and Ecology Center, OH (http://www.shepherdscorner.org/); Genesis Farm, NJ (http://www.genesisfarm.org/index.taf); Crown Point Ecology Center, OH (http://crownpt.org/); Green Gulch Farm, San Francisco Zen Center (http://www.sfzc.org/ggf/); and Isabella Freedman Jewish Retreat Center Adamah Farm Fellowship (http://isabellafreedman.org/adamah/intro).

gotten involved in soil and land stewardship.* Thus, within a given region, a particular religion can provide cultural cues, respond to cultural trends, and establish links between religious and cultural perspectives that provide the basic frame of orientation for those who work the soil. Soil stewardship can be greatly empowered by such links. And yet, religious intentions and sensibilities are usually not sufficient in themselves.

12.4 RELIGIOUS–CULTURAL PERSPECTIVE IN NEED OF SOIL SCIENCE

As we have seen, religious traditions can influence individuals, communities, and polities either through shared values or principled devotion, or through direct management. At these various levels, religious traditions and writings contribute to soil stewardship in at least two ways. In general, they foster an appreciation for soil as a gift to humans and as an essential resource for food production that sustains life (see Hinduism, Table 12.1); in some cases, they also provide more specific management instructions (Sikhism, Table 12.1; soil nutrients, Table 12.2). But the intention to be a good steward of the soil does not automatically lead to best management practices. Sound stewardship can be religiously motivated, but scientific understanding is often needed to make the stewardship effective. Consider the examples of irrigation and of nutrient management.

12.4.1 Irrigation and Salinization of Soil

Most irrigation water contains salts that accumulate in the irrigated soil as water is removed by direct evaporation and transpiration through the crops. Sustainable irrigation requires (1) flushing salts from the soil by the application of water in excess of crop needs and (2) soil drainage so that the salty flush water can be carried away without waterlogging the soil. In addition, there must be some place for the drainage water to go. Along streams and rivers, downstream water quality will deteriorate as salt concentrations increase. In a geological basin, some land will be sacrificed in low-lying ponds sterilized by the salt (Hillel 2008a; Szabolcs 1986).

For the most part, prescientific peoples have not understood this, and as a consequence, their irrigation projects often failed over time because of salinization of the soil. The classic case is in the Tigris–Euphrates valley of Mesopotamia (Hillel 2008b). The cultures of Sumer (ca. 4000 to 2500 BCE), Akkad (ca. 2500 to 2100 BCE), and Babylonia (ca. 2100 to 500 BCE) were each established further up the valley than their predecessors because the lower delta lands became too salty for crop production. Even the relatively salt-tolerant barley (compared to wheat) eventually failed because of too much salt in the soil (Artzy and Hillel 1988; Hillel 2008b), and the salinized soils remain a problem into modern times (Richardson et al. 2005).

Another example is found on the soils of the American Southwest originally irrigated by the First Americans. Modern soil scientists have uncovered evidence of salinization in those soils (Olson 1981; Palacios-Fest 1994), indicating that natural

---

* See Section 12.5.4 below for more information about some of these programs.

salts in irrigation water were accumulating in pre-Columbian times. Although subject to some controversy, Native Americans are usually credited with a religiously based land ethic that promotes good stewardship (Jostad et al. 1996). Likewise, the ancient Mesopotamians are also known to have been very religious, though their land ethic is less clear (Jacobsen 1978). Nevertheless, in both of these religious cultures that predate soil science, absence of scientific understanding led to the damage and even destruction of their soil resources and the decay of their civilizations.

12.4.2 Soil Nutrient Management

Religious communities also struggle with soil nutrient management, often in association with the high concentrations of livestock in modern times. Such concentrations make it possible that the application of livestock manure to the soil will be in excess of crop demands and that soil nutrients will accumulate in the soil to levels that cause environmental damage (Meek et al. 1982; Sharpley et al. 2002). Excess nitrogen pollutes surface waters and groundwaters (Smith et al. 1990). Excess phosphorus leads to eutrophication of surface waters (Sharpley 2003). Excess potassium leads to health problems of livestock that eat forage produced on soils very high in potassium available to the crop (Cherney et al. 1998).

In the United States, the problem of excess plant nutrients from manure has become serious on Amish farms in southeastern Pennsylvania (Lanyon et al. 2006), and problems with water quality in other Amish areas are also well documented (Lanyon et al. 2006; OEPA 2007; Widner 2010; Young et al. 1985). The spread of Amish communities across the United States is notable (Cross 2010; Donnermeyer and Cooksey 2004). High population densities among the Amish in traditionally Amish areas and the opportunity to acquire land elsewhere are important reasons for out-migration, but the deterioration of soil and water quality associated with too much livestock manure for the available land on old Amish farms is concerning. The religious views of the Amish toward community and generational transfer of farms place a high value on land stewardship (Moore et al. 2001). The Amish want their children to inherit farms with soils that are fertile and not eroded (Widner 2010). However, suspicion and avoidance of science education can jeopardize those goals. Thus, agricultural extension and soil and water conservation programs are being applied to help the Amish be better soil stewards (Drake and James 1993; Hoorman and Spencer 2001; Lanyon et al. 2006). The concepts of soil stewardship and soil science need to work together.

12.5 EXAMPLES OF RELIGIOUS–CULTURAL INTEGRATION WITH SOIL STEWARDSHIP

Promising integrations of science-based land stewardship principles can be seen in many examples throughout the world, and religiously based soil stewardship traditions have been evolving for ages. Anthropologists have documented the prevalence and multifunctional uses of religion with regard to soil fertility. In most cases, we see complex social, natural, and physical systems that are integrated (Moore 2009a). Lansing (2006) and Lansing and Kremer (1993) have argued for the emergent property of this integration. In other words, most social systems have evolved an ideological system over thousands of years, and when we examine it today, it is nearly impossible to unravel exactly how it developed. This was the case for Lansing's water temple networks of the Indonesian island of Bali, where the soil in terraced rice paddies has been preserved through the ritual processes among decentralized water temples. Rappaport (1984) noted the importance of ritual symbolically regulating the ecosystem, especially preventing the overintensification of agricultural production.

Equally, religion plays a key role in building soil. For example, the cerrado region of Brazil where ultisols and oxisols predominate is only 20% of its original size because it has been cleared along with the forests to make way for new forms of agriculture. Japanese development projects along with world development agencies have promoted and funded the development of large-scale farms in this area composed mostly of cattle grazing, coffee production, and charcoal production, as well as irrigated agriculture. Cerrado includes native plant and animal species, which are adapted to the acidity and high aluminum and low calcium content of the soil and thus have a better chance of succeeding, especially because they are good at regenerating from roots (Durigan 1997). Cerrado consists of savanna, woodland/savanna, and dry forest ecosystems, which may seem harsh by some standards but biologically have high species diversity as well as endemism (species that are found only within that ecoregion). Darrell Posey, an ethnoecologist and anthropologist, researched how the Kayapo of Brazil were able to create patches of forest island (*apete*) in the open cerrado.

12.5.1 Kayapo Sustainable Anthropogenic Landscape in Brazil

According to Cooper et al. (2005) and Giannini (1991), when classifying natural objects, the Kayapo divide nature into categories that are related to the domains of the Kayapo universe: soil (*puka*), sky (*koikwa*), and water (*ngô*). The Kayapo associate specific plants and animals with the context of particular ecological zones. Each ecological zone represents a system of interactions among plants, animals, soils, and the Kayapo themselves. There is a continuum of zones from forest to savanna. The village is the center of their universe, while the forest is an antisocial space where people can be transformed into dangerous spirits. Their cosmology revolves around the tension between social (village) and natural (forest). Accordingly, the Kayapo have rituals and chants to transform the forest into their social area.

As Posey (2000) has pointed out, "perhaps the most exciting aspect of these new data is the implication for reforestation. This indigenous example not only provides new ideas about how to build forests 'from scratch,' but also how to successfully manage what has been considered infertile." The Kayapo were able to create value to their environment symbolically, economically, and socially (Moore 2002) into one where the plants and animals were more diverse, more locally concentrated, of greater size and density, and more youthful and vigorous than they would be in a forest that lacked these indigenous resource managers. They did this by using their local knowledge of the microclimate and soils to transplant nondomesticated local varieties (endemism) into wooded concentrations of useful plants (Posey 2000). It is also related to their practice of transplanting termite mounds, ant nests, and forest litter to the site and later protecting it from fire when the savannas are burnt (Posey 2000). Historically, these *apate* were considered "natural" by researchers, but Posey discovered that this was an anthropogenic landscape (Posey 2000) only after living with the Kayapo for more than 7 years.

12.5.2 Amish Land Stewardship in Collaboration with Agroecologists

Understanding the role of religious influences on soil and land often requires more than surface investigation, and the relation between religious and other factors is often dynamic and changing. In contrast to the negative examples described above, a promising example of soil scientists and other partners leveraging Amish community religious resources to empower better land stewardship has grown out of the Sugar Creek project in Northeast Ohio. For the Amish, the belief that people are "tending the garden" (of Eden) provides the spiritual basis for values that promote biodiversity (Moore et al. 2001). The Sugar Creek Project involves researchers from Ohio State University teaming up with local communities to provide them with water quality data so that each farm family can see its own impact on the stream. One problem for Amish farmers in the North Fork of Sugar Creek near Kidron was high nitrate and bacteria levels in the stream that were causing the cows to get sick when they drank the water. The Amish farmers, with the help of the university researchers and the Wayne County Soil and Water Conservation District, fenced cows out of the stream so that the nitrate and bacteria levels decreased. They received government funds, which paid for most of the cost of the fencing. They paid for the cost-share on the fencing through church communal labor to install the fencing. They found that the somatic cell count in the cows' milk dropped dramatically, the mastitis rates decreased, and herd health improved, and they received a higher premium for their milk from the dairy as a result of the lower bacteria count. Later, two farmers heading this effort became the CEO and president of the Green Field Farms organic dairy cooperative, formed with a main goal of promoting family farming values among Amish youth.

A second case involved the Amish farmers in the South Fork of Sugar Creek who implemented conservation measures as part of the Alpine Cheese Company water quality trading plan. Most of the conservation measures improved the manure management of their dairy herd. They also recycled the milk house waste, high in phosphorus (that was previously draining into the ditch), back onto their fields. The success of the 25 Amish farmers implementing conservation measures in the Alpine Cheese project served as a model for 21 counties to come together to form a Joint Board of Soil and Water Conservation Districts in the Muskingum Watershed. In both cases, the Sugar Creek researchers coupled natural science with an understanding and appreciation of the local culture, and small successes were multiplied into larger projects.

12.5.3 Soil Stewardship Sundays and the Dust Bowl

Another example of promising complementarity between religious and conservation communities is Soil Stewardship Sundays. This program emerged from parallel concerns for land and cultural decline in the United States in the 1920s. Agencies concerned about rural life and the connection between good soil and rural livelihoods set about to help farmers adopt more sustainable soil management practices but did so using the platform of Catholic tradition. Leaders from the National Catholic Rural Life Conference (NCRLC) felt that the best way to instill a stewardship ethic among farmers was to tie it to local religious traditions and give deeper purpose to conservation practices. The NCRLC renewed attention to Rogation Days (spring rituals to bless and protect fields and crops) and Ember Days (celebrations focused on the seasons), traditions that had faded in observance as agricultural livelihoods became less common.

The NCRLC also saw these efforts as an opportunity to renew religious life and make it more present to the challenges, meaning, and practice of rural life. Allying with leaders interested in liturgical renewal, they brought elements of farming life directly into worship traditions, such as prayers related to soil and farming and serving Eucharistic bread and wine from local farms to highlight the connection of the sacramental elements to the daily lives of parishioners. In so doing, they highlighted the relevance of the liturgy, employing it as a spiritually rooted means of encouraging better practice. Soil Stewardship Sundays, first known as Rural Life Sundays, began as an early expression of these common agrarian and liturgical goals and successfully effected change toward more sustainable practices (Woods 2009).

As the movement began to grow, and as the Dust Bowl of the 1930s hit, Rural Life Sundays were expanded beyond Catholic tradition to reach communities and farmers of other denominations. The International Association of Agricultural Missions recommended observing Rural Life Sunday starting in 1929. Episcopal liturgy drew on resources similar to those found within Catholic prayers. The United Methodist Rural Fellowship also supported the movement. Lutherans broadened the theme to include Earth stewardship in all its manifestations. In 1955, the National Association of Conservation Districts adopted the program, and it now claims that Soil Stewardship Sundays and Soil Stewardship Week may be the largest religiously linked conservation effort in the world, with materials and resources distributed widely since the 1950s.*

12.5.4 Emerging Religious Approaches to Land Stewardship in Africa

More recently, various religiously affiliated development and relief agencies have begun to connect land use with religious principles to enhance prospects for farm communities in Africa. "Farming God's Way," sponsored by Caring for Creation Kenya, is a program developed in Zimbabwe that promotes science-based farming practices in concert with biblical principles to "radically transform farming practices."† Farming God's Way is also being promoted as an extension movement unto itself, offering technical and managerial instruction integrated with biblical themes.‡ Another example is The Reckoning International, a Christian nongovernmental development, microfinance, and business training agency, which is beginning to expand their work into forestry and agriculture in Africa (Melissa Herman,

---

* The National Association of Conservation Districts (http://www.nacdnet.org/stewardship/) provides many related resources on their Web site.
† The Farming God's Way program (http://kenya.careofcreation.net/what-we-do/farming-gods-way/) is part of the larger Care of Creation organization (http://www.careofcreation.net/).
‡ See for example http://www.farming-gods-way.org/home.htm.

personal communication, September 30, 2011). These emerging efforts echo the integrative approaches of long-standing organizations like the Educational Concerns for Hunger Organization (ECHO), which provides small-farm tropical agriculture development help, appropriate technology, and forest stewardship assistance for workers in 180 countries around the world,* based on core Christian values and scientifically informed sustainability principles.

The examples highlighted in this section range from complex social traditions evolved over millennia to emerging soil stewardship programs attempting to enhance the care of soil in connection with religious and theological approaches. Two other anecdotes are suggestive. The first is the case of Jewish farmers in the village of Komemiyut during the 1952 Sabbath year. According to Jewish law and its biblical sources, the principle of the Sabbath requires farmers to rest from their labors every seventh day and every seventh year, leaving their fields fallow in the Sabbath year. Komemiyut was one of the few villages that refrained from working the land that year. At the end of the Sabbath season, the farmers could find only inferior seed to plant. A Rabbi advised them to sow the seed anyway, even though it was 3 months after neighboring villages had planted their fields. They did so, and that year, the fall rains came late, the day after they sowed their poor seed. As a result, the neighboring villages had a meager harvest, while their village had a bumper crop. This experience was seen as a fulfillment of the biblical promise of bounty (Kuber 2007).

A similar experience was observed among Tangier Island watermen in the Chesapeake Bay in 1998. Following a religious revival wherein many crab fishermen covenanted to respect the law and obey environmental regulations they had traditionally ignored, an unprecedented peel run (blue crabs soon to shed their hard shell) supplied additional harvest that helped the watermen meet their quotas that year. Their new observance of environmental guidelines had diminished their catch and lost them money in the spring, but the unusually long run made up for it,[†] and they took their good fortune as a sign of divine affirmation of their faithfulness (Pohorski 1998; Emmerich 2009). These same watermen later traveled to visit Christian farmers in Pennsylvania and invited them to make a similar covenant to be good stewards of their land, which is within the Chesapeake Bay Watershed, in hopes that improved farming practices in the watershed would help stabilize the health of the fishery and the livelihoods of the watermen (Pohorski 2001). In both of these cases, stewardship behavior was motivated directly by religious principles of restraint and an ethos of obedience reinforced by the farmers' and watermens' religion.

Stories of indigenous cultures or of rural religious subcultures are sometimes set aside as quaint anecdotes unfit for contemporary agricultural or land use practices. But paying closer attention to examples like these and the principles that underlie them may be precisely the sort of corrective to modern hubris that successful Earth stewardship requires. The examples above of heroic or transformed stewardship ethics were founded, spread, and reinforced through faith communities. Though these

---

* For more information, see http://www.echonet.org/.
† As Tangier Island waterman James "Ooker" Eskridge says in the film: "I figure I made less money this spring, but God more than made up for it by what He provided toward the end of the season" (Pohorski 1998).

examples may seem to some a tentative fit for contemporary systems of control, regulation, and production, they also suggest a corrective to the sort of hubris that desires human control of everything; such an impulse to control, whether through simple or complex means, may perennially need adjustment if stewardship ethics are to be realized. Such an ethic of trusting divine or communal provision rather than controlling all means of production to assure unflinching yields suggests an ethos that empowers individuals and communities to live within the earth's means, operating in concert with natural systems rather than trying to live over the top of them. The integrated indigenous systems whose long-term success commends them as models for sustainability and the emerging programs that seek to integrate religious sensibilities in local land use practice may therefore model an embrace of human limitations. Given the complexity of the issues we face today, and the mathematical certainty that every input of complex systems cannot be controlled nor every output predicted, adding such humility into the balance may be the better part of wisdom.

12.6 CONCLUSION

Religion is no panacea or universal solution to soil stewardship issues, but to the extent that religion plays a role in shaping the lives and behaviors and cultures of humans, to understand and connect with resonant religious and spiritual values and traditions is important if soil stewardship is to be adopted by the mass of people. While religious and theological elements can thus be a key to soil stewardship, they still require integration with good science and policy to be most fruitful—negative legacies can just as easily arise if this balance is not achieved, particularly given the complexity of pressures impacting soils today.

Some may still be tempted to believe that science alone (or at least without religion) will suffice to achieve soil stewardship. Scientists often tend to think that facts about "what is going on" will lead to policies and actions to correct problems such as soil quality, but as philosopher Kathleen Dean Moore (2010) has emphasized, this is a logical error—intervening values are needed upon which people can make use of facts in guiding their choices, behavior, communities, and polities; indeed, such values are what cultures and religions are uniquely equipped to express and maintain. As such, further attention to religious and theological dimensions of soil stewardship can be a key complement to the work of soil scientists.

At the same time, religious communities need the insights of soil science to apply their beliefs and values in redeeming, enlightening, peaceful, and just ways. When the authors of the Hebrew Bible command adherents to follow God's laws (Lv 26:3–5), in the context of a covenant between humans, the divine, and the land, the promise is conditional. *If* the people follow these tenets, they will live long in the land and will enjoy abundant harvests. But if not, all bets are off (Lv 26:14–16,19–20). In the context of the unprecedented pressure on soil resources and ecosystem services on a planet of 7 billion people, this ancient synergy between sustainability and faithful adherence to principles offers strong help, and *if* such cultural, spiritual, and religious resources can complement scientifically derived best practices, be informed by good soil science, and reinforce a tradition of soil stewardship to meet the complex challenges of the present, perhaps prospects for sustainable soil management have a prayer.

REFERENCES

Artzy, M. and D. Hillel. 1988. A defense of the theory of progressive soil salinization in ancient southern Mesopotamia. *Geoarchaeology* 3:235–238.

Brady, N.C. and R.R. Weil. 2008. *The Nature and Properties of Soil*, 14th ed. Upper Saddle River, NJ: Pearson Prentice Hall.

Central Intelligence Agency (CIA). 2012. *World Factbook*. https://www.cia.gov/library/publications/the-world-factbook/.

Cherney, J.H., D.J.R. Cherney, and T.W. Bruulsema. 1998. Potassium management. In *Grass for Dairy Cattle*, eds. J.H. Cherney and D.J.R. Cherney, 137–160. New York: CABI Publishing.

Cooper, M., E.R. Teramoto, P. Vidal-Torrado, and G. Sparovek. 2005. Learning soil classification with the Kayapó Indians. *Sci Agric (Piracicaba, Braz)* 62(6):604–606.

Cross, J.A. 2010. The expanding role of the Amish in America's dairy industry. *Focus on Geography* 50:7–16.

Donnermeyer, J. and E. Cooksey. 2004. The demographic foundations of Amish society. Paper presented at the Plain Communities Conference, Elizabethtown College, Pennsylvania.

Drake, B.H. and R.E. James. 1993. Extension in religious communities. *J Extension* 31(1):18–19.

Durigan, G. 1997. Restoration of "cerrado" vegetation in degraded areas, SE Brazil. *World Forestry Congress* Vol. 2, Topic 7.

Emmerich, S.D. 2009. Fostering environmental responsibility among watermen of Chesapeake Bay: A faith and action research approach. In *Mutual Treasure: Seeking Better Ways for Christians and Culture to Converse,* eds. H. Heie and M.A. King, 73–92. Telford, PA: Cascadia Publishing House.

Fick, G.W. 2008. *Food, Farming, and Faith*. Albany, NY: SUNY Press.

Gardiner, D.T. and R.W. Miller. 2008. *Soils in Our Environment*, 11th ed. Upper Saddle River, NJ: Pearson Prentice Hall.

Gardner, G. 2006. *Inspiring Progress: Religions' Contribution to Sustainable Development.* New York: W.W. Norton & Company.

Giannini, I.V. "A ave resgatada: A impossibilidade da leveza do ser." MSc diss., University of São Paulo/FFLCH, 1991, 205.

Hatfield, J.L. and T.J. Sauer, eds. 2011. *Soil Management: Building a Stable Base for Agriculture.* Madison, WI: American Society of Agronomy and the Soil Science Society of America.

Hillel, D. 2008a. Soil chemical attributes and processes. In *Soil in the Environment: Crucible of Terrestrial Life*, ed. D. Hillel, 135–149. New York: Academic Press.

Hillel, D. 2008b. Soil water management. In *Soil in the Environment: Crucible of Terrestrial Life*, ed. D. Hillel, 175–195. New York: Academic Press.

Hitzhusen, G.E. 2007. Judeo-Christian theology and the environment: Moving beyond scepticism to new sources for environmental education in the United States. *Environ Edu Res* 13(1):55–74.

Hoorman, J.J. and E.A. Spencer. 2001. Engagement and outreach with Amish audiences. *J Higher Educ Outreach Engagement* 7:157–168.

Hughes, J.D. 1975. *Ecology in Ancient Civilizations*. Albuquerque, NM: University of New Mexico Press.

Hughes, J.D. 1994. *Pan's Travail: Environmental Problems of the Ancient Greeks and Romans.* Baltimore: The Johns Hopkins University Press.

Jacobsen, T. 1978. *The Treasures of Darkness: A History of Mesopotamian Religion.* New Haven, CT: Yale University Press.

Jostad, P.M., L.H. McAvoy, and D. McDonald. 1996. Native American land ethics: Implications for natural resource management. *Soc Nat Res* 9:565–581.

Kanagy, C.L. and H.M. Nelsen. 1995. Religion and environmental concern: challenging the dominant assumptions. *Rev Relig Res* 37(1):33–45.

Kuber, M. 2007. Shmittah for the clueless. *JA Mag Orthodox Union* 68(2):68–75. http://www.ou.org/index.php/jewish_action/article/33256/.

Lal, R. 2010. Soil quality and ethics: The human dimension. In *Food Security and Soil Quality*, eds. R. Lal and B.A. Stewart, 301–308. Boca Raton, FL: CRC Press.

Lansing, S. 2006. *Perfect Order: Recognizing Complexity in Bali* (Princeton Studies in Complexity). Princeton, NJ: Princeton University Press.

Lansing, J.S. and J.N. Kremer. 1993. Emergent properties of Balinese water temple networks—coadaptation on a rugged fitness landscape. *Am Anthropol* 95:97–114.

Lanyon, L.E., K.E. Arrington, C.W. Abdalla, and D.B. Beegle. 2006. Nutrient management on dairy farms in southeastern Pennsylvania. *J Soil Water Conserv* 61:51–58.

Meek, B., L. Graham, and T. Donovan. 1982. Long-term effects of manure on soil nitrogen, phosphorus, potassium, sodium, organic matter, and water infiltration rate. *Soil Sci Soc Am J* 46:1014–1019.

Moore, K.D. 2010. Climate change: A moral crisis (podcast video), *Changing Climates, Colorado State University*, http://changingclimates.colostate.edu/movies/kathy_moore_796_kbits.mov.

Moore, R.H. 1990. *Japanese Agriculture: Patterns of Rural Development*. Boulder, CO: Westview Press.

Moore, R.H. 2002. Consideration of value-added landscape from the viewpoint of sustainability (Jizoku-teki Seitaiken Hozen no Shiten yori Fukakachi-fukei wo Kangaeru). *Agric Survey Res (Norin Tokei Tyosa)* 52(February):50–61.

Moore, R.H. 2009a. Integrating the social and natural sciences in the Sugar Creek method. In *Sustainable Agroecosystem Management: Integrating Ecology, Economics, and Society*, eds. P. Bohlen and G. House, 21–40. Boca Raton, FL: CRC Press Taylor and Francis Group.

Moore, R.H. 2009b. Yama no kami kou: The mountain goddess fertility association in Northeast Japan. In *Pika-Pika: The Flashing Firefly. Essays to Honour and Celebrate the Life of Pauline Hetland Walker (1938–2005)*, ed. A. Walker, 277–294. New Delhi, India: Hindustan Publishing Corporation.

Moore, R.H., D. Stinner, D. Kline, and E. Kline. 2001. Honoring creation and tending the garden: Amish views of biodiversity. In *Human Values of Biodiversity*, eds. D. Posey and G. Dutfield, 305–309. Special volume for the UN Environmental Programme Global Assessment of Indigenous People, London: Intermediate Technologies.

Ohio EPA (OEPA). 2007. *Total Maximum Daily Loads for Bacteria for the Sugar Creek Watershed*. Columbus, OH: Ohio EPA. http://www.epa.ohio.gov/dsw/tmdl/SugarCreekBacteriaTMDL.aspx.

Olson, G.W. 1981. *Soils and the Environment: A Guide to Soil Surveys and Their Applications*. New York: Chapman and Hall, 130–140.

Palacios-Fest, M.R. 1994. Nonmarineostracode shell chemistry from ancient hohokam irrigation canals in central Arizona: A paleohydrochemical tool for interpretation of prehistoric human occupation in the North American Southwest. *Geoarchaeology* 9:1–29.

Palmer, M. and V. Finlay. 2003. *Faith in Conservation: New Approaches to Religions and the Environment*. Washington, DC: Worldbank Publications.

Pohorski, J. 1998. Tangier Island: Faith-based stewardship for the Chesapeake (video). *Skunkfilms*. http://skunkfilms.com.

Pohorski, J. 2001. When heaven meets Earth: How a faithful few inspired change (video). *Skunkfilms*. http://skunkfilms.com.

Pollan, M. 1991. *Second Nature: A Gardener's Education*. New York: Dell Publishing.

Posey, D. 2000. Introduction: Culture and nature—The inextricable link. In *Cultural and Spiritual Values of Biodiversity: A Complementary Contribution to the Global Biodiversity Assessment*, eds. D. Posey and G. Dutfield, 3–18. Special Volume for the UN Environmental Programme Global Assessment of Indigenous People. London: Intermediate Technologies.

Rappaport, R.A. 1984. *Pigs for the Ancestors*, 2nd ed. New Haven: Yale University Press.

Richardson, C.J., P. Reiss, N.A. Hussain, A.J. Alwash, and D.J. Pool. 2005. The restoration potential of the Mesopotamian marshes of Iraq. *Science* 307(5713): 1307–1311.

Sharpley, A.N. 2003. Agricultural phosphorus and eutrophication. USDA Research and Information Collection. http://hdl.handle.net/10113/26693.

Sharpley, A.N., J.J. Meisinger, A. Breeuwsma, J.T. Sims, T.C. Daniel, and J.S. Schepers. 2002. Impacts of animal manure management on ground and surface water quality. In *Animal Waste Utilization: Effective Use of Manure as a Soil Resource*, eds. J.L. Hatfield and B.A. Stewart, 173–241. Boca Raton, FL: CRC Press.

Smith, S.J., J.S. Schlepers, and L.K. Porter. 1990. Assessing and managing agricultural nitrogen losses to the environment. *Adv Soil Sci* 14:1–43.

Szabolcs, I. 1986. Agronomical and ecological impact of irrigation on soil and water salinity. *Adv Soil Sci* 4:189–218.

Tucker, M.E. 2007. Ecology, religion and policymaking: Survey of the field. *Bull Boston Theol Inst* 6(2/Spring):8–15.

Widner, D.E. 2010. Old Order Amish beliefs about environmental protection and the use of best management practices in the Sugar Creek watershed. M.A. thesis, Kent State University, Kent, OH, 2010.

Wirzba, N. 2006. *Living the Sabbath: Discovering the Rhythms of Rest and Delight*. Grand Rapids, MI: Brazos Press.

Woods, M.J. 2009. *Cultivating Soil and Soul: Twentieth-Century Catholic Agrarians Embrace the Liturgical Movement*. Collegeville, MN: Liturgical Press.

Young, C.E., B.M. Crowder, J.S. Shortle, and J.R. Alwang. 1985. Nutrient management on dairy farms in southeastern Pennsylvania. *J Soil Water Conserv* 40:443–445.

13 Traditional Knowledge for Sustainable Management of Soils

B. Venkateswarlu, Ch. Srinivasarao, and J. Venkateswarlu

CONTENTS

13.1 Introduction 304
13.2 Traditional Classification of Soils..305
13.3 Traditional Concept of Soil Fertility 309
    13.3.1 Sheep, Goat, and Cow Penning 310
    13.3.2 Manuring/Farmyard Manure 310
    13.3.3 Fallowing 311
    13.3.4 Crop Rotation... 312
    13.3.5 Organic Farming... 312
    13.3.6 Recycling Crop Residues and Incorporation of Weeds 313
    13.3.7 Tillage Practices.. 314
    13.3.8 Green Manuring.. 315
    13.3.9 *Beushening* in Eastern India.. 315
    13.3.10 *Maghi* Cropping—Legume-Based Sequence Cropping 315
    13.3.11 Indigenous Organic/Liquid Manures... 316
    13.3.12 Fertility Management through a Tree-Based System 316
        13.3.12.1 Alder in the Himalayas... 316
        13.3.12.2 *Khejri* in Northwest India.. 317
        13.3.12.3 *Aquilaria* and Bamboo in Meghalaya 317
13.4 Indigenous Soil Conservation Practices 318
13.5 Traditional Concepts of Integrated Natural Resource Management 321
    13.5.1 Location-Specific Runoff Control Practices................................... 322
    13.5.2 Flash Flood Control in Himalayas.. 322
    13.5.3 Terrace-Based Land Management of *Apatanis* 324
    13.5.4 Traditional Practices to Overcome Iron Toxicity........................... 324
    13.5.5 The Science of Traditional *Jhum* (Shifting Cultivation)............... 324
    13.5.6 Women and *Jhum* Cultivation... 325
    13.5.7 Slash-and-Burn Agriculture... 325
    13.5.8 Terrace or Bun Cultivation.. 326
    13.5.9 Managing Marshy Saline Soils.. 326

13.6 Traditional Land Use Systems 327
   13.6.1   Soil Depth and Crop Management 327
   13.6.2   Soil Texture and Crop Choice 328
   13.6.3   Physiography and Crop Management 329
13.7 Strengthening Modern Science by Building upon Traditional Knowledge 330
13.8 Conclusions 331
Acknowledgments 331
Abbreviations 331
References 332

13.1 INTRODUCTION

"The term traditional knowledge refers to that information which is accumulated from the wisdom of the older generation, which in turn has been gained through observation, experience and experimentation." This knowledge is often unique to a given culture, being passed on from generation to generation and, in most situations, orally (Nagnur et al. 2006).

Ever since mankind domesticated plants, and started practicing settled agriculture, it has been adopting a number of traditional practices, which have evolved over time. Many soil management practices have been followed in ancient cultures of India, China, Egypt, and Iraq, which are passed on through generations orally without any formal documentation. However, some ancient texts and stone inscriptions provide insights into these practices and their rationale. In India, the ancient texts describe four pillars of farming as *Bhoomi* (land), *Varsha* (rain), *Beej* (seed), and *Hal* (plow). Soil and its management have been given utmost importance by indigenous farmers for successful agriculture.

In *Atharva Veda* (Pant and Khanduri 1998), soil was described as an anchorage for all. Soil was the receptacle of all that lives. The following *hymn* sums up the traditional perception of Indian society to soil:

> *Important to us those vitalizing forces*
> *that come, O Earth, from deep within your body,*
> *Your Central point, your novel, purify us wholly*
> *The Earth is the mother; I am son of Earth*
> *Rain giver is my father, may he shower*

It is further said that soil is our most valuable material heritage, the basis of all terrestrial life. As an ecological factor, soil is of great significance, for it affords a medium of anchorage of plants and a depot for minerals and water.

The practices adopted by farmers for developing ethnotaxonomy and fertility management are on account of the adaptive skills of local people derived from years of accumulated experience (traditional knowledge) that have often been communicated through oral traditions and learned from the family members over the generations (Thrupp 1989). Traditional knowledge is a rich source of location-specific ecological information and also provides the key to understanding people's sociocultural conditions (Pawluck et al. 1992; Singh et al. 2002; Singh 2003). Although, India has rich traditional knowledge on soil management, very little attention has been paid to study, understand, and assess its relevance in modern agriculture. These traditional practices are ideal risk management tools and even today are relevant if one can blend indigenous knowledge with modern practices to achieve high production and yet sustain the production environment.

Indigenous knowledge may also be defined as the sum total of knowledge and practices that are based on people's accumulated experience in dealing with situations and problems in various aspects of life. Such knowledge and practices are unique to a particular culture. The interest in traditional knowledge is growing in recent times as we are unable to maintain the momentum generated by the green revolution and many unforeseen problems are arising that defy cost-effective solutions. There is undoubtedly a need to initiate systematic efforts for collecting the traditional practices from different areas and cataloguing them for validation and wider use. The indigenous knowledge would be helpful in developing ecologically compatible and socially accepted technologies in the difficult areas of crop and animal culture.

The complementary role indigenous knowledge can play in modern agriculture has been acknowledged by many (Sandor and Furbee 1996). Indigenous knowledge systems in soil characterization, fertility assessment, and land use decision making are widely recognized (Zurayk et al. 2001; Greiner 1998; Payton et al. 2003). More recently, much research has focused on the usefulness of traditional soil taxonomies as they relate to agriculture production (Dvorak 1988; Osunda 1988). The concept of sustainable agriculture was expounded on thousands of years ago in the *Vedas*, the ancient Hindu texts (Nene 2004). The objective of this chapter is therefore to understand the traditional knowledge and indigenous practices for sustaining soil management and discuss their importance in meeting current challenges facing agriculture.

13.2 TRADITIONAL CLASSIFICATION OF SOILS

Soil classification and fertility management are some of the earliest concepts that have been used by indigenous farmers since time immemorial. Farmers' wisdom on soil taxonomy and fertility management dynamics is based on trial and error, problem solving, and a group approach with an objective to meet the challenges they face in their local environment. Traditionally, soils are classified according to recognizable and easily observable attributes. Farmers' criteria for soil classification are yield; the topographical position of the field; soil depth, color, and texture; retention capacity; and the presence of stoniness. Keeping the land fallow is one of the widely followed practices by farmers to restore fertility. However, most Indian farmers are now unable to follow the traditional wisdom of keeping land fallow due to high fragmentation of land (0.2 ha per household) and the need to produce food for their families year after year. Traditional subsistence farmers throughout the tropics, however, have a deep understanding of their local ecosystem, including soil, climate, and water (Talawar and Rhodes 1998). A number of studies in India have documented the indigenous classification of soil by farmers in different regions.

Bharara and Mathur (1994) studied the indigenous systems of soil classification in Rajasthan (Table 13.1). The local names reveal the soil capacity to produce good crop, which are aligned with different classes and categories.

Dvorak (1988) described the traditional soil classification followed by farmers in semiarid regions of Andhra Pradesh and Maharashtra (Table 13.2). The terms used by the farmers describe not only the color, texture, and depth of the soil but also whether the soil is prone to drought or it can hold more water, and so forth. The cropping systems in these regions have evolved over time based on these traditional classifications.

In eastern India, there is a clear toposequence due to hilly terrain. In this region, farmers described their land not only as upland, medium, and lowland but also in terms of subcategories within each ecosystem, which govern the type of paddy to be grown (direct seeded or transplanted) and the timing of the planting. Paddy is the major crop grown on such lands in Jharkhand, West Bengal, Orissa, and Chattisgarh (Table 13.3). Even today, most of the recommendations and the package of practices evolved by agricultural scientists are made according to this traditional classification being followed by farmers. It is interesting to note that this classification is based on easily determinable characteristics.

In the northeast state of Arunachal Pradesh, Kuldip et al. (2011) found that the traditional *Nyishis* tribes identify their soils by visual properties such as color, texture, and topographic positioning of land/terrain. The five classes of land/terrain identified by them are given in Table 13.4.

TABLE 13.1
Indigenous Soil Classification in Rajasthan (Northwestern India)

Quality of Land	Class	Local Name	Color	Average Depth (cm)
Excellent	*Barani-I*	Badis Unav	Black	135
Very good	*Barani-II*	Kundala Jod Jodio	Less black	105
Good	*Barani-III*	Dabla Banwla	Yellow	75
Fair	*Barani-IV*	Bedala Dhal Tils	White	45

Source: Bharara, L.P., and Y.N. Mathur, People's perception and indigenous knowledge of land resource conservation in Rajasthan. *Proc. 8th Intl. Soil Cons. Conference on Soil and Water Conservation.* Eds. L.S. Bhushan, I.P. Abrol, and M.S. Rama Mohan Rao, 976–985, Indian Association of Soil Water Conservationists, Dehradun, India, 1994.

TABLE 13.2
Farmers' Classification of Soils in Andhra Pradesh and Maharashtra

Farmer Soil Group	Predominant Soil Type	Depth (cm)	Remarks
Andhra Pradesh (Mahabubnagar District)			
Bette	Red (Alfisols and related soils)	Shallow (<30)	*Bette* means "drought"
Dubba	Red (as above)	Shallow (30–60)	*Dubba* means "loose"
Erra	Red (as above)	Medium (45–90)	*Erra* means "red"
Nalla	Black (Vertisols and related soils)	Medium (>60)	*Nalla* means "black"
Regadi	Black (Vertisols)	Deep (90–275)	*Regadi* means "clayey"
Maharashtra (Solapur and Akola Districts)			
Kali, Morwandi	Black (Vertisols)	Medium to deep (90–135)	*Kali* means "black"
Karal	Black (Vertisols; poor infiltration)	Medium to deep (90–135)	Harder soil
Barad	Black-gray soil	Medium (<100)	*Barad* is drought prone
Marul	Black (Vertisols)	Deep (>200)	Irrigated
Tambadi	Red (Alfisols)	Lies below top black soil	It is considered as soil amendment
Bhari kali	Black (Vertisols)	Deep (90–180)	*Bhari* means "heavy"
Madhyam kali	Black (as above)	Medium to shallow (30–75)	*Madhyam* means "medium"
Halki	Mixed color (black and red)	Shallow to medium (8–45)	*Halki* means "light"

Source: Dvorak, K.A., Indigenous soil classification in semiarid tropical India, Progress Report 84, ICRISAT, India, 1988.

TABLE 13.3
Traditional Classification of Soils Followed by Farmers in Eastern India

	Traditional Classification			
Physiographic Location	Ranchi (Jharkhand)	Purulia (West Bengal)	Kandhamhal (Orissa)	Raipur (Chattisgarh)
Upland	$Tanr_1$; $Tanr_2$	Tanr	Podar	Dadar
Medium land	$Tanr_3$; Don_3	Baid, Kanali	Dhipa, Majhya	Surra
Lowland	Don_2; Don_1	Bohal	Khatia	Bahara

Source: Venkateswarlu, J., and K.P.R. Vittal, Case studies on adoption of production technologies in rainfed rice, groundnut, *rabi* sorghum and pearlmillet, Module 3, ICAR/World Bank project, CRIDA, Hyderabad, India, 1999.

TABLE 13.4
Traditional Soil Classification Followed in Arunachal Pradesh

Local Name of the Land/Terrain	Texture of the Soil	Color	Remarks
Uttola	Sandy	Pale brown	Poor fertility
Kanla	Sandy clay	Dark grayish Brown soil	High fertility
Kannan	Clayey	Dark brown	Moderate fertility
Lengehing	Loam	Red	Low in water and nutrient holding
Ponglung	Loam	Yellow brown	Good fertility

Source: Kuldip, G.A., A. Arunachalam, and B.K. Dutta, *Indian J. Tradit. Knowl.*, 10, 508–511, 2011.

In Assam, all soils were described by textural properties and some by color. Soils are either sandy (Balu), clayey (Poli), or sticky (Athail). The other two soils, Lal and Ujar, are distinguished by their color (Das and Das 2005). Table 13.5 shows the five types of soil identified by rice farmers of this region.

In central Himalayas, farmers classified lands as per their location in the forests, hills, or valley and likely use they are put to (Sanyal and Dhyani 2007). This is an excellent toposequence-based classification (Table 13.6).

TABLE 13.5
Traditional Soil Classification Followed in Assam

Farmers' Soil Name	Color	Remarks
Poli	Dark brown	Tendency to become waterlogged
Athail	Dark brown	Good water retention capacity
Lal	Reddish often mixed with dark colored soil	Water retention capacity adequate
Ujar	Whitish to grayish white	Water retention capacity adequate
Balu	Light brown	Water retention capacity very poor especially during dry season

Source: Das, T., and A.K. Das, *Indian J. Tradit. Knowl.*, 4(1), 94–99, 2005.

TABLE 13.6
Land Classification in Central Himalayas

Local Name	Location	Use
Nalies	Adjacent to the forests	Irregular cultivation
Upran	Between Nalies and valley	Rainfed agriculture
Talaon	Valley area	Irrigated by rivulets, springs, and streams

Source: Sanyal, P.K., and P. Dhyani, *Outlook Agric.*, 36, 49–56, 2007.

13.3 TRADITIONAL CONCEPT OF SOIL FERTILITY

Krishi parashara (ca. 400 BC), one of the well-known ancient Indian texts on agriculture, stressed soil management: "Farms yield gold if properly managed but lead to poverty if neglected." It also emphasized, "Crops grown without manure will not give yield." In *Krishi Gita* (ca. 1500 AD), *Parashurama* recommended deep summer plowing and green manuring for rice to obtain good yields (Nene 2009). These verses indicate that farmers in ancient India realized that good management and nurturing of soil are essential to realize good yields.

Farmers in India describe productive and fertile land as *urwar* and *upjau*, which literally means well-prepared land that has a good seedbed, is brown and plain, and is able to provide good nutrients to crops. Farmers also saw soil fertility as a multifaceted concept. It includes factors such as the soil capacity, drainage, tillage, and manure requirement, and shows how easy it is to work (Rhoades 1998). Probably the earliest form of soil fertility management originated when nomads, people who moved between grazing areas with their animals, discovered that they could grow crops readily on new land, but that if they tried to grow crops on the same piece of land for more than one season, they would obtain less food from the crops planted. Once this lesson had been learnt, the process of "shifting cultivation" gradually became more sophisticated. This form of soil management by shifting cultivation has been widely practiced throughout the world. It is an essential response to the problem of obtaining food where the soil itself is incapable of sustaining the continuous production of crops for an unlimited period. The system is sustainable as long as there is sufficient land for the soil to be allowed to recover until its productivity returns to the former level. It is also dependent on the knowledge and experience acquired by the farmers themselves (FAO 1994). In terms of fertility, three main classes are recognized by traditional Indian farmers, namely, *Bangar* (fertile), *Karail* (moderately fertile),

TABLE 13.7
Different Soil Fertility Practices as Ranked by Farmers and Constraints in Adoption in Present-Day Farming

Soil Fertility Practices	Rank	Constraints in Adoption
Sheep, goat, and cow penning	I	As animal migration came down, this practice declined. More expensive to farmers now
Crop rotation	II	Shift in preference towards cash crops
Application of crop residue	III	Use of crop residue as fodder and fuel
Fallowing	IV	Shortage of land
Incorporation of weeds	V	Shortage of labor
Manuring	VI	Manure used as fuel and decline of livestock
Tillage practices	VII	Reduction in draft power
Green manuring	III	Lack of seeds of green manuring crop and attack of wild blue cows

Source: Singh, R.K., and P.K. Singh, *Asian Agrihist.*, 9(4), 291–303, 2005.

FIGURE 13.1 Sheep penning. A flock of sheep is retained in a cage for a night and moved to the next site the following day. (Reprinted from Reddy, S., *Soil Carbon Sequestration for Climate Change Mitigation and Food Security*, November 24 to December 3, Central Research Institute for Dryland Agriculture, 322 pp., 2011. With permission.)

and *Kiyari* (least fertile). Considering quality and the importance of improving the fertility status of soil, farmers have ranked different practices (Table 13.7).

Sheep, goat, and cow penning was ranked first and crop rotation as the second-best soil management practice on account of their no-cost nature. Others are ranked according to the cost implications.

13.3.1 Sheep, Goat, and Cow Penning

There are two types of penning: cattle penning and sheep penning. There are two particular castes, *Ganderiya* and *Yadav*, in India who pen their sheep and cows on a farmer's field in return for money and grain, as these sheep owners are landless and need food grains. It is a kind of barter system. They keep the bovine animals (cattle and sheep) in the fallow land after the harvest of the last crop throughout day and/or night supplied with suitable feed and shelter (Figure 13.1). Fresh sheep dung contains 0.5%–0.7% N, 0.4%–0.6% P_2O_5, and 0.3%–1.0% K_2O, and sheep urine contains 1.5%–1.7% N, traces of P_2O_5, and 1.8%–2.0% K_2O.

13.3.2 Manuring/Farmyard Manure

FYM is an important input for maintaining and enhancing the soil fertility. FYM prepared from cattle dung has been the backbone of traditional Indian agriculture since time immemorial (Figure 13.2). It was used as a source of nutrition for plants. Cow dung is also used for seed treatment to provide protection against preemergence

FIGURE 13.2 FYM application in the fields. (Reprinted from CRIDA. With permission.)

and postemergence diseases. On average, well-decomposed FYM contains 0.5% N, 0.2% P_2O_5, and 0.5% K_2O (Reddy 2011). There are two types of FYM, the first made in the rainy season and the second made in winter. Farmers recognize the second type of manure as the best for crops due to its well-decomposed nature. Farmers in India always attempted to use manure as an alternative for chemical fertilizer. But with the increase in population, more and more dung is used as fuel cakes for cooking. Particularly in northern India, shrinking animal population due to mechanization and labor shortage to transport manure to far-flung fields have all led to rapid decline of this practice. Applying green leaf and FYM is the traditional method of improving the condition of the soil, in which green leaves and FYM are plowed into the soil. It helps in improving soil fertility through nitrogen fixation (Sarma 2002; Talukdar et al. 2002; Satapathy 2002; Laxminarayana 2002; Tiwari et al. 2002; Bregman 1993).

13.3.3 Fallowing

The indigenous method of restoring the soil fertility is fallowing. Traditionally, farmers observed that yields on poor land (*Kiyari* soils) decline if they continuously grow two or three crops a year under irrigated conditions without any fallow period. The fallow period varies from soil to soil. The *Kiyari* soil type is kept fallow for 1 year, while *Bangar* and *Karail* are kept fallow for one season. In general, farmers adopt two methods of fallowing. In the first method, land is kept fallow after taking the paddy in the *Kiyari* soil for over 1 year, and timely plowing is done after every rain. In the second method, the field is fallowed for only one season (4 months after harvesting *rabi* season crops).

13.3.4 Crop Rotation

Farmers in ancient India were aware of the importance of crop rotation to improve soil productivity (Figure 13.3). Farmers choose the crop to be grown in rotation depending upon soil, rainfall pattern, and available resources. The ancient texts describe what type of crop ought to be rotated in each soil, rainfall condition, and so forth, much before the modern-day agricultural scientists highlighted crop rotation for building soil fertility and reducing pest and disease incidence. The choice of crop rotation is heavily influenced by the desire to reduce the need for labor and intensive land preparation (basic principles of conservation agriculture as we understand today). In addition, some farmers have learnt that to prepare the nursery bed for paddy crop, it is necessary to grow berseem, which is one of the dominant fodder crops during the *rabi* season. Local farmers perceive that it helps to increase the amounts of residue in the soil as well as strengthen the soil health.

13.3.5 Organic Farming

Organic farming is not a new concept in India. Natural organic farming is the true and original form of agriculture, "the 'method less' method of agriculture—the unmoving way of Bodhidharma" (Aishwath 2007). The philosophy of sustainable agriculture was expounded millennia ago in the *Vedas*, the ancient Hindu scriptures, and consists of four components. (1) Humans are a part of the complex universe, and the intimacy of humans with nature is a matter of great joy. (2) Humans should live in harmony with natural forces and ensure that harmony with the different natural forces or *panchabhutas* (earth, water, air, fire, and *akash* [ether]) is not disturbed.

FIGURE 13.3 Cereal–legume intercropping system. (Reprinted from CRIDA. With permission.)

FIGURE 13.4 Vermicompost making—Nalgonda cluster. (Reprinted from Srinivasarao, C. et al., *Livelihood Impacts of Soil Health Improvement in Backward and Tribal Districts of Andhra Pradesh*, Central Research Institute for Dryland Agriculture, 119 pp., 2011. With permission.)

(3) Humans should respect and show gratitude to useful objects, inanimate or animate. (4) Organic matter should be recycled (Nene 2004).

In order to maintain the productivity of the soil, the organic method of farming solely depends on the use of crop residues, animal manures, green manures, off-farm organic wastes, crop rotation with legumes, and biological pest control (Palaniappan and Annadurai 1999). The philosophy of organic farming is to feed the soil rather than the crop, to maintain soil health, and to be a means of giving back to nature whatever is taken from it (Funtilana 1990). *Agnihotra* (ashes left after a *yagna* is performed) is often touted as a complete plant food (Pathak and Ram 2002). The major components of organic farming such as green manure, FYM, vermicompost (Figure 13.4), cover crops and mulching, microbial fertilizers, crop rotation, and crop management are the major sources of organic matter, which provide the humus in the soil after decomposition, which helps in alteration of the physical, chemical (Aishwath et al. 2003), and biological properties of the soil (Pettersson et al. 1992). It also helps in soil aeration by providing substrate to the microbial population as well as an improved habitat for macro fauna (worms, millipedes, spiders, etc.).

13.3.6 Recycling Crop Residues and Incorporation of Weeds

The *Rigveda* (ca. 8000 BC), the *Atharvaveda* (ca. 1000 AD), and also the *Holy Quran* specify that at least one-third of what we take out from the soil must be returned to it—the recycling of crop residues. Most farmers, even today, leave residues of crops in the field itself to mix them in the soil for increasing the fertility level. Paddy stem, wheat straw, and sugarcane leaves are generally mixed in the soil and, after irrigation, are left for decaying. Singh et al. (2002) showed that spreading the dry paddy stem in the standing crop of sugarcane in June and July (when the crop is about

4 months old) will suppress the weeds and ensures the availability of all nutrients to the crop. This practice also helps to destroy the weed and insect pests and improve the fertility of soil.

In the waterlogged areas having *Karail* soil, organic matter cannot be added to the soil due to transportation problems. The weeds are uprooted, made into small bundles, and pushed inside the soil with the roots above the ground. This practice increases the soil fertility (Singh 2003). The weeds obtained from the upland of a paddy crop are, however, used as fodder for the animals because under such condition, it takes more time for weeds to decay, and also, animals relish fodder.

13.3.7 Tillage Practices

Farmers perceive that even after applying a good amount of manures and fertilizers, if the soil is not tilled adequately, its productivity declines (Singh et al. 2002, 2003). Generally, in most of the cash crops like sugarcane and potato, farmers try to provide more and more light tillage under irrigated conditions for good aeration and destroying weeds and insect pests. Tillage and plowing are both determined by the crop variety, types of soil, and economic conditions of the farmers. Hundreds of varieties of wooden plows were used by ancient farmers in India across the country designed as per the soil type and type of bullocks, some of which are still used in many parts (Figure 13.5).

FIGURE 13.5 Traditional plows used in India for tillage and seeding. (Reprinted from www.google.com/images_traditionalploughs. With permission.)

13.3.8 GREEN MANURING

Green manuring is an age-old practice prevalent since ancient times (Sofia et al. 2006). Farmers try to grow *dhaincha* (*Sesbania acculeata*) and sunhemp (*Crotalaria juncea*) as green manuring crops for improving soil health. Just around flowering (30–45 days after sowing), the crop is cut down and mixed into the soil, after which the season's main crop is sown. These practices are more popular under irrigated conditions. Some of the farmers do incorporate okra and potato stem in the field, which are comparatively much better than other crops. Since these crops are termed as cash crops, after the harvesting of pod and tuber, remaining biomass is used for soil strengthening. So, incorporation of this green biomass after adding water in the soil has a significant role in maintaining soil health. Green manuring is beneficial in two ways—first, it fixes nitrogen, and second, the addition of biomass (around 5 to 10 tons/acre) greatly helps in improving the soil texture and water holding capacity.

13.3.9 BEUSHENING IN EASTERN INDIA

Beushening in rainfed lowland rice is an age-old practice in the shallow submerged lowland in most of the eastern states (Chandra 1999). *Beushening* is criss-cross plowing in a standing rice crop of 30 to 45 days after seeding when water 10 to 15 cm deep stands in the field. It helps in burying the weeds and excess rice seedlings. This is followed by laddering and seedling redistribution. Sometimes weeding is also taken up. The weed population decreases considerably with *beushening*. Through *beushening*, as much as 310 g/m^2 of weeds are buried in the field as green manure. This is besides the additional rice seedlings plowed in. The weeds and additional rice seedlings add at least 45 kg N/ha/year. However, weed control is needed for better yields under *beushening*. The additional yield with *beushening* even in hand-weeded situations is due to root pruning of the rice seedlings. Thus, in the absence of labor, at least *beushening* maintains good productivity. Farmers practice to *beushening* suppresses weeds and adds nutrients through green manure for the standing rice crop. Such a "green manuring" effect is ideal in lowland rice as its decomposition does not need high energy (being anaerobic). Also, *beushening* saves demand on labor, an increasingly felt constraint in modern agriculture.

13.3.10 MAGHI CROPPING—LEGUME-BASED SEQUENCE CROPPING

A unique practice of legume-based sequence cropping is carried out by farmers in the Khammam district of Andhra Pradesh. This area receives 1000-mm rainfall, with an assured moisture supply period of 180 to 210 days. Soils are predominantly Vertisols low in nitrogen and high in potassium. A unique practice of the green gram–sorghum cropping system is followed here to take care of the nutrient requirement. The green gram is sown by early June. After 65 to 70 days, the pods are harvested, the haulms are incorporated in the soil, and the sorghum crop is sown during the third week of August. This is neither a kharif nor a rabi crop. Green gram incorporation adds 35–40 kg N/ha, adequate to meet the N requirement of sorghum. This practice has been in vogue for many years, and the physical properties of the soils like aggregation and infiltration rate are maintained very well.

13.3.11 Indigenous Organic/Liquid Manures

The ancient indigenous technology of low-cost nutrient sources such as *Panchagavya* (organic liquid manure), FYM, and crop residues can improve the soil fertility, soil biological activity, and physical properties of soil, concurrently supplementing all nutrients, including micronutrients (Meena and Bhimavat 2009). Various studies have revealed that under summer plowing with green manure incorporation at 60 days and foliar application of *Panchagavya*, the water use efficiency increased, and also, remarkable improvements in yield and yield attributes were recorded due to green manure incorporation and foliar spray of *Panchagavya*. The *Panchagavya* solution (Sanjutha et al. 2008) was prepared from five cow products (i.e., cow dung, cow urine, cow milk, cow ghee, and fermented curd) and kept for 15 days before use. *Panchagavya* is widely used in Tamil Nadu in many villages. Following the success of *Panchagavya*, farmers started using *Gunapajalam*, which is a mixture of pork, beef, mutton, pig dung, cow dung, goat urine, gur/palm gur, amla, jackfruit, dates, bananas, honey, ghee, tender coconut water, toddy, *urad* (black gram) paste, and sesame oil cake. This *Gunapajalam* can be applied as a foliar and basal nutrient, using 1.5 L/acre. Field experiments showed that the application of *Gunapajalam* is more effective than *Panchagavyam* (Narayana 2006).

The other commonly used indigenous liquid organic manure/biomanure is *kunapajala* (Nene 2006). The dictionary meaning of the Sanskrit word *kunapa* is "smelling like a dead body, stinking." The manure *kunapambu* or *kunapajala* (*jala* means water), which has been prepared and used since ancient times in India, was appropriately named because it involved fermentation of animal remains, such as flesh, marrow, and so forth. Two contemporary documents have references to *kunapajala*, that is, *Vrikshayurveda* by *Surapala* and *Lokopakara* by *Chavundaraya*. The excreta, marrow of the bones, flesh, brain, and blood of a boar mixed with water and stored underground is called *kunapa*. *Surapala* also mentioned that the wastes from animals such as cow, porpoise, cat, bird, deer, elephant, and so forth can be used for the preparation of *kunapajala*; therefore, it gives flexibility to farmers in picking the animal of their choice. Experimental work carried out on *kunapajala* in Arunachal Pradesh produced an herbal *kunapa* called *Sasyagavya*, which, when applied to soil, showed good healthy-looking tea plants.

13.3.12 Fertility Management through a Tree-Based System

Trees are a common component in arable lands all over India. In low-rainfall regions, trees provide fodder for animals and fuel wood. However, farmers in India have integrated trees in the farming systems and used their biomass and nitrogen fixation properties to maintain soil fertility. The following specific examples describe how this integration has helped farmers sustain soil fertility over centuries.

13.3.12.1 Alder in the Himalayas

Alder (*Alnus nepalensis*) is grown in Nagaland, Sikkim, and other northeastern states for enhancing soil productivity. Alder grows well in altitudes between 1000 and 3000 m msl and in high-rainfall areas of Nagaland (>1500 mm). It is a nonleguminous tree but fixes atmospheric N through *Frankia* to the tune of 150 kg/ha (Sharma et al. 2002). Dhyani (1998) reported a 2.2-fold increase in the yield of cardamom under an alder canopy.

The alder trees vary in population in the arable lands. The N contribution through leaf litter varies with the population of the alder trees. A population of 60/ha adds about 48 kg/ha, which goes up to 113 kg/ha if the population is 166 (Rathore et al. 2010). These trees are pollarded at 2–2.5 m above ground level. The twigs and stems are used for fuel. The leaves are left on the field and burnt along with the stubbles of the earlier crop to add nutrients to the soil and also to oxidize Fe^2 to Fe^{3+} (an irreversible reaction), thus reducing possible Fe^{2+} toxicity in the oxisols. Then different crops are grown as a mixture by the farmer. By the time the crops come to maturity, a full canopy develops so that next season, the cycle is repeated.

13.3.12.2 *Khejri* in Northwest India

Khejri (*Prosopis cineraria*) is common tree in arid regions of northwest India (western Rajasthan, Haryana, Gujarat, dry region of Deccan Plateau). It is largely limited to rainfall below 500 m. *Khejri* is a moderate-sized evergreen leguminous thorny tree. It partly sheds leaves from mid-October to mid-February. Thus, considerable leaf litter accumulates under the trees, and light will not be limiting for crops that grow under the canopy. Agarwal et al. (1975) reported that *khejri* can fix up to 250 kg N/ha.

In the farmers' fields under a *khejri* canopy, the yield of barley was 999 kg/ha, compared to 537 kg/ha away from the canopy in the Hisar district of Haryana under rainfed conditions (Kumar et al. 1998). Such yield increases under *khejri* were reported in the case of chickpea, pearl millet, mung bean, and cluster bean. Therefore, *khejri* is considered as the *kamadhenu* (well-wisher) of the arid-zone farmer, providing sustainability to crop production. The above two examples indicate that traditional agroforestry systems in India are helpful in soil fertility management.

13.3.12.3 *Aquilaria* and Bamboo in Meghalaya

Areca nut (*Areca catechu*), bamboo (*Bambusa* sp.), banana (*Musa paradisiacal* Linn.), black pepper (*Piper nigrum* Linn.), and canes (*Calamus* sp.) are cultivated with *Aquilaria* (*Aquilaria agallocha* Roxb.) in Meghalaya. It is a common practice in all parts of this hill state. Areca nut (*A. catechu* Linn.) is considered as the economically important tree species of the Indian peninsula and northeast. Black pepper (*P. nigrum* Linn.), ginger (*Zingiber officinale* Rosc.), maize (*Zea mays* Linn.), and turmeric (*Curcuma domestica* Valeton) are cultivated in the interspaces. Farming practice based on bamboo–areca nut–betel is prominent in some isolated pockets of Meghalaya. In this system, areca nut (*A. catechu* Linn.), bamboo species (*Bambusa tulda* Roxb., *Bambusa pallida* Munro, *Bambusa balcooa* Roxb., *Dendrocalamus hamiltonii* Nees., *Neohouzeaua dulloa* A. Camus, *Melocana baccifera* Kurz.), and *Piper betel* Linn. are cultivated together (Jeeva et al. 2006). All these systems are based on internal recycling of residues and litter, which supports the soil with external inputs.

13.4 INDIGENOUS SOIL CONSERVATION PRACTICES

Traditional Indian farmers pay much attention to soil and water conservation (Figure 13.6). Much before modern science understood the loss of soil due to erosion and the need for conservation, several indigenous practices followed by farmers were documented in different parts of India. Table 13.8 gives a summary of indigenous practices adopted in different states of India to conserve soil and control runoff. In addition to the above list, there are many location-specific conservation practices that are linked to crop cultivation in different regions in India.

Silt harvesting structures: The smallholders in Peninsular India place loose rock check dams across gullies to harvest the soil eroded in the upper catchment. These structures are useful in two ways. First, they prevent widening and deepening of gullies, promote deposition of nutrient-rich fertile sediments, and reduce the velocity of the runoff in the gullies. Second, the area where such soil is accumulated is used for growing high-value crops (Sanghi et al. 1994).

FIGURE 13.6 Traditional soil management practices in India that conserve soil and improve fertility. (Reprinted from Srinivasarao, C. et al., *Soil Carbon Sequestration for Climate Change Mitigation and Food Security*, November 24 to December 3, Central Research Institute for Dryland Agriculture, 322 pp., 2011. With permission.)

TABLE 13.8
Indigenous Practices of Soil and Water Conservation in India

Indigenous Practices	Crops/Plants Grown	Purpose/Benefit	State Where Followed
Vegetative fencing/ barrier	Kiluvai (*Blasmo Dendron verii*) and agave	Reduce water runoff and increase infiltration	Karnataka
Mixed intercropping as vegetative barrier	Groundnut, pigeon pea, and pulses	Runoff control	Andhra Pradesh
Relay cropping	Onion–rabi sorghum or chickpea	Reduction in runoff and better utilization of soil moisture	Karnataka
Spur structure	Any crop	Protection of cropland from erosion by diverting the runoff	Jharkhand
Mixed cropping (*Mishrabele paddati*)	Onion, chili, and cotton	Reduction in runoff and better utilization of soil moisture	Karnataka
Preemergence soil stirring	Any crop	Removal of weeds, loosening the soil for conserving moisture	Uttar Pradesh
Compartmental bunding; loose boulder checks; peripheral stone bunding	Any crop	Soil conservation and runoff management	Andhra Pradesh, Orrisa, Maharashtra
Bunds protected with vegetal cover	Any crop	Strengthening of earthen bunds to reduce erosion and runoff	Gujarat
Bandh system of cultivation	Any rabi crop	Harvest runoff for assured rabi crops in rainfed areas	Madhya Pradesh
Live bunding	By planting cactus	Reduce runoff and check soil erosion	Uttar Pradesh
Cross plowing	Any crop	Check runoff and soil loss	Uttar Pradesh
Plantation of grasses on field bound	With *Vitex negundo* (*Nirgundi*)	Reduce runoff and soil loss	Maharashtra
Peripheral bunding; vegetative barrier across gullies	Agave spp., *Ipomea*	Gully control and runoff management	Andhra Pradesh, Karnataka
Nala plugging	General	Control of flow of water in *nalas* to minimize development of gully and lateral recharge of water	Uttar Pradesh

(*continued*)

TABLE 13.8 (Continued)
Indigenous Practices of Soil and Water Conservation in India

Indigenous Practices	Crops/Plants Grown	Purpose/Benefit	State Where Followed
Conservation furrows with traditional plow; intercropping; wide row spacing; tank silt application	Groundnut and pigeon pea	In situ moisture conservation	Karnataka, Andhra Pradesh, Orrisa
Repeated tillage during monsoon season	Rabi crops	Soil moisture conservation for sowing of winter crops	Haryana
Crop residue in the field	Rabi crops	Prevent sheet erosion and increase in situ conservation	Gujarat
Deep summer plowing	Any crop	Harvest early showers	Maharashtra
Interculturing (Hoeing) and earthening in standing crop	Any crop	Harvesting rainwater and providing soil mulch	Gujarat
Sand mulching, cultivation; stone bunding	All kharif crops	Soil and moisture conservation	Andhra Pradesh
Grassing waterways	All kharif crops in uplands	Water conservation and land degradation control	Jharkhand
Vegetative barrier supported with small section bund	All kharif crops	Soil moisture conservation and avoidance of encroachment by wild animals	Uttar Pradesh
Contour cultivation (operation across slope)	All crops	Elimination slope length, creating barriers for water flow, enhancing soil moisture status	Uttar Pradesh
Ridge and furrow planting	Kharif crops	Conservation of rainwater, modulating excess water, control of soil loss	Madhya Pradesh, Uttar Pradesh
Vegetative barriers (Munj and Khus) on field bunds	Kharif crops	Soil moisture conservation and protection from wild animals	Uttar Pradesh
Conservation furrow (Gurr)	Pearl millet	Reduction of runoff and soil moisture conservation	Uttar Pradesh

(continued)

TABLE 13.8 (Continued)
Indigenous Practices of Soil and Water Conservation in India

Indigenous Practices	Crops/Plants Grown	Purpose/Benefit	State Where Followed
Application of tank silt	All crops	Improvement of soil moisture holding capacity of soil	Andhra Pradesh Karnataka
Green manuring with sun hemp; sesbania and cowpea	Paddy	Improve organic matter for soil fertility improvement, weed control, soil and water conservation	Madhya Pradesh
Crop stubbles/residue management	Rabi crops	Improve organic matter in soil and to improve water holding capacity of soil	Madhya Pradesh
Planting trees	Acacia sp.	Reduce salinity of soil	Uttar Pradesh

Source: Srivastava, S.K., and Hema Pandey, *Indian J. Tradit. Knowl.*, 5(1), 122–131, 2006.

Jal lands: In the southern Telangana region of Andhra Pradesh, there are several gully courses that suffer from waterlogging with an underlying hardpan. These are called *Jal* lands. Such lands are terraced and put to a mixture of ragi and short-duration rice crops by tribals. The logic is, if the rainfall is normal, both the crops yield some grain. If there is low rainfall, at least ragi would provide sufficient grain. If the rainfall is high, ragi may yield less, but rice will provide food security. In larger-sized gully courses, embankments are constructed to store inflow of rainwater for (1) direct use for irrigation by gravity and (2) recharging groundwater. This recharges open wells also (Venkateswarlu 2004).

Jhola lands: In the high-rainfall Koraput district of Odisha, tribals reclaim gullies and hilly tracts between two hills and convert them to terraced rice fields. The onflowing perennial streams are the source of irrigation. They add adequate quantities of nutrients through the silt load carried from the hillocks (MSSRF 2012). This supports excellent rice crop without any need for chemical fertilizers.

13.5 TRADITIONAL CONCEPTS OF INTEGRATED NATURAL RESOURCE MANAGEMENT

Soil, water, and vegetation were three basic natural resources recognized in ancient India. The survival of God's creation depends upon them, and nature has provided them as assets to human beings (Mishra 1998). Since the pre-Vedic era, people have been managing natural resources to meet their requirements. As the farmers had the village administration in their hands, they were ranked high in the social system. Farmers gained technical knowledge suitable to the specific conditions of that region in order to manage land, water, and vegetation through experience and learning by doing. Every culture of a social system is the result of people's action to survive and their attempts to optimize the use of available resources (i.e., soil, water, and vegetation).

13.5.1 Location-Specific Runoff Control Practices

Quite early, farmers realized the need for protection of land from erosion, landslides, and floods. In most agricultural fields across the country, farmers protect their land from erosion through traditional methods such as by using bamboo culms, stones, and gunny bags filled with soil (NIC 2001). In the Meghalaya state of northeast India, a primitive form of bench terraces was in vogue even while practicing shifting cultivation. The vertical slope between the terraces is not usually more than a meter. It prevents soil erosion and retains maximum rainwater within the slopes and safely disposes of the excess runoff to the foothills.

In Peninsular India and the central plains, farmers in India followed the concept of watershed while delineating village boundaries. Village boundaries were decided upon on a watershed basis by the expert farmers in the villages. Such boundaries were socially acceptable to all the members of the system. Such age-old village boundaries are fixed at the common point of the drainage system in between two villages. It is still in vogue, and people do not go beyond the limits of their hydrological boundaries. In the western Himalayas, farmers used to carry water to their fields through small irrigation channels known as *gulas* (Figure 13.7). These go from the source of water along the slopes to the fields. In order to avoid seepage losses, pipes made of bamboos and tree trunks were also used (Sharma and Sinha 1993). In the Garhwal Himalayas, traditionally, rainwater was harvested in small dug-out ponds. Such ponds were dug at several places along the slope. Natural streams in the hills were the source of drinking water. Villagers planted vegetation all along the stream bunds in order to prevent sediment entry and pollution of water.

13.5.2 Flash Flood Control in Himalayas

In the entire Himalayas, flash floods cause landslides, and silt carried by water fills up the water bodies. Ancient Indian farmers have indicated three strategies to deal with this problem. They are mechanical, agricultural, and vegetative. The main occupation of the hill farmers is agriculture. They usually construct terraces for cultivation known as *nala*, with risers known as *pusata*. These terraces are small, but there are many. In 1 acre of land, a farmer possesses 50 *nalas*. In these, it is possible to manage rainwater. The farmers, with their expertise, are able to prepare fields for crop production in hills. Though cultivation is allowed on up to a 33% slope, hill farmers are able to make terraces from top to bottom of the mountain terrain with indigenous practices. With terraces, they construct loose boulder retention walls (risers) and grass them. These grasses keep both stones and the land intact. In making these risers, farmers simply arrange boulders of the proper size along the terrace wall. They retain the soil perfectly and gradually become stabilized. Farmers of the hill region used to make brushwood or long wood check dams across the drainage channels for controlling soil loss by means of local materials. In the Doon Valley, *Ipomea carnea* and *Arando donex* plant species are used as vegetative spurs in order to train torrents.

Traditional Knowledge for Sustainable Management of Soils 323

FIGURE 13.7 Traditional soil and water management practices in Himachal Pradesh. (Reprinted from Lal, C., and L.R. Verma. *Indian J. Tradit. Knowl.*, 7(3), 485–493, 2008. With permission.)

13.5.3 TERRACE-BASED LAND MANAGEMENT OF *APATANIS*

Apatanis in Arunachal Pradesh practiced a unique terrace-based method of land and water management system in steep hills (Chaudhary et al. 1993). The terraces are quite broad, perfectly leveled, and provided with strong bunds (risers). These risers are made of soil and supported by flattened wooden clips fixed at the base. Bamboo or boulder support is provided if the riser is tall or if there are chances of erosion due to runoff. All terraces are provided with inlet and outlet pipes for proper water management. The risers are used for finger millet cultivation. Although the yield of finger millet on risers is low to average, it checks weed growth and acts as a binding material for soil on risers. *Apatanis* do not use any chemical fertilizer in their wet terraces. Nutrient and fertility management of the terraces is done mainly by recycling agricultural wastes. All types of biomass from the rice field, cattle yards, poultry houses, domestic waste, and leaves collected from the adjoining jungles are recycled in order to replenish humus and nutrient in the soil. The paddy straw, approximately 4–5 t/ha, is allowed to decompose in the wet terraces and finally incorporated at the times of land preparation. After the rice crop is harvested, cattle are allowed free grazing in the fields from December to February, and thus, the cow dung is also recycled. Thus, the entire hills, surrounding valleys, and uplands around the villages are conserved as forests even today. Soil erosion, silting of rivers, drying of the water sources, and loss of nutrients, flora, fauna, and forest resources are negligible in this plateau.

13.5.4 TRADITIONAL PRACTICES TO OVERCOME IRON TOXICITY

In the humid hilly areas in India, lateritic soils predominate. In most cases, rice is the staple crop. In these soils, Fe^{2+} is found to be in excess/toxic levels. Viswanath (1937) pointed out the traditional knowledge systems for detoxicating excess Fe^{2+}. The Assam rice soils suffer from iron toxicity. The leaves of rice crop turn yellow. To alleviate the problem, the farmers walk up and down the rice fields at regular intervals. This facilitates aeration, and Fe^{2+} gets converted to Fe^{3+} (Ferrous to Ferric) and becomes insoluble and nontoxic. Toky and Ramakrishnan (1983) found that slash-and-burn farming (shifting cultivation) would improve soil aggregation and irreversibly oxidize Fe^{2+} to Fe^{3+}, leaving the soil less toxic or nontoxic with Fe^{2+} iron to the crops in northeast India. In the Western Ghats of Maharashtra, rain-fed rice is grown by transplanting seedlings in puddled soils. To raise the nursery, farmers clear a piece of land, place leaf litter, and burn it over the plot meant for growing the rice nursery. By doing so, farmers are converting iron from the Fe^{2+} form to the inactive Fe^{3+} form. Otherwise, the young rice seedlings will suffer from iron toxicity.

13.5.5 THE SCIENCE OF TRADITIONAL *JHUM* (SHIFTING CULTIVATION)

In the slash-and-burn system of *jhum* cultivation, Ramakrishnan (1992) pointed out that the soils cleared for shifting cultivation have steep slopes, and so spatial variability is very high. The farmers by choice grow the C_3 rice crop in lower parts of the

TABLE 13.9
Best-Performing *Jhum* Cycles in Northeast India

Jhum Cycle (Years)	Best-Performing Crops
30	Rice, maize
10	Sesame, kenaf, cassava, tuber crops
5	Banana, leafy vegetables

Source: Toky, O.P., and P.S. Ramakrishnan, *J. Ecol.*, 71, 735–745, 1983.

slope, which will be nutrient-rich microsites, while in the upper reaches, C_4 species with high nutrient use efficiency area are grown, as these areas are nutrient-poor microsites. Presently, the *jhum* cycles are reduced from 30 years to as low as 5 years. Toky and Ramakrishnan (1983) studied the performance of selected crops under *jhum* cycles of 30, 10, and 5 years. The best-performing different cycles are given in Table 13.9. Ramakrishnan (1992) also found that as the *jhum* cycle is now limited to a lesser number of years, the farmers are choosing cassava, tuber crops, banana, leafy vegetables, and early rice to meet their domestic needs.

13.5.6 WOMEN AND *JHUM* CULTIVATION

As procreators of life, women in tribal and traditional cultures over the centuries have had highly significant interactions with the environment. They have contributed to the management, conservation, and use of natural resources such as soil, water, and biodiversity. Women in Mizoram (India) practice *jhum* (shifting cultivation) (Figure 13.8). *Jhum* is a farming method where land is extensively used to cultivate crops for a few years and then left fallow for few years so that it recuperates its nutritional level, and then it is used again (Hmingthanzuali and Pande 2009). This practice is followed by majority of the population, amounting to 52%. Women carry out all kinds of operations in *jhum* cultivation, while men do not. During the traditional period, all the members of the community are involved in *jhum* cultivation. The method of *jhum* cultivation demands strenuous labor on the part of men and women in the field. According to one of the Mizo mythological stories, a woman named Lalmanga Nu knew that June 21 or 22 was the longest day in the year, so she would arrange for her villagers to work on her *jhum* land on that day in order to get as much out of them as possible (Hmingthanzuali and Pande 2009).

13.5.7 SLASH-AND-BURN AGRICULTURE

Arunachal Pradesh, in the northeastern part of India, which constitutes a large number of tribes and subtribes, depends largely on the age-old traditional slash-and-burn farming system, which has been maintained as a part of the traditional livelihood system over generations, mostly on hillslopes (Tangjang 2009). Slash-and-burn agriculture refers to any temporally and spatially cyclical agricultural system that involves slashing of forests, usually with the assistance of fire, followed by phases of

FIGURE 13.8 Agroforestry interventions for restoring *jhum* lands. (Reprinted from ICAR Research Complex, North East region, Meghalaya, Annual Report. With permission.)

cultivation and fallow periods (Thrupp 1997). This system is manmade and reflects the perception of traditional culture that has evolved over the years, which is based mainly on strong sociocultural and traditional beliefs, confounded by the economic status of people. In this extensive method of agriculture, farmers rotate land rather than crops to sustain livelihoods.

13.5.8 Terrace or Bun Cultivation

In order to maintain productivity of the system, to conserve soil moisture, and to prevent land degradation, bun cultivation has been practiced on hillslopes and valleys of the Himalayas for several decades (NIC 2001) (Figure 13.9). In this system, bench terraces are constructed on hillslopes running across the slopes. The space between two buns is leveled using the cut-and-hill method. The vertical interval between the terraces is not usually more than 1 m. Such measures help to prevent soil erosion, retain maximum rainwater within the slopes, and safely dispose of the excess runoff from the slopes to the foothills (Awasthi and Borthakur 1986; Misra et al. 1992; Sonowal and Dutta 2002; Singh 2002).

13.5.9 Managing Marshy Saline Soils

In some situations, the soils become sodic due to lack of drainage. This is common in Vertisols of central India. The rainfall is low here. Such situations are managed by

FIGURE 13.9 Traditional crop management practices in hill ecosystems of India. (Reprinted from ICAR Research Complex, North East region, Meghalaya, Annual Report. With permission.)

the farmers by flocculating the soil using saline water for irrigation. In the midwest hinterlands with low rainfall in Maharashtra, the sodic Vertisols in valley plains are irrigated with saline waters to grow salt-tolerant crops like safflower and coriander (Venkateswarlu 2004). The marshy lands in coastal Gujarat are sodic and not permeable. Viswanath (1937) reported that farmers spray the extract of a halophyte, rich in salts, along with irrigation to flocculate such marshy soils.

13.6 TRADITIONAL LAND USE SYSTEMS

Farmers in India have evolved land use systems and cropping patterns based not only on climate but also on the soil types. The entire cropping pattern and cultural traditions evolved in a given region are based on the soil type. Farmers generally considered edaphic factors in choosing the crop and the level of management. Soil depth and texture are two major factors considered by them.

13.6.1 Soil Depth and Crop Management

In the Yavatmal district of Maharashtra, the Vertisols occur in a toposequence. The depth increases as we move down the slope. The predominant cropping systems are cotton and sorghum in the kharif season (Joshi et al. 1996). The following table describes how the farmers use production inputs depending on the soil depth. Farmers apply FYM and higher fertilizer inputs in a deep soil and much less in a shallow soil (Table 13.10). Improved crop varieties with high yield potential are grown in the better soil. They also invest on pest management only on the crop grown in good soil.

Thus, the resource allocation has been adopted to suit the production capacity of the soil type. Another example is the postmonsoon rabi sorghum crop grown in the medium black soils of Maharashtra and Karnataka (Table 13.11). In these soils, the variety M-35-1 has been popular with farmers for over 50 years because no other

TABLE 13.10
Soil Type and Use of Production Inputs

Input	Good Soil (Deep)	Poor Soil (Shallow)
Tillage	Normal	More passes to loosen the soil
FYM (cartloads/ha)	5	None
Fertilizer (18-18-10) (kg/ha) (complex)	125	50
Seed	Cotton improved and hybrid Hybrid sorghum	Cotton local and improved Hybrid sorghum
Pesticide	2–4 sprays on cotton	–

Source: Joshi, P.K., S.P. Wani, and V.K. Chopdi, *Economic and Political Weekly* (June 29 issue), A-89–92, 1996.

TABLE 13.11
Production Inputs in Rabi Sorghum-Based Cropping System of Maharashtra Based on Soil Depth

Management Practice	Soil Type		
	Deep	**Medium**	**Shallow**
Variety	*Maldandi*	*Maldandi*	*Maldandi*
Land preparation	Two tractor plowings and harrowings	Two plowings with bullocks and harrowings	Two plowings with bullocks and harrowings
Seeding	One or two bowl seed drills	One or two bowl seed drills	One bowl seed drill
FYM (t/ha)	10–12	5–12	5–6
Fertilizer (kg/ha)	Basal: 100–150 (urea) or 125 (*Suphala* 15-15-15) Top dressing: 100 (urea)	50 kg (urea) –	– –
Crop rotation	Sorghum + safflower–pigeon pea or sorghum–pigeon pea/sunflower/groundnut	Sorghum–pigeon pea or sole sorghum	Sole sorghum; rarely sorghum–pigeon pea

Source: Sohan, L., and D.L. Dent, Effect of soil depth on yield of crop in vertisols. In *Land Evaluation for Land Use Planning.* Eds. D.L. Dent and S.B. Deshpande, NBSS & LUP Nagpur, 1993.

improved variety could match the soil depth of the region, and farmers still adopt the cropping pattern according to the soil depth.

13.6.2 SOIL TEXTURE AND CROP CHOICE

In the arid soils of Rajasthan in northwest India, farmers choose the cropping pattern depending on the soil texture. In heavy-texture soils, they keep the land fallow

TABLE 13.12
Soil Texture-Based Cropping Pattern Followed by Farmers in Madhya Pradesh (Central India)

Local Name	Fertility Index	Characteristics	Crops
Kali	8.5	Heavy, deep black flat lands	Rice, maize, wheat, gram, pigeon pea, sesame, vegetables
Bohra	7.0	Heavy, moderately deep black flat lands	Rice, wheat, peas, vegetables
Bharra	6.5	Light, medium red, undulating	Maize, minor millets, pulses, niger
Mooth Bharra	7.0	Loamy, medium red, undulating	Early rice, minor millets, pulses, niger
Karkariya Bharra	4.5	Light, moderately sloppy, shallow, red	Minor millets, pulses, niger
Kachal	5.0	Flat, black, deep	Rice, peas, vegetables
Sehra	8.0	Yellow to medium black, medium to deep	Rice, minor millets pulses, sesame, peas

Source: Singh, R.K., and A.K. Sureja, *Indian J. Tradit. Knowl.*, 7. 642–654, 2008.

during kharif, followed by wheat crop during rabi, while in medium- and light-textured soils, they grow only a single rabi crop of short duration.

Likewise, in the tribal region of Madhya Pradesh, farmers adopt a clear soil texture-based cropping pattern where the soils are given different names in the local language, which is matched with a fertility index, and its characters are described based on the scientific study of soils (Singh and Sureja 2008). All farmers in the region religiously follow the same cropping pattern, which generally does not result in failure of the crop even in years with subnormal rainfall (Table 13.12).

13.6.3 Physiography and Crop Management

In the eastern Indian states of Orissa, Chattisgarh, and Jharkhand, the familiar catenae of rice-growing fields have uplands, medium lands, and lowlands. The soils are lateritic with soil acidity common in most regions. The difference in the physiographic locations is texture, being light in uplands and heavier in lowlands. The moisture regime improves from uplands to lowlands. Rice is the main crop in the region. The average yield levels are 6, 9, and 17 q/ha, respectively, in uplands, medium lands, and lowlands. There is a clear distinction in crop management and input use depending on the physiographic location (Venkateswarlu and Vittal 1999). Even the sowing method differs (Table 13.13).

TABLE 13.13
Farming Practices for Rainfed Rice Grown in Different Zones of the Catena in Eastern India

Production System	Uplands Unbunded	Uplands Bunded	Medium Land	Lowlands
Sowing	Broadcasting (100–120 kg/ha)	Broadcast/line sowing (60–80 kg/ha)	Broadcast/line sowing and transplanting (random dibble)	Transplanting (random dibble)
Fertilizers	FYM	FM + 50 kg Diammonium phosphate	FYM + 20-10-10	FYM + 36-16-16
Weeding	Hand weeding	*Bueshening* after 4–5 weeks in standing water	*Bueshening* after 4–5 weeks in standing water	Hand weeding or use of tire harrow
Yield (t/ha)	0.4	0.7	0.9	1.7

Source: Venkateswarlu, J., and K.P.R. Vittal, Case studies on adoption of production technologies in rainfed rice, groundnut, *rabi* sorghum and pearlmillet, Module 3, ICAR/World Bank project, CRIDA, Hyderabad, India, 1999.

* *Bueshening* is done in fields that are broadcast.

13.7 STRENGTHENING MODERN SCIENCE BY BUILDING UPON TRADITIONAL KNOWLEDGE

Traditional knowledge has played a significant role in the development of modern science and will continue to do so in the future. This can be seen in the development of hypotheses, research designs, methods, and interpretations employed by scientists, as shown by contemporary historians of science. To sustain crop production and productivity per unit area, we need to build upon the foundations of traditional knowledge by integrating with modern science. It is important to build upon the traditional knowledge and avail of the benefits of modern innovations. It is not an "either/or" scenario; modern science must synthesize the traditional knowledge and build upon it. Overall, what we need ideally is a high-yielding, income-generating, science-based, farmer-empowering, and eco-friendly agriculture system that provides nutritional and food security first to smallholder farmers and eventually to the nation. In the northeastern Indian context, the short-term strategy for agriculture development is best formed by strengthening the agroforestry component, using traditional technology, for building up soil fertility, which otherwise is built up through natural processes of forest succession, where trees play a key role. Modern agricultural technology, through external subsidies of fertilizer, has not been able to effectively replace the traditional way of recovering soil fertility through forest regrowth under shifting agriculture in the humid tropics. Activities related to food security and nutrition can build on the development of underutilized, indigenous crops and their processing, to better prepare for food emergencies.

The traditional knowledge of indigenous peoples and local communities on ecosystem management and sustainable use of natural resources is gaining credence as a key weapon in the fight against climate change. Traditional knowledge on crops and farming practices offer huge potential for building resilience and adapting agriculture to climate change. Finally, the resources available to all farmers are the same everywhere, but the distinction between traditional and modern lies in the level of knowledge and the management of resources. The management of natural resources aims to sustain the production base so that poor farmers can continue deriving reasonable production and incomes with little external input.

13.8 CONCLUSIONS

Indigenous knowledge is closely tied to the sustainable use and maintenance of a healthy and vibrant ecosystem. Many successful examples of regenerating ecosystems and supporting local livelihoods are found in areas where users themselves have established a management structure, or management that is based upon an indigenous system. Further efforts are needed to document traditional, sustainable farming systems and best practices, and to design projects that blend traditional and new technologies. The persistence of millions of agricultural hectares under ancient, traditional management in the form of raised fields, terraces, polycultures (with a number of crops growing in the same field), agroforestry systems, and so forth documents a successful indigenous agricultural strategy and constitutes a tribute to the "creativity" of traditional farmers. These microcosms of traditional agriculture offer promising models for other areas because they promote biodiversity, thrive without agrochemicals, and sustain year-round yields. The International Assessment of Agricultural Knowledge, Science, and Technology (AKST) commissioned by the World Bank and the Food and Agriculture Organization (FAO) of the United Nations stressed that traditional and local knowledge systems enhance agricultural soil quality and biodiversity as well as nutrient, pest, and water management, and the capacity to adapt to climate change. Broadening local indigenous knowledge on soil management by including the knowledge of physical processes by linking soil knowledge with socioeconomic context and integrating traditional wisdom with scientific knowledge will open up new avenues for tackling several problems of unsustainable agriculture that the world is challenged with.

ACKNOWLEDGMENTS

The authors would like to thank Vijay Sandeep Jakkula, Sumanta Kundu, and Pushpanjalli for their valuable contributions in manuscript preparation and other technical assistance.

ABBREVIATIONS

AKST: Agricultural Knowledge, Science, and Technology
FAO: Food and Agriculture Organization
FYM: farmyard manure

REFERENCES

Agarwal, R.K., J.P. Gupta, and S.K. Saxena. 1975. Studies on soil physico-chemical and ecological changes under twelve year old five desert tree species of western Rajasthan. *Indian Forester* 102:863–872.

Aishwath, O.P. 2007. Concept, background and feasibility of organic agriculture and biodynamic agriculture. *Asian Agri-History* 11(2):119–132.

Aishwath, O.P., M.S. Dravid, and M.S. Sachdev. 2003. Radiotracer study on phosphorus utilization as influenced by bacterial seed inoculation, farmyard manure and nitrogen fertilization in *Triticum aestivum* grown on Typic Ustifluvents (saline phase). *J Nuclear Agric Biol* 3(3–4):158–166.

Awasthi, R.P. and D.N. Borthakur. 1986. Agronomic evaluation of rice varieties in upland terraces of Meghalaya. *Indian J Agric Sci* 56(10):703.

Bharara, L.P. and Y.N. Mathur. 1994. People's perception and indigenous knowledge of land resource conservation in Rajasthan. *Proc. 8th Intl. Soil Cons. Conference on Soil and Water Conservation*. Eds. L.S. Bhushan, I.P. Abrol, and M.S. Rama Mohan Rao, 976–985. Indian Assoc. Soil Water Conservationists, Dehra Dun.

Bregman, L. 1993. Comparison of the erosion controls potential of agro forestry systems in the Himalayan region. *Agroforest Syst* 21:116.

Chaman, L. and L.R. Verma. 2008. Indigenous technological knowledge on soil and water management from Himachal Himalaya. *Indian J Trad Knowl* 7(3):485–493.

Chandra, D. 1999. *Beushening*—a wonderful cultural operation of rainfed lowland rice in eastern India. *Indian Farming* July:10–12.

Chaudhary, R.G., R.N. Dwivedi, K.K. Dutta, B.K. Sarma, C.S. Patel, and R.N. Prasad. 1993. Rice based farming of Apatani-an efficient indigenous system of hill farming. *Indian J Hill Farm* 6(1):93–102.

Das, T. and A.K. Das. 2005. Local soil knowledge of smallholder rice farmers: A case study in Barak Valley, Assam. *Indian J Trad Knowl* 4(1):94–99.

Dhyani, S.K. 1998. Tribal areas: an agroforestry system in the northeastern hills of India. *Agroforestry Today* 10:14–15.

Dvorak, K.A. 1988. Indigenous soil classification in semiarid tropical India. Progress Report 84, ICRISAT, India.

FAO. 1994. *Soil Management for Sustainable Agriculture and Environmental Protection in the Tropics*. Land and Water Development Division (Food and Agriculture Organization of the United Nations), Rome, Italy.

Funtilana, S. 1990. Safe, inexpensive, profitable and sensible. International agricultural development, March–April.

Greiner, L. 1998. *Working with Indigenous Knowledge. A Guide for Researchers*. International Development Research Centre, 1-11, IDRC, Ottawa, ON, Canada.

Hmingthanzuali and R. Pande. 2009. Women's indigenous knowledge and relationship with forests in Mizoram. *Asian Agri-History* 13(2):129–146.

Jeeva, S.R.D.N., R.C. Laloo, and B.P. Mishra. 2006. Traditional agricultural practices in Meghalaya, North East India. *Indian J Trad Knowl* 5(1):7–18.

Joshi, P.K., S.P. Wani, and V.K. Chopdi. 1996. Farmer's perception of land degradation: a case study. *Econ Polit Weekly* (June 29 Issue), A-89-92.

Kuldip, G.A., A. Arunchalam, and B.K. Dutta. 2011. Indigenous knowledge of soil fertility management in the humid tropics of Arunachal Pradesh. *Indian J Trad Knowl* 10:508–511.

Kumar, A., M.S. Hooda, and R. Bahadur. 1998. Impact of multipurpose trees on productivity of barley in arid ecosystem. *Annals Arid Zone* 37:153–157.

Laxminarayana, K. 2002. Agroforestry for fertility improvement and management of soils. In *Natural Resources Conservation and Management of Mountain Development*, 331. International Book Distributor, Dehra Dun, India.

Meena, R.P. and B.S. Bhimavat. 2009. Moisture use functions and yield of rainfed maize as influenced by Indigenous technologies. *Asian Agri-History* 13(2):155–158.

Mishra, A.S. 1998. Traditional knowledge and management of natural resources. http://ignca.nic.in.

Misra, J., H.N. Pandey, R.S. Tripathi, and U.K. Sahoo. 1992. Weed population dynamics under jhum (slash and burn agriculture) and terrace cultivation in northeast India. *Agric Ecosyst Environ* 41:295.

MSSRF. 2012. Rice cultivation in Odisha—a case study, A global recognition for Odisha tribal's traditional green farming, Odisha today, MS Swaminath Research Foundation.

Nagnur, S., G. Channal, and R.A. Bharathi. 2006. Traditional knowledge of rural health security. *Asian Agri-History* 10(4):293–302.

Narayana, R.S. 2006. Application of *Gunapajalam (kunapajala)* as liquid biofertilizer in organic farms. *Asian Agri-History* 10:161–164.

Nene, Y.L. 2004. Sustainable agriculture through vedic concepts and literature. In *Souvenir of the National Conference on Organic Farming for Sustainable Production, March 23–25, 2004*, 3–4. New Delhi, India. Horticulture Society of India, New Delhi, India.

Nene, Y.L. 2006. *Kunapajala*—a liquid organic manure of antiquity. *Asian Agri-History* 10(4):315–321.

Nene, Y.L. 2009. Indigenous knowledge in conservation agriculture, *Asian Agri-History* 3(4):321–326.

NIC. 2001. Soil and water conservation, Meghalaya (National Informatics Centre, Meghalaya State Centre, Shillong). http://megsoil.nic.in/shifting_cul.htm.

Osunda, A.M.A. 1988. Soil suitability classification by small farmers. *Prof Geograph* 40(2):194.

Palaniappan, S.P. and K. Annadurai. 1999. *Organic Farming—Theory and Practice*. Scientific Publishers, Jodhpur, Rajasthan, India.

Pant, R. and R. Khanduri. 1998. Ecological degradation due to exploitation of natural resources and development. In *The Cultural Demands of Ecology*. Indira Gandhi National Centre for the Arts, New Delhi.

Pathak, R.K. and K.A. Ram. 2002. *Biodynamic Agriculture*. Central Institute of Subtropical Agriculture, Lucknow, Uttaranchal, India.

Pawluck, R.R., J.A. Sander, and J.A. Tabor. 1992. The role of Indigenous soil knowledge in agriculture development. *J Soil Water Conserv* 47(2):298–302.

Payton, R.W., J.J.F. Barr, A. Martin, P. Sillitoe, J.F. Deckers, J.W. Gowing, N. Hatibu, S.B. Naseem, M. Tenywa, and M.I. Zuberi. 2003. Contrasting approaches to integrating indigenous knowledge about soils and scientific soil survey in East Africa and Bangladesh, *Geoderma* 111(3–4):355–386.

Pettersson, B.D., H.J. Reents, and E.V. Wistinghausen. 1992. *Fertilization and Soil Properties: Results of a 32-year Field Experiment in Jarna, Sweden*. Institit fur biologisch-dynamiische Forschung Darmstadt, Germany.

Ramakrishnan, P.S. 1992. Ecology and traditional wisdom. In *Biodiversity Conservation*. Eds. C.A. Perrings et al. Kluwer Academic Publishers, Netherlands.

Rathore, S.S., K. Karunakaran, and B. Prasad. 2010. Alder based farming systems: traditional farming practices in Nagaland for amelioration of *Jhum* land. *Indian J Trad Knowl* 9:677–680.

Rhoades, R.E. 1998. Scientific and local classification and management of soils. *Agric Human Val* 15(2):3–14.

Reddy, S. 2011. Impact of sheep penning on soil organic carbon, nutrient build up and crop productivity in red soil region of Andhra Pradesh. Soil carbon sequestration for climate change mitigation and food security. Central Research Institute for Dryland Agriculture, Hyderabad, India.

Sandor, J.A. and L. Furbee. 1996. Indigenous knowledge and classification of soils in the Andes of Sothern Peru. *Soil Sci Soc Am J* 60(5):1502–1512.

Sanghi, N.K., J. Kerr, and S. Sharma. 1994. Indigenous soil and water conservation practices in Telangana region of Andhra Pradesh. *Indian Fertilizer Manag* 44:57–64.

Sanjutha, S., S. Subramanian, I. Rani, and Maheswari. 2008. Integrated nutrient management in *Andrographis paniculata*. *Res J Agric Bio Sci* 4(2):141–145.

Sanyal, P.K. and P. Dhyani. 2007. Indigenous *soil* fertility maintenance and pest control practices in traditional agriculture in the Central Himalayas: Empirical evidence and issues. *Outlook Agric* 36:49–56.

Sarma, B.K. 2002. Watershed management and biodiversity conservation. In *Integrated Watershed Management for Sustainable Development*, 202. ICAR Research Complex for NEH Region, Umiam, Meghalaya.

Satapathy, K.K. 2002. Natural resources conservation and management for mountain development in northeastern region. In *Natural Resources Conservation and Management of Mountain Development*, 322. International Book Distributor, Dehra Dun, India.

Sharma, G., R. Sharma, and E. Sharma. 2002. Performance of an age series of *Alnus*—cardamom plantation in Sikkim in Himalaya: Nutrients dynamics, *Ann Botany* 89:273–282.

Sharma, J.P. and B.P. Sinha. 1993. Traditional wisdom of hill farmers of Uttarkashi. National Seminar on Indigenous Technology for Sustainable Agriculture.

Singh, R.K. 2002. Soil conservation measures in agriculture land. In *Integrated Watershed Management for Sustainable Development*, 114. ICAR Research Complex for NEH region, Umiam, Meghalaya.

Singh, R.K. 2003. Micro farming situation based cropping systems. The ICAR Mission Mode Project on "Collection, Identification, Documentation and Validation of Indigenous Technological Knowledge," ZC Unit, JNKVV Campus, Jabalpur, MP. Indian Council of Agricultural Research, New Delhi, India.

Singh, R.K., B.S. Dwivedi, R. Tiwari, and P.K. Singh. 2002. Ethno-taxonomy and fertility management dimension: An appraisal of farmers' wisdom. In *Food and Environment: Second International Agronomy Congress on Balancing Food and Environment Security—A Continuing Challenge held at IARI, Pusa, New Delhi, India*. An extended summary, Vol II. Indian Agricultural Research Institute, New Delhi, India.

Singh, R.K. and A.K. Sureja. 2008. Indigenous knowledge and sustainable agricultural management under rainfed conditions. *Indian J Trad Knowl* 7:642–654.

Singh, R.K. and P.K. Singh. 2005. Fertility management dynamics of soil: Exploration of farmers' hidden wisdom. *Asian Agri-History* 9(4):291–303.

Sofia, P.K., Rajendra Prasad, and V.K. Vijay. 2006. Organic farming—traditional reinvented. *Indian J Trad Knowl* 5(1):139–142.

Sohan, L. and D.L. Dent (1993). Effect of soil depth on yield of crop in Vertisols. In *Land Evaluation for Land Use Planning*. Eds. D.L. Dent and S.B. Deshpande. NBSS & LUP Nagpur.

Sonowal, D.K. and K.K. Dutta. 2002. Planning soil and water conservation works. In *Integrated Watershed Management for Sustainable Development*, 80. ICAR Research Complex for NEH region, Umiam, Meghalaya.

Srivastava, S.K. and Hema Pandey. 2006. Traditional knowledge for agro-ecosystem management. *Indian J Trad Knowl* 5(1):122–131.

Talawar, S. and R.E. Rhoades. 1998. Scientific and local classification and management of soils. *Agriculture and Human Values* 15:3–14.

Talukdar, N.C., P. Hazarika, D. Dutta, and Y.P. Singh. 2002. Restoration of degraded soils using organic amendment and bio-inoculant. In *Natural Resources Conservation and Management of Mountain Development*, 303. International Book Distributor, Dehra Dun, India.

Tangjang, S. 2009. Traditional slash and burn agriculture as a historic land use practice: A case study from the ethnic noctes in Arunachal Pradesh, India. *World J Agric Sci* 5(1):70–73.

Thrupp, L.A. 1989. Legitimizing knowledge: Scientized packages or environment for Third

World People. In *Indigenous Knowledge Systems: Implications for Agriculture and International Development*, Eds. M. Warren, J. Slikkerverer, and O. Titlolo, 15–20. Studies on Technology and Social Change, No. 11. Iowa State University, Ames, Iowa, USA.

Thrupp, L.A. 1997. *The Diversity and Dynamics of Shifting Cultivation: Myths, Realities and Policy Implications*. World Resources Institute, Washington DC.

Tiwari, S.C., J.K. Das, B. Loktongbam, and S.S. Singh. 2002. Potential forest tree species for amelioration of soil properties in humid tropics. In *Natural Resources Conservation and Management of Mountain Development*, 381. International Book Distributor, Dehra Dun, India.

Toky, O.P. and P.S. Ramakrishnan. 1983. Secondary succession following slash and burn agriculture in northeastern India. I. Biomass, litter fall and productivity. *J Ecol* 71:735–745.

Tripathi, K.P. 2002. Rehabilitation of degraded lands. In *Integrated Watershed Management for Sustainable Development*, 114. ICAR Research Complex for NEH region, Umiam, Meghalaya.

Venkateswarlu, J. 2004. *Wisdom in Traditional Farming Systems*. Centre for Sustainable Agriculture, Hyderabad.

Venkateswarlu, J. and K.P.R. Vittal. 1999. Case studies on adoption of production technologies in rainfed rice, groundnut, *rabi* sorghum and pearlmillet. Module 3, ICAR/World Bank project, CRIDA, Hyderabad.

Viswanath, B. 1937. Presidential Address, Indian Science Congress, Hyderabad.

Zurayk, R., A. Faraj, S. Hamadeh, S. Talhouk, C. Sayegh, A. Chehab, and K. Al Shab. 2001. Using indigenous knowledge in land use investigations: a participatory study in semi-arid mountainous region of Lebanon. *Agric Ecosyst Environ* 86(3):247–262.

14 Sustainable Soil Management Is More than What and How Crops Are Grown

Amir Kassam, Gottlieb Basch, Theodor Friedrich, Francis Shaxson, Tom Goddard, Telmo J. C. Amado, Bill Crabtree, Li Hongwen, Ivo Mello, Michele Pisante, and Saidi Mkomwa

CONTENTS

14.1 Introduction 338
14.2 Agricultural Soil Degradation: Definitions and Extent 339
14.3 Causes of Degradation of Agricultural Soils 341
    14.3.1 Soil Degradation from an Ecological Perspective 343
    14.3.2 Agricultural Intensification Based on the "Interventionist" Paradigm ...343
    14.3.3 Examples of Large-Scale Agricultural Soil Degradation 346
        14.3.3.1 Brazil 346
        14.3.3.2 Australia 349
        14.3.3.3 China 350
14.4 Nurturing Soils and Landscapes as Living Biological Systems 352
14.5 Sustainable Soil Management Based upon Agroecological Principles 354
    14.5.1 Conservation Agriculture as a Base for Sustainable Soil Management and Production Intensification 355
    14.5.2 Linkage with Landscape Health 360
    14.5.3 Restoring Degraded Agricultural Soils and Landscapes 362
        14.5.3.1 Brazil 363
        14.5.3.2 Australia 367
        14.5.3.3 China 370
14.6 Integrating Sustainable Soil Management Principles into Farming Systems 371
    14.6.1 Crop Management Practices and Sustainable Soil Management 372
    14.6.2 Sustainable Soil Management with Intercropping as an Alternative in Permanent No-Till Systems 373
    14.6.3 Crop–Livestock Integration for Sustainable Soil Management 374
    14.6.4 Farm Power and Mechanization for Sustainable Soil Management ...375

14.7 Large-Scale Landscape-Level Benefits from Sustainable Soil Management....376
    14.7.1 Canada: Carbon Offset Scheme in Alberta..377
    14.7.2 Brazil: Watershed Services in the Paraná Basin 378
    14.7.3 Spain: Soil Conservation in Olive Groves...379
14.8 Policy, Institutional, Technology, and Knowledge Implications 381
    14.8.1 Policy and Institutional Support 381
    14.8.2 Technology and Knowledge Support...383
14.9 Concluding Comments 385
Abbreviations..387
References..387

14.1 INTRODUCTION

Soil management in agricultural landscapes should deploy production practices that are in harmony with soil-mediated ecosystem functions if they are to deliver a broad range of ecosystem services. Such services include edible and nonedible biological products, clean drinking water, processes that decompose and transform organic matter, and cleansing processes that maintain air quality. Several categories of ecosystem services are recognized: provisioning, regulating, cultural, and supporting (Millennium Ecosystem Assessment [MEA] 2005). In agricultural landscapes, provisioning ecosystem services can be delivered effectively and efficiently when the linked regulatory and supporting services are allowed to operate normally. Ecosystem functions that protect and enhance regulatory and supporting ecosystem services in the soil and landscape in which crops are grown appear, in general, to offer an effective way of harnessing the best productivity, ecological, and economic performances.

Thus, agricultural soil management can only be considered sustainable if field soil health and productive capacity are kept at an optimum to provide ecosystem services such as provision of clean water, hydrologic and nutrient cycling, habitats for microorganisms and mesofauna, carbon sequestration, and climate regulation. Across agricultural and mixed land use landscapes, such ecosystem services form the necessary conditions for society to be able to sustainably harness the biological potentials of the altered agroecosystems and the associated provisioning services of food, vegetation, water, etc.

In general, over the past several millennia, agricultural land use globally has led to soil physical, chemical, biological, and hydrological degradation, and this state of affairs continues unabated in most farmlands (MEA 2005; Montgomery 2007; FAO 2011a). This is true on small and large farms, on farms using mechanized or manual farm power, in developing and in industrialized countries, in the tropics, and outside the tropics. The dominant farming systems paradigm globally is based on mechanical tillage of various types to control weeds (often along with herbicides), soften the seedbed for crop establishment, and loosen compacted subsoil. At the center of this paradigm, there are farming practices for crop, soil, nutrient, water, and pest management that are considered by most agricultural stakeholders to be "modern, good, and normal." However, the same farming practices have also forced farmers to accept that, supposedly, any accompanying soil degradation and loss of ecosystem services are inevitable and "natural" consequences of farming—consequences that can be kept under control but not avoided altogether. This view is increasingly being challenged and considered to be outdated, and inherited farming practices are considered unable to deliver the multifunctional objectives of productivity with ecosystem services now being demanded from agricultural land and producers who use it for farming.

In the past three decades, ideas and concepts, as well as an ecosystem approach to sustainable production intensification, have led to the emergence of an alternative approach to farming across all continents. The title of this chapter is "Sustainable Soil Management Is More Than What and How Crops Are Grown." Not only how and what crops are grown matters but also the interactions of the two in space and time lead to effects and consequences that influence system performance and delivery of ecosystem services. Some ecosystem services involve processes such as hydrological, carbon, and nutrient cycling that operate at the level of the fields on farms, landscapes, watersheds, and beyond. In addition, agricultural soil management is undertaken within different farming systems for the purpose of producing biological products for markets, and a range of production inputs, equipment and machinery, and management skills are needed to operate successfully. Thus, the topic of sustainable soil management has a wide and complex scope as reflected in the list of 10 tenets proposed by Lal (2009).

This chapter is about soil degradation in agricultural land, its root causes, and what solutions are being implemented in different parts of the world to integrate sustainable soil management into sustainable farming and landscape management. Section 14.2 describes what is meant by agricultural soil degradation and its extent. Section 14.3 provides an explanation of some of the major causes of soil degradation in agricultural land use and illustrates three cases of widespread soil degradation in contrasting environments. This is followed, in Section 14.4, by a discussion on the elements of sustainable soil management. Section 14.5 provides an elaboration of sustainable soil management based on the agroecological paradigm that is increasingly being promoted internationally, including how sustainable soil management has been able to restore degraded soils in different agricultural environments. Section 14.6 illustrates the kind of contributions crop management, intercropping, crop–livestock integration, and farm power that can make to sustainable soil management objective. Section 14.7 presents three examples of large-scale landscape level ecosystem service benefits that are being harnessed from sustainable soil management systems. This is followed by Section 14.8 on policy and institutional implications for sustainable soil management. Section 14.9 offers some concluding remarks regarding the current trend toward sustainable soil management and what policy makers can do to support the trend.

14.2 AGRICULTURAL SOIL DEGRADATION: DEFINITIONS AND EXTENT

Soil is considered to be a nonrenewable resource that ensures crucial environmental, social, and economic functions, and it has a central role in any approach aimed at defining the principles and practices of sustainable agriculture (Warkentin 1995). To identify the causes of agricultural soil degradation, it is necessary to agree on signs that clearly characterize this phenomenon and its degree. However, the "definition" of what is considered soil degradation has been regarded as a rather relative term, because an objective or quantitative evaluation of the evolution of soil quality and productivity is quite a complex undertaking. Further, similarly to what has been proposed by many authors with regard to the process of soil erosion (Verheijen et al. 2009), the extent of soil degradation, which may be considered "acceptable" or "tolerable" (i.e., which is not understood as such), is far from being clear.

Agricultural soil degradation is generally understood as loss in the quality or productivity of soil as a result of human activities, leading as a consequence to less productivity or even its abandonment for agricultural use. In the *Guidelines for General Assessment of the Status of Human-Induced Soil Degradation* (Oldeman 1988), the different forms of human-induced soil degradation are distinguished comprehensively between two main categories: (1) displacement of soil material through water and wind erosion and (2) chemical and physical deterioration, such as depletion of soil nutrients and organic matter, salinization, acidification, and pollution, but also compaction, sealing and crusting, truncation of the soil profile, or waterlogging. Despite this distinction between the two categories, there is a strong relationship between them once occurrence and degree of soil displacement are appreciated as being a consequence of chemical and physical deterioration of the soil. In addition, both categories of agricultural soil degradation may lead to severe off-site effects such as sedimentation of reservoirs, harbors, or lakes; flooding; riverbed filling and riverbank erosion; and eutrophication of water bodies.

In these earlier definitions and descriptions of agricultural soil degradation, soil is treated mainly as a physical entity. In reality, however, a productive agricultural soil is a living system in which biological processes carried out by soil microorganisms and mesofauna are key elements in the creation, maintenance, and enhancement of soil health and its productive capacity. Soil health represents the soil's physical, chemical, hydrological, and biological status and its ability to respond to agricultural production inputs and to climatic variability including extreme weather events. For example, soil physical and chemical characteristics such as soil structure and porosity, soil aeration, water infiltration and drainage, soil water and nutrient holding capacity, total exchange capacity, and pH are greatly influenced by soil biological properties such as soil organic matter (SOM) turnover and the dynamics of soil biodiversity, which has an intimate relationship with plant roots, affecting its phenotypic expression below and above the ground. Deterioration of soil biological health, and consequent loss in soil productive capacity, is often not given much prominence in agricultural soil management and degradation research or in farming system management. Thus, the role of soil microorganisms and mesofauna and the SOM they require in order to function effectively and self-sustainably in the maintenance of soil health and the important role they play in crop phenotypic expression and crop performance are overlooked. This includes diverse kinds of symbiotic relationships that exist between soil biodiversity and plants about which we know very little (Uphoff et al. 2006), presumably because of difficulty in establishing, through scientific experimentation, the causal relationships with productivity and ecosystem services.

According to The Global Assessment of Human-Induced Soil Degradation (GLASOD), up to half the world's agricultural land is degraded to some degree (Oldeman et al. 1991). Degradation of cropland is most extensive in Africa, affecting 65% of cropland, compared with 51% in Latin America and 38% in Asia (CA 2007). Loss of organic matter and physical degradation of soil not only reduce nutrient availability but also have significant negative impacts on infiltration and porosity, which consequently impacts local and regional water productivity; the resilience of agroecosystems; and global carbon cycles. Accelerated on-farm soil erosion leads to substantial yield losses and contributes to downstream sedimentation and degradation of water bodies and infrastructure (Vlek et al. 2010). Nutrient depletion and chemical degradation of soil are a primary cause of decreasing yields and result in low on-site water productivity and off-site water pollution. Globally, agriculture is the main contributor to nonpoint-source water pollution. Water quality problems can often be as severe as those of water availability. Secondary salinization and water logging in irrigated areas threaten productivity gains. According to the MEA (2005), some two-thirds of our ecosystems are degraded. According to FAO (2011a), only some 10% of the global agricultural land is considered to be under improving condition; the rest has suffered some degree of degradation, with 70% characterized as being moderately to highly degraded.

Unfortunately, the problem of agricultural soil degradation is often considered to be unique to tropical and subtropical regions (Greenland and Lal 1977) or in developing regions, which is now recognized to be not so. Soil mismanagement and the traditional physical view of soils have led to serious soil degradation in temperate agroecologies in the industrialized countries (Pretty 2002; Montgomery 2007). For example, in 2002, the European Union initiated the so-called "Thematic Strategy for Soil Protection," as it recognized that "Soil is a vital and non-renewable resource and had not been the subject of comprehensive EU action." At that time, the Commissioner of Environment even said that "for too long, we have taken soil for granted. However, soil erosion, the decline in soil quality and the sealing of soil are major problems across the EU." The ensuing discussion in the frame of this strategy identified eight major soil threats: soil erosion; decline in SOM; soil contamination; soil sealing; soil compaction; decline in soil biodiversity; salinization; and floods and landslides. Notwithstanding this, the approach to understanding root causes of soil degradation in any agricultural environment has remained relatively narrow, lacking the fuller appreciation of the role of soil biology in the maintenance of soil health, the role of symbiotic relationships between soil microorganisms and crop performance, and the disruptive effect of mechanical soil disturbance on soil health and productive capacity, and on production system resilience (Kassam et al. 2009).

14.3 CAUSES OF DEGRADATION OF AGRICULTURAL SOILS

The root cause of soil degradation in agricultural land use and of decreasing productivity—as seen in terms of loss of soil health—is the low soil-carbon and soil-life disrupting paradigm of mechanical soil tillage, which, in order to create conditions for improved crop performance, debilitates many important soil-mediated ecosystem functions. For the most part, agricultural soils are becoming destructed, our landscape is exposed and unprotected, and soil life is starved of organic matter, reduced in biological activity, and deprived of habitat. The loss of soil biodiversity, damaged structure, and its self-recuperating capacity or resilience, increased topsoil and subsoil compaction, runoff and erosion, and greater infestation by pests, pathogens, and weeds indicate the current poor state of the health of many of our soils. In the developing regions, this is a major cause for inadequate food and nutrition security.

In industrialized countries, the poor condition of soils due to excessive disturbance through mechanical tillage is being exacerbated by

1. Overreliance on application of mineral fertilizers, as the main source of plant nutrients, onto farmland that has been losing its ability to respond to nutrient inputs due to degradation in biological soil health—related to declining stocks of soil carbon—including loss/destruction of adequate soil porosity and reduced soil moisture storage and increased runoff, leading to poor root system, nutrient loss, and decrease in nutrient uptake
2. Reducing or doing away with crop diversity and rotations including legumes and pastures (which were largely in place around the time of World War II [WWII]) facilitated by high levels of agrochemical inputs, standardized fixed agronomy, and commodity-based market forces that are insensitive to on-farm and landscape ecosystem functions.

The situation in industrialized nations is now leading to further problems of increased threats from insect pests, diseases, and weeds against which farmers are forced to apply even more pesticides and herbicides, and which further damages biodiversity and pollutes the environment.

It seems that with mechanical tillage (intensive or otherwise) and with low soil input of atmospheric carbon and nitrogen and exposed soil surfaces as a basis of the current agriculture production and intensification paradigm, we have now arrived at a "dangerous" point in soil and agroecosystem degradation globally, including in the industrialized North. However, we also know that the solution for sustainable soil management for farming has been known for a long time, at least since the mid-1930s when the Midwest of the United States suffered massive dust storms and soil degradation due to intensive plowing of the prairies. Dust bowls and large-scale soil degradation continue to occur in vast regions and in developed and developing countries (Baveye et al. 2011), despite the recognition of soil health being critical to life on earth.

For instance, in 1945, Edward Faulkner wrote a book *Ploughman's Folly* in which he provocatively stated that there is no scientific evidence for the need to plow. More recently, David Montgomery in his well-researched book *Dirt: The Erosion of Civilizations* shows that generally with any form of tillage, including noninversion tillage, the rate of soil degradation (loss of soil health) and soil erosion is generally by orders of magnitude greater than the rate of soil formation, rendering agroecosystems unsustainable. According to Montgomery's research, tillage has caused the destruction of the agricultural resource base and of its productive capacity nearly everywhere, and continues to do so (Montgomery 2007).

For these natural science writers as far back as 1945, tillage, regardless of type and intensity, is not compatible with sustainable agriculture. We only have to look at the various international assessments of the large-scale degradation of our land resource base and the loss of productivity globally to reach a consensus as to whether or not the further promotion of any form of tillage-based agriculture is a wise development strategy. We contend that to continue with intensive tillage agriculture now verges on irresponsibility toward society and nature. Thus, we maintain that with tillage-based agriculture in all agroecologies, no matter how different and unsuitable they may seem for no-till farming, crop productivity (efficiency) and output *cannot be optimized* to the full potential. Further, agricultural land under tillage is not fully able to deliver the needed range and quality of environmental services that are mediated by ecosystem functions in the soil system. Obviously, something must change.

14.3.1 Soil Degradation from an Ecological Perspective

Agricultural land is derived from natural forest, savanna, and grassland ecosystems in which topsoil formation processes are driven by the natural bio-chemo-physical environment. The attending ecosystem functions mediated by soil, terrain, climate, and vegetation are driven by nature. Human-induced changes of the land by removing original vegetation, tilling and cultivating, burning, introducing new species of plants and animals, and adding agrochemicals are significant changes that can equal in their effect to rare catastrophic changes during geological time that set off sequences of erosion and reshaping of the topography. The altered hydrology, limited crop residue input, and long periods when bare soil is exposed to effects of sun, wind, and rain are the basic causes of land degradation. This has been the traditional view held by many experts during much of the past century, which led to large-scale (though, as we now see, insufficiently effective) soil conservation measures that were developed after the North American "dust bowl" disaster in the 1930s. The first measures involved practices such as contour plowing, terracing, and/or strip cropping to reduce runoff and soil erosion. However, they did not specifically target damage to soil aggregation, depletion of SOM, and loss of porosity by pulverization and compaction—which are significant factors in changing the balance between infiltration and runoff.

Tillage results in accelerated oxidation of carbon-rich organic matter by soil biota, faster than it may be being replaced, leading to progressive depletion of carbon-rich SOM. The common belief is that tillage accelerates crop residue breakdown, leading to increase in soil biota and nutrient flushes when residue is mixed with soil. Any positive effect is of very short duration and with little positive effect on soil quality and function. Rapid breakdown of crop residues starves soil organisms of their future source of energy for life processes, with consequent decline in their effectiveness in maintaining/improving the health and quality of the soil as a medium for plants' rooting and functioning.

14.3.2 Agricultural Intensification Based on the "Interventionist" Paradigm

The post-WWII agricultural intensification placed increasing reliance upon breeding "new" high yielding seeds and more intensive tillage of various types pulled by heavy and more powerful machines, combined with even more chemical fertilizers, pesticides, and herbicides, supposedly making crop rotations superfluous and promoting apparent efficiency through specialization with monocropping. According to our reading (e.g., Perkins 1997; Helvarg 2001; Posner 2005), factories producing nitrates and ammonia for manufacturing explosives needed for WWII had to find an alternate market once the war ended. The crop production sector was susceptible to nitrate and ammonia salesmen who went around convincing farmers, government officials, and scientists that high yields and more profit could be obtained with mineral nitrogen and that there was presumably no real need for crop diversification and rotations with legumes or for adding plant sources of nutrients or animal manure. Crop production could be decoupled from livestock production. This was complemented with the notion that with more mineral nitrogen input comes the need for new more responsive cultivars because traditional cultivars are not capable of responding to higher doses of mineral nitrogen. A slogan of that era, coined by DuPont, was "Better Living through Chemistry." Agroindustry and the Land Grant Colleges joined forces in promoting an industrial model for agriculture that was based on the use of chemical inputs and large volumes of output. Even FAO launched in 1961 the Freedom from Hunger Campaign (FFHC), which was partly financed by the world fertilizer industry. The FFHC's main target was to encourage the use of fertilizers by small-scale farmers through education, effective means of distribution, and credit. The overall idea was that agricultural production cannot be significantly increased in developing countries of the world without improving the nutrient status of most soils. In the late 1970s, the FFHC was replaced by FAO's Fertiliser Programme. Concurrently, rapid urbanization and land consolidation in industrialized countries forced agriculture "labor" to be substituted by "capital," particularly in the form of agricultural equipment and machinery. Large tractors with large plows became common in the 1980s and symbolized modern farming. This technological "interventionist" approach became the accepted paradigm for production intensification and was promoted globally including in the developing regions—referred to as the Green Revolution paradigm of the 1950s, 1960s, and 1970s—and that, despite boosting crop yields, increased the likelihood of

- Loss of SOM, porosity, aeration, and biota (corresponding to decline in soil health) leading to collapse of soil structure, which in turn results in surface sealing, often accompanied by mechanical compaction, decrease in infiltration, waterlogging, and flooding (Figure 14.1).
- Loss of water as runoff, as well as of soil microorganisms, of soil particles, and of organic matter in top soil as sediment.
- Loss of time, seeds, fertilizer, and pesticide (erosion, leaching).
- Less capacity to capture and slowly release water and nutrients.
- Less efficiency of mineral fertilizer.
- Loss of biodiversity in the ecosystem, below and above soil surface.
- More pest problems (breakdown of food webs for microorganisms and natural pest control).
- Falling input efficiency and factor productivities, declining yields.
- Reduced resilience and reduced sustainability.

Soil Management Is More Than What and How Crops Are Grown

FIGURE 14.1 Soil compaction and loss in water infiltration ability caused by regular soil tillage leads to impeded drainage and flooding after a thunderstorm in the plowed field (b) and no flooding in the no-till field (a). Photograph taken in June 2004 in a plot from a long-term field trial "Oberacker" at Zollikofen, close to Berne, Switzerland, started in 1994 by SWISS NO-TILL. The three water-filled "cavities" in the no-till field were derived from soil samples taken for "spade tests" prior to the thunderstorm. (Courtesy of Wolfgang Sturny.)

- Poor adaptability to climate change and its mitigation.
- Higher production costs, lower farm productivity and profit, minimal ecosystem services, and abandoned and desertified farmland and landscapes.

Smallholder farmers in developing regions using manual labor to till the land and burning or removing all crop residues from the field also experience the above consequences and remain trapped in a degrading vicious cycle that cannot be broken just by applying mineral fertilizer and replacing traditional varieties with the latest breeding results. This also applies to farms in industrialized regions where the voices demanding more sustainable farming practices, both environmentally and economically, are getting louder. As soil degradation advances, the need for purchased inputs increases until the point where compensatory effect is no longer possible, forcing farmers to use even higher inputs with equally higher environmental impact.

According to Derpsch (2004), research on "conservation" or reduced tillage with early versions of a chisel plow was initiated in the Great Plains in the 1930s to alleviate wind erosion. Stubble mulch farming was also developed and can be seen as a forerunner of no-tillage farming. This collection of practices led to what became known as conservation tillage, which includes a range of tillage practices from high soil disturbance tillage to low soil disturbance that maintains at least 30% soil cover.

The book *Ploughman's Folly* by Edward Faulkner (1945) was an important milestone in the development of sustainable soil management for agriculture. Faulkner questioned the wisdom of plowing and explained the destructive nature of soil tillage. Further research in the United Kingdom, the United States, and elsewhere during the late 1940s and 1950s, and the development of herbicide technology made no-tillage farming possible, and the practice began to spread in the United States in the 1960s, and in Brazil, Argentina, Paraguay, and Australia in the 1970s. In 1973, Shirley Phillips and Harry Young published the book *No-Tillage Farming*, the first of its kind in the world, and this was followed in 1984 by the book *No-Tillage Agriculture: Principles and Practices* by Phillips and Phillips (1984).

The modern successor of no-till farming—now generally known as conservation agriculture (CA)—goes much further as elaborated in Section 14.4. It involves simultaneous application of three practical principles based on locally formulated practices (Friedrich et al. 2009; Kassam et al. 2011a): minimizing soil disturbance (no-till seeding); maintaining a continuous soil cover of organic mulch of crop residues and plants (main crops and cover crops including legumes); and cultivation of diverse plant species that, in different farming systems, can include annual or perennial crops, trees, shrubs, and pastures in associations, sequences, or rotations, all contributing to enhance system resilience.

14.3.3 Examples of Large-Scale Agricultural Soil Degradation

Examples of large-scale agricultural soil degradation from different parts of the world appear to share several common experiences as can be seen from the cases of South America, China, and Australia presented in the following sections. These cases reflect contrasting agricultural environments ranging from the tropical and subtropical environment with summer rainfall in Brazil, to subtropical environment with winter precipitation in Western Australia, to temperate environment with winter precipitation in northern China from east to west.

14.3.3.1 Brazil

Although in South America there is a diversity of soil types and ecologies, the dominant croplands are found on Oxisols, Ultisols, and Alfisols situated for the most part in tropical and subtropical climates. Usually, in undisturbed conditions, these soils have good physical properties (deep, well-drained, stable aggregates, and rapid water infiltration), but they have low natural soil fertility as reflected by low activity clay, acidity, high aluminum content, high phosphorus fixation, and low base saturation. These soils represent one of the world's biggest agricultural soil reserves. Therefore, understanding the risk of soil degradation associated with mismanagement is crucial. The dominant weather characteristics result in high intensity rainfall, especially in the spring and summer seasons, which lead to high risk of water erosion and nutrient leaching. Other processes associated with the weather characteristics are the fast organic matter turnover due to higher soil temperature and moisture, which favors microbial activity year-around. Further, there is also the potential to produce high amounts of biomass due to the high solar radiation reaching the land surface. In most humid ecologies, it is possible to design intensive cropping systems with at least two crops per year accompanied with a diversity of cover crops to fill up the autumn season window (Amado et al. 2006).

Until the 1960s, the agricultural features in South America were those of a predominantly subsistent agriculture with land use change from native vegetation (natural forest and grass pasture) to grain crop and sown pastureland. The slash-and-burn and conventional tillage with human and animal traction were dominant practices in this agricultural expansion. The agricultural inputs were mainly any available organic material and very few inorganic fertilizers, thus an imbalance in the input–output status. This soil management was introduced by European settlers and was mainly based upon experiences from a temperate environment. The main aim of agriculture production was to supply the increasing local and regional market demand. Although the soil management practice could be classified as poor, the limited capacity to expand the cropland area due to the associated high labor demand resulted in a relatively low environmental impact. During this period, soil degradation was more intense in mountainous and steep areas due to high erosion rates (Bernoux et al. 2006). Shifting cultivation was a common practice among smallholders in response to rapid soil degradation and loss of soil productivity. The growing use of technology and inputs (investment) and increasing land value induced farmers to stay longer on the same land parcels.

The adoption of mechanization in South Brazil that occurred in the late 1960s resulted in a huge impact on land use change. Mielniczuk (2003) reported that until 1969, the cropland in Rio Grande do Sul State was lower than 1 Mha, but 8 years later, in 1977, it reached 4 Mha. The main cash crops were wheat and soybean associated with long fallow periods. The large-scale application of lime was an important tool to improve soil fertility in the acid Oxisols. Also, phosphorus application resulted in the amelioration of low soil fertility. The improvement in soil fertility was not followed by better production stewardship. Thus, practices such as intensive soil tillage, crop residue burning, low crop intensity, and bare soil were widely adopted by farmers. The high intensive soil tillage system accompanied with high intensive rainfall resulted in unprecedented erosion and contamination of water reservoirs and rivers (Cogo et al. 1978; Gianluppi et al. 1979; Mielniczuk and Schneider 1984). Frequent tillage was used as a tool to control weeds, reduce diseases in wheat, increase water infiltration, incorporate lime and fertilizers, make a seedbed, loosen the soil, break the soil crust, incorporate herbicides, accelerate the decomposition of roots and residues of native vegetation, decrease surface roughness, and eliminate/disguise rill erosion. Thus, the conventional tillage was typically composed of two plow and four to six disc operations per year. During the 1970s, it was estimated that for each kilogram of soybean harvested, approximately 10 kg of fertile soil was lost (Gianluppi et al. 1979; Mielniczuk 2003; Amado et al. 2006). In Brazil, the annual values of erosivity ranged from 3116 to 20,035 MJ mm ha^{-1} h^{-1} year^{-1}. Highest erosivity values were observed in November to February, a period that can constitute more than 70% of total annual erosivity (Cogo et al. 2003). This period is coincident with the main summer cash crop (soybean and maize) establishment, increasing the risk of soil erosion (Cogo et al. 1978).

The record soil erosion documented in South Brazil occurred on November 1978. This event was known as red November because of the amount of Oxisol sediments that was carried out to the waterways, changing the color of the rivers from blue to red. During this event, 90% of cropland was managed under conventional tillage (Mielniczuk 2003), and the soil was bare or recently disturbed with soybean plantings. There was a precipitation event of approximately 200 mm (8 in.) in just 4 days resulting in 192,200 ha that had lost at least 10 cm of topsoil (truncated) by rill and gully erosion (Gianluppi et al. 1979). The loss of seeds, fertilizers, and agrochemical from croplands resulted in US$33 million of damage (Gianluppi et al. 1979). Another environmental indicator of the intensity of soil erosion verified during this period was at the Passo Real Dam, which had 1.6 kg of soil per 1000 m^3 of water, resulting in a total of 6 M t of suspended soil sediments in the water.

The estimated soil erosion in South Brazil during the conventional tillage period was approximately 25 t ha^{-1} year^{-1}. After 15 years, adoption of conventional tillage practice resulted in two-thirds of the agricultural land in Southern Brazil showing soil degradation, expressed by the depletion of SOM, reduced water infiltration rate, structural degradation, soil compaction, and an increase in plant susceptibility to short duration droughts. The social consequences of high erosion and soil degradation were as follows: sedimentation of rivers, smallholders forced to migrate to cities increasing the unemployed population, sales of small farms, and interregional migration of farmers (south to central and north) (Cassol 1984; Amado and Reinert 1998; Pöttker 1977; Bolliger et al. 2006). During the time period that conventional tillage was the prevalent soil management practice, increases in crop yields were very modest regardless of the increase in inputs and germplasm improvement.

In Paraguay, the semideciduous subtropical forest was replaced by agricultural land use, which, along with conventional tillage practice, promoted soil degradation (Riezebos and Loerts 1998), similar to that verified in South Brazil. Prior to deforestation, SOM ranged from 2.09% to 2.42% but decreased to 1.59% under conventional tillage management. Mechanically tilled fields appear to have a more rapid decline in organic matter than manually tilled fields (1.59% vs. 1.89%) suggesting more severe soil degradation in mechanized agriculture. In South Brazil, a decline in SOM in conventional tillage pulled by tractors also was noted, although the effect of poor management in reducing soil carbon was more pronounced in soils with lower content of clay and iron oxides and under high soil erosion rates (Fabrizzi et al. 2009). Séguy et al. (1996) reported that in degraded soils of Brazil, the SOM stocks were depleted by as much as 30% to 50%.

Conventional tillage causes the physical destruction of crop residues, increases the soil-residue contact, promotes higher aeration and higher soil temperature, and increases soil N mineralization (Amado et al. 2006; Aita and Giacomini 2007). These processes in combination cause a sharp increase in microbial biomass activity that consume crop residues and labile SOM resulting in an exponential rate of decay (Pes et al. 2011). Soil tillage causes the disruption of soil aggregates and exposes particulate SOM to microbial biomass attack (Amado et al. 2006; Pes et al. 2011).

In summary, the main causes of soil degradation in South America were associated with the cumulative effects of the reduction of plant biomass being returned to the soil, reduction of crop diversity, soil erosion, soil disturbance by tillage, maintenance of bare soil or limited soil cover in periods of high rainfall erosivity, depletion of SOM, depletion of soil fertility by unbalanced input–output agroecosystems, deterioration of soil structure, soil compaction, loss of microbial biomass diversity, and decrease in soil quality.

14.3.3.2 Australia

Historically, Australian farmers had pasture as an alternative "crop." This ley pasture farming system was common up until 1990. It enabled farmers to control weeds with animals and thereby reduce their reliance on herbicides. However, the profitability of this farming system was challenged with poor wool prices in the late 1980s, and it was largely replaced with continuous cropping. Running livestock in dry regions also created soil degradation concerns with compaction common in wet heavy soils and wind erosion common on the sandier soils.

The most obvious and concerning soil degradation issues in dryland Australian agriculture have been wind erosion, followed by water erosion. The emergence of saline soil in Western Australia, about 30 years after clearing of the native vegetation, is a serious threat to some areas of the landscape (George et al. 1997). On the other hand, other areas experience more subtle soil degradation such as nutrient export, compaction, waterlogging, sodicity, water repellence, and acidity.

The degree of concern for each of these issues varies across regions and states in accordance with soil type, soil slope, geological parent material, proximity to the coast, and the local climate. Other temporal issues also had a strong influence, including intensity and duration of wind and rainfall events, level of soil cover, grazing pressure, the level of tillage used, and the level of knowledge of techniques capable of mitigating against degradation issues.

Australia is known as a "land of drought and flooding plains." The last 12 years have seen about 7 years of widespread drought and 3 years of widespread flooding plains. Such contrasting climatic conditions present soil management challenges. The climate across southern Australia is classical Mediterranean with winter wet (June–August) and summer dry (December–February). Toward northern NSW, rainfall becomes more evenly distributed throughout the year, with summer the dominant rainfall period in Queensland.

The strongest Mediterranean climate is found in the southwestern area of Australia. This area has received 40% less winter rainfall since the early 1970s. In contrast, the northern third of Australia, during the same time, has had more rainfall. However, there is limited cropping activity in these northern regions—though there is grazing of livestock, mostly cattle. Therefore, the focus of this article is on southern Australia where cropping is common.

Australian soil is reported to be part of the most ancient and weathered landscape of anywhere in the world (McArthur 2004). Large areas have a very sandy surface—some have almost no clay in the topsoil. When sandy soil is combined with the often dry climate, it creates a recipe for significant land degradation potential. The clearing of the native vegetation of mostly mallee, or Eucalypt trees for agricultural purposes, has predisposed these surface soils to wind (Crabtree 1990) and water erosion (Bligh 1989, 1991). The majority of this vegetation clearing in Western Australia occurred during the 1950s and 1960s. Over 400,000 ha was cleared each year during that decade.

The most profound and obvious forms of soil degradation in Australia were wind and water erosion. Immediately after the land was cleared, soil erosion (caused by wind and water) occurred. Sandy soils, associated with the mallee vegetation of southern Australia, occupy large regions of Victoria and South Australia. Soil erosion began on these soils soon after clearing in the late 1800s. Similar erosion occurred in Western Australia when its sandy soils were released in the early 1960s.

Prior to the availability of herbicides in the 1980s, tillage was essential for controlling weeds. However, the burial of surface organic matter, through tillage, exposed the soil to the erosive forces of the wind. During wetter years, in the 1960s, the level of erosion was small due to the rapid soil cover from regreening of annual pastures or weeds near the "break of the season." In contrast, the poor ground cover during successive drought years caused serious wind erosion, which could persist for much of the year. The mallee area of Victoria had regularly horrifying dust storms in the 1930s. In Western Australia, similar erosion occurred in the 1969 drought and regularly thereafter during the dry 1970s.

During this time, pasture was often overgrazed and the soil was left bare, and this also predisposed soil to wind erosion. The common practice, at the time, was two preseeding tillages to control weeds and soften the soil, for even seed placement. Similarly to sheep grazing, this tillage removed surface vegetation that could protect the soil against erosion (Robertson 1987).

During autumn (March–May), and before the pasture or crop could fully cover the ground with new growth, strong prefrontal winds would blow the topsoil against the seedlings, often cutting the plants off and blowing the soil off-site. Both emerging crops and pastures were damaged. On other sandy soils, on more hilly terrain, and in higher rainfall areas, water erosion was more of a concern to farmers. Similarly, the sandy soil was loosened with tillage and was also left bare, providing little soil cover to protect the soil from water erosion.

14.3.3.3 China

In Asia, population pressure on natural resources is already high, and it is expected to increase further. However, based on past trends, as population continues to grow toward a plateau level of 9 to 10 billion people, the expansion of land will become increasingly modest. The growth in commodity production in South Asia is now almost completely (94%) based on increases in yields and cropping intensities (FAO Agriculture Towards 2050), and available water resources are the limiting factors there. In East and Southeast Asia, there is still a lot of water that could be used for irrigation, but the agricultural land resources are becoming scarce (Pisante et al. 2010).

China is one of the Asian countries that have been seriously endangered and affected by soil degradation. The area of land degradation is estimated to be 370,000 km^2 corresponding to a direct economic loss of 54 billion yuan ($8.5 billion) each year. Soils in dryland areas have suffered severe degradation and desertification through water and wind erosion impacting the main grain-producing area of the country.

The threat of water erosion in dryland areas is affected by the amount and intensity of rainfall, the type of irrigation, the erodibility of the soil, cropping and management factors, and erosion control practices. The impact of raindrops or the flood irrigation on the soil surface is the beginning, and the most important part, of the erosion process. In recent decades, sand storms in China have also done great harm to the farmland. As affected by all the reasons mentioned above, the degradation of farmland finally caused the decline of productivity.

Annual rainfall ranges from 200 to 600 mm in the Loess Plateau, which is a one crop per year region. Soil in the Loess Plateau is easily eroded and is intensively cropped with dryland winter wheat. Limited crop-available water is one of the major factors constraining agricultural production on the Loess Plateau, and severe erosion has resulted in degradation of soil properties, such as water retention (Zha and Tang 2003).

In cold and semiarid Northeast China, spring maize is one of the most important grain crops in terms of area and output (Liu et al. 2002). The annual rainfall here varies from 400 to 1000 mm, and the average cumulative evaporation is ~1800 mm, which is about four times higher than the average total rainfall received during the growing stage of spring maize. Therefore, the low status of soil moisture in the root zone usually limits productivity of spring maize in this region. Conserving moisture accumulated in the root zone during the rainfall season can increase productivity of spring maize in the dry Northeast China.

In annual double cropping areas of the North China Plain, the annual rainfall is 450 to 800 mm, and the annual cumulative evaporation hugely exceeds the annual rainfall. Since the 1980s, the cropping system in this region has changed from a single- to a double-cropping system (winter wheat–summer maize) (Liu 2004). Therefore, the demand for plant available water has jumped and water scarcity is a serious issue.

In the pastoral ecology of Inner Mongolia, the annual rainfall is 450 to 500 mm, and the annual cumulative evaporation is 1300 to 1880 mm, hugely exceeding the annual rainfall. In some parts of the pastoral areas, the annual rainfall is even less than 50 mm (He et al. 2009a). In the last 100 years, large areas of grassland have been converted into cropland due to an increased population and food demand (Zhang et al. 1998). The agriculture–pasture transition region has about 32.8 Mha land, representing 27.8% of the total land area of Inner Mongolia (LZU 2005). In this region, conversion of grassland to cropping combined with insufficient rainfall and wind erosion has resulted in serious soil nutrient depletion and structural deterioration (Liu et al. 2007).

In Northwest China, water shortage is one of the major constraints to the production of agricultural crop. The average annual precipitation varies from 40 to 200 mm (Xie et al. 2005), and the annual potential evaporation in this region exceeds 1500 mm resulting in a moisture deficit of at least twice the growing season requirements of spring wheat for the area (>600 mm).

The dryland areas of China have soil that are easily eroded and intensively cropped with dryland crops (wheat, maize, etc.), which occupy 56% of the arable land (Zhu 1989). Over the past 20 years, crop yields have increased through fertilizer application and increasing water consumption; however, soil water is often not fully replenished during the fallow period (Huang and Zhong 2003). Since crop yield varies strongly with rainfall (Li 2001), water shortage becomes the greatest threat to crop production. Some 80% to 90% precipitation is lost through evaporation or runoff, and only 10% to 20% can infiltrate into the soils. Thus, the soil water storage capacity is a crucial indicator for increasing production (Li et al. 2007; Zhang et al. 2009).

Conventional tillage practices based on moldboard plowing and preparing fine seedbeds with residue removed or burned have resulted in poor soil fertility and degraded soil structure as indicated by soil surface sealing, low mesoporosity (pores of diameter <60 μm), unstable soil aggregates, and low SOM content, all of which reduce water infiltration and soil water retention (Elliott 1986; Fabrizzi et al. 2005), creating a harsh environment for crop growth. Notably, after a long period with conventional tillage, a hard plow pan forms, which prevents water infiltration and results in a lower soil water storage capacity, increased runoff, and erosion. Dust storms have increased considerably in recent years (Zhang et al. 2004; Wang et al. 2006).

CA using no-till can improve soil water storage once the hard plow pan is broken through subsoiling or ripping. Soil residues cover and no or minimum tillage can reduce evaporation and promote soil water infiltration by mitigating the direct attack of rainfall and decreasing soil crusting. The decomposed roots can form the channels in the soils, thereby reducing runoff and increasing soil water infiltration. A positive effect of CA in conserving soil water has been proved in demonstration sites established in dryland areas of China (Wang et al. 2008; He et al. 2008, 2009a,b, 2011).

14.4 NURTURING SOILS AND LANDSCAPES AS LIVING BIOLOGICAL SYSTEMS

Alongside the concern for soil erosion and the destruction of soil structure and soil life caused by frequent and intensive tillage has been the growing understanding of the important role soil life and soil biology play in the maintenance of soil health. In the 1940s, Eve Balfour referred to this in terms of "the living soil" as being a necessary condition for healthy crops, environment, and people (Balfour 1943; Primavesi 1984). According to Doran and Zeiss (2000):

> Soil health is the capacity of soil to function as a living system with ecosystem and land use boundaries, to sustain plant and animal productivity, maintain or enhance water and air quality, and promote plant and animal health.

According to Peter Trutmann, quoted in FAO (2008), this emphasizes a unique property of biological systems, since inert components cannot be sick or healthy. Management of soil health thus becomes synonymous with management of the living portion of the soil to maintain essential functions of soil to sustain plant and animal productivity, maintain or enhance water and air quality, and promote plant and animal health. According to David Wolfe, quoted in FAO (2008), healthy soils maintain a diverse community of soil organisms that help to control plant disease, insect, and weed pests; form beneficial symbiotic associations with plant roots (e.g., nitrogen-fixing bacteria and mycorrhizal fungi); recycle essential plant nutrients; improve soil structure (e.g., aggregate stability) with positive repercussions or soil water and nutrient holding capacity; and ultimately improve crop production.

In this context, ongoing supply of carbon-rich organic matter for soil organisms is essential, from which they source both energy and nutrients. Examples of management practices for maximizing soil health include maintaining vegetative cover of the land year-round to increase organic matter input and minimize soil erosion, more reliance on biological as opposed to chemical approaches to maintain crop productivity (e.g., rotations with legumes and disease- and weed-suppressive cover crops), and avoiding physical (mechanical) interventions that might compact, alter, or destroy the biologically created porous structural arrangements of soil components (FAO 2008).

A key factor for sustainability in any production system, in contrast to sustainable intensification, was described by Uphoff et al. (2006) as follows:

> Of particular importance for sustainable agriculture is the enhancement of soil water-holding capacity and drainage. This is very dependent on the kinds of biological activity that lead to better particle aggregation, creating soil that can be both better aerated and infused with water at the same time. …Improving soil characteristics through biological activity and management will store water, the most essential source for agriculture, in soil horizons and root zones where it is most needed…

Similarly, in FAO (2008), it was described as follows:

> Sustainability of land's capacities to continue yielding both plant products and water year after year depends primarily on maintaining the soil in fit condition for active life processes of the whole soil/plant system. This relates to the ongoing generation and re-generation of the porous soil architecture—the soil's 'self-recuperation capacity'—with respect to the repair of damaged soil and to its physical resilience in the face of adverse shocks of weather and/or of poor management.

It is now recognized more widely that a productive agricultural soil, together with its inhabiting plants and other biota, is a living biological system (Tikhonovich and Provorov 2011; Doran and Zeiss 2000; Doran 2002) that is made up of a complex web of interactions between a large diversity of microorganism and mesofauna and between microorganisms and plant roots as well as aboveground parts. Relatively little is known about this complex agrobiodiversity or soil biota and its ecosystem functions as its role in crop productivity has been generally ignored, even during the recent decades.

For example, four main aggregate ecosystem functions are performed by the belowground soil biota (Swift et al. 2008): (1) decomposition of organic matter brought about by the enzymatic activity of bacteria and fungi, and facilitated by soil animals such as mites, millipedes, earthworms, and termites; (2) nutrient cycling, which is closely associated with biological nitrogen fixation, uptake of various nutrients from lower soil horizons, organic matter turnover, and organic decomposition, with transformations mediated through microorganisms; (3) bioturbation through the activities of plant roots, earthworms, termites, ants, and some other soil mesofauna and macrofauna that form channels, pores, aggregates, and mounds, and physically moving particles from one horizon to another; and (4) disease and pest control through, for example, the regulations of activities of pathogens by the microbivore and micropredator portions of the soil biota that feed on microbial and animal pests, respectively.

The above-described soil biological processes and ecosystem functions cannot be performed adequately in soils that are mechanically disturbed by tillage and whose structure and porosity are repeatedly impaired as a result. Soil biological health is further hindered by the inadequate amount of organic substrate being supplied to feed and maintain soil microorganisms and their functions at rates equal to, or faster than, its rate of oxidation following tillage.

In addition, we are discovering the importance and significance of the fact that living organisms including plants and animals each have coevolved with a large number of symbiotic endophytes and nonendophytes that form mutually beneficial relationships with plants and animals that can lead to a superior phenotypic performance from the same genotype. In other words, the G × E (genotype × environment) equation can work differently depending on whether certain microorganisms are present or not in the soil, in the rhizosphere, and within the plants. In some cases, microorganisms such as the Rhizobia, which are well known for their ability to fix atmospheric nitrogen in legumes, have recently been shown to behave as a symbiotic endophyte in rice plants, where it has been shown to penetrate through the root system all the way into the leaves, increasing unit leaf photosynthesis rate by some 15% (Mishra et al. 2006). Similarly, in the case of mobile phosphorus level in the soil, values as high as 50 to 60 ppm have been recorded in soils with phosphobacteria, which would otherwise show phosphorus deficiency (Turner and Haygarth 2001).

What is being discovered is that a living soil has a different productive capacity and resilience when farming practices encourage and facilitate soil life to play its important role in maintaining soil health and quality. Such soils respond differently and more efficiently to farming practices that are applied to intensify production, and there is increasing evidence that the phenomenon of "more from less," which is often observed with biologically active soils, is due to the role soil microorganisms play in the various ecosystem processes and functions in the soil (Uphoff et al. 2006).

14.5 SUSTAINABLE SOIL MANAGEMENT BASED UPON AGROECOLOGICAL PRINCIPLES

Evidence from different parts of the world suggests that it may not be possible to separate sustainable soil management from sustainable production system management. Both are inextricably linked in ways that sustainable crop production systems must first be ecologically sustainable. This means that any production system that permits the mechanical disruption of soil life and biology and soil structure and quality, and therefore ecosystem services, cannot be considered to be sustainable ecologically. The aim of "sustainable soil management" should be to reverse the trends indicated by the items listed above, via the inducing of improvements in the quality of the soil as a rooting environment for plants. Also, an agricultural soil system is of no value if the crops grown are attacked by weeds, insects, and pathogens. In other words, sustainable soil management is not enough for sustainable production as an outcome, and certainly not where sustainable production intensification is the objective in which crop, soil, nutrient, water, pest, and farm power management in space and time must be taken care of to remain ecologically and economically viable.

The agroecological principles that underpin sustainable production systems for small and large farmers from an eco-commercial viewpoint relate to resource conservation and efficiency of resource use, both natural and purchased, while profitably managing sustainable production intensification and ecosystem services. At the core, and based upon large amount of empirical evidence from farmers themselves in all continents, we can say that sustainable production derives from a number of practical principles that can be applied simultaneously through combined crop–soil–water–nutrient–pest–ecosystem management practices. These practices are locally devised and adapted to capture a range of productivity, socioeconomic, and environmental co-benefits of agriculture and ecosystem services at the farm, landscape (watershed), and provincial or national scale (Pretty 2008; Kassam et al. 2009; Godfray et al. 2010; FAO 2011b; Pretty et al. 2011).

However, different from the tillage-based *interventionist approach* to farming described above, there are now many production systems with a predominantly *ecosystem or agroecological approach* generally characterized by minimal disturbance of the ecosystem, with both natural and managed biodiversity in order to provide food, raw materials, and ecosystem services. Biologically healthy soils underpin these systems. Thus, in order to achieve sustainable intensification, a production system must be able to support and maintain the ecosystem functioning, and services derived from it, by limiting interventions (which may appear necessary for intensifying the production) to levels that do not disrupt these functions.

Sustainable production systems based on ecosystem approaches offer a range of productivity, socioeconomic, and environmental benefits to producers and to society at large. To achieve the increased productivity required to meet 2050 food demands and the range of ecosystem services expected by society, sustainable production systems should be based on five technical principles:

- Simultaneous achievement of increased agricultural productivity and enhanced ecosystem services.
- Enhanced input-use efficiency, where key inputs include water, nutrients, pesticides, energy, land, and labor.
- Reduced dependency from external inputs derived from fossil fuels (such as mineral fertilizer and pesticides) and preference for alternatives (such as biological nitrogen fixation and integrated pest management).
- Protection of soil, water, and biodiversity through use of minimum disturbance of natural systems; interventions must not have accumulative effects but must have an impact and frequency lower than the natural recovery capacity of the ecosystem.
- Use of managed and natural biodiversity to build and/or rebuild system resilience to abiotic, biotic, and economic stresses.

Over time, systems following these principles will show increasing production levels and decreasing levels of input use. In many degraded situations, better retention of incoming water—its capture, infiltration, and in-soil storage at plant-available tensions—is an important achievement, which makes possible the optimum functioning of the entire soil/plant system.

14.5.1 Conservation Agriculture as a Base for Sustainable Soil Management and Production Intensification

The farming practices required to implement the above-mentioned key principles will differ according to local conditions and needs but will have the following required characteristics, based on optimizing conditions in the root zone as being essential to (1) biotic activity; (2) provision of water and crops; and (3) assurance of self-sustainability of soil structure and porosity.

These include capacities for achieving the following: maximum rain infiltration/minimum runoff and optimum water storage; minimum compaction; reduced diurnal temperature ranges in upper soil layers; regular supply of C-rich organic matter to the surface; minimal loss of SOM by oxidation; N levels in soil maintained; and optimized P availability. Such are best achieved by incorporating the following three main tenets of CA as a base or a foundation for sustainable soil management (see www.fao.org/ag/ca):

1. *Minimizing soil disturbance by mechanical tillage.* Whenever possible, seeding or planting directly into untilled soil, in order to maintain SOM, soil structure, and overall soil health.
2. *Enhancing and maintaining permanent mulch cover on the soil surface.* Use of crops, cover crops, or crop residues to protect the soil surface conserves water and nutrients, promotes soil biological activity, and contributes to integrated weed and pest management.
3. *Diversification of species.* Utilize both annuals and perennials in associations, sequences, and rotations that can include trees, shrubs, pastures, and crops (some or all of which may be N-fixing legumes). All will contribute to enhanced crop nutrition and improved system resilience.

CA practices related to the above-described principles are now widely used in a range of farming systems in all continents on nearly 10% of the global crop land. They add to sustainability of production and soil systems and generate a range of ecosystem services (Table 14.1). They also improve soil conditions (Table 14.2a) and result in beneficial outcomes for production, ecosystem services, and socioeconomic conditions (Table 14.2b). However, to achieve the sustainable *intensification* necessary to meet future food requirements, these CA practices need to be complemented by additional best management practices:

- Use of well-adapted, high-yielding varieties, and good-quality seeds
- Enhanced crop nutrition, based on healthy soils
- Integrated management of pests, diseases, and weeds
- Efficient water management
- Careful management of machines and field traffic to avoid soil compaction

Sustainable crop production intensification (SCPI) is the combination of all of these improved practices applied in a timely and efficient manner. For this, the ensuring of soil stability and the favoring of self-recuperation of appropriate soil structural conditions are essential (see Table 14.1 and Figure 14.1a and b). Thus, sustainable soil management depends on how and what crops are grown. However, for sustainable production *intensification* to occur, the core or foundation CA practices must integrate with other complementary practices that allow the intensification of output and the optimization of the production inputs. Such sustainable production systems, and the associated sustainable soil management practices, are knowledge

TABLE 14.1
Effects of Production System Components Fully Applied Together on Sustainability and Ecosystem Services

To Achieve	Mulch Cover[a]	No Tillage[b]	Legumes[c]	Crop Rotation[d]
Simulate optimum "forest floor" conditions	√	√		
Reduce evaporative loss of moisture from soil surface	√			
Reduce evaporative loss from upper soil layers	√	√		
Minimize oxidation of SOM, CO_2 loss		√		
Minimize compactive impacts by intense rainfall, passage of feet, machinery	√	√		
Minimize temperature fluctuations at soil surface	√			
Provide regular supply of organic matter as substrate for soil organisms' activity	√			
Increase and maintain nitrogen levels in root zone	√	√	√	√
Increase CEC of root zone	√	√	√	√
Maximize rain infiltration, minimize runoff	√	√		
Minimize soil loss in runoff and wind	√	√		
Permit and maintain natural layering of soil horizons by actions of soil biota	√	√		
Minimize weeds	√	√		√
Increase rate of biomass production	√	√	√	√
Speed up soil porosity's recuperation by soil biota	√	√	√	√
Reduce labor input		√		
Reduce fuel–energy input		√	√	√
Recycle nutrients	√	√	√	√
Reduce pest pressure of pathogens				√
Rebuild damaged soil conditions and dynamics	√	√	√	√
Pollination services	√	√	√	√

Source: Friedrich T. et al., Conservation agriculture, In: *Agriculture for Developing Countries*, Science and Technology Options Assessment (STOA) Project, European Technology Assessment Group, Karlsruhe, Germany, 2009.

[a] Crop residues, cover crops, green manures.
[b] Minimal or no soil disturbance.
[c] As crops for fixing nitrogen and supplying plant nutrients.
[d] For several beneficial purposes.

TABLE 14.2a
How CA Improves Soil Conditions

Components of Soils' Productive Capacity	Key Features of Conservation Agriculture			
	No-Till	Mulch	Rotations	Legumes
Hydrological	1	4		
Physical	2	5	7	10
Biological	3	6	8	11
Chemical			9	12

Note: Key: 1 = Water percolation; 2 = Varied soil porosity; 3 = Favors biological soil layering; 4 = Buffers impacts of rainfall, wide diurnal ranges of surface temperature; 5 = Prevents soil crusting; 6 = Source of energy and nutrients; 7 = Augments root channels—distribution and depth; 8 = Favors biodiversity in soil; 9 = Beneficial root exudates; 10 = Favors development of optimum soil architecture (solids × spaces); 11 = Nitrogen + C-rich organic matter; 12 = Nitrogen.

TABLE 14.2b
Some Resulting Beneficial Outcomes with CA

For Agricultural Production	For Ecosystem Services	For Socioeconomic Conditions
Greater security of output under varying weather conditions	Diminished water pollution by agrochemicals and eroded soil; reduced costs of water treatment	Greater efficiencies of use of labor and financial resources
Greater efficiency of rainwater use, leading to more stable yields	Less frequency, depth, and duration of flooding after unit storms of equal severity	Better health and nutrition
No/minimal soil erosion; smaller losses of applied energy, fertilizers, seeds, etc.	Longer duration of streamflow; recharge of groundwaters	Reduced frequency of flooding and severity of damages to roads, bridges, etc.
Improved soil health provides better biological controls of weeds and pests	Reduced loss of SOM by tillage-induced oxidation to CO_2	More time for diverse activities on-farm (technical)
Recirculation of carbon, micronutrients, and macronutrients	Maintenance/improvement of soil carbon content	More time for diverse activities off-farm (social)
Lesser effects of climatic drought events	Lesser damage to normal multiple functioning of soil in wider ecosystem	
Etc.	Etc.	Etc.

and management intensive and relatively complex to learn and implement. They are dynamic systems, offering farmers many possible combinations of practices to choose from and adapt, according to their local production conditions and constraints (Kassam et al. 2009; Godfray et al. 2010; FAO 2011b; Pretty et al. 2011).

The development of SCPI requires building on the core principles and practices outlined above as the production base and finding ways to support and self-empower producers to implement them all, through participatory approaches and stakeholder engagement. In addition, SCPI must be supported by coherent policies, institutional support, and innovative approaches to overcome any barriers to adoption. Monitoring and evaluating the progress of change in production system practices and their outcomes at the farm and landscape levels are critical.

One of the main criteria for ecologically sustainable production systems such as CA is the maintenance of an environment in the root zone to optimize conditions for soil biota, including healthy root function to the maximum possible depth. Roots are thus able to function effectively and without restrictions to capture plant nutrients and water as well as interact with a range of soil microorganisms beneficial for soil health and crop performance (Uphoff et al. 2006; Pretty 2008). In such systems with the above attributes, there are many similarities to resilient "forest floor" conditions (Kassam et al. 2009). Maintenance or improvement of SOM content and soil structure and associated porosity are critical indicators for sustainable production and other ecosystem services.

A key factor for maintaining soil structure and organic matter is to limit mechanical soil disturbance in the process of crop management. For this reason, no-tillage production methods—as practiced, for example, in CA—have in many parts of the world been shown to improve soil conditions, reduce degradation, and enhance productivity. However, as a stand-alone practice, the elimination of tillage would not necessarily lead to a functioning sustainable production system. This requires a set of complementary practices to enable a functioning soil system as well as the whole agroecosystem to deliver a range of ecosystem services.

The contribution of practices that implement the technical principles of CA—including mulch cover, no-tillage, legume crops, and crop rotations—in important ecosystem services is shown in Table 14.1 and Figure 14.1a and b. Even where it is not possible to install all desirable practical aspects in the production system at the same time, progressive improvements toward those goals should be encouraged. However, for any agricultural system to be sustainable in the long term, the rate of soil erosion and degradation (loss of organic matter) must never exceed the rate of soil formation (though the steeper the slope, the greater the danger that this could happen). In the majority of agroecosystems, this is not possible if the soil is mechanically disturbed (Montgomery 2007). For this reason, the avoidance of mechanical soil disturbance can be seen as a starting point for sustainable production. Once it has been brought into good physical condition, no further tilling of the soil is therefore a necessary condition for sustainability but not a sufficient condition. For SCPI, including ecosystem services, other complementary techniques are required as mentioned already, of which the practices related to the above three CA principles constitute the bare minimum for ecological sustainability (FAO 2011b).

To achieve and sustain the necessary intensification of these production systems to meet the increasing demand for food and other ecosystem services, productivity needs to be optimized by applying best management practices such as good-quality adapted seeds, adequate nutrition, and protection from pests and diseases (weeds, insects, and pathogens) and avoiding soil compaction. In addition, efficient water management and timely operations are required within suitable cropping systems to achieve desirable and acceptable outcomes.

In light of the above, it is clear that sustainable soil management depends on both what and how crops are grown, as well as on additional aspects of soil and landscape management, which includes the horizontal integration of other production sectors such as livestock and forestry. The special role of deep-rooted legumes such as pigeon pea (*Cajanus cajan*), lablab (*Dolichos lablab*), and *Mucuna* (*Stizolobium cinereum*) in building soil structure and biopores for drainage and aeration, in contributing biologically fixed nitrogen to improved nitrogen stocks in soils, and in generating both biomass and edible products is a case in point. Beneficial effects of cover crops on soil and water quality, ecological sustainability, and crop and livestock productivity have been known for many years (e.g., Hargrove 1991). Similarly, species diversification as the third principle of CA is related to integrated management of insect pests, pathogens, and weeds, and the effectiveness of control of pests, pathogens, and weeds depends on both what and how crops are grown. Species diversification involving crops of different durations and complementarity is also related to the use and management of resources of different crops in space and time to maximize and optimize the production during the growing season every year to its fullest potential in an increasingly variable and unpredictable climate. Furthermore, in order to establish diversity of soil biological activity, it is necessary to include in the cropping system a diversity of crops instead of monocropping or reduced crop diversity.

CA is now adopted on about 125 million ha of arable land worldwide, which corresponds to nearly 10% of the total cropland (Friedrich et al. 2012). Some 50% of this area is located in the developing regions. During the past decade, it has been expanding with an average rate of more than 6 million ha/year. The highest adoption levels, exceeding 50% of the cropland, are found in the southern part of South America, the Canadian prairies, and Western Australia. Fast adoption rates are now being seen in Central Asia and China, alongside increasing policy support and early large-scale adoption taking place across Africa, particularly in Zambia, Zimbabwe, South Africa, Tanzania, Kenya, Morocco, and Tunisia. Europe now has some few pockets of adoption, particularly in Finland, Spain, France, Italy, the United Kingdom, and Switzerland (Kassam et al. 2010; Derpsch and Friedrich 2009; Friedrich et al. 2012).

14.5.2 Linkage with Landscape Health

Soil forming factors include topography, climate (microclimates), and parent materials, all of which vary by landscape type and magnitude (Jenny 1980). Soils are variable according to their positions in the landscape. Landscapes distribute water and energy according to landform characteristics. In the northern hemisphere, the north-facing side of a hill, in contrast to the south-facing side, will receive less radiation and be cooler and moister, have more organic matter, and be less drought prone. The top or crown of a hill or hummock will catch less rainfall, and a shallow or more weakly developed soil profile will be found. By contrast, a depression or foot slope position will receive more water and have a deeper soil profile.

Soils formed on different landform facets will have different risks and fragility characteristics related to crop conditions. Soil biologic processes will occur differentially as well by landscape position because of the variable microclimate conditions and soil development (or degradation). Land managers need to recognize the range of soil health and functional characteristics associated with landscapes in order to develop conservation agriculture systems as well as monitoring and evaluating performance and risks.

Soil quality strongly affects agricultural land use and thus the shaping of the landscape. Any change in soil quality, whether through degradation processes or soil health improvement, will have consequences not only on the field or farm level but also on a greater scale, the landscape. In addition, landscape normally consists of a combination of different ecosystems that are interlinked more or less closely with each other. The healthier the soil is under agricultural use, the lesser the off-site effects that can be expected upon adjacent ecosystems of the same landscape.

> Good land husbandry is the active process of implementing and managing preferred systems of land use and production in such ways that there will be an increase—or, at worst, no loss—of productivity, of stability, or of usefulness for the chosen purpose. Also, in particular situations, existing uses or management may need to be changed so as to halt rapid degradation and to return the land to a condition where good land husbandry can have fullest effect (Shaxson et al. 1977).

If a production system, as represented by the features of the type of land use and those of its management characteristics, is imposed on an area of fragile or hazardous land (e.g., sandy soil, steep slope, and/or shallow depth, etc.), any erosional degradation arising from inadequacy of management will occur more rapidly toward a condition of lower productive potential than if the enterprise were located on flatter and/or less fragile land; the land itself will "wear out" toward a condition of lower productivity.

This has two implications:

- If the enterprise cannot be transferred to another "safer" or suitable location, then a more protective production system such as CA or agroforestry (Saha et al. 2010) would provide increased security and prolong the soils' usefulness (better management systems).
- If a choice of sites on a landscape is possible, then the safest strategy will be to locate the physical production system(s) on a (varied) landscape in such ways that there is rational matching of "hazardous" land uses onto the "safer" land units and of the "safer" uses onto the land units of greater hazard (site-specific management).

  To achieve any such rationalization, due attention needs to be given to catchment-oriented land resource survey, assessment, and mapping, followed by physical planning of layout of fields and infrastructure items in catchment-related patterns, to facilitate effective management of any runoff that may occur in consequence of excessive rainstorms (Carver 1981; Shaxson et al. 1977). This is of particular significance where "new" land is being opened to cropping. This is because a physical allocation of proposed land uses that is sensitive to the physical characteristic of the chosen landscape is more forgiving of mistakes in management than where land use allocations have not taken account of such realities.

Achieving this effectively represents the achievement of good land husbandry (Shaxson et al. 1989).

14.5.3 Restoring Degraded Agricultural Soils and Landscapes

A sustainable approach to soil management in rainfed and irrigated production cannot be a single technology but rather a range of mutually reinforcing practices. For both tillage and no-tillage systems, their best performances can be achieved only when the production systems are supported by effective plant nutrition, soil moisture provision, and best agronomic practices. Production systems are most sustainable and function best when all three key soil, crop, and environmental management principles listed in Section 14.5.1 are applied simultaneously. CA is a good example of progress in this regard as it is based on no-till and maintenance of soil cover and has now spread across all continents and ecologies (Hobbs 2007; Friedrich et al. 2009; Kassam et al. 2009, 2010). There are other complementary ecosystem-based approaches, which together form lead to SCPI, that have also proven to be successful as a basis for sustainable intensification in all continents under a wide range of circumstances (Uphoff et al. 2011; Kassam et al. 2011b). The responses of rice plants to aerobic soil environment suggest the possibility of discovering comparable positive responses in other crops also and establishing the scientific knowledge that can explain the effects of the symbiotic interactions between root systems and their coevolved soil microorganisms on the crop's phenotypic performance.

Sustainable production systems also mobilize plant nutrients through biological transformations of organic matter, providing micronutrients that may not otherwise be available (Flaig et al. 1977). For example, mulch-based no-till production systems can retain and mimic the soil's original desirable characteristics ("forest floor conditions") on land being first opened for agricultural use. Throughout the transformation to agricultural production, sustainable systems based on an agroecological no-tillage approach can safeguard desirable soil characteristics, sustain the health of long-opened farmland that is already in good condition, and regenerate land that has reached poor condition due to past misuse (Doran and Zeiss 2000).

Such types of information from soils and ecosystems in good condition under CA systems provide a range of "yardsticks" against which to compare the benefits of CA and the health of the soil and the ecosystem, as against the "classical" tillage agriculture. Tillage agriculture with monocropping and no organic cover represents the most vulnerable and detrimental production system, whereas CA represents a more sustainable option (Montgomery 2007).

Sustainable soil management as practiced in CA systems has resulted in the enhancement or rehabilitation of the soil resource base and its agroecological potentials, thus enabling the avoidance of soil degradation and repair of lands, leading to sustainable intensification and the harnessing of ecosystem services. This is illustrated in the examples for Brazil, Australia, and China given in the next sections.

14.5.3.1 Brazil

The soil degradation in South Brazil was reverted initially by reducing tillage intensity. This involved the use of a chisel in substitution of the moldboard plow and the reduction in the number of disc operations. The first no-till experimental plots were set up in the early 1970s in Rio Grande do Sul and Paraná States. However, the successful diffusion of no-till systems on a broader scale remained erratic until late 1980s. The first obstacles that had to be overcome were control of weeds without soil tillage or a hoe, as well as unavailability of planters able to work with crop residues. There was also the need to select appropriate cover crop options to intensify the cropping system in substitution of fallow, to produce enough crop residue to protect the soil and deal with the scarce technical assistance, high price of herbicides, and many technical doubts about the efficiency of lime and fertilizer surface broadcast instead of soil placement (Amado and Reinert 1998; Bernoux et al. 2006; Bolliger et al. 2006).

In the 1980s, farmers began to organize themselves into no-till-promoting associations, such as the "Clube da Minhoca" ("Earthworm Clubs") and the "Clubes Amigos da Terra" ("Friends of the Soil" clubs or "Earth" clubs), as well as private research institutions, such as the "Fundacão ABC" (ABC Conglomerate of Farmers' Cooperatives) to promote the adoption and diffusion of no-till (Borges Filho 2001; Dijkstra 2002).

The initial drive to expand the adoption of no-till was led by pioneer farmers, who also organized the first Brazilian no-till conference in 1981 (Steiner et al. 2001). No-till technologies and systems subsequently spread fairly rapidly from Paraná to other Southern Brazilian states and neighboring Paraguay, where similar environmental conditions existed.

A steady interregional migration of farmers from Southern Brazil to tropical Brazil brought a transfer of the basic zero-till principles in its wake, but the different agroecological conditions of humid subtropical Southern Brazil compared to those of frost-free, seasonally dry, tropical Brazil, as well as the different scales of large cerrado farms compared to generally smaller farms in the South, meant that no-till systems had to undergo scale and regional adaptation (Spehar and Landers 1997; Bolliger et al. 2006). The first records of mechanized no-till in South America were in the Brazilian state of Goiás dating from 1981/1982 (Landers et al. 1994). In Brazil, especially, no-till-type land management expanded from an estimated less than 1000 ha in 1973/1974 to nearly 26 million ha in 2010/2011 (Bolliger et al. 2006; Kassam et al. 2010).

More than 45% of total cultivated land in Brazil is now estimated to be managed with no-till (Scopel et al. 2004), although in Southern Brazil, this figure is reported to exceed 80% (Amado et al. 2006; Denardin and Kochhann 1999; Bolliger et al. 2006). Among the leading no-till nations, Brazil is purportedly the only one with both substantial no-till in the tropics as well as, importantly, a significant amount of smallholder no-till farms (Ralisch et al. 2003; Wall and Ekboir 2002; Bolliger et al. 2006). The latter is perhaps of particular significance, as, contrary to no-till spread in general, the adoption of true (permanent rather than sporadic) no-till systems by smallholder farmers worldwide has been poor, remaining, as yet, relatively marginal outside Brazil, Paraguay (where appropriate systems have spread from Southern Brazil), and small parts of Central America, where similar systems were already traditional (Buckles et al. 1998). Berton (1998) suggests that the main reasons for smallholder farmers in Southern Brazil to adopt no-till practices include labor and time savings, erosion control, greater income, and higher yields. Ribeiro and Milléo (2002) concur that once plowing and mechanical weeding are discontinued, labor savings and less drudgery are the major incentives expressed by smallholder farmers. Some Brazilian farmers are now into their third decade of practicing no-till land management.

In regions that experience high-intensity rainfall and support undulating terrain and/or erodible soils, protecting the soil from erosive raindrop impact through sufficient vegetative mulch is conceivably the best strategy against excessive runoff and erosion (Amado 1985; Calegari 2000, 2002; Erenstein 2003; Wildner 2000). Only not plowing, in turn, means that a protective biomass cover or mulch from previous crops is maintained on the soil surface.

The main advantages of mulch agriculture include reducing evaporation from bare soil (Stone and Moreira 1998), mediating soil temperature extremes (Derpsch 2001), providing a buffer against compaction under the weight of heavy equipment (Séguy et al. 2003), smothering weeds (Darolt 1997; Kumar and Goh 2000), creating a favorable environment for beneficial soil fauna and flora (Balota et al. 1996), and preventing soil and water contamination from pesticides and nutrient leaching (Scopel et al. 2004). However, the practice may also make the planting process more complicated, allow pests and pathogens to reproduce and spread longer in close proximity to crops (Forcella et al. 1994), protract the warming up of soil after cold periods, induce erratic crop germination, and decrease the efficiency of fertilizers and herbicides (Banks and Robinson 1982; Rodrigues 1993). Nevertheless, no-till in itself, without soil cover (e.g., if residues are burnt, grazed, or otherwise exported from the field) or under an unbalanced nutrient budget, can lead to similar soil degradation and reduced crop productivity issues as conventional tillage system.

Rather than rely purely on crop residues from a main crop to provide adequate and permanent soil cover, especially in regions where the climate favors fast decomposition of residues, one of the major Brazilian adaptations of no-till has been the strong emphasis on integrating fast-growing winter cover crops and summer crop rotations into no-till cropping systems. Such crops can be intercropped prior or planted immediately after the harvest of the main crop and rapidly produce abundant mulch, consequently allowing a succession of enhanced, year-round biomass accumulation. This can compensate for fast residue decomposition, as well as offsetting any potential lack of soil cover (Séguy et al. 1996).

Due to the high amount of mulch left on the soil surface at seeding time, Brazilian farmers hence commonly refer to no-till as "plantio direto na palha" or "planting directly into straw" (Amado et al. 2006). Derpsch (2001) and Steiner et al. (2001) argue that the complete integration of cover crops into no-till cropping systems is probably the single most fundamental key to the success of such systems in Brazil. Two decades of farm experience with cover crop management in fully integrated no-till systems result in good improvement. Seed quality and genetic material of cover crop are key points. Cover crops in cropping systems deserve the same attention in the quality of management as do cash crops. Cover crops sometimes need fertilizer input such as nitrogen and phosphorus or can even be used after chiseling or ripping the soil when the soil resistance is too high. The mixture of cover crops is a very ecological approach, and many mixtures used include black oat and vetch, black oat and oil radish, rye, black oat, vetch, and so forth. Some mineral fertilization can be split between cover crops and cash crops; farmers call this a crop rotation fertilization instead of cash crop fertilization. The main advantage of this system is the avoidance of applying too much fertilizer in a single application, which increases the environmental impact, increases the cover crop biomass, stimulates nutrient cycling, and stimulates biological activity.

Functions of cover crops broadly include the following: (1) providing additional fodder, forage, food, biofuel, and secondary commercial or subsistence products for livestock and humans; (2) directly adding or sparing nitrogen to/from the soil through symbiotic N_2 fixation from the atmosphere; (3) converting otherwise unused resources, such as sunlight and residual soil moisture, into additional biomass and, concomitantly, upon the breakdown of their residues, increasing the buildup of SOM; (4) capturing and recycling easily and moderately easily leachable nutrients (NO_3, S, K, Mg, and Ca) that would otherwise be lost beyond the rooting zone of commercial crops; (5) ameliorating soil structure and buffering against compaction by creating and stabilizing additional root channels that differ from those of the main crops and by stimulating soil biological activity through, *inter alia*, the release of root exudates; (6) improving the management of acidic soils by releasing various products that can mobilize lime movement through the soil profile, decarboxylize organic anions, function in ligand exchange, and add basic cations to the soil; (7) facilitating weed management by competing against or smothering weeds that would otherwise become noxious in the main crop cycle; and (8) breaking the cycle of, or repelling or suppressing, certain pests and diseases that could otherwise build up in continuous monocropping systems. On the other hand, integrating cover crops into existing cropping systems generally incurs extra costs of seed and agrochemicals (e.g., herbicides to terminate the crop before the next main crops or nitrogen and phosphorus fertilization), extra labor and managerial skill required to establish and maintain the crop, as well as the opportunity cost of the land and equipment, while the rewards of cover crops may well take time to properly manifest themselves.

Tropical soils have a mineralogy that is dominated by low-activity clays and sesquioxide material, making soil fertility and functionality integrity much more SOM dependent than temperate soils. In some tropical Brazilian soils, 70% to 95% of cation exchange capacity (CEC) is dependent on the SOM (Bayer et al. 2000a). In such soils, SOM status is crucial to ensuring good crop productivity and is often postulated as the single most important element of the soil restoration process associated with Brazilian no-till regimes. In principle, both the decreased erosion losses of SOM-rich topsoil (Lal 2002; Rasmussen and Collins 1991) and the slower SOM mineralization rates in zero-till soil compared to plowed soil suggest that no-till provides more favorable conditions for SOM buildup than conventional tillage. Not turning the soil, for example, means the following: (1) less soil macroaggregates are disrupted, consequently leading to the increased formation of stable microaggregates that occlude and protect particulate organic matter (POM) from microbial attack (Amado et al. 2006; Feller and Beare 1997; Lal et al. 1999; Six et al. 1998, 1999, 2000; Fabrizzi et al. 2009); (2) there is less stimulation of sharp increase in microbial activity and concomitant release of CO_2 in response to enhanced soil aeration (Bayer et al. 2000a,b; Bernoux et al. 2006; Kladivko 2001); and (3) there is less mixing of residues deeper into the soil where conditions for decomposition are often more favorable than on the soil surface (Blevins and Frye 1993; Karlen and Cambardella 1996). In this context, Mielniczuk (2003) estimated the rate of SOM mineralization under conventional tillage regimes in Southern Brazil to be, on average, 5% to 6% per year compared to an average of about 3% per year in no-till soils. Although the actual amount of SOM storage potential in a given soil is in turn largely determined by climate and the capability of soils to stabilize and protect SOM, this in itself generally is largely determined by soil texture, soil mineral surface area, and soil mineralogy, with soil parameters such as water-holding capacity, pH, and porosity acting as rate modifiers (Six et al. 2000). The large majority of Brazilian literature does indeed suggest that SOM accumulation in no-till soils exceeds that of plowed soils and that this is the case over a range of soil textures, from sandy loams (Amado et al. 1999, 2000, 2001, 2002, 2006; Bayer et al. 2000a,b, 2002) to heavy clay (>60% clay) soils (Amado et al. 2006; De Maria et al. 1999; Perrin 2003), both in Southern Brazil (Muzilli 1983; Sá et al. 2001a,b; Zotarelli et al. 2003) as well as in the degraded savanna cerrado region further north (Corazza et al. 1999; Freitas et al. 1999; Resck et al. 1991, 2000; Scopel et al. 2003). Bernoux et al. (2006) reviewed some 25 published and unpublished data sets on the rate of C accumulation in Brazilian no-till soils and observed that reported C accumulation rates in excess of those found in comparable plowed soils vary from around 0.4 to 1.7 t C ha^{-1} year^{-1} for the 0- to 40-cm soil layer in the cerrado region and between 0.5 and 0.9 t C ha^{-1} year^{-1} in Southern Brazil, with an overall average accumulation of 0.6 to 0.7 t C ha^{-1} year^{-1}.

Brazilian research data also indicate that the composition and quality of SOM in no-till soils differ from those of plowed soils. Various studies have also found that the relative amount of free labile or more recent (e.g., POM) rather than humified and occluded SOM fractions is higher in no-till soils compared to plowed soils, which in turn has important ramifications for soil structure and nutrient cycling and as a source of energy for soil microbial biomass. Other studies suggest that SOM responds linearly to increasing rates of residue input over a variety of soils and climates (Bayer 1996; Black 1973; Burle et al. 1997; Rasmussen and Collins 1991; Testa et al. 1992; Teixeira et al. 1994). For example Burle et al. (1997) obtained a close relationship between SOC in the 0- to 17.5-cm soil layer and residue quantity added by 10 different no-till cropping systems. Sisti et al. (2004) and Amado et al. (2006) further studied the role of N additions in SOM buildup under no-till in

Brazil, and both found that where rotations with N_2-fixing legumes were included, much more SOM was accumulated, hence highlighting the fact that for there to be an accumulation of SOM, there must be not only a C input from crop residues but also a net input of N. They further postulated that where net N balance was close to zero over the whole crop rotation, little SOM accumulation was to be expected. Amado et al. (2006) reported that pigeon pea and *Mucuna* cover crops integrated into no-till maize cropping systems had the highest C accumulation rates under no-till than that of the intensive cropping systems, including mixtures of black oat with hairy vetch in winter and maize with cowpea in summer. Sá et al. (2001a) suggests that the immobilization process is most intense during the first years of no-till but, after 5 or more years, gradually diminishes due to the increased surface concentration of SOM acting as an N source, thereby effectively counteracting N limitations induced by residues input on the soil surface.

Both in tropical and subtropical Brazil, legume residues left on the soil surface decompose rapidly and provide a prompt N release, sometimes so fast that it causes asynchronies with maize demand (Acosta 2005; Aita and Giacomini 2003; Vinther 2004). Common vetch residue left on the soil surface in Santa Maria, for example, released 60 kg of N per hectare in only 15 days (Acosta 2005).

14.5.3.2 Australia

Australian farmers became increasingly concerned about soil degradation during a dry period in the 1970s. They saw how plowing of the soil was not sustainable in such an unforgiving and harsh climate (Crabtree 2010). There were also other dry periods where soil erosion was a serious concern, particularly in the 1930s and then potentially again in the first decade of the 2000s. However, during this recent decade, Australian farmers were prepared! They had widely adopted no-till farming, and this greatly mitigated the severely damaging effects on the soil and maintained financial viability during such droughts.

No-till solved most degradation: An in-depth Australian experience with land degradation and the usefulness of no-tillage techniques to manage these concerns are well documented in Crabtree (2010). It was wind erosion concerns that initiated farmers' determination to find a better way to farm. There was no other soil degradation concern that motivated farming practice change. As two broad-spectrum herbicides, SpraySeed (paraquat:diquat) and Roundup (glyphosate), became available in the early 1980s, farmers began reducing their reliance on tillage. Through trial and error, both farmers and researchers gained confidence in the technique (Crabtree 1983, 2010; Flower et al. 2008).

While the initial adoption was slow, the technique of spraying herbicides, and then planting the crop, with little soil disturbance, was the beginning of no-tillage in Western Australia. The experience of farmers revealed many other soil benefits. In fact, most of the concerns with soil degradation were significantly mitigated with no-tillage through time.

No-till both improved soil structure with less vehicle compaction and increased the steady state of microbial activity, which gave the soils some biological structure. Waterlogging became less common due to better infiltration (and some dryer years). Soil salinity was somewhat mitigated as soil water runoff was less common and there was less water running to low-lying areas where it contributed to elevated water table levels. Perhaps three soil degradation challenges are not improved by no-tillage: nutrient removal, acidity, and water repellence.

Nutrient removal: The soils of South West Australia are highly weathered and generally have coarse-textured surfaces with low soil fertility and acidity, limiting crop and pasture production (Moore 2001; McArthur 2004). Soil availability of macronutrients nitrogen (N), sulfur (S), potassium (K), and phosphorus (P) and micronutrients copper (Cu) and zinc (Zn) have the potential to limit crop and pasture growth (Moore 2001). While calcium (Ca), magnesium (Mg), manganese (Mn), and molybdenum (Mo) are important nutrients, they are generally not considered to be limiting plant production. Soil fertility of Australian soils has been increased by the application of fertilizers (Weaver and Wong 2011). Nevertheless, cropping results in significant removal of nutrients, and continued application of nutrients is required to maintain long-term sustainability and productivity of cropping systems.

Soil acidity: Soil acidification is a natural process enhanced by agriculture. Each crop that is harvested is essentially alkaline material. Since no-tillage increases whole farm yields, it is effectively removing more alkaline material from the paddocks. Also, the use of nitrogenous fertilizers and the growing of pulse crops cause acidification. Consequently, soils are becoming more acid through time, especially when cropped. Lime application is required to maintain the productivity of most soils, the exception being soils with an alkaline base. Acidification happens more rapidly in slightly acid sandy soils and where leaching rains are common. Also, these soils have low levels of organic carbon, less than 1.5%, giving the soil low capacity for the prevention of soil pH decline.

Some native Australian plants have adapted to these acidic conditions, over many thousands of years, and they can also fix atmospheric N, further acidifying the soil. The most common of these species comes from the *Acacia* genus. After many years, their N fixation results in the soil becoming very acidic at depth. Such soils in Western Australia are known as Wodgil soils; however, the area affected by these naturally very acidic soils is less than 5% of Western Australia's agricultural land (Gazey and Davies 2009). The result of such strong acidification is severe soil degradation, making the soil unproductive.

The solution to this form of soil degradation is the addition of large amounts of lime. Under a no-tillage system, the movement of this lime into a 20-cm soil profile depth can take 4 years (Flower and Crabtree 2011). For a more rapid amelioration of these acid subsoils, farmers have also used deep tillage or plowing. By doing so, they are exposing themselves to soil erosion risk. However, large yield responses have been achieved immediately (Davies 2011), and this has encouraged farmer adoption of this technique.

Water repellence: Native vegetation can induce water repellence (McGhie 1980). Nevertheless, Australian sandy soils that contain less than 3% clay are also capable of becoming water repellent within 10 years of agriculture practices (Crabtree 1983). The sands develop a wax coat around individual sand particles (Mashum et al. 1988). The wax is the remnant of plant residue decomposition, and it causes water to run to the lowest-lying hollows, causing wetting in preferred pathways. Such a phenomenon exists across several countries. However, Australia has the largest area, with about

5 million ha naturally predisposed to the problem (Summers 1987; Crabtree and Henderson 1999; Cann 2000).

This problem is debilitating to farming. King (1981) discusses crops and weeds germinating over a 3-month period and how water repellence makes weed control very difficult. Nutrients are tied up in the dry topsoil, and microbial activity is restricted. Insects can become established on the first flush of emerging weeds that typically grow in the hollows, making later-emerging weeds, in colder conditions, more exposed to insect attack.

Several solutions have been adopted to overcome the problem. The most common and successful technique is to apply clay to the topsoil and physically mix the sand into the clay such that the top 15 cm of soil now contains an average of 3% to 5% clay (Cann 2000). The hydrophilic nature of the clay overcomes the hydrophobic nature of the waxed sand. The technique is called "claying," and it is considered a likely permanent solution. After claying, farmers typically revert back to no-tillage.

In some environments, no-till can reduce the impact of water repellence. The technique requires disc seeders, continuous no-till, full stubble retention, no sheep in the farming system, and no pulse crops in the rotation (Margaret Roper, personal communication). The less tillage, the better, and indeed work by Roper at Munglinup (Western Australia) has shown that such a system creates biopores that assist in soil wetting. It is not clear if this option has broad applicability, though.

Compaction: Soil compaction is a real constraint, although subtle and often unseen. The driving of vehicles across paddocks causes compaction at up to 50-cm depth (Ellington 1986). This compaction restricts root growth. Farming with livestock can also cause surface compaction. Some soils, with shrink-and-swell clay characteristics, can self-heal, while others, like loamy sands, do not and may require deep tillage to ameliorate them (Jarvis 2000). Improved microbial activity, as a result of no-tillage and stubble retention, also helps to soften soils.

A combination of controlled traffic and no-till has been shown to give strong yield improvements and enables soils to soften (Tullberg et al. 1998). The technique has been readily enabled by GPS guided farming machinery with matching implement widths and is becoming increasingly adopted.

Waterlogging and sodicity: These two degradation issues are restricted to small areas of the Australian cropping landscape. No-till enables more diverse crop rotations, which help manage waterlogging. Permanent raised beds (PRBs) are also used with good effect (Bakker et al. 2005). Sodicity is improved with no-till and further improved with the addition of gypsum.

Australia is a harsh and unforgiving countryside, due to some poor soils and erratic weather where drought and floods are common. Australian farmers have to make radical adaptations to their agricultural practices to minimize the extent of soil degradation. The standout and single most successful soil management strategy, which has had almost complete adoption in Western Australia, is no-till. No-till adoption was largely and proudly farmer led, with minimal government support or investment. The Australian farmer groups, the universities, and the Australian and state governments each play a necessary part in monitoring and providing insight into the best practices to help overcome soil degradation and to even improve Australian soils over their natural state.

14.5.3.3 China

Conservation agriculture for soil conservation: China is paying more attention to tilled soils being more susceptible to water and wind erosion. The best solution to control water and soil erosion is to eliminate tillage. Practices that improve water use efficiency and natural resource management by reducing runoff and erosion are of great importance. Therefore, the adoption of CA practices, providing more residue cover and less soil disturbance, has received considerable attention. Since their beginnings in response to issues from the American "dust bowl" era, several decades of development have demonstrated that CA systems are a valuable means of reducing erosion by both water and wind (Uri et al. 1998) because of low soil disturbance and soil surface protection with crop residues.

Conventional tillage (CT) in dryland farming areas of northern China includes moldboard plowing to a depth of about 20 cm, followed by harrowing, hoeing, rolling, and leveling. All the residues in the fields are removed for animals or as fuel before plowing. In some parts of northern China, particularly in the North China Plain, burning crop residue has increased during the last decades.

Long-term moldboard plowing and residue removal/burning have increased the risks of wind and water erosion and the formation of hardpan in the deep soil layer. It has also resulted in poor soil physical and chemical properties, as well as high inputs of energy and labor, which apparently lead to low farmer incomes. To address these

TABLE 14.3
Magnitudes of Soil Sediment Transport in Comparisons of CA and CT

Region	Testing Site	Collection Time	CA (g)	CT (g)	Reduction (%)
Loess Plateau of China	Yanggao, Shanxi	March 25, 2004– April 3, 2004	8.4	15.1	44.7
Northeast ridge tillage areas	Lingyuan, Liaoning	March 25, 2004– April 3, 2004	16.3	10.2	37.3
North China Plain	Fengning, Hebei	March 22, 2002– April 21, 2002	12.7	42.5	70.0
	Zhangbei, Hebei	April 8, 2002– May 8, 2002	12.7	42.5	70.0
	Changping, Beijing	March 28, 2005– April 17, 2005	16.7	19.0	12.1
	Yanqing, Beijing	March 16, 2005– March 20, 2005	4.2	5.0	17.0
Farming— pastoral areas	Chifeng, Inner Mongolia	April 22, 2003– May 3, 2003	4.7	7.1	34.2
	Zhenglanqi, Inner Mongolia	March 23, 2003– April 27, 2003	11.3	25.0	54.8
	Wuchuan, Inner Mongolia	March 26, 2003– April 6, 2003	2.9	7.4	61.6
Northwest China	Hetian, Xinjiang	March 16, 2004– April 27, 2004	7.4	105.5	92.9

problems, various kinds of CA practices have been developed in northern China, such as no-till, controlled traffic, and PRBs, leading to a range of beneficial effects on soil quality, including increase in SOM, decrease in bulk density and improvement in soil structure, higher infiltration rate, greater soil moisture holding capacity, and reduced runoff and erosion (Bai et al. 2008; Gao et al. 2008; Chen et al. 2008).

Effect of CA on wind erosion: Research measuring springtime wind erosion losses in the Yanggao region of the Loess Plateau has shown that CA treatments reduced topsoil loss by 44.7% compared to CT (Table 14.3). At nine other sites across northern and north central China, from the dry, windy conditions of the far west to the relatively temperate plains in the Beijing area, CA treatments consistently reduced springtime wind erosion losses from 12% to 93% depending upon the measurement duration, ambient conditions, and erosive winds (Table 14.3).

14.6 INTEGRATING SUSTAINABLE SOIL MANAGEMENT PRINCIPLES INTO FARMING SYSTEMS

Sustainable soil management and crop production principles of CA can be integrated into most if not all types of production or farming systems. This is because they provide the ecological underpinnings to production and farming systems to generate greater productivity and environmental benefits. Below are some examples.

Organic agriculture based on CA can lead to greater soil health and productivity, increased efficiency of use of organic matter, and reduction in use of energy. Organic CA farming is already being practiced on a smaller scale in the United States, Brazil, and Germany, as well as by subsistence CA farmers in Africa and elsewhere. Tillage-based organic farming is often characterized more by what practices it excludes from its production systems than by what it actually does to harness sustainable production intensification and ecosystem services. Introducing CA principles into organic farming would reduce soil disturbance, improve weed control with mulch cover and crop diversification, and generate greater amounts of organic matter from in situ sources within a more diversified cropping system involving legumes (Altieri et al. 2011).

Agroforestry systems involve the cultivation of woody perennials and annual crops together in a sustainable manner and are increasingly practiced in degraded areas with perennial legumes (Saha et al. 2010). CA works well with trees and shrubs and within agroforestry and related systems. In fact, several tree crop systems in the developing and developed regions already practice some form of CA, but these systems can be further enhanced with improved crop associations including legumes and integration with livestock. Alley cropping has been one innovation in this area that is beginning to offer productivity, economic, and environmental benefits to producers (Sims et al. 2009).

CA with trees has now become an important option for many farming situations, particularly in the tropics. These CA systems incorporate varying densities of fertilizer trees in order to enhance biological nitrogen fixation, increase biomass production for surface residue, and conserve moisture. They have become the basis for major scaling-up programs with hundreds of thousands of farmers in Zambia, Malawi, Niger, and Burkina Faso (Garrity et al. 2010; Garrity 2011). The incorporation of the indigenous acacia species *Faidherbia albida* into maize-based conservation agriculture in Zambia on a large scale is a noteworthy example. These programs have demonstrated the practical opportunities for combining fertilizer trees with CA in both small-scale and commercial-scale farming systems.

Shifting agriculture, also referred to as "swidden" or "slash and burn," entails the clearing of land to prepare a cultivation plot and subsequently returning this to regrowth and eventual natural reforestation, during which damaged soil structure and depleted "indigenous" plant nutrients are restored. Shifting cultivation has acquired a negative connotation, particularly because of the burning of vegetation. However, for sustainable intensification, such systems can be adapted to follow CA principles, changing from slash-and-burn systems into *slash-and-mulch* systems with diversified cropping (including legumes and perennial crops) that reduce the need for extra land clearing.

The *System of Rice Intensification* (SRI) has taken root on an international scale in more than 40 countries across all developing regions, including China, India, Indonesia, and Vietnam, moving beyond its origins in Madagascar (De Laulanié 1993). Trained farmers have shown SRI to offer higher income and productivities (use efficiencies) of inputs of labor, nutrients, and water, and to require less seeds, water, energy, fertilizer, and labor compared with conventional irrigated or rainfed flooded rice production systems. SRI advantages have been shown to apply to traditional as well as modern cultivars. As with crops in CA systems, SRI phenotypes are widely reported by farmers to be less susceptible to pest and disease damage. The SRI production concept has been defined on the basis of a set of practices (i.e., seedlings 10 days of age for transplanting, or direct seeding; single plant; wide spacing; mainly moist, not saturated and flooded, soil water regimes; regular weeding to also facilitate soil aeration; and liberal use of organic fertilizers) (Uphoff et al. 2011; Kassam et al. 2011c; Uphoff and Kassam 2011). An SRI system based on CA principles is being practiced on permanent nontilled raised beds as well as in unpuddled paddies in Asian countries, thus eliminating puddling and the soil-disturbing ways of weeding (Sharif 2011). The wheat–rice cropping system in the Indo-Gangetic Plains involves the production of no-till wheat over some 3 million ha with residues from the previous rice crop providing soil cover. It would now seem appropriate to introduce no-till SRI rice in the wheat–rice cropping system and manage the cropping system based on the CA principles.

14.6.1 CROP MANAGEMENT PRACTICES AND SUSTAINABLE SOIL MANAGEMENT

Standard agronomic crop management practices comprise crop and cultivar choice, crop establishment and yield response to water, crop genetic improvement, pest management, fertilizer and nutrient management, and crop rotation and intensification. Individual crop management practices that form a constituent part of good integrated production systems are often interrelated. The interactions among practices can work synergistically to produce outcomes in terms of productivity via improvements in conditions of the soil as a rooting medium, enabling the better expression of plants' genetic and epigenetic potentials. For example, for a given amount of rainfall, soil moisture availability to plants depends on how the soil surface, SOM, soil structure, and plant root systems are managed. Also, high water productivities under good soil moisture supply are possible only when plant nutrition is adequate. Similarly, no amount of fertilizer application and use of modern varieties will improve water use efficiency and water productivity if the soil has a hardpan in the rooting zone or if the soil has little organic matter to build and maintain good soil structure and porosity for maximum moisture storage and root growth. Equally, without the maintenance of good water infiltration and without soil cover to minimize evaporation from the soil surface, it is not possible to fully optimize and maximize water use and water productivity. Another example is the above-described SRI system: the interrelation of the soil characteristics, providing an optimal rooting environment, allowing different plant spacings, which can lead to different phenotypic plant development as compared to conventional practices.

Thus, agricultural soils maintained in good health and quality will offer the possibility of making optimum soil moisture and nutrients available for crop production over the period of the crops' development and of optimal input use efficiencies through good agronomic manipulation or good crop management. However, good crop management is not an independent variable but a function of how sustainably the production system as a whole is being managed in order to maintain or intensify production while harnessing the desired ecosystem services.

14.6.2 Sustainable Soil Management with Intercropping as an Alternative in Permanent No-Till Systems

In tropical regions, the high rate of organic material decomposition associated with warm and wet climate conditions is a challenge to meeting the prerequisite of permanent soil cover required by CA. Most of the straw input, even when maintained on the soil surface, is decomposed in 20 to 60 days according to the C/N ratio, N content, and lignin content of plant material. This fact results in bare soil and risk of soil erosion and degradation. Also, the weed infestation, depletion of SOM, nutrient leaching, and soil compaction are processes associated with bare soils in the tropics. The decrease in soil productivity as a consequence of deterioration in soil quality is a threat to permanent no-till in the tropics. In order to overcome this situation, the farmers try to increase the amount of crop residue input and select pearl millet as a grass-type cover crop in order to maintain soil cover for a longer period.

The use of perennial forage plants, such as *Brachiaria*, intercropped with grain crops is a promising alternative to providing greater soil sustainability in no-till systems in tropical Brazil. The large-scale success of *Brachiaria* in strengthening soil and production sustainability in Brazil provides a specific example of why participating crops in no-till cropping systems are important to both sustainable soil management as well as sustainable production. There are many species of *Brachiaria* that were introduced from Africa into Central Brazil in the early 1960s, the most common being *Brachiaria brizantha*, *B. decumbens*, and *B. ruziziensis* (Landers 2007). The best *Brachiaria* intercrop alternative with corn has been investigated with N fertilization. The straw of *Brachiaria* in combination with corn stalks can input as much as 17 tons of dry mass per hectare and provide soil cover for more than 100 days. *Brachiaria* pastures on cerrado soils can last up to 5 years and can raise average livestock carrying capacity from 0.3 to 1.0 AU/ha (Machedo 1997). It has been estimated that some 85% of pastures in the cerrado are *Brachiaria* (Landers 2007).

Brachiaria has a deep, well-developed root system that can penetrate depths of more than 1 m, with at least 20% of the total root system present below 0.30 m. Intercropping increases soil aggregation and stability of aggregates, lessens bulk density, and increases macroporosity and water infiltration.

The total dry biomass of a *Brachiaria* root system can reach 1.7 t ha^{-1}. This fact is important for cycling nutrients like potassium, magnesium, sulfur, and nitrogen that are subject to leaching in tropical agriculture soils. The *Brachiaria* mulch decreases the soil temperature, keeping the soil environment cool and wet, thus increasing soil biological activity. Therefore, the intercrop system is very efficient in nutrient cycling and reducing nutrient losses by runoff and leaching.

This intercrop system can sequester soil carbon in the range of 0.5 to 1.0 t C ha^{-1} year^{-1}. These rates are double those for regular no-till carbon sequestration with systems that have only a tillage change with grain crops held constant. The soil loss then reduced to the range of 0 to 3 t ha^{-1} year^{-1}, which is around three times lower than other no-till grain systems, and it is in equilibrium with soil formation.

The large amount of aboveground *Brachiaria* biomass is important to reduce weed infestation, especially with *Conyza bonariensis, Commelina benghalensis, Euphorbia heterophyla*, and *Cenchurs echinatus*. Weed infestations are one of the most serious threats to continuous no-till in the tropics. The total weed reduction provided by *Brachiaria* generally is in the range of 30% to 70%.

Brachiaria can suppress important diseases of soybean and black beans such as *Fusarium solani* infestation by approximately 60%. Also, *Rhizoctonia solani* can be reduced by *Brachiaria* intercropped with grain crop production. One of the most common pathogens in the South American tropics is *Sclerotinia sclerotiorum*, and the combination of a grain crop with *Brachiaria* is one of the best options to reduce this disease. Intercropped *Brachiaria* and maize provides competitive maize yields and forage to cattle during an otherwise fallow period, providing income diversification. This system is an important option to sustain no-till for the long term in tropical environments. It has restored degraded pastureland and degraded forestland in Central Brazil.

14.6.3 Crop–Livestock Integration for Sustainable Soil Management

Pastureland has important ecological functions. It often contains a high percentage of perennial grasses, which have the ability to sequester and safely store high amounts of carbon in the soil at rates that exceed by far those of annual crops. This capacity can be enhanced with appropriate management, for example, replacing exported nutrients, maintaining diversity in plant species, and allowing for sufficient recovery periods between use by grazing or cutting. In conventional farming systems, there is a clear distinction between arable crops and, mostly permanent, pastureland. Under CA-based farming, this distinction does not exist anymore, since annual crops may rotate into pasture and vice versa without the destructive intervention of soil tillage; it is just as additional element of cropping diversity.

Integrated crop–livestock systems including trees and pasture have long been a foundation of agriculture. In recent decades, there have been practical innovations that harness synergies between the production sectors of crops, livestock, and agroforestry that ensure economic and ecological sustainability while providing a flow of valued ecosystem services. System integration increases environmental and livelihood resilience through increased biological diversity, effective/efficient nutrient cycling/recycling, improved soil health, and enhanced forest preservation and contributes to adaptation and mitigation of climate change. The integration of production sectors can enhance livelihood diversification and efficiency through optimization of production inputs including labor, offer resilience to economic stresses, and reduce risks (Landers 2007; FAO 2010).

Integration can be on-farm as well as on an area-wide/catchment (three-dimensional) basis. Successful crop–livestock integration should be seen through the lens of nutrient use efficiency and nutrient cycling benefits, of ecosystem health advantages, and of positive biosecurity outcomes, all of which are strong public goods. Successful integration can also halt and reverse land degradation. In many fragile ecosystems, livestock is the mainstay of livelihoods, but at the same time, uncontrolled grazing can lead to land degradation. Under such cases, the issue of mutually beneficial area integration between the primary and secondary production sectors must be addressed at the community and regional levels. Issues to be addressed include dynamic grazing and functional biomass management, species composition for feed quality and ecosystem services, and matching stocking rate to carrying capacity in the context of the prevailing climatic and landscape variability in space and time. In extensive rangeland systems, greater precision in matching stocking rate with feed availability and the exposure time to the recovery requirements of vegetation is possible with satellite-guided overhead remote-sensing systems (FAO 2010).

14.6.4 Farm Power and Mechanization for Sustainable Soil Management

One of the most important yet commonly overlooked inputs in agricultural production systems is farm power. Lack of sufficient farm power in many countries is a bottleneck to increasing and intensifying production, especially where it depends on manual or animal traction power.

Farmers working manually on average can feed only three other persons. With animal traction, one farmer can feed six other persons, and with a tractor, the number further increases to 50 or more persons (Legg et al. 1993). Labor output levels vary widely according to the mechanization level and climatic conditions, and there is a clear correlation between the production levels and the farm power input (Giles 1975; Wieneke and Friedrich 1988), but they also depend on the kind of farming system used (Zweier 1985; Doets et al. 2000). At each of these levels, the energy for the respective farm power needs to be supplied, either through human food, animal feed, or tractor fuel, which could also be biofuel. Bearing in mind the pressure to produce more food for an increased population, increasingly concentrated in urban centers (already now about 50% of the population no longer lives in rural areas), the need for increased mechanization of agricultural crop production becomes obvious (Mrema 1996). It is worth noting that suitable mechanization options can lead to improved energy efficiency in crop production, leading to better sustainability, higher productive capacity, and lower environmental damage (Baig and Gamache 2009; Lindwall and Sonntag 2010).

Suitable CA mechanical technologies are commercially available for all technology levels, from the small farmer using exclusively manual power to the large-scale mechanized farmer applying precision farming with satellite guidance. However, small-scale hand and animal traction tools and equipment for CA so far are easily accessible only to farmers in Southern Brazil and Paraguay, while single-axle tractors with CA attachments can be found on the market only in Bangladesh and Brazil. The actual challenge is to improve the accessibility and commercial availability of such tools and equipment for the smallholder farmer in Africa and Asia, as well as in parts of Latin America. In several developing and middle-income countries in Africa and Asia, small workshops and manufacturers are now starting to produce manual and animal traction no-till planters as well as tractor-drawn direct seeding equipment (Friedrich and Kassam 2011; Sims et al. 2011).

Modern technologies do allow a much more efficient use of energy and other production inputs, and they have also been instrumental for allowing ecologically oriented crop production concepts, such as CA, to develop. A crucial input into the development and increased adoption of CA is direct seeding technology, which enables the establishment of crops in undisturbed soils. These modern mechanized technologies have contributed to the success and area spread of CA, which also facilitates the improved delivery of ecosystem services and allows the development toward sustainable agriculture through the reduction of waste and an increased input efficiency (Baker et al. 2007). Yet, in addition, agricultural mechanization can also directly—with more precise application equipment for agricultural inputs and the additional use of precision farming tools—contribute to a reduction in input use. GIS technologies further allow control of traffic of agricultural machinery, so as to minimize areas of soil compaction and, with this, facilitate the development of a functioning soil ecosystem, increasing at the same time the energy efficiency of crop production systems (Tullberg 2007; Wang et al. 2009).

14.7 LARGE-SCALE LANDSCAPE-LEVEL BENEFITS FROM SUSTAINABLE SOIL MANAGEMENT

Benefits from sustainable production systems are scale independent. They do occur at the field-point scale, but benefits accrue to landscapes, farms, communities, and regions. The four major sets of benefits from sustainable soil management and production systems are as follows:

1. Higher stable production output, productivity, and profitability
2. Adaptation to climate change and reduced vulnerability
3. Enhanced ecosystem functioning and services
4. Reduced greenhouse gas (GHG) emissions and "carbon footprint" of agriculture

Soil Management Is More Than What and How Crops Are Grown 377

All these are of direct benefit to producers and society as a whole. The relevant socioeconomic indicators include the following: farm profit, factor productivity (efficiency), amount of pesticides applied per unit of output, yield per unit area and per farmer practicing sustainable intensified systems, and stability of production. The relevant ecosystem service indicators include the following: clean water provisioning from catchment areas under an intensive agriculture area; reduced erosion, both wind and water (Mello and Raij 2006; Laurent et al. 2011); increased biodiversity/wildlife within agricultural landscapes; and increase in carbon sequestration and reduction in carbon footprint and GHG emissions of methane and nitrous oxide (Baig and Gamache 2009; Kassam et al. 2011c; FAO 2012).

It is important to identify key indicators that would detect changes in the desired direction at the field, farm, and physiographic landscape level within whose boundary the farm is located and whose management has an impact on the aggregate behavior of the landscape as a whole. CA-based ecosystem services operate in different parts of the world and include the following: the agricultural carbon offset scheme in Alberta, Canada; the hydrological services from the Paraná III Basin in Brazil; the control of soil erosion in Andalusia, Spain; the controlling of water erosion and dust storms and combating of desertification and drought in the Loess Plateau of the Yellow River basin in China; and reducing susceptibility/increasing resilience to land degradation in Western Australia. Controlling land degradation, particularly soil erosion, caused by tillage, exposed soils, and depletions of SOM, has been a main objective of most of such initiatives. Such landscape schemes are possible only when the landscape has a contiguous network of sustainable soil management that mediates such large-scale environmental and economic benefits to the producers and rural as well as urban society. With sustainable soil management practices being applied over large areas, it is then possible to overlay landscape-level development programs to harness large-scale ecosystem services such as carbon sequestration in Alberta, Canada; water-related services in Paraná Basin, Brazil; or erosion control in Andalusia, Spain.

14.7.1 CANADA: CARBON OFFSET SCHEME IN ALBERTA

The province of Alberta has operated a GHG offset system since 2007 that allows regulated companies to offset their emissions by purchasing verified tonnes from a range of approved sources including agriculture projects (Haugen-Kozyra and Goddard 2009). This compliance system for large emitters has provided a rich venue for learning on behalf of all players—the regulated companies, government, scientists, consultants, aggregator companies, and farmers. Climate change legislation was amended in 2007 to require regulated companies to reduce their emissions to a set target below their 2003 to 2005 baseline. If they could not achieve their target in any year, they could settle their accounts with any of three options: pay into a research fund at a fixed rate of C$15 per tonne CO_2e; trade emission performance credits if they were generated by any company reducing emissions beyond their target; or purchase verified offsets generated within Alberta using Alberta government-approved protocols. The latter option triggered interest and activities in developing protocols across all industrial sectors including agriculture. Offset tonnes trade at a discount to the C$15 fund payment option in order to cover the aggregation and transaction costs.

The Alberta government provides the enabling legislation and regulations. They also provide oversight of protocol development and approvals. Beyond that, the private sector invests in development of protocols, aggregation of offsets and assembly of projects, third-party verification of projects, and the bilateral sales to the regulated emitters. A nongovernment agency, Climate Change Central, also plays a role of facilitator and is the designated operator of the registry of the offsets. All verified tonnes are serialized and tracked by the registry through to the retirement (used for a compliance year) of a particular tonne. The regulator/government ministry holds annual review meetings with the players in the market to review performance, new developments, regulatory changes, and guidance. The amount of offsets used by companies for compliance has been relatively consistent at about 36% of the total annual accounts (CCC 2011). Agricultural offsets have contributed about 36% to 40% of all offsets. The most popular protocol has been the Tillage System protocol, which acknowledges the soil carbon sequestration through implementation of no-till practices. The Tillage System protocol has contributed over 8 M t of offsets worth C$100 million over the last 5 years of the offset system.

The offset system has had many cobenefits beyond reducing GHG emissions and reducing the C footprint of industries. Scientists come together in helping to develop protocols and share a systems view of the production system under review. Science and policy come together and integrate to form protocols and develop a market. The private sector of aggregator and verification companies have integrated efforts and developed streamlined systems to bring offsets to market efficiently. Farmers have developed improved production and record systems. Very often, the financial benefits to the farmer by adopting a protocol far exceed any offset payment for the GHG savings portion. All players are now further along the capacity curve to be in a better position to see and take advantage of other ecosystem goods and services opportunities.

14.7.2 BRAZIL: WATERSHED SERVICES IN THE PARANÁ BASIN

As part of a strategy for improvement, conservation, and sustainable use of natural resources, the Itaipú Dam *Programa Cultivando Água Boa* ("cultivating good water") has established a partnership with farmers to achieve their goals in the Paraná III Basin located in the western part of Paraná State on the Paraguay border (ITAIPU 2011; Mello and van Raij 2006). The dam's reservoir depends on the sustainable use and management of soil and water in the watershed/catchment for efficient electricity generation. Sediments and nutrients entering the reservoir resulting from inappropriate land use pollute the water used by the turbines to generate electricity. This phenomenon shortens the reservoir's useful life and increases the maintenance costs of power-generating turbines, increasing therewith electricity generation costs. Thus, in principle, payments could be made through a program to improve the conditions of electricity generation. The spatial unit covered by this program is the whole watershed/catchment. Functioning as a community, joining many farmers in the watershed, they reach a scale where environmental impact can be monitored with suitable indicators to establish a system of payment for environmental services.

One of the partnerships established in the *Cultivando Água Boa* program and developed through an agreement with the Brazilian No-Till Federation (FEBRAPDP) is the Participatory Methodology for Conservation Agriculture Quality Assessment (Laurent et al. 2011), based upon former positive experiences with catchment development in Brazil. The first phase in the program is that the partners plan to measure the impacts of farm management through a scoring system indicating how much each farm is contributing to the improvements of the water conditions. (The system is available online in Portuguese at http://plantio.hidroinformatica.org/.) In this regard, a scoring index model for rating the quality of no-till systems has been devised. The model relies on expert knowledge and is being applied to identify soil erosion and land degradation risks arising from any weakness in the adopted CA practices, and possible action needed to address the weakness (Roloff et al. 2011). Consolidating this phase and adapting the principles established for the "water producer" by the National Water Agency, the partners will assign values to ecosystem services generated from farms participating in the program (ANA 2011). Considering the polluter/payer and provider/receiver principles set in the Brazilian Water Resources Policy, farmers with good scores will be paid for their proactive action to deliver watershed services once the Paraná Watershed Plan is established. This will be a new framework for services provided by farmers as compensation for their proactive approach to improve the reservoir water quality and reduce costs for electricity generation by the Itaipú Dam.

14.7.3 SPAIN: SOIL CONSERVATION IN OLIVE GROVES

Olive orchards are an important agroecosystem in the Mediterranean. Andalusia, the southernmost region of Spain, is the main olive cultivation area in the world as it produces a third of the world's olive oil, and around 1.5 Mha or 17% of the surface area is covered with olive groves (Gomez et al. 2009a), which account for 60% of the Spanish olive growing area. Historically, olive cropping has been concentrated on hilly lands, where soil erosion happens to be a very severe and widespread problem. Locally, historical soil loss rates have been reported to reach up to 184 Mg ha^{-1} year^{-1} (Vanwalleghem et al. 2010). Erratic but high-intensity rainfall especially during winter, but also the management of the orchards, lies at the origin of soil erosion. Commercial olive orchards, mainly grown under rainfed conditions, are characterized by extremely adverse management conditions as farmers tend to till intensively to avoid competition of weeds with tree water and nutrient uptake. Therefore, simple conservation strategies, such as no-till with natural vegetation or the establishment of cover crops, are not easily adopted by farmers. Conventional tillage has been the dominant management system in olive orchards over the last decades. The combination of this human-induced low vegetation cover with the steep slope gradients on which these orchards are located, together with the high-intensity rainfall events that characterize the Mediterranean climate, explains why high soil erosion rates have been associated with olive oil production (Beaufoy 2001).

Despite these alarming erosion rates that have been reported in olive groves on sloping and mountainous land, there are authors questioning the severity and extent of water erosion in olive orchards in southern Spain (Fleskens and Stroosnijder 2007). Other authors, however, insist on soil erosion being a widespread threat to the sustainable land use through olive production (Gomez et al. 2008; Vanwalleghem et al. 2010, 2011). Moreover, land use change and the abandonment of the terraced slopes, functioning as anthropic hydrological infrastructures, which protected the soil and preserved the natural vegetation in the recent past, have been progressively collapsing, mainly due to the rapid removal of the soil, causing important land degradation problems (Dunjó et al. 2003).

Despite the gradual introduction of no-till as the soil management system in olive groves, a first agrienvironmental measure scheme was introduced in Andalucía in the late 1990s, aiming to fight soil erosion in olive orchards mainly by vegetation cover between trees and natural vegetation on the land borders. Other soil erosion control practices were also promoted such as soil tillage along contour lines and the maintenance of pruning residues in the interrow space (Franco and Calatrava 2006). The adoption especially of no-till increased tremendously from 1995 onward and covers today, depending on the study region, between 50% and 95% of the area under olive production (Franco and Calatrava 2006; Leyva et al. 2007; Martinez 2009).

Despite this notable progress in terms of adoption of soil conservation measures in the case of perennial crop production in Spain (Table 14.4), there are still regions where the adoption of soil conservation practices is very low, and, in general, there is much room left for the extension of policy measures to mitigate and invert soil degradation (Calatrava et al. 2011). In addition, the findings of Gomez et al. (2009b) that

TABLE 14.4
Evolution of the Area Under Cover Crop Soil Management Systems in Total Woody Crops and Olives in Spain

	2011	%	2010	%	2009	%	2006	%
Total woody crops (ha)	4.932.002	100	4.986.046	100	5.043.896	100	5.039.440	100
With cover crops (ha)	1.178.297	23.9	1.218.726	24.4	1.066.182	21.1	836.731	16.6
With no-till (bare soil) (ha)	453.219	92	443.309	8.9	431.472	8.6	347.449	6.9
Olives, total area (ha)	2.580.577	100	2.572.793	100	2.568.383	100	2.476.540	100
Olives with cover crops (ha)	680.510	26.4	683.363	26.6	627.1668	24.4	438.828	17.7
Olives no-till (bare soil) (ha)	341.674	13.2	328.716	12.8	299.711	11.7	225.998	9.1

Source: Encuesta sobre Superficies y Rendimientos Cultivos (ESYRCE) 2006, 2009, 2010, 2011.

bare soil, though untilled, is capable of providing more runoff and sediment yield in olive groves under certain conditions should be seriously taken into account while designing conservation strategies. These have to be driven by their real delivery of ecosystem services and not just by cost-effective minimal conservation approaches. The faster adoption of cover crops compared with no-till as a soil conservation measure in perennial woody crops, and especially in olives (Table 14.4), can therefore be considered an important step toward the mitigation of soil erosion and degradation.

14.8 POLICY, INSTITUTIONAL, TECHNOLOGY, AND KNOWLEDGE IMPLICATIONS

An enabling policy and institutional environment is needed to promote sustainable soil management for agriculture development, which in practice entails a change in process in which interested stakeholders become engaged to produce, in nondestructive ways, based on available and affordable resources, agricultural products desired by the producers, individual groups, and society. However, it is necessary to implement an enabling environment to promote farmers' interest in undertaking sustainable soil management and production intensification and maintenance of ecosystem services. For this, given the necessary understanding, the requirements include effective and integrated development planning and policies backed up by relevant research and advisory/extension systems, and the mobilization of concerned stakeholders in all sectors.

14.8.1 POLICY AND INSTITUTIONAL SUPPORT

Principles of sustainable soil management for agriculture production based on an ecosystem approach form the basis for good agricultural land use and management. It indicates the urgent need for a significant change in "mindset" concerning care of the soil and landscape, after the realization that erosion of soil (deemed a major and continuing problem) is a *consequence* rather than a prime cause of land degradation, in as much as loss of stable soil aggregates and their counterpart spaces in the soil precedes the accumulation of runoff. This understanding has major implications for how best to encourage and achieve sustainability of productive land uses. It indicates the need to respect and make best and careful use of agroecosystem processes, rather than try to usurp their functions by use of technologies that prove to be inimical to soil life and therefore not suitable or ecologically sustainable.

Policy coherence and cohesion are critical as all governments already have a number of institutions involved in caring for the development of their natural resources. However, the fragmented nature of their organizational arrangement across several ministries (e.g., Agriculture, Forestry, National Parks, Energy, Water), the disconnection from production sectors, and nonworkable relationships within a government, often inhibit their full effectiveness.

At national and state levels, the adoption of CA policies is often congruent and supportive of other policies related to the environment, natural resources, energy efficiencies, and more recently, climate change. Policy makers need to both align and document the support, compatibilities, and synergies that may arise from suggested

CA policies. Since CA is a systems approach, policy impacts are numerous and interrelated.

At an agriculture sector level, CA is compatible with robust policies for innovation, technologies, diversity, resource conservation, enterprise risk management, and community development. Policy research and analysis is needed to identify sector policies and institutions that publically fund policies that are counter to CA adoption, societal values, and government directions. For example, some governments have historically had fuel subsidies for farmers to reduce costs (and lower income risks). With increasing fuel costs, the burden on society is projected to increase. CA realizes fossil fuel savings, so the argument for a subsidy diminishes. Crop insurance programs are another example where historic policies favor conventional cultivation systems over CA.

The private sector is another rapidly emerging champion of CA systems. Large retailers have adopted sustainability policies and are starting to require simple certification or proof of production practices. Practices favored are often components of CA systems, or conversely, full CA is the optimization of the desired production characteristics. The early work in life cycle analysis focused upon carbon, GHGs, or energy. More recently, work has moved toward more comprehensive or encompassing approaches such as environmental footprints. Again, CA profiles more favorably than conventional production systems. Financial institutions are other players in the private sector that are increasingly looking at production practices of clients, including agriculture, from an environmental risk perspective and innovation in market opportunities. CA receives high marks. The private sector is unique in that it can formulate and implement policies much more quickly than governments. The private sector has leading players in CA policy that governments need to pay attention to.

Thus, it is necessary to ensure that all relevant institutions in both private and public sectors and at all scales (international to local) have a clear awareness of the basic agroecological and socioeconomic principles upon which sustainable land use is based, and of the ways in which each institution's particular interests and responsibilities may be able to support and embody the CA principles. This commonality of underlying concern with the care of land, underpinning policy cohesion, will facilitate the needed interdisciplinary collaborations to be undertaken with farmers and other land users, and the alignment and linkages of new progressive policies.

Agricultural development policy can and should therefore have a clear commitment to sustainable soil management and production intensification. Best sustainable systems cannot be devised based on high-soil-disturbance agriculture. Where agriculture development is maintained by tillage systems, it will generally not be possible to maintain production intensification as well as to continue to deliver ecosystem services because of suppression of soil biotic capacities for self-repeating soil structure regeneration. Hence, all agricultural development activities dealing with crop production intensification should be assessed for their compatibility with dynamic ecosystem functions and their desired services. Any environmental management schemes in agriculture, including certification protocols and payments for environmental services that do not promote the emulation of CA principles and practices as a basis for sustainable soil management, are unlikely to be economically and environmentally sustainable in the long run. This does not mean that non-CA alternatives based on tillage agriculture cannot be considered in new developments, but when they are being planned for deployment, the results in terms of output, productivity, and ecosystem services will generally not match those from agroecological low-disturbance systems in terms of sustainability, and the decision makers and policy planners must be made aware of this.

Regardless of which institution is developing or revising policies, the policies need to be adaptable to changing societies, changing markets, and developing farming practices. This is a case for creating adaptive policies (Swanson and Bhadwal 2010). Analysis must be integrative and forward looking, not a reinvention of the past. Collective and collaborative discussions are needed to ensure that concepts and understanding are consistent, and that all points of view result in common values and agreement on direction. Ultimately, adaptive policies have automatic adjustments that arise because the system is well understood, and policies adjust when anticipated conditions arise. Such is not the case with CA, and a more conservative approach of formal policy review and continuous learning should be favored. A key component of adaptive policy is to enable self-organization and social networking. Successes in CA are associated with these developments, and further support is needed for CA organizations. Finally, because CA is complex, an integrated promotion of variation in policy should be analyzed. If a variety of policies are directed at an issue from different directions or sources, and if one fails, then the others may succeed.

14.8.2 Technology and Knowledge Support

Current crop production systems vary widely. There are many production systems that take a predominantly ecosystem approach and are not only productive but also more sustainable in terms of environmental impacts (FAO 2011b; Pisante et al. 2012). Such sustainable production systems, when fully developed, are based on sustainable soil management. They are, by definition, management and knowledge intensive and relatively complex systems to learn and implement as they must work with nature and integrate as much as possible with the natural ecosystem processes into the design and management of the production systems. This is a continuing task with many possible permutations for farmers to choose from so as to suit their local production circumstances and constraints. They cannot be reduced to a simple standard technology, and thus, pioneers and early adopters face many hurdles before the full benefits of such systems can be realized. Indeed, the upscaling of no-tillage systems to achieve national impact requires a dynamic complement of enabling policies and institutional support to producers and supply-chain service providers.

One bottleneck is often insufficient knowledge about the new soil management and production system. Site-specific research is needed to assist farmers in responding to no-till soil management and production system changes, such as in nutrient requirements and pest, disease, and weed problems, as well as for options of green manure cover crops to be incorporated into the crop rotations.

Farmers are not alone in the need for education. Across the countryside, farm consultants and input suppliers also need to learn about and understand CA systems. They are important partners and contributors to local clubs, farmer associations, and CA conferences. If field staff of consulting and retail companies understand

CA systems and see both what is needed at the field level for adoption and how their corporate culture and policies are supportive or not, they can then serve both sides by becoming catalysts for policy changes at the corporate level. Corporate executives will appreciate advice from their own staff to compare against what they are hearing from farmers and farm associations. Farm organizations will appreciate informed dialogues with consultants who are at the forefront of knowledge supporting CA. Consulting and farm input retailers can develop win–win situations with their clients and companies.

Academic institutions (universities, colleges) and large research agencies funded by governments or commodity commissions also need to catch up to CA through their policies and programs. Universities focus on training in reductionist science and place little effort on integrative science. CA is an integrative discipline and, thus, likely unfamiliar territory to those developing curricula and lecturing in the agriculture sciences. Similarly, large research institutions have inertia that is difficult to alter. They may see CA as only a deviation from conventional, intensive production systems and study only components rather than the system. Indeed, in western Canada, innovative agronomic scientists with the federal agriculture department pushed for agronomic treatments to be imposed on no-till rather than conventional-till plots/fields. Only in the last decade has all agronomic research at all prairie federal research stations been carried out on no-till land. Sadly, plant breeders in many places (Canada and elsewhere) still conduct trials and selections on tilled plots. One speculates as to what traits are being selected for those that favor CA cropping. Disease or pest issues in residues have been cited as concerns as has a lack of plot-sized equipment.

A particular bottleneck for wide adoption of CA is the availability of suitable equipment. While small-scale CA can be undertaken without special tools by just using a narrow hoe or planting stick, the full benefits of labor saving and work precision can be achieved only using special tools or equipment. These tools all exist at manual, animal traction, and tractor power mechanization levels, yet their local availability for the farmers in most parts of the world is a real challenge. Even where this equipment, such as no-till planters, is on the market, it is often more expensive than corresponding conventional equipment and constitutes a considerable initial investment for the farmer. These bottlenecks can be overcome, for example, by facilitating input supply chains and local manufacturing of the equipment, where feasible, and by offering contractor services or sharing equipment among farmers in a group to reduce the cost for a single farmer. In most small-farm scenarios, even animal traction no-till planters have a working capacity that would exceed the requirements of a single farmer.

Knowledge, information, and technology are increasingly generated, diffused, and applied through the private sector. Exponential growth in information and communications technology (ICT), especially the Internet, has transformed the ability to take advantage of knowledge developed in other places or for other purposes. The knowledge structure of the agricultural sector in many countries is changing markedly (OECD 2011), incorporating a greater awareness in education, research, and development of the need for ecological sustainability of agricultural production systems and landscape management.

14.9 CONCLUDING COMMENTS

Essentially, we have two farming paradigms operating, and both aspire to manage the soil and landscape sustainably. The two paradigms are as follows: (1) The tillage-based farming systems, including intensive tillage with inversion plowing during the last century, aim at modifying soil structure to create a clean seedbed for planting seeds and to bury weeds or incorporate residues. This is the *interventionist paradigm* in which most aspects of crop production are controlled by technological interventions such as soil tilling; genetically engineered varieties; protective or curative pest, pathogen, and weed control with agrochemicals; and the application of mineral fertilizers for plant nutrition. This is still the predominant cropping system around the world. (2) With the development of no-tillage technologies from the 1940s onward, and the discovery of specific farming systems since the 1970s, many farmers have taken a predominantly ecosystem approach and are productive and ecologically sustainable. This is the *agroecological paradigm* characterized by minimal disturbance of the soil and the natural environment; the use of traditional or modern adapted varieties; plant nutrition from organic and nonorganic sources including biological nitrogen fixation, feeding first of all the soil from which crops derive a balanced nutrition; and the use of both natural and managed biodiversity to produce food, raw materials, and other ecosystem services. Crop production based on an ecosystem or agroecological approach can sustain the health of farmland already in use and can regenerate land left in poor condition by past misuse.

The core agroecological elements of sustainable soil management, and of production intensification based on it, are the practices that implement the three principles—minimum mechanical soil disturbance, permanent organic soil cover, and species diversification—plus other best practices dealing with crop management, including integrated pest control, plant nutrition, water management, and so forth, as well as the integration of pastures, trees, and livestock into the production system, supported by adequate and appropriate farm power and equipment. This concept and associated practical implications must be placed at the center of any effort to intensify production at any farm scale.

With suitable forms of feeding, protection, and management, the living components of the plant/soil ecosystems integrate and energize the other key components of agricultural production systems—chemical, physical, hydrological—effectively almost free of charge. Through its capacity to reproduce itself, the soil biota sustains the land's potentials and their outcomes. Damaging these capacities within agricultural systems of land use, through poor husbandry of these resources, should be avoided since it reduces the resilience, sustainability, and potentials for intensification of the current systems, with results that in fact can be foreseen and can therefore be avoided (FAO 1982).

The development of sustainable soil management and intensification systems requires the acceptance of these principles and finding ways to support and empower the producers to implement them through participatory approach/stakeholder engagement, policy cohesion, coherent policy and institutional support, innovative approaches to overcome equipment bottlenecks, and monitoring progress of change in production system practices and their outcomes at the farm as well as at the landscape level.

There are three nested levels of economic, social, and environmental impacts that can be recognized for identifying, monitoring, and measuring progress by different stakeholders, including farmers. At level 1, it is the change in people's concepts and mind-set as well as production system practices that is the goal. (For example, to monitor progress in the case of sustainable production systems and practices based on CA principles, the indicators would be the specification of effectiveness and stability of the production system, the number of farmers practicing and the area covered, and the rate of innovation.) At level 2, it is the outcomes resulting from the change in mindset and practices that are being sought (e.g., yield, income, stability, and productivity [efficiency], as well as ecosystem services such as soil health and quality, SOM, biological nitrogen fixation, water infiltration, soil biota (especially earthworms), erosion/runoff, crop health, and specific components of biodiversity such as pollinator bees or natural enemies of pests or specific soil microorganisms). There would be outcomes on the social side such as increases in innovative farm business management, on-farm experimentation, and social capital development in terms of farmers coming together to innovate and capture economies of scale. At the third level, it is the change in the state of the economic, social, and environmental conditions of the target group and their area that is being sought. (For example, in the case of the environment, four parameters are important for monitoring progress—physical state of landscape and soil quality, of functional biodiversity, and of water resources in quantity and quality, and climate change mitigation.) In terms of the change in social and economic conditions, social benefits can be decreased stress in the community, increased institutional innovation, stable incomes, and greater resilience. This includes the target groups' own perceptions of type and degree of change.

Our overall conclusion is that sustainable soil management as a basis for sustainable agricultural production is essential and practicable, but depends on both how and what crops are grown, as well as on the engagement of all stakeholders who are aligned toward transforming the unsustainable tillage-based farming systems to conservation agriculture systems regardless of soil, climate, and farmers' economic capacity to invest. It is possible to develop a sustainable production system based on how and what crops are grown but always following CA principles. This would allow the maintenance of the underpinnings of ecological sustainability of production systems in good order so that sustainable production of food and other ecosystem services becomes the norm. This transformational change is now occurring worldwide on all continents and ecologies and covers nearly 10% of the global arable land.

To enable the reduction or elimination of soil degradation on all agricultural soils as a basis for sustainable agriculture, the following policy and institutional action points for policy makers and institutional decision makers are suggested:

- Establish clear and verifiable guidelines and protocols for agricultural production systems, which would qualify as sustainable intensification based on conservation agriculture and other good practices from a socioeconomic and environmental point of view.
- Institutionalize the new way of farming with sustainable soil management in public-sector education and advisory services as an officially endorsed policy.

- Establish the conditions for a conducive environment to support this new kind of agriculture involving sustainable soil and landscape management, including research and technology adoption and the provision of suitable technologies and inputs through the commercial supply markets.
- Establish incentive mechanisms such as payments for environmental or community services, based on the adherence to the established protocols for sustainable soil management and production intensification, and align any eventually existing payments to farmers to such a service-based approach.
- As adoption levels of sustainable soil management increase and the sustainable production intensification becomes an accessible option to every farmer, introduce penalties for polluting or degrading ways of agricultural land use and landscape management as additional incentive for late adopters.

ABBREVIATIONS

Ca:	calcium
CA:	conservation agriculture
CEC:	cation exchange capacity
CT:	conventional tillage
Cu:	copper
FEBRAPDP:	Brazilian No-Till Federation
FFHC:	Freedom from Hunger Campaign
GIS	geographic information system
GPS	global positioning system
K:	potassium
Mg:	magnesium
Mn:	manganese
Mo:	molybdenum
N:	nitrogen
P:	phosphorus
POM:	particulate organic matter
PRBs:	permanent raised beds
S:	sulfur
SCPI:	sustainable crop production intensification
SOM:	soil organic matter
SRI:	System of Rice Intensification
Zn:	zinc

REFERENCES

Acosta, J. A. A. 2005. Improving the fertilizer recommendations for nitrogen in maize, adapted for use in the crop production systems of conservation agriculture. Alban Programme Final Report, Copenhagen, and M.Sc. Thesis. Department of Soil Science, Federal University of Santa Maria, RS, Brazil.

Aita, C. and S. H. Giacomini. 2003. Crop residue decomposition and nitrogen release in single and mixed cover crops. *Rev Bras Cienc Solo* 27(4): 601–612.

Aita, C. and S. J. Giacomini. 2007. Matéria orgânica do solo, nitrogênio e enxofre nos diversos sistemas de exploração agrícola. In *Nitrogênio e enxofre na agricultura brasileira*, eds. T. Yamada et al., 1–41. Piracicaba, IPNI Brasil.

Altieri, M. A, M. A. Lana, H. V. Bittencourt, A. S. Kieling, J. J. Comin, and P. E. Lovato. 2011. Enhancing Crop Productivity via Weed Suppression in Organic No-Till Cropping Systems in Santa Catarina, Brazil. *J Sustain Agric* 35: 1–15.

Amado, T. J. C. 1985. Relações da erosão hidrica dos solos com doses e formas de manejo do residuo da cultura da soja. Porto Alegre. Thesis (Mestrado Agronomía), 104. Fac. Agronomía, Universidade Federal do Río Grande do Sul, Porto Alegre.

Amado, T. J. C. and D. J. Reinert. 1998. Zero tillageage as a tool for sustainable agriculture in South Brazil. In *Conservation Tillage for Sustainable Agriculture. Proceedings of the International Workshop, Harare, Zimbabwe, Annexe III*: Background Papers (International) eds. J. Benites, E. Chuma, R. Fowler, J. Kienzle, K. Molapong, I. Manu, Nyagumbo, K. Steiner, and R. van Veenhuizen, June 22–27, 1998. GTZ, Eschborn.

Amado, T. J. C., J. Mielniczuk, S. B. V. Fernandes, and C. Bayer. 1999. Culturas de cobertura, acúmulo de nitrogênio total no solo e produtividade de milho. *Rev Bras Cienc Solo* 23: 679–686.

Amado, T. J. C., J. Mielniczuk, and S. B. V. Fernandes. 2000. Leguminosas e adubaçaõ mineral como fontes de nitrogênio para o milho em sistemas de preparo do solo. *Rev Bras Cienc Solo* 24: 179–189.

Amado, T. J. C., C. Bayer, F. L. F. Eltz, and A. C. R. Brum. 2001. Potencial de culturas de cobertura em acumular carbono e nitrogênio no solo no plantio direto e a melhoria daqualidade ambiental. *Rev Bras Cienc Solo* 25: 189–197.

Amado, T. J. C., J. Mielniczuk and C. Aita. 2002. Recomendaçaõ de adubaçaõ nitrogenada para o milho no RS e SC adaptado ao uso de culturas de cobertura, sob sistema plantio. *Rev Bras Cienc Solo* 26: 241–248.

Amado, T. J. C., C. Bayer, P.C. Conceicaõ, E. Spagnollo, B. C. Campos, and M. da Veiga. 2006. Potential of carbon accumulation in zero tillage soils with intensive use and cover crops in Southern Brazil. *J Environ Qual* 35: 1599–1607.

ANA. 2011. Programme for Water Producers (Programa Produtor de Água). Available at http://www2.ana.gov.br/Paginas/imprensa/noticia.aspx?id_noticia=9304.

Bai, Y. H., F. Chen, H. W. Li, H. Chen, J. He, Q. J. Wang, J. N. Tullberg, and Y. S. Gong. 2008. Traffic and tillage effects on wheat production on the Loess Plateau of China: 2. Soil physical properties. *Aust J Soil Res* 46: 652–658.

Baig, M. N. and P. M. Gamache. 2009. *The Economic, Agronomic and Environmental Impact of No-Till on the Canadian Prairies*. Alberta Reduced Tillage Linkages. Canada

Baker, C. J., K. E. Saxton, W. R. Ritchie, W. C. T. Chamen, D. C. Reicosky, M. F. S. Ribeiro, S. E. Justice, and P. R. Hobbs. 2007. *No-Tillage Seeding in Conservation Agriculture* (2nd edn). Rome: CABI and FAO.

Bakker, D. M., G. Hamilton, D. Houlbrooke, and C. Spann. 2005. The effect of raised beds on soil structure, waterlogging, and productivity on duplex soils in Western Australia. *Aust J Soil Res* 43: 575–585.

Balfour, E. B. 1943. *The Living Soil*. London: Faber & Faber Ltd.

Balota, E. L., M. Kanashiro, and A. Calegari. 1996. Adubos verdes de inverno na cultura do milho e a microbiologia do solo. In *I Congresso Brasilero de Plantio Direto para uma Agricultura Sustenavel*, March 18–22, 1996, 12–14. Ponta Grossa, PR, Brazil, Resumos expandidos.

Banks, P. A. and E. L. Robinson. 1982. The influence of straw mulch on soil reception and persistence of metribuzin. *Weed Sci* 30: 164–168.

Baveye, P. C., D. Rangel, A. R. Jacobsen, M. Laba, C. Darnault, W. Otten, R. Radulovich, and F. A. O. Camargo. 2011. From dust bowl to dust bowl: soils are still very much a frontier of science. *SSAJ* 75: 2037–2048.

Bayer, C. 1996. Dinâmica da matéria orgânica em sistemas de manejo de solos. PhD thesis. Universidade Federal do Rio Grande do Sul, Porto Alegre, Brazil.

Bayer, C., J. Mielniczuk, T. C. Amado, L. Martin Neto, and J. V. Fernández. 2000a. Effect of zero tillage cropping systems on soil organic matter storage in a clay loam Acrisol from southern Brazil monitored by electron spin resonance and nuclear magnetic resonance. *Soil Till Res* 53: 95–104.

Bayer, C., J. Mielniczuk, T. C. Amado, L. Martin-Neto, and J. V. Fernández. 2000b. Organic matter storage in a clay loam Acrisola Vected by tillage and cropping systems in southern Brazil. *Soil Till Res* 54: 101–109.

Bayer, C., D. P. Dick, G. M. Ribeiro, and K. K. Scheuermann. 2002. Estoques de carbono em fraçõ es da matéria orgânica afetados pelo uso e manejo do solo, com ênfase ao plantio direto. *Ciénc Rural* 32: 401–406.

Beaufoy, G., 2001. The Environmental Impact of Olive Oil Production in the European Union: Practical Options for Improving the Environmental Impact. European Forum on Nature Conservation and Pastoralism. Available at http://ec.europa.eu/environment/agriculture/pdf/oliveoil.pdf (last accessed September 7, 2012).

Bernoux, M., C. C. Cerri, C. E. P. Cerri et al. 2006. Cropping systems, carbon sequestration and erosion in Brazil, a review. *Agron Sust Dev* 26: 1–8.

Berton, A. L. 1998. Viabilidade do plantio direto na pequena propriedade. In *Conferência anual de plantio direto 3*, 43–48. Aldeia Norte Editoria, Passo Fundo, Brazil.

Black, A. L. 1973. Soil property changes associated with crop residue management in a wheat–fallow rotation. *Soil Sci Soc Am J* 37: 943–946.

Blevins, R. L. and W. W. Frye. 1993. Conservation tillage: An ecological approach to soil management. *Adv Agron* 51: 33–78.

Bligh, K. 1989. Narrow points. Western Australia Dept of Agric, DRM report.

Bligh, K. J. 1991. Narrow-winged seeder points reduce water erosion and maintain crop yields. *J Agric Western Australia* 32: 62–65.

Bolliger, A., J. Magid, T. J. C. Amado et al. 2006. Taking stock of the Brazilian "Zero-Till Revolution": a review of landmark research and farmers' practice. *Adv Agron* 91: 47–110.

Borges Filho, E. P. 2001. O desenvolvimento do plantio direito no Brasil: A conjunção de interesses entre agricultores, indú strias e estado, MSc Thesis, UNICAMP/IE, Campinas, Brazil.

Buckles, D., B. Triomphe, and G. Sain. 1998. Cover crops in hillside agriculture: Farmer innovation with Mucuna, IRDC/CIMMYT, Canada.

Burle, M. L., J. Mielniczuk, and S. Focchi. 1997. Effect of cropping system on soil chemical characteristics, with emphasis on soil acidification. *Plant Soil* 190: 309–316.

CA. 2007. *Water for Food, Water for Life: A Comprehensive Assessment of Water Management in Agriculture*. London: Earthscan and Colombo: International Water Management Institute.

Calatrava, J., G. G. Barbera, and V. M. Castillo. 2011. Farming practices and policy measures for agricultural soil conservation in semi-arid Mediterranean areas: the case of the Guadalentin basin in southeast Spain. *Land Degrad Dev* 22 (1): 58–69.

Calegari, A. 2000. Adubação Verde eRotação de Culturas. In *Feijão: Tecnologia de Producão*, 29–34. Instituto Agronômica do Paraná (IAPAR) Circular 135, Londrina, PR, Brazil.

Calegari, A. 2002. The spread and benefits of no–till agriculture in Paraná State, Brazil. In *Agroecological Innovations: Increasing Food Production with Participatory Development*, ed. N. Uphoff, 187–202. London: Earthscan.

Cann, M. A. 2000. Clay spreading on water repellent sands in the south east of South Australia—promoting sustainable agriculture. *J Hydrol* 231: 333–341.

Carver, A. J. 1981. Air Photography for Land Use Planners. Dept. of Conservation & Extension, Salisbury, Rhodesia.

Cassol, E. A. 1984. Erosao do solo-influencia do uso agricola, do manejo e preparo do solo. Publicaçao IPRNR 15. Instituto de Pesquisas de Recursos Naturais Renovaveis "AP," 40. IPRNR, Porto Alegre.

CCC. 2011. Specified Gas Emitters Regulation Results for the 2010 Compliance Year. Climate Change Central, Alberta, Canada. Available at http://carbonoffsetsolutions. climatechangecentral.com/policy-amp-regulation/alberta-offset-system-compliance-a-glance/compliance-review-2010.

Chen, H., Y. H. Bai, Q. J. Wang, F. Chen, H. W. Li, J. N. Tullberg, J. R. Murray, H. W. Gao, and Y. S. Gong. 2008. Traffic and tillage effects on wheat production on the Loess Plateau of China: 1. Crop yield and SOM. *Aust J Soil Res* 46: 645–651.

Cogo, N. P., C. R. Drews, and C. Gianello. 1978. Í ndice de erosividade das chuvas dos municípios de Guaíba, Ijuí, e Passo Fundo, no Estado do Rio Grande do Sul. In *Encontro nacional de pesquisa sobre conservação do solo, II.* Passo Fundo, 1978. Anais, 145–152. CNPT, Passo Fundo, Brazil.

Cogo, N. P., R. Levine, and R. A. Schwarz. 2003. Soil and water losses by rainfall erosion influenced by tillage methods, slope-steepness classes, and soil fertility levels. *Rev Bras Ciênc Solo* 27 (4): 743–753.

Corazza, E. J., J. E. Silva, D. V. S. Resck, and A. C. Gomes 1999. Comportamento de diferentes sistemas de manejo como fonte e depó sito de carbono em relação a vegetação de cerrado. *Rev Bras Ciênc Solo* 23: 425–432.

Crabtree, W. L. 1983. The effect of cultivation on soil fertility. University of Western Australia, Honours Project.

Crabtree, W. L. 1990. Toward better minimum tillage for south-coastal sandplain soils. The role of minimum tillage on wind erosion prone south coast sandy soil. Western Australian Department of Agriculture Technical Report, Division of Resource Management 1110.

Crabtree, W. L. 2010. *Search for Sustainability with no-till Bill in Dryland Agriculture.* Crabtree Agricultural Consulting, 204. Available at www.no-till.com.au.

Crabtree, W. L. and C. W. Henderson 1999. Furrows, press wheels and wetting agents improve crop emergence and yield on water repellent soils. *Plant Soil* 214: 1–8.

Darolt, M. R. 1997. Manejo do sistem de plantio direto na pequena propriedade. In *Plantio direto: o caminho para uma agricultura sustentavel*, eds. R. T. G. Peixoto, D. C. Ahrens, and M. J. Samaha, 72–83. Instituto Agronomico do Parana (IAPAR), Ponta Grossa, PR, Brazil.

Davies, S. 2011. Wheat response to rotary spading of water repellent sand at Marchagee. Dept. Ag & Food of W. Australia (DAFWA) and Liebe Group.

De Laulanié, H. 1993. Le système de riziculture intensive malgache. *Tropicultura* (Belgium) 11: 110–114.

De Maria, I. C., P. C. Nabude, and O. M. Castro. 1999. Long-term tillage and crop rotation effects on soil chemical properties of a Rhodic Ferralsol in southern Brazil. *Soil Till Res* 51: 71–79.

Denardin, J. E. and R. A. Kochhann. 1999. Fast zero tillage adoption in Brazil without subsidies: A successful partnership. In *Northwest Direct Seed Cropping Systems Conference Proceedings*, January 5–7, 1999, Spokane, Washington. Available at http://pnwsteep.wsu.edu/directseed/conf99/dspropBr.htm.

Derpsch, R. 2001. Frontiers in conservation tillage and advances in conservation practice. In *Sustaining the Global Farm. Selected papers from the 10th International Soil Conservation Organization Meeting held May 24–29, 1999 at Purdue University and the USDA National Soil Erosion Laboratory*, eds. D. E. Stott, R. H. Mohtar, and G. C. Steinhardt.

Derpsch, R. 2004. History of crop production, with and without tillage. *Leading Edge* 3: 150–154.

Derpsch, R. and T. Friedrich. 2009. Development and Current Status of No-till Adoption in the World, Proceedings on CD, 18th Triennial Conference of the International Soil Tillage Research Organization (ISTRO), Izmir, Turkey, June 15–19, 2009.

Dijkstra, F. 2002. Conservation tillage development at the ABC Cooperatives in Paraná, Brazil. In *Making Conservation Tillage Conventional: Building a Future on 25 Years of Research. Proceedings of the 25th Annual Southern Conservation Tillage Conference for Sustainable Agriculture held June 24–26, 2002 in Auburn*, ed. E. van Santen, 12–18. Alabama Agricultural Experimental Station and Auburn University, AL, USA.

Doets, C. E. M., G. Best, and T. Friedrich. 2000. *Energy Conservation Agriculture*, Occasional Paper, FAO SDR Energy Program, Rome

Doran, J. W. 2002. Soil health and global sustainability: translating science into practice. *Agric Ecosyst Environ* 88 (2): 119–127.

Doran, J. W. and M. R. Zeiss. 2000. Soil health and sustainability: managing the biotic component of soil quality. *Appl Soil Ecol* 15: 3–11.

Dunjó, G., G. Pardini, and M. Gispert. 2003. Land use change effects on abandoned terraced soils in a Mediterranean catchment, NE Spain. *Catena* 52 (1): 23–37.

Ellington, A. 1986. Effects of deep ripping, direct drilling gypsum and lime on soils, wheat growth and yield. *Soil Till Res* 8: 29–49.

Elliott, E. T. 1986. Aggregate structure and carbon, nitrogen and phosphorous in native and cultivated soils. *Soil Sci Soc Am J* 50: 627–633.

Encuesta sobre Superficies y Rendimientos Cultivos (ESYRCE). 2006, 2009, 2010, 2011. Ministerio de Agricultura, Alimentación y Medio Ambiente, Madrid.

Erenstein, O. 2003. Smallholder conservation farming in the tropics and sub-tropics: a guide to the development and dissemination of mulching with crop residues and cover crops. *Agric Ecosyst Environ* 100: 17–37.

Fabrizzi, K. P., F. O. Garcia, J. L. Costa, and L. I. Picone. 2005 Soil water dynamics, physical properties and corn and wheat responses to minimum and no-tillage systems in the southern Pampas of Argentina. *Soil Till. Res* 81: 57–69.

Fabrizzi, K. P., C. W. Rice and T. J. C. Amado. 2009. Protection of soil organic C and N in temperate and tropical soils: effect of native and agroecosystems. In *3rd International Conference on Mechanisms of Organic Matter Stabilization and Destabilization in Soils and Sediments*, Glenelg, September 23–26, 2007 (also in *Biochemistry* 92 (1–2): 129–143, January 2009; DOI: 10.1007/s10533-008-9261-0).

FAO. 2008. *Investing in Sustainable Crop Intensification: The Case for Soil Health*. Report of the International Technical Workshop, FAO, Rome, July. Integrated Crop Management, Vol. 6. Rome: FAO. Available at http://www.fao.org/ag/ca/.

FAO. 2010. *An International Consultation on Integrated Crop-Livestock Systems for Development—The Way Forward for Sustainable Production Intensification*. Integrated Crop Management, Vol. 13. Rome: FAO.

FAO. 2011a. *The State of the World's Land and Water Resources for Food and Agriculture (SOLAW): Managing Systems at Risks*. Summary Report. Rome: FAO.

FAO. 2011b. *Save and Grow. A Policymakers' Guide to the Sustainable Intensification of Smallholder Crop Production*. Rome: FAO. 98. Available at www.fao.org/ag/save-and-grow/.

FAO. 2012. Soil organic carbon accumulation and greenhouse gas emission reductions from conservation agriculture: a literature review. *Integr Crop Manage* 16, 89 p. Rome: FAO.

Faulkner, E. H. 1945. *Ploughman's Folly*. London: Michael Joseph.

Feller, C. and M. H. Beare. 1997. Physical control of soil organic matter dynamics in the tropics. *Geoderma* 79: 69–116.

Flaig, W., B. Nagar, H. Sóchtig, and C. Tietjen. 1977. *Organic Materials and Soil Productivity*. FAO Soils Bulletin no. 35. Rome: FAO.

Fleskens, L. and L. Stroosnijder. 2007. Is soil erosion in olive groves as bad as often claimed? *Geoderma* 141 (3–4): 260–271.

Flower, K., B. Crabtree, and G. Butler. 2008. No-till cropping systems in Australia. In *No-Till Farming Systems,* eds. T. Goddard, M. Zoebisch, Y. Gan, W. Ellis, A. Watson, and S. Sombatpanit, 457–467 Special Pub. No. 3. Bangkok: World Association of Soil and Water Conservation (WASWAC).

Flower, K. C. and W. L. Crabtree. 2011. Soil pH change after surface application of lime related to the levels of soil disturbance caused by no-tillage seeding machinery. *Field Crops Res* 121: 75–87.

Forcella, F., D. D. Buhler, and M. E. McGriffn. 1994. Pest management in crop residues. In *Advances in Soil Science: Crop Residue Management,* eds. J. L. Hatfield and B. A. Stewart, 173–189. London: Lewis Publishers.

Franco, J. A. and J. Calatrava. 2006. Adoption of soil erosion control practices in Southern Spanish olive groves. *Proceedings of the International Association of Agricultural Economists,* Gold Coast, Australia, August 12–18.

Freitas, P. L., P. Blancaneaux, E. Gavinelli, M. C. Larré Larroy, and C. Feller. 1999. Nível e natureza do estoque orgânico de Latossolos sob diferentes sistemas de uso e manejo. *Pesq Agropec Bras* 35: 157–170.

Friedrich, T. and A. H. Kassam. 2011. Mechanization and the Global Development of Conservation Agriculture. 23rd Annual SSCA Conference, January 13, 2011, Saskatoon, Canada.

Friedrich, T., A. H. Kassam, and T. F. Shaxson. 2009. Conservation Agriculture. In *Agriculture for Developing Countries.* Science and Technology Options Assessment (STOA) Project. Karlsruhe, Germany: European Technology Assessment Group.

Friedrich, T., R. Derpsch, and A. Kassam. 2012. Overview of the global spread of conservation agriculture. *Field Actions Sci Rep Special Issue* 6. Available at http://factsreports.revues.org/1941.

Gao, H. W., H. W. Li, and W. Y. Li. 2008. Development of conservation tillage. *Transact Chin Soc Agric Mach* 9: 43–48 (in Chinese).

Garrity, D. P. 2011. Making conservation agriculture ever green. 5th World Congress of Conservation Agriculture incorporating 3rd Farming Systems Design Conference, September 2011, Brisbane, Australia.

Garrity, D. P., F. K. Akinnifesi, and A. Oluyede. 2010. Evergreen agriculture: a robust approach to sustainable food security in Africa. *Food Sec* 2: 197–214.

Gazey, C. and S. Davies. 2009. Soil acidity: a guide for WA farmers and consultants. Bulletin 4784, Department of Agriculture and Food, Western Australia, 47. Available at http://www.agric.wa.gov.au/objtwr/imported_assets/content/lwe/land/acid/liming/bn_soil_acidity_guide.pdf.

George, R., D. McFarlane, and R. A. Nulsen. 1997. Salinity threatens the viability of agriculture and ecosystems in Western Australia. *Hydrogeol J* 5 (1): 6–21.

Gianluppi, D., I. Scopel, and J. Mielniczuk. 1979. Alguns prejuizos da erosao do solo no RS. In *Congresso Brasileiro de Ciencia do Solo, XVII, Manaus, 1979. Resumos,* 92 SBCS, Campinas, Brazil.

Giles, G. W. 1975. *The Reorientation of Agricultural mechanization for Developing Countries: Politics and Attitudes for Action Programmes.* Report on the Meeting of the FAO/OECD Expert Panel on the Effects of Farm Mechanization on Production and Employment, Rome.

Godfray, C., J. R. Beddington, and I. R. Crute. 2010. Food security: the challenge of feeding 9 billion people. *Science* 327: 812–818.

Gomez, J. A., J. V. Giraldez, and T. Vanwalleghem. 2008. Comments on "Is soil erosion in olive groves as bad as often claimed?" by L. Fleskens and L. Stroosnijder. *Geoderma* 147 (1–2): 93–95.

Gomez, J. A., S. Alvarez, and M.-A. Soriano. 2009a. Development of a soil degradation assessment tool for organic olive groves in southern Spain. *Catena* 79 (1): 9–17.

Gomez, J. A., T. A. Sobrinho, J. V. Giraldez, and E. Fereres. 2009b. Soil management effects on runoff, erosion and soil properties in an olive grove of Southern Spain. *Soil Till Res* 102 (1): 5–13.

Greenland, D. and R. Lal. 1977. *Soil Conservation and Management in the Humid Tropics.* Chichester: John Wiley & Sons.

Hargrove, W. L. (ed.). 1991. *Cover Crops for Clean Water.* The Proceedings of an International Conference. West Tennessee Experiment Station, April 9–11, 1919, Jackson, Tennessee. Soil and Water Conservation Society, Ankeny, Iowa.

Haugen-Kozyra, K. and T. Goddard. 2009. Conservation agriculture protocols for green house—gas offsets in a working carbon markets. Paper presented at the *IV World Congress on Conservation Agriculture*, February 3–7, 2009, New Delhi, India.

He, J., H. W. Li, A. D. McHugh et al. 2008. Spring wheat performance and water use efficiency on permanent raised beds in arid northwest China. *Aust J Soil Res* 46: 659–666.

He, J., Q. J. Wang, H. W. Li et al. 2009a. Soil physical properties and infiltration after long-term no-tillage and ploughing on the Chinese Loess Plateau. *NZ J Crop Horticult Sci* 37: 157–166.

He, J., N. J. Kuhn, X. M. Zhang, X. R. Zhang, and H. W. Li. 2009b. Effect of 10 years of conservation tillage on soil properties and productivity in the farming–pastoral ecotone of Inner Mongolia, China. *Soil Use Manag* 25: 201–209.

He, J., H. W. Li, R. G. Rasaily et al. 2011. Soil properties and crop yields after 11 years of no tillage farming in wheat-maize cropping system in North China Plain. *Soil Till Res* 113: 48–54.

Helvarg, D. 2001. *Blue Frontier.* New York: W.H. Freeman & Co.

Hobbs, P. R. 2007. Conservation agriculture: what is it and why is it important for future sustainable food production? *J Agric Sci* 145: 127–137.

Huang, M. B. and L. P. Zhong. 2003. Evaluating the EPIC model to simulate soil water content of the Loess Plateau, China. *J Exp Botany* 54: 25–26.

ITAIPU. 2011. Cultivando Agua Boa (Growing Good Water). Available at http://www2.itaipu.gov.br/cultivandoaguaboa/.

Jarvis, R. 2000. Deep tillage. In *The Wheat Book: Principles and Practice*, eds. W. K. Anderson and J. R. Garlinge, 185–187. Bulletin 4443. Department of Agriculture, Western Australia.

Jenny, H. 1980. *The Soil Resource.* New York: Springer.

Karlen, D. L. and C. A. Cambardella. 1996. Conservation strategies for improving soil quality and organic matter storage. In *Structure and Organic Matter Storage in Agricultural Soils. Advances in Soil Science,* eds. M. Carter and B. A. Stewart, 395–420. CRC Press, Boca Raton.

Kassam, A. H., T. Friedrich, T. F. Shaxson, and J. N. Pretty. 2009. The spread of conservation agriculture: justification, sustainability and uptake. *Int J Agric Sustain* 7 (4): 292–320.

Kassam, A. H., T. Friedrich, and R. Derpsch. 2010. Conservation Agriculture in the 21st Century: A Paradigm of Sustainable Agriculture. European Congress on Conservation Agriculture, October 4–6, 2010, Madrid, Spain.

Kassam, A., T. Friedrich, T. F. Shaxson, T. Reeves, J. Pretty, and J. C. de Moraes Sà. 2011a. Production systems for sustainable intensification: integrated productivity with ecosystem services. *Technikfolgenabschatzung Theorie Praxis* 2: 39–45.

Kassam, A., W. Stoop, and N. Uphoff. 2011b. Review of SRI modifications in rice crop and water management and research issues for making further improvements in agricultural and water productivity. *Paddy Water Environ Special Issue* 9 (1): 163–180.

Kassam, A. H., I. Mello, T. Goddard, T. Friedrich, F. Laurent, T. Reeves, and B. Hansmann. 2011c. Harnessing Ecosystem Services with Conservation Agriculture in Canada and Brazil. 5th World Congress of Conservation Agriculture incorporating 3rd Farming Systems Design Conference, September 2011 Brisbane, Australia. Available at www.wcca2011.org.

King, P. M. 1981. Comparison of methods for measuring severity of water repellency of sandy soils and assessment of some factors that affect its measurement. *Aust J Soil Res* 19: 275–285.

Kladivko, E. 2001. Tillage systems and soil ecology. *Soil Till Res* 61: 61–76.

Kumar, K. and K. M. Goh. 2000. Crop residues and management practices: Effects on soil quality, soil nitrogen dynamics, crop yield, and nitrogen recovery. *Adv Agron* 68: 197–319.

Lal, R. 2002. Soil carbon dynamic in cropland and rangeland. *Environ Pollut* 116: 353–362.

Lal, R. 2009. Ten tenets of sustainable soil management. *J. Soil Water Conserv* 64 (1): 20–21A.

Lal, R., R. F. Follet, J. Kimble, and C. V. Cole. 1999. Managing U.S. cropland to sequester carbon in soil. *J. Soil Water Conserv* 54: 374–381.

Landers, J. 2007. *Tropical Crop-Livestock Systems in Conservation Agriculture: The Brazilian Experience.* Integrated Crop Management, Vol. 5. Rome: FAO.

Landers, J. N., S. M. Teixeira, and A. Milhomen. 1994. Possíveis impactos da técnica dePlantio Direto sobre a sustentabilidade da produção de grãos na região dos cerrados. In *Congresso brasileiro de economia e sociologia, 32, Brasília.* Desafio do Estado diante de uma agricultura em transformação. Anais. 2: 799–820. SOBER, Brasília.

Laurent, F., G. Leturcq, I. Mello, J. Corbonnois, and R. Verdum. 2011. La diffusion du semis direct au Brésil, diversité des pratiques et logiques territoriales: l'exemple de la région d'Itaipu au Paraná. *Confins* 12. Available at http://confins.revues.org/7143.

Legg, B. J., D. H. Sutton, and E. M. Field. 1993. Feeding the world. Can engineering help? Fourth Erasmus Darwin Memorial Lecture, November 17, 1993, Silsoe, UK.

Leyva, J. C., J. A. F. Martinez, and M. C. G. Roa. 2007. Analysis of the adoption of soil conservation practices in olive groves: the case of mountainous areas in southern Spain. *Spanish J Agric Res* 5 (3): 249–258.

Li, H. W., H. W. Gao, H. D. Wu, W. Y. Li, X. Y. Wang, and J. He. 2007. Effects of 15 years of conservation tillage on soil structure and productivity of wheat cultivation in northern China. *Aust J Soil Res* 45: 344–350.

Li, Y. 2001. Variation of crop yield and soil drying under high fertility conditions. *Acta Pedol Sin* 38: 353–356 (in Chinese with English abstract).

Lindwall, C. W. and B. Sonntag (eds.). 2010. Landscape Transformed: The History of Conservation Tillage and Direct Seeding, Knowledge Impact in Society, Saskatoon, University of Saskatchewan, Canada.

Liu, L. J. 2004. Systematic experiments and effect analysis of all year conservation tillage in two crops a year region. Ph.D. Dissertation. China Agricultural University, Beijing, China (in Chinese).

Liu, L. Y., X. Y. Li, P. J. Shi et al. 2007. Wind erodibility of major soils in the farming–pastoral ecotone of China. *J Arid Environ* 68: 611–623.

Liu, X. E., H. A. Guo, and L. C. Li. 2002. The question and developmental counter measure of breeding for maize in Northeast China. *J Jilin Agric Sci* 27: 20–23 (in Chinese).

LZU. 2005. The Chinese map for agriculture—pasture transition region based on GIS. Lanzou University, Lanzou, China (in Chinese).

Machedo, M. C. M. 1997. Sustainability of pasture production in the savannas of tropical America. International Grassland Conference, Canadian Society of Agronomy, Canadian Society of Animal Science, Winnipeg and Saskatoon.

Martinez, J. A. F. 2009. Impacto de la política agroambientaleuropea de lucha contra la erosión sobre la olivicultura en Andalucía. *Ecol Apl* 8 (2): 37–45.

Mashum, M., M. E. Tate, G. P. Jones, and J. M. Oades. 1988. Extraction and characterization of water-repellent materials from Australian soils. *J Soil Sci* 39: 99–110.

McArthur, W. M. 2004. *Reference Soils of South-western Australia.* Western Australia, Perth: Department of Agriculture.

McGhie, D. A. 1980. The origins of water repellence in some Western Australian soils. Ph.D. thesis, University of Western Australia.

MEA. 2005. *Ecosystems and Human Well-Being: Synthesis.* Millennium Ecosystem Assessment. Washington, DC: Island Press.
Mello, I. and B. van Raij. 2006. No-till for sustainable agriculture in Brazil. *Proc World Assoc Soil Water Conserv* 1: 49–57.
Mielniczuk, J. 2003. Manejo do solo no Rio Grande do Sul: Uma sintesehistorica. *In Curso de fertilidade do solo em plantio direto, VI. Passo Fundo, 2003. Resumo de palestras*, 5–14. Aldeia Norte Editora Ltd., Ibiruba.
Mielniczuk, J., and P. Schneider. 1984. Aspectos socio-economicos do manejo de solos no sul do Brasil. In *Anais do I Simposio de Manejo de Solo e Plantio Direto no Sul do Brasil e II Simposio de Conservaçao de Solo do Planalto*, 3–19. Passo Fundo, RS, Brasil.
Mishra, R. P. N., R. K. Singh, H. K. Jaiswal, V. Kumar, and S. Maurya. 2006. Rhizobium-mediated induction of phenolics and plant growth promotion in rice (*Oryza sativa* L.). *Curr Microbiol* 52 (5): 383–389.
Montgomery, D. R. 2007. *Dirt: The Erosion of Civilizations.* Berkeley and Los Angeles: University of California Press.
Moore, G. 2001. *Soil Guide: A Handbook for Understanding and Managing Agricultural Soils.* Agriculture Western Australia, Bulletin No. 4343, Perth.
Mrema, G. C. 1996. Agricultural development and the environment in Sub-Saharan Africa: an engineer's perspective. Keynote paper presented at the First International Conference of SEASAE, October 2–4, 1996, Arusha, Tanzania.
Muzilli, O. 1983. Influência do sistema de plantio direto, comparado ao convencional, sobre a fertilidade da camada arável do solo. *Rev Bras Cien Solo* 7: 95–102.
OECD. 2011. *Fostering Productivity and Competitiveness in Agriculture.* Paris: OECD Publishing. DOI: 10.1787/9789264166820-en.
Oldeman, L. R. 1988. *Guidelines for General Assessment of the Status of Human-Induced Soil Degradation.* International Soil Reference and Information Centre (ISRIC), Wageningen, the Netherlands.
Oldeman, L. R., R. T. A. Hakkeling, and W. G. Sombroek. 1991. *World Map of the Human-Induced Soil Degradation: An Explanatory Note.* International Soil Reference and Information Centre (ISRIC), Wageningen, the Netherlands.
Perkins, J. H. 1997. *Geopolitics and the Green Revolution: Wheat, Genes, and the Cold War.* New York: Oxford University Press.
Perrin, A. S. 2003. EVets de diVérents modes de gestion des terres agricoles sur la matie`reorganique et la biomasse microbienne en zone tropicale humide au Brésil, M.Sc. Thesis, Ecole Polytechnique Fédérale de Lausanne, Lausanne, Switzerland.
Pes, L. Z., Amado, T. J. C., and Scala N. L. Jr. 2011. The primary sources of carbon loss during the crop-establishment period in a subtropical Oxisol under contrasting tillage systems. *Soil Till Res* 117: 163–171.
Phillips, E. R. and S. H. Phillips. 1984. *No-Tillage Agriculture: Principles and Practices.* New York: Van Nostrand Reinhold Co.
Phillips, S. H. and H. M. Young. 1973. *No-Tillage Farming.* Milwaukee: Reiman Associates.
Pisante, M., F. Stagnari, and C. Grant. 2012. Agricultural innovations for sustainable crop production intensification. *Italian J Agron* 7 (4): 300–311.
Pisante, M., S. Corsi, A. Kassam, and T. Friedrich. 2010. The challenge of agricultural sustainability for Asia and Europe. *Trans Studio Rev* 17 (4): 662–667.
Posner, J. 2005. Mitigation of Ecosystem Damage by Good Agricultural Stewardship Valuation of Ecosystems in Agriculture Workshop, Augusta, MI, October 26–28, 2005 Preliminary briefing paper.
Pöttker, D. 1977. Efeito do tipo de solo, tempo de cultivo e da calagem sobre a minerlizaçao da material organica em solos do Rio Grande do Sul. M.Sc. thesis. Universidade Federal do Rio Grande do Sul (UFRGS), Porto Alegre.
Pretty, J. 2002. *Agri-Culture: Reconnecting People, Land and Nature.* London: Earthscan.

Pretty, J. 2008. Agricultural sustainability: concepts, principles and evidence. *Philos Transact R Soc Lond B* 363 (1491): 447–466.
Pretty, J. N., C. Toulmin, and S. Williams. 2011. Sustainable intensification in African agriculture. *Int J Agric Sustain* 9 (1): 5–24.
Primavesi, A. 1984. *Manejo Ecologico del Suelo: La Agricultura en Regiones Tropicales.* 5th ed. Buenos Aires: Libreria "El Ateneo" Editorial.
Ralisch, R., O. J. G. Abi-Saab, and R. L. P. Cainzos. 2003. Bright Spots Research Project: Drivers Effecting No-Till System in Brazilian Small Farms, Report. CIAT. San Salvador.
Rasmussen, P. E., and H. P. Collins. 1991. Long-term impacts of tillage, fertilizer, and crop residue on soil organic matter in temperate semiarid regions. *Adv Agron* 45: 93–134.
Resck, D. V. S., J. Pereira, and J. E. Silva. 1991. Dinâmica da matéria orgânica na região dos cerrados. (Série Documentos, 36). Empresa Brasileira de Pesquisa Agropecuária, Planaltina, Brazil.
Resck, D. V. S., C. A. Vasconcellos, L. Vilela, and M. C. M. Macedo. 2000. Impact of conversion of Brazilian cerrados to cropland and pastureland on soil carbon pool and dynamics. In *Global Climate Change and Tropical Ecosystems, Advances in Soil Science*, eds. R. Lal, J. M. Kimble, and B. A. Stewart, 169–196. CRC Press, Boca Raton.
Ribeiro, M. F. S. and R. D. S. Milléo (eds.). 2002. Referências em Plantio Direto paraAgricultur Familiar do Centro Sul do Paraná. Instituto Agronômica do Paraná (IAPAR), Londrina, PR, Brazil.
Riezebos, H. Th. and A. C. Loerts. 1998. Influence of land use change and tillage practice on soil organic matter in southern Brazil and eastern Paraguay. *Soil Till Res* 49 (3): 271–275.
Robertson, G. 1987. Soil Management for Sustainable Agriculture. Department of Agriculture, Western Australia. Resource Management Technical Report No. 95.
Rodrigues, B. N. 1993. Influência da cobertura morta no comportamento dos herbicidasimazaquin e clomazone. *Planta Dahina* 11: 21–28.
Roloff, G, R. A. T. Lutz, and I. Mello. 2011. An index to rate the quality of no-till systems: a conceptual framework. 5th World Congress of Conservation Agriculture incorporating 3rd Farming Systems Design Conference, September 2011 Brisbane, Australia.
Sá, J. C. M., C. C. Cerri, W. A. Dick et al. 2001a. Organic matter dynamics and carbon sequestration rates for a tillage chronosequence in a Brazilian oxisol. *Soil Sci Soc Am J* 65: 1486–1499.
Sá, J. C. M., C. C. Cerri, W. A. Dick et al. 2001b. Carbon sequestration in a plowed and zero tillage chronosequence in a Brazilian oxisol. In *Sustaining the Global Farm. Selected papers from the 10th International Soil Conservation Organization Meeting held May 24–29, 1999 at Purdue University and the USDA National Soil Erosion Laboratory*, eds. D. E. Stott, R. H. Mohtar, and G. C. Steinhardt, 266–271. West Lafayette, IN: Purdue University.
Saha, R., P. K. Ghosh, V. K. Mishra, B. Majumdar, and J. M. S. Tomar. 2010. Can agroforestry be a resource conservation tool to maintain soil health in the fragile ecosystem of northeast India? *Outlook Agric* 39 (3): 191–196.
Scopel, E., E. Doucene, S. Primot, J. M. Douzet, A. Cardoso, and C. Feller. 2003. Diversity of direct seeding mulch based cropping systems (DMC) in the Rio Verde region (Goias, Brazil) and consequences on soil carbon stocks. In *Book of Extended Summaries of the II World Congress on Conservation Agriculture—Producing in Harmony with Nature*, 286–289. August 11–15, 2003, Foz do Iguaçu, Brazil, Volume II. Food and Agriculture Organization (CD).
Scopel, E., B. Triomphe, M. F. S. Ribeiro, L. Séguy, J. E. Denardin, and R. A. Kochann. 2004. Direct seeding mulch-based cropping systems (DMC) in Latin America. In *New Directions for a Diverse Planet: Proceedings for the 4th International Crop Science Congress*, eds. T. Fischer, N. Turner, J. Angus, L. McIntyre, M. Robertsen, A. Borrell, and D. Llyod Brisbane. Australia, September 26–October 1, 2004. Available at www.cropscience.org.au.

Séguy, L., S. Bouzinac, A. Trentini, and N. A. Cortez. 1996. L'agriculturebrésilienne des front-pionniers. *Agric Dév* 12: 2–61.

Séguy, L., S. Bouzinac, E. Scopel, and M. F. S. Ribeiro. 2003. New concepts for sustainable management of cultivated soils through direct seeding mulch based cropping systems: The CIRAD experience, partnership and networks. In *Proceedings of the II World Congress on Conservation Agriculture—Producing in Harmony with Nature*, August 11–15, 2003, Foz do Iguaçu, Brazil. Food and Agriculture Organization (CD).

Sharif, A. 2011 Technical adaptations for mechanised SRI production to achieve water saving and increased profitability in Punjab, Pakistan. *Paddy Water Environ, Special Issue*: 9 (1): 111–119.

Shaxson, T. F., N. D. Hunter, T. R. Jackson, and J. R. Alder. 1977. *A Land Husbandry Manual: Techniques of Land-Use Planning and Physical Conservation*. Zomba, Malawi: Govt. Printer.

Shaxson, T. F., N. W. Hudson, D. W. Sanders, E. Roose, and W. C. Moldenhauer. 1989. *Land Husbandry: A Framework for Soil and Water Conservation*. Ankeny (USA): Soil & Water Conservation Society.

Sims, B., T. Friedrich, A. H. Kassam, and J. Kienzle. 2009. Agroforestry and Conservation Agriculture: Complementary practices for sustainable agriculture. 2nd World Congress on Agroforestry, August 2009, Nairobi, Kenya.

Sims, B., C. Thierfelder, J. Kienzle, T. Friedrich, and A. Kassam. 2011. Development of the Conservation Agriculture Equipment Industry in sub-Saharan Africa. 5th World Congress of Conservation Agriculture incorporating 3rd Farming Systems Design Conference, September 2011, Brisbane, Australia.

Sisti, C. P. J., H. P. dos Santos, R. Kohhann, B. J. R. Alves, S. Urquiaga, and R. M. Boddey. 2004. Change in carbon and nitrogen stocks in soil under 13 years of conventional or zero tillage in southern Brazil. *Soil Till Res* 76: 39–58.

Six, J., E. T. Elliot, K. Paustian, and J. W. Doran. 1998. Aggregation and soil organic matter accumulation in cultivated and native grassland soils. *Soil Sci Soc Am. J* 62: 1367–1377.

Six, J., E. T. Elliot, and K. Paustian. 1999. Aggregate and soil organic matter dynamics under conventional and no-tillage systems. *Soil Sci Soc Am J* 63: 1350–1358.

Six, J., E. T. Elliot, and K. Paustian. 2000. Soil macroaggregate turnover and microaggregate formation: a mechanism for C sequestration under no-tillage agriculture. *Soil Biol Biochem* 32: 2099–2103.

Spehar, C. R. and J. N. Landers. 1997. Características, limitações e futuro do Plantio Direto nos cerrados. In *Seminário Internacional Do Sistema Plantio Direto*, 2, 1997, Passo Fundo. Anais, 127–131. EMBRAPA-CNPT, Passo Fundo.

Steiner, K., R. Derpsch, G. Birbaumer, and H. Loos. 2001. Promotion of Conservation Farming by the German Development Cooperation. In *Conservation Agriculture: A Worldwide Challenge. Proceedings of the First World Congress on Conservation Agriculture*, eds. L. Garcia-Torres, J. Benites, and A. Martinez-Vilela, Madrid, October 1–5, 2001, Vol. 2 60–65. XUL, Cordoba, Spain.

Stone, L. F. and J. A. L. Moreira. 1998. A irrigação no plantio direto. *J Plantio Direto no Cerrado* 8: 5–6 (Maio).

Summers, R. N. 1987. The incidence and severity of non-wetting soils on the south coastal sandplain of Western Australia. Master of Science Thesis, University of Western Australia.

Swanson, D. and S. Bhadwal (eds.). 2010. *Creating Adaptive Policies—A Guide for Policy-Making in an Uncertain World*. New Delhi: Int'l Institute for Sustainable Dev. and SAGE Publications.

Swift, M. J., D. E. Bignell, F. M. S. Moreira, and E. J. Huising. 2008. The inventory of soil biological diversity: concepts and general guidelines. In *A Handbook of Tropical Soil Biology: Sampling and Characterization of Below-ground Biodiversity*, eds. F. M. S. Moreira, E. J. Huising, and D. E. Bignell, 1–16. London: Earthscan.

Teixeira, L. A. J., V. M. Testa, and J. Mielniczuk. 1994. Nitrogenio no solo, nutrico e rendimento de milho afetados por sistemas de cultura. *Rev Bras Cien Solo* 18: 207–214.

Testa, V. M., L. A. J. Teixeira, and J. Mielniczuk. 1992. Caracterýsticas quý micas de um Podzolico Vermelho-escuro afetadas por sistemas de cultura. *Rev Bras Cienc Solo* 16: 107–114.

Tikhonovich, I. A. and N. A. Provorov. 2011. Microbiology is the basis of sustainable agriculture: an opinion. *Annals Appl Biol* 159 (2): 155–168.

Tullberg, J. N., D. F. Yule, and T. Jensen. 1998. Introduction. Attachment to Proceedings of Second National Controlled Traffic Conference, eds. J. N. Tullberg and D. F. Yule, The University of Queensland, Gatton College, August 26–27.

Tullberg, J. N. 2007. Special section: soil management for sustainability–Introduction. *Soil and Tillage Research* 97 (2): 247–248.

Turner, B. L. and P. M. Haygarth. 2001. Biochemistry: phosphorus solubilization in rewetted soils. *Nature* 411: 258.

Uphoff, N. and A. Kassam (eds.). 2011. Paddy and water management with the System of Rice Intensification (SRI). *Paddy Water Environ, Special Issue* 9 (1): 182.

Uphoff, N. et al. 2006. Issues for more sustainable soil system management. In *Biological Approaches to Sustainable Soil Systems*, 715–716. CRC Press, Boca Raton, FL.

Uphoff, N., A. Kassam, and R. Harwood. 2011. SRI as a methodology for raising crop and water productive adaptations in rice agronomy and irrigation water management. *Paddy Water Environ Special Issue* 9 (1): 3–11.

Uri, N. D., J. D. Atwood, and J. Sanabria. 1998. An evaluation of the environmental costs and benefits of conservation tillage. *Environ Impact Assess Rev* 18: 521–550.

Vanwalleghem, T., A. Laguna, J. V. Giraldez, and F. J. Jimenez-Hornero. 2010. Applying a simple methodology to assess historical soil erosion in olive orchards. *Geomorphology* 114 (3): 294–302.

Vanwalleghem, T., J. Amate, M. G. de Molina, D. S. Fernandez, and J. A. Gomez. 2011. Quantifying the effect of historical soil management on soil erosion rates in Mediterranean olive orchards. *Agric Ecosyst Environ* 142 (3–4): 341–351.

Verheijen, F. G. A., R. J. A. Jones, R. J. Rickson, and C. J. Smith. 2009. Tolerable versus actual soil erosion rates in Europe. *Earth-Sci Rev* 94 (1–4): 23–38.

Vinther, M. 2004. Hairy vetch a green manure and cover crop in conservation agriculture. The Royal Veterinary and Agricultural University. Master Science Thesis. Denmark.

Vlek, P. L. G., Q. B. Le, and L. Tamene. 2010. Assessment of land degradation, its possible causes and threat to food security in sub-Saharan Africa. In *Advances in Soil Science—Food Security and Soil Quality*, eds. R. Lal and B. A. Stewart, 57–86. CRC Press, Boca Raton.

Wall, P. C. and J. Ekboir. 2002. Conservation Agriculture for small farmers: Challenges and possibilities, ASA, CSCA, SSSA Meeting, Indianapolis, USA, November 10–14.

Wang, Q. J., Y. H. Bai, H. W. Gao et al. 2008. Soil chemical properties and microbial biomass after 16 years of no-tillage farming on the Loess Plateau, China. *Geoderma* 144: 502–508.

Wang, Q. J., H. Chen, H. W. Li et al. 2009. Controlled traffic farming with no tillage for improved fallow water storage and crop yield on the Chinese Loess Plateau. *Soil Till Res* 104 (1): 192–197.

Wang, X. B., O. Oenema, W. B. Hoogmoed, U. D. Perdok, and D. X. Cai. 2006. Dust storm erosion and its impact on soil carbon and nitrogen losses in northern China. *Catena* 66: 221–227.

Warkentin, B. P. 1995. The changing concept of soil quality. *J Soil Water Conserv* 50: 226–228.

Weaver, D. M. and M. T. F. Wong. 2011. Phosphorus balance efficiency and P status in crop and pasture soils with contrasting P buffer indices: scope for improvement. *Plant Soil* 349: 37–54.

Wieneke, F. and T. Friedrich. 1988. Agricultural Engineering in the Tropics and Subtropics, CENTAURUS, Pfaffenweiler.
Wildner, L. 2000. Soil cover. In *Manual on Integrated Soil Management and Conservation Practices*. FAO Land and Water Bulletins 8. Rome: IITA and FAO.
Xie, Z. K., Y. J. Wang, and F. M. Li. 2005. Effect of plastic mulching on soil water use and spring wheat yield in arid region of northwest China. *Agric Water Manag* 75: 71–83.
Zha, X. and K. Tang. 2003. Change about soil erosion and soil properties in reclaimed forestland of loess hilly region. *Acta Geogr Sin* 58: 464–469 (in Chinese).
Zhang, G. Y., S. X. Zhao, and J. H. Sun. 2004. Analysis of climatological characteristics of severe dust storms in recent years in the northern China. *Clim Environ Res* 9 (1): 101–115.
Zhang, Q., X. Zhao, and H. L. Zhao. 1998. *Sandy Grassland in China*. Beijing: China Weather Press (in Chinese).
Zhang, X. R., H. W. Li, J. He, Q. J. Wang, and M. H. Golabi. 2009. Influence of conservation tillage practices on soil properties and crop yields for maize and wheat cultivation in Beijing, China. *Aust J Soil Res* 47: 362–371.
Zhu, X. 1989. *Soil and Agriculture in the Loess Plateau*. Beijing: Agricultural Science Press (in Chinese).
Zotarelli, L., B. J. R. Alves, S. Urquiaga, E. Torres, K. Paustian, R. M. Boddey, and J. Six. 2003. Efeito do preparo do solo nos agregados do solo e no conteúdo de matéria orgânica. XXXIV Congresso Brasileiro de Ciência do Solo, Ribeirão Preto, Brazil.
Zweier, K. 1985. Energetische Beurteilung von Verfahren und Systemen in der Landwirtschaft der Tropen und Subtropen—Grundlagen und Anwendungsbeispiele; Ph.D. Thesis, University Göttingen; Forschungsbericht Agrartechnik des Arbeitskreises Forschung und Lehre der Max-Eyth-Gesellschaft (MEG) (Agricultural engineering report of the workgroup research and teaching of the Max-Eyth-Society for agricultural engineering) no. 115, 337, Göttingen.

15 Mining of Nutrients in African Soils Due to Agricultural Intensification

Eric T. Craswell and Paul L. G. Vlek

CONTENTS

15.1 Introduction 401
    15.1.1 Soils in Africa..403
    15.1.2 Crop Intensification 404
15.2 Rates of Nutrient Depletion 406
    15.2.1 Methodology..406
    15.2.2 Nutrient Balances at International and Regional Scales 408
    15.2.3 Nutrient Balances at National Scale 409
    15.2.4 Nutrient Balances at Farm and Plot Scales..................................... 411
15.3 Drivers of Nutrient Mining.. 412
    15.3.1 Population Pressure 412
    15.3.2 Low External Nutrient Inputs 413
    15.3.3 Climatic Risk.. 414
15.4 Addressing the Challenge of Nutrient Mining 414
    15.4.1 Recycling Nutrients in Peri-Urban and Urban Agriculture............ 414
    15.4.2 Conservation Agriculture 415
    15.4.3 Integrated Soil Fertility Management... 416
    15.4.4 Policy Issues 416
15.5 Conclusions.. 417
Abbreviations 418
References... 418

15.1 INTRODUCTION

In the twenty-first century, food security in many African countries remains a major concern. A recent United Nations report estimates that in sub-Saharan Africa (SSA—this region is the focus of the chapter, but some reference is made to the rest of the continent), 218 million people, some 30% of the population, suffer from chronic malnutrition (FAO 2009). Average cereal yields in the region of 1.2 Mg/ha contrast with average yields of 3 Mg/ha in the developing world as a whole. Many economic and technological factors have been cited to explain these low crop yields, including the fact that 80% of the total farm area in SSA is made up of the 33 million farms of less than 2 ha. Whereas smallholder farms also dominate agriculture in most areas of Asia, the high yields there were built on intensification utilizing the Green Revolution packages of irrigation, modern crop varieties, and external inputs. Apart from the lack of irrigation in SSA, the most obvious technological factor contributing directly to the low yields is the inadequate use of external inputs. The Food and Agriculture Organization (FAO 2009) estimates that fertilizer use on arable land in SSA in 2002 amounted to only 13 kg/ha compared with 190 kg/ha in Asia. Continuing low yields will not meet the demand for food as the population increases from 770 million in 2005 to between 1.5 and 2 billion in 2050. The consequent intensification through continuous cropping to produce more food will exacerbate land degradation, which is already widespread (Vlek et al. 2010).

Human impacts on nutrient cycling and budgets vary widely across the globe. Vlek et al. (1997) estimate that 230 Tg of plant nutrients are removed yearly from agricultural soils, whereas global fertilizer consumption of N, P_2O_5, and K_2O is 130 Tg. (This chapter utilizes the International System of Units as follows: Mg = 1000 kg [1 metric ton]; Gg = 1,000,000 kg [1000 t]; Tg = 1,000,000,000 kg [1 million t]; Pg = 1,000,000,000,000 kg [1 billion t]. Unless specified as the oxide forms P_2O_5 or K_2O, amounts of P and K are expressed as uncombined elements.) Biological nitrogen fixation (BNF) additionally contributes an estimated 90 Tg. Tan et al. (2005) estimated the global average rates of soil nutrient deficit in 2000 as 19 kg N/ha, 5 kg P/ha, and 39 kg K/ha. Many of the vast cereal-producing food bowl areas of the United States, Canada, Australia, Brazil, and Argentina relied for many years on the mining of native soil fertility before farmers purchased external inputs to maintain yields. More than half a century ago, Martin and Cox (1956) documented the progressive loss of up to 40% of the total nitrogen in fertile Vertisols that had been repeatedly fallowed and cropped without fertilizer inputs in the northern grain belt of Australia. Continued cropping in these areas was eventually possible only with major investments in nitrogen fertilizer production and distribution. The global dimensions of widespread soil fertility decline due to the transformation of land for agriculture are indicated by the decline in soil organic carbon (SOC) levels, estimated to have released 55 to 90 Pg to the atmosphere as carbon dioxide and methane (Lal 2006).

Continuing cropping, which exports nutrients from the farm as harvested product, combined with inadequate fertilizer use, clearly has serious repercussions for the productivity of the soils as well as the release of greenhouse gases (GHGs) as organic matter decomposes. Hartemink (2006) differentiates between nutrient depletion or nutrient decline (larger removal than addition of nutrients) and nutrient mining, which involves no inputs to replace the removal of large quantities of nutrients. While this paper focuses on the implications of nutrient mining due to crop intensification, the term is generally used in the ensuing discussion with the knowledge that it is clearly difficult to differentiate between it and Hartemink's other less severe categories. The important food security implications in SSA of this problem have led to extensive literature on the nutrient balance at a range of scales (see Vlek 1990; Stoorvogel and Smaling 1990; Stoorvogel et al. 1993; Smaling et al. 1993; Van den

Bosch et al. 1998; Henao and Banaante 1999, 2006; Cobo et al. 2010). This chapter examines the causes and extent of long-term nutrient mining in African cropping systems, focusing particularly on the impacts on the soils. The conclusion considers ways to arrest the decline in soil fertility while increasing crop yields.

15.1.1 Soils in Africa

The initial soil fertility levels in the soils of Africa, as elsewhere, are largely determined by the soil organic matter (SOM) content, which in turn depends strongly on the type of climax vegetation. The climax vegetation depends on the climate, which in Africa ranges from arid, with a length of growing period of 0 days, to humid, which has a length of growing period of >270 days. Vegetation types vary accordingly from desert to tropical forests. Sanchez et al. (1982) suggested that some soil orders in the tropics (Ultisols, Alfisols, and Mollisols) have similar levels of organic matter as the same soil orders in temperate regions. However, Vlek and Koch (1992) point out that published studies for Africa indicate that median SOM contents vary greatly between regions, with 4% in East Africa, 0.5% in West Africa, and a range from 0.12% to 0.17% in the Sahelian zone. They concluded that overall, the natural endowment of African soils appears limited compared to other parts of the world (Table 15.1).

Eswaran et al. (1997) estimate that with the low inherent fertility of African soils and subsequent degradation of the land, only 16% of the land can be classified as having soil of high quality, whereas 13% has soil of medium quality (Table 15.2). A total of 9 million km² of high- and medium-quality arable land supports 400 million people, making up 45% of the total population, and this area is the main focus

TABLE 15.1
Distribution of Major Soil Orders in Tropical Africa

Soil Order	Area (Million ha)	Percentage
Oxisol	550	22
Aridisols	840	34
Alfisols	550	23
Ultisols	100	4
Inceptisols	70	3
Entisols	300	12
Vertisols	40	2
Total	**2450**	**100**

Source: Vlek, P.L.G., and H. Koch, Göttinger Beiträge zur Land—und Forstwirtschaft in den Tropen und Subtropen, 71, 139–160, 1992.

TABLE 15.2
Proportion of Land (%) of Continental Africa in Each Population Density/Soil Quality Class

Soil Quality	<10% Land Area	10–100% Land Area	100–200% Land Area	200–400% Land Area	>40% Land Area	Total % Land Area
High	3.7	10.8	1.1	0.5	0.1	16.3
Medium	6.1	6.1	0.4	0.2	0.1	12.8
Low	6.2	8.0	0.9	0.3	0.1	15.5
Unsustainable	44.3	9.4	0.6	0.1	0.1	54.6
Total	60.3	34.4	3.0	0.1	0.4	99.1[a]

Population Density—Persons per km²

Source: Eswaran, H. et al., *J. Sustain Agric*, 10, 75–94, 1997.
[a] An additional 0.9% of the land area is made up of water bodies—total land area is 30,650.2 km².

for the nutrient mining discussed in this chapter. As much as 55% of the land area is nonarable but supports a further 250 million people or 30% of the population, made up mainly of nomadic herders.

15.1.2 Crop Intensification

Sustainable agriculture is not a new concept to African farmers cultivating food crops. Traditionally, their practice of shifting cultivation continued for millennia and involved hand clearing of forest and savanna land to plant crops for 2 or several years, after which the sites were abandoned and revert to bush for a lengthy (at least 10 years) fallow period. The hand clearing left the largest trees in place, thus stabilizing the soil against erosion, whereas the use of fire to clear ground vegetation and smaller trees provided charcoal and ash that enriched the soil by returning part of the nutrients stored in the fallow vegetation to the soil, providing nutrients to crops (Nye and Greenland 1960). During the long fallow periods, regrowth of vegetation recycled nutrients from the root zone and leaf litter and recharged the SOM pool. However, the cost of vegetation burning in volatilized nutrients is substantial (Kugbe et al. 2012). Provided the population density was limited, land was abundant, and consequently, bush fallow periods were long, the shifting cultivation system was sustainable.

Figure 15.1 shows the classic path for the intensification of agriculture from shifting cultivation to repeated cropping of land. The shifting cultivation system practiced by subsistence farmers intensifies through reduced fallow length until the land is cropped every year. This practice by subsistence farmers leads to the declining soil fertility. Consequently, the reduced yields and vegetation biomass expose the land to wind and water erosion, the latter of which is a particular concern in the low-potential steeplands of humid tropical areas (Craswell 2000).

The mining of soil fertility with annual cropping can be arrested by the use of organic and mineral fertilizer inputs. However, in most countries of SSA, these are

Mining of Nutrients in African Soils Due to Agricultural Intensification

```
                    Increasing intensity of crop cultivation
    ----------------------------------------------------->
    |    Subsistence systems            |    Cash cropping systems        |
    |    Poor farmers, labor out-migration    |    Land value high, peri-urban  |
    |---------------------------------|---------------------------------|
    |  Shifting cultivation  | Annual cropping |  2–3 crops  |  Continuous  |
    |  Long fallow–short fallow |              |  per year   |  vegetables  |
    ---------------------------------->---------------------------------->
     Declining soil fertility and organic matter    Increasing inputs and income
            Increasing erosion risk                 Increased pollution risk
```

FIGURE 15.1 Impacts of the intensification of cropping.

often unavailable or unaffordable. In areas with access to produce markets, a shift to cash cropping can make purchased inputs affordable. In the most intensive cash cropping areas, two or even three crops are grown annually, but careful management of fertilizer inputs becomes critical in order to reduce the risks of pollution of surface water and groundwater, especially in urban vegetable production systems. According to Drechsel et al. (2008), the 20 million farmers in urban areas of West Africa could productively use wastewater for irrigation or recycle nutrients in animal manures provided that mechanisms for reducing health risks are put in place.

The highest rates of depletion of soil fertility occur when subsistence farmers, who cannot afford inputs, increase cropping intensity to one or more crops per year. McCown and Jones (1992) summarized the problem and depicted it as a spiral poverty trap (Figure 15.2). The main drivers are population growth and low use of

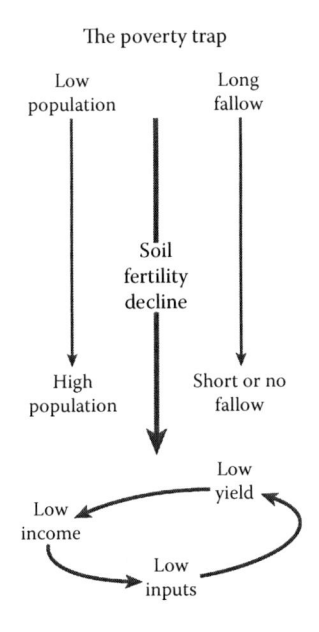

```
                        The poverty trap

        Low                              Long
     population                          fallow
         |                                 |
         |                                 |
         |         Soil                    |
         |       fertility                 |
         |        decline                  |
         |          |                      |
         ↓          ↓                      ↓
        High                         Short or no
     population                         fallow
                              Low
                         ←--  yield  ←--
                    Low                    
                   income                  
                         --→  Low    --→
                              inputs
```

FIGURE 15.2 Poverty trap. (Adapted from McCown, R.L., and R.K. Jones, Agriculture of semi-arid eastern Kenya: Problems and possibilities, In *A Search for Strategies for Sustainable Dryland Cropping in Semi-arid Eastern Kenya*, ed. M.E. Probert, 8–14, Proceedings No. 41, Canberra, Australian Centre for International Agricultural Research, 1992.)

inputs, where low income leads to low inputs, low yields, and so on. While the number and distribution of African subsistence farmers caught in the poverty trap is not well known, the widespread incidence of the problem in various countries can be surmised from the calculation of soil nutrient balances using models and available national level data as well as data on other scales.

15.2 RATES OF NUTRIENT DEPLETION

15.2.1 Methodology

Nutrient balance calculations can be made at a range of scales from plot to farm to region to nation and beyond. At the global level, Grote et al. (2005) and Craswell et al. (2003) assessed the net flows of nutrients in commodities transported internationally in trade. The total Nitrogen, Phosphorus and Potassium (NPK) in the different food commodities can be calculated from the amount of the commodities multiplied by average values for their N, P, and K contents. One problem is that these studies have not been able to source and include data on plantation crops, which commonly receive much of the fertilizer used in many African countries. Furthermore, this approach does not provide information about within-country nutrient flows, which in an increasingly urbanized Africa may largely be in the direction from rural to urban areas (Craswell et al. 2003; Drechsel and Zimmermann 2005). Nevertheless, whether derived from nationally or regionally aggregated data, the international trade balance provides an indication of the massive scale of NPK movement in food commodities globally.

At the national and farm scale, NPK mining can be calculated from the following (Henao and Baanante 1999):

$$Rn_t = \sum{}^t (AP_t + AR_{\Delta t} - RM_{\Delta t} - L_{\Delta t}) \quad (15.1)$$

where Rn_t is the amount of nutrient coming from organic and inorganic sources remaining available in the soil after a period of time t; AP_t is the inherent soil nutrient available at time t; $AR_{\Delta t}$ is the total of the nutrient in mineral and organic compounds added or returned to the soil during time interval Δt; $RM_{\Delta t}$ is an estimate of the plant nutrients removed in crop harvest and residues during time interval Δt; and $L_{\Delta t}$ is the amount of nutrient lost via various pathways during time interval Δt. The nutrient balance calculation generally is based on estimates of the extent of five inputs—mineral fertilizers, organic inputs, BNF, wet/dry deposition, and sedimentation—and of five losses, such as crop product, crop residues, gaseous losses, leaching, and soil erosion. Partial nutrient balances can be estimated from data on fertilizer use and crop yields, utilizing databases such as FAOSTAT. Sheldrick et al. (2002) combined FAOSTAT data with nutrient efficiency ratios to develop nutrient audits without resorting to more detailed modeling. More detailed analysis tools such as the International Fertilizer Development Center (IFDC) model and NUTMON (Henao and Baanante 1999; Stoorvogel et al. 1993) applied at the farm or supranational level (Figure 15.3) require estimates of transfer functions for soil erosion and other components of nutrient losses and gains. Of the three major

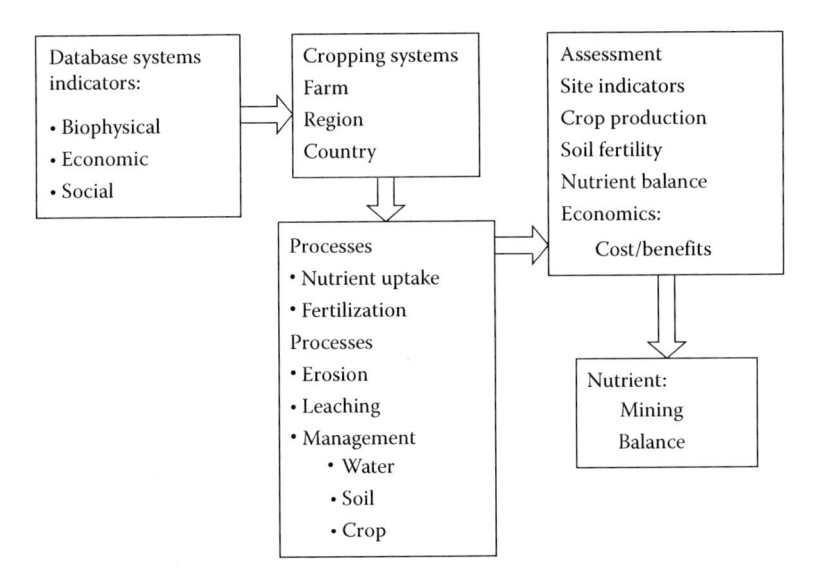

```
┌─────────────────┐   ┌──────────────────┐   ┌──────────────────┐
│ Database systems│   │ Cropping systems │   │ Assessment       │
│ indicators:     │──▶│ Farm             │   │ Site indicators  │
│ • Biophysical   │   │ Region           │   │ Crop production  │
│ • Economic      │   │ Country          │   │ Soil fertility   │
│ • Social        │   └──────────────────┘   │ Nutrient balance │
└─────────────────┘            │             │ Economics:       │
                               ▼             │   Cost/benefits  │
                       ┌──────────────────┐  └──────────────────┘
                       │ Processes        │           │
                       │ • Nutrient uptake│──▶        │
                       │ • Fertilization  │           ▼
                       │ Processes        │  ┌──────────────────┐
                       │ • Erosion        │  │ Nutrient:        │
                       │ • Leaching       │  │   Mining         │
                       │ • Management     │  │   Balance        │
                       │    • Water       │  └──────────────────┘
                       │    • Soil        │
                       │    • Crop        │
                       └──────────────────┘
```

FIGURE 15.3 Schema used in nutrient balance calculations. (Adapted from Henao, J., and C.A. Baanante, *Agricultural Production and Soil Nutrient Mining in Africa: Implications for Resource Conservation and Policy Development*, International Fertilizer Development Center, Muscle Shoals, Alabama, 2006.)

nutrients—NPK—nitrogen is the most dynamic, being subject to gaseous losses as well as leaching and runoff. The necessity to utilize such estimates of nitrogen loss and other processes of NPK loss in soil particles eroded by wind or water opens the nutrient balance models to criticism. From a review of the transfer functions used in NUTMON, and their impact on the nutrient balance calculations, Faerge and Magid (2004) concluded that the transfer functions used have a strong tendency to overestimate losses. Lesschen et al. (2007) refined the methodology using transfer functions based on more up-to-date information and adding a spatially explicit dimension to a nutrient balance calculation for Burkina Faso. Clearly, any method used to calculate nutrient balances requires a leap of faith as data are estimated across landscapes, farming enterprises, social equity dimensions, and time.

A recent comprehensive review of literature on nutrient balance in Africa collected data sets from 57 published studies, which used a wide range of study types, nutrient types, spatial units for balance data, types of balance, and time frames, as shown in Table 15.3 (Cobo et al. 2010). Most of the balances were calculated at the plot and farm scale and mostly in East Africa. The high proportion of studies showing negative balances (>75% for N and K and 56% for P) support the general conclusion that soil nutrients are being mined. On the other hand, positive balances, particularly for P (44%), were also reported, indicating the use of inputs, particularly by wealthier farmers who presumably were producing cash crops. Only 40% of the studies linked the balance to soil nutrient stocks, and 79% did not show the variability of the balance data. Cobo et al. (2010) point out the many pitfalls of nutrient

TABLE 15.3
Selected Characteristics of 57 Nutrient Balance Studies Reviewed by Cobo et al. (2010)

Characteristic	Number of Studies	% of Studies
Country		
Kenya	19	33
Ethiopia	8	14
Mali	7	12
Uganda	6	11
Study Type		
Agroecosystem assessment	42	74
Experiment	13	23
Scenario/simulation	8	14
Nutrients		
N	55	96
P	47	82
K	36	63
Time Frame		
1 year	23	40
1 season	11	19
2 years	8	14
Results Showing Negative Balances		
N	48	85
P	32	56
K	43	76
Studies Showing Positive Balances		
N	14	24
P	25	44
K	9	15

Source: Cobo, J.G. et al., *Agric Ecosyst Environ*, 136, 1–15, 2010.

balance estimations especially when results are scaled up spatially and temporally. These limitations suggest that many published nutrient balance papers seek to provide an indication of trends rather than precise estimates.

15.2.2 Nutrient Balances at International and Regional Scales

Net international trade of agricultural food commodities leads to major nutrient transfers (Figure 15.4). The analysis by Grote et al. (2005) shows that in 1997, through net trade, SSA imported the equivalent of 0.266 Tg of NPK, projected to

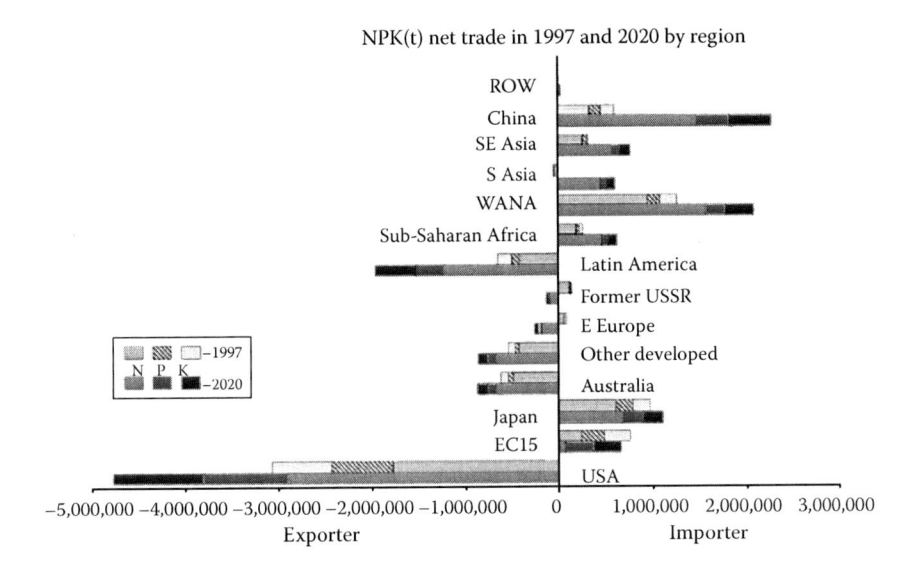

FIGURE 15.4 Nutrient flows in net trade of agricultural commodities in 1997 and projected to 2020. (From Craswell, E.T. et al., Nutrient flows in agricultural production and international trade: Ecological and policy issues, ZEF Discussion Paper on Development Policy No. 78, Center for Development Research, University of Bonn, Bonn, 2003.)

increase to 0.662 Tg of NPK in 2020. This positive side of the net trade balance is much less than that in less populous, net food importing, regions such as West Asia and North Africa, reflecting the subsistent nature and self-sufficiency of many countries in SSA. It is also closely associated with differences in the availability of water between the regions and the need for virtual water trade (Grote et al. 2007). At first sight, the positive nutrient stream due to food imports to SSA may appear to contradict concerns about soil nutrient mining in rural areas due to low rates of fertilizer inputs. However, the data presented are averages across and within countries. If, as appears likely, the nutrients imported in food are consumed and end up as wastes in the major cities, distant rural lands will not benefit. The imported nutrients represent a useful resource, considering that the 1997 level represents 26% of the 1.102 Tg of NPK fertilizer used in that year. As mentioned above, an issue in SSA is that the fertilizer data used by Grote et al. (2005) include nutrients applied to plantation crops, whereas the NPK net trade data exclude plantation and industrial crops. Vlek (1993) estimated that in 1987, the net export from Africa of N, P_2O_5, and K_2O in agricultural commodities, mainly cotton (*Gossypium* spp.), tobacco (*Nicotiana tobacum*), sugar (*Saccharum officinarum*), coffee (*Coffea* spp.), cocoa (*Theobroma cacao*), and tea (*Camellia sinensis*), was 0.296 Tg.

15.2.3 NUTRIENT BALANCES AT NATIONAL SCALE

Several major studies have utilized NUTMON or similar formulae to calculate nutrient balances for arable land on an average basis for different African countries. Table 15.4 shows a comparison of the data on nutrient mining from two such

TABLE 15.4
Average Country Nutrient Balances in Selected African Countries

	Year (Average)					
	Henao and Baanante			Stoorvogel et al.		
	(N + P$_2$O$_5$ + K$_2$O kg/ha)			(N + P$_2$O$_5$ + K$_2$O kg/ha)		
Country	1981–1985	2002–2004	% Change	1982–1984	2000	% Change
Benin	−37	−44.0	19	−27	−33	22
Botswana	−1	−47.0	460	−1	−4	300
Cameroon	−25	−44.0	76	−34	−41	21
Ethiopia	−56	−49.0	−12	−73	−102	40
Ghana	−32	−58.0	81	−50	−68	36
Kenya	−48	−68.0	42	−74	−91	23
Malawi	−42	−72.0	71	−144	−148	3
Mali	−34	−49.0	44	−18	−27	50
Nigeria	−35	−57.0	63	−72	−83	15
Rwanda	−101	−77.0	−24	−131	−158	21
Senegal	−38	−41.0	8	−26	−37	42
Tanzania	−44	−61.0	39	−58	−68	17
Zimbabwe	−12	−53.0	342	−62	−62	0

Source: Henao, J., and C.A. Baanante, *Estimating Rates of Nutrient Depletion in Soils of Agricultural Lands of Africa*, International Fertilizer Development Center, Muscle Shoals, Alabama, 1999; Stoorvogel, J.J. et al., *Fert Res*, 35, 227–235, 1993.

Note: All data converted to N, P$_2$O$_5$, and K$_2$O so that estimates could be directly compared.

studies—Henao and Baanante (1999, 2006) and Stoorvogel et al. (1993). The countries listed have been selected on the basis of the list published by Stoorvogel et al. (Henao and Baanante had a more complete list). Both studies show consistently high annual rates of nutrient depletion both in the early 1980s and 20 years later. In most countries listed, both data sets show that over the 20-year periods shown, the rates of nutrient mining continued to increase significantly. Several countries show inconsistent trends in one or the other data set. Botswana, which has a relatively high income, shows the lowest nutrient depletion between 1981 and 1985. This increases significantly to 147 kg ha^{-1} year^{-1} in the 2002 to 2004 data of Henao and Baanante but not Stoorvogel et al. Zimbabwe, on the other hand, had a low level of depletion between 1981 and 1985 of 12 kg ha^{-1} year^{-1}, but in the equivalent period, Stoorvogel et al. estimated a nutrient mining rate of 62 kg ha^{-1} year^{-1}. Finally, the data of Henao and Baanante show a decline in nutrient mining rates in Rwanda from 101 to 77 kg ha^{-1} year^{-1}, whereas the data of Stoorvogel et al. show the opposite trend and higher rates of depletion. These inconsistencies aside, the major conclusion from the country-level studies in Table 15.4 is that NUTMON-type national level balances show high and continuing average rates of nutrient mining in most SSA countries listed.

TABLE 15.5
SSA Nations Grouped According to On-Site Costs of Restoring Mined Nutrients as a Percentage of the AGDP

Country	% of AGDP
Benin, Botswana, Cameroon, Central African Republic, Democratic Republic of the Congo, Republic of the Congo, Gabon, Ghana, Guinea, Kenya, Mauritania, Mauritius, Sierra Leone, Swaziland, Zambia, Zimbabwe	≤5
Angola, Burkina Faso, Burundi, Chad, Côte d'Ivoire, Ethiopia, Lesotho, Madagascar, Malawi, Mali, Nigeria, Senegal, Togo, Uganda	6–11
Mozambique, Niger, Rwanda, Tanzania	≥11
SSA (Average)	7

Several studies have considered the economic dimensions of nutrient mining, but the nutrient replacement costs at farm-gate prices probably capture the national-level impact better than other more detailed analyses. Drechsel et al. (2001) based their national and regional economic analysis on the nutrient balance data published by Stoorvogel and Smaling (1990). Data on nutrient mining in 37 African countries were used to calculate the replacement costs as a proportion of the agricultural gross domestic product (AGDP). The results summarized in Table 15.5 show that countries such as Rwanda, Mozambique, Niger, and Tanzania have nutrient replacement costs of >25% of AGDP. Across SSA, the average nutrient replacement costs are 7% of AGDP, indicating that nutrient mining is draining the natural resource capital of most countries at an unsustainable rate.

15.2.4 Nutrient Balances at Farm and Plot Scales

Studies of nutrient mining at a detailed level exhibit the wide variability under the influence of localized conditions such as resource endowment, soil type, erosion rates, input use, cropping systems, and so forth, as well as the conditions under which studies are made, such as spatial–temporal boundaries (see comprehensive review by Cobo et al. 2010). To illustrate the extent of this variability, the results of several diverse detailed studies are summarized in Table 15.6. The study area for Harris (1998) in Nigeria was an intensively cropped area where legume nitrogen fixation and the recycling of N through ruminant animals were a major factor limiting the mining of soil fertility. Large negative balances in the Ethiopian studies can be attributed to high rates of soil erosion, whereas the balances in Uganda varied widely amongst different land use types, in some cases reflecting the transfer of nutrients to high-value crops such as bananas (*Musa paradisiaca*). The degree of recycling of nutrients from crop residues and manure in mixed crop–livestock systems was clearly a determinant of eventual farm balances. In most studies, the negative balances of N exceeded those of P and K, reflecting the dynamic and loss-prone features of this nutrient.

TABLE 15.6
Nutrient Balance Data from Selected Published Studies in Different Countries

Country	Land Use	Scale of Study	Nutrient Balance (kg ha^{-1} year^{-1}) N	P	K	Authors
Nigeria	Intense cropping	Farm	−28 to +3	−3 to +3	−	Harris (1998)
Ethiopia	Grain cropping	Farm	−102 to 112	−4 to −3	−70 to −65	Okumu et al. (2004)
Ethiopia	Crop–livestock	Field/farm	−102 to +1	+1 to +12		Elias et al. (1998)
Uganda	Crop–livestock	Crop/field/farm	−125 to +18	−8 to +4	−19 to +10	Wortmann and Kaizzi (1998)
Kenya	Crop–livestock	Farm	−71	+3	−9	Van den Bosch et al. (1998)
Mozambique	Mixed cropping	Land use system	−122 to −13	−27 to 0	−97 to −14	Folmer et al. (1998)
Mali	Crop–livestock	Farm	−25	−20	−5	Pol and Traore (1993)

15.3 DRIVERS OF NUTRIENT MINING

15.3.1 POPULATION PRESSURE

As discussed above, shifting cultivation systems with extended fallow periods allows time for the bush fallow to restore a degree of soil fertility following a short period of cropping. This is changed as increased population pressure drives crop intensification, leading to soil nutrient mining. Table 15.2 shows the levels of population pressure on land with different soil quality, but this has been taken a step further; Drechsel et al. (2001) combined population density data with nutrient balance estimates to show how the degree to which cropping mines the soil N is directly related to population (Figure 15.5). They concluded that Malthusian mechanisms were at work, making the population–agriculture–environment nexus unsustainable. This finding runs counter to the Boserup-type success story of Tiffen et al. (1994), who coined the phrase "more people, less erosion." This concept is based on the idea that high population density leads to greater care being taken by farmers in conserving their soil resource. Unfortunately, as pointed out by Simpson et al. (1996), Tiffen et al. (1994) did not document the mining of soil nutrients in the Machakos district, which they studied from a retrospective point of view, focusing on the implementation of soil conservation measures. Simpson et al. (1996) speculated that the reduced soil erosion would have increased soil water availability, increasing yields and thus

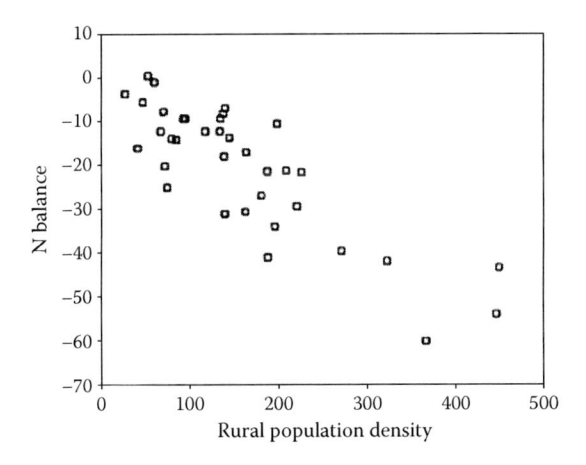

FIGURE 15.5 Relationship between nitrogen balance in soils and rural population density. (From Drechsel, P. et al., *Popul Environ*, 22, 411–423, 2001.)

the rate of nutrient mining in harvested products. As demonstrated by Kaizzi et al. (2004) in Uganda, the increase in production of maize (*Zea mays*) due to fertilizer use did not pay for the investment in all but the most favorable production environments, even when the yields doubled or tripled in the less favorable sites. Without fertilizer subsidies, farmers will have to generate income from cash crops for the purchase of fertilizer inputs or face continued declining soil productivity as cropping intensifies.

15.3.2 Low External Nutrient Inputs

The most striking and tangible feature of the nutrient mining problem in SSA is the continuing low level of fertilizer use in the region over the past 25 years. Figure 15.6 shows

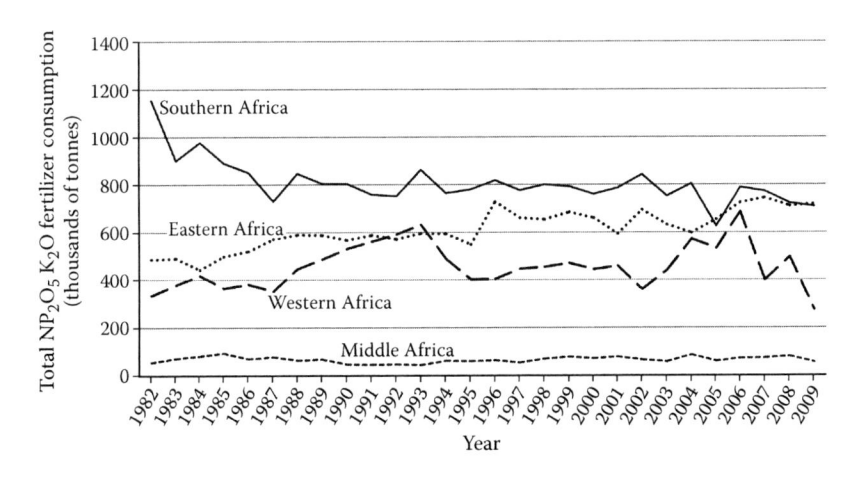

FIGURE 15.6 Total fertilizer consumption in different regions of Africa, 1982–2008. (From FAOSTAT, Fertilizers, http://faostat.fao.org/site/575/default.aspx#ancor (accessed July 14, 2012).)

the NPK fertilizer consumption in four regions between 1982 and 2008. The lowest consumption is in middle or central Africa, and there appears to be no upward trend in recent years. Southern Africa shows a declining consumption of fertilizer nutrients over the period covered, whereas the data for Western Africa show considerable variability with time. The only consistent upward trend is shown in Eastern Africa. Considering that during the period shown, the total population of SSA increased from 400 million to over 700 million (FAO 2009), the lack of a similar rate of growth in fertilizer consumption helps explain the stagnation of food crop yields at 1.2 t ha^{-1}. Looking to the future, Tenkorang and Lowenberg-DeBoer (2009) forecast fertilizer nutrient use in SSA to grow from a total of 1.4 Tg in 2005 to only 2.7 Tg in 2030, only 1.1% of global consumption. According to these authors, the resulting continued soil nutrient drawdown will, in the long run, exacerbate food shortages and undermine plans to produce biofuels.

15.3.3 Climatic Risk

In SSA, only 4% of arable land is irrigated, compared with 20% globally and 38% in Asia. Many ascribe the lack of adoption of fertilizer inputs by farmers in Africa—that is, a "Green Revolution" such as Asia experienced—to the risk associated with dryland farming (Cooper and Coe 2011). This problem is exacerbated by climate change, as the variability and unreliability of rainfall increases. Crop simulation models are valuable tools for assessing the risk of fertilizer use if climate change models generate plausible rainfall patterns for future seasons. Dixit et al. (2011) showed that 50 years of simulations by the Agricultural Production Systems Simulator (APSIM) model gave maize yield responses to N fertilizer that agreed with observed data published in the literature. They also found that across seasons, the rates of return to investment in N fertilizer use varied from 1.1 to 6.2. This level of variability in risk must be a serious deterrent to smallholder farmers. Farmers are well aware of the increasing risk due to climate change (Fosu-Mensah et al. 2012), and such risks are likely to further hamper the adoption of fertilizers in the future.

15.4 ADDRESSING THE CHALLENGE OF NUTRIENT MINING

Irrespective of the temporal and spatial variability in the data, the published nutrient balances at the national scale together with the statistics showing low average rates of fertilizer use by farmers indicate clearly that nutrient mining in African soils continues unabated as cropping continues to intensify to meet the demands of a burgeoning population. The obvious remedy to this problem, based on the experience of farmers in industrialized countries, is to replace or to at least slow down the loss of soil nutrients by applying external organic and/or mineral sources. Our consideration of this topic forms the concluding section of the chapter.

15.4.1 Recycling Nutrients in Peri-Urban and Urban Agriculture

Tracing the nutrients mined from rural areas, the trail leads to urban areas, which are increasingly also a magnet for rural people looking for employment and where population grows at an estimated 4.6% per year (Drechsel et al. 2008). One of the important features of income growth in cities is the increased demand for animal protein, leading to the establishment of intensive poultry and other livestock production systems. The resultant manure resources provide a valuable source of nutrients for urban agriculture that can thus maintain the productivity of soils that previously were subject to nutrient mining (Drechsel et al. 2007). In addition to the nutrients mined from rural areas, nutrients imported in feed and food grains may contribute significantly to the resources available to urban agriculture. Grote et al. (2005) suggested that the positive nutrient balance in the international trade in agricultural commodities—estimated at 0.268 Tg in 1997, rising to 0.63 Tg in 2020—may largely be utilized in urban areas rather than rural areas. African cities have serious problems in disposing of the resulting solid and liquid wastes, but Drechsel and Zimmermann (2005) found that while manures and wastewater were not used by subsistence farmers, farmers producing high-value vegetables for the urban inhabitants were open to such methods of soil fertility enhancement. Drechsel and Zimmermann contend that, as land availability reduced the bush fallow periods, farmers in their study area near Kumasi in Ghana had three courses of action: (1) look for more land to continue bush fallow at a more remote area; (2) change their farming system to meet specific demands from urban areas; and (3) abandon farming and seek employment in the city. The cutoff point in terms of the length of bush fallow period is an important area for further research.

15.4.2 Conservation Agriculture

Conservation agriculture (CA) largely based on minimum tillage or no-till (NT) has been widely adopted in the major cropping areas of North and South America and Australia, as a substitute for mechanized plowing. The three main elements of CA are minimum soil disturbance, permanent soil cover, and species diversification in the cropping system (FAO 2008). In formerly mechanized systems, the use of herbicides with NT reduces the use of fossil fuels, reduces soil erosion, and helps reduce the rate of SOC depletion (Vlek and Tamene 2009). It could be said that some elements of CA are already practiced by many subsistence farmers in SSA, since they do not disturb the soil, having not "progressed" to mechanized tillage, and their farming systems are generally based on mixed cropping as their method for the management of risk. From a nutrient perspective, the retention of high-carbon residues may require increased use of N fertilizer to counteract N immobilization (Kihara et al. 2011). Given the cost and availability constraints to N fertilizer use, legume crops that have effective symbioses with root nodule bacteria may provide the African farmer with an important and cheap source of nitrogen. The potential role of biologically fixed N from including legumes in rotations has been well documented (Peoples and Craswell 1992), and recent experiments show the benefits to maize of including crops like velvet bean (*Mucuna pruriens*) in the cropping systems (Kaizzi et al. 2004). Nevertheless, many soils need P fertilizer inputs to ensure effective legume growth and N fixation, so general economic constraints to fertilizer supply and affordability may constrain adoption of these solutions to the nutrient balance problem. Regarding attitudes to CA in Africa, a recent paper by Giller et al. (2009) has listed some of the problems with CA in Africa, including decreased yields, increased labor requirements when herbicides are not used, increased labor for women, and the high priority given to feeding crop residues to livestock. Added to a lack of access to, and use of, external inputs, these constraints take the shine off CA as a panacea for the problem of poor yields. More research involving smallholder participation is needed to assess practical ways of implementing the principles of CA in the context of African farms.

15.4.3 Integrated Soil Fertility Management

Integrated soil fertility management (Sanginga and Woomer 2009; Vanlouwe and Giller 2006) and low-external-input agriculture (Graves et al. 2004) are both based on the principle that farmers should use available and affordable organic and mineral nutrient sources judiciously to enhance crop productivity. By increasing the recycling of nutrients within the farm, the decline in soil nutrient levels can be partially re-dressed and crop yields improved. Since many smallholder farms combine livestock and cropping, ways have been sought to increase the recovery, recycling, and synchronization of manure with external nutrient inputs (Saleem 1998). In practice, labor and other constraints affect the allocation of manures and other nutrient resources by smallholder farmers, who may favor some fields, often those closer to their houses, and in the process pass up some potential benefits in crop yield (Rowe et al. 2006). However, simply redistributing nutrients within a farm is not a long-term solution; external inputs are needed to balance the output in harvested products. The inputs can come as purchased fertilizer or as N fixed by leguminous tree/crops, which can be planted as companion, rotation, or fodder species. Agroforestry systems based on legume trees and shrubs have been advocated for many years. The best-known example is alley cropping, in which crops are planted between rows of legume shrubs, the foliage of which is used as a nutrient supplement to the crops. Adoption of alley cropping has been limited by problems with the effects on alley crop yields of competition with the legume for soil water and labor costs (Akinnifesi et al. 2010). The labor factor depends on the ratio of the cost of labor to the value of the crop. In fragile steep land areas, planting the legume shrubs in hedgerows on the contour has the important additional benefit of reducing soil erosion and may be attractive to farmers if the harvest of some trees planted in the hedgerows brings cash that can also be used to purchase fertilizers (Craswell et al. 1998). Legumes can be used to improve fallows and as green manures, but the opportunity costs of labor and of leaving land fallow, combined with immediate concerns for providing food, limit the adoption of these technologies (Mafongoya et al. 2006). Garrity et al. (2010) have proposed another model called evergreen agriculture, which utilizes woody legume trees, such as *Faidherbia albida*, grown at a spacing of 25 to 100 per hectare and managed to transfer biomass and nutrients to food crops grown in association with the trees. This approach builds on the early research on *F. albida* by French scientists (CTFT 1988; Pieri 1989).

15.4.4 Policy Issues

The sustainable intensification of food crop production in Africa demands a solution to the problem of soil nutrient mining. Pretty et al. (2011) reviewed 40 relevant projects in 20 countries during the 1990s and 2000s. The lessons learned included the need to scale up and spread science and farmer inputs into technologies and practices that combine crops and animals with agroecological and agronomic management. Additionally, the private sector must be engaged; capacities and knowledge of farmers should be enhanced. The microfinance and technology needs of women should be a focus, and public sector support for agriculture should be ensured. An indicator of progress with the last issue is the Abuja Declaration from the African Union Ministers of Agriculture at the 2006 Africa Fertilizer Summit, which declared fertilizer, from both inorganic and organic sources, to be a strategic commodity without borders (African Development Bank 2012). The stated goal was to increase the level of fertilizer use from the current average of 8 kg ha^{-1} to at least 50 kg ha^{-1} by 2015. Achieving such a goal would no doubt effectively resolve the nutrient mining problem. As a directly related case study, the Malawi subsidy scheme was devised to supply improved seed and heavily subsidized fertilizer to maize producers during the period from 2005 and 2006 to 2008 and 2009 (Dorward and Chirwa 2011). Maize productivity was increased as were food availability, real wages, and poverty reduction. The economic impact of the program was clear, but high international fertilizer prices and higher maize prices undermined the poverty reduction impacts. Subsequent commodity price reductions from the 2008 spike provide a scope for building on the original program achievements. An alternative or complementary form of subsidy may arise from carbon pricing, which Marenya et al. (2012) estimate may provide a means for farmers to break through the resource barriers constraining their adoption of productivity-enhancing nutrient management technologies. This approach, driven by global concern about the effects of GHGs from SOM decomposition on climate change, may provide valuable support to African farmers, who themselves face serious environmental, economic, and social impacts from global warming (Fosu-Mensah et al. 2012).

15.5 CONCLUSIONS

With half the population of Africa now living in cities, it is unavoidable that large quantities of nutrients mined from the farms are concentrating there and in need of disposal. With many cities located along rivers or near the coastline, it should be no surprise that many of these nutrients are finding their way into the oceans, which makes them almost impossible to retrieve, and which causes untold environmental damage. This review indicates that concerns about nutrient mining of SSA soils are well justified. The evidence for this conclusion is founded on the following: farmers intensify their crop cultivation and reduce fallow periods as population grows; statistics show the continuing low rates of average fertilizer use in most countries; and research data show negative nutrient balances at a range of scales. There is spatial and temporal variation in the rate and impact of nutrient mining, but the law of conservation of mass dictates that nutrients repeatedly removed in harvests must eventually be replaced, or else crop yields will suffer and farming will become uneconomic. The social impacts of declining yields, then, are manifested in the migration of rural populations to urban areas, driving additional demands for food and further nutrient exports from the land.

In its simplest terms, the problem of nutrient mining calls for minimizing the export of nutrients by returning as much of the residues to the field as possible. Local processing of foodstuffs would favor recycling as nutrients in residues cannot be economically returned to the field when distances exceed tens of kilometers. The remaining deficit will require the replacement/supplementation of soil nutrient reserves by nutrient additions. In the case of N, this can be achieved by the use of legumes that fix N from the atmosphere (BNF), and various models such as intercropping and relay cropping with leguminous ground cover, *Azolla* culture, and agroforestry systems have been proposed. Soil scientists and agronomists have proposed many novel systems, but there is no magic bullet. The research base on the various systems is extensive, but the rate of adoption has been disappointing. The overarching lesson of the past is that development of such complex cropping systems needs participation of farmers as a way of ensuring that innovations have a reasonable chance of adoption.

Ultimately, SSA farmers will need to expand their use of fertilizers beyond the current rate of 10 kg ha^{-1} to guarantee food security over the long term. This will require farmers to have liquidity for investment and to move away from purely subsistence farming and introduce cash crops, preferably on the same farms. Government policies and programs that facilitate or even subsidize fertilizer marketing and distribution are needed, as well as efforts to educate farmers on the use of fertilizers. Governments will also have to take measures to avoid the market distortions due to importation of heavily subsidized food grains from industrialized countries, against which SSA farmers cannot compete. Progress may ultimately depend on whether sufficient political will can be mobilized in Africa and the industrialized countries to remedy an unsustainable agricultural sector in SSA. If as a consequence, the nutrient balance on the farm in SSA has been brought into equilibrium, there remains only the problem of disposing of nutrients accrued in cities in an environmentally acceptable fashion. That problem may be resolved in part by the recycling of the waste nutrients in peri-urban agriculture.

ABBREVIATIONS

AGDP: agricultural gross domestic product
BNF: biological nitrogen fixation
CA: conservation agriculture
GHGs: greenhouse gases
NT: no-till
SOC: soil organic carbon
SOM: soil organic matter
SSA: sub-Saharan Africa

REFERENCES

African Development Bank. 2012. Abuja declaration on fertilizer for the African green development. http://www.afdb.org/en/topics-and-sectors/initiatives-partnerships/african-fertilizer-financing-mechanism/abuja-declaration/ (accessed July 14, 2012).

Akinnifesi, F.K., O.C. Ajayi, G. Sileshi, P.W. Chirwa, and J. Chianu. 2010. Fertilizer trees for sustainable food security in the maize-based production systems of East and Southern Africa. A review. *Agron Sustain Dev* 30:615–629.

Cobo, J.G., G. Dercon, and G. Cadisch. 2010. Nutrient balances in African land use systems across different spatial scales: A review of approaches, challenges and progress. *Agric Ecosyst Environ* 136:1–15.

Cooper, P.J.M. and R. Coe. 2011. Assessing and addressing climate-induced risk in sub-Saharan rainfed agriculture. *Exp Agric* 47:179–184.

Craswell, E.T. 2000. Save our soils–Research to promote sustainable land management. In *Food and Environment Tightrope*, Proceedings of Seminar, 24 November 1999, at Parliament House, Canberra. Melbourne: Crawford Fund for International Agricultural Research, pp. 85–95.

Craswell, E.T., U. Grote, J. Henao, and P.L.G. Vlek. 2003. *Nutrient Flows in Agricultural Production and International Trade: Ecological and Policy Issues*. ZEF Discussion Paper on Development Policy No. 78. Bonn: Center for Development Research, University of Bonn.

Craswell, E.T., A. Sajjapongse, D.J.B. Howlett, and A.J. Dowling. 1998. Agroforestry in the management of sloping lands in Asia and the Pacific. *Agrofor Syst* 38:121–137.

CTFT. 1988. *Faidherbia Albida Del. Monographie*. Nogent-sur-Marne, France: Centre Technique Forestier Tropical.

Dixit, P.N., P.J.M. Cooper, K.P. Rao, and J. Dimes. 2011. Adding value to field-based agronomic research through climate risk assessment: A case study of maize production in Kitale, Kenya. *Exp Agric* 47:317–338.

Dorward, A. and E. Chirwa. 2011. The Malawi agricultural input subsidy programme 2005–06 and 2008–09. *Int J Agric Sustain* 9:232–247.

Drechsel, P., O. Cofie, and S. Niang. 2008. Sustainability and resilience of the urban agricultural phenomenon in Africa. In *Conserving Land, Protecting Water*, eds. D. Bossio and K. Geheb, 120–128. Wallingford: CABI International.

Drechsel, P., S. Graefe, and M. Fink. 2007. *Rural-Urban Food, Nutrient and Virtual Water Flows in Selected West African Cities*, IWMI Research Report 115. Colombo, Sri Lanka: International Water Management Institute.

Drechsel, P., D. Kunze, and F. Penning de Vries. 2001. Soil nutrient depletion and population growth in sub-Saharan Africa: A Malthusian nexus? *Popul Environ* 22:411–423.

Drechsel, P. and U. Zimmermann. 2005. Factors influencing the intensification of farming systems and soil-nutrient management in the rural-urban continuum of SW Ghana. *J Plant Nutr Soil Sci* 168:694–701.

Elias, E., S. Morse, and D.G.R Belshaw. 1998. Nitrogen and phosphorus balances of Kindo Koisha farms in southern Ethiopia. *Agric Ecosyst Environ* 71:93–113.

Eswaran, H., R. Almaraz, P.F. Reich, and P.F. Zdruli. 1997. Soil quality and soil productivity in Africa. *J Sustain Agric* 10:75–94.

Faerge, J. and J. Magid. 2004. Evaluating NUTMON nutrient balancing in sub-Saharan Africa. *Nutr Cycl Agroecosyt* 69:101–110.

FAO. 2008. *Conservation Agriculture*. 2008-07-08 http://www.fao.org/ag/ca/index.html (accessed 7/14/2012).

FAO. 2009. *High Level Expert Forum–How to Feed the World in 2050*. Office of the Director, Agricultural Development Economics Division, Economic and Social Development Department.

FAOSTAT. 2012. *Fertilizers*. http://faostat.fao.org/site/575/default.aspx#ancor (accessed 7/14/2012).

Folmer, E.C.R., P.M.H. Geurts, and J.R. Francisco. 1998. Assessment of soil fertility depletion in Mozambique. *Agric Ecosyst Environ* 71:159–167.

Fosu-Mensah, B.Y., P.L.G. Vlek, and D.S. MacCarthy. 2012. Farmers' perception and adaptation to climate change: a case study of Sekyedumase district in Ghana. *Environ Dev Sustain* 14:495–505.

Garrity, D.P., F.K. Akinnifesi, O.C. Ajayi, S.G. Weldesemayat, J.G. Mowo, A. Kalinganire, M. Larwanou, and J. Bayala. 2010. Evergreen agriculture: a robust approach to sustainable food security in Africa. *Food Sec* 2:197–214.

Giller, K.E., E. Witter, M. Corbeels, and P. Tittonel. 2009. Conservation agriculture and smallholder farming in Africa: The heretic's view. 2009. *Field Crop Res* 114:23–34.

Graves, A., R. Matthews, and K. Waldie. 2004. Low external input technologies for livelihood improvement in subsistence agriculture. *Adv Agron* 82:473–555.

Grote, U., E.T. Craswell, and P.L.G. Vlek. 2005. Nutrient flows in international trade: Ecology and policy issues. *Environ Sci Policy* 8:439–451.

Grote, U., E.T. Craswell., and P.L.G. Vlek. 2007. Nutrient and virtual water flows in traded agricultural commodities. In *Land Use and Soil Resources*, eds. A.K. Braimoh and P.L.G. Vlek, 121–143. Berlin: Springer Science+Business Media B.V.

Harris, F. 1998. Farm-level assessment of the nutrient balance in northern Nigeria. *Agric Ecosyst Environ* 71:201–214.

Hartemink, A.E. 2006. Soil fertility decline: Definitions and assessment. In *Encyclopedia of Soil Science*, ed. R. Lal, 1618–1621. New York: Marcel Dekker.

Henao, J. and C.A. Baanante. 1999. *Estimating Rates of Nutrient Depletion in Soils of Agricultural Lands of Africa*. Muscle Shoals, Alabama: International Fertilizer Development Center.

Henao, J. and C.A. Baanante. 2006. *Agricultural Production and Soil Nutrient Mining in Africa: Implications for Resource Conservation and Policy Development*. Muscle Shoals, AL: International Fertilizer Development Center.

Kaizzi, C.K., H. Ssali, and P.L.G. Vlek. 2004. The potential of velvet bean (*Mucuna pruriens*) and N fertilizers in maize production on contrasting soils and agro-ecological zones of East Uganda. *Nutr Cycl Agroecosyst* 69:59–72.

Kihara, J., A. Bationo, D.N. Mugendi, C. Martius, and P.L.G. Vlek. 2011. Conservation tillage, local organic resources and nitrogen fertilizer combinations affect maize productivity, soil structure and nutrient balances in semi-arid Kenya. *Nutr Cycl Agroecosyst* 90:213–225.

Kugbe, J., M. Fosu, L.D. Tamene, M. Derich, and P.L.G. Vlek. 2012. Annual vegetation burns across the northern savanna region of Ghana: period of occurrence, area burns, nutrient losses and emissions. *Nutr Cycl Agroecosyst* 93:265–284.

Lal, R. 2006 Enhancing crop yields in the developing countries through restoration of the soil organic carbon pool in agricultural lands. *Land Degrad Dev* 17:197–209.

Lesschen, J.P., J.J. Stoorvogel, E.M.A. Smaling, G.B.M. Heuveling, and A. Veldkamp. 2007. A spatially explicit methodology to quantify soil nutrient balances and their uncertainties at the national level. *Nutr Cycl Agroecosyst* 78:111–131.

Mafongoya, P.L., A. Bationo, J. Kirhara, and B.S. Waswa. 2006. Appropriate technologies to replenish soil fertility in southern Africa. *Nutr Cycl Agroecosyst* 76:137–151.

Marenya, P., E. Nkonya, W. Xiong, J. Deustua, and E. Kato. 2012. Which policy would work better for improved soil fertility management in sub-Saharan Africa, fertilizer subsidies or carbon credits? *Agric Syst* 110:162–172.

Martin, A.E. and J.E. Cox. 1956. Nitrogen studies on black soils from the Darling Downs, Queensland. I. Seasonal variations in moisture and mineral nitrogen fractions. *Aust J Agric Res* 2:169–183.

McCown, R.L. and R.K. Jones. 1992. Agriculture of semi-arid eastern Kenya: Problems and possibilities. In *A Search for Strategies for Sustainable Dryland Cropping in Semi-Arid Eastern Kenya*, ed. M.E. Probert, 8–14. Proceedings No. 41. Canberra: Australian Centre for International Agricultural Research.

Nye, P.H. and D.J. Greenland. 1960. *The Soil Under Shifting Cultivation*. Tech Commun No. 51. Harpenden: Commonwealth Agricultural Bureau of Soils.

Okumu, B.M., M.A. Jabbar, D. Colman, D. Russell, M.A.M. Saleem, and J. Pender. 2004. Technology and policy impacts on economic performance, nutrient flows and soil erosion at watershed level: The case of Ginchi in Ethiopia. *J Agric Econ* 55:503–524.

Peoples, M.B. and E.T. Craswell. 1992. Biological nitrogen fixation: Investments, expectations and actual contributions to agriculture. *Plant Soil* 141:13–39.

Piéri, C. 1989. *Fertilité des terres de savanes. Trente ans de recherche et de développement agricoles au sud du Sahara.* Montpellier, France: Agridoc-International, Centre de coopération internationale en recherche agronomique pour le développement (CIRAD).

Pol, F. and B. Traore. 1993. Soil nutrient depletion by agricultural production in southern Mali. *Nutr Cycl Agroecosyst* 36:79–90.

Pretty, J., C. Toulmin, and S. Williams. 2011. Sustainable intensification in African agriculture. *Int J Agric Sustain* 9:5–24.

Rowe, E.C., M.T. van Wijk, N. de Ridder, and K.E. Giller. 2006. Nutrient allocation strategies across a simplified heterogeneous African smallholder farm. *Agric Ecosyst Environ* 116:60–71.

Saleem, M.A.M. 1998. Nutrient balance patterns in African livestock systems. *Agric Ecosyst Environ* 71:241–254.

Sanchez, P.A., M.P. Gichuru, and L.B. Katz. 1982. Organic matter in major soils of the tropical and temperate regions. *Trans 12th Int Congr Soil Sci New Delhi* I:99–114.

Sanginga, N. and P.L. Woomer. 2009. *Integrated Soil Fertility Management in Africa: Principles, Practices and Developmental Process.* Nairobi: Tropical Soil Biology and Fertility Institute, CIAT.

Sheldrick, W.F., J.K. Syers, and J. Lingard. 2002. A conceptual model for conducting nutrient audits at national, regional and global scales. *Nutr Cycl Agroecosyst* 62:61–67.

Simpson, J.R., J.R. Okalebo, and G. Lubulwa. 1996. *The Problem of Maintaining Soil Fertility in Eastern Kenya: A Review of Relevant Research.* Monograph No. 41. Canberra: Australian Centre for International Agricultural Research.

Smaling, E.M.A., J.J. Stoorvogel, and P.N. Windmeijer. 1993. Calculating soil nutrient balances in Africa at different scales II. District scale. *Fert Res* 35:237–250.

Stoorvogel, J.J. and E.M.A. Smaling. 1990. *Assessment of Soil Nutrient Depletion in Sub-Saharan Africa: 1983–2000.* Winand Staring Centre Report 28. Wageningen, The Netherlands: Winand Staring Centre, 137 pp.

Stoorvogel, J.J., E.M.A Smaling, and B.H. Jansen. 1993. Calculating soil nutrient balances at different scales: I. Supra-national scale. *Fert Res* 35:227–235.

Tan, Z.X., R. Lal, and K.D. Wiebe. 2005. Global nutrient depletion and yield reduction. *J. Sustain Agric* 26:123–146.

Tenkorang, F. and J. Lowenberg-DeBoer. 2009. Forecasting long-term global fertilizer demand. *Nutr Cycl Agroecosyst* 83:233–247.

Tiffen, D.D., M. Mortimer, and F. Gichuki. 1994. *More People, Less Erosion. Environmental Recovery in Kenya.* Chichester: John Wiley and Sons.

Van den Bosch, H., J.N. Gitari, V.N. Ogaro, S. Maobe, and J. Vlaming. 1998. Monitoring nutrient flows and economic performance in African farming systems (NUTMON): III. Monitoring nutrient flows and balances in three districts in Kenya. *Agric Ecosyst Environ* 71:63–80.

Vanlauwe, B. and K.E. Giller. 2006. Popular myths around soil fertility management in sub-Saharan Africa. *Agric Ecosyst Environ* 116:34–46.

Vlek, P.L.G. 1990. The role of fertilizers in sustaining agriculture in sub-Saharan Africa. *Fert Res* 26:327–339.

Vlek, P.L.G. 1993. *Strategies for Sustaining Agriculture in Sub-Saharan Africa: The Fertilizer Technology Issue. Technologies for Sustainable Agriculture in the Tropics.* ASA Special Publication No. 56. Madison, Wisconsin, USA, pp. 265–276.

Vlek, P.L.G. and H. Koch. 1992. The soil resource base and food production in the developing world: Special focus on Africa. In *Göttinger Beiträge zur Land- und Forstwirtschaft in den Tropen und Subtropen*, Erich Toltze-Verl., Göttingen, Germany. 71:139–160.

Vlek, P.L.G. and L. Tamene. 2009. Conservation agriculture: Why? *Proceedings of the 4th World Congress on Conservation Agriculture, February 4–7, 2009, New Delhi, India.* http://www.fao.org/ag/ca/doc/wwcca-leadpapers.pdf#page=17 (accessed 14 July, 2012).

Vlek, P.L.G., R.F. Kühne, and M. Denich. 1997. Nutrient resources for crop production in the tropics. *Phil Trans Soc Lond B* 352:975–985.

Vlek, P.L.G., Q.B. Le, and L. Tamene. 2008. *Land Decline in Land-Rich Africa: A Creeping Disaster in the Making.* Rome: CGIAR Science Council Secretariat.

Vlek, P.L.G., Q.B. Le, and L. Tamene. 2010. Assessment of land degradation, its possible causes and threat to food security in Sub-Saharan Africa. In *Advances in Soil Science—Food Security and Soil Quality*, eds. R. Lal and B.A. Stewart, 57–86. Boca Raton: CRC Press.

Wortmann, C.S. and C.K. Kaizzi. 1998. Nutrient balances and expected effects of alternative practices in farming systems of Uganda. *Agric Ecosyst Environ* 71:115–129.

16 Carbon Sink Capacity and Agronomic Productivity of Soils of Semiarid Regions of India

*Ch. Srinivasarao, B. Venkateswarlu,
Rattan Lal, A. K. Singh, Sumanta Kundu,
and Vijay Sandeep Jakkula*

CONTENTS

16.1	Introduction	424
16.2	Soils of Rainfed Regions	425
16.3	Constraints in Rainfed Production Systems	426
	16.3.1 Uncertain Rainfall	426
	16.3.2 Degraded Soils	426
	16.3.3 Low-Input Application	426
	16.3.4 Biotic Stresses	428
	16.3.5 Untapped Water–Nutrient Synergy	429
	16.3.6 Poor Crop Management	429
	16.3.7 Lack of Focused Extension Program	431
	16.3.8 Lack of Appropriate Policy Support	432
16.4	Yield Gaps in Rainfed Agriculture	432
16.5	Poor Soil Health and Multinutrient Deficiencies	434
16.6	Declining NUE: A Serious Concern	437
16.7	Importance of SOC	437
	16.7.1 Carbon Stocks in Different Soil Types of Rainfed Agroecosystems of India	438
	16.7.2 Soil Carbon Stocks in Relation to Production System	440
	16.7.3 Carbon Stocks in Relation to Rainfall	441
16.8	Reasons for Soil Carbon Depletion in Rainfed Tropical Systems	443
16.9	Carbon Sequestration	444
16.10	Carbon Management Options	446
	16.10.1 Site-Specific RMPs	446
	16.10.1.1 Improvement of Crop Yield through *Gliricidia* Green Leaf Manuring	447

16.11 Enhanced NUE with Manuring 448
16.12 SOC in Relation to Water Retention and Water Productivity 449
16.13 Carbon Management in Rainfed Production Systems—Experiences
         from Long-Term Experiments in Semiarid Tropics 449
         16.13.1 Mean Yield and Sustainable Yield Index Influenced by
                    Different Nutrient Management Treatments in Diverse
                    Cropping Systems 452
         16.13.2 Soil Carbon and Yield Sustainability 458
                    16.13.2.1 Groundnut Production System 458
                    16.13.2.2 Finger Millet-Based Production System 459
                    16.13.2.3 Winter Sorghum-Based Production System 459
                    16.13.2.4 Pearl Millet-Based Production System 459
                    16.13.2.5 Safflower- and Soybean-Based Production System 459
                    16.13.2.6 Rice-Based Production System 465
         16.13.3 Minimal Carbon Input Requirements for Arresting Carbon
                    Depletion 465
16.14 Conclusions 468
16.15 Future Research Needs 469
Abbreviations 469
References 470

16.1 INTRODUCTION

The global rainfed croplands were estimated at 1.132 billion hectares (B ha) at the end of the last millennium (Biradar et al. 2009). This is 2.78 times the net irrigated areas (407 M ha) of the world (http://www.iwmigiam.org/info/main/aboutGMRCA2.asp). Rainfed agroecosystems occupy a considerable place in Indian agriculture too, covering 80 million ha (M ha), in arid, semiarid, and subhumid climatic zones, constituting nearly 57% of the net cultivated area. India has 18% of the world's population, 15% of the world livestock with only 2.3% of the geographical area, 4.2% of freshwater resources, 1% of forests, and 0.5% of pastureland (Srinivasarao 2011; Srinivasarao et al. 2011a). Rainfed areas in India contribute almost 100% of forest products, 84% to 87% of coarse-grain cereals and pulses, 80% of horticulture, 77% of oilseeds, 60% of cotton (*Gossypium hirsutum*), and 50% of fine cereals like rice (*Oryza sativa*), wheat (*Triticum aestivum*), maize (*Zea mays*), sorghum (*Sorghum bicolor*), and so forth (Srinivasarao et al. 2010, 2011b). Rainfed regions support 60% of livestock and 40% of the human population and contribute 40% of food grains and several special-attribute commodities such as seed spices, dyes, herbs, gums, and so forth. (Srinivasarao et al. 2011a). However, the double-cropped area in rainfed farming is negligible. Rainfed agroecoregions are complex, diverse, fragile, risky, and underinvested and require regionally differentiated investments and management strategies. Achieving high production potential is difficult in these rainfed areas due to vagaries of rainfall.

The above-mentioned statistics indicate the high pressure on finite natural resources. The net sown area in India has remained constant for many years at 141 M ha, but the human and livestock population has been steadily increasing. The slowdown in agricultural productivity growth has been attributed to resource degradation problems and declining factor productivity in major production systems. With the skewed and intense rainfall and concurrent rise in temperatures, the fertile topsoil is prone to degradation. Low factor productivity in Indian agriculture has raised serious concerns in terms of economy as well as environmental quality. Progressively more and more inputs are needed for unit production of food. The depletion of soil organic matter (SOM) is leading to declining response to fertilizers and widespread deficiencies of secondary nutrients and micronutrients. Therefore, improving SOM concentration may be an important option. This chapter discusses the soil-related constraints of arid and semiarid India, carbon (C) stocks, potential of C sequestration in different cropping systems, as well as different C management options under rainfed conditions.

16.2 SOILS OF RAINFED REGIONS

India's diverse soil types can be grouped into eight orders. Taxonomically, soils of India fall under Entisols (80.1 M ha), Inceptisols (95.8 M ha), Vertisols (26.3 M ha), Aridisols (14.6 M ha), Mollisols (8.0 M ha), Ultisols (0.8 M ha), Alfisols (79.7 M ha), Oxisols (0.3 M ha), and miscellaneous types (23.1 M ha). In terms of the distribution of rainfall, 15 M ha of the land receives annual rainfall of <500 mm, 15 M ha receives 500 to 750 mm, 42 M ha receives 750 to 1150 mm, and 25 M ha receives >1150 mm. In terms of land use and management, alluvial soils are the most dominant (93.1 M ha), followed by red soils (79.7 M ha), black soils (55.1 M ha), desert soils (26.2 M ha), and lateritic soils (17.9 M ha) (Table 16.1). Both the

TABLE 16.1
Major Soil Groups and Their Moisture Storage Capacities in Rain-Dependent Areas of India

Broad Soil Group	Subgroup (Based on Soil Depth)	Moisture Storage Capacity (mm)
Vertisols and related soils	Shallow to medium (up to 45 cm)	135–145/45 cm
	Medium to deep (45–90 cm)	145–270/90 cm
	Deep (>90 cm)	300/m
Alfisols and related soils	Shallow to medium (up to 45 cm)	40–70/45 cm (sandy loam)
	Deep (>90 cm)	70–100/45 cm (loam)
		180–200/90 cm
Aridisols	Medium to deep (up to 90 cm)	80–90/90 cm
Inceptisols	Deep	90–100/m (loamy sand)
		110–140/m (sandy loam)
		140–180/m (sandy loam)
Entisols	Deep	110–140/m (sandy loam)
		140–180/m (loam)

Source: Srinivasarao, Ch., *Indian J Fertil*, 7(4), 12–25, 2011.

ecological (supporting biomass, forests, buffering action) and the economic (food, fodder, and fiber production) functions of these soils are important as the crop and livestock production systems in India are highly integrated in a farming systems mode.

16.3 CONSTRAINTS IN RAINFED PRODUCTION SYSTEMS

Unlike irrigated agriculture, the productivity of rainfed crops has remained low with some exceptions. Low yields under rainfed conditions are attributed to several constraints ranging from cultivation to marketing. Such constraints lead to lower biomass production, lesser root biomass, and less or no residue recycling with the attendant low levels of soil organic C (SOC) concentration in rainfed soils. Principal constraints to achieving higher productivity of rainfed crops are briefly discussed below.

16.3.1 UNCERTAIN RAINFALL

Due to uncertainty in rainfall, agriculture in rainfed regions depends on the amount and distribution of monsoons. Increased frequency of droughts is a major challenge to rainfed agriculture (Figure 16.1). Prolonged moisture stress at critical growth stages results in lower productivity.

16.3.2 DEGRADED SOILS

Degraded soils with high risks of accelerated erosion resulting in loss of fertile surface soil and SOC are major factors. The magnitude of soil loss ranges from 5 to 150 Mg ha^{-1} year^{-1} depending upon soil type, vegetation, and slope of the land. There have been several estimates of degraded lands/wastelands reported by different organizations. These estimates have been harmonized by adopting spatial data integration in a geographical information system (GIS) environment (Maji 2007) (Figures 16.2 and 16.3). The total degraded area is estimated at 120.72 M ha, of which 104.19 M ha (86.3%) is arable land and 16.53 M ha (13.7%) is open forestland. Of the total degraded land area, 73.27 M ha (60.7%) is affected by water erosion, 12.40 M ha (10.3%) by wind erosion, 5.44 M ha (4.5%) by salinity, and 5.09 M ha (4.2%) by soil acidity. Some areas are affected by more than one degradation process (Maji 2007).

16.3.3 LOW-INPUT APPLICATION

Use of production inputs like fertilizers, supplemental irrigation, good-quality seeds, pesticides, and herbicides is lower in rainfed than irrigated crops. Thus, productivity has remained low. Though it has been amply demonstrated that soils in rainfed regions are multinutrient deficient, balanced use of these inputs in rainfed crops is rarely achieved. There are wide disparities of inputs between irrigated and rainfed regions due to numerous uncertainties in the crop production. Statewise consumption of fertilizers under

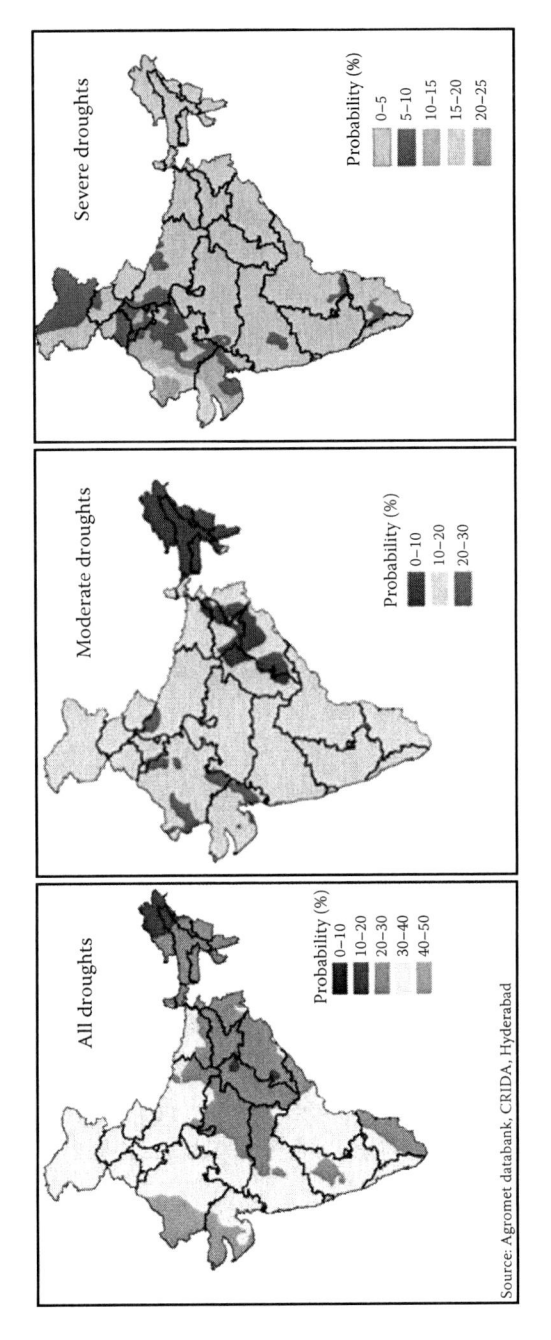

FIGURE 16.1 Drought scenario of India. (From Rao, G.G.S.N. et al., *Indian Journal of Dryland Agricultural Research and Development*, 24, 10–20, 2009.)

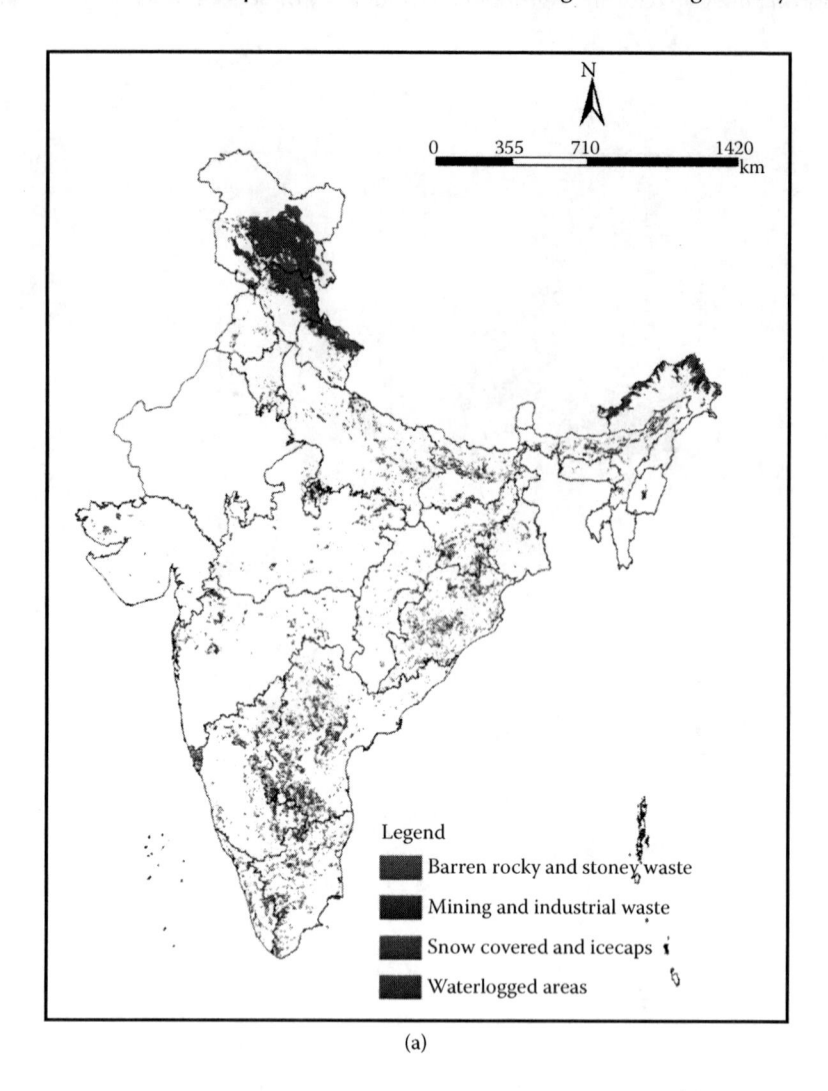

FIGURE 16.2 (a) Physical land degradation in India. (b) Soil loss by water erosion (>10 Mg ha^{-1} year^{-1}). (c) Wind erosion in India (>10 Mg ha^{-1} year^{-1}). (From Venkateswarlu, B. et al., *Natural Resource Management for Accelerating Agricultural Productivity*, Studium Press (India) Pvt. Ltd., New Delhi, India, 2012.)

irrigated and rainfed conditions and consumption of nitrogen, phosphorus, and potassium (NPK) in major rainfed crops in India are shown in Figures 16.4 and 16.5.

16.3.4 Biotic Stresses

Yield losses of several rainfed crops in general and of pulses in particular are high due to severe incidence of diseases and insect pests. In some cases, an entire crop is lost because of severe infestation. Though technological and management options are well established, they are seldom practiced because of economic and other human dimensions.

(b)

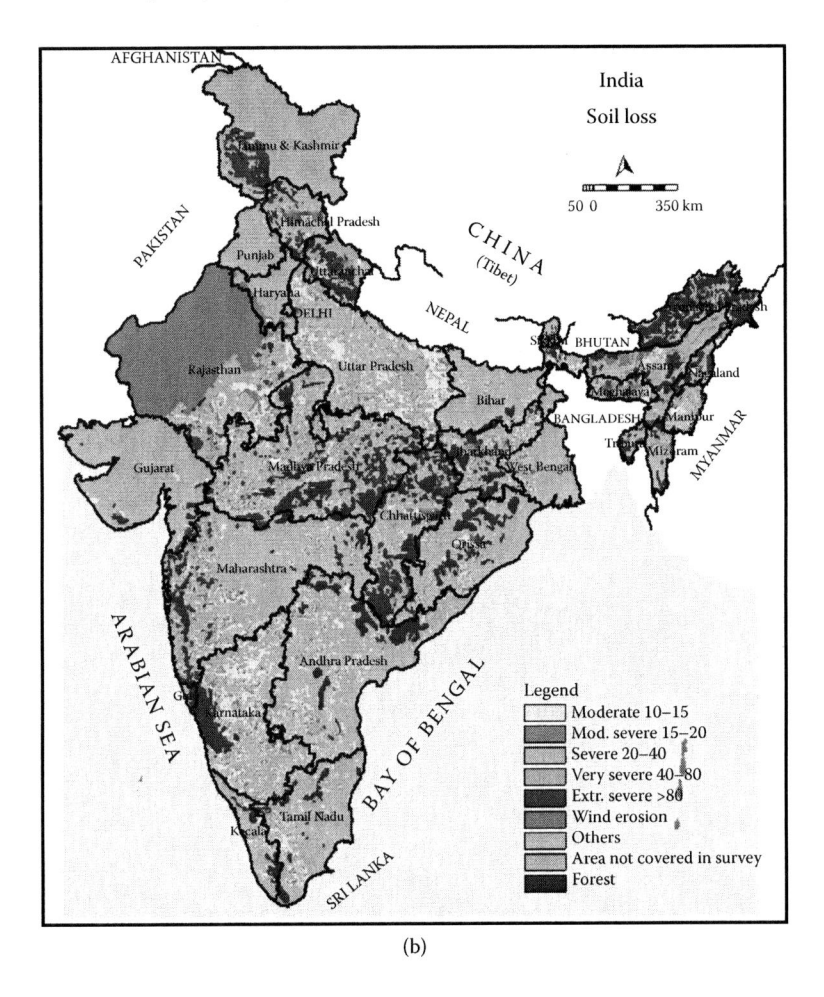

FIGURE 16.2 (Continued) (a) Physical land degradation in India. (b) Soil loss by water erosion (>10 Mg ha^{-1} year^{-1}). (c) Wind erosion in India (>10 Mg ha^{-1} year^{-1}). From Venkateswarlu, B. et al., *Natural Resource Management for Accelerating Agricultural Productivity*, Stadium Press (India) Pvt. Ltd., New Delhi, India, 2012.)

16.3.5 Untapped Water–Nutrient Synergy

Rainfed regions often suffer from water scarcity and are characterized by multinutrient-deficient soils. However, applying plant nutrients in synergy with profile moisture storage is crucial in improving crop productivity and enhancing nutrient and water use efficiencies.

16.3.6 Poor Crop Management

Untimely sowing, lack of weeding and supplemental irrigation, and suboptimum plant population result in poor crop stand. Excessive weed infestation can smother growth of the crops and severely reduce their productivity. Yet, a judicious crop

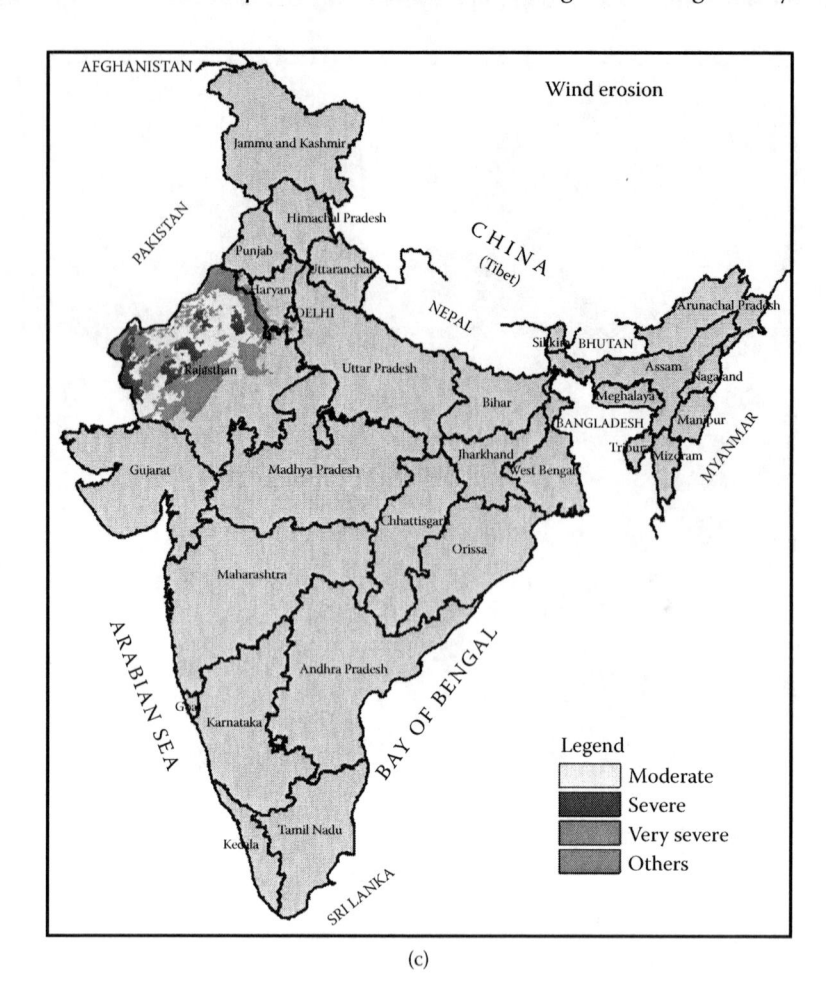

FIGURE 16.2 (Continued) (a) Physical land degradation in India. (b) Soil loss by water erosion (>10 Mg ha^{-1} year^{-1}). (c) Wind erosion in India (>10 Mg ha^{-1} year^{-1}). (From Venkateswarlu, B. et al., *Natural Resource Management for Accelerating Agricultural Productivity*, Studium Press (India) Pvt. Ltd., New Delhi, India, 2012.)

FIGURE 16.3 Degraded soils: a predominant feature of rainfed regions.

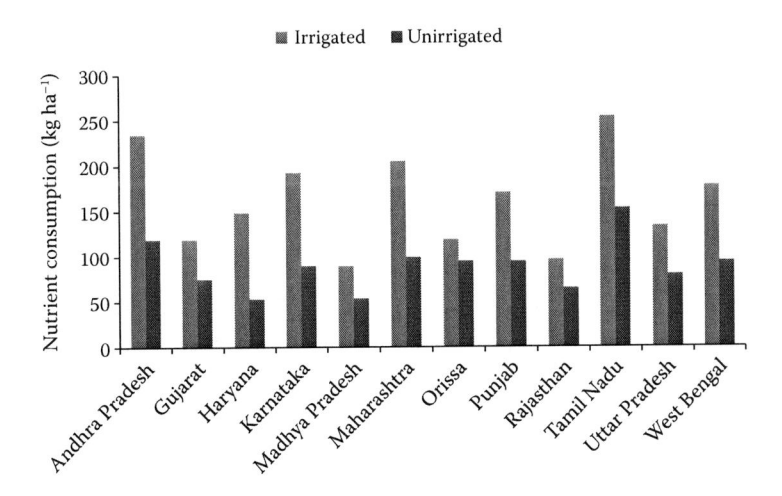

FIGURE 16.4 Statewise consumption of fertilizers under irrigated and rainfed conditions. (From Venkateswarlu, B. et al., *Natural Resource Management for Accelerating Agricultural Productivity*, Studium Press (India) Pvt. Ltd., New Delhi, India, 2012.)

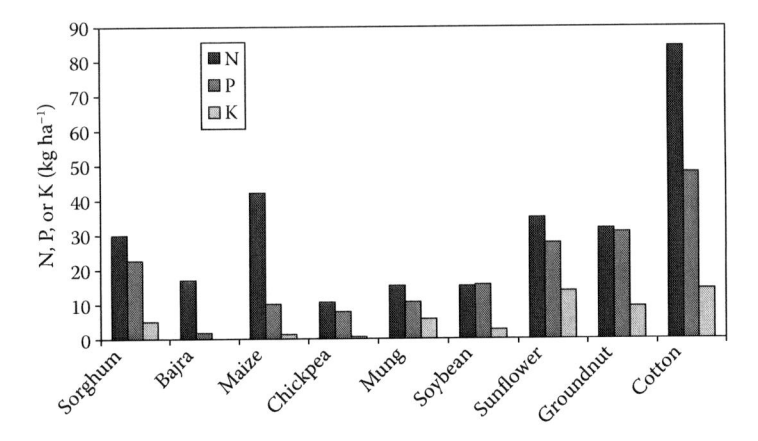

FIGURE 16.5 Consumption of NPK in major rainfed crops in India. (From Venkateswarlu, B. et al., *Natural Resource Management for Accelerating Agricultural Productivity*, Studium Press (India) Pvt. Ltd., New Delhi, India, 2012.)

management can substantially enhance productivity and improve agronomic/economic returns.

16.3.7 Lack of Focused Extension Program

Improved and short-duration varieties and matching production technologies are available for rainfed cropping. However, the lack of a focused extension program specially aimed at rainfed regions and adoption of improved technologies are among the major constraints.

16.3.8 LACK OF APPROPRIATE POLICY SUPPORT

Many coarse grains like finger millet (*Eleusine coracana*), sorghum (*Sorghum bicolor*), pearl millet (*Pennisetum glaucum*), and others do not have a fair price policy. Yet, these crops are highly nutritious and important components of diet.

16.4 YIELD GAPS IN RAINFED AGRICULTURE

The factors described above are among major constraints to achieving high productivity levels of rainfed crops in the farmers' fields. Thus, the yield gaps between potential, on-station, and farm yields and adopted crop management are high. Unless better water and nutrient management technologies are adopted, along with improved seed combination, harnessing the benefits of modern technologies to enhancing crop production cannot be realized.

The large yield gap between demonstration plots managed by scientists and actual national averages exists in most cereals as well as oilseed crops (Table 16.2). Crop yield directly or indirectly is the numerator and the determinant of the nutrient use efficiency (NUE) of the crop and of the soil and agronomic factors that increase crop yield and improve the NUE. Prior to any discussion on management factors, it is appropriate to assess the potential and attainable yields of crops in different regions of India. Some examples of data on potential yield, on-station yield, and on-farm yield for the regions where research stations are located are summarized in Table 16.3.

There exists a yield gap of 37% to 52% between potential and on-station yields and 35% to 70% between potential and on-farm yields. The gap between on-station and on-farm yields ranges between 6% and 44%. In general, the yield gaps are wider in

TABLE 16.2
Yield Gaps in Dryland Agriculture in India

	Average Yield (Mg ha^{-1})		
Crop	National Demonstration	National Average	Yield Gap (Mg ha^{-1})
Cereals			
Kharif	1.40	0.78	0.62
Rabi	1.73	0.92	0.81
Millets	0.92	0.61	0.31
Pulses	0.75	0.35	0.40
Oilseeds			
Edible oilseeds	0.60	0.34	0.26
Castor bean	0.52	0.23	0.29

Source: Srinivasarao, Ch., *Indian J Fertil*, 7(4), 12–25, 2011.

TABLE 16.3
Potential, On-Research-Station and On-Farm Yields of Rice and Wheat in Different Zones of the Indo-Gangetic Plains

Site	Potential Yield (A) (Mg ha^{-1})	On-Station Yield (B) (Mg ha^{-1})	On-Farm Yield (C) (Mg ha^{-1})	Yield Gap (%) 100 (A–B)/A	Yield Gap (%) 100 (A–C)/A	Yield Gap (%) 100 (B–C)/A
			Rice			
Ludhiana	10.7	5.6	5.6	48	48	0
Karnal	10.4	6.8	3.8	35	64	44
Kanpur	9.5	4.5	2.8	52	71	38
Pantnagar	9.0	5.5	4.2	39	53	24
Varanasi	9.2	4.1	3.2	55	65	22
Falzabad	9.1	4.2	2.8	54	69	33
			Wheat			
Ludhiana	7.9	4.7	4.3	41	46	6
Karnal	7.3	4.6	3.6	37	51	22
Kanpur	7.0	4.6	2.8	34	60	39
Pantnagar	6.5	3.9	4.2	40	35	–0.1
Varanasi	7.0	3.8	3.2	46	54	16
Falzabad	6.7	3.4	2.8	49	58	18

Source: Srinivasarao, Ch., *Indian J Fertil*, 7(4), 12–25, 2011.

rice than wheat. The adoption of recommended management practices (RMPs) can reduce the on-station versus on-farm yield gap, which can increase rice and wheat production by 15% to 20%. The fact remains that the yield gap between on-station versus on-farm is zero in the Ludhiana region of Punjab for rice and in Pantnagar, Uttaranchal, for wheat. These trends show that farmers have already adopted RMPs in these regions. Thus, with good extension programs, the yield gap can also be abridged in other regions of the country. Information on benefits of adopting RMPs vis-à-vis the farmers' practices in increasing crop yield is also available for oilseeds (Hegde and Babu 2008) and pulses (Ali et al. 2002).

Rainfed crops are largely grown on marginal and submarginal lands and receive the least attention from farmers in application of fertilizers and manures. Thus, nutrient deficiencies are emerging as the most important constraint to achieving the yield targets in these crops. In most of the rainfed crops, for example, in pulses (Table 16.4), average on-farm yields are about one-third of the potential on research plots and half of the yield in on-farm demonstration plots. Therefore, the rainfed agriculture in India has a vast but an untapped potential. Adoption of RMPs, synchronizing nutrient application with moisture availability or rainfall, and timely weed control can improve crop yields by as much as 30% to 60%.

TABLE 16.4
Critical Yield Gaps (q ha^{-1}) in Major Rainfed Pulse Crops

Crop	Yield Potential on Research Plots	Yield in FLDs at Farmers Fields	National Average
Chickpea	20–22	15–18	8.06
Pigeon pea (early)	15–17	12–15	7.97
Pigeon pea (late)	20–25	20–22	–
Mung bean	11–12	9–10	3.81
Urd bean	10–12	8–9	4.40
Field pea	20–22	15–18	10.34
Lentil	15–18	12–14	7.32

Source: Chatuverdi, S.K. and M. Ali, Poor man's meat needs fresh fillip. In *The Hindu Survey of Indian. Agriculture 2002*, 63–69. Chennai, India, 2002.

Note: FLD, front-line demonstration.

16.5 POOR SOIL HEALTH AND MULTINUTRIENT DEFICIENCIES

Therefore, soils under rainfed agriculture are not only thirsty but also hungry. Soil quality is severely degraded because of low levels of SOM in dry regions of India. Soils of arid and semiarid regions in India are highly diverse and comprise Vertisols and Vertic subgroups, Alfisols, Oxisols, Inceptisols, Aridisols, Entisols, and so forth. Furthermore, there exists a large variation in rainfall, which ranges between 400 and 1500 mm per annum. Accordingly, the length of the growing season ranges between 60 and 180 days and sometimes even lower. Above all, soils are highly degraded, extremely depleted of their SOM concentration (Figure 16.6), and are multinutrient deficient (Table 16.5). The SOC concentrations in some regions of rainfed agriculture

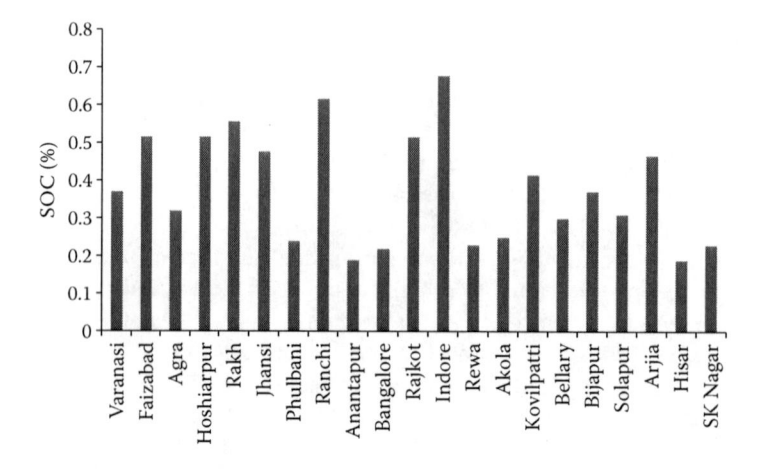

FIGURE 16.6 Rainfed regions in arid and semiarid tropical India are low in SOC. (From Srinivasarao, Ch., and K.P.R. Vittal, *Indian J Fertil*, 3, 37–46, 2007.)

TABLE 16.5
Emerging Nutrient Deficiencies in Dryland Soils (0–15 cm) under Diverse Rainfed Production System of India

Location	Rainfall (mm)	Soil Type	Production System	Limiting Nutrient Needs to Supplied
Varanasi	1080	Inceptisol	Upland rice–lentil	N, Zn, B
Faizabad	1060	Inceptisol	Upland rice	N
Phulbani	1400	Oxisol	Upland rice–horse gram	N, Ca, Mg, Zn, B
Ranchi	1300	Alfisol	Upland rice	Mg, B
Rajkot	615	Vertisol	Groundnut	N, P, S, Zn, Fe, B
Anantapur	590	Alfisol	Groundnut	N, K, Mg, Zn, B
Indore	950	Vertisol	Soybean	–
Rewa	900	Vertic Inceptisol	Soybean	N, Zn
Akola	825	Vertisol	Cotton	N, P, S, Zn, B
Kovilpatti	750	Vertic Inceptisol	Cotton	N, P
Bellari	500	Vertisol	Rabi sorghum	N, P, Zn, Fe
Bijapur	680	Vertisol	Rabi sorghum	N, Zn, Fe
Jhansi	1020	Inceptisol	Kharif sorghum	N
Solapur	720	Vertisol	Rabi sorghum	N, P, Zn
Agra	665	Inceptisol	Pearl millet	N, K, Mg, Zn, B
Hisar	412	Inceptisol	Pearl millet	N, Mg, B
SK Nagar	550	Aridisol	Pearl millet	N, K, S, Ca, Mg, Zn, B
Bangalore	925	Alfisol	Finger millet	N, K, Ca, Mg, Zn, B
Arjia	650	Vertisol	Maize	N, Mg, Zn, B
Ballowal-Saunkri	1000	Inceptisol	Maize	N, K, S, Mg, Zn
Rakh-Dhiansar	1200	Inceptisol	Maize	N, K, Ca, Mg, Zn, B

Source: Srinivasarao, Ch., and K.P.R. Vittal, *Indian J Fertil*, 3, 37–46, 2007.

are as low as 0.15%. In addition to a widespread deficiency of nitrogen (N) and phosphorus (P), deficiencies of potassium (K), magnesium (Mg), sulfur (S), zinc (Zn), and boron (B) are also among major constraints to crop production in arid and semiarid regions (Srinivasarao et al. 2006, 2008a,b, 2009a,b, 2011a,b).

Declining soil quality and nutrient imbalance are major controls of agronomic productivity. The SOM levels have declined sharply in intensively cropped Indo-Gangetic plains, leading to stagnant yields of rice–wheat cropping systems. In addition to the widespread deficiency of N, those of K, S, and micronutrients are also increasing. Deficiency of Zn has become most acute, followed by that of S and B (Motsara 2002). It is estimated that 29.4 M ha of soils in India is experiencing decline in fertility, with a net negative balance of 8–10 Tg (1 Tg = teragram = 10^{12} g) of nutrients per annum (Lal 2004). Thus, low NUE is another cause of concern. The SOM level of 63% and 26% arable lands remains low and medium, respectively; and Indian soils are low in organic matter with 63% low, 26% medium, and 11% high status. About 80% of the soils tested are low to medium in available P and 50% in K (Table 16.6). Similarly among micronutrients, the deficiency is most severe in Zn, followed by that in iron (Fe), manganese (Mn), and copper (Cu) (Table 16.7). Thus far, soil fertility issues are mainly focused on irrigated agriculture, but nutrient deficiencies are also severe in rainfed agriculture (Srinivasarao et al. 2006, 2008a,b; Srinivasarao and Vittal 2007; Sharma et al. 2002, 2004), which are exacerbated by low SOC concentrations and multinutrient deficiencies.

TABLE 16.6
Fertility Status of N, P, and K in Some States of India

State	No. of Samples Analyzed	N L	N M	N H	P L	P M	P H	K L	K M	K H
Madhya Pradesh	1,38,553	40	41	19	39	38	23	10	32	58
Uttar Pradesh	8,07,424	80	15	5	71	26	3	12	55	33
Maharashtra	93,142	67	26	7	86	12	2	8	18	74
Andhra Pradesh	3,12,521	62	21	17	57	29	14	9	30	61
Karnataka	3,17,213	29	37	34	31	48	21	7	32	61
Orissa	2,51,196	60	23	17	59	28	13	33	41	26
Tamil Nadu	4,91,657	75	16	9	24	41	35	12	36	52
India	36,50,004	63	26	11	42	38	20	13	37	50

Source: Motsara, M.R., *Fertil News*, 47(8), 15–21, 2002.

TABLE 16.7
Micronutrient Deficiencies in Some States of India

State	No. of Samples (Range of Samples Analyzed for Different States)	Zn	Fe	Cu	Mn
Andhra Pradesh	5219–6563	51	2	1	2
Bihar	17,802–19,078	54	6	3	2
Gujarat	29,532	24	8	4	4
Madhya Pradesh	11,204–12,000	63	3	1	3
Tamil Nadu	19,559–20,580	53	15	3	8
Uttar Pradesh	24,425–25,122	45	6	1	3
Karnataka	24,411–25,542	78	39	5	19

Source: Takkar, P.N., *Indian Soc Soil Sci*, 44, 563–581, 1996.

16.6 DECLINING NUE: A SERIOUS CONCERN

Fertilizer consumption increased 322 times in India between 1950 and 1951 and 2007 and 2008. Yet, the fertilizer use efficiency is extremely low. The yield curve in most irrigated crops has plateaued because of the stagnated response to added inputs. Nonetheless, because of a large yield gap, there exists a vast potential to enhance the crop yields in rainfed agriculture. The potential yield of sorghum in research plots is about 3970 kg ha^{-1} compared with the average yield of 2090 kg ha^{-1} in verification trials. However, the average yield levels in rural farms are as low as 370 kg ha^{-1}. Data from on-farm trials conducted under the Simple Fertilizer Trials Scheme of the Indian Council of Agricultural Research (ICAR) show that the response ratio is the highest for N (11.6 to 16.7 kg grain/kg N), followed by that of P (5.5 to 12.5 kg grain/kg P_2O_5), and the least for K (3.6 to 6.2 kg grain/kg K_2O). Furthermore, response to nitrogen-phosphorus (NP), nitrogen-potassium (NK), or NPK is not additive of their individual responses, which leads to an excessive use of N. The rate of nutrient application in these trials is high (120 kg N + 60 kg P_2O_5 + 60 kg K_2O), and although the increase in yield of wheat by fertilizer use is much higher (1.1 to 2.6 Mg ha^{-1} compared to 0.47 to 1.25 Mg ha^{-1} for tall varieties), the response ratio to NPK application is low and ranges from 4.7 to 10.9 kg grain/kg of nutrients.

16.7 IMPORTANCE OF SOC

World soils are among the Earth's largest terrestrial reservoirs of C and have a large potential of C sequestration. Thus, C sequestration in agricultural soils is an important strategy of reducing atmospheric concentration of CO_2 (Lal 2004). Further, sequestration of C in agricultural soils has numerous cobenefits in terms of increased soil fertility and environmental quality. Because soils under rainfed cropping are severely depleted of their SOC stock and are strongly degraded, there is a large potential for C sequestration (Wani et al. 2003). Low SOM level in tropical soils, particularly in those of arid, semiarid, and subhumid climates, is a major determinant of low productivity (Syers et al. 1996; Katyal et al. 2001). Therefore, proper management of SOM is important to sustaining productivity, ensuring food security, and restoring marginal lands (Scherr 1999). With a low fertilizer input in dryland agriculture, mineralization of SOM is a major source of plant nutrients. Yet, maintaining or improving SOM levels in soils of the tropics is difficult because of a high rate of oxidation under prevailing high temperatures (Lal 1997; Lal et al. 2003). Nonetheless, maintaining or improving SOM level is a necessary prerequisite to improving soil quality, productivity, and sustainability.

The SOC balance of terrestrial ecosystems is changed markedly by human activities—including deforestation, biomass burning, and land use change, which result in the emission of trace gases that enhance the "greenhouse effect" (Bolin 1981; Trabalka and Reichle 1986; IPCC 1990; Batjes 1996; Bhattacharya et al. 2000; Lorenz and Lal 2005). Routinely, soil surveys estimate SOC pool to 0.5- to 1-m depth. However, SOC can be sequestered in subsoil (>1-m depth) by selecting plants/cultivars with deeper and thicker root systems that are high in recalcitrant compounds like suberin and lignin (Wani et al. 2003; Lorenz and Lal 2005).

16.7.1 CARBON STOCKS IN DIFFERENT SOIL TYPES OF RAINFED AGROECOSYSTEMS OF INDIA

Organic C stocks, inorganic C stocks, and TCSs differ between and within soil types (Table 16.8). Vertisols and associated soils contain higher SOC stocks, followed by Inceptisols > Alfisols > Aridisols (Figure 16.7). In general, the SOC concentration is greater than that of soil inorganic carbon (SIC) in Alfisols and Aridisols, while SIC > SOC in Vertisols and Inceptisols. The SOC stocks (Mg ha^{-1}) range from 26.7

TABLE 16.8
Organic C Stocks, Inorganic C Stocks, and TCSs in Topical Soils of India

Soil Location	Production System	Organic	Inorganic	Total
		C Stocks (Mg ha^{-1})		
Inceptisols				
Varanasi	Rice	32.54	112.36	144.90
Faizabad	Rice	29.81	22.30	52.11
Agra	Pearl millet	26.69	26.76	63.45
Ballowal-Sauntri	Maize	56.73	72.21	128.94
Rakh-Dhiansar	Maize	59.71	45.75	105.46
Jhansi	Rabi sorghum	56.97	135.11	192.08
Alfisols/Oxisols				
Parbhani	Rice	23.28	8.81	32.10
Ranchi	Rice	49.83	23.35	73.18
Anantapur	Groundnut	25.41	57.02	82.43
Bangalore	Finger millet	24.75	17.88	42.63
Vertisols/Vertic Groups				
Rajkot	Groundnut	58.02	154.77	212.79
Indore	Soybean	95.90	88.33	184.24
Rewa	Soybean	28.71	16.03	44.74
Akola	Cotton	28.60	367.63	396.23
Kovilpatti	Cotton	48.20	183.05	231.26
Bellary	Rabi sorghum	34.67	298.53	333.20
Bijapur	Rabi sorghum	36.60	326.06	362.67
Solapur	Rabi sorghum	49.73	106.70	156.42
Arjia	Maize	36.93	62.07	99.01
Aridisols				
Hisar	Pearl millet	20.10	14.27	34.38
SK Nagar	Pearl millet	27.36	20.50	47.86

Source: Srinivasarao, Ch. et al., Carbon sequestration strategies under rainfed production systems of India, Central Research Institute for Dryland Agriculture, Hyderabad (ICAR), India, 2009.

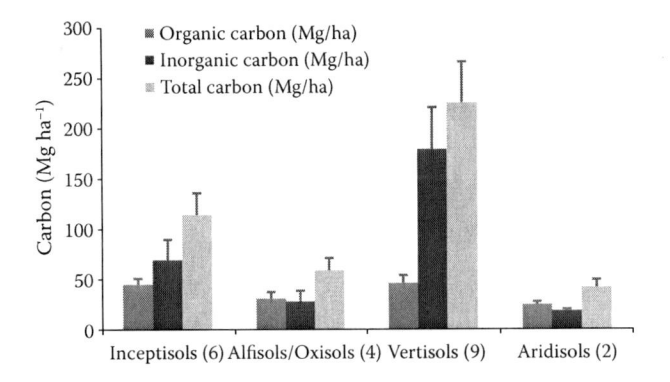

FIGURE 16.7 Carbon stocks in diverse soil types under rainfed systems. (From Srinivasarao, Ch. et al., Carbon sequestration strategies under rainfed production systems of India, Central Research Institute for Dryland Agriculture, Hyderabad (ICAR), India, 2009. Srinivasarao, Ch. et al., *Comm Soil Sci Plant Anal*, 40(15), 2338–2356, 2009.)

to 59.7 with a mean of 43.7 in Inceptisols, 23.3 to 49.8 with a mean of 30.8 in Alfisols, 28.6 to 95.9 with a mean of 46.4 in Vertisols, and 20.1 to 27.4 with a mean of 23.7 in Aridisols. The stabilizing effect of clay particles on SOC in aggregation decreases in the following sequence: allophone > amorphous minerals > smectite > illite > kaolinite (Van Breemen and Feijtel 1990). In arid and semiarid regions, Vertisols with smectite as a dominant mineral have larger SOC stocks than illitic Inceptisols and kaolinitic Alfisols.

Similar to SOC, the SIC concentration also varies widely among the soil types. Vertisols contain larger SIC, followed by Inceptisols, Alfisols, and Aridisols. The SIC stocks (Mg ha^{-1}) range from 22.3 to 135.1 (mean of 69.1) in Inceptisols, 8.8 to 57.0 (mean of 26.8) in Alfisols, 16.0 to 367.6 (mean of 178.1) in Vertisols, and 14.3 to 20.5 (mean of 17.4) in Aridisols. In most cases, surface soil storage of SOC is greater than in subsoil layers, while the reverse is the trend in the case of SIC. TCSs (Mg ha^{-1}) range from 52.1 to 192.1 (mean of 112.8) in Inceptisols, 32.1 to 82.4 (mean of 57.6) in Alfisols, 44.7 to 396.2 (mean of 224.5) in Vertisols, and 34.4 to 47.9 (mean of 41.1) in Aridisols. The TCS is also greater in Vertisols, followed by that in Inceptisols, Alfisols, and Aridisols.

The SOC concentration and stock depend on climate, soil type, and land use (Dalal and Mayer 1986). Wani et al. (2003) reported an increase in C sequestration in Vertisols by adopting RMPs (32 kg SOC ha^{-1} year^{-1}) in pigeon pea-based systems as compared to sorghum-based systems under on-farm management. The SOC concentrations reported are lower in soils of tropical regions in India than those of Australia (Dalal and Mayer 1986; Dalal 1989; Murphy et al. 2002; Young et al. 2005). Significantly lower concentrations of SOC in these soils are attributed to high rates of oxidation of SOM due to prevailing high temperatures and frequent cultivation (Dalal and Chan 2001; Wani et al. 2003). Young et al. (2005) reported that Vertisols with high clay content contain higher SOC stocks than other soils with lower clay contents. Sahrawat (2003) reported that calcium carbonate is a common mineral in soils of the dry regions of the world, stretching from subhumid to arid zones, as soils of these regions are calcareous in nature. According to an estimate by the National Bureau of Soil Survey and Land Use Planning, Nagpur, India, calcareous soils occupy about 230 M ha and constitute 69% of the total geographical area of India. Furthermore, SIC stocks consist of primary inorganic carbonates or lithogenic inorganic carbonates (LICs) and secondary/pedogenic inorganic carbonates (PICs). The reaction of atmospheric CO_2 with water (H_2O) and calcium (Ca^{2+}) in the upper horizons of the soil, leaching into the subsoil, and subsequent reprecipitation results in formation of secondary carbonates and in the sequestration of atmospheric CO_2. This is a reason why deeper layers have higher SIC than surface soils in most profiles (Sahrawat 2003).

16.7.2 Soil Carbon Stocks in Relation to Production System

The TCSs vary with production system and soil type. The data on organic C stocks, inorganic C stocks, and TCSs under different production systems are presented in Figure 16.8. In general, soybean-based production systems contain higher SOC stocks (62.3 Mg C ha^{-1}), followed by maize-based (47.6 Mg ha^{-1}) and groundnut-based (41.7 Mg ha^{-1}) systems. Soils under pearl millet- and finger millet-based systems contain lower SOC stocks. On the other hand, cotton-based systems (275.3 Mg ha^{-1}) and postrainy (*rabi*) sorghum production systems (243.7 Mg ha^{-1}), primarily on Vertisols and associated soils, contain higher SIC stocks, while the SIC is the lowest in soils under lowland rice systems (18.2 Mg ha^{-1}). The highest level of TCSs is reported under cotton-based production systems, followed by summer sorghum-based and the lowest in pearl millet-based systems. However, relative contribution of

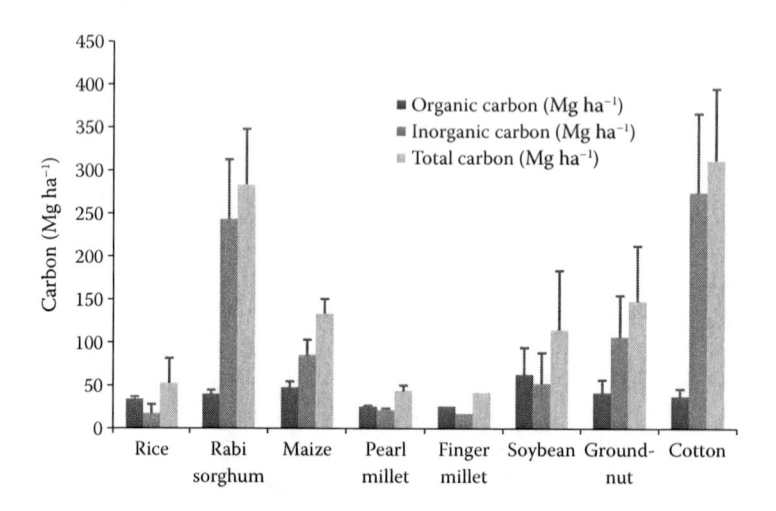

FIGURE 16.8 Soil C stocks under rainfed production systems across soil types. (From Srinivasarao, Ch. et al., Carbon sequestration strategies under rainfed production systems of India, Central Research Institute for Dryland Agriculture, Hyderabad (ICAR), India, 2009. Srinivasarao, Ch. et al., *Comm Soil Sci Plant Anal*, 40(15), 2338–2356, 2009.)

SOC to TCSs is higher under rice-based systems, while the highest SIC contribution to TCSs is in soils supporting the cotton-based system.

On a regional scale, aboveground and below-ground biomass production is probably the major determinant of the relative distribution of SOC with depth (Jobbagy and Jackson 2000). The aboveground biomass has probably only a limited effect on SOM levels compared to the below-ground biomass as has been demonstrated by several long-term residue management studies (Clapp et al. 2000; Reicosky et al. 2002). The dominant role of root C and a greater relative contribution of root versus shoot biomass to the SOC stock is widely recognized (Rasse et al. 2004). Root-to-shoot ratio is 0.21 to 0.25 for corn, 0.23 for soybean, and 0.50 for barley (Allmaras et al. 2004; Bolinder et al. 1997). In addition to the root biomass, its composition also has a strong impact on SOC sequestration. In general, leafy plants decompose faster than the woody plants, and leaves decompose faster than roots (Wang et al. 2004). The second most abundant compounds after proteins are lignins, which largely contribute to terrestrial biomass residues. These compounds exhibit a higher resistance to microbial degradation as compared to celluloses (Martin and Haider 1986). Suberin is mostly found in root tissues and is a major contributor to SOM (Nierop et al. 2003). Among dryland crops, concentration of lignin is 4% in sorghum, 8% in soybean, 10% in maize, 9% to 13% in millet, 11% to 13% in rice, and 6% to 16% in legumes like alfalfa (Scheffer 2002; Fernandez et al. 2003; Bilbro et al. 1991; Devevre and Horwath 2000; Clement et al. 1998). Corn roots also contain a wide range of fatty acids beside carbohydrates, lignin, lipids, and alkyl-aromatics (Gregorich et al. 1996).

The effect of a cropping system on SOC concentration varies with soil type. In Vertisols and associated soils, cotton- and sorghum-based systems have larger SIC stocks, while soybean and groundnut systems have higher SOC stock. In general, legume-based systems have higher SOC stocks than cereal-based systems practiced on Vertisols in the tropics (Wani et al. 1995, 2003). In Inceptisols, maize-based systems have high SOC and SIC concentrations. In Alfisols, rice-based systems (Ranchi and Phulbani) have relatively higher SOC concentration, while groundnut-based (Anantapur) systems have larger SIC concentrations. These trends may be due to larger carbonate deposits observed in the subsoil layers of the profile and frequent addition of gypsum to groundnut along with differences in rainfall, parent material, and other management practices adapted at these locations. In Aridisols, pearl millet-based systems at Sant Kabir (SK) Nagar have higher TCSs than those in Hisar, Haryana.

16.7.3 Carbon Stocks in Relation to Rainfall

In general, SOC stocks increase with an increase in the mean annual rainfall (Figures 16.9 and 16.10, $r = 0.59$, $P < 0.05$). In contrast, however, the SIC stocks decrease with the increase in mean annual rainfall, from 156.4 Mg ha^{-1} for rainfall of <550 mm to 26.0 Mg ha^{-1} for >1100 mm. As the SIC stocks exceed those of SOC, TCS decreases with increase in mean annual rainfall from 183.8 Mg ha^{-1} in the arid environment (<550 mm) to 70.2 Mg ha^{-1} in subhumid regions (>1100 mm). However, cation exchange capacity (CEC) is significantly and positively correlated ($r = 0.81$ [highly

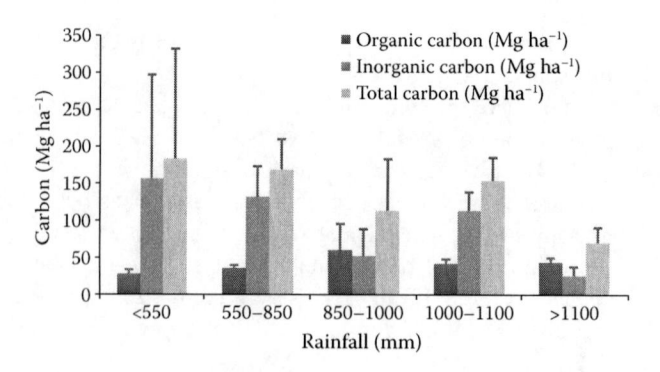

FIGURE 16.9 Soil carbon stocks in rainfed production systems in relation to rainfall. (From Srinivasarao, Ch. et al., Carbon sequestration strategies under rainfed production systems of India, Central Research Institute for Dryland Agriculture, Hyderabad (ICAR), India, 2009.)

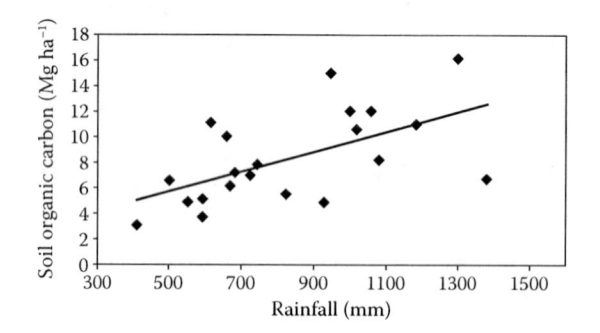

FIGURE 16.10 Relationship between mean annual rainfall (mm) and SOC in surface layer (0–15 cm) under rainfed conditions (OC stock = 1.91 + 0.008x, $r = 59$ [significant at $P \leq 0.05$ level]). (From Srinivasarao, Ch. et al., Carbon sequestration strategies under rainfed production systems of India, Central Research Institute for Dryland Agriculture, Hyderabad (ICAR), India, 2009.)

significant at $P \leq 0.01$ level]) (Figure 16.11), while clay content is only slightly correlated with the SOC stocks (Figure 16.12). This trend indirectly indicates that the specific type of clay mineral with a larger surface area is largely responsible for higher potential of C sequestration.

It has been postulated that climatic aridity is responsible for the formation of PIC, and this is a reverse process to the enhancement in SOC stock. Yet, there exists a strong synergism between increase in SOC and SIC stocks (Sahrawat 2003). The present scenario of changing climatic parameters (e.g., temperature and annual rainfall in some regions of the country) may reduce the potential of C sequestration in soils of the Indian subcontinent. Therefore, the arid climate will continue to remain as a bane for Indian agriculture because it will deplete SOC stock but lead to formation of PIC along with the concomitant development of sodicity and/or salinity (Eswaran et al. 1993; Bhattacharya et al. 2000).

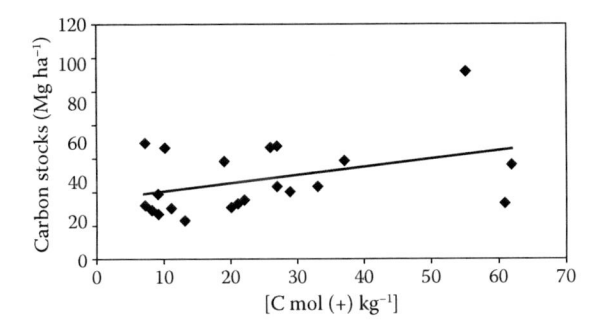

FIGURE 16.11 Relationship between CEC and organic carbon stocks in soils ($r = 0.81$ [highly significant at $P \leq 0.01$ level]). (From Srinivasarao, Ch. et al., Carbon sequestration strategies under rainfed production systems of India, Central Research Institute for Dryland Agriculture, Hyderabad (ICAR), India, 2009.)

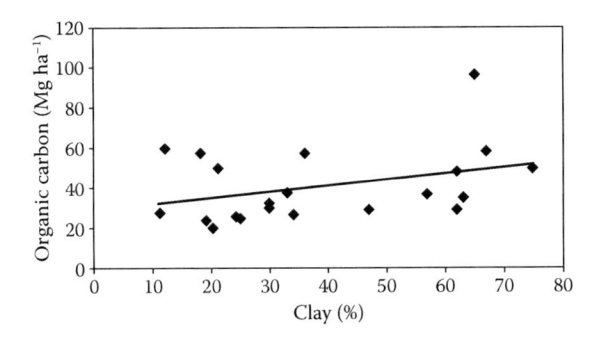

FIGURE 16.12 Relationship between clay and organic carbon stocks in soils ($r = 0.35^{NS}$ [NS means nonsignificant]). (From Srinivasarao, Ch. et al., Carbon sequestration strategies under rainfed production systems of India, Central Research Institute for Dryland Agriculture, Hyderabad (ICAR), India, 2009.)

16.8 REASONS FOR SOIL CARBON DEPLETION IN RAINFED TROPICAL SYSTEMS

India, similar to China and other countries, is facing a dual challenge of reducing CO_2 emissions and enhancing the gross domestic product (GDP) by 20% to 25% by 2020 compared with the 2005 baseline. In this context, the importance of sustainable management of soils of agroecosystems to enhance SOC stocks by sequestering atmospheric CO_2 cannot be overemphasized. Both the magnitude and quality of SOC stock are critical to improving soil quality, increasing crop productivity, and offsetting CO_2 emissions (Lal 2004; Smith 2007). Optimum levels of SOC can be managed through the adoption of RMPs such as appropriate crop rotations (Wright and Hons 2005), soil fertility management, using inorganic fertilizers and organic amendments (Schuman et al. 2002; Mandal et al. 2007; Majumder et al. 2008), and use of conservation tillage methods (Lal 2009). In rainfall-scarce environments of tropical and subtropical regions characterized by arid and semiarid climates, soils are inherently low in SOC stock, and agronomic productivity is strongly related to soil quality. Therefore, curtailing depletion and enriching SOC stock are essential to adapting and mitigating climate change, buffering agroecosystems in harsh climates against extreme events (drought), and stabilizing agronomic productivity by ensuring some returns even during the bad seasons.

Crop cultivation adversely affects the distribution and stability of soil aggregates and reduces SOC stock (Kong et al. 2005). The magnitude of reduction in SOC due to cropping, however, varies among climates and cropping systems (Lal 2004). Because of the prevailing high temperatures, soils of the tropics generally emit more oxidative products (i.e., CO_2) per unit SOC stock than those of temperate and cooler regions. However, crop species also play an important role in maintaining quantity and quality of SOC stock despite the diverse nature of crop residues (CRs) with highly variable turnover or the mean residence time (MRT) in the soil (Mandal et al. 2007).

The duration and timing of "fallowing" within a cropping system can also affect the magnitude of SOC stock (Halvorson et al. 2002) because of the differences in cropping intensity and specific management practices. Once the pathways of C sequestration in soils are identified, suitable agricultural strategies may be identified that have the potential to enhance SOC stocks, offset CO_2 loading into the atmosphere, and mitigate global warming (Lal 2009). A large proportion of the research done thus far on SOC sequestration in soils of agroecosystems is confined to cold and temperate regions. The effects of soil moisture regime on SOC decomposition (and hence, long-term SOC storage) are highly variable. In general, the rate of SOC decomposition is high when precipitation equals evapotranspiration (ET) (Parton et al. 1987). Thus, the SOC dynamics must be different in arid and semiarid rainfed conditions where precipitation always exceeds the ET.

Soils of the rainfed regions are characterized by low SOC and N stocks despite large variations in the cropping system, soil type, rainfall, temperature, and soil/crop/water management practices such as manuring and fertilization. Low crop yields and low or no biomass residue retention, coupled with long fallow periods, which extend up to 9 months in a year, result in adverse environments that do not sustain SOC levels. However, the magnitude of change of SOC due to continuous cultivation depends on the balance between the loss by oxidative forces during tillage, the quantity and quality of CRs that are returned, and the amount of organic amendments added to the soils. Therefore, crop and soil management practices have to be fine-tuned to ensure long-term sustainability. The use of plant nutrients and organic amendments and the inclusion and cultivation of legumes support SOC and its sustainability.

16.9 CARBON SEQUESTRATION

It is widely recognized that atmospheric CO_2 concentrations are steadily increasing and leading to global climate change (IPCC 2001, 2007). The Kyoto Protocol negotiated a framework for reducing the emissions of greenhouse gases (GHGs) in December 1997. The protocol also recognized that some terrestrial ecosystems have the potential to sequester large amounts of C and thus slow down the increase of atmospheric CO_2 concentrations. An increase in SOM and the biomass pools could buy time while reducing fossil fuel-related emissions of CO_2. The SOM contains ~50% C on a weight basis.

The term "soil carbon sequestration" implies transferring atmospheric CO_2 into long-lived pools and storing it securely so that it is not immediately remitted. Thus, soil C sequestration means increasing SOC and SIC stocks through judicious land and crop management practices (Lal 2004) (Table 16.9). The potential soil C sink capacity of managed ecosystems approximately equals the cumulative historic C loss estimated at 55 to 78 Pg (1 Pg = 10^9 Mg = 10^{15} g). The attainable soil C sink capacity is only 50% to 66% of the potential capacity. Atmospheric CO_2 is one of the principal GHGs, and its concentration has been steadily increasing from 280 μmol mol^{-1} in 1750 to about 370 μmol mol^{-1} by 2000 (+32%) and 390 μmol mol^{-1} by 2010 (+39%). The atmospheric CO_2 is currently increasing at the rate of 0.5% per year (3.4 Pg C $year^{-1}$).

The development of agriculture during the last two centuries and particularly since the 1990s has entailed depletion of SOC stocks. World soils are among the planet's largest reservoirs of C and hold potential for expanded C sequestration, and thus provide a prospective way of mitigating the increasing atmospheric concentration of CO_2. The United Nation Convention to Combating Desertification (UNCCD) is concerned that extensive areas of formerly productive land, in the arid and semi-arid regions in particular, have been rendered unsuitable for crop production due to land degradation. Land degradation not only reduces crop yields but also often reduces the C content of agroecosystems, which is of concern in the context of global warming.

TABLE 16.9
Comparison between Traditional Methods and RMPs in Relation to SOC Sequestration

Traditional Methods	RMPs
1. Biomass burning and residue removal	Residue returned as surface mulch
2. CT and clean cultivation	Conservation tillage, NT, and mulch farming
3. Bare/idle fallow	Growing cover crops during the off-season
4. Continuous monoculture	Crop rotations with high diversity
5. Low-input subsistence farming and soil fertility mining	Judicious use of off-farm input
6. Intensive use of chemical fertilizers	INM with compost, biosolids, and nutrient cycling; precision farming
7. Intensive cropping	Integrating trees and livestock with crop production
8. Surface flood irrigation	Drip irrigation, furrow irrigation, or subirrigation
9. Indiscriminate use of pesticides	Integrated pest management
10. Cultivating marginal soils	Conservation, restoration of degraded soils through land use change

Source: Lal, R., *Geoderma*, 123, 1–22, 2004.

Principal reservoirs of C (Pg) are the following (Kimble et al. 1990): oceans = 39,000; soils (worldwide) = 1576; atmosphere = 800; tropical soils = 506, and soil litter = 3. The oceanic pool is the largest, followed by the geologic, pedologic (soil), biotic, and atmospheric pools (Lal 2004). All these pools are interconnected, and C circulates among them. The pedologic or soil C pool comprises SOC and SIC components. The SIC stock is especially important in soils of dry regions. The SOC concentration ranges from low in soils of the arid regions to high in soils of the temperate regions and extremely high in organic/peat soils (Eswaran et al. 2000).

16.10 CARBON MANAGEMENT OPTIONS

The SOM is the storehouse of many plant nutrients, and it strongly influences the biological activity and productive capacity of soils. Over the years, efforts have been made to improve SOM status in continuously cropped soils by fertilization, manuring, and residue management practices. However, maintaining and increasing SOM is a major challenge in dryland soils due to high temperature and moisture stress. A study of 21 locations across rainfed regions of India covering eight production systems revealed that most soils are low in SOC and available N, and low to high in available P, K, and S. Many soils are deficient in available Mg, Zn, and B (Srinivasarao et al. 2006, 2008a,b). Therefore, crop and soil management practices have to be adapted to ensure long-term crop/cropping systems. Application of plant nutrients (Paustian et al. 1997) and organic amendments and inclusion/cultivation of legumes favor improvement of soil fertility and sustainability. This is directly related to maintaining the quantity of SOM, which is a critical component of soil productivity. However, resource-poor farmers in dryland regions apply a meager quantity of nutrients, and thus, crops suffer from multinutrient deficiencies (Srinivasarao et al. 2003). The best option seems to be an integrating farm-generated organic manure with inorganic fertilizers to enhance SOM, improving productivity and advancing sustainability of dryland agriculture. Crop rotation, residue management, and fertilization can maintain the level of SOM (Campbell et al. 1991). Ali et al. (2002) stated that by improving physical, chemical, and biological properties of soils, food legume cultivation could arrest the declining trend in productivity of cereal–cereal systems. The availability of CR is a major problem in India due to its competing uses. In some crops, by-products (e.g., groundnut shells) are available for soil application, as they do not have major alternate uses. Though chemical fertilization has a demonstrable and great effect on crop yields, poor farmers in rainfed regions mostly rely on farmyard manure (FYM) and other organic manures because of the high cost of chemical fertilizers. Thus, nutrient management and residue incorporation can make a major contribution to soil C sequestration.

16.10.1 SITE-SPECIFIC RMPS

Adoption of RMPs can enhance the rate of SOC sequestration (Lal 2004), thereby sequestering atmospheric CO_2 and mitigating climate change. West and Post (2002) observed that conversion from conventional tillage (CT) to no-till (NT) can sequester 57 ± 14 g C m^{-2} $year^{-1}$, excluding wheat (*Triticum aestivum* L.)–fallow systems, which may not sequester SOC. Enhancing rotation complexity can sequester an average of 20 ± 12 g C m^{-2} year^{-1}, excluding a change from continuous corn (*Zea mays* L.) to corn–soybean (*Glycine max* L.) rotation, which may not result in a significant accumulation of SOC. The rate of SOC sequestration upon conversion from CT to NT can peak in 5 to 10 years, with SOC reaching a new equilibrium in 15 to 20 years. Changes in agricultural practices for the purpose of increasing SOC must either increase inputs of biomass C to the soil or decrease decomposition of SOM and oxidation of SOC, or a combination thereof (Follett 2001; Paustian et al. 2000).

A menu of management options is being promoted by the Central Research Institute for Dryland Agriculture (CRIDA) in farmers' participatory action mode diverse physical and ethnic regions of Andhra Pradesh. Some RMPs being promoted include CR retentions, FYM, biofertilizers, inclusion of legumes in the cropping sequence or as intercrops, green manure crops, green leaf manuring, tank silt addition, and vermicomposting along with chemical fertilizers (Srinivasarao et al. 2011a,b,c,d). Among them, green leaf manure (GLM) with *Gliricidia sepium* is the most promising and climate-smart technology. *Gliricidia*, commonly known as *Kakawate*, used as insecticide, repellant, and rodenticide, can thrive in dry, moist, acidic soils or even on degraded soils under rainfed conditions. The GLM is one of the important practices for increasing SOM concentration. Most degraded and depleted soils lack sufficient amount of N. Growing *Gliricidia* plants on farm bunds serves a dual purpose of producing GLM rich in N and conserving soil (Srinivasarao et al. 2011c). Farm bunds could be productively used for growing N-fixing shrubs and trees to generate N-rich loppings. For example, growing *Gliricidia sepium* at a close sparing of 75 cm on farm bunds could provide 28 to 30 kg N ha^{-1} year^{-1} in addition to being a valuable source of organic matter (Wani et al. 2009). Indeed, the integrated nutrient management (INM) strategy is important to improving soil quality for enhancing water use efficiency (WUE) and increasing farmers' incomes. Farmers should, therefore, be encouraged to grow *Gliricidia* on farm bunds and border of fields for use in crop production, which is a sustainable means of maintaining soil fertility along with being a source of nutrients.

16.10.1.1 Improvement of Crop Yield through *Gliricidia* Green Leaf Manuring

Using GLM of *Gliricidia* can enhance soil productivity and increase crop yields of several rainfed crops. The impact of mulch-cum-manuring with *Gliricidia* on the productivity of castor showed a threefold increase compared with that of the control and one-and-a-half-fold increase in comparison with application of FYM and fertilizer (Table 16.10). Yield response of different rainfed crops to *Gliricidia* GLM has been positive in many regions including that of finger millet in red soils of Karnataka, groundnut in red soils of Andhra Pradesh, pearl millet in light-textured soils of Gujarat, and sorghum in medium to deep black soils of Maharashtra. At Bhubaneswar, in acid red and lateritic soils, maize yield improved from 1.7 to 2.1 Mg ha^{-1} with *Gliricidia* GLM equivalent to 20 kg N ha^{-1} (AICRPDA 2008–2010). In a long-term manurial experiment on Alfisols of a CRIDA farm near Hyderabad, sorghum and green gram yields were significantly higher with the addition of 1 to 2 Mg

TABLE 16.10
Impact of Mulch-cum-Manuring with *Gliricidia* on the Productivity of Castor

Treatments	Castor Equivalent (kg ha^{-1})	Net Income (Rs. ha^{-1})
T1: No FYM and no fertilizers	328	2035
T2: FYM at 5 t ha^{-1} + 40:30:0 kg NPK ha^{-1}	691	5493
T3: T2 + *Gliricidia*	984	8307

Source: Srinivasarao, Ch. et al., Carbon sequestration strategies under rainfed production systems of India, Central Research Institute for Dryland Agriculture, Hyderabad (ICAR), India, 2009.

ha^{-1} of *Gliricidia* along with 10–20 kg N ha^{-1}. Sharma et al. (2004) reported that the two INM treatments, 2 Mg *Gliricidia* loppings + 20 kg N and 4 Mg compost + 20 kg N, were the most effective in increasing the sorghum grain yield by 85% and 78% over control, respectively. However, the highest amount of SOC concentration (0.74%) was observed in 100% organic treatment (4 Mg compost + 2 Mg *Gliricidia* loppings). Some of these options of managing nutrients by using farm-based organics can be a component of organic farming. Based on a long-term experiment, Sharma et al. (2002) reported that the conjunctive use of urea and organics such as loppings of *Leucaena* and *Gliricidia* (1:1 ratio on N equivalent) increased sorghum grain yield to 1.69 and 1.72 Mg ha^{-1}, respectively, and thus revealed that a minimum of 50% N requirement of sorghum can be easily met from farm-based organic sources. This research information can be used to supplement fertilizer N up to 50% by using green loppings of *Gliricidia maculata* and *Leucaena leucocephala* and will be useful while planning for the cultivation of organic produce. Furthermore, the SOC concentration was significantly increased by application of CRs such as sorghum stover and *Gliricidia* at the rate of 2 Mg ha^{-1} under minimum-tillage and CT systems for rainfed sorghum–castor rotations on Alfisols. Sharma and colleagues also reported that an increase in N levels from 0 to 90 kg N ha^{-1} significantly improved the SOC level in these soils over a period of 8 years.

16.11 ENHANCED NUE WITH MANURING

Efficiency of applied nutrients can be substantially improved by using an INM strategy. The data from a 32-year-long manurial trial at Bangalore under rainfed conditions showed that partial factor productivity (PFP) can be sustained only with INM involving organic manures. In contrast, application of chemical fertilizers alone reduced PFP gradually in finger millet (Figure 16.13) (Srinivasarao 2011). Therefore, using INM is essential to sustainability of agricultural production on degraded soils with low SOM and multinutrient deficiencies.

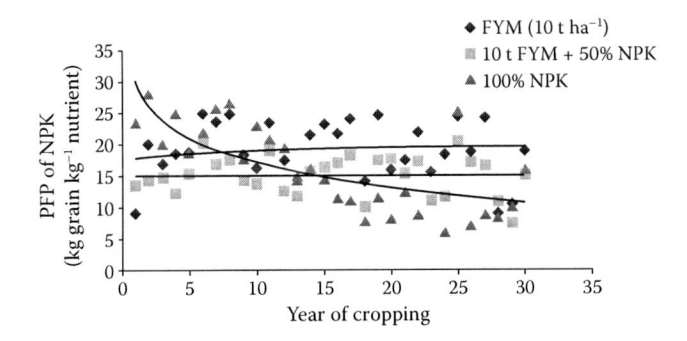

FIGURE 16.13 PFP of NPK in finger millet with various nutrient management options under rainfed conditions during 1978–2007. (From Srinivasarao, Ch., *Indian J Fertil*, 7(4), 12–25, 2011.)

16.12 SOC IN RELATION TO WATER RETENTION AND WATER PRODUCTIVITY

Water productivity (grain yield per unit amount of rainfall) has been assessed for six production systems in long-term manuring experiments of 13- to 27-year duration in different agroecological regions of India (Srinivasarao et al. 2009a). The data indicated that INM involving use of chemical fertilizers along with organic manure (FYM/CRs/GLM) increased WUE compared to unfertilized control or chemical fertilizers (Table 16.11).

16.13 CARBON MANAGEMENT IN RAINFED PRODUCTION SYSTEMS—EXPERIENCES FROM LONG-TERM EXPERIMENTS IN SEMIARID TROPICS

Seven ongoing long-term experiments involving permanent manurial trials and INM studies established under the auspices of All India Co-ordinated Research Project for Dryland Agriculture (AICRPDA) are sited in six ecoregions in semiarid climates at Anantapur (Andhra Pradesh), Bangalore (Karnataka), Solapur (Maharashtra), Indore (Madhya Pradesh), Sardar Krushinagar (Gujarat), and Varanasi (Uttar Pradesh) (Figure 16.14). Details of the soil type, cropping systems, and ecoregional characteristics are outlined in Table 16.12, and the cropping systems and manurial treatments are listed in Table 16.13. Common treatments across seven experiments are control (no fertilizer or organics), 100% recommended dose of fertilizers (RDFs), 50% RDF+ 50% organics, and 100% organics (Srinivasarao et al. 2011e, 2012a,b,c,d,e,f). All treatments at each site have been laid out in triplicate according to a randomized block design. Crop response and soil quality are briefly described below.

TABLE 16.11
Water Productivity of Different Crops as Influenced by Different Organic and INM Practices in Various Rainfed Production Systems across Different Soil Types and Climates in India

Location/State	Production System	Soil Type	Climate	Average Annual Rainfall (mm)	Duration of Permanent Manurial Trial	Crop	Average Water Productivity (kg ha^{-1} mm^{-1})				
Anantapur (Andhra Pradesh)	Groundnut	Alfisols	Arid	566	1985–2005 (20 years)	Groundnut	Control	100% RDF (20:40:40 N, P$_2$O$_5$, K$_2$O)	50% RDF + 4 t groundnut shells per hectare	50% RDF + 4 t FYM per hectare	100% organic (5 t FYM per hectare
							1.38	1.80	1.73	1.80	1.66
Bangalore (Karnataka)	Finger millet	Alfisols	Semiarid	768	1978–2005 (26 years)	Groundnut	Control	FYM at 10 t ha^{-1}	FYM at 10 t ha^{-1} + 50% NPK	FYM at 10 t ha^{-1} + 100% NPK	Recommended NPK (25:50: 25 kg ha^{-1}, groundnut; 50: 50:25 kg ha^{-1}, finger millet
							0.53	1.72	1.53	1.39	0.94
						Finger millet	0.99	4.02	4.76	4.94	3.19
Solapur (Maharashtra)	Rabi Sorghum	Vertisols	Semiarid	723	1985–2006 (21 years)	Sorghum	Control	50 kg N ha^{-1}, urea	25 kg N ha^{-1}, CR, + 25 kg N ha^{-1}, urea	25 kg N ha^{-1}, FYM, + 25 kg N ha^{-1}, urea	25 kg N ha^{-1}, CR, + 25 kg N ha^{-1}, *Leucaena*
							0.84	1.44	1.45	1.47	1.67

Location	Crop	Soil	Climate	Rainfall	Period	Crop	Control	100% RDN	50% RDN (FYM)	50% RDN (fertilizer) + 50% RDN (FYM)	Farmers' method (5 t of FYM ha^{-1} once in 3 years)
SK Nagar (Gujarat)	Pearl millet	Entisols	Arid	550	1988–2006 (18 years)	Pearl millet	0.75	1.41	1.01	1.49	1.05
						Cluster bean	0.56	0.81	0.86	1.04	0.76
						Castor	0.80	1.45	1.15	1.50	1.03
Indore (Madhya Pradesh)	Soybean	Vertisols	Semiarid	900	1992–2007 (15 years)		Control	40 kg N + 26 kg P	FYM 6 t ha^{-1} + N$_{20}$P$_{13}$	Soybean residue 5 t ha^{-1} + N$_{20}$P$_{13}$	CRs of soybean at 5 t ha^{-1}
						Soybean	1.40	2.10	2.33	1.99	1.81
						Safflower	0.72	1.34	1.72	1.39	1.17
Varanasi (Uttar Pradesh)	Rice	Inceptisols	Subhumid	1080	1986–2007 (21 years)		Control	100% RDF	100% organic (FYM)	50% organic (FYM) + 50% RDF	Farmers' practice
						Rice	1.04	1.73	1.40	1.77	1.29
						Lentil	0.46	0.69	0.65	0.91	0.47

Source: Srinivasarao, Ch. et al., Carbon sequestration strategies under rainfed production systems of India, Central Research Institute for Dryland Agriculture, Hyderabad (ICAR), India, 2009.

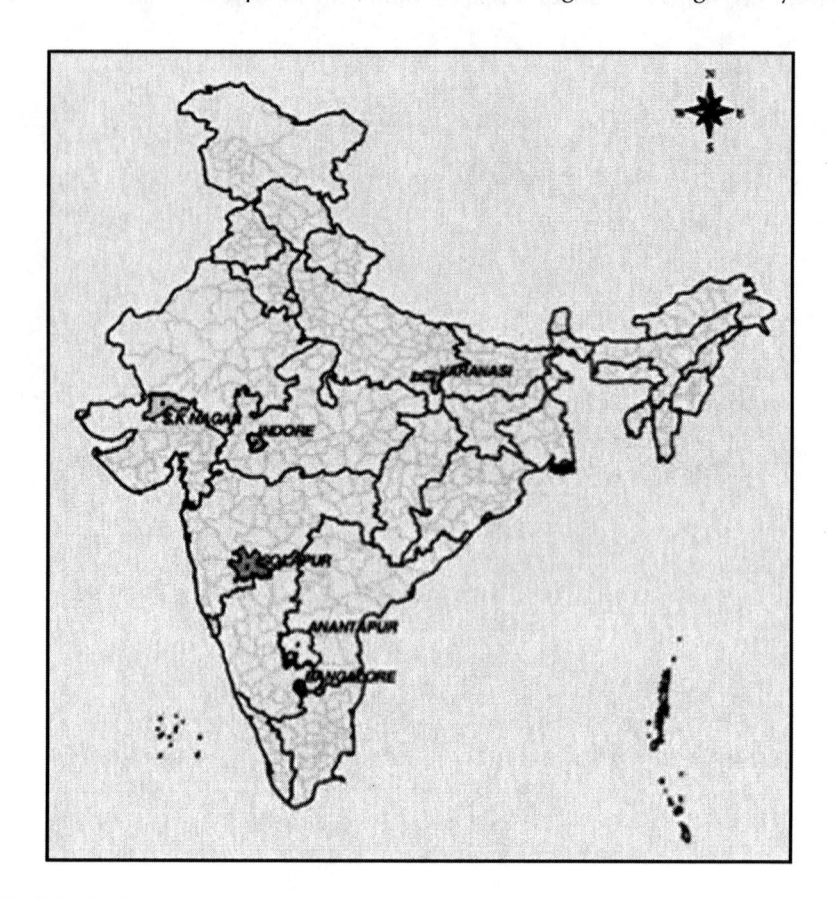

FIGURE 16.14 Location of seven long-term experiments representing predominant rainfed production systems in India. (From Srinivasarao, Ch. et al., *Comm Soil Sci Plant Anal*, 40(15), 2338–2356, 2009.)

16.13.1 Mean Yield and Sustainable Yield Index Influenced by Different Nutrient Management Treatments in Diverse Cropping Systems

Soil fertility management treatments produced significantly higher mean yield than control (Table 16.14). Higher yields were obtained by using INM involving organic manure used in combination with chemical fertilizers. For example, a higher mean pod yield of groundnut (Mg ha^{-1}) was obtained in 50% RDF + 4 Mg ha^{-1} FYM (1.03), followed by 50% RDF + 4 Mg ha^{-1} groundnut shells (GNS) (1.02), 100% RDF (0.98), and the least in the control (0.78) (Srinivasarao et al. 2012a).

Higher mean yield (Mg ha^{-1}) of pods (1.34) of groundnut was obtained with the application of FYM followed by FYM + 50% NPK (1.22), FYM + 100% NPK (1.10), recommended NPK (0.72), and the least in control (0.40) over the seven cropping seasons. In contrast, higher grain yield (Mg ha^{-1}) of finger millet was obtained through the application of FYM + 100% NPK (3.96) followed by FYM + 50% NPK (3.77), FYM (3.25), recommended NPK (2.58), and the least in the control (0.82)

TABLE 16.12
Details of Location, Soil Type, and Production System Included in the Present Study

Location	Crop/Cropping System	State	Latitude, Longitude, and Altitude	Soil Type	Climate	Average Annual Rainfall (mm)
Anantapur	Monocropping of groundnut	Andhra Pradesh	14°42′ N, 77°40′ E, 350 m	Alfisols	Semiarid	566
Bangalore	Finger millet–finger millet	Karnataka	12°46′ N, 77°11′ E, 810 m	Alfisols	Semiarid	768
Bangalore	Groundnut–finger millet rotation	Karnataka	12°46′ N, 77°11′ E, 810 m	Alfisols	Semiarid	768
Solapur	Winter sorghum	Maharashtra	17°51′ N, 75°32′ E, 480 m	Vertisols	Semiarid	723
SK Nagar	Pearl millet–cluster bean–castor rotation (once in 3 years)	Gujarat	24°30′ N, 72°13′ E, 152.5 m	Entisols	Semiarid	550
Indore	Soybean–safflower sequence	Madhya Pradesh	22°51′ N, 75°51′ E, 530 m	Vertisols	Semiarid	900
Varanasi	Upland rice–lentil sequence	Uttar Pradesh	25°11′ N, 82°51′ E, 480 m	Inceptisols	Subhumid	1080

Source: Srinivasarao, Ch. et al., Carbon sequestration strategies under rainfed production systems of India, Central Research Institute for Dryland Agriculture, Hyderabad (ICAR), India, 2009.

over the six cropping seasons in a groundnut–finger millet rotation (Srinivasarao et al. 2012b). In finger millet monocropping, higher mean yields (Mg ha^{-1}) of grains (3.28 over 27 cropping seasons) were obtained through use of RDF along with 10 Mg FYM ha^{-1}, followed by 2.92 with 10 Mg FYM ha^{-1} + 50% NPK (Srinivasarao et al. 2012c). Even with the application of 10 Mg ha^{-1} FYM alone, sustained and significantly higher grain yield was obtained compared to sole application of chemical fertilizer or unfertilized control. For the entire 22-year period, higher grain yield was obtained through the application of 25 kg N from CRs + 25 kg N (*Leucaena*) (1.2 Mg ha^{-1}). It was on par with 25 kg N (FYM) + 25 kg N (urea, 1.06 Mg ha^{-1}); 25 kg N (CR) + 25 kg N (urea, 1.05 Mg ha^{-1}); and 50 kg N (urea, 1.04 Mg ha^{-1}) and was the least in control (0.61 Mg ha^{-1}) (Srinivasarao et al. 2012d). The higher average seed yields (Mg ha^{-1}) of pearl millet (0.81), cluster bean (0.58), and castor (0.83) over six

TABLE 16.13
Selected C Management Options for Different Rainfed Productions in Seven Permanent Manurial Trials in the Studied Locations

Production Systems/ Location	Duration/No. of Years of Permanent Manurial Trial	Treatment Details
Groundnut (Anantapur)	1985–2004 (20 years)	T_1 = control (no fertilizer), T_2 = 100% RDF (20:40:40 N, P_2O_5, K_2O), T_3 = 50% RDF + 4 Mg groundnut shells (GNS) per hectare, T_4 = 50% RDF + 4 Mg FYM ha^{-1}, T_5 = 100% organic (5 Mg FYM ha^{-1})
Finger millet (Bangalore; finger millet–finger millet)	1978–2004 (27 years)	T_1 = control, T_2 = 10 Mg FYM ha^{-1}, T_3 = 10 Mg FYM ha^{-1} + 50% NPK, T_4 = 10 Mg FYM ha^{-1} + 100% NPK, T_5 = recommended NPK (50:50:25 kg NPK ha^{-1}, finger millet)
Finger millet (Bangalore; finger millet–groundnut)	1991–2004 (13 years)	T_1 = control, T_2 = 10 Mg FYM ha^{-1}, T_3 = 10 Mg FYM ha^{-1} + 50% NPK, T_4 = 10 Mg FYM ha^{-1} + 100% NPK, T_5 = recommended NPK (25:50:25 kg NPK ha^{-1}, groundnut; 50:50:25 kg NPK ha^{-1}, finger millet)
Rabi sorghum (Solapur)	1985–2006 (22 years)	T_1 = control, T_2 = 25 kg N ha^{-1} (urea), T_3 = 50 kg N ha^{-1} (urea), T_4 = 25 kg N ha^{-1} through sorghum residue (CR), T_5 = 25 kg N ha^{-1} through FYM, T_6 = 25 kg N ha^{-1} (CR) +25 kg N ha^{-1} (urea), T_7 = 25 kg N ha^{-1} (FYM) + 25 kg N ha^{-1} (urea), T_8 = 25 kg N ha^{-1} (CR) + 25 kg N ha^{-1} (*Leucaena* clippings), T_9 = 25 kg N ha^{-1} (*Leucaena*), T_{10} = 25 kg N ha^{-1} (*Leucaena*) +25 kg N ha^{-1} (urea)
Pearl millet (SK Nagar)	1988–2006 (18 years)	T_1 = control, T_2 = 100% recommended dose of N through mineral fertilizer (RDN (F)), T_3 = 50% RDN (F), T_4 = 50% recommended N (FYM), T_5 = 50% recommended N (fertilizer) + 50% recommended N (FYM), T_6 = farmers' method (5 Mg of FYM ha^{-1} once in 3 years)
Soybean (Indore)	1992–2007 (15 years)	T_1 = control, T_2 = 20 kg N + 13 kg P, T_3 = 30 kg N + 20 kg, T_4 = 40 kg N + 26 kg, T_5 = 60 kg N + 35 kg P, T_6 = 6 Mg FYM ha^{-1} + $N_{20}P_{13}$, T_7 = 5 Mg soybean residue per hectare + $N_{20}P_{13}$, T_8 = 6 Mg FYM ha^{-1}, T_9 = 5 Mg soybean residue per hectare
Rice (Varanasi)	1986–2007 (21 years)	T_1 = control, T_2 = 100% RDF (inorganic), T_3 = 50% RDF (inorganic), T_4 = 100% organic (FYM), T_5 = 50% organic (FYM), T_6 = 50% RDF + 50% (foliar), T_7 = 50% organic (FYM) + 50% RDF, T_8 = farmers' practice

Source: Srinivasarao, Ch. et al., Carbon sequestration strategies under rainfed production systems of India, Central Research Institute for Dryland Agriculture, Hyderabad (ICAR), India, 2009.

cropping seasons were obtained through INM followed by 0.78, 0.45, and 0.80 Mg ha^{-1}, respectively, with 100% RDF (Srinivasarao et al. 2011d). For the entire 15-year period, higher grain yield was obtained through the application of 6 Mg FYM + N$_{20}$P$_{13}$ ha^{-1} (2.10 and 1.49 Mg ha^{-1} for soybean and safflower, respectively), and it was the least in the control (1.04 and 0.63 Mg ha^{-1} for soybean and safflower, respectively) (Srinivasarao et al. 2012e). Both crops responded well to a higher dose of fertilizer application and also to INM. Significantly higher yields were obtained through the application of 60 kg N + 35 kg P ha^{-1} (1.99, 1.21 Mg ha^{-1}) compared to that with application of 5 Mg CR + N$_{20}$P$_{13}$ ha^{-1} (1.79, 1.16 Mg ha^{-1}); 6 Mg FYM ha^{-1} (1.86, 1.22 Mg ha^{-1}); and 5 Mg CR ha^{-1} (1.63, 0.97 Mg ha^{-1}) (Srinivasarao et al. 2012e). Pooled data of 21 years suggest a higher grain yield (Mg ha^{-1} of rice and lentil, respectively) through the application of 50% organic (FYM) + 50% RDF (1.95, 1.04) followed by 100% RDF (mineral; 1.85, 0.77). Yield was on par with 100% organic (FYM; 1.75, 0.82), and the least yield was measured in control (1.08, 0.48) (Srinivasarao et al. 2012f).

The sustainable yield index [(SYI = $Y - \sigma$)/Ym, where Y is the estimated average yield of a practice across the years, σ is its estimated standard deviation, and Ym is the observed maximum yield in the experiment during the years of cultivation] also followed a trend similar to that of agronomic yield. Significantly higher SYI was observed with the application of organic amendments either alone or in combination with chemical fertilizers compared to control or sole application of chemical fertilizers. The SYI of groundnut was also higher in 50% RDF + 4 Mg ha^{-1} GNS (0.48) followed by 50% RDF + 4 Mg ha^{-1} FYM (0.46), 5 Mg ha^{-1} FYM (0.38), 100% RDF (0.32), and the lowest in control (0.25) (Srinivasarao et al. 2012a). The highest SYI was obtained with the application of FYM + 50% NPK (0.24 for groundnut; 0.80 for finger millet), followed by FYM + 100% NPK (0.21 for groundnut, 0.76 for finger millet) in groundnut–finger millet rotation (Srinivasarao et al. 2012b). A similar trend in SYI was obtained in the case of finger millet monocropping (Srinivasarao et al. 2012c). In the case of winter sorghum, the highest SYI was obtained with 25 kg N (CR) + 25 kg N (*Leucaena*, 0.48), followed by 25 kg N (CR) + 25 kg N (urea, 0.45), and the lowest in control (0.38) (Srinivasarao et al. 2012d). In the case of pearl millet-based systems, the SYI of all three crops was higher in 50% RDN (F) + 50% RDN (FYM) (0.30 for pearl millet, 0.69 for cluster bean, and 0.46 for castor), followed by 100% RDN (F) (0.24, 0.46, 0.40); 50% RDN (FYM) (0.25, 0.46, 0.39); 50% RDN (F) (0.22, 0.36, 0.38); farmers' practice (0.16, 0.34, 0.42); and lowest in control (0.14, 0.23, 0.32), respectively (Srinivasarao et al. 2011e). Similarly, significantly higher SYI was obtained with the application of organics either alone or in combination with chemical fertilizers compared either to the control or with the sole application of chemical fertilizers at a lower dose in the case of safflower (Srinivasarao et al. 2012e). A significantly higher residual effect was observed in FYM compared to that with the use of residues of soybean and safflower. The highest SYI was measured with 6 Mg FYM + N$_{20}$P$_{13}$ ha^{-1} (0.48 and 0.45 for soybean and safflower, respectively) and the lowest in the control (0.41 and 0.22). In the case of rice-based systems, the highest SYI was observed in 50% organic (FYM) + 50% RDF (0.29, 0.30) followed by 100% organic (FYM; 0.26, 0.26), and the lowest was in control (0.17, 0.14) (Srinivasarao et al. 2012f).

TABLE 16.14
Mean Crop Yield and SYI of Different Crops in Seven Long-Term Experiments under Rainfed Production Systems

Location/Production System	Treatment					
Anantapur (Groundnut)	Control	100% RDF	50% RDF + 4 Mg ha⁻¹ GNS	50% RDF + 4 Mg ha⁻¹ FYM	100% organic (5 Mg ha⁻¹ FYM)	
Mean yield	0.78[D]	0.98[B]	1.02[A]	1.03[A]	0.92[C]	
SYI	0.25[D]	0.32[C]	0.48[A]	0.46[A]	0.38[B]	
Bangalore (groundnut–finger millet)	Control	FYM at 10 Mg ha⁻¹	FYM at 10 Mg ha⁻¹ + 50% NPK	FYM at 10 Mg ha⁻¹ + 100% NPK	Recommended NPK	
Mean yield						
Groundnut	0.40[E]	1.10[C]	1.22[B]	1.34[A]	0.72[D]	
Finger millet	0.82[E]	3.25[C]	3.77[B]	3.96[A]	2.58[D]	
SYI						
Groundnut	0.09[D]	0.23[A]	0.24[A]	0.21[B]	0.16[C]	
Finger millet	0.24[E]	0.71[C]	0.80[A]	0.76[B]	0.62[D]	
Bangalore (finger millet)	Control	FYM at 10 Mg ha⁻¹	FYM at 10 Mg ha⁻¹ + 50% NPK	FYM at 10 Mg ha⁻¹ + 100% NPK	Recommended NPK	
Mean yield	0.84[E]	2.48[C]	2.92[B]	3.28[A]	2.16[D]	
SYI	0.04[D]	0.58[B]	0.62[A]	0.59[B]	0.36[C]	
Solapur (winter sorghum)	Control	50 kg N ha⁻¹, urea	25 kg N ha⁻¹, CR + 25 kg N ha⁻¹, urea	25 kg N ha⁻¹, FYM, + 25 kg N ha⁻¹, urea	25 kg N ha⁻¹, CR + 25 kg N ha⁻¹, Leucaena	25 kg N ha⁻¹, Leucaena, + 25 kg N ha⁻¹, urea
Mean yield	0.61[D]	1.04[B]	1.06[B]	1.05[B]	0.85[C]	1.19[A]
SYI	0.38[D]	0.41[C]	0.45[B]	0.44[B]	0.48[A]	0.44[B]
SK Nagar (pearl millet)	Control	100% recommended dose of N through mineral fertilizer RDN (F)	50% RDN (F)	50% recommended N (FYM)	50% RDN (F) + 50% RDN (FYM)	Farmers' method (5 Mg of FYM ha⁻¹ once in 3 years)

	Control	20 kg N + 13 kg P	30 kg N + 20 kg P	40 kg N + 26 kg P	60 kg N + 35 kg P	FYM 6 Mg ha⁻¹ + $N_{20}P_{13}$	Soybean residue 5 Mg ha⁻¹ + $N_{20}P_{13}$	FYM at 6 Mg ha⁻¹	Soybean residues at 5 Mg ha⁻¹
Mean yield									
Pearl millet	0.43[E]	0.78[B]	0.62[C]	0.55[D]	0.81[A]	0.60[C]			
Cluster bean	0.32[E]	0.45[C]	0.43[D]	0.48[B]	0.58[A]	0.42[D]			
Castor	0.44[F]	0.80[B]	0.67[C]	0.63[D]	0.83[A]	0.57[E]			
SYI									
Pearl millet	0.14[E]	0.24[B]	0.22[C]	0.25[B]	0.30[A]	0.16[D]			
Cluster bean	0.23[D]	0.46[B]	0.36[C]	0.46[B]	0.69[A]	0.34[C]			
Castor	0.32[D]	0.40[B]	0.38[C]	0.39[C]	0.46[A]	0.42[B]			
Indore (soybean)	Control	20 kg N + 13 kg P	30 kg N + 20 kg P	40 kg N + 26 kg P	60 kg N + 35 kg P	FYM 6 Mg ha⁻¹ + $N_{20}P_{13}$	Soybean residue 5 Mg ha⁻¹ + $N_{20}P_{13}$		
Mean yield									
Soybean	1.04[E]	1.62[D]	1.76[C]	1.89[B]	1.99[A]	2.10[A]	1.79[C]	1.86[B]	1.63[D]
Safflower	0.63[G]	0.80[F]	1.02[D]	1.16[C]	1.21[B]	1.49[A]	1.16[C]	1.22[B]	0.97[E]
SYI									
Soybean	0.41[C]	0.47[A]	0.46[A]	0.48[A]	0.48[A]	0.48[A]	0.38[D]	0.42[B]	0.44[B]
Safflower	0.22[E]	0.29[D]	0.29[D]	0.31[C]	0.31[C]	0.45[A]	0.36[B]	0.38[B]	0.32[C]
Varanasi (upland rice)	Control	100% RDF (inorganic)	50% RDF (inorganic)	100% organic (FYM)	50% RDF + 50% (foliar)	50% organic (FYM) + 50% RDF	Farmers' practice		
Mean yield									
Rice	1.08[F]	1.85[B]	1.51[D]	1.75[C]	1.57[D]	1.95[A]	1.37[E]		
Lentil	0.48[D]	0.77[B]	0.54[D]	0.82[B]	0.66[C]	1.04[A]	0.51[D]		
SYI									
Rice	0.17[D]	0.25[B]	0.21[C]	0.26[B]	0.22[C]	0.29[A]	0.20[C]		
Lentil	0.14[F]	0.24[C]	0.16[E]	0.26[B]	0.19[D]	0.30[A]	0.15[F]		

Sources: Srinivasarao, Ch. et al., *Land Degrad Develop*, doi: 10.1002/ldr.1158, 2011. Srinivasarao, Ch. et al., *Eur J Agron*, doi: 10.1016/j.eja.2012.05.001, 2012. Srinivasarao, Ch. et al., *Int J Agric Sustain*, 10(3), 1–15, 2012. Srinivasarao, Ch. et al., *J Plant Nutrition Soil Sci*, doi: 10.1002/jpln.201000429, 2012. Srinivasarao, Ch. et al., *Geoderma*, 175–176, 90–97, 2012. Srinivasarao, Ch. et al., *Can J Soil Sci*, 92, doi: 10.4141/cjss2011-098, 2012. Srinivasarao, Ch. et al., *Soil Sci Soc Am J*, 76(1), 168–178, 2012.

Note: Different capital letters within rows are significantly different at $P = 0.05$.

The general trend of mean yield and SYI of all the crops irrespective of production system and soil types confirms that improvement of SOC concentration of the soil profile largely influenced the agronomic yield and sustainability of the system. For example, the highest SOC concentration (4.6 g kg^{-1}) was measured in 50% RDF + 4 Mg ha^{-1} GNS, followed by that in 50% RDF + 4 Mg ha^{-1} FYM (4.3 g kg^{-1}), 5 Mg ha^{-1} FYM (4.2 g kg^{-1}), 100% RDF (3.3 g kg^{-1}), and the least in control (2.8 g kg^{-1}) in groundnut production systems (Table 16.15). In finger millet production system, significantly higher SOC concentration was observed in 10 Mg FYM ha^{-1} + 100% NPK treatment. Similarly, the SOC concentration was also increased in other production system for treatments receiving organics along with mineral fertilizers. For example, the following were observed: in winter sorghum, 25 kg N ha^{-1} (*Leucaena*) + 25 kg N (urea, 5.6 g kg^{-1} profile mean SOC); in pearl millet, 50% RDN (F) + 50% RDN (FYM) (2.2 g kg^{-1}); in soybean, 6 Mg FYM ha^{-1} + N$_{20}$P$_{13}$ (5.4 g kg^{-1}); in rice, 100% organic (FYM, 2.1 g kg^{-1}). There was a significant reduction in SOC concentration with the sole application of inorganic fertilizers (100% RDF) compared to the treatments receiving organics, which was indirectly reflected by lower SYI values (Srinivasarao et al. 2011d, 2012d,e).

16.13.2 Soil Carbon and Yield Sustainability

Crop yield sustainability under rainfed conditions depends on improvement in soil quality through increase in SOC stock. The relationships between SYI and SOC concentration and inputs of biomass C indicate that maintenance of SOC concentration through regular organic or mineral inputs determines the sustainability of rainfed production systems. The improvement in SOC stock is also related to an enhanced water holding capacity of the soil profile (Du et al. 2009), which alleviates intermittent droughts, a major constraint in dryland agriculture.

16.13.2.1 Groundnut Production System

There exists a positive relationship between SOC stock to 1-m depth and cumulative yield of the groundnut pod (15 kg ha^{-1} year^{-1} Mg^{-1} of SOC) (Srinivasarao et al. 2012a). A highly significant correlation was observed between SYI and internal C inputs (CR; $Y = 0.0002X + 0.17$, $r = 0.95$, $R^2 = .91$, $P < 0.005$), total C inputs ($-0.0001X + 0.222$, $r = 0.97$, $R^2 = 0.95$, $P < 0.05$), profile mean SOC concentration ($Y = 0.18X - 0.17$, $r = 0.99$, $R^2 = 0.98$, $P < 0.05$), and SOC stock ($Y = 0.008X + 0.051$, $r = 0.70$, $R^2 = 0.48$, $P < 0.05$) (Table 16.16). Application of chemical fertilizers alone did not sustain agronomic productivity on a long-term basis. Significantly higher SYI was obtained with the application of organics either alone or in combination with chemical fertilizers compared to control or sole application of chemical fertilizers. Further, application of FYM/CR/GLM significantly improved SYI of both crops compared to use of the recommended rate of NPK, probably due to a high moisture retention capacity in treatments receiving organic amendments. Thus, increase in soil moisture storage by application of FYM is important to obtaining high yields under rainfed conditions. Ghosh et al. (2006) reported that total system productivity was 130% higher in a groundnut-based [rainy-season groundnut followed by other post–rainy season crops, viz., groundnut, chickpea (*Cicer arietinum*), wheat (*Triricum aestivum*), mustard (*Brasicca compestriss*), sunflower (*Carthamus tinctorius*)] than in a fallow-based (rainy-season fallow followed by other post–rainy season crops) system and was in the following order: groundnut–groundnut > groundnut–chickpea > groundnut–wheat > groundnut–mustard > groundnut–sunflower, though SYI was the highest in the groundnut–groundnut system in Vertisols (Typic Haplustert).

16.13.2.2 Finger Millet-Based Production System

In the case of groundnut–finger millet rotations, the SYI was in accord with the SOC pool, and there existed a significant correlation between the annual input of CRs ($r = 0.98–0.99$, $R^2 = 0.97$, $P < 0.05$), total cumulative C input ($r = 0.85–0.94$, $R^2 = 0.73–0.88$, $P < 0.05$), C buildup ($r = 0.96–0.98$, $R^2 = 0.93–0.97$, $P < 0.05$), profile SOC ($r = 0.96–0.98$, $R^2 = 0.93–0.97$, $P < 0.05$), and C sequestrated ($r = 0.96–0.98$, $R^2 = 0.93–0.97$, $P < 0.05$) (Table 16.17). In case of finger millet monocropping, a significant relationship was observed between SYI and annual CR C inputs ($r = 0.97$, $R^2 = 0.94$, $P < 0.05$), total cumulative C input ($r = 0.93$, $R^2 = 0.87$, $P < 0.05$), C buildup ($r = 0.95$, $R^2 = 0.90$, $P < 0.05$), profile SOC ($r = 0.95$, $R^2 = 0.90$, $P < 0.05$), and C sequestrated ($r = 0.95$, $R^2 = 0.90$, $P < 0.05$) (Table 16.18). Application of chemical fertilizers alone could not sustain productivity of either a cereal- or legume-based system on a long-term basis. Manna et al. (2005) reported similar positive yield trends in the NPK and NPK + FYM treatments in soybean–wheat systems on Alfisols at Ranchi, Bihar.

16.13.2.3 Winter Sorghum-Based Production System

The SYI was also in accord with the SOC stock, and a significant correlation existed between SYI and the total C input ($r = 0.87*$ [one asterisk means significant at $P \le 0.05$ level], $R^2 = 0.76$, $P < 0.05$), % SOC sequestration ($r = 0.90*$, $R^2 = 0.81$, $P < 0.05$), profile SOC stock ($r = 0.90*$, $R^2 = 0.81$, $P < 0.05$), and SOC sequestered ($r = 0.90*$, $R^2 = 0.81$, $P < 0.05$) (Table 16.19). However, the correlation of SYI with the annual input of CR was weak ($R^2 = 0.31$). A beneficial effect of application of organic materials as a source of N on sorghum grain yield has been reported earlier by Bellakki and Badanur (2000).

16.13.2.4 Pearl Millet-Based Production System

In general, there exists a strong relationship between SOC concentration/stock and grain yield (Bronson et al. 1998; Yadav et al. 2000; Regmi et al. 2002; Singh et al. 2004). There also exists a linear relationship between the SYI and SOC fractions (Srinivasarao et al. 2011e). These trends indicate the importance of SOC fractions on crop yield through improvement in soil quality (Srinivasarao et al. 2011b). A significant relationship was also observed between SYI and total C inputs ($r = 0.67–0.75$, $R^2 = 0.45–0.56$, $P < 0.05$) and the profile SOC stock ($r = 0.78–0.88$, $R^2 = 0.61–0.77$, $P < 0.05$) (Table 16.20). Thus, the maintenance of SOC stock through regular addition of organics determines the sustainability of rainfed conditions involving pearl millet–based production systems (Srinivasarao et al. 2011e).

16.13.2.5 Safflower- and Soybean-Based Production System

The SYI of safflower is in accord with the SOC concentration and stock, and a significant correlation exists between SYI and the total C input ($R^2 = 0.53*$ [one asterisk means significant at $p \le 0.05$ level], $P < 0.05$), % SOC sequestration ($R^2 = 0.63*$, $P < 0.05$),

TABLE 16.15
Profile Mean Soil Organic and Microbial Biomass C Concentration in Seven Long-Term Experiments Under Rainfed Production Systems

Location				Treatments		
Anantapur (Groundnut)	Control	100% RDF	50% RDF + 4 Mg ha^{-1} GNS	50% RDF + 4 Mg ha^{-1} FYM	100% organic (5 Mg ha^{-1} FYM)	
SOC	2.8^D	3.3^C	4.6^A	4.3^B	4.2^B	
Bangalore (groundnut–finger millet)	Control	FYM at 10 Mg ha^{-1}	FYM at 10 Mg ha^{-1} + 50% NPK	FYM at 10 Mg ha^{-1} + 100% NPK	Recommended NPK	
SOC	3.3^D	5.2^B	5.7^A	5.7^A	4.6^C	
Bangalore (finger millet)	Control	FYM at 10 Mg ha^{-1}	FYM at 10 Mg ha^{-1} + 50% NPK	FYM at 10 Mg ha^{-1} + 100% NPK	Recommended NPK	
SOC	4.2^E	5.9^C	6.1^B	6.4^A	4.9^D	
Solapur (winter sorghum)	Control	50 kg N ha^{-1}, urea	25 kg N ha^{-1}, CR, + 25 kg N ha^{-1}, urea	25 kg N ha^{-1}, FYM, + 25 kg N ha^{-1}, urea	25 kg N ha^{-1}, CR, + 25 kg N ha^{-1}, *Leucaena*	25 kg N ha^{-1}, *Leucaena*, + 25 kg N ha^{-1}, urea
SOC	3.9^D	4.5^C	5.0^B	5.5^A	4.4^C	5.6^A

	Control	100% recommended dose of N through mineral fertilizer (RDN (F))	50% RDN (F)	50% recommended N (FYM)	50% RDN (F) + 50% RDN (FYM)	Farmers' method (5 Mg of FYM ha^{-1} once in 3 years)		
SK Nagar (pearl millet)								
SOC	1.3^D	1.5^C	1.4^C	1.9^B	2.2^A	1.4^C		
Indore (soybean)	Control	20 kg N + 13 kg P	30 kg N + 20 kg P	40 kg N + 26 kg P	60 kg N + 35 kg P	FYM 6 Mg ha^{-1} + N$_{20}$P$_{13}$	Soybean residue 5 Mg ha^{-1} + N$_{20}$P$_{13}$	FYM at 6 Mg ha^{-1}	Soybean residue at 5 Mg ha^{-1}
SOC	3.7^G	4.1^F	4.4^E	4.5^D	4.6^D	5.4^A	5.3^A	5.1^B	4.8^C
Varanasi (upland rice)	Control	100% RDF (inorganic)	50% RDF (inorganic)	100% organic (FYM)	50% RDF + 50% (foliar)	50% organic (FYM) + 50% RDF	Farmers' practice		
SOC	1.3^D	1.7^B	1.5^C	2.1^A	1.5^C	2.0^A	1.5^C		

Source: Srinivasarao, Ch. et al., *Land Degrad Develop*, doi: 10.1002/ldr.1158, 2011. Srinivasarao, Ch. et al., *Eur J Agron*, doi: 10.1016/j.eja.2012.05.001, 2012. Srinivasarao, Ch. et al., *Int J Agric Sustain*, 10(3), 1–15, 2012. Srinivasarao, Ch. et al., *J Plant Nutrition Soil Sci*, doi: 10.1002/jpln.201000429, 2012. Srinivasarao, Ch. et al., *Geoderma*, 175–176, 90–97, 2012. Srinivasarao, Ch. et al., *Can J Soil Sci*, 92, doi: 10.4141/cjss2011-098, 2012. Srinivasarao, Ch. et al., *Soil Sci Soc Am J*, 76(1), 168–178, 2012.

Note: SOC is expressed in g kg^{-1}. Different capital letters within rows are significantly different at $P = 0.05$.

TABLE 16.16
Relationships of C Inputs, Profile C Content, Stocks, Sequestration Rate, and Different Forms of C with SYI in a 20-Year Long-Term Experiment

Parameters		Regression Equation	Correlation (r)	(R^2)
Internal C inputs (X)	SYI (Y)	Y = 0.0002X + 0.17	0.95	0.91***
Total C inputs (X)		Y = –0.00007X + 0.222	0.97	0.95***
Profile mean C content (X)		Y = 0.18X – 0.17	0.99	0.98***
Profile C stocks (X)		Y = 0.008X + 0.051	0.70	0.48*
CSR (X)		Y = 0.015X + 0.30	0.99	0.98***

Source: Srinivasarao, Ch. et al., *Eur J Agron*, doi: 10.1016/j.eja.2012.05.001, 2012.
Note: * and *** denote significance at $P < 0.05$ and 0.001, respectively.

TABLE 16.17
Relationships of SYI with C Input, C Buildup, Profile SOC, and C Sequestration in a 13-Year Long-Term Experiment

Parameters		Regression Equation	Correlation (r)	(R^2)
Annual CR C input (X)	SYI (Y)	Y_G = 0.2X + 0.03	0.99	0.97
		Y_F = 0.72X + 0.05	0.98	0.97
Total cumulative C input (X)		Y_G = 0.003X + 0.1	0.94	0.88
		Y_F = 0.01X + 0.35	0.85	0.73
C buildup % (X)		Y_G = 0.003X + 0.1	0.96	0.93
		Y_F = 0.01X + 0.27	0.98	0.97
Profile SOC (X)		Y_G = 0.01X – 0.25	0.96	0.93
		Y_F = 0.02X – 1.02	0.98	0.97
C sequestrated (X)		Y_G = 0.01X + 0.17	0.96	0.93
		Y_F = 0.02X + 0.57	0.98	0.97

Source: Srinivasarao, Ch. et al., *Int J Agric Sustain*, 10(3), 1–15, 2012.

profile SOC stock ($R^2 = 0.63*$, $P < 0.05$), and SOC sequestered ($R^2 = 0.63*$, $P < 0.05$) (Table 16.21). However, the correlation of SYI with the annual input of CR is weak ($R^2 = 0.26$). But these relationships may not be always significant in the case of soybean. Thus, the maintenance of SOC stock through application of FYM and chemical fertilizers is essential to the sustainability of rainfed production systems, particularly in cropping systems like soybean–safflower, which need high external inputs. Any improvement in SOC enhances available water capacity of the soil profile (Du et al. 2009), which reduces frequency and intensity of the drought stress. Although the application of organics may not drastically increase the SYI of soybean, it can sustain the

TABLE 16.18
Relationships of C Input, C Buildup, Profile SOC, and C Sequestration with SYI in a 27-Year Long-Term Experiment

Parameters		Regression Equation	Correlation (r)	(R^2)
Annual CR C input (X)	SYI (Y)	Y = 0.82X − 0.24	0.97	0.94
Total cumulative C input (X)		Y = 0.01X + 0.09	0.93	0.87
C buildup % (X)		Y = 0.02X + 0.11	0.95	0.90
Profile SOC (X)		Y = 0.03X − 1.53	0.95	0.90
C sequestrated (X)		Y = 0.03X + 0.29	0.95	0.90

Source: Srinivasarao, Ch. et al., *J Plant Nutrition Soil Sci*, doi: 10.1002/jpln.201000429, 2012.

TABLE 16.19
Relationships of SYI to CR C Input, Total Cumulative C Input, % Change in SOC Stock, and the Magnitude of SOC Sequestered after 22 Years of Cropping

Parameters	Regression Equation	R^2
Annual CR C input (X)	SYI = 0.16X + 0.30	0.31
Total cumulative C input (X)	SYI = 0.001X + 0.39	0.76*
C buildup % (X)	SYI = 0.002X + 0.39	0.81*
Profile SOC (X)	SYI = 0.004X + 0.2	0.81*
C sequestrated (X)	SYI = 0.004X + 0.41	0.81*

Source: Srinivasarao, Ch. et al., *Geoderma*, 175–176, 90–97, 2012.
Note: * Significant at P = 0.05 level.

agronomic productivity of the soybean–safflower system on a long-term basis. Besides including a legume in sequence, combined use of available organic sources along with chemical fertilizers might prove beneficial for long-term productivity and sustainability of the system (Behera et al. 2007). A significantly higher SYI in soybean has been obtained by the sole application of chemical fertilizers. This trend is mainly attributed to resilience of a soil system to intermittent droughts, with high moisture retention capacity in plots receiving organic amendments compared with those receiving inorganic fertilizers in case of postrainy season crop of safflower. Such a positive response underlines the importance of organics in enhancing soil resilience under harsh climatic conditions during the cropping period, a common feature of rainfed agriculture. In comparison, soybean, being a nutrient-exhaustive (i.e., P) crop, requires more nutrients during the growing period. Manna et al. (2005) reported a negative yield trend of soybean with unbalanced use of N and NP application in Vertisol.

TABLE 16.20
Relationships between Different Forms of C (C Inputs, Soil C Sequestration Rate, and SYI) of Pearl Millet-Based Systems in Semiarid Tropical Conditions

Independent Variable		Regression Equation	Coefficient of Determination (R^2)
C sequestration rate	SYI	Y(pearl millet) = 0.02X + 0.34	0.67*
		Y(cluster bean) = 0.05X + 0.84	0.77**
		Y(castor) = 0.01X + 0.5	0.56*
Total C inputs		Y(pearl millet) = 0.003X + 0.17	0.45
		Y(cluster bean) = 0.008X − 0.29	0.56*
		Y(castor) = 0.002X + 0.36	0.45
Profile mean SOC content		Y(pearl millet) = 2.01X − 0.07	0.61*
		Y(cluster bean) = 5.99X − 0.44	0.77**
		Y(castor) = 1.64X + 0.16	0.64*

Source: Srinivasarao, Ch. et al., *Can J Soil Sci*, 92, doi: 10.4141/cjss2011-098, 2012.
Note: * and ** denote significance at $P < 0.05$ and 0.01, respectively.

TABLE 16.21
Relationships of SYI to CR C Input, Total Cumulative C Input, % Change in SOC Stock, and the Magnitude of SOC Sequestered after 15 Years of Cropping

Parameters	Regression Equation	R^2
Annual CR C input (X)	$SYI_{Soybean}$ = 0.001X + 0.37	0.21^{NS}
	$SYI_{Safflower}$ = 0.002X + 0.17	0.26^{NS}
Total cumulative C input (X)	$SYI_{Soybean}$ = −0.005X + 0.48	0.10^{NS}
	$SYI_{Safflower}$ = 0.002X + 0.20	0.53*
C buildup % (X)	$SYI_{Soybean}$ = −0.001X + 0.46	0.03^{NS}
	$SYI_{Safflower}$ = 0.004X + 0.24	0.63*
Profile SOC (X)	$SYI_{Soybean}$ = −0.001X + 0.51	0.03^{NS}
	$SYI_{Safflower}$ = 0.01X − 0.19	0.63*
C sequestrated (X)	$SYI_{Soybean}$ = −0.001X + 0.45	0.03^{NS}
	$SYI_{Safflower}$ = 0.008X + 0.29	0.63*

Source: Srinivasarao, Ch. et al., *Can J Soil Sci*, 92, doi: 10.4141/cjss2011-098, 2012.
Note: NS, nonsignificant; *, significant at $P = 0.05$ level.

TABLE 16.22
Relationships of C Input, C Buildup, Profile SOC, and C Sequestration with SYI in a 21-Year Long-Term Experiment

Parameters		Regression Equation	(R^2)
Annual CR C input (X)	SYI (Y)	$Y_{rice} = 0.09X^{**} + 0.07$	0.99
		$Y_{lentil} = 0.14X^{**} - 0.03$	0.97
Total cumulative C input (X)		$Y_{rice} = 0.001X^* + 0.18$	0.62
		$Y_{lentil} = 0.001X^* + 0.13$	0.68
C buildup % (X)		$Y_{rice} = 0.002X^* + 0.20$	0.65
		$Y_{lentil} = 0.003X^* + 0.16$	0.69
Profile SOC (X)		$Y_{rice} = 0.01X^* + 0.04$	0.65
		$Y_{lentil} = 0.01X^* - 0.09$	0.69
C sequestrated (X)		$Y_{rice} = 0.01X^* + 0.22$	0.65
		$Y_{lentil} = 0.02X^* + 0.19$	0.69

Source: Srinivasarao, Ch. et al., *Soil Sci Soc Am J*, 76(1), 168–178, 2012.
Note: * and ** indicate significance at $P < 0.05$ and 0.01 levels, respectively.

16.13.2.6 Rice-Based Production System

A significant relationship has been observed between SYI and the annual input of CR C ($R^2 = 0.97$–0.99^*, $P < 0.01$), total cumulative C input ($R^2 = 0.63$–0.68^*, $P < 0.05$), C restored ($R^2 = 0.65$–0.69^*, $P < 0.05$), profile SOC stock ($R^2 = 0.65$–0.69^*, $P < 0.05$), and total SOC sequestrated ($R^2 = 0.65$–0.69^*, $P < 0.05$) (Table 16.22).

16.13.3 MINIMAL CARBON INPUT REQUIREMENTS FOR ARRESTING CARBON DEPLETION

Long-term cropping without using any organic amendment and/or mineral fertilizers causes depletion of the SOC stock. This depletion ranges from 0.15 Mg C ha^{-1} year^{-1} in rice-based systems to 0.92 Mg C ha^{-1} year^{-1} in groundnut–finger millet systems (Table 16.23). The highest rate of depletion observed in groundnut–finger millet systems was 0.92 Mg C ha^{-1} year^{-1} in semiarid Alfisols, followed by 0.67 Mg C ha^{-1} year^{-1} in pearl millet-based systems in Entisols and 0.47 Mg C ha^{-1} year^{-1} in soybean-based system in Vertisols. The SOC depletion has been the least in rice-based systems. There is a negative relationship between mean annual C inputs and mean SOC depletion rate across locations, soil types, and production systems (Figure 16.15). However, rainfall plays a significant role in SOC depletion. Thus, a reciprocal relationship is expected between rainfall and SOC depletion rate (Figure 16.16). In a similar study, the SOC sequestration potential [C sequestration potential (CSP)], defined as the rate of increase in the SOC stock over the antecedent stock in the 0- to 0.2-m depth, ranged from −0.178 Mg C ha^{-1} year^{-1} (unfertilized control) to 0.572 Mg C ha^{-1} year^{-1} (50% RDF + 4 Mg groundnut shells per hectare) (Bhattacharyya et

TABLE 16.23
Mean Annual C Input, C Depletion Rate, and Critical C Input Requirement in Different Rainfed Production Systems

Location/Production System	Mean Annual C Input (Mg C ha^{-1} year^{-1})	Mean C Depletion Rate (Mg C ha^{-1} year^{-1})	Critical C Input Requirement (Mg ha^{-1} year^{-1})
Anantapur (groundnut)	0.5–3.5	0.18	1.12
Bangalore (groundnut–finger millet rotation)	0.3–3.0	0.92	1.62
Bangalore (finger millet)	0.3–3.1	0.25	1.13
Solapur (winter sorghum)	0.6–3.4	0.23	1.10
SK Nagar (pearl millet)	0.2–1.9	0.67	3.30
Indore (soybean)	1.9–7.0	0.47	3.47
Varanasi (upland rice)	1.1–5.6	0.15	2.47

Sources: Srinivasarao, Ch. et al., *Land Degrad Develop*, doi: 10.1002/ldr.1158, 2011. Srinivasarao, Ch. et al., *Eur J Agron*, doi: 10.1016/j.eja.2012.05.001, 2012. Srinivasarao, Ch. et al., *Int J Agric Sustain*, 10(3), 1–15, 2012. Srinivasarao, Ch. et al., *J Plant Nutrition Soil Sci*, doi: 10.1002/jpln.201000429, 2012. Srinivasarao, Ch. et al., *Geoderma*, 175–176, 90–97, 2012. Srinivasarao, Ch. et al., *Can J Soil Sci*, 92, doi: 10.4141/cjss2011-098, 2012. Srinivasarao, Ch. et al., *Soil Sci Soc Am J*, 76(1), 168–178, 2012.

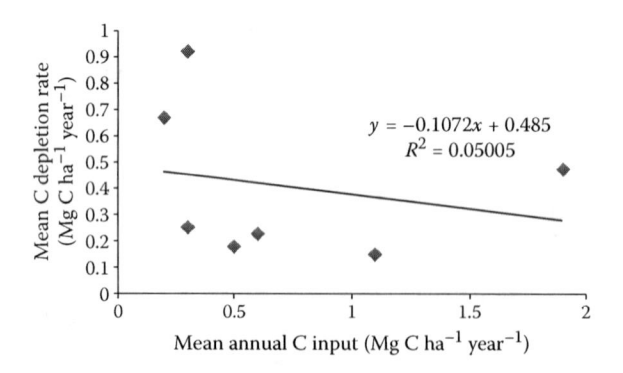

FIGURE 16.15 Relationship between mean annual C inputs and mean C depletion rate in seven long-term experiments.

al. 2009). Globally, rates of C sequestration by different types of management range from 0.11 to 3.04 Mg C ha^{-1} year^{-1}, with a mean of 0.54 Mg C ha^{-1} year^{-1}, and are highly influenced by biome type and climate (Conant et al. 2001). The positive linear relationship between the changes in SOC stock and the total cumulative C inputs to the soils (external organics plus CR) over the years has been highly significant and indicates that even after 13 to 27 years of C input, ranging from 0.2–1.9 to 1.9–7.0 Mg C ha^{-1} year^{-1} in the pearl millet-based systems, the unsaturated C sink capacity is

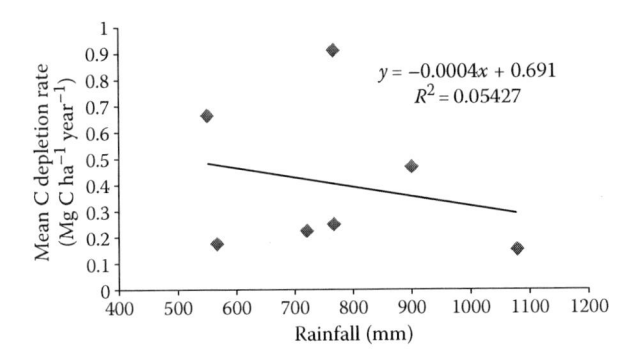

FIGURE 16.16 Relationship between mean annual rainfall and mean C depletion rate in seven long-term experiments.

not filled. Therefore, these soils have a high C sink capacity. However, sink capacity and/or storage rate cannot continue indefinitely (Six et al. 2002). Each soil with a different C loading may reach a new steady state of SOC stock over time. Assessment of SOC stock for these treatments at periodic, perhaps decadal, intervals may provide insights into the strategies of C management in these soils. Lal et al. (2007) estimated that the rate of SOC sequestration in the United States, ranging from 100 to 1000 kg ha^{-1} year^{-1}, depends on climate, soil type, and site-specific management. The global potential of SOC sequestration and restoration of degraded/desertified soils is estimated at 0.6 to 1.2 Pg C year^{-1} for about 50 years, with a cumulative sink capacity of 30 to 60 Pg (Lal et al. 2003, 2007), comprising 0.4 to 0.8 Pg C year^{-1} through adoption of RMPs on cropland (1350 Mha); 0.01 to 0.03 Pg C year^{-1} on irrigated soils (275 Mha); and 0.01 to 0.3 Pg C year^{-1} through improvements of rangelands and grasslands (3700 Mha). Maintaining a constant level of SOC stock (zero change) requires C input of 1.10 Mg C ha^{-1} year^{-1} in Vertisols under a winter sorghum system to 3.47 Mg C ha^{-1} year^{-1} under a soybean-based system. The rate of C input required for groundnut, finger millet, and winter sorghum systems is much lower than those reported by Kong et al. (2005) (3.1 Mg ha^{-1} year^{-1}) for Davis, CA; Majumder et al. (2007) (4.59 Mg ha^{-1} year^{-1}) for rice–wheat–jute systems; Majumder et al. (2008) (3.56 Mg ha^{-1} year^{-1}) for irrigated rice–wheat systems of the Indo-Gangetic plains; and Mandal et al. (2007) (2.92 Mg ha^{-1} year^{-1}) for rice-based systems in subtropical India. The lower input of C needed to maintain a constant level in this study may be due to lower initial SOC levels (1.4–3.9 g kg^{-1} soil) (Srinivasarao et al. 2006). In the studies referred to above, the initial SOC concentrations were approximately three to six times higher (>6–15 g kg^{-1} soil) than those in the present study. But in the case of soybean, pearl millet, and upland rice, this rate is comparatively higher (Srinivasarao et al. 2011e, 2012e,f). In the case of soybean–safflower sequences cultivated in Vertisols, initial SOC concentration of the soil was quite higher (6.2 g kg^{-1}). To maintain or improve this level, a significant amount of biomass C is required. In the case of pearl millet-based systems in Entisol, soil was degraded and had low fertility. Summer temperature of the experimental location is also quite high compared to other locations. This may be the reason for requirement of higher critical C input.

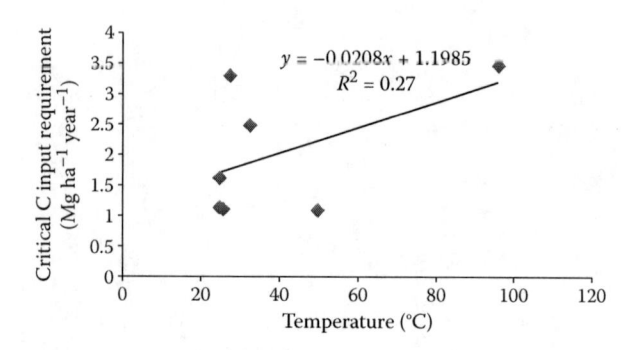

FIGURE 16.17 Relationship between critical C input requirement and mean temperature of the location in seven long-term experiments.

A positive relationship has been observed between temperature and critical C input requirement (Figure 16.17). Information regarding the critical C input requirement for maintaining the SOC stock of the location at antecedent levels is strategically very important in terms of C sequestration, soil quality management, and overall food security in semiarid regions (Lal 2004; Mandal et al. 2007).

16.14 CONCLUSIONS

Poor soil quality with multiple nutrient deficiencies, low NUE, and declining PFP are among the major concerns in semiarid and rainfed agriculture. The SOC stock plays a crucial role in the soil's quality, availability of plant nutrients, environmental functions, and global carbon cycle. Drylands are generally low in fertility and low in organic matter and hence have a high carbon sink capacity. Carbon storage in the soil profile not only improves fertility but also abates global warming. Several soil, production, and management factors influence carbon sequestration, and it is important to identify production and management factors that enhance carbon sequestrations in dryland soils. SOC stocks in the soil profiles across the country show wide variations and follow the order Vertisols > Inceptisols > Alfisols > Aridisols. Inorganic carbon and TCSs are larger in Vertisols than in other soil types. SOC stocks decrease with depth in the profile, while inorganic carbon stocks increase with depth. Among the production systems, soybean-, maize-, and groundnut-based systems maintain higher organic carbon stocks than other production systems. Data from experiments involving long-term cropping, fertilization, and manuring show that mineral fertilizer application alone can neither sustain the productivity nor maintain the SOC stock. A strong depletion of carbon occurs irrespective of production systems and soil type. Application of organics along with mineral fertilizer can sustain productivity and sequester a significant amount of soil carbon. Depletion of SOC can range from 0.15 Mg C ha^{-1} year^{-1} in rice-based systems to 0.92 Mg C ha^{-1} year^{-1} in groundnut–finger millet system. To arrest this depletion, C input of 1.10–3.47 Mg C ha^{-1} year^{-1} is required as a maintenance dose. Further potential of tropical soils to sequester more C in soil can be harnessed by identifying appropriate production systems and management practices for sustainable development and improved livelihoods in the tropics.

16.15 FUTURE RESEARCH NEEDS

- There is an urgent need to launch long-term (>10 years) field experiments to quantify the influence of RMPs on the soil carbon sequestration in various ecosystems and crop production systems.
- There is a need for strengthening the existing data on critical C input requirement as well as for developing this type of data set for other crop production systems in diverse soil types.
- Quantification is required for C stabilization (%) from different types of residues and organic manures.
- Monitoring and verification of the rate of SOC sequestration in a transparent, cost-effective, and credible manner are also required for developing a user-friendly carbon trading system.
- Quantification of the country's soil CSP for land use and management scenarios is needed in a spatial perspective by using soil carbon models.
- Research information is needed for the recommended land use and management practices for soil C sequestration and on the corresponding rates under "on-farm" conditions.
- The same relevant processes of soil C sequestration (e.g., aggregation, humification, formation of secondary carbonates) must be identified, and the residence time of C thus sequestered must be estimated.
- There is a strong need for developing cost-effective, credible, transparent, and simple methods of measuring the rate of soil C sequestration, along with their accuracy and precision.
- Because sustainable land use/land use change and soil management are a net sink for C, it is important to identify and implement policy instruments that facilitate realization of this sink.
- Based on the data on the soil C pool in relation to tillage and cropping systems, a holistic package must be evolved for creating a positive ecosystem C budget and a sustained and high rate of C sequestration.

ABBREVIATIONS

B: boron
B ha: billion hectares
C: carbon
CRIDA: Central Research Institute for Dryland Agriculture
CT: conventional tillage
Cu: copper
ET: evapotranspiration
Fe: iron
FYM: farmyard manure
GDP: gross domestic product
GHGs: greenhouse gases
GIS: geographical information system
GLM: green leaf manure

INM:	integrated nutrient management
K:	potassium
LICs:	lithogenic inorganic carbonates
M ha:	million hectares
Mg:	magnesium
Mn:	manganese
MRT:	mean residence time
N:	nitrogen
NT:	no-till
NUE:	nutrient use efficiency
P:	phosphorus
PFP:	partial factor productivity
PICs:	pedogenic inorganic carbonates
RDF:	recommended dose of fertilizer
RMPs:	recommended management practices
S:	sulfur
SIC:	soil inorganic carbon
SOC:	soil organic carbon
SOM:	soil organic matter
TCSs:	total C stocks
UNCCD:	United Nations Convention to Combat Desertification
WUE:	water use efficiency
Zn:	zinc

REFERENCES

This article is not included in your organization's subscription. However, you may be able to access this article under your organization's agreement with Elsevier.

AICRPDA Annual Report, 2008–2010. All India Coordinated Research Project on Dryland Agriculture, CRIDA, Hyderabad.

Ali, M., A.N. Ganeshamurthy, and C. Srinivasarao. 2002. Role of pulses in soil health and sustainable crop production. *Indian J Pulses Res* 15(2):1–14.

Allmaras, R.R., D.R. Linden, and C.E. Clapp. 2004. Corn residue transformation into root and soil carbon as related to nitrogen, tillage and stover management. *Soil Sci Soc Am J* 68:1366–1375.

Batjes, N.H. 1996. Total carbon and nitrogen in the soils of the world. *Eur J Soil Sci* 47:151–163.

Behera, U.K., A.R. Sharma, and H.N. Pandey. 2007. Sustaining productivity of wheat–soybean cropping system through integrated nutrient management practices on the Vertisols of central India. *Plant Soil* 297:185–199.

Bellakki, M.A. and V.P. Badanur. 2000. Residual effect of crop residues in conjunction with organic, inorganic and cellulolytic organisms on chickpea grown on Vertisol. *J Indian Soc Soil Sci* 48(2):393–395.

Bhattacharya, T., D.K. Pal, C. Mandal, and M. Velayutham. 2000. Organic carbon stock in Indian soils and their geographical distribution. *Curr Sci* 79:655–660.

Bhattacharyya, R., Ved Prakash, S. Kundu, A.K. Srivastava, and H.S. Gupta. 2009. Soil properties and their relationships with crop productivity after 30 years of different fertilization in the Indian Himalayas. *Arc Agron Soil Sci* 55(6):641–661.

Bilbro, J.D., D.J. Undersander, D.W. Fryrear, and C.M. Lester. 1991. A survey of lignin, cellulose and acid detergent fiber ash contents of several plants and implications for wind erosion control. *J Soil Water Conserv* 46:314–316.

Biradar, C.M., P.S. Thenkabail, P. Noojipady, Y. Li, D. Venkateswarlu, H. Turral, M. Velpuri, M.K. Gumma, O.R.P. Gangalakunta, X.L. Cai, M.A. Schull, R.D. Alankara, S. Gunasinghe, and S. Mohideen. 2009. A global map of rainfed cropland areas at the end of last millennium using remote sensing. *Int J App Earth Observ Geoinform* 11(2):114–129.

Bolin, B. (ed). 1981. *Carbon Cycle Modeling*. SCOPE 16, John Wiley & Sons, Chichester.

Bolinder, M.A., D.A. Angers, and J.P. Dubuc. 1997. Estimating shoot to root ratios and annual carbon inputs in soils for cereal crops. *Agric Ecosyst Environ* 63:61–66.

Bronson, K.F., K.G. Cassman, R. Wassmann, D.C. Olk, M. Noordwijk, and D.P. van Garrity. 1998. Soil carbon dynamics in different cropping systems in principal eco-regions of Asia. In *Management of Carbon Sequestration in Soil*, (eds) R. Lal, J.M. Kimble, R.F. Follett, and B.A. Stewart, 35–57. CRC, Boca Raton, FL.

Campbell, C.A., K.E. Bowren, M. Schnitzer, R.P. Zentner, and L. Townley-Smith. 1991. Effect of crop rotations and fertilization on soil biochemical properties in a thick Black Chernozem. *Can J Soil Sci* 71:377–387.

Chaturvedi, S.K. and M. Ali. 2002. Poor man's meat needs fresh fillip. In *The Hindu Survey of Indian. Agriculture 2002*, 63–69. Chennai, India.

Clapp, C.E., R.R. Allmaras, M.F. Layese, D.R. Linden, and R.H. Dowdy. 2000. Soil organic carbon and ^{13}C abundance as related to tillage, crop residue and nitrogen fertilization under continuous corn management in Minnesota. *Soil Till Res* 55:127–142.

Clement, A., J.K. Ladha, and F.P. Chalifour. 1998. Nitrogen dynamics of various green manure species and the relationship to lowland rice production. *Agron J* 90:149–154.

Conant, R.T., K. Paustian, and E.T. Elliott. 2001. Grassland management and conversion into grassland: Effects on soil carbon. *Ecol Appl* 11(2):343–355.

Dalal, R.C. 1989. Long-term effects of no-tillage, crop residue and nitrogen application on properties of a Vertisol. *Soil Sci Soc Am J* 53:1511–1515.

Dalal, R.C. and K.Y. Chan. 2001. Soil organic matter in rainfed cropping systems of the Australian cereal belt. *Austr J Soil Res* 39:435–464.

Dalal, R.C. and R.J. Mayer. 1986. Long-term trends in fertility of soils under continuous cultivation and cereal cropping in southern Queensland II. Total organic carbon and its rate of loss from the soil profile. *Austr J Soil Res* 24:281–292.

Devevre, O.C. and W.R. Horwath. 2000. Decomposition of rice straw and microbial carbon use efficiency under different soil temperatures and moisture. *Soil Biol Biochem* 32:1773–1785.

Du, Z., S. Liu, K. Li, and T. Ren. 2009. Soil organic carbon and physical quality as influenced by long-term application of residue and mineral fertilizer in the North China Plain. *Austr J Soil Res* 47:585–591.

Eswaran, H., E. Vanden Berg, and P. Reich. 1993. Organic carbon in soils of the world. *Soil Sci Soc Am J* 57:192–194.

Eswaran, H., P.F. Reich, J.M. Kimble, F.H. Beinroth, E. Padmanabhan, and P. Moncharoen. 2000. Global carbon stocks. In *Global Change and Pedogenic Carbonate*, (eds) R. Lal, J.M. Kimble, H. Eswaran, and B.A. Stewart, 15–25. CRC Press, Boca Raton, FL.

Fernandez, I., N. Mahieu, and G. Cadisch. 2003. Carbon isotopic fractionation during decomposition of plant materials of different quality. *Glob Biogeochem Cycle* 17:1075–1087.

Follett, R.F. 2001. Soil management concepts and carbon sequestration in cropland soils. *Soil Till Res* 61:77–92.

Ghosh, P.K., M.C. Manna, D. Dayal, and R.H. Wanjari. 2006. Carbon sequestration potential and sustainable yield index for groundnut- and fallow-based cropping systems. *J Agric Sci* 144:249–259.

Gregorich, E.G., C.M. Monreal, M. Schnitzer, and H.R. Schulten. 1996. Transformation of plant residues into soil organic matter: Chemical characterization of plant tissue, isolated soil fractions, and whole soils. *Soil Sci* 161:680–693.

Halvorson, A.D., B.J. Wienhold, and A.L. Black. 2002. Tillage, nitrogen and cropping system effects on soil carbon sequestration. *Soil Sci Soc Am J* 66:906–912.

Hegde, D.M. and S.N.S. Babu. 2008. Agro-economic analysis of oilseed based cropping system. *Indian J Fertil* 4(4):93–120.

http://www.iwmigiam.org/info/main/aboutGMRCA2.asp.

IPCC. 1990. *Climate Change*. (eds) J.T. Houghton, G.J. Jenkinson, and J.J. Ephraums, Cambridge University Press, Cambridge.

IPCC. 2001. *Climate Change 2001: Mitigation—Contribution of Working Group III to the Third Assessment Report of the Intergovernmental Panel on Climate Change (IPCC)*, (eds) B. Metz, O. Davidson, R. Swart, and J. Pan, Cambridge University Press, Cambridge.

IPCC. 2007. *Climate Change 2007: Impacts, Adaptation and Vulnerability. Contribution of Working Group II to the Fourth Assessment Report of the Intergovernmental Panel on Climate Change*, (eds) M.L. Parry, O.F. Canziani, J.P. Palutikof, P.J. van der Linden, and C.E. Hanson. Cambridge University Press, Cambridge, United Kingdom.

Jobbagy, E.G. and R.B. Jackson. 2000. The vertical distribution of soil organic carbon in relation to climate and vegetation. *App Ecol* 10:423–436.

Katyal, J.C., N.H. Rao, and M.N. Reddy. 2001. Critical aspects of organic matter management in the tropics: The example of India. *Nutr Cycling Agroecosyst* 61:77–88.

Kimble, J.M., H. Eswaran, and T. Cook. 1990. Organic carbon on a volume basis in tropical and temperate soils. In *Transactions XIV Congress of the International Society of Soil Science*, 248–253. International Society of Soil Science, Kyoto, Japan.

Kong, A.Y.Y., J. Six, D.C. Bryant, R.F. Denison, and C. Van Kessel. 2005. The relationship between carbon input, aggregation, and soil organic carbon stabilization in sustainable cropping systems. *Soil Sci Soc Am J* 69:1078–1085.

Lal, R. 1997. Residue management, conservation tillage and soil restoration for mitigating green house effect by CO_2-enrichment. *Soil Till Res* 43:81–107.

Lal, R. 2004. Soil carbon sequestration to mitigate climate change. *Geoderma* 123:1–22.

Lal, R. 2009. Soil carbon sequestration for climate change mitigation and food security. Souvenir, Platinum Jubilee Symposium on Soil Science in Meeting the Challenges to Food Security and Environmental Quality. Indian Society of Soil Science, New Delhi.

Lal, R. 2010. Carbon sequestration potential in rainfed agriculture. *Indian J Dryland Agric Develop* 25(1):1–16.

Lal, R., R.F. Follett, and J.M. Kimble. 2003. Achieving soil carbon sequestration in the U.S.: A challenge to the policy makers. *Soil Sci* 168:827–845.

Lal, R., R.F. Follett, B.A. Stewart, and J.M. Kimble. 2007. Soil carbon sequestration to mitigate climate change and advance food security. *Soil Sci* 172(12):943–956.

Lorenz, K. and R. Lal. 2005. The depth distribution of soil organic carbon in relation to land use and management and the potential of carbon sequestration in subsoil horizons. *Adv Agron* 88:35–66.

Maji, A.K. 2007. Assessment of degraded and wastelands of India. *J Indian Soc Soil Sci* 55(4):427–435.

Majumder, B., B. Mandal, P.K. Bandyopadhyay, and J. Chaudhury. 2007. Soil organic carbon pools and productivity relationships for a 34 year old rice–wheat–jute agroecosystem under different fertilizer treatments. *Plant Soil* 297:53–67.

Majumder, B., B. Mandal, P.K. Bandyopadhyay, A. Gangopadhyay, P.K. Mani, A.L. Kundu, and D. Majumder. 2008. Organic amendments influence soil organic carbon pools and crop productivity in a 19 years old rice–wheat agroecosystems. *Soil Sci Soc Am J* 72:775–785.

Mandal, B., B. Majumder, P.K. Bandyopadhyay, G.C. Hazra, A. Gangopadhyay, R.N. Samantaray, A.K. Mishra, J. Chaudhury, M.N. Saha, and S. Kundu. 2007. The potential of cropping systems and soil amendments for carbon sequestration in soils under long-term experiments in subtropical India. *Global Change Biol* 13:1–13.

Manna, M.C., A. Swarup, and R.H. Wajari. 2005. Long-term effect of fertilizer and manure application on soil organic carbon storage, soil quality and yield sustainability under sub-humid and semi-arid tropical India. *Field Crops Res* 93(2–3):264–280.

Martin, J.P. and K. Haider. 1986. Influence of mineral colloids on turnover rates of soil organic carbon. In *Interactions of Soil Minerals with Natural Organics and Microbes*, (eds.) P.M. Huang and M. Schnitzer, 283–304. Soil Science Society of America, Madison.

Motsara, M.R. 2002. Available nitrogen, phosphorus and potassium status of Indian soils as depicted by soil fertilizer maps. *Fertilizer News* 47(8):15–21.

Murphy, B., A. Rawson, L. Ravenscroft, M. Rankin, and R. Millard. 2002. Paired sit sampling for soil organic carbon estimation-NSW. Australian Greenhouse Office, Technical Report, Canberra.

Nierop, G.J.L., D.F.W. Haafs, and J.M. Verstraten. 2003. Occurrence and distribution of ester-bound lipids in Dutch coastal dune soils along a pH gradient. *Organic Geochem* 34:719–729.

Parton, W.J., D.S. Schimel, C.V. Cole, and D.S Ojima. 1987. Analysis of factors controlling soil organic matter levels in Great Plains grasslands. *Soil Sci Soc Am J* 51:1173–1179.

Paustian, K., H.P. Collins, and E.A. Paul. 1997. Management controls in soil carbon. In *Soil Organic Matter in Temperate Ecosystems: Long-term Experiments in North America*, (eds) E.A. Paul, K. Paustian, and E.T. Elliot, 15–49. CRC Press, Boca Raton, FL.

Paustian, K., J. Six, E.T. Elliott, and H.W. Hunt. 2000. Management options for reducing CO_2 emissions from agricultural soils. *Biogeochem* 48:147–163.

Rao, G.G.S.N., A.V.M. Rao, V.U.M. Rao, M. Vanaja, M. Srinivasarao, K.V. Rao, S. Desai, and Ch. Srinivasarao. 2009. Impact, adaptation and vulnerability of climate change on rainfed agriculture. *Indian Journal of Dryland Agricultural Research and Development* 24:10–20.

Rasse, D.P., C. Rumpel, and M.F. Dignac. 2004. Is soil carbon mostly root carbon? Mechanisms for specific stabilization. *Plant Soil* 269:341–356.

Regmi, A.P., J.K. Ladha, H. Pathak, E. Pasuquin, D. Dawe, P.R. Hobbs, D. Joshy, S.L. Maskey, and S.P. Pandey. 2002. Analysis of yield and soil fertility trends in a 20 years old rice wheat experiment in Nepal. *Soil Sci Soc Am J* 66:857–867.

Reicosky, D.C., S.D. Evans, C.A. Cambardella, R.R. Allmaras, A.R. Wilts, and D.R. Huggins. 2002. Continuous corn with moldboard tillage. Residue and fertility effects on soil carbon. *J Soil Water Conserv* 57:277–284.

Sahrawat, K.L. 2003. Importance of inorganic carbon in sequestering carbon in soils of the dry regions. *Curr Sci* 84:864–865.

Scheffer, F. 2002. *Lehrbuch der Bodenkunde/Scheffer/Schachtschabel*. Spektrum Akademischer Verlag, Heidelberg.

Scherr, S. 1999. Soil degradation. A threat to developing-country food security by 2020? IFPRI Discussion Paper 27.

Schuman, G.E., H.H. Janzen, and J.E. Herrick. 2002. Soil carbon dynamics and potential carbon sequestration by rangelands. *Environ Pollut* 116:391–396.

Sharma, K.L., K. Srinivas, S.K. Das, K.P.R. Vittal, and J. Kusuma Grace. 2002. Conjunctive use of inorganic and organic sources of nitrogen for higher yield of sorghum in dryland alfisol. *Indian Dryland Agric Res Develop* 17(2):79–88.

Sharma, K.L., U.K. Mandal, K. Srinivas, K.P.R. Vittal, B. Mandal, J. Kusuma Grace, and V. Ramesh. 2004. Long-term soil management effects on crop yields and soil quality in a dryland Alfisol. *Soil Till Res* 83(2):246–259.

Singh, H.P., K.D. Sharma, G. Subba Reddy, and K.L. Sharma. 2004. Dryland agriculture in India. In *Challenges and Strategies for Dryland Agriculture*. Crop Science of America and American Society of Agronomy, Madison, USA. Publication No 32:67–92.

Six, J., C. Feller, and K. Denef. 2002. Soil organic matter, biota, and aggregation in temperate and tropical soils-effects of no-tillage. *Agro* 22:755–775.

Smith, P. 2007. Carbon sequestration in croplands: the potential in Europe and the global context. *Eur J Agron* 20:29–236.

Srinivasarao, Ch. 2011. Nutrient Management Strategies in Rainfed Agriculture: Constraints and Opportunities. *Indian J Fertil* 7(4):12–25.

Srinivasarao, Ch. and K.P.R. Vittal. 2007. Emerging nutrient deficiencies in different soil types under rainfed production systems of India. *Indian J Fertil* 3:37–46.

Srinivasarao, Ch., A.N. Ganeshamurthy, and M. Ali. 2003. Nutritional constraints in pulse production. Bulletin, Indian Institute of Pulses Research, Kanpur.

Srinivasarao, Ch., K.P.R. Vittal, G.R. Chary, P.N. Gajbhiye, and B. Venkateswarlu. 2006. Characterization of available major nutrients in dominant soils of rainfed crop production systems of India. *Indian J Dryland Agric Res Dev* 21:105–113.

Srinivasarao, Ch., K.P.R. Vittal, P.N. Gajbhiye, S. Kundu, and K.L. Sharma. 2008a. Distribution of micronutrients in soils in rainfed production systems of India. *Indian J Dryland Agric Res Dev* 23:29–35.

Srinivasarao, Ch., S.P. Wani, K.L. Sahrawat, T.J. Rego, and G. Pardhasaradhi. 2008b. Zinc, boron and sulphur deficiencies are holding back the potential of rainfed crops in semi-arid India: Experiences from participatory watershed management. *Int J Plant Prod* 2:89–99.

Srinivasarao, Ch., G. Ravindra Chary, B. Venkateswarlu, K.P.R. Vittal, J.V.N.S. Prasad, S. Kundu, S.R. Singh, G.N. Gajanan, R.A. Sharma, A.N. Deshpande, J.J. Patel, and G. Balaguravaiah. 2009a. *Carbon Sequestration Strategies under Rainfed Production Systems of India*. Central Research Institute for Dryland Agriculture, Hyderabad (ICAR).

Srinivasarao, Ch., K.P.R. Vittal, B. Venkateswarlu, S.P. Wani, K.L. Sahrawat, S. Marimuthu, and S. Kundu. 2009b. Carbon stocks in different soils under diverse rainfed production systems in tropical India. *Comm Soil Sci Plant Anal* 40(15):2338–2356.

Srinivasarao, Ch., S.P. Wani, K.L. Sahrawat, and B.K. Rajasekhara Rao. 2009c. Nutrient management strategies in participatory watersheds in semi arid tropical India. *Indian J Fertil* 5(12):113–128.

Srinivasarao, Ch., B. Venkateswarlu, S. Dixit, S.P. Wani, K.L. Sahrawat, S. Kundu, K. Gayatri Devi, C. Rajesh, and G. Pardasaradhi. 2010. Productivity enhancement and improved livelihoods through participatory soil fertility management in tribal districts of Andhra Pradesh. *Indian J Dryland Agric Res Dev* 25(2):23–32.

Srinivasarao, Ch., B. Venkateswarlu, S. Dixit, S. Kundu, and K. Gayatri Devi. 2011a. *Livelihood Impacts of Soil Health Improvement in Backward and Tribal Districts of Andhra Pradesh*. Central Research Institute for Dryland Agriculture, Hyderabad, Andhra Pradesh, India.

Srinivasarao, Ch., B. Venkateswarlu, M. Dinesh Babu, S.P. Wani, Sreenath Dixit., K.L. Sahrawat, and Sumanta Kundu. 2011b. *Soil health improvement with Gliricidia green leaf manuring in rainfed agriculture*. Central Research Institute for Dryland Agriculture, Hyderabad, India.

Srinivasarao, Ch., B. Venkateswarlu, K. Srinivas, S. Kundu, and A.K. Singh. 2011c. *Soil Carbon Sequestration for Climate Change Mitigation and Food Security*, (ed) Ch. Srinivasarao et al., 372. Central Research Institute for Dryland Agriculture, Hyderabad, India.

Srinivasarao, Ch., B. Venkateswarlu, R. Lal, A.K. Singh, S. Kundu, K.P.R. Vittal, J.J. Patel, and M.M. Patel. 2011d. Long-term manuring and fertilizer effects on depletion of soil organic carbon stocks under pearl millet–cluster bean–castor rotation in western India. *Land Degrad Dev* doi: 1002/ldr.1158.

Srinivasarao, Ch., B. Venkateswarlu, R. Lal, A.K. Singh, S. Kundu, K.P.R. Vittal, G. Balaguruvaiah, M. Vijaya Shankar Babu, G. Ravindra Chary, M.B.B. Prasadbabu, and Y. Reddy. 2012a. Soil carbon sequestration and agronomic productivity of an Alfisol for a groundnut based system in a semi-arid environment in South India. *Eur J Agron* 43:40–48.

Srinivasarao, Ch., B. Venkateswarlu, A.K. Singh, K.P.R. Vittal, S. Kundu, G.N. Gajanan, and B.K. Ramachandrappa. 2012b. Yield sustainability and carbon sequestration potential of groundnut–fingermillet rotation in Alfisols under semi-arid tropical India. *Int J Agric Sustain* 10(3):1–15.

Srinivasarao, Ch., B. Venkateswarlu, A.K. Singh, K.P.R. Vittal, S. Kundu, G.N. Gajanan, B. Ramachandrappa, and G.R. Chary. 2012c. Critical carbon inputs to maintain soil organic carbon stocks under long term finger millet (*Eleusine coracana* (L.) Gaertn) cropping on Alfisols in semi-arid tropical India. *J Plant Nutr Soil Sci* 175(5):681–688.

Srinivasarao, Ch., A.N. Deshpande, B. Venkateswarlu, R. Lal, A.K. Singh, S. Kundu, K.P.R. Vittal, P.K. Mishra, J.V.N.S. Prasad, U.K. Mandal, and K.L. Sharma. 2012d. Grain yield and carbon sequestration potential of post monsoon sorghum cultivation in Vertisols in the semi arid tropics of central India. *Geoderma* 175–176:90–97.

Srinivasarao, Ch., B. Venkateswarlu, R. Lal, A.K. Singh, S. Kundu, K.P.R. Vittal, S.K. Sharma, R.A. Sharma, M.P. Jain, and G. Ravindra Chary. 2012e. Sustaining agronomic productivity and quality of a Vertisolic Soil (Vertisol) under soybean-safflower cropping system in semi-arid central India. *Can J Soil Sci* 92(5):771–785.

Srinivasarao, Ch., B. Venkateswarlu, R. Lal, A.K. Singh, K.P.R. Vittal, Sumanta Kundu, S.R. Singh, and S.P. Singh. 2012f. Long-term effects of soil fertility management on carbon sequestration in a rice-lentil cropping system of the Indo-Gangetic plains. *Soil Sci Soc Am J* 76(1):168–178.

Syers, J.K., J. Lingard, J. Pieri, and G. Ezcurra. 1996. Sustainable land management for the semiarid and sub-humid tropics. *Ambio* 25:484–491.

Takkar, P.N. 1996. Micronutrient research and sustainable agricultural productivity in India. *J Indian Soc Soil Sci* 44:563–581.

Trabalka, J.R. and D.E. Reichle (eds.). 1986. *The Changing Carbon Cycle: A Global Analysis*. Springer-Verlag, New York.

Van Breemen, N. and T.C.J. Feijtel. 1990. Soil processes and properties involved in the production of greenhouse gases with special reference to soil taxonomic systems. In *Soils and the Greenhouse Effect*, ed. A.F. Bouwman, 195–223. John Wiley & Sons, Chichester, UK.

Venkateswarlu, B., A.K. Singh, Ch. Srinivasarao, G. Kar, A. Kumar, and S.M. Virmani. 2012. *Natural Resource Management for Accelerating Agricultural Productivity*, Studium Press (India) Pvt. Ltd., New Delhi, India.

Wang, W.J., J.A. Baldock, R.C. Dalal, and P.W. Moody. 2004. Decomposition dynamics of plant materials in relation to nitrogen availability and biochemistry determined by NMR and wet-chemical analysis. *Soil Biol Biochem* 36:2045–2058.

Wani, S.P., T.J. Rego, S. Rajeswari, and K.K. Lee. 1995. Effect of legume-based cropping systems on nitrogen mineralization potential of Vertisol. *Plant Soil* 175:265–274.

Wani, S.P., P. Pathak, L.S. Jangawad, H. Eswaran, and P. Singh. 2003. Improved management of Vertisols in semiarid tropics for increased productivity and soil carbon sequestration. *Soil Use Manag* 19:217–222.

Wani, S.P., T.K. Sreedevi, J. Rockström, and Y.S. Ramakrishna. 2009. Rainfed agriculture—Past trend and future prospectus. In *Rainfed Agriculture: Unlocking the Potential. Comprehensive Assessment of Water Management in Agriculture Series*, (eds.) S.P. Wani, J. Rockström, and T. Oweis, 1–35. CAB International, Wallingford, UK.

West, T.O. and W.M. Post. 2002. Soil organic carbon sequestration rates by tillage and crop rotation: A global analysis. *Soil Sci Soc Am J* 66:930–1046.

Wright, A.L. and F.M. Hons. 2005. Tillage impacts on soil aggregation and carbon and nitrogen sequestration under wheat cropping sequences. *Soil Till Res* 84:67–75.

Yadav, R.L., B.S. Dwivedi, K. Prasad, O.K. Tomar, N.J. Shurpali, and P.S. Pandey. 2000. Yield trends, and changes in soil organic-C and available NPK in a long-term rice–wheat system under integrated use of manures and fertilisers. *Field Crops Res* 68:219.

Young, R., B.R. Wilson, M. Mcleod, and C. Alston. 2005. Carbon storage in the soils and vegetation of contrasting land uses in northern New South Wales, Australia. *Austr J Soil Res* 43:21–31.

17 Soil Renewal and Sustainability

Richard M. Cruse, Scott Lee, Thomas E. Fenton, Enheng Wang, and John Laflen

CONTENTS

17.1 Introduction	477
17.2 Soil Resources and Agriculture Production	478
17.3 Climate Implications	479
17.4 Agriculture and Soil Erosion	480
17.5 Soil Development and Agriculture	480
17.6 Soil Renewal Rates	481
17.7 History of Tolerable Soil Loss (T)	485
17.8 Erosional Impact on Productivity	488
17.9 Erosional Impact on Productivity—China Perspective	492
17.10 Soil Sustainability	495
Abbreviations	496
References	497

17.1 INTRODUCTION

"The nation that destroys its soil destroys itself"—Franklin D. Roosevelt, 32nd president of the United States. Volumes of evidence, including failed civilizations, real-time observations, and extensive refereed research testify to the absolute necessity of productive soils to meet the growing demands of a world struggling to feed its people. Human wisdom and the desire to sustain or improve standards of living or even insure survival of future generations should offer sufficient incentive to make sustaining the world's soil resource base a top priority. Too often, however, soil sustainability has been a lower priority than the short-term goal of maximizing production and/or profit at the lowest possible cost, which has led to risky soil management choices, resulting in widespread soil erosion, soil structure deterioration, salinization, and/or organic matter depletion. Most agricultural soils have been degraded such that agricultural production potential has been compromised to some extent; as much as 25% of the world's agricultural soils have been damaged such that they cannot be reliably used for agricultural purposes (Food and Agriculture Organization [FAO] 2011). Yet elevated crop yields seem increasingly necessary as increasing agricultural productivity on existing lands is viewed as necessary to meet the growing world demand for food (Cassman et al. 2011). This demand will continue to grow, with projections that food production must increase 50% by 2030 to meet human dietary needs (United Nations Secretary-General's High-Level Panel on Global Sustainability 2012), meaning that yields will have to increase as the per capita land area available for food production is steadily decreasing. Yields must increase on already-degraded soil, soil that will experience increasing production pressure, and with this pressure comes a high probability of continued and even increased degradation.

Soil erosion is arguably the most important land degradation process associated with farming. Pimentel (2006) and Lal (1995) estimated independently that globally, 10 million and 3 million ha, respectively, of cropland are lost annually due to soil erosion. The total area of productive land destroyed by erosion since the beginning of settlement agriculture may be as high as 130×10^6 ha (Lal 1995) or approximately 9% of today's arable land area. Continued erosion based on current estimated soil degradation, and potential future accelerated erosion rates, have dangerous consequences. The goal of this chapter is to address the role and implications of soil erosion in light of estimated soil regeneration rates on sustained production such that agriculture's ability to meet the future demand for agricultural products might be sustained.

17.2 SOIL RESOURCES AND AGRICULTURE PRODUCTION

The implication of lost production due to soil degradation looms large. This is especially disconcerting because degradation often results from human choices regarding land care and soil management. Humans also make choices regarding use of land capable of producing agricultural commodities. Land use choice is often determined by the market value associated with various potential uses of the property, especially in more developed countries. Often, agriculture and crop production do not offer the greatest financial return, at least in the short term, and therefore, agricultural land is converted to uses such as urban development, housing, or roads/infrastructure. Essentially, none of the converted agricultural land will be, or can be, recovered for agricultural uses. The Food and Agriculture Organization (2002) estimates that 7% of the world's agricultural land of 2002 will be lost to other uses by 2030.

The world's soil resource base suitable for food, feed, fiber, and fuel production is fixed. That is, land area suitable for agricultural production, both farmed and potentially farmable, is finite. Expansion of agriculture into unfarmed areas that have acceptable production potential remains a possibility and a distinct probability considering growing global demand for agricultural commodities. Fischer et al. (2002) indicated that about 57% of the world's rain-fed agricultural land with production potential (considering existing technology) was being farmed in the mid-1990s. The potential for expansion of rain-fed agricultural land to meet growing food demand and to replace degraded/lost agricultural land seems reassuring; however, one must look a bit deeper to gain a calculated understanding of what expansion of rain-fed agriculture realistically means for potential increases in agricultural production. This estimate of potentially farmable area is the gross expansion potential. From this area, roads, communities, and other legally protected areas must be reserved, which would remove 10% to 30% of the gross estimated land from producing crops. Fischer et al. (2002) cautioned that substantial expansion areas would have relatively low productivity and would require extensive fertilizer applications to bring production to acceptable levels. They further indicated that large areas of Asia not suitable for rain-fed agriculture were being farmed at the time of the study, suggesting that these areas will likely contribute relatively minor rain-fed production amounts in the future. As much of this potential rain-fed agricultural area is in developing countries in the lower latitudes, that is, in areas that have climate change predictions somewhat unfavorable for agriculture (Intergovernmental Panel on Climate Change 2007), agriculture production increases will not likely be proportional to the potential agricultural expansion area. Expansion into areas with potential rain-fed productivity holds some promise, but considering these areas as replacements for past and future degraded land and for land taken out of production seems more than cautiously optimistic, especially in considering the need to meet the rising global food, feed, and fuel feedstock demand.

17.3 CLIMATE IMPLICATIONS

Climate change amplifies the food/feed production challenges previously identified. When considered globally, the overwhelming evidence portrays future atmospheric conditions leading to reduced potential for crop yields (Intergovernmental Panel on Climate Change 2007). Recent studies further suggest that yields of major crops have already been reduced significantly (Lobell et al. 2011); even more disconcerting, research suggests that previous climate change crop yield reduction projections may have been substantially too conservative (Lobell et al. 2012). That is, temperature extremes may reduce selected crop yields to a greater extent than models have previously predicted.

The connection between changing climatic conditions and crop yield potential has been made with at least one oversimplification—that soil quality remains static. Because most agricultural soils have been degraded and will very likely continue to degrade (FAO 2011), crop stress associated with less favorable soil conditions coupled with a more variable climate must be included in crop modeling efforts if a realistic picture of future crop production potential is to be obtained. Degraded soil typically has lower rainfall infiltration rates, lower water holding capacity, lower soil organic matter content, poorer drainage, and lower fertility than a comparable soil in an undegraded condition. Not only must these conditions be considered, but observed and predicted increased frequency of extreme rainfall events will escalate soil erosion pressures, aggravating existing degraded soil conditions found in many locations (Soil and Water Conservation Society 2006; Nearing et al. 2004). In fact, erosion rates are expected to increase disproportionately relative to rainfall by a factor of about 1.7 (Nearing et al. 2004). The soil degradation drag on food production, agricultural land conversion to other uses, limits to agricultural land expansion, climate change-related production reductions, and accelerating demand for agricultural products portray a food security system margin of error that continually narrows. Food security will rely increasingly on technological advances, that is, advanced genetics, but technology's contribution to food, feed, fiber, and fuel feedstock production will be met only if the soil in which crops grow can supply the required (and timely) quantities of water and nutrients. Continued degradation of soils will limit technology's contribution to global food security.

17.4 AGRICULTURE AND SOIL EROSION

Humans have altered the relationship between soils, climate, and plants. Through agriculture, we have replaced native species, in most situations, with species that are not as well adapted to a specific location, but that yield products favorable for human consumption or human use. As would be expected, the most productive soils (those with relatively high native fertility and favorable physical condition), Mollisols and Alfisols, account for a major portion of farmed soils globally (Eswaran et al. 2012). That is, rain-fed agriculture is practiced in areas of the world having a combination of climate and parent material properties that historically favored relatively high amounts of plant growth. In part, because crops selected for agricultural production normally are not the most naturally productive/competitive species for the soil and climate combination where farming is practiced, humans manipulate soil physical and chemical properties through practices such as tillage and/or fertilization to enhance production of their chosen crop. In doing so, the soil is often exposed to rain and wind and left vulnerable to soil erosion, which removes the upper layer(s) of soil material. The physical and chemical soil environment is changed; usually, plant rooting is shifted downward into less weathered materials lower in both organic matter and fertility and with poorer soil physical conditions.

Accelerated soil erosion is a degradation process; however, concurrently, soil physical and chemical changes occur, producing new soil from parent material. Soil renewal, or development, involves physical, chemical, mineralogical, and biological processes. In reality, the net soil change at a given location on the landscape is the combined effect of soil detachment, transport, and deposition through erosion and soil addition through soil development.

17.5 SOIL DEVELOPMENT AND AGRICULTURE

Plant biochemistry associated with growth and food production (seed, fruit, and tubers) relies on chemical elements (nutrients) and water obtained from the soil. Without these materials, selected required plant chemical processes are limited or do not occur, plant productivity decreases, and in worst-case scenarios, the plant simply cannot survive. Further, plants must obtain quantities of these elements in proportion to their needs, that is, elements have a very limited capacity to substitute one element for another in basic plant physiological processes. If one element is limiting, even when others are in adequate supply, plant function is disrupted. Also, and equally important, different plants require different amounts of the necessary elements for survival and maximum production. Certain soil conditions are better suited for certain plants.

The amount and proportion of the different plant-required elements vary spatially and are related to the chemical and mineral characteristics of the parent material. These factors also vary vertically in the soil profile depending on a variety of factors, largely influenced by the chemistry of the parent material making up the plant rooting medium (soil), climate, topography, time of weathering, and even the native plants growing in this soil material. The unique combination of nutrient availability, physical characteristics of the developing soil, and climate leads to a condition that typically favors growth of selected plants or plant types over others. Plant characteristics, in turn, impact soil development through root and shoot contributions to soil organic matter, soil structure development, soil–plant–water dynamics, and nutrient cycling. Soil (and climate conditions) favoring highly productive plant systems, prairies, for example, develop with high soil organic matter contents and favorable soil structure because of massive plant root systems; because of relatively fertile parent material, uptake of nutrients and incorporation in the growing plant tissue, and efficient nutrient cycling, these soils also tend to develop and maintain highly fertile conditions *if not disturbed*. In contrast, less favorable climate and/or soil materials that have poor physical and/or nutrient conditions (hereafter termed fertility) support only plants that can tolerate such conditions. These plants typically have low total production and/or have low nutrient content; as such, these plants typically have low or little food/feed value.

The synergistic relationship between soil and plants develops very slowly relative to a human life span. In fact, most of the world's highly productive soils have developed over thousands of years. When this relationship is disrupted or changed, the soil development process changes. Not only do development processes change when agriculture husbandry replaces natural ecosystem processes, soil itself is often heavily manipulated, a process normally unfavorable for sustained productivity. With these changes occurring, and often leading to soil erosion and alterations of the soil profile, soil development rates become very important, as once a soil's productive capacity is reduced, or even lost, new development is key to regaining productivity levels.

17.6 SOIL RENEWAL RATES

To understand soil renewal rates, the ecosystem, of which soils are a part, must be understood to the best of our ability. For the past 60+ years, the qualitative method of understanding soil formation has been based on the soil forming factors proposed by Jenny (1941): climate, vegetation, relief, parent material, and time. Most of these factors were first proposed by Dokuchaev (1883) but were developed into a state factor model by Jenny. This model has been used to provide a broad conceptual framework for understanding soil formation, but Jenny's original assumption of the independence of the factors is not generally accepted by soil scientists. However, these factors set the stage for the processes of additions, removals, transfers, and transformations outlined by Simonson (1959) to occur. These models have been used to explain soil properties and horizon differentiation. Subsequently, man was added as a sixth soil forming factor (Jenny 1980). An objective of this section is to review rates of soil formation and renewal as they relate to erosion rates, soil loss tolerance, and sustainability.

Soils are classified and evaluated based on horizons present and their degree of differentiation. In the US classification system, soil taxonomy, a distinction is made between genetic horizons and diagnostic horizons. Genetic horizons, for example, A, B, and C horizons, reflect a qualitative judgment about soil forming processes. Diagnostic horizons such as mollic epipedon, cambic horizon, and argillic horizons have defined quantitative properties that may include two or more genetic horizons. It seems that diagnostic horizons provide a more quantitative method of evaluating soil properties that contribute to sustainability. However, the majority of data in the literature are reported based on genetic horizons.

The time factor is important in understanding formation rates of soil. The aging/development of a soil in the simplest case can be thought of as starting at time zero with the exposure of unweathered parent material to physical, chemical, and biological processes. Most cases are not this simple. Methods of determining the absolute age related to time zero include radioisotope dating and historical records. Chemical composition data have also been used on a watershed basis. Alexander (1988) used outputs of silica and major cations to estimate rates of formation. Wakatusi and Rasyidin (1992) used geochemical mass balance equations to estimate rates of weathering and soil formation. Geomorphic surfaces can also aid in refining the time factor in a relative way. A geomorphic surface is a portion of the land surface that can be defined in space and time and is mappable (Ruhe 1969). It must be defined in relation to other geomorphic surfaces to place it in the proper spatial and time sequence. Parsons et al. (1970) and Ruhe et al. (1975) used this approach to develop time lines for formation of surface and subsurface horizons.

Another method of evaluating the time factor utilizes the degree of development or horizon differentiation of soil profiles. However, this approach has many problems. For example, combining data from Hutton (1947) and Ruhe (1969) demonstrates that on stable geomorphic surfaces dated at about 14,000 years before present (YBP), soils developed in Wisconsinan loess and classified as Mollisols range from those with minimal cambic horizons to those with argillic horizons having clay contents in excess of 55%. Differences in profile differentiation are related to distance from loess source and sorting, loess thickness, depth to restrictive layer, and internal drainage and are not necessarily controlled exclusively, or even dominantly, by the time factor.

Today's soils result from a myriad of processes occurring over hundreds to thousands (and even millions) of years. Some components of soil development such as soil organic matter additions can occur in a relatively short time—a period of years. Chemical weathering of parent materials into clay minerals may take centuries or even longer. Leaching of materials, such as clays, forming different soil horizons will take even longer as this process will occur subsequent to conversion of parent materials to clay minerals. Thus, clearly articulating soil renewal rates must involve integration of multiple soil development processes that occur at different rates and give different results based on other supporting soil formation factors. Identifying soil development rates is not an exact science; however, it is one that is critical in evaluation of soil erosion impacts on soil sustainability.

Many studies report both soil loss and soil formation in units of soil depth, while arguably the world's most recognized soil conservation agency, the Natural Resources Conservation Service of the United States Department of Agriculture (NRCS), bases soil erosion and conservation programs on mass of soil eroding or

Soil Renewal and Sustainability

developing per unit area. As a guide for the following discussion, the NRCS typically recognizes soil loss rates of 3 to 5 tons/acre (6.7–11.2 Mg/ha) as tolerable (discussed in more detail later in this chapter). A soil loss or development rate of 11.2 Mg/ha equates to a depth of soil (assuming a bulk density of 1.3 Mg/m^3) of 0.85 mm/year. For the subsequent discussions, development rates or soil erosion rates expressed on a mass-per-unit basis have been converted to depth assuming a soil material bulk density of 1.3 Mg/m^3.

Soil renewal rates vary widely between different studies and, as might be expected, depend heavily on the component of the soil profile being addressed and methods used. Alexander (1988), based on data from 18 watersheds, concluded that rates of soil formation ranged from 0.002 to 0.09 mm/year, while Wakatuski and Rasyidin (1992) estimated soil formation on a worldwide basis of 0.056 mm/year. Using present morphological descriptions to determine soil thickness for four well-drained loess derived Mollisols (Tama, Marshall, Otley, Galva) with cambic horizons or minimal argillic horizons in Iowa on stable geomorphic surfaces dating from 14,000 YBP, the average rate of soil formation is about 0.11 mm/year, with a range of 0.11 to 0.13 mm/year. Reported soil formation rates for these three studies varied by approximately two orders of magnitude, illustrating the challenge associated with relating soil loss to soil sustainability and the implication of methods used in estimating formation rates.

The surface soil horizon seems the most rapidly/easily modified soil profile component but also is the most sensitive to degradation through mismanagement and/or soil erosion. Other profile components change considerably slower. For a 100- and a 50-year period in the Central and North Central United States, surface horizon formation rates of 3 mm/year were observed by Hallberg et al. (1978) and Simonson (1959). Ruhe et al. (1975) observed an average A horizon formation rate of approximately 6.3 mm/year for soil developed on the Missouri River floodplain, but for a relatively short time of less than 100 years. When additional subsurface horizons are considered, the formation rate decreases. Crocker (1960) concluded that organic carbon increased rapidly in the early stages of all sequences he studied. In one sequence, he reported a rapid buildup of organic carbon in the first 60 years and an apparent decreasing accumulation rate after that time. Parsons (1962) reported that the formation of a 25.4-cm A + A2 (now E) horizon took 1000 years, which equates to 0.254 mm/year. Using data from Parson et al. (1962) for three B horizons with an estimated formation time of 2500 years, the formation rate for the total soil (A, E, and B) was 0.31 mm/year. Correcting for a soil organic matter loss from the surface horizon might be relatively quickly fixed, but that does not necessarily equate to soil formation rates in other horizons or correcting other interrelated properties and processes impacted by soil loss.

Soil formation contributes to sustaining the soil resource but is counterbalanced by soil loss through erosion processes. Recently, global cropland erosion rates summarized by Montgomery (2007) averaged 0.64 mm/year, more than an order of magnitude greater than the global soil formation rates given by Wakutuski and Rasyidin (1992) or Alexander (1988). In contrast, if the T value of 0.85 mm/year recognized for many US soils by the NRCS appropriately reflects sustainable soil loss rates globally (a very big assumption), soil degradation through erosion might not be as serious an issue as many suggest. Using these two evaluation approaches leads to vastly different conclusions regarding soil erosion impacts on sustainability.

Montgomery's (2007), or any, average global soil erosion value is likely a poor indicator of soil erosion influences on land degradation. Soil loss rates averaged over large areas simply give a distorted view of soil erosion impact on sustainability (Cox et al. 2011). To exemplify on a smaller scale than that used by Montgomery, the latest United States Department of Agriculture—National Resource Inventory (USDA-NRI) report estimated average soil erosion rates for Iowa in 2007 to be 0.88 mm/year, very close to the T value of 0.85 mm/year. However, Fenton (2010), using land use data and the prime farmland definition given by the USDA and NRI, estimated that the majority of the soil loss in Iowa was coming from about 27% of the cropland. Using this assumption, the soil loss from those areas would be over 3.14 mm/year. These data suggest that relatively large areas in Iowa would be losing soil at an unsustainable rate, even though based on the reported average, one would assume that soil loss and T are balanced, leading to statewide sustained production potential. This soil loss trend is supported by data from the five most extensive soil series (2,223,467 ha and all Mollisols) in Iowa that have slope and erosion phases (Fenton 2012). Soil map unit summaries from these soils show that on slope gradients of 5% to 9% or greater, only 8 to 18 cm of surface soil remains on 75% to 100% of the slope groups. Original depths were approximately 40 to 45 cm.

The Iowa Daily Erosion Project (IDEP) (Cruse et al. 2006), using the Water Erosion Prediction Project (WEPP) soil erosion model, estimated average statewide 2007 soil erosion rates, based on spatial and temporal rainfall characteristics, similar to those of the NRI; yet greater than 10.9 mm of erosion was estimated in isolated areas of Iowa (Cox et al. 2011). This estimate is 12 times greater than the estimated average annual erosion rate of 0.88 mm/year and also an order of magnitude greater than the value of T used for most soils. The Losing Ground Report (Cox et al. 2011) states that statewide erosion averages provide a poor representation of actual erosion rates and that many areas are eroding faster than the "sustainable" rate. In fact, they estimated that approximately 20% of the state's agricultural land was eroding at twice the T level (0.85 mm/year). If the soil renewal rates of Wakatuski and Rasyidin (1992) and/or Alexander (1988) are considered the best metric for determining allowable soil loss rates, then 20% of the state's agricultural land eroded at more than an order of magnitude greater than what should be considered sustainable. In the more heavily impacted areas, erosion was two orders of magnitude greater than what sustainability would dictate acceptable. Further, when focusing on specific geographical areas, the combination of erosion rates and crop production can generate relatively large impacts not factored in averaged yields (Den Biggelaar et al. 2001).

Erosion rates may even be worse than current estimates when focusing on short-duration, high-intensity rainfall events. Soil movement is most noticeable during such rainfall events, and particularly when the soil is not protected by cover (Larson et al. 1983). When studying individual storms and their spatial distribution, more detailed and accurate estimates of soil erosion can be obtained than those obtained by relying on an average annual estimate, especially over a large area, as illustrated above. To further exacerbate the issue, areas of the world, exemplified by watersheds in the Upper Mississippi River Basin and other parts of the United States, are experiencing increasing frequency of high-intensity rainfall events (Karl et al. 2009), strongly suggesting that elevated soil erosion rates have been occurring and could be an increasing threat. An even greater complication involves limitations of current soil erosion modeling technology. The universal soil loss equation (USLE) and its derivative technologies and the WEPP model estimate only sheet and rill erosion. Sediments lost in ephemeral gullies, small channels, and ditches are not included in their estimates. Thus, erosion values given in numerous projects that use these approaches for estimating soil erosion rates are likely very conservative. In considering the wide array of research methods used, interpretations made of data presented, and seriousness of the issue, multiple authors have articulated that globally, topsoil is thinning at an alarming rate (Pimentel 2006; Brown 1984; Lal 1995). This condition poses serious threats to future food, feed, fiber, and fuel feedstock security.

17.7 HISTORY OF TOLERABLE SOIL LOSS (*T*)

The United States' national interest in soil erosion can be traced to the 1930s Dust Bowl. Wind erosion generated nationwide attention and resulted in the formation of the Soil Erosion Service in the Department of Interior in 1933 under the leadership of Hugh Bennett as chief. In 1935, the Soil Erosion Service was transferred to the Department of Agriculture and combined with other units to form the Soil Conservation Service (SCS, now known as the NRCS).

Recognizing the damage caused by soil erosion was the first step in addressing an existing and newly recognized serious national issue; however, quantifying the soil loss occurring and damage was a prerequisite to developing a long-term strategy to address the growing problem. Not only was there a need to quantify soil loss; quantifying soil renewal rates was necessary to determine the net change in soil depth and crop production potential. What ensued is arguably the largest historical human conservation impact on soil resources, agricultural management, water quality, and food/feed/fiber/fuel feedstock production known to man. The multiple-decade research programs' goal was to determine the acceptable level of soil erosion and what management practices on the country's farmed fields ensured that the identified acceptable soil erosion level would not be exceeded.

The "*T* factor," as it has become known, is the soil loss tolerance (in mass of soil per unit area). It is defined as the maximum amount of erosion at which the quality of a soil medium for plant growth can be maintained. Maintaining soil quality has three focus areas. It includes maintaining (1) the surface soil as a seedbed for plants; (2) the atmosphere–soil interface to facilitate the entry of air and water into the soil while protecting the underlying soil from wind and water erosion; and (3) the total soil volume as a reservoir for water and plant nutrients preserved by minimizing soil loss. The following discussion outlines the US developmental steps to determine *T* for every soil and thus to nationally combat soil erosion based on quantitative science. This same soil erosion issue plagues international soil resources and threatens human capability to meet growing food demands, especially in the face of changing climate (Soil and Water Conservation Society 2006).

In 1940, Austin W. Zingg published the first equation relating soil erosion to land slope and length:

$$A = CS^m L^{n-1} \quad (17.1)$$

where A is the average soil loss per unit area from a land slope of unit width, S is the land slope in percent, L is the horizontal length of the land slope, and C is a constant of variation. Data evaluated by Zingg were from work reported by Duley and Hays on two Kansas soils: by Diseker and Yoder on a Cecil clay soil in Alabama and from soil erosion plots located at Tyler, TX; Guthrie, OK; Clarinda, IA; Bethany, MO; and LaCrosse, WI. The work of Duley and Hays was conducted using a sprinkling-can rainfall simulator. At Bethany, MO, a rainfall simulator was used in addition to the soil erosion plots. Zingg recommended values of 1.4 for m and 1.6 for n.

Zingg's work ushered in a long period of research that ultimately led to the development of the USLE and provided direction in soil erosion research that soon led to the concept of an allowable, tolerable, or limiting soil loss. Further, the equations, methods, and factors developed were used to select and design practices that limited soil loss to levels supporting sustainable land use.

In 1941, Dwight D. Smith, building on the work of Zingg, extended Equation 17.1 by the introduction of a factor P to provide for the effect of practices on soil loss. The factor P is the ratio of A_1/A, where A_1 is the soil loss with a given practice and A is given by Equation 17.1. He further developed the concept of designing soil conservation systems by substituting A_1/P for A in Equation 17.2:

$$A_1/P = CS^{7/5} L^{3/5} \quad (17.2)$$

Equation 17.2 was rewritten to develop an equation to determine a maximum acceptable slope length based on the allowable soil loss:

$$L = (A_1/PC)^{5/3} S^{-7/3} \quad (17.3)$$

where L would be the maximum length of slope allowable for a specific site with values of C, S, A_1, and P.

Smith applied this approach to conditions for the Shelby soil at the soil conservation experiment station at Bethany, MO. Based on a 3-year rotation of corn, wheat, clover, and timothy, he computed a C value for the Shelby surface soil.

The value of A_1 (allowable average annual soil loss) Smith recommended for the Shelby soil was 9 Mg/ha. Smith expressed the view that the allowable soil loss limit should be based upon that rate of soil loss that permits a constant, or preferably an increasing, time gradient of soil fertility.

Smith indicated that the 9-Mg/ha soil loss per year may be too great a loss for maintenance of soil fertility when erosion has progressed to the point that plowing is diluting the surface soil with thin layers of subsoil. He also stated that the rate of allowable loss may be even greater than 9 Mg/ha/year, with fertility being maintained because certain nutrients had larger concentrations in the Shelby subsoil than in the surface soil.

Smith developed recommendations for maximum length and degree of slope combinations for different conservation practices including strip cropping, contouring, and terracing. In his summary, Smith pointed to the equation's development that "provides for the effect of soil–climate–crop–treatment, length and degree of slope, and mechanical conservation practices," and "a soil loss of not more than 4 tons per acre per year (9 Mg/ha) is suggested as a rate which would allow maintenance of fertility with recommended cropping practices."

Smith was an engineer, and he seemed to approach the design of conservation systems as an engineer. In his paper, he stated, "Mathematical formulas have been used in practically all design work. Early formulas were largely empirical, and as additional knowledge was gained, they have been modified or replaced by theoretical relationships. Design formulas, even though empirical, should be of assistance to soil conservationists, particularly to the inexperienced technician, in making more correct field applications."

Browning et al. (1947), published a paper on a method for determining the use and limitations of rotation and conservation practices in the control of soil erosion in Iowa. This paper gave permissible annual soil loss limitations for 13 important Iowa soils. These rates ranged from 4.5 to 13.5 Mg/ha. Equation 17.3 and a permissible annual soil loss of 11.2 Mg/ha were used to demonstrate how a table might be developed for various practices, a permissible soil loss, and a range of slopes. Equation 17.3 was also used to evaluate slope lengths for different permissible soil loss rates.

One of the most-quoted papers related to soil erosion prediction was published by Musgrave (1947). The title of the paper was "The Qualitative Evaluation of Factors in Water Erosion—A First Approximation." This work was "the result of the findings of a group of workers showing the relationship between the major casual factors and the resulting rate of erosion." Included in this group were D.D. Smith and George Browning. This paper was a major work in the development of the USLE, which has been widely used, along with allowable soil erosion, to select practices and management strategies to keep average annual soil erosion below a tolerable limit. Yet, this 1947 paper does not mention tolerable or allowable soil loss limits.

Smith and Whitt (1947, 1948) presented a method of predicting soil loss they referred to as the "factor system," which drew heavily on the previous work of Smith and Browning. In these papers, they discussed "allowable soil loss." They indicated that the ultimate objective of soil conservation is to maintain soil fertility, and hence crop production, indefinitely. Smith and Whitt used organic matter as their criterion of soil fertility and, based on plots of organic matter versus erosion rate, suggested a limit of 9 Mg/ha for the Marshall and Shelby soils and 6.7 Mg/ha for the Putnam soils. They recommended a limit of 6.7 Mg/ha for the claypan soils and 4.5 Mg/ha for the Ozark Region soils. All soils were in Missouri.

The USLE was first published in 1961. It is clear that the major driving force for the USLE development was the perceived need for a technology that could be used to select conservation practices and cropping and management that would limit soil losses from a particular field "to a level that will not reduce its future production potential." The USLE could be used "to select practices that will permit a farmer to make the most profitable use of the field and still protect and improve the soil."

A series of regional workshops were held in 1961 to introduce the equation to SCS personnel. The agenda for the regional workshop in Little Rock, AR, for SCS personnel in Mississippi, Louisiana, Arkansas, Texas, and Oklahoma included the history of soil loss prediction by D. D. Smith; the rainfall erosion index and the cropping and management factor by W. H. Wischmeier; soil erodibility values by T. C. Olson; and slope length and steepness and conservation practice factors by D. D. Smith. A session on soil factor and soil loss tolerance and a 2-h session on actual soil factors and soil loss tolerances were held. At an evening session, state soil scientists assigned soil factor values to important soils based on soil factor values that were derived from plot studies on major US soils. While no workshop objective was explicitly stated, the agenda clarifies the expectation that attendees were in a position to apply soil loss prediction to the land in those states. Attendees were led through soil loss computations using the USLE. There was also a discussion of implementing soil loss prediction in the SCS field programs (Laflen, Agenda—Personal communication from D. D. Smith, evening session based on personal attendance).

In 1965, *Agricultural Handbook 282*, the first complete publication of the USLE (Wischmeier and Smith 1965) became available. While it was well known that the erosion prediction equations were for sheet and rill erosion, this was explicitly stated in *Agricultural Handbook 282*, and there was discussion of estimating sediment delivery from watersheds. There was a separate section on soil loss tolerance, which was defined as the "maximum rate of soil erosion that will permit a high level of crop productivity to be sustained economically and indefinitely."

In 1978, *Agricultural Handbook 537* was released; it presented a revision of the USLE, expanding *Agricultural Handbook 282* to include new tillage systems, improved yields, and improved science. It also included an enlarged treatment of soil loss tolerances—including a discussion of considerations such as water quality standards and sediment delivery.

In 1997, the revised universal soil loss equation (RUSLE) was released (Renard et al. 1996), which also discussed soil loss tolerance. As in earlier works, T values were recommended to remain as originally defined and intended. If issues of water quality, economics, and policy were to be addressed for erosion control, T values were recommended to be designated as T_{wq}, T_{ep}, and T_{pol}.

The concept of soil loss tolerance emerged when it became apparent that one could predict sheet and rill erosion on agricultural lands using relationships between slope gradient, length, and factors representing cropping and management. The primary focus, in terms of soil loss, since Zingg first published Equation 17.1 expressing the effect of slope and length on soil loss, has been to develop better technology with a wider applicability for predicting soil loss, rather than on soil tolerance values.

17.8 EROSIONAL IMPACT ON PRODUCTIVITY

The negative effects of erosion on soil properties have been well documented by researchers (Follett and Stewart 1985; Lal and Stewart 1990; Pimentel 1993; Cleveland 1995; Loch and Silburn 1997): (1) reduction in soil depth and potential rooting depth; (2) reduction in soil organic matter content; (3) reduction in nutrient availability; (4) nonuniform removal of topsoil within a field; (5) exposure of, and/or mixing of topsoil with, subsoil of poorer physical, biological, and chemical properties; (6) changes in soil physical properties (such as changes in bulk density, water infiltration, water holding capacity, texture, or structure); or (7) some combination of the above factors (Den Biggelaar et al. 2001).

The detrimental effects of soil erosion associated with agricultural production have also been demonstrated through numerous experiments conducted where erosion was simulated either by artificial desurfacing (Mbagwu et al. 1984; Dormaar et al. 1986; Gollany et al. 1992; Malhi et al. 1994; Tanaka 1995; Larney et al. 2000) or by comparing yield on strongly eroded areas with yield on less eroded areas (Bramble-Brodahl et al. 1984; Busacca et al. 1984; White et al. 1984; Mielke and Schepers 1986; Olson and Carmer 1990; Kosmas et al. 2001; Bakker et al. 2007). Study results suggest that yield reductions at the field scale are approximately 4% for each 0.1 m of soil loss, where yield reductions could generally be attributed to a reduction in rooting depth (Bakker et al. 2007).

Erosion rates alone are not necessarily good indicators of damage to productivity. There is little establishment of the direct cause–effect relationship, and most studies of erosional processes are lacking in quantification of structural attributes (Lal 1998). Further, there is a need for more credible data on soil erosion rates, soil formation rate, soil loss tolerance, and impact of erosion on productivity (Lal 1998). To better estimate the effects of erosion on productivity, the development of more sophisticated and reliable crop simulation models may be required (Schumacher et al. 1994).

The relationship between soil erosion and productivity is complex. Although soil erosion is thought to be occurring at unsustainable rates, portions of the United States have experienced unprecedented increases in crop yields in recent years. The complexity in predicting impact on yield may occur because soil properties are not the only factors influencing crop yields. It is well documented that technology has greatly impacted yields. The increased use of higher-yielding corn hybrids and increased nitrogen fertilizer inputs has masked the effects of erosion on yields, and in many cases, yields have increased over time (Fenton et al. 2005). Climatic factors such as changes in growing season, precipitation, and temperature further complicate this relationship (Schumacher et al. 1994).

Many believe that the reliance on technology is masking the long-term damage being done to soil resources. While soil fertility can be improved through technological advances, the impacts of erosion on soil texture and function cannot be readily corrected by technology (Craft et. al 1992). Soil degradation effects can be grouped into two categories: (1) reversible, with such components as nutrient levels, pH, organic matter, and biological activity, and (2) irreversible, such as occurs with changes in rooting depth, water holding capacity, structure, and texture. The reversibility or irreversibility of soil degradation is a function not only of available technology but also of economic return (Den Biggelaar et al. 2001).

C, or the A horizon, can be considered an irreversible impact of soil erosion based on current estimates of soil formation. The A horizon plays a critical role in soil fertility. Plant roots and available nutrients are concentrated in this layer, and it is critical for nutrient retention and water holding capacity. In many cultivated soils, the A horizons have significantly decreased in thickness or been removed by erosion, leaving soils in the B horizon that are unfavorable for plant growth (Larson et al. 1983).

The importance of organic matter in soils has long been recognized. Stevenson (1972) summarized the important contributions of soil organic matter as it impacts nutrients, nutrient holding capacity, microbial growth, water holding capacity, buffering capacity, rainfall energy absorption, cementing agent for soil structural units, and possible plant growth stimulation. Verity and Anderson (1990) reported that 5 cm of organic matter–rich topsoil added to severely eroded knolls increased grain yields by more than 50%, which is in sharp contrast to the results of Bakker et al. (2007) cited above. Fenton et al. (2005) studied the impact of erosion on organic matter content and productivity of some Iowa soils. They concluded that organic matter content was significantly correlated with erosion phase.

Organic matter content for well- and moderately drained till-derived Mollisols with clay contents ranging from 18% to 40% was 3.65%, 2.46%, and 1.82% for slight, moderately, and severely eroded phases, respectively. For loess-derived soils with similar parameters, organic matter contents were 3.7%, 2.9%, and 2.0%. Models developed in the same study showed that for till-derived soils, corn yields decreased by 1.37 Mg/ha from slightly eroded to severely eroded soil phases. For loess-derived soils, corn yields decreased by 0.67 Mg/ha for the same erosion phase change. Decreased organic matter content and associated change in properties were major factors in the decreased productivity of the eroded soils.

Field studies and simulation models confirm large variations in soil degradation impacts to soil properties and quality (Maetzold and Alt 1986). It may not directly be the decrease in the A horizon depth that impacts yields but rather the corresponding change to other soil properties. Hoag (1998) concluded that "depth is not generally an adequate measure of productivity and that the most profitable management of a soil will depend on the quality and distribution of soil layers." Comparisons are also difficult to make among sites because of differences in soil properties and weather conditions (Schumacher et al. 1994). During degradation, some soils experience consistent yield reductions, while others suffer no impact until a critical soil loss level is reached. Once this level is reached, significant yield losses occur with further erosion (Hoag 1998).

Numerous studies have aimed at quantifying the relationship between soil erosion and productivity through relating yield declines to loss in Topsoil Depth (TSD). The relationship is typically nonlinear with greater yield declines at thinner TSD, but also with diminishing increased returns in relation to increasing TSD in areas of eroded sediment deposition (Walker and Young 1983). This relationship was also studied by Fenton et al. (2005) for medium-textured loess-derived and till-derived soils in Iowa, with corn yield expressed as a function of the A horizon thickness (Figure 17.1). The slope of yield curves increases as soil transitions from slightly to moderately to severely eroded, indicating increased yield sensitivity for more severely eroded soils. This relationship was further developed through yield response curves to increased nitrogen input (Figure 17.2). The slope of the curves decreases with increased nitrogen fertilizer input but also indicates that the decrease in yield was not corrected totally by the additional fertilizer (Fenton et al. 2005).

Soil Renewal and Sustainability

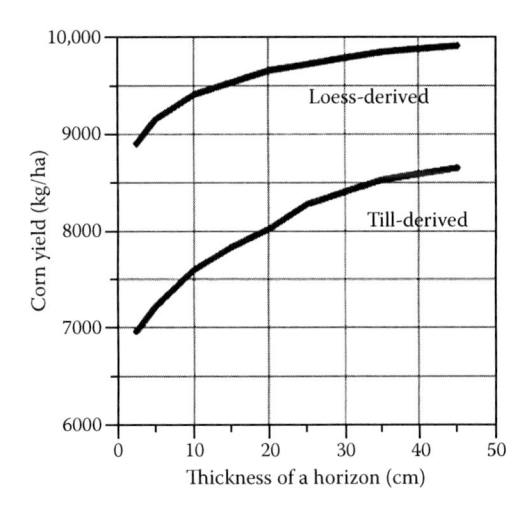

FIGURE 17.1 Effect of thickness of the A horizon on corn yields for loess- and till-derived soils. (Reprinted from *Soil Tillage Res*, 81, Fenton T.E. et al. Erosional impact on organic matter content and productivity of selected Iowa soils, 163–171, Copyright 2005, with permission from Elsevier.)

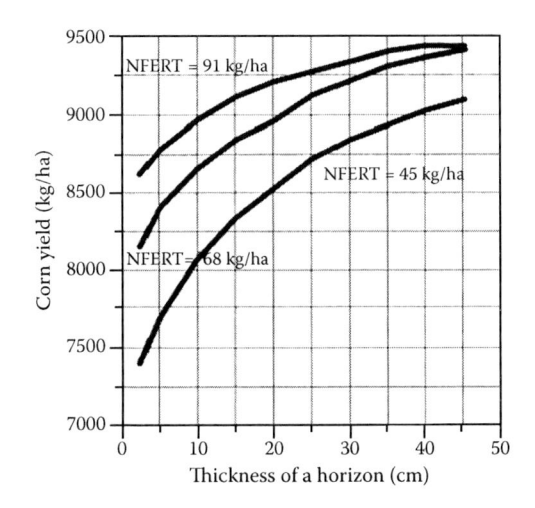

FIGURE 17.2 Effect of thickness of the A horizon on corn yields for loess- and till-derived soils combined with 45, 68, and 91 kg/ha of nitrogen fertilizer. (Reprinted from *Soil Tillage Res*, 81, Fenton T.E. et al. Erosional impact on organic matter content and productivity of selected Iowa soils, 163–171, Copyright 2005, with permission from Elsevier.)

The impact of technology on the yield curve must be better understood in relation to the true cost of lost "potential" production. Evidence suggests (Figure 17.3) that technology shifts the function to greater yields but in a skewed manner (Walker et al. 1983). There is a greater influence of technology with increasing TSD, or less eroded soils; yet the greatest need for recovery through technological influence needs to

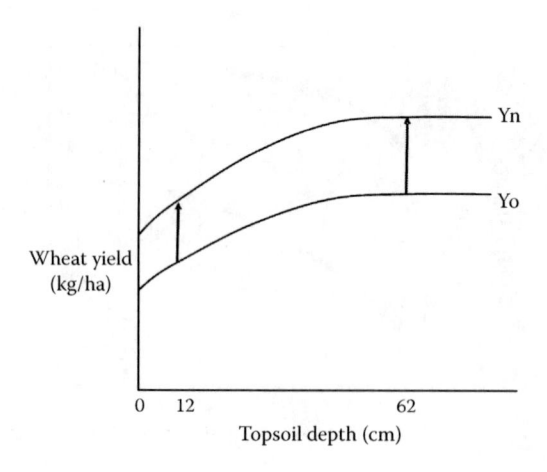

FIGURE 17.3 Technology shifts yield function. (Redrawn from Walker, D. and D. Young, A perspective that technology may not ease the vulnerability of US agriculture to erosion, NRE Staff Report, "Perspectives on the Vulnerability of US Agriculture to Soil Erosion," 1983.)

occur in more severely eroded soils. While increased N input has some impact for thinner TSD, it can only partially shift the relationship toward the maximum yield potential.

17.9 EROSIONAL IMPACT ON PRODUCTIVITY— CHINA PERSPECTIVE

The black soil (Mollisols) region in Northeast China has some of the most fertile soil in the world and is a critical area for commodity (soybean and corn) production. Average TSD in the region has decreased from approximately 70 to 100 cm to 20 to 70 cm (National Soil Survey Office 1994), and some areas of intense cultivation have no remaining topsoil. Average erosion rates of cultivated soil in this area range from 0.316 to 0.433 mm annually (Yan and Tang 2005), with reported losses of up to 1 cm in 1 year (Zhang et al. 2006). The reduction of TSD has resulted in lower soybean and corn production.

Topsoil thinning in this region, consistent with other locations on the globe, impacts crop performance. However, the impact of topsoil thinning is dependent on the depth of the topsoil being modified and the crop being produced, which seems intuitively logical. It further seems logical that the relationship between change of TSD and crop performance would be nonlinear. That is, as the TSD decreases, successive removal of a given depth increment has an increasing impact on crop performance, at least until the parent material influence dominates. This is supported by field data (Zhang 2006, 2007), which additionally suggest that Mollisol TSD for corn and soybeans of about 20 cm is required to approach full production potential.

Because most soils are tilled annually, soil in the top 5–20 cm (or sometimes deeper) is mixed on a yearly basis. This process, associated with surface soil loss through erosion, creates a surface plant growth medium composed of topsoil essentially diluted with underlying soil materials. One might expect that the "dilution" process would modify the typical yield loss trends normally observed in soil erosion/surface soil removal studies. However, this process still seems to result in an exponential yield decrease, at least for soybeans grown on a Mollisol with an average topsoil layer of 50 to 60 cm (Wang et al. 2009). Fertilizer application seems to modify the yield values observed for a given topsoil condition but fails to negate the erosion/crop yield relationship (Figure 17.4). Yields were reduced from 1.68 to 0.61 Mg/ha when topsoil loss increased from 0 to 70 cm with fertilizer addition (N: 46.4 kg/ha²; P_2O_5: 60.0 kg/ha²; K_2O: 13.8 kg/ha²) and reduced from 1.35 to 0.36 Mg/ha without fertilizer addition under corresponding topsoil thicknesses. The average yield reductions per 10 cm of soil loss when fertilized were 28.8%, 18.9%, 14.5%, 13.2%, 10.6%, 8.8%, and 9.2% at the erosion depths of 10, 20, 30, 40, 50, 60, and 70 cm compared to the yield at an erosion depth of 0 cm; but when unfertilized, they were 32.6%, 21.1%, 17.7%, 13.5%, 13.2%, 11.6%, and 10.4%, respectively. Inorganic fertilizer additions

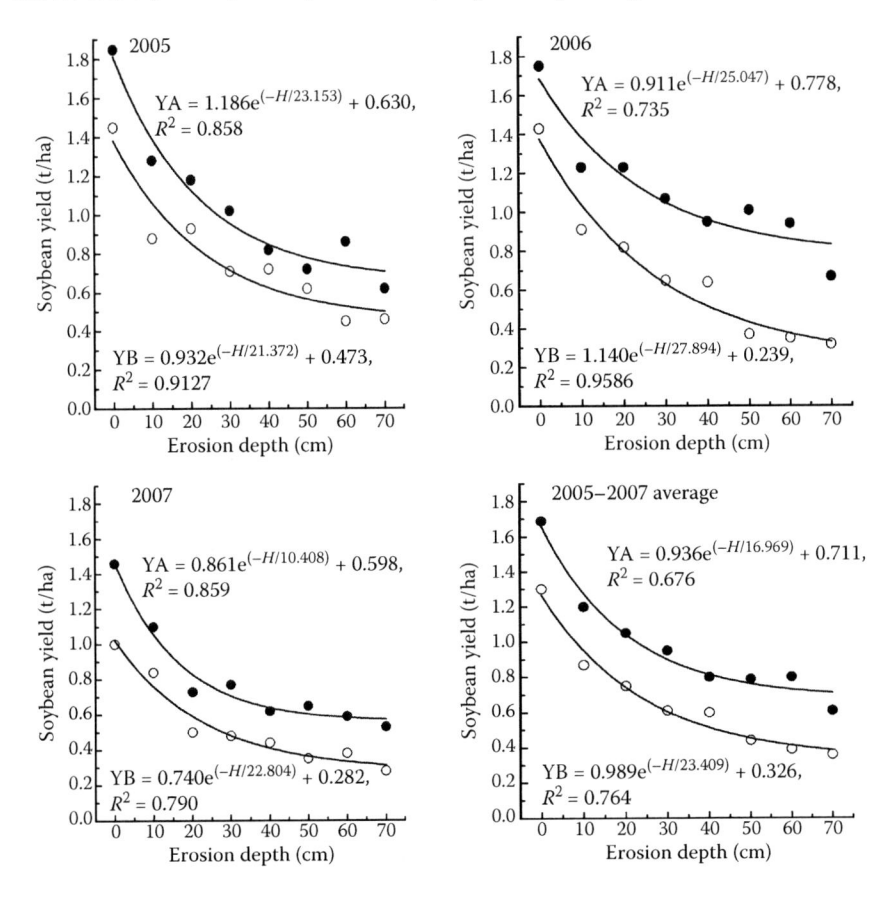

FIGURE 17.4 Soybean yield as a function of thickness of soil loss under fertilized condition (a, solid points) and unfertilized condition (b, circles). (From Wang, Z. et al. *Sci. China D*, 52, 1005–1021, 2009.)

seem to enhance the yields of eroded soil to some extent but fail to bring yields to preerosion levels (Figure 17.4).

The Loess Plateau, located in Southwest China, is one of the most eroded regions in the world. With sloping topography, approximately 70% of the area has suffered soil erosion to different extents; 36% of these eroded areas lose more than 50 Mg/ha annually (Zhu 1956). Soil erosion has significantly reduced crop production and yield potential in the southern region of the Loess Plateau, which is also a critical wheat production area of China.

The potential impact of the Loess Plateau soil loss on wheat production might be inferred from erosion and wheat yield research. Jia et al. (2004) examined the impact of soil erosion on wheat production and yield of slightly sloping loess farmland (Inceptisol) without fertilizer application. Wheat yield showed a positive incremental exponential relationship with increases in topsoil depth. Yield decreased by 39.1% and 69.1% when simulated erosion removed 10 and 20 cm, respectively, of topsoil compared to uneroded soil conditions, and increased by 53.1% when 10 cm of topsoil was added to the field surface. For each 1 cm loss of topsoil, wheat yield reduction ranged from 2.3% to 4.0%. In contrast, wheat yield increased by 5.3% for each 1-cm increase in topsoil depth. A linear relationship between topsoil thickness and grain weight was found. The study indicated that soil erosion not only reduced wheat yield but also resulted in degradation of grain quality. Based on wheat sensitivity to TSD from this study, erosion rates identified by Zhu (1956) in Southwest China could have very severe impacts on wheat production in that region.

While maize, soybean, and wheat are among the most important commodities globally and considering that soil erosion impacts to these crops is critical, other crops play key roles in feeding the world population as well, and erosional influences are also critically important. Limited evidence shows a negative relationship between thinning TSD and yields of potato, flack, and pea (Chen et al. 2003), with pea being the most sensitive of these three crops. Spring wheat was also part of that study, and similar to Jia et al.'s (2004) study, it was highly sensitive to TSD change. When removing topsoil depths of 5, 10, and 20 cm, reductions in spring wheat yield equaled 15.68%, 22.34%, and 37.92%, while reductions in pea were 9.79%, 26.20%, and 41.38%, respectively. Adding 10 cm of topsoil to the field brought an increase of 16.86% and 12.21% to spring wheat and pea yield, respectively.

The Sichuan Province in Southwest China consists of purple soils (Entisols) and is considered an agriculture center of China. In the region, 68% of the land is farmed, of which 72% is defined as low to medium yielding, with average topsoil depths of 20 to 60 cm (He et al. 1990; Zhu et al. 2009). Thinner topsoil induced by soil erosion has restricted crop growth and yield in the region. As displayed in Figure 17.5, corn yield showed an exponential increase, while wheat displayed a positive logarithmic relationship when increasing the topsoil thickness from 20 to 100 cm (Zhu et al. 2009). For wheat, the largest impacts in yield were seen at TSD, ranging from 20 to 60 cm, while there was no significant difference in yield when the thickness exceeded 80 cm.

Winter wheat and corn yields also increased with increasing topsoil thickness, although the yield impact varied during different growth periods. The effects of topsoil depth on crop yield could be partially explained by examining root growth under

Soil Renewal and Sustainability

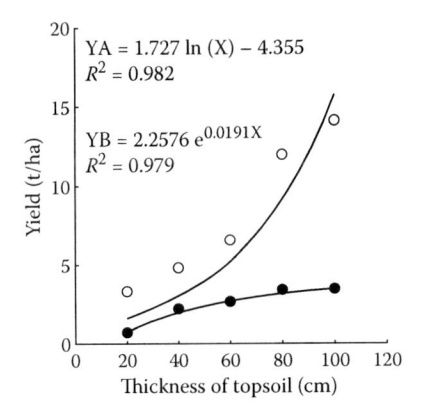

$$YA = 1.727 \ln(X) - 4.355$$
$$R^2 = 0.982$$

$$YB = 2.2576\, e^{0.0191X}$$
$$R^2 = 0.979$$

FIGURE 17.5 Relationship between crop yield and thickness of topsoil. (From Zhu, B. et al. *J Mountain Sci*, 27, 735–739, 2009.)

different thicknesses. Root weight of winter wheat and corn gradually increased as a function of topsoil depth. For both crops, 20 cm of topsoil resulted in the lightest root weight when compared to the other topsoil thicknesses. In normal rainfall years, the study concluded that the critical thickness of topsoil for both crops was 40 to 60 cm.

17.10 SOIL SUSTAINABILITY

Sustainability is defined as "managing soil and crop cultural practices so as not to degrade or impair environmental quality on- or off-site, and without eventually reducing yield potential as a result of the chosen practice though exhaustion of either on-site resources or nonrenewable inputs" (Soil Science Society of America 2008). The 1990 Farm Bill has a more expansive definition and states: "the term sustainable agriculture means an integrated system of plant and animal production practices having a site-specific application that will, over the long term (i) satisfy human food and fiber needs, (ii) enhance environmental quality and the natural resource base upon which the agricultural economy depends, (iii) make the most efficient use of nonrenewable resources and on-farm resources and integrate, where appropriate, natural biological cycles and controls, (iv) sustain the economic viability of farm operations, and (v) enhance the quality of life for farmers and society as a whole."

Questions have been raised regarding the consideration of soil as a renewable resource. Friend (1992) states, "The leading edge of academic thinking now is that a majority of world soils, apart from the deepest and most rapidly forming, are non-renewable within a human lifetime." Thus, time becomes an important factor in any discussion of both sustainability and renewability. Kummerer et al. (2010) proposed several general rules in order to use soils in a sustainable manner. They called one of these the Rule of Transition: resources that are not renewable within human time scales should be used only within the time needed for the transition from an unsustainable to a sustainable economy. All their rules involve varying time scales and vary in importance at different points in time and in their interconnectedness in soil formation and soil development. They conclude that an interdisciplinary temporal view is needed as a basis for the sustainable use of soils. The focus on temporal features will allow improved understanding of various forms of soil degradation and the full range of sustainable and unsustainable use of soil resources. We agree with the conclusion made by Hall et al. (1982) that the highest priority should be given to maintaining a favorable rooting volume. This automatically places those soils with restrictive subsurface horizons in the category of high risk for maintaining a sustainable system of crop production.

The irreversible impact of reduced rooting depth on production may be slowed in more fertile areas of the world with naturally deep A horizons. Areas such as those containing Mollisols or Alfisols may show greater resistance to initial soil erosion where the long-term severe impacts of topsoil removal may not yet have been fully experienced. However, there are many regions of the world where disappearing A horizons have greatly altered agriculture production and yield to unrecoverable levels (FAO 2011).

Soil and crop management practices that increase soil erosion are often driven by economic considerations. The impacts of erosion must be considered through both long-term physical impacts to the land and economic impacts associated with lost production. These impacts include the costs incurred to reduce or offset the yield penalty associated with soil loss (Crosson 1997) as well as fertilizer input use increases to compensate for declines in soil quality due to erosion (ERS 1997; Den Biggelaar et al. 2001). Nutrient losses, direct damage to plants, and offsite damages must also be considered (Larson et al. 1983). Despite the impacts of soil erosion, the world has managed to maintain and even increase production levels through technological advances. However, to take advantage of technological advances leading to greater productivity, especially those related to plant genetics, soil quality must be sufficient to supply increasing amounts of water and nutrients, as deficiencies in either of these will negate increased plant production potential.

Globally, evidence strongly suggests that erosion of agricultural soils is occurring at a rate considerably greater than soil renewal rates. Soil erosion-based land degradation varies greatly in space and time. The temporal component of soil erosion is being aggravated by increased frequencies of heavy rainfall events attributed to a changing global climate and increased production pressure placed on land due to elevated demand for agricultural products. As elevated demand draws additional land into production, spatial soil erosion will almost certainly increase as well, especially erosion associated with farming environmentally sensitive lands that offer agriculture production potential.

The imbalance between soil erosion and soil renewal presents a dilemma often centered on human choices. Minimizing soil loss through appropriate management, while maintaining existing production, is often a choice avoided for financial, political, or cultural reasons. As previous civilizations have failed due to soil misuse and/or abuse, as inferred by Franklin D. Roosevelt, our choice to sustain or degrade our soil resources is a choice of self-preservation or self-destruction.

ABBREVIATIONS

IDEP: Iowa Daily Erosion Project
NRCS: Natural Resources Conservation Service of the United States Department of Agriculture

RUSLE: revised universal soil loss equation
SCS: Soil Conservation Service
T: tolerable soil loss
USDA-NRI: United States Department of Agriculture–National Resource Inventory
USLE: universal soil loss equation
WEPP: Water Erosion Prediction Project
YBP: years before present

REFERENCES

Alexander, E.B. 1988. Rates of soil formation: Implications for soil-loss tolerance. *Soil Sci* 145:37–45.

Bakker, M.M., G. Govers, R.A. Jones, and D.D.A. Rounsevell. 2007. The effect of soil erosion on Europe's crop yields. *Ecosystems* 10(7):1209–1219.

Bramble-Brodahl, M., M.A. Fosberg, D.J. Walker, and A.L Falen. 1984. Changes in soil productivity related to changing topsoil depth on two Idaho Palouse soils. In *Erosion and Soil Productivity*, ed. D.K. McCool, 18–27. American Society of Agricultural Engineering, St. Joseph, MI.

Brown, L.R. 1984. Conserving soil. In *State of the World*, ed. L.R. Brown, 53–75. Norton, New York.

Browning, G.M., C.L. Parish, and J. Glass. 1947. A method for determining the use and limitations of rotation and conservation practices in the control of soil erosion in Iowa. *J Am Soc Agron* 39:65–73.

Busacca, A.J., D.K. McCool, R.I. Papendick, and D.L. Young. 1984. Dynamic impacts of erosion processes on productivity of soils in the Palouse. In *American Society of Agricultural Engineers*, eds. Erosion and soil productivity, 152–69. American Society of Agricultural Engineers, St. Joseph, MI.

Cassman, K.G., P. Grassini, and J. van Wart. 2011. Crop yield potential, yield trends, and global food security in a changing climate. In *Handbook of Climate Change and Agroecosystems: Impacts, Adaptation, and Mitigation*, eds. D. Hillel and C. Rosenzweig. Imperial College Press, London.

Chen, Q.B., K.Q. Wang, S. Qi, and D.L. Sun. 2003. Soil and water erosion in its relation to slope field productivity in hilly gully area of the Loess Plateau. *Acta Ecol Sin* 23:1463–1469.

den Biggelaar, C., R. Lal, K. Wiebe, and V. Breneman. 2001. Impact of soil erosion on crop yields in North America. *Adv Agron* 72:1–52.

Cleveland, C.J. 1995. Resource degradation, technical change, and the productivity of energy use in U.S. agriculture. *Ecol Econ* 13:185–201.

Cox, C., A. Hug, and N. Bruzelius. 2011. Losing Ground. Environmental Working Group. Available at: http://static.ewg.org/reports/2010/losingground/pdf/losingground_report.pdf.

Craft, E.M., R.M. Cruse, and G.A. Miller. 1992. Soil erosion effects on corn yields assessed by potential yield index model. *Soil Sci Soc Am J* 56:878–883.

Crocker, R.L. 1960. The plant factor in soil formation. Proc. 9th Pacific Sci., Congr. of Pacific *Sci Assn* 18:84–90.

Crosson, P.R. 1997. The on-farm economic costs of soil erosion. In *Methods for Assessment of Soil Degradation*, eds. R. Lal, W.H. Blum, C. Valentin, and B.A. Stewart, 495–511. CRC Press, Boca Raton, FL.

Cruse, R., D. Flanagan, J. Frankenberger, B. Gelder, D. Herzmann, D. James, W. Krajewski et al. 2006. Daily estimates of rainfall, water runoff, and soil erosion in Iowa. *J Soil Water Conserv* 61:191–198.

Dokuchaev, V.V. 1883. Russian Chernozems (Russkii Chernozem). Translated by N. Kaner, Israel *Prog. Sci Trans,* Jerusalem, Israel.

Dormaar, J.F., C.W. Lindwall, and G.C. Kozub. 1986. Restoring productivity to an artificially eroded dark brown chernozemic soil under dry land conditions. *Can J Soil Sci* 66:273–285

Economic Research Service, Natural Resources and Environment Division, 1997. Land and soil quality. In *Agricultural Resources and Environmental Indicators 1996–97*, pp. 41–49. USDA Economic Research Service, Washington, D.C.

Eswaran, H., P.F. Reich, and E. Padmanabhan. 2012. World soil resources opportunities and challenges. In *World Soil Resources and Food Security*, eds. R. Lal and B.A. Stewart. CRC Press, Boca Raton, FL, USA.

Fenton, T.E. 2010. Soil erosion in Iowa. In *Getting Into Soil and Water*, ed. R. Unger, Iowa Water Center, Ames, IA.

Fenton, T.E. 2012. The impact of erosion on the classification of Mollisols in Iowa. *Can J Soil Sci* 92:413–418.

Fenton, T.E., M. Kazemi, and M.A. Lauterbach-Barrett. 2005. Erosional impact on organic matter content and productivity of selected Iowa soils. *Soil Tillage Res* 81:163–171.

Fischer, G., H. van Velthuizen, M. Shah, and F. Nachtergaele. 2002. *Global Agro-ecological Assessment for Agriculture in the 21st Century: Methodology and Results*. International Institute for Applied Systems Analysis. FAO. Rome, Italy.

Follett, R.F. and B.A. Stewart. 1985. *Soil Erosion and Crop Productivity*. ASA-CSSASSSA, Madison, WI.

Food and Agriculture Organization. 2002. *World Agriculture: Towards 2015/2030*. FAO, Rome.

Food and Agriculture Organization. 2011. *State of the World's Land and Water Resources for Food and Agriculture*. Summary Report. FAO. Rome.

Friend, J.A. 1992. Achieving soil sustainability. *J Soil Water Conserv* 47(2):157–167.

Gollany, H.T., T.E. Schumacher, M.J. Lindstrom, P.D. Evenson, and G.D. Lemme. 1992. Topsoil depth and desurfacing effects on properties and productivity of a Typic Agriustoll. *Soil Sci Soc Am J* 56:220–225.

Hall, G.F., R.B. Daniels, and J.E. Foss. 1982. Rate of soil formation and renewal in the USA. In *Determinants of Soil Loss Tolerance*. ASA Publication Number 45. American Society of Agronomy and Soil Science Society of America, Madison, WI.

Hallberg, G.R., N.C. Wollenhaupt, and G.A. Miller. 1978. A century of soil development in spoil derived from loess in Iowa. *Soil Sci Soc Am J* 42:339–343.

He, Y.Y., F.J. Zhang, L.H. Pan, and A.B. Wen. 1990. Research on the degradation of purple soils in the hilly areas, Sichuan Basin I. The physical properties and the degradation of the purple soils. *Resource Dev Protect* 6:3–7, 11.

Hoag, D.L. 1998. The intertemporal impact of soil erosion on non-uniform soil profiles: A new direction in analyzing erosion impacts. *Agric Sys* 56(4):415–429.

Hutton, C.E. 1947. Studies of loess-derived soils in Southwestern Iowa. *Soil Sci Soc Am Proc* 12:424–431.

Intergovernmental Panel on Climate Change. 2007. Climate Change 2007: Impacts, Adaptation, and Vulnerability. In *Contribution of Working Group II to the Third Assessment Report of the Intergovernmental Panel on Climate Change*, eds. M.L. Parry, O.F. Canziani, J.P. Palutikof, P.J. van der Linden, and C.E. Hanson, 1000. Cambridge University Press, Cambridge, UK.

Jenny, H. 1941. *Factors of Soil Formation*. McGraw-Hill, New York.

Jenny, H. 1980. *The Soil Resource*. Springer-Verlag, New York.

Jia, R. Y., X.G. Zhao, and C.P. Du. 2004. Evaluation on productivity decreased by soil and water loss in the southern Loess Plateau. *J Northwest Forestry Univ* 19:77–81.

Kosmas, C., S. Gerontidis, M. Marathianou, B. Detsis, T. Zafiriou, W. Muysen, G. Govers et al. 2001. The effects of tillage displaced soil on properties and wheat biomass. *Soil Tillage Res* 58:31–44.

Karl, T.R., J.M. Melillo, and T.C. Peterson. 2009. *Global Climate Change Impacts in the United States*. Cambridge University Press, Cambridge, UK, 196 pp.

Kummerer, K., M. Held, and D. Pimentel. 2010. Sustainable use of soil and time. *J Soil Water Conserv* 65:141–149.

Lal, R. 1995. Global soil erosion by water and carbon dynamics. In *Soils and Global Change. Adv. in Soil Science*, eds. R. Lal et al., pp. 131–141. CRC Press.

Lal, R. 1998. Soil erosion impact on agronomic productivity and environment quality. *Crit Rev Plant Sci* 17(4):319–464.

Lal, R. and B.A. Stewart. 1990. *Soil Degradation: Advances in Soil Science 11*. Springer-Verlag, New York.

Larney, F.J., B.M. Olson, H.H. Janzen, and C.W. Lindwall. 2000. Early impacts of topsoil removal and soil amendments on crop productivity. *Agron J* 92:948–956.

Larson, W.E., F.J. Pierce, and R.H. Dowdy, 1983. The threat of soil erosion to long-term crop production. *Science* 219(4584):458–465. Article Stable URL: http://www.jstor.org/stable/1690845.

Lobell, D.B., W.S. Schlenker, and J. Costa-Roberts. 2011. Climate trends and global crop production since 1980. *Science* 333:616–620. doi:10.1126/science.1204531.

Lobell, D.B., A. Sibley, and J.I. Ortiz-Monasterio. 2012. Extreme heat effects on wheat senescence in India. *Nat Clim Change* 2:186–189.

Loch, R.J. and D.M. Silburn. 1997. Soil erosion. In *Sustainable Crop Production in the Subtropics: An Australian Perspective*, eds. A.L. Clarke and P.B. Wylie, 27–63. Department of Primary Industries, Brisbane, Qld.

Maetzold, J. and K. Alt. 1986. Forum on Erosion Productivity Impact Estimators: Assessment and Planning Report. Soil Conservation Service, Appraisal and Program Development Division, Washington, DC.

Malhi, S.S., R.C. Izaurralde, M. Nyborg, and E.D. Solberg. 1994. Influence of topsoil removal on soil fertility and barely growth. *J Soil Water Conserv* 49:96–101.

Mbagwu, J.S.C., R. Lal, and T.W. Scott. 1984. Effects of desurfacing of Alfisols and Ultisols in Southern Nigeria: I. Crop performance. *Soil Sci Soc Am J* 48:828–833.

Mielke, L.N. and J.S. Schepers. 1986. Plant response to topsoil thickness on an eroded loess soil. *J Soil Water Conserv* 41:59–63.

Montgomery, D.R. 2007. Soil erosion and agricultural sustainability. *Proc Natl Acad Sci* 104:13268–13272.

Musgrave, G.W. 1947. The quantitative evaluation of factors in water erosion—a first approximation. *J Soil Water Conserv* 2:133–138, 170.

National Soil Survey Office. 1994. *Chinese Soil Census Records*. China Agricultural Press, Beijing.

Nearing, M.A., F.F. Pruski, and M.R. O'Neal. 2004. Expected climate change impacts on soil erosion rates: a review. *Soil Water Conserv J* 59:43–50.

Olson, K.R. and S.G. Carmer. 1990. Corn yield and plant population differences between eroded phases of Illinois soils. *J Soil Water Conserv* 45:562–566.

Parsons, R.B., C.A. Balster, and A.O. Neas. 1970. Soil development and geomorphic surfaces, Willamette Valley, Oregon. *Soil Sci Soc Am Proc* 34:485–491.

Parsons, R.B., W.H. Scholtes, and F.F. Riecken. 1962. Soils of Indian mounds in northeastern Iowa as bench marks for studies of soil genesis. *Soil Sci Soc Am Proc* 26:491–496.

Pimentel, D. 2006. Soil erosion: a food and environmental threat. *Environ Dev Sustain* 8:119–137.

Pimentel, D. (ed.). 1993. *World Soil Erosion and Conservation*. Cambridge Univ. Press, Cambridge, UK.

Ruhe, R.V. 1969. *Quaternary Landscapes in Iowa*. Iowa State Univ. Press, Ames, IA.

Ruhe, R.V., T.E. Fenton, and L.L. Ledesma. 1975. Missouri River history, floodplain construction, and soil formation in southwestern Iowa. *Iowa Agric Home Econ Exp Sta Res Bull* 580:783–791.

Schumacher, T.E., M.J. Lindstrom, D.L. Mokma, and W.W. Nelson. 1994. Corn yield: Erosion relationships of representative loess and till soils in the North Central United States. *J Soil Water Conserv* 49(1):77–81.

Simonson, R.W. 1959. Outline of a generalized theory of soil genesis. *Soil Sci Soc Am Proc* 23:152–156.

Smith, D.D. 1941. Interpretation of soil conservation data for field use. *Agr Eng* 22:173–175.

Soil and Water Conservation Society. 2006. *Planning for Extremes: A Report from a Soil and Water Conservation Society Workshop*. Soil and Water Conservation Society. Ankeny, Iowa.

Soil Science Society of America. 2008. *Glossary of Soil Science Terms*. Madison, WI.

Stevenson, F.J. 1972. *Humus Chemistry: Genesis, Composition, Reactions*. Wiley Press, New York.

Tanaka, D.L. 1995. Spring wheat straw production and composition as influenced by topsoil removal. *Soil Sci Soc Am J* 59:649–654.

United Nations Secretary-General's High-Level Panel on Global Sustainability. 2012. *Resilient People, Resilient Planet: A Future Worth Choosing*. United Nations, New York.

Verity, G.E. and D.W. Anderson. 1990. Soil erosion effects on soil quality and yield. *Can J Soil Sci* 70:471–484.

Wakatuski, T. and A. Rasyidin. 1992. Rates of weathering and soil formation. *Geoderma* 52:251–262.

Walker, D. and D. Young, 1983. A Perspective that Technology May Not Ease the Vulnerability of US Agriculture to Erosion. NRE Staff Report, "Perspectives on the Vulnerability of US Agriculture to Soil Erosion."

Wang, Z.Q., B.Y. Liu, X.Y. Wang, X.F. Gao, and G. Liu. 2009. Erosion effect on the productivity of black soil in Northeast China. *Sci China D Earth Sci* 52:1005–1021.

White, A.W., R.R. Bruce, A.W. Thomas, G.W. Langdale, and H.F. Perkins. 1984. Characterizing productivity of eroded soils in the Southern Piedmont. In *Erosion and Soil Productivity*. American Society of Agricultural Engineers, New Orleans, Louisiana.

Wischmeier, W.H. and D.D. Smith. 1965. Predicting rainfall-erosion losses from cropland east of the Rocky Mountains—Guide for selection of practices for soil and water conservation. U.S. Dept. of Agric., Agr. Handbook No. 282.

Yan, B.X. and J. Tang. 2005. Study on black soil erosion rate and the transformation of soil quality influenced by erosion. *Geogr Res* 24:499–506.

Zhang, X.Y., X.B. Liu, Y.Y. Sui, S.L. Zhang, J.M. Zhang, H.J. Liu, and J.J. Stephen. 2006. Effects of artificial topsoil removal on soybean dry matter accumulation and yield in Chinese Mollisols. *Soybean Sci* 25:123–126.

Zhang, X.Y., L.Q. Meng, X.B. Liu, Y.Y. Sui, S.L. Zhang, and J.H. Stephen. 2007. Effects of soil erosion on corn dry matter accumulation and yield in black soil are. *China Water Res* 22:47–49.

Zhu, B., F.H. Kuang, M.R. Gao, T. Wang, X.G. Wang, and J.L. Tang. 2009. Effects of soil thickness on productivity of sloping cropland of purple soil. *J Mountain Sci* 27:735–739.

Zhu, X.M. 1956. Classification of soil erosion in loess plateau. *Acta Pedol Sin* 4:17–21.

Zingg, A.W. 1940. Degree and length of land slope as it affects soil loss in runoff. *Agr Eng* 21:59–64.

18 Organic Carbon Sequestration Potential and the Co-Benefits in China's Cropland

Genxing Pan, Kun Cheng, Jufeng Zheng, Lianqing Li, Xuhui Zhang, and Jinwei Zheng

CONTENTS

18.1 Role of SOC in China's Agriculture 501
18.2 Status of SOC Storage of China's Croplands 505
    18.2.1 Background SOC Stock in the 1980s 505
    18.2.2 Historical Changes in Soil Carbon Stocks of China's Croplands 506
    18.2.3 Carbon Stock Dynamics and Sequestration between
        1980 and 2006 507
18.3 Sequestration of Organic Carbon in Croplands: Drivers, Potential, and
    Technical Feasibility 509
18.4 Co-Benefits of SOC Sequestration for Crop Production and Ecosystem
    Health in Croplands 512
18.5 Conclusion 516
Acknowledgment 516
Abbreviations 516
References 516

18.1 ROLE OF SOC IN CHINA'S AGRICULTURE

Soil organic matter (SOM) is a key natural pool of energy and capital in a holistic assessment of global terrestrial ecosystem services (Brown and Ulgiati 1999). It is also a vital material for soil fertility of croplands and has been proposed as a key soil parameter for characterizing soil quality for productivity and ecosystem functioning (Tiessen et al. 1994). In a global perspective of key soil quality parameters, topsoil SOM content has been adopted as a more appropriate overall soil quality indicator than other properties for the European Union's (EU) agricultural and forestry sectors (European Commission 2002). It has been widely reported that contents of SOM in croplands have significant control on crop productivity and functioning (Dawea et al. 2003), and this control has been well addressed by Manlay et al. (2007).

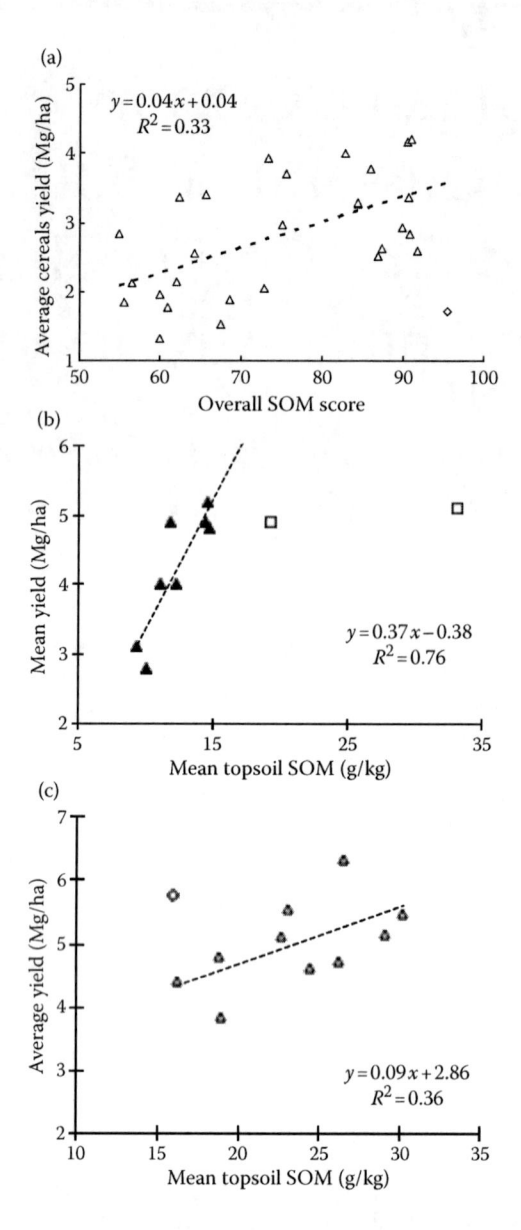

FIGURE 18.1 Cropland topsoil SOM related to the averaged cereals yield of China's provinces. (a) Overall correlation of provincial mean cereals yield over 1949–1999 with the mean topsoil SOM content surveyed in the early 1980s. (b) Correlation of the mean cereals yield over 1994–1999 for North China provinces, with rainfed crops in black triangles and the white squares representing Jilin and Inner Mongolia. (c) Correlation of the mean cereals yield over 1994–1999 of South China's provinces, with mainly rice in open triangles. The white square represents Jiangsu. (Data synthesized from the Yearbook of Agricultural Statistics of China 2001, China Agricultural Press, Beijing, 2002.)

By a statistical integration of SOM contents of croplands, Pan et al. (2009) reported linear response of provincial mean cereal productivity to the average SOM content of croplands of China, though varying in incremental response depending upon climatic conditions and socioeconomical and technological factors (Figure 18.1). The mean provincial cereal yields of the major crop production provinces of North China (solid triangles) from 1950 to 1999 are strongly related to their mean topsoil SOM contents. However, such a relationship is weak for the provinces in southeast China, and there was no response in other provinces such as Jiangsu and Shanghai. The linear correlations suggest an approximate productivity of 8 Mg/ha for rainfed uplands and 6 Mg/ha for rice paddies corresponding to a SOM content of 10 g/kg (Figure 18.1).

Furthermore, data from long-term monitoring and field experiments with different management practices revealed also significant SOM control on crop productivity and sustainability (Table 18.1). In general, a change in SOM content of 0.1% could reduce the variability by 1% to 2%. However, the overall changes in soil organic carbon (SOC storage under agricultural development for the last decades and the future sequestration capacity of the croplands have experienced a large C debt since the early twenty-first century. The SOC storage has been a major issue in relation to the highly variable crop production since the end of the last century (Table 18.1).

In addition, there was also evidence that SOM had prominent impacts on rural economic development, even though it was not entirely based on agriculture. A regression analysis of data compiled from the second Chinese soil survey showed that the mean agriculture production of a municipality in Jiangsu in the early 1980s at the time of the survey was correlated with the mean SOM score (a ranking value of croplands determined by the proportion of cropland area with high SOM category, corresponding to a relative SOM abundance in croplands) (Figure 18.2).

TABLE 18.1
Temporal Variability in Crop Yield of Long-Term Experiments in Relation to SOC Contents across China

Agro/Ecoregion	Crop	Yield Variability (%)	Topsoil SOC (g/kg)
Central China plain	Wheat–maize	30–45	7–8
Tai Lake plain	Rice	8–30	16–20
Northern Zhejiang plain	Rice	12–30	15–17
Rolling area of red soils in Hunan	Winter wheat	30–40	9–11
Purple basin of Sichuan	Winter wheat	15–30	15–18
	Summer rice	13–20	
Rolling area of red soil, Jiangxi	Double rice	8–17	13–20

Source: Yield and SOC data from the long-term experiment sites between the late 1980s and early 2000s, courtesy of Dr. Xu Mingang.

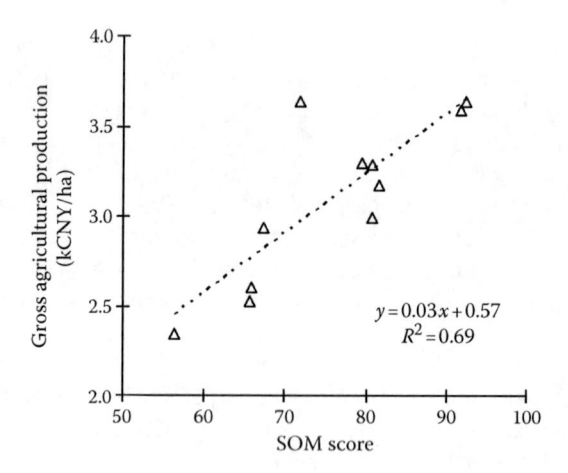

FIGURE 18.2 Gross agricultural production (1982) as a function of SOM ranking values of 11 municipalities of Jiangsu as per the second soil survey conducted between 1979 and 1982.

This relationship may be explained by the impact of SOM on crop yield and stability that helped in ensuring food supply and income from agriculture for the local people before industrialization. Such an impact is also indicated by a correlation of mean GDP with the SOM scores of their croplands (Figure 18.3). Nevertheless, such a role is becoming less important because the local economy is progressively less dependent on crop production (Figure 18.3a and b).

Therefore, SOM plays an important role in China's crop production and, consequently, in the sustainable food supply of China. This role may lead further to a strong impact on the local economy and, in turn, the livelihood of the local people. For a region with agriculture as the predominant economic sector, the role of SOM is critical and must not be underestimated.

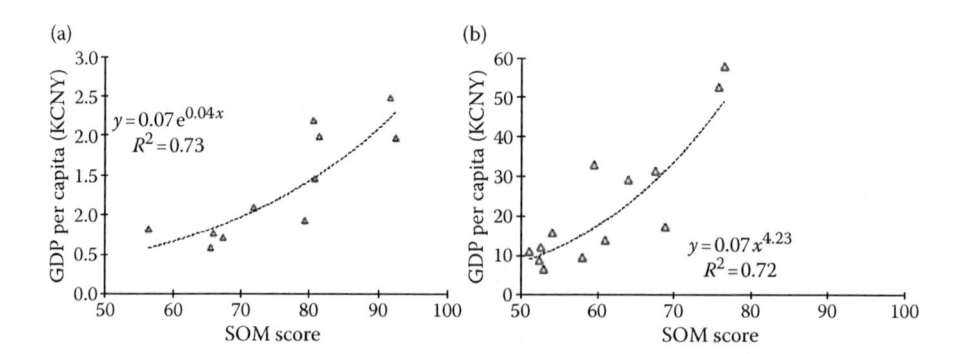

FIGURE 18.3 Gross domestic product (GDP) as a function of cropland SOM ranking values for the municipalities of Jiangsu Province. (a) GDP data for the year of 1982 and SOM data from the second soil survey finalized in 1982. (b) GDP data for the year 2004 and SOM data from the geological survey of the province completed in 2004.

18.2 STATUS OF SOC STORAGE OF CHINA'S CROPLANDS

18.2.1 Background SOC Stock in the 1980s

The multiple types of terrestrial ecosystems existing in large areas of China have a significant role in preserving C stocks and reducing greenhouse gas (GHG) emissions through land use and land cover change. There have been numerous studies on estimating total SOC stock among land use types. A total SOC stock to 1-m depth for soils of China has been estimated at 90 Pg by a national panel conference on SOC (Xiangshan Science Conference Series No. 362 in 2004). Of the total stock, 15 Pg C is in China's croplands. Detailed data of topsoil SOC measurements from the second national soil survey allowed a reliable quantitative assessment of background topsoil SOC stock. A statistical analysis by Song et al. (2005) estimated that China's cropland soils contain 5.1 Pg of SOC stock to 0.2-m depth, of which 1.3 Pg is contained in soils of rice paddies cultivated on 30 Mha, and another 3.8 Pg in soils of dry croplands cultivated on about 100 Mha. In comparison, soils under China's forestland contain 5.9 Pg C in topsoil in an area of 1.42 Mha, and soils under China's grassland contain merely 1.15 Pg with a land area of 3.31 Mha (Fang et al. 2007). In addition, there exists a relatively small stock of 0.9 Pg in wetlands covering an area of 0.4 Mha (Zhang et al. 2008). The distribution of SOC stocks in China's croplands is presented in Table 18.2.

As a vital part of soil, topsoil is rich in SOM, which supports soil fertility and biodiversity and is crucial to ecosystem functioning. In agriculture, the SOM content of topsoil largely determines the cropland productivity. China's croplands were depleted of SOM stock because of intensive cultivation (Song et al. 2005), widespread degradation (Lal 2002), and improper management (Feng et al. 2011). An official report released by the Ministry of Agriculture of China in 2004 indicated that topsoil SOC content was as low as 1% in 65% of China's croplands and was no higher than 1.5% in croplands from all the major production regions including Henan, Hebei, Shandong, Hubei, Hunan, Jiangxi, and Sichuan provinces. In contrast,

TABLE 18.2
Estimates of the Baseline SOC Stock in China's Croplands in the Early 1980s

Land Use Type	SOC Stock (Pg) Whole Soil	SOC Stock (Pg) Topsoil	Topsoil SOC (Mg C/ha)
Uncultivated	75	33	49.84 ± 46.69
Cultivated soil	15	5.1	35.87 ± 32.77
Rice paddy	/	1.3	43.98 ± 19.07
Dry cropland	/	3.8	33.44 ± 36.89

Source: Recalculated from the data by Song et al. (2005). Pan et al. (2003). Topsoil refers to the depth of 0–20 cm for croplands and of 0–30 cm for uncultivated soils, while whole soil implies a depth of 0–100 cm.

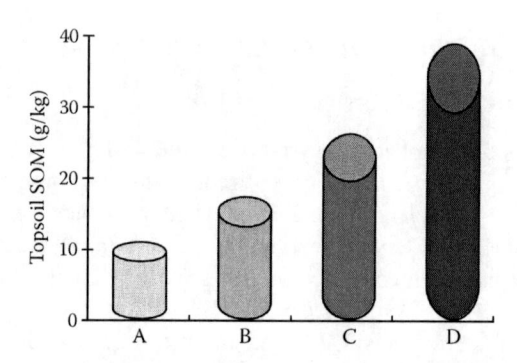

FIGURE 18.4 Comparison of China's cropland SOM level with those of the United States and EU. (A) SOM level under 10 g/kg for the majority of croplands of China. (Data from the Ministry of Agriculture and State Commission of Development and Reform, China. *A National Planning of Conservation Tillage Project.* [AO/BO]. Available at www.gov.cn/gzdt/2009-08/28/content140367.html, 2004.) (B) SOM level under 15 g/kg for all croplands from the major agricultural production areas of China. (Data from the Ministry of Agriculture and State Commission of Development and Reform, China. *A National Planning of Conservation Tillage Project.* [AO/BO]. Available at www.gov.cn/gzdt/2009-08/28/content140367.html, 2004.) (C) Mean level of SOM from the United States and EU countries. (D) Critical SOM level for the United Kingdom. (From Pan, G., *Adv Clim Change Res* 5, 11–18, 2009.)

the mean topsoil SOC content in croplands from the United States and EU countries has generally been reported to be between 2% and 3%. In addition, Loveland and Webb (2003) suggested a critical SOM level of 3.5% for temperate zone agriculture. Thus, in comparison with soils of the United States and the EU, soils of China are severely depleted of their SOC stock (Figure 18.4).

The low SOC stock in soils of China is also attributed to the increasing dependence on chemicals and decreasing sustainability of China's crop production since the end of the last century (Pan and Zhao 2005). The small per capita cropland area, approaching the Food and Agriculture Organization's (FAO's) (2001) critical limit of 0.05 ha, and the low topsoil SOC content, which is lower than the world average by 30% (Huang and Sun 2006), are among the major constraints to China's agricultural sustainability. It is this fact that urged the creation of a national climate change policy in agriculture to promote SOC enhancement through increasing input of biomass C and protecting soils against degradation caused by intensified agricultural use of croplands. This policy has been addressed in a national panel meeting of Xiangshan Science Conference No. 362 (Pan 2009).

18.2.2 Historical Changes in Soil Carbon Stocks of China's Croplands

It is widely accepted that the historic land use reduced the global SOC stock by 5% (Lal 1999), with total loss of SOC stocks due to land use change estimated to be 55 Pg by agricultural land use (Intergovernmental Panel on Climate Change 1996). However, there exists much uncertainty in the estimates of historical loss of SOC stocks in China's historical agricultural development. Lal (2002) estimated a total loss of 3.5 Pg since the agricultural development in China, including about 2 Pg from land desertification caused by improper land use (Lal 2002). Lindert et al. (1996) reviewed SOC data from early agricultural literature published before 1960 and reported severe SOC losses from a wide range of China's ecosystems. Using the DeNitrification-DeComposition (DNDC) model, Li (2000) addressed a serious and continuing decline of SOC storage in China's croplands since the 1950s, resulting in a total SOC stock loss of up to 70 Tg since the 1970s. Wu et al. (2003) conducted a statistical analysis of SOC storage from the archived data from the second national soil survey for cultivated soil compared with natural soils and estimated a loss of whole soil SOC stock of China's croplands at 7 to 8 Pg, primarily due to historical degradation in SOC storage from cultivation of the native soils. Drastic loss in SOC stock occurred mainly in the north and other arid and semiarid regions of China. Nevertheless, it has been argued that there have been large areas and soil and land use associations where changes in SOC stocks were either minimal or even where increases in SOC storage had occurred, especially in irrigated areas (Pan et al. 2003). Song et al. (2005b) used data for cultivated and uncultivated soils from the national soil survey in a comparative analysis and revealed that cultivation-induced loss of topsoil SOC stock could be up to 14.8 ± 15.1 Mg/ha, giving a total stock loss of 2 Pg for cultivated topsoil. Song and colleagues also observed that over 60% of this loss occurred in soils of Northeast China, Northwest China, and Southwest China. Therefore, the significant loss of SOC stock from soils under intensive cultivation and increased inorganic fertilizer inputs may account for, at least partly, the low level of SOC in croplands, but also indicates a significant role of CO_2 emission to the atmosphere from China's historical land use/land cover changes.

18.2.3 CARBON STOCK DYNAMICS AND SEQUESTRATION BETWEEN 1980 AND 2006

The SOC stock of croplands, among other C reservoirs in the terrestrial ecosystem, is prone to anthropogenic activities, especially with progressive advances in agricultural technology. China's agriculture experienced consecutive increases in crop production capacity for ensuring food supply with technology transfer for improving crop management since the earlier 1980s, and consequently, the SOC followed an increasing trend in cropland soils. Thus, several studies have reported adverse impacts of China's intensive agriculture on natural resources and the environment, including increase in desertification, extension of arid areas, secondary salinization, N overuse, and water quality deterioration, as well as historical SOC loss. However, an increase in overall topsoil SOC stock has also been well documented at a county level in Jiangsu (Zhang et al. 2004b), a village level in Jiangxi (Zhang 2004), and a regional level in the North China plain and Northeastern China's black soil region (Xu et al. 2004).

The increase in topsoil SOC stock has been validated by a range of long-term agroecosystem experiments as well as by modeling. Data from long-term agroecosystem experiments show topsoil SOC enhancement over the last decades. For example, increases in topsoil SOC stocks have been reported with rational fertilization, mainly via combined use of organic and inorganic fertilizers, over 10 to 20 years either in irrigated rice paddies (Pan et al. 2003; Zhou et al. 2006; Chen et al. 2007) or in unirrigated dry croplands (Meng et al. 2005; Huang et al. 2006a; Wang et al. 2005). An experiment with tillage treatments in a purple sand-derived rice paddy also showed a SOC enhancement since 1987 under conventional rice/rape rotation farming systems from Southwest China (Huang et al. 2006b). Consequently, an overall SOC increase could be reasonably validated in China's croplands over the last two decades (between 1990 and 2010), though with a wide range of 8 to 27 Tg/year. A large uncertainty could be due to variable data size (Xie et al. 2007) and/or the different accounting procedures used in these studies (Liu et al. 2004). For example, a statistical analysis of the data from 26 sites of long-term agroecosystem experiments in China showed the annual gain of topsoil SOC stock for all of the country's croplands to be between 0.05 and 0.29 g/kg, using scenarios of balanced, compound chemical fertilizer and a combined inorganic and organic fertilization. Therefore, an overall annual SOC increase can be estimated to be in the range of 0.2 to 1.6 Pg between 1985 and 2005 (Wu and Cai 2007).

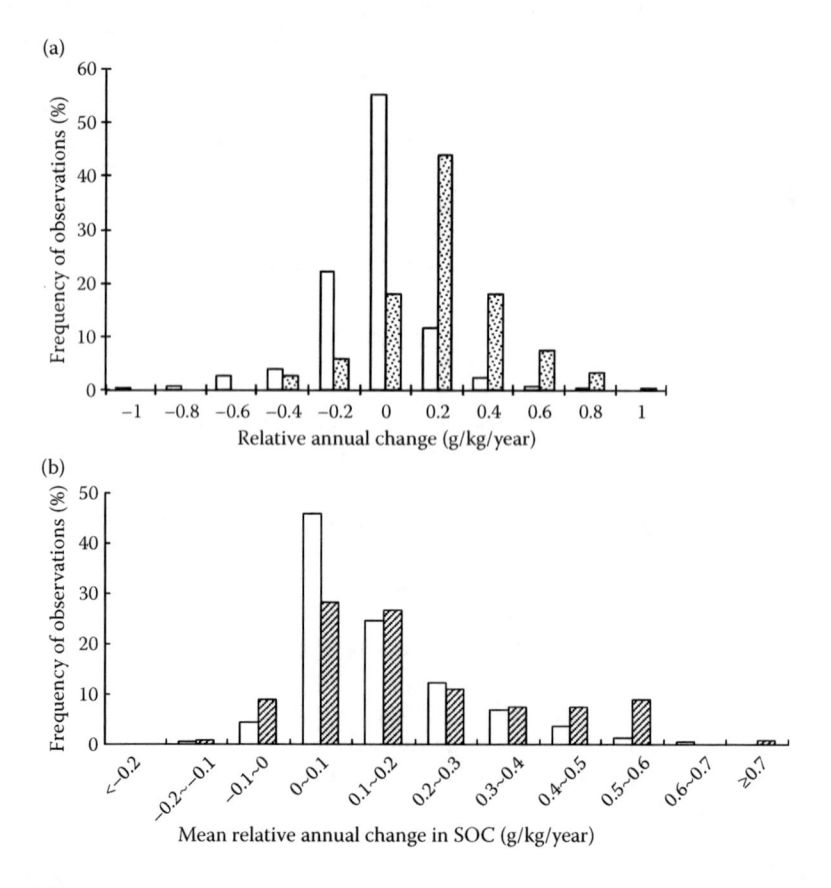

FIGURE 18.5 Topsoil SOC dynamics of China's croplands between 1982 and 2006. (a) Estimated using data from soil monitoring. (b) Estimated using data from long-term experiments. Blank, dry croplands; shaded, rice paddies.

Toward an attempt to reduce the uncertainty, Huang and Sun (2006) conducted a metaanalysis of topsoil SOC data from 200 publications, covering 60,000 single measurements over the period 1993 to 2005, and observed an increase in SOC at a frequency of over 50% and a decrease of 30%, while the other 20% remained unchanged. Huang and colleagues reported a semiquantitative estimate of increase in SOC stock of the topsoil by 0.3 to 0.4 Pg for the period from 1985 to 2005. They updated the estimates by using soil monitoring survey data available from 146 publications and calculated the SOC sequestration rate in China's croplands at 21.9 Tg per year with a total of 0.44 Pg for the period between 1980 and 2000 (Sun et al. 2010). Pan et al. (2009) developed a methodology to estimate the relative increase in SOC compared with the baseline year when soil monitoring was initiated, and they performed a metaanalysis of the mean annual rate of SOC change (Figure 18.5) over different time durations for 1099 single SOC measurements from the soil monitoring system from the published literature up until 2006. Based on this analysis, Pan and colleagues estimated a mean annual overall rate of increase in SOC in the topsoil of 0.06 ± 0.20 g/kg for unirrigated croplands and 0.11 ± 0.24 g/kg for rice paddies. Therefore, the total annual increase in SOC stock was estimated at 25.5 Tg for the period between 1985 and 2006. However, there was wide variation in the rate of SOC increase across different regions of China. Thus, Pan and colleagues concluded that the topsoil (0–20 cm) of China's croplands had sequestered as much as 0.58 ± 0.38 to 0.65 ± 0.53 Pg C over this period, which could have been a significant contribution to offsetting the nation's GHG emissions from energy consumption. Summarizing these indicates the carbon sequestration rates in the range of 20 to 25 Tg per annum for China's croplands over the two decades between 1990 and 2010.

18.3 SEQUESTRATION OF ORGANIC CARBON IN CROPLANDS: DRIVERS, POTENTIAL, AND TECHNICAL FEASIBILITY

Topsoil is the most sensitive to climate change and human interference. An analysis of SOC increase in 966 monitored cropland sites revealed a wide range of SOC sequestration rates varying with land use and management practices (Xu 2009). Tai et al. (2011) showed that garden soils had a much higher SOC storage than cropland soils due to high organic matter (OM) input and solicited management. Thus, soils of the rice paddies also have topsoil SOC storage higher than those of the dry croplands by almost 10 Mg/ha, because crop C input is higher in rice paddies than in dry croplands by 170% (Xu et al. 2009; Cheng et al. 2009). A statistical analysis of a cropland quality survey sponsored by the Ministry of Agriculture of China revealed a significant change with land use in topsoil SOC storage of a county from Jiangsu Province for the two decades between 1990 and 2010. This study also indicated a general SOC decline with the conversion of rice paddies to garden soils, dry croplands, nurseries, and forestlands in the short term but a continuous SOC increase in the rice paddies since the 1980s (Hou et al. 2007). However, in a case study of a similar area, maize cultivation in a rice paddy caused loss of topsoil SOC by 30% of the original SOC stock over 3 to 5 years, and the loss was documented by ^{13}C isotope abundance changes (Li et al. 2007b). The loss in SOC by cultivation of maize may be attributed to enhanced SOC decomposition by plowing for maize production and the loss of biophysical protection of SOC in the paddy. Land ownership and management system changes can also impact on cropland SOC dynamics. Fragmentation of land holdings, as affected by household ownership, has prevailed for a long time in China (Tan et al. 2006) and has caused adverse effects on land quality, resource efficiency, and profitability of agriculture (Tan et al. 2008; Rahman and Rahman 2009). The potential impact of the land fragmentation on SOC had received much attention elsewhere but not in China (De Costa et al. 2006). A pilot study of soil quality and household farming was conducted in a rural area of Jiangxi Province where the local people had been living on the land (Feng et al. 2011). They concluded that the topsoil SOC stock varied widely from 1.7 to 25.2 g/kg, depending on the size of the total cropland area and the land tenure status. The amount of SOC in plots of <0.1 ha was significantly lower (by 20%) than in those of >0.1 ha, and the owned croplands had a higher SOC level by almost 100% on average than those leased or contracted out. Therefore, improving land tenure and land management systems could play a key instrumental role in enhancing SOC sequestration in China's croplands.

Therefore, the SOC sequestration in cropland soils of China could be accounted for mostly by the increasing amount of straw or residues being returned, and these increased crop inputs have been affected by fertilizer application as well as enhanced humification with increased N status. However, a low rate of SOC sequestration and even loss of SOC in some cropland soils could occur in some regions of China (Figure 18.6). While SOC sequestration could be prominent in East, North, and Northwest China, large losses have happened in Northeast and also in Southwest China. The region of Southwest China had been recognized as an ecotone vulnerable to climate changes due to the shallow soils on limestone terrains and a widespread problem of

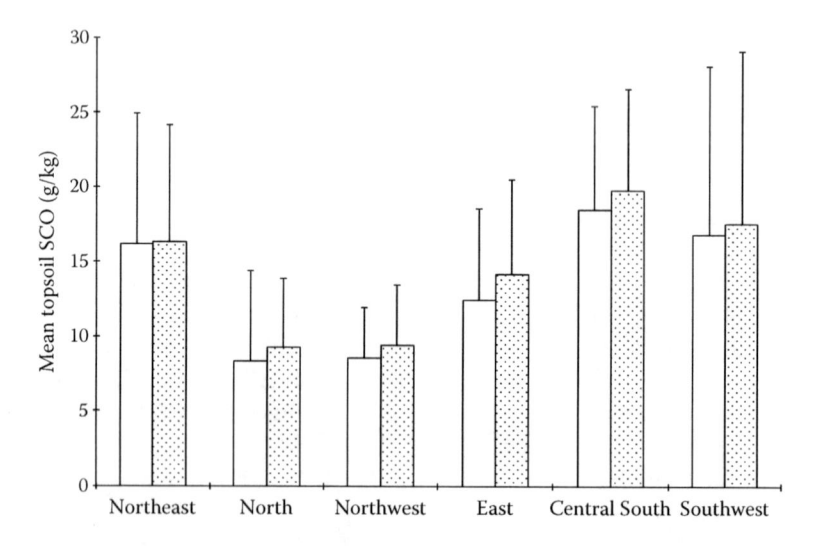

FIGURE 18.6 Mean background value and the value after t years of monitoring of croplands from major crop production regions of China. (From Xu, X.W., *Regional Distribution and Variation of SOC Storage in Agricultural Soils at Different Scales* (in Chinese), Nanjing: Ph.D. Dissertation of Nanjing Agriculture University, 2009.)

soil degradation occurring in a range of landforms. Ren et al. (2006) showed a SOC loss of over 30% in degraded limestone sloping lands compared to a preserved terrain from Guizhou, Southwest China. Climate change could be another driver for SOC loss in Northeast China, where organic soils were very common before the 1960s, either in the peatlands of Heilongjiang province or in the grass-meadow lands of the northeast great plain. Warming and drying have been common features since the 1980s in this area, which accelerated the SOC decomposition and loss of originally SOC-rich croplands. In addition, cultivation of wetlands for rice and maize/soybean production in the lower reach of Sanjiang valley of Heilongjiang could have resulted in a significant loss of SOC stock in China, which was considered as a major cause for the national total loss of 1.5 Pg C from cultivated wetlands (Zhang et al. 2008). Therefore, soil degradation and climate change are important factors in aggravating SOC sequestration in China's croplands. In other words, protection of soils against desertification and adaptation to climate change would be a win–win strategy for attaining a high SOC sequestration potential and for ensuring the ecological safety of China (Lal 2002).

SOC stock and its sequestration capacity are key parameters for evaluating the climate change mitigation potential of soils of some regions. Thus, several studies on SOC dynamics focused on the magnitude of the SOC sequestration potential. However, the technical attainability should also be addressed in soliciting practical options for SOC sequestration in mitigating climate change as well as ensuring food production in agriculture (Lal 2004).

As estimated by the FAO, the SOC sequestration capacity of global cropland soils may be as much as 20 Pg, with an expected mean annual sequestration rate of 0.9 ± 0.3 Pg in the first 25 years (Lal 2004). As first proposed by Lal (2002), China's soils could have an overall C sequestration capacity of 11 Pg until 2050, including 105 to 198 Tg/year for SOC and 7 to 138 Tg/year for soil inorganic carbon (SIC). Recently, other workers estimated the total sequestration capacity to be in the range of 2 to 2.5 Pg C up to 2050 (Wu and Cai 2007). However, Li (2004) argued for a sequestration potential of 500 Tg by reclamation of low-yielding croplands over a long time.

There are several approaches to assess an acceptable estimation of topsoil SOC sequestration of China's croplands in terms of both biophysical and technological attainability. First, an estimation of the sequestration capacity could be assessed by determining the recovery of SOC losses from cultivation, the maximum or ultimate biophysical potential for SOC sequestration in croplands. By this means, a SOC sequestration capacity of as high as 2 Pg C could be potentially reached in the long run when good management and conservation are maintained. A second approach may be by synthesizing the SOC sequestration rates against the initial SOC level, where the gap between the attainable level and the present level could be considered as the sequestration potential (Xu 2009). This approach led to an estimation of total SOC sequestration potential of China of 0.8 ± 0.2 and 1.2 ± 0.5 Pg, respectively, for rice paddies and dry croplands (Cheng et al., revision submitted). Alternatively, the sequestration potential of croplands may also be considered as the capacity to store all of C the input, which would be the case with the adoption of conservation tillage with full straw return, already known as the best means for enhancing SOC sequestration in croplands of China (Wang et al. 2009). Thus, the gap between the attainable SOC saturation level under the best management practice (BMP) scenario of conservation tillage with straw return and the present SOC level (Pan et al. 2010) would be the sequestration potential under BMP. By such an approach, Cheng et al. (revision submitted) estimated a potential of 0.91 and 1.01 Pg SOC, respectively, for rice paddies and dry croplands of China. The estimated overall potential of 2 Pg is also similar to that by Lal (2002). However, it is also argued that a technically attainable SOC sequestration potential may be much smaller than these estimates mainly via biophysical saturation methods (Cheng et al., revision submitted). Constraints of technology adoption and potential area for BMPs will still exist; for example, conservation tillage with crop residue return will not be adopted in all of China's croplands. Therefore, an assessment of technically attainable SOC sequestration potential would be more appropriate to identifying the useful approaches for mitigation of climate change by improving China's agriculture.

18.4 CO-BENEFITS OF SOC SEQUESTRATION FOR CROP PRODUCTION AND ECOSYSTEM HEALTH IN CROPLANDS

Several studies conducted on field soils have indicated that SOC sequestration may have a number of co-benefits for crop production and agroecosystem functioning. Those conducted at the site level or plot level have already demonstrated a strong benefit of SOC sequestration for crop productivity and sustainability. The high SOC sequestration rate found in two sites of long-term agroecosystem experiments from South China (Zhou et al. 2006, 2009; Pan et al. 2003) could be partly attributed to enhanced chemical stabilization through binding with iron oxyhydrates (Zhou et al. 2009; Song et al. 2011). In both of these sites, much lower yield variability over the years was observed in plots with high SOC contents under combined organic/inorganic fertilization (Pan and Zhao 2005). The high productivity is more evidence of a major benefit of SOC sequestration for crop production, suggesting that SOC sequestration can substantially enhance food production sustainability of agriculture in croplands that are low in SOC stock. Data from long-term experiments on rice paddies from South China (Figure 18.7) and from cropland productivity monitoring sites (Figure 18.8) revealed that higher SOC sequestration coincided with high rice yield and is also indicated by a high N use efficiency under combined organic/inorganic fertilization (Pan et al. 2009). While there were few studies on SOM effect on crop growth, field experiments with biochar soil amendment to enhance the SOM pool support a stimulating effect of biochar added SOM on root system development of dry crops, including maize, in soils already depleted of SOM in arid regions (Figure 18.9). This would be an indispensible reason for SOM's role in crop production.

There has been increasing evidence that SOC sequestration could lead to a healthy soil microbial community. With SOC sequestration, enhanced microbial abundance and gene diversity are observed under combined organic/inorganic fertilization in rice paddies (Zhang et al. 2004a). Jin et al. (2012) demonstrated that soil microbial community and diversity and soil enzyme activity were greatly enhanced with an increase in SOC storage in rice paddies converted from sandy wetlands in the Hubei Province. This could be the intrinsic mechanism whereby SOC accumulation enhances soil productivity through a buildup of an active and healthy soil microbial community in rice paddies.

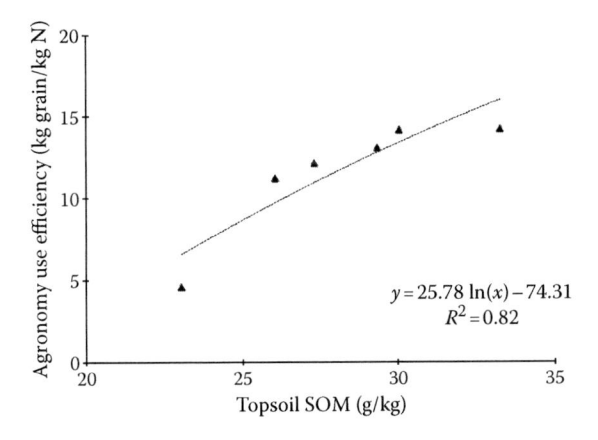

FIGURE 18.7 Correlation of mean agronomy use efficiency of N with SOM under long-term fertilization trials in a rice paddy in Jiangxi, China.

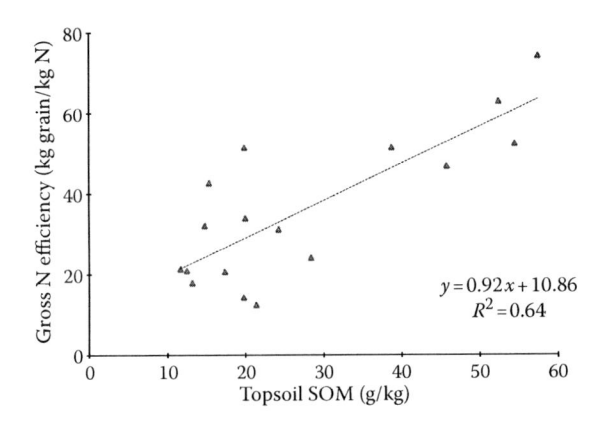

$y = 25.78 \ln(x) - 74.31$
$R^2 = 0.82$

FIGURE 18.8 Observed mean gross output of N applied as a function of topsoil OM in rice paddies by the monitoring system of the Ministry of Agriculture, data of 2004.

$y = 0.92x + 10.86$
$R^2 = 0.64$

The health of an agroecosystem could also be addressed by moderating the efflux of GHG emission, especially in terms of C intensity of crop production, which has been studied in detail by Zheng et al. (2011). For example, increase in N efficiency could help to decrease fertilizer-induced C emission per unit of rice production, which, in turn, results in a higher net C sink (Li et al. 2009a,b). And this has also been true for a rice paddy in a site at Jiangxi, where a higher rate of SOC sequestration had been observed (Li et al. 2009b). In a case study by Zheng et al. (2006), the metabolic quotient of soil microbial community and the respiratory quotient of SOC were both reduced, while soil CO_2 evolution may be higher under SOC accumulation under good fertilization. In many field studies, the correlation coefficient of soil respiration with SOC content in rice soils without N being limiting was generally low. Instead, soil respiratory activity and respired CO_2 flux were often shown to

FIGURE 18.9 Well-developed root system (left, dark colored, amended with biochar) in OM-rich soil in contrast to OM-depleted soil (right, bright color) of maize in an Ochrept from Shanxi, China. (Courtesy of Genxing Pan, May 2012.)

be lower in plots rich in SOC under a well-designed fertilizer scheme when compared with relatively SOC-poor plots under long-term agroecosystem experiments, as was reported by Williams et al. (2007). Decreases in soil respiratory activity with increases in SOC have been observed in dry croplands under organic fertilization, unlike those under nonorganic amendments (Meng et al. 2005; Yin and Cai 2006). An integrated field study using a number of long-term experiments from rice paddies in South China has demonstrated an increasing dominance of the fungal over the bacterial community with increasing SOC accumulation, which in turn supports a reduction both in soil respiratory quotient and in microbial metabolic quotient under good agromanagement (Liu et al. 2011). A laboratory incubation study of methane production from rice paddies with different SOC contents also showed a reduction in C intensity from methane emission in rice paddies containing high SOC under combined organic/inorganic fertilization, and this is further proven by another study with enhanced diversity of methanotrophs, which is responsible for methane exhaustion in rice paddies (Zheng et al. 2008). A new insight into SOC sequestration and GHG emission was that a total global warming potential calculated from all the GHG fluxes in a plot continuously receiving compound fertilizers seemed smaller than from the one receiving only the chemical fertilizer (Li et al. 2009a,b; Liu et al. 2009). Some studies have shown that the net C sink can be 1.5 to 3 times more under a combined organic/inorganic fertilizer regime than that under chemical fertilization in rice paddies from Jiangxi and Jiangsu, China (Li et al. 2009a,b). A similar study indicated that there was a higher net C sink (by 1.1- to 1.7-fold) under organic amendments compared to under chemical fertilization only (Peng et al. 2009).

Meanwhile, there have also been other studies with long-term experiments under well-managed practices that indicated that SOC accumulation would exert additional positive effects on ecosystem health. In coincidence with the findings of microbial abundance and diversity, Xiang et al. (2006) documented enhanced soil fauna community and diversity, especially those of soil earthworms in a long-term experiment site from the Tai Lake region. Conducting the same experiment used by Xiang et al.

(2006), Feng et al. (2006) and Li et al. (2007) demonstrated also enhanced diversity of weeds in the field and the seed bank in soil with SOC accumulation under the good fertilization schemes. Using soil samples from the same experiment, a study by Han et al. (2009) indicated decreased content but increased degradation of PAHs with SOC accumulation under different fertilization practices. Thus, SOC sequestration in croplands, particularly in rice paddies, would offer multiple win–win effects for crop productivity, mitigation of GHG emission, as well as ecosystem health in agriculture, thus ensuring food security and sustainability of agriculture for China (Pan and Zhao 2005).

We have developed a hypothesis as a concept model for characterizing the mutual co-benefits of SOM's role in enhancing productivity and ecosystem functioning (Figure 18.10). The input of OM to soil may at first act as single particulate material, which could have chemical reactivity during the first time. With time, OM gets into soil organomineral complexes through physical and chemical binding, which could further form microaggregates within which soil microbes could colonize. The microaggregates could then be transformed into macroaggregates, which offer a wide range of microhabitats for different species and communities of soil organisms. In this way, a functional entity is developed with mutual interaction of mineral, chemical, biological, and ecological forces. Therein, the soil processes and functioning could be magnified to enhance soil capacity for nutrient, moisture, and biotic conservation. Therefore, the soil activity with accumulation of SOM in topsoil would be a linear response not to the content of SOM but to the interaction of SOM with soil attributes. In such a framework, soil crop productivity and ecosystem functioning can be harmonized for ensuring sustainability.

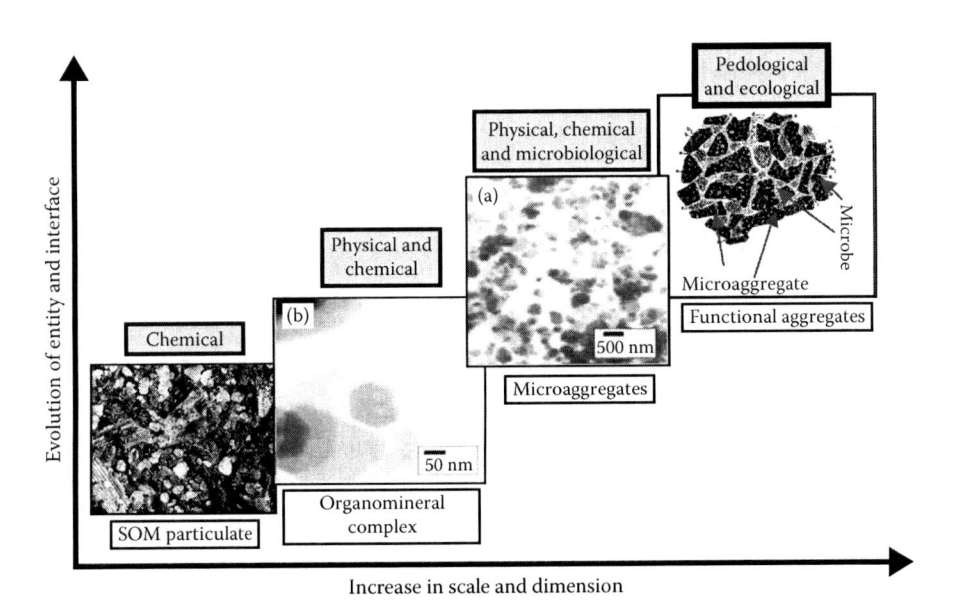

FIGURE 18.10 Concept model of the role of SOM in enhancing ecosystem functioning.

In conclusion, SOC sequestration is a matter not of an increase in soil's OM content but of the increase in soil's capacity to act in an integrated and interactive manner for cropland productivity and ecosystem health. Therefore, SOC sequestration in China's croplands should be given much more attention than ever before when food safety and ecological sustainability are at stake.

18.5 CONCLUSION

There have been many studies on estimating the SOC pools, storage, and the changes due to land use and land use changes of China's soils. It has become clear that China's soils have been relatively depleted of OM and thus have a great potential to sequester carbon for climate change mitigation in agriculture. However, it is still poorly understood to what extent the biophysical SOC sequestration potential could be realized or technically attainable with BMPs even under rational incentives within the climate change policy framework of the state. There have been increasing evidences that SOC sequestration benefits mitigation and also crop productivity and health, as well as ecosystem functioning also though the coherent processes and mechanisms are still unclear. Finally, there has been an urgent need to enhance and stabilize SOM in China's croplands for ensuring food production and sustainability of farming systems of the country.

ACKNOWLEDGMENT

This work was partly funded by the Natural Science Foundation of China under grants 40830528 and 40270010092.

ABBREVIATIONS

BMP: best management practice
SOC: soil organic carbon
SOM: soil organic matter

REFERENCES

Brown, M.T. and S. Ulgiati. 1999. Energy evaluation of biosphere and natural capita. *Ambio* 28(6): 486–491.
Chen, A. L., K.R. Wang, X.L. Xie et al. 2007. Responses of microbial biomass P to the changes of organic C and P in paddy soils under different fertilization systems (in Chinese). *Chin J Appl Ecol* 18: 2733–2738.
Cheng, K., J.F. Zheng, D. Nayak, P. Smith, and G. Pan. (in press). Re-evaluating biophysical and technologically attainable potential of carbon sequestration by China's croplands. *Soil Use Manag.*
Cheng, K., G.X., Pan, Y.G. Tian et al. 2009. Changes in topsoil organic carbon of China's cropland evidenced from the national soil monitoring network (in Chinese). *J Agro-Environ Sci* 28: 2476–2481.
Dawea, D., A. Dobermannb, and J.K. Ladhaa. 2003. Do organic amendments improve yield trends and profitability in intensive rice systems? *Field Crops Res* 83: 191–213.

De Costa, W.A.J.M. and U.R. Sangakkara. 2006. Agronomic regeneration of soil fertility in tropical Asian smallholder uplands for sustainable food production. *J Agric Sci Cambridge* 144: 111–133.

European Commission. 2002. Communication of 16 April 2002 from the Commission to the Council, the European Parliament, the Economic and Social Committee and the Committee of the Regions—Towards a Thematic Strategy for Soil Protection [COM (2002) 179 final, 35 pp. http://europa.eu.int/scadplus/printversion/en//lvb/l28122.htm].

Fang, J.Y., Z.D. Guo, and S.L. Piao. 2007. Terrestrial vegetation carbon sinks in China, 1981–2000. *Sci China Ser D Earth Sci* 37: 804–812.

Feng, S.Y., S.H. Tan, A.F. Zhang et al. 2011. Effect of household land management on cropland topsoil organic carbon storage at plot scale in a red earth soil area of South China. *J Agr Sci.* doi:10.1017/S0021859611000323.

Feng, W., G.X. Pan, S. Qiang et al. 2006. Influence of long-term fertilization on soil seed bank diversity of a paddy soil under rice/rape rotation (in Chinese). *Biodiversity Sci* 14: 461–469.

Food and Agriculture Organization (FAO). 2001. *Soil carbon sequestration for improved land management.* World Soil Resources Reports, No. 96, ISSN 0532-0488. Rome, Italy: FAO.

Han, X.J., G.X. Pan, and L.Q. Li. 2009. Effects of the content of organic matter on the degradation of PAHs: A case of a paddy soil under a long-term fertilization trial from the Tai Lake Region, China (in Chinese). *J Agro-Environ Sci* 28: 2533–2539.

Hou, P.C., X.D. Xu, and G.X. Pan. 2007. Influence of land use change on topsoil organic carbon stock—a case study of Wujiang Municipality (in Chinese). *J Nanjing Agr Univ* 30: 68–72.

Huang, B., J.G. Wang, H.Y. Jin et al. 2006a. Effects of long-term application fertilizer on carbon storage in a calcareous meadow soil (in Chinese). *J Agro-Environ Sci* 25: 161–164.

Huang, X.X., M. Gao, C.F. Wei et al. 2006b. Tillage effect on organic carbon in a purple paddy soil. *Pedosphere* 16: 660–667.

Huang, Y. and W.J. Sun. 2006. Changes in topsoil organic carbon of croplands in China over the last two decades. *Chin Sci Bull* 51: 1785–1803.

Intergovernmental Panel on Climate Change. 1996. Climate Change 1995. *Impact Adaptations and Mitigation of Climate Change Scientific Technical Analysis [C].* Working Group I, Cambridge: Cambridge University Press.

Lal, R. 1999. Soil management and restoration for C sequestration to mitigate the accelerated greenhouse effect. *Progress Environ Sci* 1: 307–326.

Lal, R. 2002. Soil C sequestration in China through agricultural intensification and restoration of degraded and desertified soil. *Land Degrad Dev* 13: 469–478.

Lal, R. 2004. Soil C sequestration impacts on global climatic change and food security. *Science* 304: 1623–1627.

Li, C.S. 2000. Loss of soil carbon threatens Chinese agriculture: A comparison on agroecosystem carbon pool in China and the U.S. (in Chinese). *Quater Sci* 20: 345–350.

Li, J.J., G.X. Pan, X.H. Zhang et al. 2009a. An evaluation of net carbon sink effect and cost/benefits of a rice-rape rotation ecosystem under long-term fertilization from Tai Lake region of China (in Chinese). *Chin J Appl Ecol* 20: 1664–1670.

Li, J.J., G.X. Pan, L.Q. Li et al. 2009b. Estimation of net carbon balance and benefits of rice-rice cropping farm of a red earth paddy under long term fertilization experiment from Jiangxi, China (in Chinese). *J Agro-Environ Sci* 28: 2520–2525.

Li, L., H.A. Xiao, and J. Wu. 2007a. Decomposition and transform of organic substrates in upland and paddy soils red earth region (in Chinese). *Acta Pedol Sin* 44: 669–674.

Li, Z.P. 2004. Density of soil organic carbon pool and its variation in hilly soil region (in Chinese). *Soils* 36: 292–297.

Li, Z.P., G.X., Pan, and X.H. Zhang. 2007b. Top soil organic carbon pool and 13C natural abundance changes from a paddy after 3 years corn cultivation (in Chinese). *Acta Pedol Sin* 44: 244–251.

Lin, F., D.Y. Li, G.X. Pan et al. 2008. Organic carbon density of soil of wetland and its change after cultivation along the Yangtze River in Anhui Province, China (in Chinese). *Wetland Sci* 6(2): 192–197.

Lindert, P.H., I. Lu, and W. Wu. 1996. Trends in the soil chemistry of South China since the 1930s. *Soil Sci* 161: 329–342.

Liu, D.W., X.Y. Liu, Y.Z. Liu et al. 2011. SOC accumulation in paddy soils under long-term agro-ecosystem experiments from South China. VI. Changes in microbial community structure and respiratory activity. *Biogeosci Discuss* 8: 1–26.

Liu, J.Y., S. Wang, J.M. Chen et al. 2004. Storages of soil organic carbon and nitrogen and land use changes in China 1990–2000. *Acta Geogr Sin* 59: 483–496.

Liu, X.Y., G.X. Pan, L.Q. Li et al. 2009. CO_2 emission under longterm different fertilization during rape growth season of a paddy soil from Tai Lake region, China. *J Agro-Environ Sci* 28(12): 2506–2511.

Loveland, P. and J. Webb. 2003. Is there a critical level of organic matter in the agricultural soils of temperate regions: a review. *Soil Tillage Res* 70(1): 1–18.

Manlay, R.J., C. Feller, and M.J. Swift. 2007. Historical evolution of soil organic matter concepts and their relationships with the fertility and sustainability of cropping systems. *Agric Ecosyst Environ* 119: 117–233.

Meng, L., Z.C. Cai, and W.X. Ding. 2005. Carbon contents in soils and crops as affected by long-term fertilization (in Chinese). *Acta Pedol Sin* 42: 769–776.

Ministry of Agriculture and State Commission of Development and Reform, China. 2004. *A National Planning of Conservation Tillage Project*. [AO/BO]. Available at www.gov.cn/gzdt/2009-08/28/content140367.html.

Pan, G. 2009. Stock, dynamics of soil organic carbon of china and the role in climate change mitigation. *Adv Clim Change Res* 5(Suppl.): 11–18.

Pan, G.X. and Q.G. Zhao. 2005. Study on evolution of organic carbon stock in agricultural soil of China: facing the challenge of global change and food security (in Chinese). *Adv Earth Sci* 20(4): 384–392.

Pan, G.X., L.Q. Li, L.S. Wu et al. 2003. Storage and sequestration potential of topsoil organic carbon in China's paddy soils. *Glob Change Biol* 10: 79–92.

Pan, G.X., X.W. Xu, P. Smith et al. 2010. An increase in topsoil SOC stock of China's croplands between 1985 and 2006 revealed by soil monitoring. *Agric Ecosyst Environ* 136: 133–138.

Pan, G., P. Smith, and W. Pan. 2009. The role of soil organic matter in maintaining the productivity and yield stability of cereals in china. *Agric Ecosyst Environ* 129: 344–348.

Peng, H., X.H. Ji, B. Liu et al. 2009. Evaluation of net carbon sink effect and economic benefit in double rice field ecosystem under long-term fertilization (in Chinese). *J Agro-Environ Sci* 28: 2526–2532.

Rahman, S. and M. Rahman. 2009. Impact of land fragmentation and resource ownership on productivity and efficiency: the case of rice producers in Bangladesh. *Land Use Policy* 26: 95–103.

Song, G.H., L.Q. Li, and G. Pan. 2005. Topsoil organic carbon storage of China and its loss by cultivation. *Biogeochem* 74: 47–62.

Song, X.Y. 2011. *Sequestration of Crop Carbon in Soil: a Laboratory Incubation Study Using Four Different Types of Soils* (in Chinese). Nanjing: Ph.D. dissertation of Nanjing Agriculture University.

Sun, W.J., Y. Huang, S.T. Chen et al. 2007. Dependence of wheat and rice respiration on tissue nitrogen and the corresponding net carbon fixation efficiency under different rates of nitrogen application. *Adv Atmos Sci* 24: 55–64.

Sun, W.J., Y. Huang, W. Zhang et al. 2010. Carbon sequestration and its potential in agricultural soils of China. *Glob Biogeochem Cycles* 24: 1302–1307.

Tai, J.C., Z.J. Jin, L.Q. Ciu et al. 2011. Changes in soil organic carbon fractions with land uses in soils reclaimed from wetland of Jianghan plain, Hubei province (in Chinese). *J Soil Water Conserv* 25(6): 124–128.

Tan, S., N. Heerink, G. Kruseman, and F. Qu. 2006. Land fragmentation and its driving forces in China. *Land Use Policy* 23: 272–285.

Tan, S., N. Heerink, G. Kruseman et al. 2008. Do fragmented landholdings have higher production costs? Evidence from rice farmers in Northeastern Jiangxi province, P.R. China. *Econ Rev* 19: 347–358.

Tiessen, H., E. Cuevas, and P. Chacon. 1994. The role of soil organic matter in sustaining soil fertility. *Nature* 371: 783–785.

Wang, B.R., M.G. Xu, and S.L. Wen. 2005. Effect of long time fertilizers application on soil characteristics and crop growth in red soil upland (in Chinese). *J Soil Water Conserv* 19: 97–100.

Wang, C.J., G.X. Pan, and Y.G. Tian. 2009. Characteristics of cropland topsoil organic carbon dynamics under different conservation tillage treatments based on long-term agroecosystem experiments across mainland China (in Chinese). *J Agro-Environ Sci* 28: 2464–2475.

Williams, M.A., D.D. Myrolda, and P.J. Biolltomley. 2007. Carbon flow from ^{13}C-labelled clover and ryegrass residues into a residue-associated microbial community under field conditions. *Soil Biol Biochem* 39: 819–822.

Wu, H.B., Z.T. Guo, and C.H. Peng. 2003. Land use induced changes of organic carbon storage in soils of China. *Glob Change Biol* 9: 305–315.

Wu, Y.Z. and Z.C. Cai. 2007. Estimation of the change of topsoil organic carbon of croplands in China based on long-term experimental data (in Chinese). *Ecol Environ* 16(6): 1768–1774.

Xiang, C.G., P.J. Zhang, G.X. Pan et al. 2006. Changes in diversity, protein content and amino acid composition of earthworms from a paddy soil under long-term different fertilizations in the Tai Lake Region, China (in Chinese). *Acta Ecol Sin* 26: 1667–1674.

Xie, Z.B., J.G. Zhu, G. Liu et al. 2007. Soil organic carbon stocks in China and changes from 1980s to 2000s. *Glob Change Biol* 13: 1989–2007.

Xu, X.W. 2009. *Regional Distribution and Variation of SOC Storage in Agricultural Soils at Different Scales* (in Chinese). Nanjing: Ph.D. Dissertation of Nanjing Agriculture University.

Xu, X.W., G.X. Pan, Y.L. Wang et al. 2009. Research of changing characteristics and control factors of farmland topsoil organic carbon in China (in Chinese). *Geogr Res* 28: 601–612.

Xu, Y., F.R. Zhang, J.K. Wang et al. 2004. Temporal changes of soil organic matter in Ustic Cambisols and Udic Isohumosols of China in recent twenty years (in Chinese). *Chin J Soil Sci* 35: 102–105.

Yin, Y.F. and Z.C. Cai. 2006. Effect of fertilization on equilibrium levels of organic carbon and capacities of soil stabilizing organic carbon for Fluvo-aquic soil. *Soils* 38: 745–749.

Zeng, J., T. Guo, and X.G. Bao. 2008. Effect of soil organic carbon and soil inorganic carbon under long-term fertilization (in Chinese). *China Soils Fertil* 2: 11–14.

Zhang, P.J., L.Q. Li, G.X. Pan et al. 2004a. Influence of long-term fertilizer management on topsoil microbial biomass and genetic diversity of a paddy soil from the Tai Lake region, China (in Chinese). *Acta Ecol Sin* 24: 2819–2824.

Zhang, Q. 2004. Change in organic carbon in paddy soils over the last two decades—a study at scales of county and village level (in Chinese). Nanjing: MS thesis, Nanjing Agricultural University.

Zhang, Q., L.Q. Li, G.X. Pan et al. 2004b. Dynamics of topsoil organic carbon of paddy soils at Yixing over the last 20 years and the driving factors (in Chinese). *Quarter Sci* 24: 236–242.

Zhang, X.H., D.Y. Li, G.X. Pan et al. 2008. Conservation of wetland soil C stock and climate change of China (in Chinese). *Adv Clim Change* 4: 202–208.

Zheng, J.F. 2008. SOC mineralization and CO2, CH4 production under long-term different fertilizations from two typical paddy soils of South China [D]. Nanjing, China: Ph.D. Dissertation of Nanjing Agriculture University, (2008031538).

Zheng, J.F., X.H. Zhang, G.X. Pan et al. 2006. Diurnal variation of soil basal respiration and CO2 emission from a typical paddy soil after rice harvest under long-term different fertilizations (in Chinese). *Plant Nutr Fertil Sci* 12: 485–494.

Zheng, J.F., G.X. Pan, L.Q. Li et al. 2008. Effect of long-term different fertilization on methane oxidation potential and diversity of methanotrophs of paddy soil (in Chinese). *Acta Ecol Sinica* 28: 4864–4872.

Zheng, J.F., G.X. Pan, and X.X. Wu. (in press). CO_2–C emission flux and stability of organic carbon of soil in beach in Sengjin Lake, Anhui province, China in dry season (in Chinese). *Wetland Sci* 9.

Zhou, P., X.H. Zhang, and G.X. Pan. 2006. Effect of long-term fertilization on content of total and particulate organic carbon and their depth distribution of a paddy soil: an example of huangnitu from the Tai Lake region, China (in Chinese). *Plant Nutr Fertil Sci* 12: 765–771.

Zhou, P., G.H. Song, and G.X. Pan. 2009. SOC enhancement in three major types of paddy soils in a long-term agro-ecosystem experiment in south china III. Structural variation of particulate organic matter of two paddy soils (in Chinese). *Acta Pedol Sin* 4: 263–273.

19 Soil Management for Sustaining Ecosystem Services

Rattan Lal and Bobby A. Stewart

CONTENTS

19.1 Introduction 521
19.2 Processes, Factors, and Causes of Soil Degradation 522
19.3 Soil Erosion and Other Degradation Processes 523
19.4 Soil Degradation and ESs 524
19.5 Technological Options 526
    19.5.1 Management of Cropland Soils 526
    19.5.2 Pastureland Soils 528
    19.5.3 Management of Soils under Forestland Use 529
    19.5.4 SQ and Water Security 529
19.6 Soil Quality Index and Sustainable Management 531
19.7 Conclusions 531
Abbreviations 533
References 533

19.1 INTRODUCTION

Soil degradation of cropland (Miao et al. 2011), pastureland (Muller et al. 2004; Martinez and Zinck 2004), forestland (Kasel and Bennett 2007), and the land under villages in tropical environments including footpaths (De Meyer et al. 2011) is a serious global issue. The problem is also exacerbated by changing and uncertain climate with an increasing frequency of extreme events. Climate change affects soil moisture and thermal regimes, with strong impacts on soil productivity potential on a global scale (Mueller et al. 2010). With a projected increase in food demand of 70% between 2005 and 2050 (Lele 2010), sustainable management of soil resources following the basic principles discussed in this volume is crucial. As the critical review of the Rio+20 declaration clearly states, there is a striking lack of awareness among policy makers regarding the importance of soil resource management. World soil resources support numerous ecosystem services (ESs) or a multitude of benefits that humans derive from the upper layer of the earth's crust (e.g., food production, freshwater supply, decomposition of waste). Thus, sustainable management of world soils is essential to the wellbeing of humans and other habitants of the planet Earth. Soil management involves the manipulation of physical, chemical, and biological properties, such that soil processes underpinning ESs are sustained and risks of degradation of soil quality (SQ) are minimized. The goal of soil management is to optimize and sustain the positives and minimize the risks of negatives (e.g., erosion, salinization, compaction). Thus, the objective of this chapter is to describe the rationale for science-based management of soils and outline the basis of priorities for research, development, and outreach, and the need to enhance awareness about the importance of sustainable management of the world's soil resources.

19.2 PROCESSES, FACTORS, AND CAUSES OF SOIL DEGRADATION

Processes refer to mechanisms or biological, physical, chemical, ecological, and geological reactions and transformations underpinning changes in SQ, ecosystem functions, and services. Principal soil degradation processes include decline in soil structure, crusting, compaction, erosion, depletion of nutrients and soil organic carbon (SOC), salinization, acidification, elemental imbalance, pollution and contamination, anaerobiosis, and drought or the hydrologic imbalance (Figure 19.1). Factors

Processes
- Organic carbon and nutrients
- Decline of soil structure
- Erosion
- Salinization
- Depletion of soil
- Acidification
- Compaction
- Crusting
- Anaerobiosis
- Pollution
- Contamination
- Drought

Factors
- Parent material
- Terrain and slope climate and climate change
- Climax vegetation
- Drainage
- Landscape position
- Soil type

Soil degradation

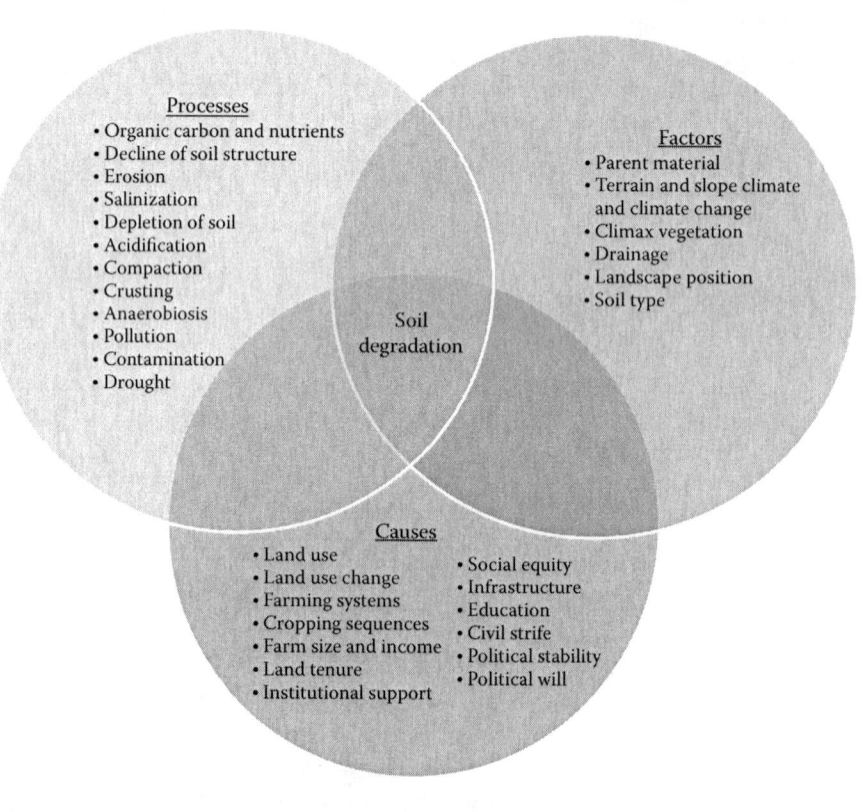

Causes
- Land use
- Land use change
- Farming systems
- Cropping sequences
- Farm size and income
- Land tenure
- Institutional support
- Social equity
- Infrastructure
- Education
- Civil strife
- Political stability
- Political will

FIGURE 19.1 Interactive effects of processes, factors, and causes of soil degradation.

of soil degradation are biophysical environments such as terrain characteristics; slope (gradient, length, shape, aspect); landscape position (summit, shoulder slope, foot slope); parent material; soil type; drainage (surface and profile); climax vegetation; climate (mean annual rainfall/precipitation, mean annual temperature, evapotranspiration); and so forth. Causes of soil degradation may be natural (terrain, parent material, soil type, groundwater, climate) and anthropogenic. Important among anthropogenic causes are land use change, deforestation, biomass burning, drainage of wetlands/peat soils, farming/cropping systems, soil and crop management, soil tillage, nutrient management, and the human dimensions. The latter consists of farm size, farm income, education, social/ethnic/gender equity, land tenure, access to market and credit, infrastructure, civil strife, and political stability (Figure 19.1). The extent and severity of soil degradation also depends on the interaction among processes, factors, and causes.

19.3 SOIL EROSION AND OTHER DEGRADATION PROCESSES

Accelerated soil erosion and the associated land degradation are among the major environmental and economic issues, especially in subhumid, semiarid, and arid environments. It aggravates the challenge of achieving food security, alleviating poverty, and improving the environment. Accelerated erosion is exacerbated by the tragedy of the common lands (Kabubo-Mariara 2006). About 75 billons tons of soil may be eroded annually from the world's terrestrial ecosystems, with a rate of soil loss of 13 to 40 Mg/ha/year from some agroecosystems (Pimentel and Kounang 1998). Because of its preferential removal by erosion, SOC concentration is a reliable means of monitoring soil degradation by accelerated erosion (Rajan et al. 2011). The soil erosion hazard is accentuated by increased/excessive and inappropriate mechanization, simplification of crop rotations, reduction in landscape diversity, and loss of noncrop features. Among several methods of assessment of soil erosion at different scales (Lal 1994, 2001), ^{137}Cs is a widely used technique (Junge et al. 2010; Alewell et al. 2009). Using the ^{137}Cs technique, Junge et al. (2010) estimated rates of 14.4 Mg/ha/year for gross and 13.3 Mg/ha/year for net erosion on cultivated farmland in Nigeria. A study conducted in the Lake Victoria Basin in Uganda indicated that household compounds, unpaved roads, and footpaths are a major source of sediments into the lake. De Meyer et al. (2011) reported that the average soil loss rate in compounds could be as much as 107 Mg/ha/year per unit compound and 207 Mg/ha/year per unit landing site. The mean soil loss rate was estimated at 34 Mg/ha/year from footpaths and 35 Mg/ha/year on unpaved roads. In the Caribbean island of St. Lucia, Cox et al. (2006) reported soil loss of 10 Mg/ha/year from agricultural land and 0.5 Mg/ha/year from forestland.

Accelerated soil erosion remains to be a serious problem in northern Ethiopia (Tesfahunegn et al. 2012) and is a serious constraint to achieving food scarcity. In comparison with the traditional management for the Mai-Negus catchment in northern Ethiopia, Tesfahunegn and colleagues reported that conservation measures decreased water runoff by 70% (168 vs. 50 mm), erosion by 77% (42,000 vs. 9215 Mg/year), total N loss by 72% (22,400 vs. 6284 kg/year), and total P loss by 75% (1330 vs. 341 kg/year). In the central rift valley of Ethiopia, which is characterized by arid and semiarid climate, soil erosion by water is a serious problem during the rainy season. Meshesha et al. (2012) analyzed soil erosion rates from 1973 to 2006. Because of the decline in vegetation cover and deforestation, soil erosion rates were 31 Mg/ha/year in 1973, 38 Mg/ha/year in 1985, and 56 Mg/ha/year in 2006. Restoring degraded lands by installing exclosures and planted vegetation, and constructing stone bunds in cropland reduced erosion by 12.6% and 63.8%, respectively. Identifying hot spots (erosion rate >20 Mg/ha/year) by integrated management is a conservation-effective approach (Meshesha et al. 2012).

Projected climate change may lead to a more vigorous hydrological cycle and to an increase in climatic erosivity of wind, rain, and runoff. An increase in frequency of extreme events caused by the climate change is another factor. A simulation study conducted in the Midwestern region of the United States indicated that relative to 1990–1999 as the baseline, erosion-related risks will increase by +10% to +310% for runoff and +33% to +277% for soil loss (O'Neal et al. 2005). Wherever the rainfall increases, erosion and runoff will increase drastically in the order of 1:1.7 (Nearing et al. 2004). Reduction in vegetative cover in case of a decrease in rainfall would also exacerbate soil erosion hazard in an uncertain and changing climate.

In addition to the adverse impacts on agronomic productivity and food security, soil degradation by erosion also reduces biodiversity. Gilroy et al. (2008) linked soil penetrability and SOC concentration (compaction) to the abundance of yellow wagtails (*Motacilla flava*) in arable fields. This link was the strongest during the latter part of the breeding season and indicated a significant relationship between soil degradation and population decline. Soil erosion also affects the C dynamics and the global C cycle (Lal 2005). It can be a major source of CO_2 and CH_4, depending on the site-specific pathways of C transported by erosion (Van Oost et al. 2007; Lal 2003).

There also exists a strong link between erosion and desertification. Removal of vegetative cover of trees by deforestation can affect water and energy balance. Further, runoff harvesting practices become less effective with an increase in a site's aridity (Safriel et al. 2011). Thus, effective erosion control through appropriate land use and judicious soil management is essential to sustaining ESs.

19.4 SOIL DEGRADATION AND ESS

Principal global challenges facing humanity are food insecurity, abrupt climate change with a high frequency of extreme events (e.g., drought in the United States in 2012), scarcity of freshwater resources, and high energy demands for a rapidly expanding world population. Soil degradation affects ESs both directly and indirectly (Figure 19.2). Direct effects include those on provisional, regulatory, and aesthetical ESs. Indirect changes include adverse effects on water security, food security, energy security, and soil security (Figure 19.2).

Food insecurity remains to be a major concern, especially in developing countries of sub-Saharan Africa (SSA) and South Asia (SA). China is also facing a great challenge of increasing agronomic production. In China, the annual cereal production must be increased to 600 Tg (Tg = 1 million Mg) by 2030 on a shrinking cropland area and limited water resources (Miao et al. 2011). In SSA, 28% to 30% of the

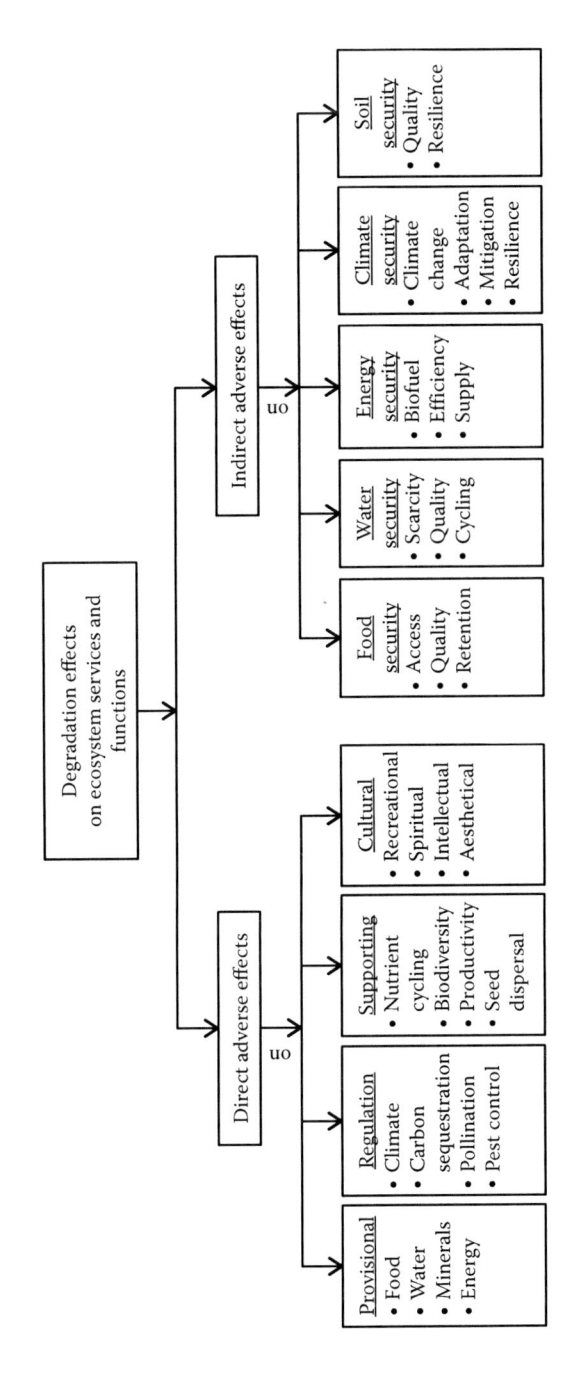

Soil Management for Sustaining Ecosystem Services 525

FIGURE 19.2 Soil degradation and ecosystem services.

population is vulnerable to food insecurity (Bationo et al. 2007). Rice import into SSA increased from 0.5 Tg of milled rice in 1961 to 6.0 Tg in 2003 (Balasubramanian et al. 2007). In Bangladesh, and elsewhere in SA, water management in rice is linked with accumulation of arsenic (As) in both soil and rice under wetland conditions, with severe health hazard (Khan et al. 2010). Productivity of the rice–wheat system, the basis of the Green Revolution of the 1970s in SA, is stagnating because of soil degradation and excessive withdrawal of groundwater. The tube well irrigation in the Indo-Gangetic Plains is closely linked with the energy security. Therefore, production of biofuels (both the first and second generation) can increase competition for grains, biomass, land area, water, and nutrients. The payback period for the so-called carbon debt created by conversion of primary forest (or drainage of peatland) may range from 100 to 1000 years depending on the specific ecosystem involved in the land use change event (Kim et al. 2009; Fargione et al. 2008). Energy consumption in farmland operations also involves the use of fertilizers (especially of N), herbicides, and pesticides (Lal 2004a). About 300 Tg of CO_2 is released annually from the manufacture of 100 Tg of fertilizer N by the Haber–Bosch processes (Jensen et al. 2012). Therefore, integration of fertilizer trees within cropland can supply as much as 60 kg N/ha/year (Akinnifesi et al. 2010), and use of legumes is considered crucial to mitigation of human-induced climate change (Jensen et al. 2012). Thus, enhancing use efficiency of fertilizer, water, pesticides, and other fossil fuel-based farm operations is critical to sustainability of agroecosystems. For example, use of legumes in the rotation cycle can reduce the use of nitrogenous fertilizer and decrease emission of N_2O. Jensen et al. (2012) estimated that globally, between 350 and 500 Tg CO_2 emission is avoided as a result of 33 to 46 Tg N that is biographically fixed each year. With increased energy conservation and improved use efficiency, the US economy could save an estimated 33% of its energy consumption (Pimentel et al. 2004).

19.5 TECHNOLOGICAL OPTIONS

The strategy is to adopt innovative technology, enhance SQ, and support diverse ESs and functions (e.g., food security, water security, and energy security). Thus, there is a range of best management practices (BMPs) of soil, crop, water, and nutrient management, and for adaptation to changing and uncertain climate. Some generic BMPs listed in Figure 19.3 must be adapted, fine-tuned, and validated under site-specific situations.

19.5.1 Management of Cropland Soils

Arable land use and its intensification can have strong ecological impacts (Stoate et al. 2001), including changes in the SOC pool and its dynamics. With regards to management and restoration of degraded/desertified cropland soils, therefore, the SOC pool is a key parameter that must be enhanced/maintained above the threshold level (Powlson et al. 2011). In this context, measurement and modeling of the SOC pool on arable lands is a priority (Tang et al. 2006). In general, conversion of natural ecosystems to agroecosystems on land use changes can lead to reduction in the SOC

Soil Management for Sustaining Ecosystem Services 527

Adaption to climate change

Soil management
- No-till farming
- Mulching
- Cover cropping
- Nano-enhanced fertilizers
- Precision farming
- Nitrogen sensors
- Nutrient cycling
- Disease-suppressive soils

Crop management
- Complex rotations
- Improved varieties/species
- Legume-based systems
- Agroforestry systems
- Agropastoral systems
- Species that generate molecular-based signals when under stress
- Species with a high HI and prolific root system

Sustainable systems

- Timing of farm operations
- Choice of species and varieties
- Drought avoidances
- Flood tolerance
- Heat tolerance

Water management
- Water conservation in the root zone
- Microirrigation (fertigation)
- Water harvesting and recycling
- Improving WUE

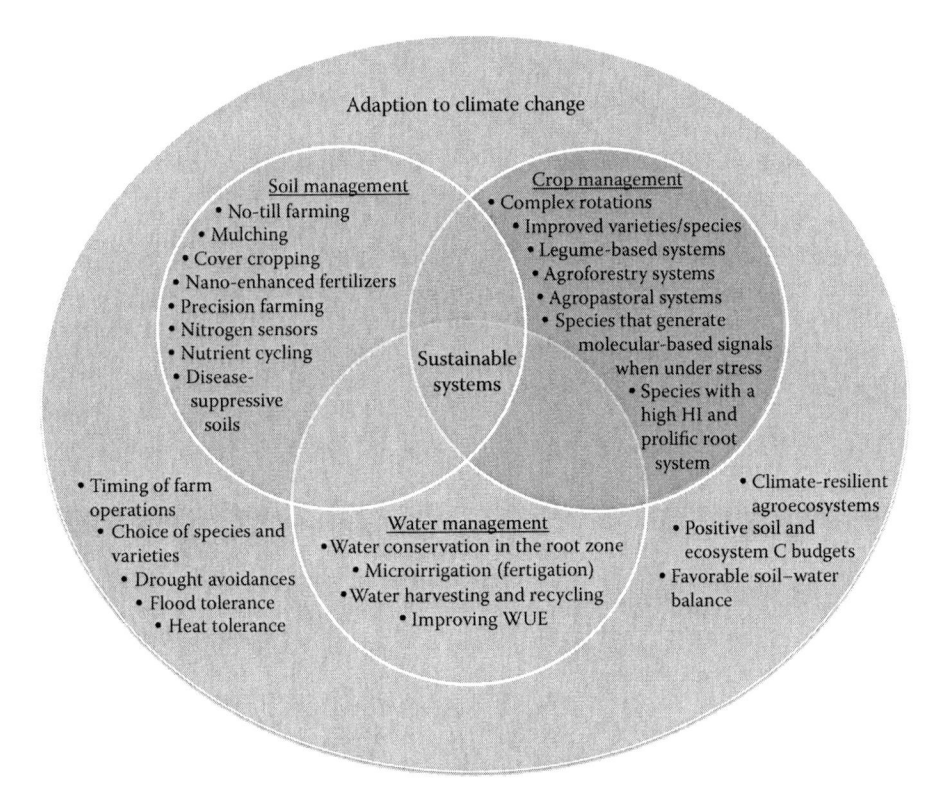

- Climate-resilient agroecosystems
- Positive soil and ecosystem C budgets
- Favorable soil–water balance

FIGURE 19.3 Technological options for sustaining agricultural productivity.

pool (Bationo et al. 2007). Management of cropland soils by NT farming, mulching, cover cropping, manuring, and so forth can exert a strong influence on the SOC pool (Batjes 2006), which depends on the difference between input and losses of biomass C. The input of root biomass has a strong influence on SOC dynamics (Rees et al. 2005). The SOC pool influences several soil properties, and even small changes in the SOC pool can impact the use efficiency of inputs and agronomic productivity. Soils with a favorable level of the SOC pool support high and rigorous biotic activity. Bioturbation by earthworms, which flourish in soil with mulch and in those managed by NT farming, improves aggregation and enhances soil tilth.

Land use conversion affects the SOC pool, with negative effects on conversion of natural ecosystems to agroecosystems and vice versa. In southeastern Australia, Kasel and Bennett (2007) observed a 30% decrease in SOC concentration with conversion of native forest to pine plantation. In New Zealand, deforestation caused a reduction of 3.0 Mg C/ha in the top 0.1 m (Oliver et al. 2004). Conversion of perennial grassland to cropland can reduce the SOC pool (Qiu et al. 2012). In contrast, conversion of cropland to forest and grassland systems can enhance/restore the SOC pool (Qiu et al. 2012). Data from a long-term experiment (1945–2004) in the Lahn-Dill Highlands of Germany indicated that care is required when interpreting results from paired site surveys, especially when soils are dissimilar. Thus, it is pertinent to assess periodically SOC pool changes in agroecosystems (e.g., cropland) at a national level (Sleutel et al. 2003). In addition to SOC, national inventories are also needed for emission of N_2O (Li et al. 2001) and CH_4. While NT farming can be a source of N_2O, it can be a sink for CH_4 (Lal 2004a). Long-term adoption of NT farming increased SOC concentration in the top 0.2-m layer by 21.4% in some soils of Inner Mongolia, China (He et al. 2009). Furthermore, 10-year crop yields increased by 14.0% and water use efficiency (WUE) by 13.5% compared with traditional tillage. Despite numerous advantages of an NT system in soil and water conservation and improving the SOC pool in the surface layer, producers are reluctant to adopt NT in semiarid Mediterranean soils (and elsewhere) because of a potential increase in soil compaction as indicated by high bulk density and strong penetration resistance (Imaz et al. 2010). Adoption of appropriate rotations and use of cover crops can reduce the risks of soil compaction. In the rainfed farmlands of Northeast China, a soybean (*Glycine max* L.)–corn (*Zea mays*) rotation is recommended for SQ enhancement and agronomic productivity improvement (Kou et al. 2012). An integrated soil–crop system management approach can increase the productivity (Chen et al. 2011). Adoption of agroforestry systems also enhances earthworm activity and reduces loss of the SOC pool (Fonte et al. 2010). Use of integrated nutrient management (INM), through a judicious combination of inorganic fertilizers and organic amendments, is needed to enhance the SOC pool and sustain SQ.

19.5.2 Pastureland Soils

Globally, pasture soils represent a large managed ecosystem and have a strong impact on the global C cycle and emissions of greenhouse gases (GHGs). In the Amazon region, expansion of pastureland is the principal cause of tropical deforestation. Conversion of forests to pastures increases bulk density and exacerbates soil compaction (Martinez and Zinck 2003). Soil compaction hazard increases with the increase in the stocking rate, due to the trampling effects of animals (Houlbrooke et al. 2009). Intensive grazing on wet soils can severely degrade soil structure and adversely impact soil physical quality. In comparison with forests, pasture soils may have a lower CH_4 uptake and lower N_2O flux (do Carmo et al. 2012). In New South Wales, Australia, Wilson et al. (2011) reported that C, N, and C/N ratio in the surface layer were in the following order: woodland > unimproved pastures = improved pastures > cultivated soil, and soil bulk density observed an opposite trend. Wilson and colleagues observed that conversion from croplands to pastures would sequester C in soil at the rate of 0.06 to 0.15 Mg/ha/year. In Coshocton, Ohio, Owens and Shipitalo (2011) reported that sediments from a watershed under pasture had a C enrichment ratio of 1.2 to 1.5 compared with the 0 to 2.5 cm layer and that pasture sediment C concentrations were more than twice the concentration on sediments from a nearby row crop watershed.

Conversion of pasture to croplands can deplete the SOC pool. Experiments conducted in a southern Mediterranean highland of Turkey indicated that conversion of pasture into cropland reduced SOC concentration of the 0 to 20 cm layer by 49%, increased soil bulk density from 1.19 to 1.33 Mg/m^3, and decreased the mean weight diameter of aggregates from a dominant size class of 4.0 to ≤0.5 mm (Celik 2005). However, conversion of croplands to pasture can enhance the SOC pool and improve SQ (Breuer et al. 2006) and also influences the hydrological balance (Qiu et al. 2011).

Understanding the processes and causes of pasture degradation is important, especially in the neotropics of the Amazon and the cerrado/llanos of Brazil and Columbia. Pasture degradation forces farmers to open new forest areas and may alter the nutritional quality of forages without dramatically altering soil properties (Muller et al. 2004). Numata et al. (2007) assessed soil physical and chemical properties of Oxisols and Alfisols under pastures in the southwest Amazon region of Brazil. Pasture degradation caused more drastic changes in chemical properties of Alfisols than those of Oxisols, with little change in those of Ultisols.

Therefore, judicious management of soils under pastures is needed for maintaining productivity and reducing the risks of environmental degradation. On a Southland dairy farm in New Zealand, Houlbrooke et al. (2009) indicated that intrusive grazing when soil wetness is greater than the plastic limit contributed to both soil trampling damage and severe compaction. In an alpine region of China, Li et al. (2007a,b) observed that total SOC concentration decreased by 29% to 41% in 0 to 30 cm depth in annually cultivated pastures (oats). Both annual and perennial pastures decreased soil WHC in comparison with the native pasture. Conversion of dry forests into pasture (and cropland) in Mexico has increased the soil erosion hazard (Cotler and Ortega-Larrocea 2000). Management of soil N content and the stocking rate are key determinants of the quality of soil under pastoral land use (Monaghan et al. 2005).

19.5.3 Management of Soils under Forestland Use

Forested ecosystems have a large SOC sink capacity (Lorenz and Lal 2010). Afforestation and reforestation of marginal/degraded soils enhance the SOC pool (Six et al. 2002) and strengthen numerous ESs. Soil and water conservation are among the important ESs of forestlands. In St. Lucia, Cox et al. (2006) observed that soil erosion from an extreme event was merely 0.2 Mg/ha from a forested site compared with 3–78 Mg/ha from a cropland site. An afforestation study in Finland on 220,000 ha of arable land indicated that bulk density of mineral soils tended to be lower and SOC concentration higher than those of a soil under old forest sites (Wall and Hytonen 2005). High variability in SOC concentration, however, limits the precision of measurements of changes by land use conservation and management practices (Conant et al. 2003). Therefore, the accuracy of measurements and understanding of the basic principles governing SOC dynamics in forested ecosystems need to be improved (Six et al. 2002).

19.5.4 SQ and Water Security

Water scarcity and drought are serious constraints that need to be addressed. It is widely recognized that as much as 60% of the world population may be vulnerable to water scarcity by 2025 (Qadir et al. 2007). Further, water-scarce countries may not be able to advance food security with conventional sources of water. While the excessive use of water (traditional flood irrigation) must be avoided, nonconventional sources must also be identified. For example, growing a rice paddy under flooding condition in hot and arid environments is wasteful use of scarce water resources. Such an excessive use of water in cropland must be avoided by judicious water use, water conservation in the root zone, water harvesting and recycling, and microirrigation (e.g., drip subirrigation and fertigation). Rice, a semiaquatic food staple of a large population, consumes a large amount of water when grown under flooded conditions. There is a need to improve water management practices in rice-based cropping systems. The so-called aerobic rice can be grown, with appropriate varieties and weed control measures, under nonflooded soil environments (Gaydon et al. 2012).

Asia, including China and countries within the Indian subcontinent, is faced with numerous challenges of high population, limited arable land area, and scarcity of water resources. The problem of finite natural resources is confounded by the degradation of soil, pollution and contamination of water, and loss of forest and other vegetation cover. Under these scenarios, the high potential of elite varieties can be realized only if grown under optimal soil conditions. Water scarcity is a serious issue, and policy makers and researchers must realize that adoption of BMPs is futile when water availability in the root zone is scarce (Tyagi et al. 2012). Furthermore, the key to effective water resource management is the strong understanding of the close link between the hydrologic cycle and the soil management and that the choice of any BMP for soil management implies specific water requirements (Bossio et al. 2010). The water productivity can be enhanced by restoration of degraded soil and improvement of SQ in relation to SOC concentration, nutrient availability, and soil structure and tilth. Effectiveness of NT farming also lies in its capacity to conserve water in the root zone and enhance WUE (Lafond et al. 2006).

The global economy and the attendant increase in food trade have also highlighted the importance of nonconventional water resources and the so-called virtual water. Important issues among nonconventional water resources for crop production include desalination of sea and brackish groundwater, rainwater harvesting (from land and roofs of the buildings), use of agricultural drainage water, wastewater from industrial and urban uses, and so forth. Linking rivers to physically transfer water from one watershed to another is also an option (Qadir et al. 2007) that involves serious engineering, and political and logistical considerations. These practices involve conversion of blue and gray water into green water.

The term "virtual water" implies the water used to produce the food that is imported into water-scarce countries. It refers to the amount of water needed to locally produce the food that has been imported into the country. Thus, "virtual water" plays an important role in the international trade for food among countries with water surplus and deficit. Because of involved subsidies, any assessment of the magnitude of virtual water must also consider policies and incentives of the food-exporting countries. Globally, 131 out of 146 countries are involved in some virtual water trade, with an attendant increase in the gross arable land areas (Kumar and Singh 2005). It is also recognized that virtual water often flows out of water-poor, land-rich countries to land-poor, water-rich countries, accentuating the challenges and exacerbating difficulties of "distribution of scarcity" and achieving "global water use efficiency." This also implies that virtual water import for a water-poor, but land-rich, country is not a viable option toward sustainable water management (Kumar and Singh 2005). The global strategy of improving land use efficiency cannot be achieved by the ever-increasing trade in virtual water.

19.6 SOIL QUALITY INDEX AND SUSTAINABLE MANAGEMENT

Sustainable management of SQ is crucial to advancing and achieving regional and global food security. In this regard, key soil parameters have been identified and used to develop a soil quality index (SQI), of which there are numerous examples (Lal 1994; Doran and Jones 1996; Karlen et al. 1997, 2001; Andrews et al. 2004; Moebius et al. 2007; Gugino et al. 2009; USDA-NRCS 1998). Despite notable advances in the development of computer-based techniques to assess the SQI, it is the one that scientists can quantify and farmers can comprehend and relate to. Thus, it has been argued that scientific SQ knowledge must be relevant to farmer SQ knowledge. Tesfahunegn et al. (2011) observed for the Mai-Negus catchment in Northern Ethiopia, that scientifically measured soil attributes significantly differed among SQ categories identified by local farmers. Tesfahunegn and colleagues opined that farmer-derived SQ status (e.g., low, medium, high) may be crucial in providing a rational basis to decision makers and other stakeholders. Policy makers are well advised to be aware of emerging possibilities from this rapidly developing theme of sustainable soil management (Powlson et al. 2011). Judicious policy interventions are needed toward application of BMPs and proven technological options, through incentivization or education for enhancing awareness, to address emerging issues. Translation of scientific knowledge into an action plan necessitates a continued dialogue between the scientists on the one hand and policy makers and land managers on the other.

19.7 CONCLUSIONS

Research and development priorities must be identified and implemented to promote adoption of BMPs based on basic principles of soil science for sustainable management of cropland, pastureland, forestland, peatland, urban lands, and other natural and managed ecosystems (Figure 19.4). Important among basic principles are the following: soils must never be taken for granted; nutrients removed must be replaced; soil degradation is exacerbated by desperateness, and it makes a society vulnerable to civil strife and political unrest; soil can be a source or a sink of GHGs depending on land use and management; and sustainable soil management is an engine of economic development. While advancing and strengthening the basic science, it is important to establish and improve channels of communication between scientists, policy makers, and land managers. Scientific principles must be used to develop decision support systems for policy makers and for incentivization of land managers toward adoption of BMPs. Myths about its importance to human wellbeing and environment quality must be replaced by facts through doing good science, developing better curricula, and improving communication.

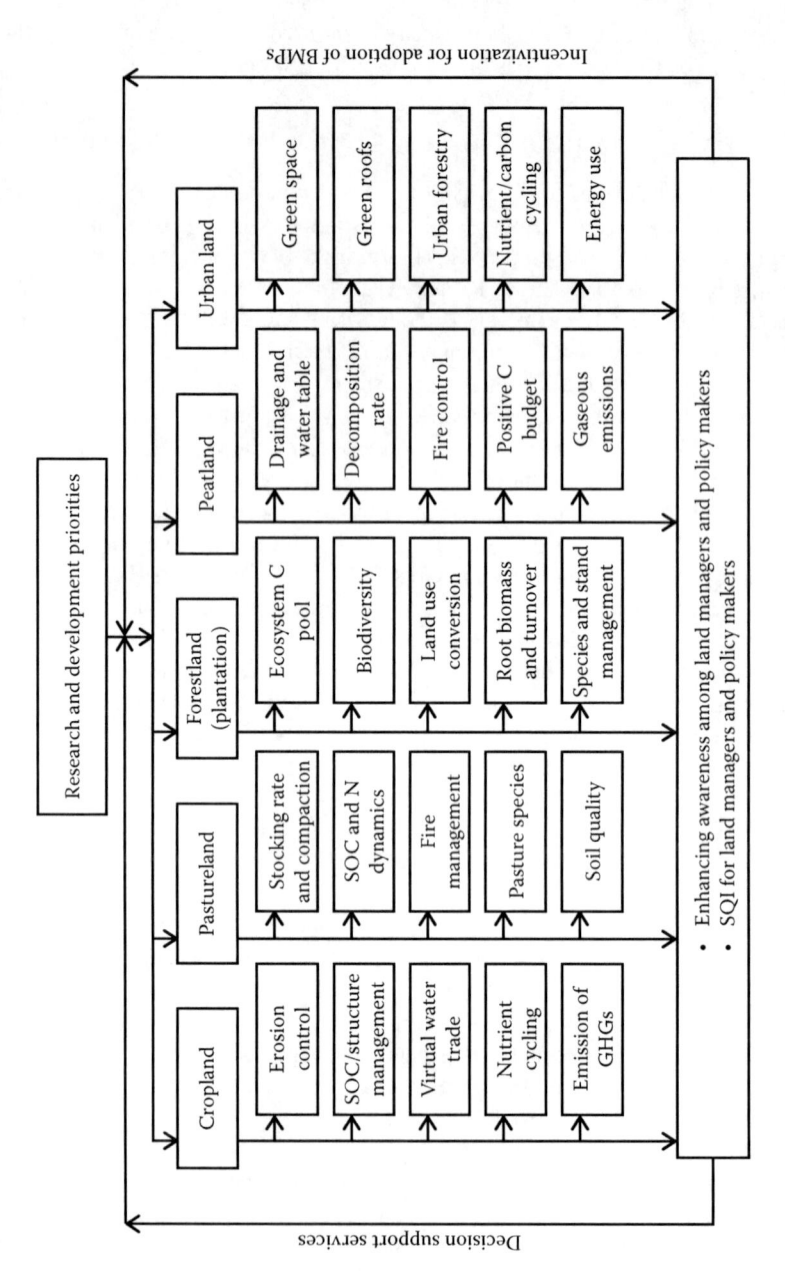

FIGURE 19.4 Research and development priorities for understanding and implementing management options based on basic principles of soil science.

ABBREVIATIONS

As: arsenic
BMPs: best management practices
ESs: ecosystem services
GHGs: greenhouses gases
HI: harvest index
INM: integrated nutrient management
N: nitrogen
NT: no-till
NUE: nutrient use efficiency
SA: South Asia
SOC: soil organic carbon
SQ: soil quality
SQI: soil quality index
SSA: sub-Saharan Africa
WUE: water use efficiency
WHC: water holding capacity

REFERENCES

Akinnifesi, F.K., O.C. Ajayi, G. Sileshi, P.W. Chirwa, and J. Chianu. 2010. Fertiliser trees for sustainable food security in the maize-based production systems of East and Southern Africa. A review. *Agron Sustain Develop* 30:615–629.

Alewell, C., M. Schaub, and F. Conen. 2009. A method to detect soil carbon degradation during soil erosion. *Biogeoscience* 6:2541–2547.

Andrews, S.S., D.L. Karlen, and C.A. Cambardella. 2004. The soil management assessment framework: a quantitative evaluation using case studies. *Soil Sci Am J* 68:1945–1962.

Balasubramanian, V., M. Sie, R.J. Hijmans, and K. Otsuka. 2007. Increasing rice production in Sub-Saharan Africa: challenges and opportunities. In *Advances in Agronomy*, vol. 94, ed. D.L. Sparks, 55–133. Elsevier, Newark, USA.

Bationo, A., J. Andre, B. Vanlauwe, B. Waswa, and J. Kimetu. 2007. Soil organic carbon dynamics, functions and management in West African agro-ecosystems. *Agric Syst* 94:13–25.

Batjes, N.H. 2006. Soil carbon stocks of Jordan and projected changes upon improved management of cropland. *Geoderma* 132:361–371.

Bossio, D., K. Geheb, and W. Critchley. 2010. Managing water by managing land: addressing land degradation to improve water productivity and rural livelihoods. *Agric Water Manag* 97:536–542.

Breuer, L., J.A. Huisman, T. Keller, and H.G. Frede. 2006. Impact of conversion from cropland to grassland on C and N storage and related spoil properties: analysis of a 60-year chronosequence. *Geoderma* 133:6–18.

Celik, I. 2005. Land-use effects on organic matter and physical properties of soil in a southern Mediterranean highland of Turkey. *Soil Till Res* 83:270–277.

Chen, X., Z. Cui, P.M. Vitousek et al. 2011. Integrated soil–crop system management for food security. *Proc Natl Acad Sci USA* 108:6399–6404.

Conant, R.T., G.R. Smith, and K. Paustian. 2003. Spatial variability of soil carbon in forested and cultivated sites: implications for change detection. *J Environ Quality* 32:278–286.

Cotler, H. and M.P. Ortega-Larrocea. 2006. Effects of land use on soil erosion in a tropical dry forest ecosystem, Chamela watershed, Mexico. *Catena* 65:107–117.

Cox, C.A., A. Sarangi, and C.A. Madramootoo. 2006. Effect of land management on runoff and soil losses from two small watersheds in St. Lucia. *Land Degrad Dev* 17:55–72.

De Meyer, A., J. Poesen, M. Isairye, J. Deckers, and D. Raes. 2011. Soil erosion rates in tropical villages: a case study from Lake Victoria Basin, Uganda. *Catena* 84:89–98.

do Carmo, J.B., N. de Sousa, R. Eraclito et al. 2012. Conversion of the coastal Atlantic forest to pasture: consequences for the nitrogen cycle and soil greenhouse gas emissions. *Agric Ecosyst Environ* 148:37–43.

Doran, J.W. and A.J. Jones. 1996. Methods for assessing soil quality. *Soil Sci Am J*, Madison.

Fargione, J., J. Hill, D. Tilman, S. Polasky, and P. Hawthorne. 2008. Land clearing and the biofuel carbon debt. *Science* 319:1235–1238.

Fonte, S.J., E. Barrios, and J. Six. 2010. Earthworm impacts on soil organic matter and fertilizer dynamics in tropical hillside agroecosystems of Honduras. *Pedobiologia* 53:327–335.

Gaydon, D.S., M.E. Probert, R.J. Buresh et al. 2012. Rice cropping systems—Modeling transitions between flooded and non-flooded soil environments. *Eur J Agron* 39:9–24.

Gilroy, J.J., G.Q.A. Anderson, P.V. Philip, J.A. Vickery, I. Bray, P. Watts, P. Nicholas, and W.J. Sutherland. 2008. Could soil degradation contribute to farmland bird declines? Links between soil penetrability and the abundance of yellow wagtails *Motacilla flava* in arable fields. *Biol Conserv* 141:3116–3126.

Gugino, B.K., O.J. Idowu, P.R. Schindelbeck et al. 2009. *Cornell soil health assessment training manual*, 2nd edn. Cornell University, Geneva.

He, J., N.J. Kuhn, X.M. Zhang, X.R. Zhang, and H.W. Li. 2009. Effects of 10 years of conservation tillage on soil properties in the farming–pastoral ecotone of Inner Mongolia, China. *Soil Use Manag* 25:201–209.

Houlbrooke, D.J., J.J. Drewry, R.M. Monaghan et al. 2009. Grazing strategies to protect soil physical properties and maximize pasture yield on a Southland dairy farm. *NZ J Agric Res* 52:323–336.

Imaz, M.J., I. Virto, P. Bescansa et al. 2010. Soil quality indicator response to tillage and residue management on semi-arid Mediterranean cropland. *Soil Till Res* 107:17–25.

Jensen, E.S., M.B. Peoples, R.M. Boddey et al. 2012. Legumes for mitigation of climate change and the provision of feedstock for biofuels and biorefineries. A review. *Agron Sustain Dev* 32:329–364.

Junge, B., L. Mabit, G. Dercon et al. 2010. First use of the Cs-137 techniques in Nigeria for estimating medium term soil redistribution rate on cultivated farmland. *Soil Till Res* 110:211–220.

Kabubo-Mariara, J., G. Mwabu, and P. Kimuyu. 2006. Farm productivity and poverty in Kenya: the effect of soil conservation. *J Food Agric Environ* 4:291–297.

Karlen, D.L., M.J. Mausbach, J.W. Doran et al. 1997. Soil quality: a concept, definition, and framework for evaluation. *Soil Sci Am J* 61:4–10.

Karlen, D.L., S.S. Andrews, and J.W. Doran. 2001. Soil quality: current concepts and applications. *Adv Agron* 74:1–40.

Kasel, K. and L.T. Bennett. 2007. Land-use history, forest conversion, and soil organic carbon in pine plantations and native forests of south eastern Australia. *Geoderma* 137:401–413.

Khan, M., I. Asaduzzaman, M. Rafiqul et al. 2010. Accumulation of arsenic in soil and rice under wetland condition in Bangladesh. *Plant Soil* 333:263–274.

Kim, H., S. Kim, and B.E. Dale. 2009. Biofuels, land use change, and greenhouse gas emissions: some unexplored variables. *Environ Sci Technol* 43:961–967.

Kou, T.J., P. Zhu, S. Huang et al. 2012. Effects of long-term cropping regimes on soil carbon sequestration and aggregate composition in rainfed farmland of Northeast China. *Soil Till Res* 118:132–138.

Kumar, M.D. and O.P. Singh. 2005. Virtual water in global food and water policy making: is there a need for rethinking? *Water Resour Manag* 19:759–789.

Lafond, G.P., W.E. May, F.C. Stevenson, and D.A. Derksen. 2006. Effects of tillage systems and rotations on crop production for a thin Black Chernozem in the Canadian Prairies. *Soil Till Res* 89:232–245.

Lal, R. 1994. Soil erosion research methods. 2nd Edn. *Soil Water Conserv Soc,* Ankeny, IA, 340 pp.

Lal, R. 2001. Soil degradation by erosion. *Land Degrad Dev* 12:519–539.

Lal, R. 2003. Soil erosion and the global carbon budget. *Env Intl* 29:437–450.

Lal, R. 2004a. Soil carbon sequestration impacts on global climate change and food security. *Science* 304:1623–1627.

Lal, R. 2004b. Carbon emissions from farm operations. *Environ Int* 30:981–999.

Lal, R. 2005. Soil erosion and carbon dynamics. *Soil Till Res* 81:137–142.

Lele, U. 2010. Food security for a billion poor. *Science* 327:1554.

Li, X., F. Li, R. Zed, and Z. Zhan. 2007a. Soil physical properties and their relations to organic carbon pools as affected by land use in an alpine pastureland. *Geoderma* 139:98–105.

Li, X., F. Li, R. Zed, and Z. Zhan. 2007b. Soil management changes organic carbon pools in alpine pastureland soils. *Soil Till Res* 93:186–196.

Li, C.S., Y.H. Zhuang, M.Q. Cao et al. 2001. Comparing a process-based agro-ecosystem model to the IPCC methodology for developing a national inventory of N(2)O emissions from arable lands in China. *Nutr Cycl Agroecosyst* 60:159–175.

Lorenz, K. and R. Lal. 2010. *Carbon Sequestration in Forest Ecosystems.* 277, Springer, Holland.

Martinez, L.J. and J.A. Zinck. 2004. Temporal variation of soil compaction and deterioration of soil quality in pasture areas of Columbian Amazonia. *Soil Till Res* 75:3–17.

Meshesha, D.T., A. Tsunekawa, M. Tsubo, and N. Haregeweyn. 2012. Dynamics and hotspots of soil erosion and management scenarios of the Central Rift Valley of Ethiopia. *Int J Sediment Res* 27:84–99.

Miao, Y., B. Stewart, and F. Zhang. 2011. Long-term experiments for sustainable nutrient in China. A review. *Agron Sustain Dev* 31:397–414.

Moebius, B.N., H.M. van Es, R.R. Scchindelbeck et al. 2007. Evaluation of laboratory-measured soil physical properties as indicators of soil quality. *Soil Sci* 172:895–910.

Monaghan, R.M., R.J. Paton, L.C. Smith, J.J. Drewry, and R.P. Littlejohn. 2005. The impacts of nitrogen fertilization and increased stocking rate on pasture yield, soil physical condition and nutrient losses in drainage from a cattle-grazed pasture. *NZ J Agric Res* 48:227–240.

Muller, M.M.L., M.F. Guimaraes, T. Desjardins, and D. Mitja. 2004. The relationship between pasture degradation and soil properties in the Brazilian amazon: a case study. *Agric Ecosyst Environ* 103:279–288.

Mueller, L., U. Schindler, W. Mirschel et al. 2010. Assessing the productivity function of soils. A review. *Agron Sustain Dev* 30:601–614.

Nearing, M.A., F.F. Pruski, and M.R. O'Neal. 2004. Expected climate change impacts on soil erosion rates: a review. *J Soil Water Conserv* 59:43–50.

Numata, I., O.A. Chadwick, D.A. Roberts et al. 2007. Temporal nutrient variation in soil and vegetation of post-forest pastures as a function of soil order, pasture age, and management, Rondonia, Brazil. *Agric Ecosyst Environ* 118:159–172.

O'Neal, M.R., M.A. Nearing, R.C. Vining, J. Southworth, and R.A. Pfeifer. 2005. Climate change impacts in soil erosion in Midwest United States with change in crop management. *Catena* 61:165–184.

Oliver, G.R., P.N. Beets, L.G. Garrett, S.H. Pearce, M.O. Kimberly, J.B. Ford-Robertson, and K.A. Robertson. 2004. Variation in soil carbon in pine plantations and implications for monitoring soil carbon stocks in relation to land-use and forest site management in New Zealand. *Forest Ecol Manag* 203:283–295.

Owens, L.B. and M.J. Shipitalo. 2011. Sediment-bound and dissolved carbon concentration and transport from a small pastured watershed. *Agric Ecosyst Environ* 141:162–166.

Pimentel, D. and N. Kounang. 1998. Ecology of soil erosion in ecosystems. *Ecosyst* 1:416–426.

Pimentel, D., A. Pleasant, J. Barron et al. 2004. US energy conservation and efficiency: benefits and costs. *Environ Dev Sustain* 6:279–305.

Powlson, D.S., P.J. Gregory, W.R. Whalley et al. 2011. Soil management on relation to sustainable agriculture and ecosystem services. *Food Policy* 36:S72–S87.

Qadir, M., B.R. Sharma, A. Bruggeman, R. Choukr-Allah, and F. Karajeh. 2007. Non-conventional water resources and opportunities for water augmentation to achieve food security in water scarce countries. *Agric Water Manag* 87:2–22.

Qiu, L., X. Wei, X. Zhang et al. 2012. Soil organic carbon losses due to land use change in semiarid grassland. *Plant Soil* 355:299–309.

Qiu, G.Y., J. Yin, F. Tian, and S. Geng. 2011. Effects of the "conversion of cropland to forest and grassland program" on the water budget of the Jinghe River catchment in China. *J Environ Qual* 40:1745–1755.

Rajan, K., A. Natarajan, A.K.S. Kumar, M.S. Badrinath, and R.C. Gowda. 2010. Soil organic carbon—the most reliable indicator for monitoring land degradation by soil erosion. *Curr Sci* 99:823–827.

Rees, R.M., I.J. Bingham, J.A. Baddeley, and C.A. Watson. 2005 The role of plants and land management in sequestering soil carbon in temperate arable and grassland ecosystems. *Geoderma* 128:130–154.

Safriel, U.N., P. Berliner, A. Novoplansky, J.B. Laronne, A. Karnieli, M.I. Arnon, A. Kharabsheh, A. Mohammad, A. Ghaleb, and G. Kusek. 2011. Soil erosion-desertification and the Middle Eastern anthroscapes. In *Sustainable Land Management: Learning from the Past for the Future*, eds. S. Kapur, H. Eswaran, W.E.H. Blum, 57–124. Springer-Verlag, Berlin.

Six, J., P. Callewaert, S. Lenders et al. 2002. Measuring and understanding carbon storage in afforested soils by physical fractionation. *Soil Sci Am J* 66:1981–1987.

Sleutel, S., S. De Neve, and G. Hofman. 2003. Estimates of carbon stock changes in Belgian croplands. *Soil Use Manag* 19:166–171.

Stoate, C., N.D. Boatman, R.J. Borralho, C.R. Carvalho, G.R. de Snoo, and P. Eden. 2001. Ecological impacts of arable intensification in Europe. *J Environ Manag* 63:337–365.

Tang, H., J. Qiu, E. Van Ranst, and C. Li. 2006. Estimations of soil organic carbon storage in cropland of China based on DNDC model. *Geoderma* 134:200–206.

Tesfahunegn, G.B., L. Tamene, and P.L.G. Vlek. 2011. Evaluation of soil quality identified by local farmers in Mai-Negus catchment, northern Ethiopia. *Geoderma* 163:209–218.

Tesfahunegn, G.B., P.L.G. Vlek, and L. Tamene. 2012. Management strategies for reducing soil degradation through modeling in a GIS environment in northern Ethiopia catchment. *Nutr Cycl Agroecosyst* 92:255–272.

Tyagi, S.K., P.S. Datta, and R. Singh. 2012. Need for proper water management for food security. *Curr Sci* 102:690–695.

USDA-NRCS. 1998. Soil quality test kit guide. In *Section 1. Test Procedures, and Section 2. Background and Interpretive Guide for Individual Tests*. UNDS-NRCS, Soil Quality Institute, Ames.

Van Oost, K., T.A. Quine, G. Govers, S. De Gryze, J. Six, J.W. Harden, J.C. Ritchie, G.W. McCarty, G. Heckrath, C. Kosmas, J.V. Giraldez, J.R. Marquesda Silva, and R. Merckx. 2007. The impact of agricultural soil erosion on the global carbon cycle. *Science* 318:626–629.

Wall, A. and J. Hytonen. 2005. Soil fertility of afforested arable land compared to continuously. *Plant Soil* 275:247–260.

Wilson, B.R., T.B. Koen, P. Barnes, S. Ghosh, and D. King. 2011. Soil carbon and related soil properties along a soil type and land-use intensity gradient, New South Wales, Australia. *Soil Use Manag* 27:437–447.

Index

Page numbers followed by f and t indicate figures and tables, respectively.

A

Abrupt climate change (ACC), 2
Abuja Declaration, 417
ACC (abrupt climate change), 2
Access to land and credit, poverty and, 174–175
Africa
    average cereal yields, 401–402
    crop intensification, 404–406, 405f
    emerging religious approaches to land stewardship in, 297–299
    nutrient balance
        calculations, 406–408
        at farm and plot scales, 411, 412t
        at international and regional scales, 408–409, 409f
        at national scale, 409–411, 410t, 411t
    nutrient mining, challenge of
        conservation agriculture, 415–416
        integrated soil fertility management, 416
        nutrients recycling in peri-urban and urban agriculture, 414–415
        policy issues, 416–417
    nutrient mining, drivers of
        climatic risk, 414
        low external nutrient inputs, 413–414, 413f
        population pressure, 412–413, 413f
    soils in, 403–404, 403t–404t
Agnihotra, 313
Agricultural gross domestic product (AGDP), 411
Agricultural growth model, soil and land degradation and (case study), 235–239
Agricultural Handbook 282, 488
Agricultural Handbook 537, 488
Agricultural intensification
    based on "interventionist" paradigm, 343–346
    as driver of land degradation, 215
    effects of, 272–275
    soil degradation and, 172–173
Agricultural Production Systems Simulator (APSIM) model, 414
Agricultural soil degradation
    cause of, 341–352
        agricultural intensification based on "interventionist" paradigm, 343–346
        conventional tillage, 346
        ecological perspective, 343
        mechanical tillage, 342, 346
    definitions, 339–341
    extent of, 339–341
    large-scale (examples)
        Australia, 349–350
        Brazil, 346–348
        China, 350–352
    soil microorganisms and mesofauna, role of, 340, 352–354
        ecosystem functions, 353
    sustainable soil management. *See* Sustainable soil management
Agriculture
    soil development and, 480–481
    soil erosion and, 480
    spirituality and, 266–268
Agriculture production, soil resources and, 478–479
Agroecological principles, for sustainable production systems, 354–355
    conservation agriculture, 355–360
    degraded agricultural soils and landscapes restoration, 362–371
        Australia, 367–369
        Brazil, 363–367
        CA systems, 362–371
        China, 370–371, 370t
    linkage with landscape health, 360–362
Agroforestry practices
    soil marginality and, 37–38
Agroforestry systems
    CA in, 371–372
Agronomic yields
    global trends and resource use, 2–6, 3f, 3t
    yield gap, 6–8
AICRPDA (All India Co-ordinated Research Project for Dryland Agriculture), 449
Alder (*Alnus nepalensis*)
    in Himalayas, 316–317
Alfisols, 19, 480
All India Co-ordinated Research Project for Dryland Agriculture (AICRPDA), 449
Alnus nepalensis (Alder)
    in Himalayas, 316–317
Amazon basin, 58, 60
American Indian, 290t
Ancient technologies *vs.* modern issues, 14

537

Andhra Pradesh, farmers' classification of soils in, 306, 307t
Andisol, 19
Animal agriculture, intensification of, 102–103
  nutrients translocation and, 102–103
Annual crops
  Cerrado, 73–76
Anthroposophy, 266
Apatanis
  terrace-based land management of, 324
APSIM (Agricultural Production Systems Simulator model), 414
Aquilaria trees, in Meghalaya, 317
Argissolos, 67
Aridisol, 19
Arunachal Pradesh, traditional soil classification in, 306, 308t
Assam, traditional soil classification in, 308t
Atharvaveda, 304, 313
Australia
  agricultural soil degradation in, 349–350
  degraded agricultural soils and landscapes restoration in, 367–369
    compaction, 369
    no-till farming, 367–369
    nutrient removal, 368
    soil acidification, 368
    waterlogging and sodicity, 369
    water repellence, 368–369
Available water capacity (AWC), 28

B

Baha'i, 290t
Balfour, Lady Eve, 266, 352
Bamboo (*Bambusa* sp.) trees, in Meghalaya, 317
Bangar (fertile) soil, 309
Beef animals, 102
Beej (seed), 304
Bequest value, defined, 233
Berry, Wendell, 265
Best management practices (BMPs), 6, 13, 43–44, 171, 526
  determinants of adoption, 44–45
"Better Living through Chemistry," 344
Beushening, in Eastern India, 315
Bhoomi (land), 304
Biodynamic farming, 266
Biofuel crops
  Cerrado, 78–79
Biological degradation, 24–26
Biological nitrogen fixation (BNF), 402
Biological processes
  integrating use of, 103–105
Biotic stresses, in rainfed production systems, 428
BMPs. *See* Best management practices (BMPs)

BNF (biological nitrogen fixation), 402
Borlaug, Norman, 97, 102
Bosch, Carl, 97
Brachiaria pastures, in Cerrado, 73
Brazil
  agricultural soil degradation in, 346–348
  degraded agricultural soils and landscapes restoration in, 363–367
  Kayapo sustainable anthropogenic landscape in, 295–296
  watershed services in Paraná Basin, 378–379
Brazilian Cerrado (*Cerrados*)
  future trends (land use, research, and environment), 80–81
    climate change, 80
    natural resources assessment, 80
    nature conservation, 80–81
    physical and economical infrastructure, 81
  main soils of, 60–64
    others, 67–69
    Oxisols, 65–66
    Quartzipsamments and sandy Oxisols, 66–67
    Ultisols, 67
  overview, 58–60, 59f
  principal land use and soil management systems
    annual crops, 73–76
    biofuel crops, 78–79
    crop/livestock integrations, 79
    crop rotations, 71–72
    fertilization, 70–71
    native vegetation clearance and soil acidity correction, 70
    pastures, 72–73
    perennial crops, 76–77
    plantation forests, 77–78
    soil tillage and seedbed preparation, 71
  soil formation factors, 60–64
    climate, 62, 63t
    organisms, 62, 64, 64f
    parent materials, 60
    time of, 60
    topography, 60–62
    vegetation, 62, 64
Brazilian Forest Act, 81
Brazilian Soil Classification System, 65, 67–68
Broadbalk plot, 91
  mean yields of wheat grain on, 92f
Browning, George, 487
Buddhism, 290t

C

CAFOs (Concentrated Animal Feeding Operations), 102

Index

Canada
  carbon offset scheme in Alberta, 377–378
Carbon management, in rainfed production systems, 446–448
  long-term experiments in semiarid tropics, 449, 452–468, 452f
  mean yield and sustainable yield index, 452–458, 456t–457t
  minimal carbon input requirements, carbon depletion and, 465–468
  soil carbon and yield sustainability
    finger millet-based production system, 459
    groundnut production system, 458–459
    pearl millet-based production system, 459
    rice-based production system, 465
    safflower and soybean-based production system, 459, 462–463
    winter sorghum-based production system, 459
Carbon offset scheme in Alberta (Canada), 377–378
Carbon sequestration. *See also* Soil organic carbon (SOC)
  rainfed agroecosystems of India, 444–446, 445t
    site-specific RMPs, 446–448
    traditional methods *vs.* RMPs and, 445t
Carbon sink capacity, and agronomic productivity of soils of India. *See* India, rainfed agroecosystems of
Carson, Rachel, 101
Carter, Vernon Gill, 278
Cation exchange capacity (CEC), 62, 365
Cattle penning, 310
CBD (UN Convention on Biological Diversity), 204
CEC. *See* Cation exchange capacity (CEC)
Central Himalayas, land classification in, 308t
Centro Internacional de Agricultura Tropical (CIAT), 58
Cerrado. See Brazilian Cerrado (*Cerrados*)
Chavundaraya, 316
Chemical degradation, 26
Chemical fertilizers
  integrating use of, 103–105
  negative effects of, 172
  organic crops *vs.* crops produced with quality of, 94–96
  safety of, 92–93
China. *See also* North China Plain (NCP)
  agricultural soil degradation in, 350–352
  agriculture, SOC role in, 501–504, 502f, 503t, 504f
  annual cereal production, 524
  degraded agricultural soils and landscapes restoration in, 370–371, 370t
  erosional impact on productivity, 492–495, 493f, 495f
  SOC sequestration in croplands, 509–512, 510f
    co-benefits of, for crop production and ecosystem health, 512–516, 513f–515f
  status of SOC storage in croplands
    background SOC stock in 1980s, 505–506, 505t, 506f
    C stock dynamics and sequestration between 1980 and 2006, 507–509, 508f
    historical changes in stocks, 506–507
Chisel plow (CP) tillage system, 28
Christianity, 290t
CIAT (Centro Internacional de Agricultura Tropical), 58
Climate change
  crop yield potential and, 479–480
  future trends in Cerrado, 80
  mitigation of, 13
The Closing Circle, 101
Coefficient of variation (CV)
  of SOC concentration, 193, 194
Commoner, Barry, 101
Concentrated Animal Feeding Operations (CAFOs), 102
Conscious self, 261
Conservation
  basic principle, 270
  indigenous soil conservation practices, 318–321, 318f, 319t–321t
  limitations of, 487
  smaller-scale methods, 271–272
  spirituality and, 270–272
Conservation agriculture (CA), 10, 125, 346, 352. *See also* No-till (NT) farming
  adoption levels, 360
  agroforestry systems, 371–372
  as base for sustainable soil management and production intensification, 355–360, 357t–358t
  degraded agricultural soils and landscapes restoration by, 362–363
  China, 370–371
    effect on wind erosion, 371
  integration into farming systems, 371–376
  nutrient mining in African soils and, 415–416
  organic agriculture based on, 371
  technical principles, 359
  tenets of, 356
Conservation Reserve Enhancement Program (CREP), 247
Conservation Reserve Program (CRP), 43, 44, 228

Conservation Stewardship Program (CSP), 44
Conservation tillage, 43, 345
Contingent valuation, defined, 234
Conventional tillage, 43
   agricultural soil degradation and, 346, 351–352
   for annual crops in the Cerrado, 74
   China, 370
Conversion of land cover
   soil degradation and, 173
Cost–benefit analysis, 227–250
   bequest value, 233
   direct current use value, 233
   DR Valdesia hydro reservoir sedimentation (case study), 239–243
   economic surplus and technological externality, concepts of, 232, 232f
   environmental economic value, types and measures of, 233–235, 234f
   existence value, 233
   external values, 233
   magnitude and economic impacts of soil erosion, 227–229
   measurements, 230–233
   Ohio evidence of soil erosion off-site economic impacts (case study), 243–245
   option value, 233
   revenues, 230
   social costs and benefits, 231
   soil and land degradation and Ag growth model (case study), 235–239
   SU studies on Ohio downstream benefits of CRP (case study), 245–250
County soil maps, 20–21, 20f–21f
Cover crop biomass, functions of, 365
Cover cropping, 10
CREP (Conservation Reserve Enhancement Program), 247
Cropland soils management, 526–528
Crop/livestock integrations
   Cerrado, 79
   for sustainable soil management, 374–375
Crop management
   physiography and, 329, 330t
   soil depth and, 327–328, 328t
   sustainable soil management and, 372–373
Crop productivity
   principles to maximize, 32
   variation in water availability and, 30–33, 31f
Crop residues, recycling of, 313–314
Crop rotation, 312, 312f
   in Cerrado, 71–72
   limitations of, 487
CRP (Conservation Reserve Program), 228
Culture
   environmental decision making and, 179
   religion and. *See* Religious–cultural integration, with soil stewardship; Religious–cultural perspective, in soil science

D

Dale, Tom, 278
Daoism, 290t
Decision-making, 21, 117
   community-level, 113
   factors affecting land management
      culture and religion, 179
      dual processing, 182
      globalization and other economic forces, 175–177
      government policies, 177–178
      heuristics in, 181–182
      population growth and urbanization, 173–174
      poverty and access to land and credit, 174–175
      psychological factors, 179–182
      rationality in, 180–181
      social risk tolerances, 180
      technology, 178–179
      worldviews, 180
Decision Support System for Agrotechnology Transfer (DSSAT)-CENTURY crop simulation model, 205–206
   integrated soil fertility management (ISFM) practices, 206
   land degrading practice, 206
Deep ecology
   defined, 261
   spirituality and, 259–261
Deep Self, 261
Deforestation, 173
Degraded soils, of rainfed agroecosystems, 426, 428f–430f
Desertification
   causes of, 11
   control of, 13
Direst current use value, defined, 233
Dirt: The Erosion of Civilizations, 342
Dominant social paradigm (DSP), 180
DR Valdesia hydro reservoir sedimentation (case study), 239–243
DSP (dominant social paradigm), 180
DSSAT-CENTURY model. *See* Decision Support System for Agrotechnology Transfer (DSSAT)-CENTURY crop simulation model
Dual process models
   land management decision making and, 181–182
Dust Bowl, 39, 296–297, 342, 343, 485

Index

Dynamism, as attribute of TK, 115
Dystrustepts, 69

E

Eastern India
　Beushening in, 315
　farmers' classification of soils in, 306, 307t
Ecological intensification
　spirituality and, 272–275
Economic analysis, 230–231
Economic growth (GDP)
　as driver of land degradation, 215–216
Economic policies
　environmental decision making and, 175–177
Economic surplus and technological externality, concepts of, 232, 232f
Ecosystem resilience, 121, 122–123. *See also* Resilience
Ecosystem services (ESs), sustainable management of, 521–531
　overview, 521–522
　processes, factors, and causes of degradation, 522–523
　soil degradation and, 524–526, 525f
　soil erosion and other degradation processes, 523–524
　soil quality index (SQI) and, 531
　technological options, 527f
　　cropland soils management, 526–528
　　pastureland soils, 528–529
　　soils under forestland use, 529
　　SQ and water security, 529–531
Endophytes
　root, beneficial effects of, 151–152
　seed, effects on root emergence and growth, 154–157, 155f, 156f
　soil microbial, effects in plant canopies, 152–154, 153t
Energy use (global), 5–6, 6t
Entisols, 19
　with lithic properties, 69
Environmental economic value, types and measures of, 233–235
Environmental Kuznet curve (EKC), 207, 209, 215
Environmental quality, 33–35
　rainfall intensity and, 33
Environmental Quality Incentives Program (EQIP), 44
EPA. *See* US Environmental Protection Agency (EPA)
Erosion. *See* Soil erosion
ESs. *See* Ecosystem services (ESs)
Existence value, defined, 233
External values, defined, 233

F

Fallowing, 311–312
FAOSTAT data, 406
1985 Farm Bill, 46
1990 Farm Bill, 495
Farm power and mechanization, for sustainable soil management, 375–376
Farmyard manure (FYM), 310–311, 311f, 316, 327, 446, 447
Faulkner, Edward, 342, 345
Fertiliser Programme (FAO), 344
Fertilization
　Cerrado soils, 70–71
Fertilizer
　global use, 2, 4, 4t
FFHC (Freedom from Hunger Campaign), 344
Financial analysis, 230–231
Finger millet-based rainfed production system, 459
Flash floods, control in Himalayas, 322
Focused extension program, lack of
　in rainfed regions, 431
Food and Agriculture Organization (FAO), 402, 477, 478
Food quality
　classification, 95
Food security
　land degradation impact on, 205–207
　crop yield gap and, 218–219, 218f
　maize and rice yield (case of Mali and Nigeria), 219–221, 220t–221t
Food Security Act of 1985, 245
Forested ecosystems, management of soils, 529
Forest plantations
　in Cerrado, 77–78
Freedom from Hunger Campaign (FFHC), 344
Fresh organic foods, 94
Fusarium culmorum, 154
FYM. *See* Farmyard manure (FYM)

G

Ganderiya, 310
Gelisol, 19
Gender connectedness, as attribute of TK, 114
Germplasm
　improved, natural resources *vs.*, 14–15
GLASOD (Global Assessment of Human-Induced Soil Degradation), 341
Gleissolos, 69
Gliricidia sepium
　GLM with, for crop yield improvement, 447–448, 448t
Global Assessment of Human-Induced Soil Degradation (GLASOD), 341
Globalization

environmental decision making and, 175–177
Global warming
  soil degradation and, 13
GNP. *See* Gross national product (GNP)
Government effectiveness
  as driver of land degradation, 216–217, 217f
Government policies
  environmental decision making and, 177–178
Grazing management, 10
Greenhouse gasses (GHGs), 13, 73, 75, 402
"Greening power," 260
Green leaf manure (GLM)
  with *Gliricidia sepium*, for crop yield improvement, 447–448, 448t
Green manuring, 315
  Beushening in Eastern India, 315
  Maghi cropping (legume-based sequence cropping), 315
Green Revolution, 2, 7, 258, 344–345, 414, 526
"Green revolution" model, of intensive agriculture, 172
Gross domestic product (GDP), 204, 207, 209, 229, 443
  land degradation and, 215–216
Gross national product (GNP), 229
Groundnut rainfed production system, 458–459
Guidelines for General Assessment of the Status of Human-Induced Soil Degradation, 340
"Guidman's Ground," 260
Gunapajalam, 316

H

Haber, Fritz, 97
Haber–Bosch process, 97–98, 98f, 99
Hal (plow), 304
Haplustepts, 69
Haplustoxes, 65, 74
Haplustults, 67
Heuristics, in land management decision making, 181–182
Himalayas
  alder trees in, 316–317
  flash flood control in, 322
Hinduism, 290t
Histosol, 19
Holism, 266, 267
Holy Quran, 313
Human dimensions, 167–168
  concept of, 169
  forms of soil degradation and land management
    conversion of land cover, 173
    intensification of agricultural management, 172–173
    lack of investment in conservation, 171–172
  land management decision-making, factors affecting
    culture and religion, 179
    dual processing, 182
    globalization and other economic forces, 175–177
    government policies, 177–178
    heuristics in, 181–182
    population growth and urbanization, 173–174
    poverty and access to land and credit, 174–175
    psychological factors, 179–182
    rationality in, 180–181
    social risk tolerances, 180
    technology, 178–179
    worldviews, 180
  systems models of, soil degradation and
    direct and indirect drivers, 168–169, 169f, 170f
    variables, 171f
Human needs
  stewardship of natural resources and, 11–12

I

ICAR (Indian Council of Agricultural Research), 437
IDEP (Iowa Daily Erosion Project), 484
IFRI (International Forest Research Institute), 216
IFRLP (Iowa Farm and Rural Life Poll), 46
Inceptisols, 19, 69
India
  rainfed agroecosystems of
    carbon management options, 446–448, 449, 452–468
    carbon sequestration, 444–446
    constraints in, 426–432
    declining NUE, as serious issue, 437
    enhanced NUE with manuring, 448, 449f
    poor soil health and multinutrient deficiencies, 434–436, 434f, 435t, 436t
    SOC, importance of, 437–443
    SOC, water retention and water productivity and, 449, 450t–451t
    SOC depletion in tropical systems, reasons for, 443–444
    soils of, 425–426, 425t
    statistics, 425
    yield gaps in, 432–433, 432t–434t
  traditional knowledge, 304–305. *See also* Traditional knowledge
Indian Council of Agricultural Research (ICAR), 437

Index

Indigenous knowledge. *See also* Traditional knowledge
  defined, 305
Indigenous organic/liquid manures, 316
Indigenous soil and water conservation (ISWC), 112, 129
Indigenous soil conservation practices, 318–321, 318f, 319t–321t
  Jal lands, 321
  Jhola lands, 321
  silt harvesting structures, 318, 318f
Induced innovation, theory of, 178
Industrial agriculture, 257
Industrial Revolution, 269
Infant mortality rate (IMR)
  as indicator of poverty, 209–210, 211f
INM (integrated nutrient management), 447, 448, 528
"Inner soil," 260
Inorganic nitrogen
  negative effects of, 172
Inorganic sources
  vs. organic, of plant nutrients, 14
Integrated natural resource management, traditional concepts, 321
  flash flood control in Himalayas, 322
  iron toxicity, practices to overcome, 324
  jhum cultivation, 324–325, 325t
  location-specific runoff control practices, 322, 323f
  marshy saline soils management, 326–327
  slash-and-burn agriculture, 325–326
  terrace-based land management of *Apatanis*, 324
  terrace/bun cultivation, 326, 327f
  women and *jhum* cultivation, 325, 326f
Integrated nutrient management (INM), 10, 447, 448, 528
Integrated soil fertility management (ISFM)
  nutrient mining in African soils and, 416
  practices, 206, 219–220
Intercropping
  as alternative in permanent no-till systems, 373–374
Intergovernmental Panel on Climate Change Synthesis Report, 168
International Fertilizer Development Center (IFDC) model, 406
International Forest Research Institute (IFRI), 216
Interventionist approach, 355
  agricultural intensification based on, 343–346
Investment, 21, 175, 233, 402, 414, 418
  global, for land degradation prevention, 204–205, 213
  lack of
    for advanced training of TK in SSA, 111
  in conservation, 171–172
Iowa Daily Erosion Project (IDEP), 484
Iowa Farm and Rural Life Poll (IFRLP), 46–47
Iron toxicity, traditional practices to overcome, 324
Irrigation
  religious–cultural perspectives, 293–294
Islam, 290t
ISWC (indigenous soil and water conservation), 112, 129

J

Jainism, 290t
Jhum cultivation, 324–325, 325t
  women and, 325, 326f
Judaism, 290t

K

Kandiustoxes, 65
Kandiustults, 67
Kanhaplustults, 67
Karail (moderately fertile) soil, 309
Kayapo sustainable anthropogenic landscape (Brazil), 295–296
Khejri (*Prosopis cineraria*)
  in northwest India, 317
Kiyari soil, 310, 311
Knowledge transfer, 275–278
Krishi Gita, 309
Krishi parashara, 309
Kunapajala, 316
Kyoto Protocol, 444

L

Land cover, conversion of
  soil degradation and, 173
Land degradation
  annual data, 204
  defined, 204
  drivers of, 214–215, 214t
    agricultural intensification, 215
    economic growth (GDP), 215–216
    government effectiveness, 216–217, 217f
    population density, 215
    relationship between, 207–209, 208t
  DSSAT-CENTURY model, 205–206
  extent of, 209
  food security, impact on, 205–207
    crop yield gap and, 218–219, 218f
    maize and rice yield (case of Mali and Nigeria), 219–221, 220t–221t
  global cost of, 204
  global investment, 204–205
  linear trend models, 206

policy implications, 221–222
poverty and
   correlation between, 209–213, 210f, 211f, 212t
   at household level (case studies), 213–214
Shapiro–Wilk normality test, 206–207
Land management
   factors affecting decision-making about
      culture and religion, 179
      dual processing, 182
      globalization and other economic forces, 175–177
      government policies, 177–178
      heuristics in, 181–182
      population growth and urbanization, 173–174
      poverty and access to land and credit, 174–175
      psychological factors, 179–182
      rationality in, 180–181
      social risk tolerances, 180
      technology, 178–179
      worldviews, 180
Landscape health, linkage with, 360–362
Land use systems, traditional
   physiography and crop management, 329, 330t
   soil depth and crop management, 327–328, 328t
   soil texture and crop choice, 328–329, 329t
Laterite, 68
Latosols, 65, 68
"law of maximum," 11
"law of the diminishing returns," 11
"law of the minimum," 11
Legume-based sequence cropping, in India, 315
Liming, 70
Linear trend models, 206
Livelihood connectedness, as attribute of TK, 115
Llanos, neotropical savannas, 58
Location-specific runoff control practices, 322, 323f
Logical positivism, 257
Lokopakara, 316
Low-input application, in rainfed production systems, 426–428

M

Macroeconomic policies
   environmental decision making and, 177
Madagascar
   high crop productivity from poor soils in, 144–148
MAE. *See* Mean annual evaporation (MAE)
Maghi cropping (legume-based sequence cropping), in India, 315

Magruder Plot, 91
Maharashtra, farmers' classification of soils in, 306, 307t
Maize
   yields in Mali and Nigeria, land degradation impact on, 219–221, 220t–221t
Mali
   land degradation impact on maize and rice yields in, 219–221, 220t–221t
Manuring, 310–311
   enhanced NUE with, 448, 449f
MAP. *See* Mean annual precipitation (MAP)
Maputo Declaration, 204, 213
Marginality, 19–47. *See also* Soil quality
   agroforestry practices and, 37–38
   concept of, 21
   defining, 22–23
   dimensions of, 23
   multifunctionality and (optimum performance of soil resource), 35–38
   scales of variation
      crop productivity, 30–33, 31f
      environmental quality, 33–35
      water availability, 30
   social values of soil functions, 38–47
      concept of risk and, 45
      determinants of BMPs adoption and, 44–45
      responsibility to protect soil resources, 45–47
      social definitions of marginality, 40–42
      willingness and capacity to protect soil resource, 42–45
   soil degradation. *See* Soil degradation
   soil management to reduce, 35–36
Marshy saline soils
   traditional management of, in India, 326–327
MAT. *See* Mean annual temperature (MAT)
Maximum sustainable yield (MSY), 234, 235
Mean annual evaporation (MAE)
   Cerrado, 62, 63t
Mean annual precipitation (MAP)
   Cerrado, 62, 63t
Mean annual temperature (MAT)
   Cerrado, 62, 63t
Mean yield, rainfed production systems, 452–458, 456t–457t
Mechanical tillage
   agricultural soil degradation, 342, 346
Meghalaya, *Aquilaria* and bamboo trees in, 317
Metaproduction function, 236
Millennium Ecosystem Assessment (MEA), 117, 168, 338
Mining, 172
   of nutrients in African soils, 401–417
      challenge of, 414–417
      drivers of, 412–414

Index

Modern issues *vs.* ancient technologies, 14
Modern science/modernization. *See also* Scientific knowledge (SK)
    policy aggressively promoting, 110–111
    traditional knowledge and development of, 330–331
Moldboard plow (MP) tillage system, 28, 74
Mollisols, 19, 480, 492
Monitoring
    integration of TK and SK, 117
Montgomery, David, 342
Morrow Plots, 91
"Mother Earth," 260
MSY (maximum sustainable yield), 234, 235
Multinutrient deficiencies
    in rainfed regions, 434–436, 434f, 435t, 436t

N

NAFTA, GATT and WTO, 176
National Organic Program (NOP), 94
Natural resources
    assessment, in Cerrado, 80
    stewardship of, 11–12
    vs. improved germplasm, 14–15
Natural Resources Conservation Service of the United States Department of Agriculture (NRCS), 482–483
Nature conservation
    in Cerrado, 80–81
NCP. *See* North China Plain (NCP)
NDVI. *See* Normalized difference vegetation index (NDVI)
Neotropical savannas, 58–81
    area of, 58
    Brazilian Cerrado. *See* Brazilian Cerrado (*Cerrados*)
    Llanos (Venezuela and Colombia), 58
    overview, 58–60
NEP (new environmental paradigm), 180
New environmental paradigm (NEP), 180
N fertilizers, 90, 91, 96
    need for better management of, 99–101, 100f
    use of, 97
Nigeria
    land degradation impact on maize and rice yields in, 219–221, 220t–221t
Nitossolos, 67
Nitrogen fixation
    Haber–Bosch process, 97–98, 98f
NOP (National Organic Program), 94
Normalized difference vegetation index (NDVI), 205, 207, 208t, 210, 212, 212t
    change in, and major drivers of land degradation, 214t, 215
North China Plain (NCP)
    agricultural soil degradation in, 351
    SOC concentration changes in different cropping systems in, 189–198
    data analysis, 193
    data collection, 191–192, 191f, 191t
    different application rates of fertilizers and, 197–198, 197t
    overview, 189–190
    results and discussions, 193–198
    study area, 190–191
    from 1980 to 1999, 193–195, 193t, 194t
    from 1999 to 2006, 195–197, 195t–196t
Northwest India
    khejri in, 317
No-Tillage Agriculture: Principles and Practices, 346
No-Tillage Farming, 346
No-till (NT) farming, 10, 28, 29, 74, 345–346. *See also* Conservation agriculture (CA)
    degraded agricultural soils and landscapes restoration by
    in Australia, 367–369
    in Brazil, 363–367
    intercropping as alternative to, 373–374
NRCS (Natural Resources Conservation Service of the United States Department of Agriculture), 482–483
NUE. *See* Nutrient use efficiency (NUE)
NUTMON, 406, 407, 409
Nutrient bank, 13
Nutrient management
    religious–cultural perspectives, 294
Nutrient use efficiency (NUE), 432
    declining, in rainfed agroecosystems, 437
    enhanced, with manuring, 448, 449f

O

OFPA (Organic Foods Production Act), 94
Ohio
    soil erosion off-site economic impacts in (case study), 243–245
    studies on downstream benefits of CRP (case study), 245–250
Olson, T.C., 488
Option value, defined, 233
Organic animal products, 94
Organic farming, 87–105
    advantages, 274–275
    agri-food-environment problems, 93
    animal agriculture intensification and nutrients translocation, 102–103
    based on CA, 371
    crops *vs.* crops produced with chemical fertilizers
    quality of, 94–96
    safety of, 92–93

defined, 88
Haber–Bosch process, 97–98, 98f
integrating use of, 103–105
limitations, 96–99
N fertilizers, need for better management of, 99–101
organic-to-conventional yield comparisons, 89–91
overview, 88
polyface farming, 273–274
principles, 88, 266
soil productivity of, 89–92
spirituality and, 266–268
as traditional method in India, 312–313, 313f
Organic fertilizers, 91
Organic Foods Production Act (OFPA), 94
Organic foods/products, 94–95
IFOAM definition of, 94
quality of, 95
Organic plant products, 94, 95
Organic sources
vs. inorganic, of plant nutrients, 14
Organic-to-conventional yield, comparisons of, 89–91
Organossolos, 69
Ostrom, Elinor, 168
Oxisols, 19
in Cerrado, 65–66
sandy, 66–67

P

Panchagavya, 316
Panentheism, 259
Paraná Basin (Brazil), watershed services in, 378–379
Partial factor productivity (PFP), 448
Particulate organic matter (POM), 366
Pastureland soils management, 528–529
Pastures
in Cerrado, 72–73
Pearl millet-based rainfed production system, 459
Penning, types of, 310, 310f
Perennial crops
Cerrado, 76–77
Permanent raised beds (PRBs), 369
Pesticides
harmful effects of, 172–173
use of, 93
Petraquepts, 67
PFP (partial factor productivity), 448
Phillips, Shirley, 346
Physical and economical infrastructure
of Cerrado, 81
Physical degradation, 25t
Planning, 21, 117
integration of TK and SK, 117

Plantation, forests
in Cerrado, 77–78
Plant nutrients
organic *vs.* inorganic sources, 14
Plinthic Quartzipsamments, 67
Plinthic soils, 68
Plintossolo, 68
Ploughman's Folly, 342, 345
Plowing, 314
Polyface farming, 273–274
POM. *See* Particulate organic matter (POM)
Population growth/density
as driver of land degradation, 215
environmental decision making and, 173–174
nutrient mining in African soils and, 412–413, 413f
Potential Pareto improvement (PPI), 231
Poultry animals, 102
Poverty
access to land and credit and, 174–175
environmental decision making and, 174–175
IMR as indicator of, 209–210, 211f
land degradation and
correlation between, 209–213, 210f, 211f, 212t
at household level (case studies), 213–214
size of landholdings and, 175
soil degradation and, 13
PPI (potential Pareto improvement), 231
Practicability, as attribute of TK, 116
Precision farming, 11
Processed organic foods, 94
Protection
integration of TK and SK, 117
Protectionism, 177

Q

Quartzipsamments, 66–67, 74

R

Raccoon River, 34
Rainfall intensity
environmental quality and, 33
Rainfed agroecosystems
constraints in
appropriate policy support, lack of, 432
biotic stresses, 428
degraded soils, 426, 428f–430f
focused extension program, lack of, 431
low-input application, 426–428
poor crop management, 429
uncertain rainfall, 426, 427f
untapped water–nutrient synergy, 429
soils of, 425–426, 425t

Index

Rajasthan (Northwestern India), indigenous soil classification in, 306t
Randriamiharisoa, Robert, Prof., 146
Rationality, in land management decision making, 180–181
Recommended management practices (RMPs), 197, 433, 443
   SOC sequestration in rainfed regions
      site-specific methods, 446–448
      traditional methods *vs.*, 445t
Relative yield
   global, 4–5, 5t
Religion, 259, 285–286. *See also* Spirituality
   environmental decision making and, 179
   principles, soil stewardship and, 286
      humans and soil in biblical creation narratives, 287–288
      humans and soil in world religious texts, 289, 290t
      Shinto and ecological integrity of Tohoku Region, Japan, 288–289, 288f
      soil themes, soil management principles and, 289, 291–292, 291t
      recognizing agricultural roots, 292–293
Religious–cultural integration, with soil stewardship (examples), 294
   Amish land stewardship in collaboration with agroecologists, 296
   Kayapo sustainable anthropogenic landscape in Brazil, 295–296
   religious approaches to land stewardship in Africa, 297–299
   Soil Stewardship Sundays and Dust Bowl, 296–297
Religious–cultural perspective, in soil science
   irrigation and salinization, 293–294
   nutrient management, 294
Renaissance, of TK, 129–132
Resilience
   concept of, 120
   conventional view of, 125–127, 125t, 126f
   defined, 123
   at ecological scale, 122–123
   managing for, 123–125
      alternative view, 127–129
   at socioecological scale, 122
   soil management dimensions, 121, 122f
   at soil scale, 123
   vulnerability and, 120–121, 123
Restoration
   cost–benefit analysis of. *See* Cost–benefit analysis
   spirituality and, 269–270
Revenues, 230
Revised universal soil loss equation (RUSLE), 488
Rhizobia, 354
Rhizospheres
   increases in microbial populations in, 148–151, 149t
Rice
   yields in Mali and Nigeria, land degradation impact on, 219–221, 220t–221t
Rice-based rainfed production system, 465
Rigveda, 313
RMPs. *See* Recommended management practices (RMPs)
Rodhustults, 67
Roosevelt, Franklin D., 477
Root emergence and growth, endophyte-induced, 154–157, 155f, 156f
Root endophytes, beneficial effects of, 151–152
Runoff control practices, location-specific, 322, 323f
RUSLE. *See* Revised universal soil loss equation (RUSLE)

S

Safe minimum standard (SMS), 234, 235
Safflower and soybean-based rainfed production system, 459, 462–463
Salinization, 172
   religious–cultural perspectives, 293–294
Sanborn Field plots, 91
Sandy oxisols, 66–67
Sasyagavya, 316
Savanna
   defined, 58
   neotropical. *See* Neotropical savannas
Scales of variation, marginality
   crop productivity, 30–33, 31f
   environmental quality, 33–35
   water availability, 30
Scientific knowledge (SK). *See also* Traditional knowledge (TK)
   lack of integration of TK and, 110–120
      idealization of TK in past studies, 112–116
      institutional failures, 110–111
      justification of promotion of TK, 117–120
      lack of framework/mechanism for, 117–119
   union of TK and, 132–133
SCPI. *See* Sustainable crop production intensification (SCPI)
Seedbed preparation
   Cerrado soils, 71
Seed endophytes, effects on root emergence and growth, 154–157, 155f, 156f
Self, types of, 261
Sensitivity to context, as attribute of TK, 115–116
Shadow self, 261
Shallow ecology, 261

Shapiro–Wilk normality test, 206–207
Sheep penning, 310
Shifting cultivation, 209, 347, 372, 404. *See also Jhum* cultivation
Shinto and ecological integrity of Tohoku Region, Japan, 288–289, 288f
Shintoism, 290t
Sikhism, 290t
Silent Spring, 101
SK. *See* Scientific knowledge (SK)
Slash-and-burn agriculture, 325–326, 372. *See also Jhum* cultivation
SLWM. *See* Sustainable land and water management (SLWM)
Smith, Dwight D., 486–488
SMS (safe minimum standard), 234, 235
Social capital, concept of, 169
Social change
    sustainability and, 278–281
Social costs and benefits, 231
Social risk tolerances
    land management decision making and, 180
Social values, of soil functions
    marginality and, 38–47
Socioecological resilience, 121, 122. *See also* Resilience
Soil(s)
    classification, 19
    functions, 20
        social values of, 38–47
    as metaphor of human behavior, 261–263
    moisture holding capacity, 40, 41
    of rainfed regions, 425–426, 425t
    traditional classification of, 305–308
Soil acidity
    correction in Cerrado, 70
Soil capability classification system, 21–22, 22t
Soil degradation
    agricultural. *See* Agricultural soil degradation
    biological, 24–26
    as cause of global warming, 13
    causes of, 11
    chemical, 26
    conversion of land cover and, 173
    cost-benefit analysis of. *See* Cost–benefit analysis
    ESs and, 524–526, 525f
    extent of, 27–30
    human dimensions that drive. *See* Human dimensions
    human-induced, categories of, 340
    intensification of agricultural management and, 172–173
    lack of investment in conservation and, 171–172
    natural and anthropogenic factors, 7–8, 8f
    physical, 25t
    poverty and, 13
    process of, 23–24, 24f
    soil management practices, 28–30
    types of, 24–26, 25t
Soil depth
    crop management and, 327–328, 328t
Soil development
    agriculture and, 480–481
Soil erosion, 26, 263, 478. *See also* Agricultural soil degradation; Soil degradation
    agriculture and, 480
    ESs and, 523–524
    factors affecting, 8
    history of tolerable soil loss (T factor), 485–488
    impact on productivity, 488–492, 491f, 492f
        China perspective, 492–495, 493f, 495f
    magnitude and economic impacts of, 227–229
    off- site economic impacts in Ohio (case study), 243–245
    soil renewal rates and, 481–485
Soil fertility, 141, 171
    as emergent and contingent property, 141–161
    traditional knowledge, 309–317
        classes, 309–310
        crop residues recycling and incorporation of weeds, 313–314
        crop rotation, 312, 312f
        fallowing, 311–312
        green manuring, 315
        indigenous organic/liquid manures, 316
        management through tree-based system, 316–317
        manuring/farmyard manure, 310–311, 311f
        organic farming, 312–313, 313f
        sheep, goat, and cow penning, 310, 310f
        tillage practices, 314, 314f
Soil health, 352–354
Soil management
    principles of, 2–15
    to reduce marginality, 35–36
Soil management, laws of
    land managers
        being proactive, 14
        masters of their own destiny, 14
        modern issues and ancient technologies, 14
        nutrient bank, 13
        organic and inorganic sources of plant nutrients, 14
    policy makers
        desertification control and mitigation of climate change, 13
        human needs and stewardship of natural resources, 11–12
        poverty and soil degradation, 13

Index

soil degradation and desertification, causes of, 11
soil degradation as cause of global warming, 13
researchers
  global issues and, 15
  lack of technology adoption, 15
  natural resources *vs.* improved germplasm, 14–15
  soil quality improvement, indicators of, 15
  training of, 15
Soil microbial endophytes, effects in plant canopies, 152–154, 153t
Soil microorganisms and mesofauna, 142–143
  endophytes, effects of, 151–157
  gene expression in plant canopy and, 157–160
  increases, in rhizospheres, 148–151, 149t
  role of, 340, 352–354
Soil organic carbon (SOC), 8, 9, 24, 64
  concentration changes in different cropping systems in NCP, 189–198
  coefficient of variation (CV), 193, 194
  data analysis, 193
  data collection, 191–192, 191f, 191t
  different application rates of fertilizers and, 197–198, 197t
  overview, 189–190
  results and discussions, 193–198
  study area, 190–191
  from 1980 to 1999, 193–195, 193t, 194t
  from 1999 to 2006, 195–197, 195t–196t
  depletion in rainfed tropical systems, 443–444, 465–468
  importance of, 72
  role in China's agriculture, 501–504, 502f, 503t, 504f
  sequestration in China' croplands, 509–512, 510f
  for crop production and ecosystem health, co-benefits of, 512–516, 513f–515f
  sequestration in rainfed agroecosystems of India, 444–446. *See also* Carbon management, in rainfed production systems
  site-specific RMPs, 446–448
  traditional methods *vs.* RMPs and, 445t
  stocks in rainfed agroecosystems of India
  different soil types, 438–440, 438t, 439f
  in relation to production system, 440–441, 440f
  in relation to rainfall, 441–442, 442f–443f
  storage status, of China's croplands
  background SOC stock in 1980s, 505–506, 505t, 506f
  C stock dynamics and sequestration between 1980 and 2006, 507–509, 508f
  historical changes in stocks, 506–507
  water productivity and, in rainfed regions of India, 449, 450t–451t
Soil organic matter (SOM), 13, 23, 340, 366
  rainfed regions of India, 446
  role in China's crop production, 501–504
Soil quality (SQ), 522. *See also* Marginality
  defined, 22
  indicators of improvement, 15
  1993 Iowa survey, 39
  key parameters, 8–10, 9t
  social meanings of, 40–42
  water security and, 529–531
Soil quality index (SQI), 531
Soil quality management, 7–8
Soil renewal, and sustainability, 477–496
  agriculture and soil erosion, 480
  climate implications, 479–480
  erosional impact on productivity, 488–492, 491f, 492f
  China perspective, 492–495, 493f, 495f
  history of tolerable soil loss (T factor), 485–488
  overview, 477–478
  rates of, 481–485
  soil development and agriculture, 480–481
  soil resources and agriculture production, 478–479
Soil resilience, 121, 123. *See also* Resilience
Soil resources
  agriculture production and, 478–479
Soil stewardship
  religious–cultural integration with (examples), 294
    Amish land stewardship in collaboration with agroecologists, 296
    Kayapo sustainable anthropogenic landscape in Brazil, 295–296
    religious approaches to land stewardship in Africa, 297–299
    Soil Stewardship Sundays and Dust Bowl, 296–297
  religious principles and, 286
    humans and soil in biblical creation narratives, 287–288
    humans and soil in world religious texts, 289, 290t
    Shinto and ecological integrity of Tohoku Region, Japan, 288–289, 288f
Soil Stewardship Sundays, 296–297
Soil sustainability, 495–496
  defined, 495
  key parameters, 8–10, 9t
  laws of soil management, 11–15
    land managers and, 13–14
    policy makers and, 11–13
    researchers and, 14–15

soil renewal and. *See* Soil renewal, and
    sustainability
technological options for, 9–10
Soil texture
    crop choice and, 328–329, 329t
South America
    agricultural soil degradation in, 346–348
South Asia (SA), 524
Spain, soil conservation in olive groves in,
    379–381
Spatial–temporal attributes, of TK, 114
Spirituality
    agriculture and, 266–268
    characteristics, 260
    connection to soil, 263–265
        dig a soil pit, 264
        hearing, 265
        smell, 265
        touching and handling the soil, 264–265
        walking on soil, 264
    deep ecology and, 259–261
    defined, 260
    engaging the spirit, 268
        ecological intensification, 272–275
        soil conservation, 270–272
        soil restoration, 269–270
    "essential Ground of Being," 260
    knowledge and wisdom exchange, 275–278
    organic farming and, 266–268
    "presence," experiencing, 259
    social change and sustainability, 278–281
    soil as metaphor of human behavior, 261–263
Spodosol, 19
Sprengel, Carl, 11
SQ. *See* Soil quality (SQ)
SQI (soil quality index), 531
SRI. *See* System of Rice Intensification (SRI)
SRI (System of Rice Intensification), 372
SSA. *See* Sub-Saharan Africa (SSA)
Steiner, Rudolf, 266
Stewardship, of natural resources, 11–12
Sub-Saharan Africa (SSA), 204, 401, 524
    crop intensification, impacts of, 404–406,
        405f
    land degradation and crop yield gap in, 218
    low crop yields in, 401–402
    nutrient balances
        at international and regional scales,
            408–409
        at national scale, 409–411, 410t, 411t
    nutrient mining, drivers of
        climatic risk, 414
        low external nutrient inputs, 413–414, 413f
        population pressure, 412–413, 413f
    renaissance of TK, 129–132
    resilience approach to soil management,
        120–129. *See also* Resilience

TK and SK, lack of integration of
    idealization of TK in past studies,
        112–116
    institutional failures, 110–111
    justification of promotion of TK, 117–120
    lack of framework/mechanism for,
        117–119
    union of SK and TK, 132–133
Sugarcane plantations
    in Cerrado, 78–79
Sugar Creek Project, 296
Surapala, 316
Sustainability. *See* Soil sustainability
Sustainable agriculture, 404
    defined, 258, 268, 353
Sustainable crop production intensification
    (SCPI), 356, 362
    development of, 359
Sustainable farming, 38
Sustainable land and water management
    (SLWM), 217
Sustainable soil management
    agroecological principles, 354–355
        conservation agriculture, 355–360
        degraded agricultural soils and
            landscapes restoration, 362–371
        linkage with landscape health, 360–362
    crop–livestock integration for, 374–375
    crop management practices and, 372–373
    farm power and mechanization for, 375–376
    integrating principles into farming systems,
        371–376
    with intercropping as alternative in
        permanent no-till systems, 373–374
    large-scale landscape-level benefits from,
        376–377
        carbon offset scheme in Alberta (Canada),
            377–378
        soil conservation in olive groves (Spain),
            379–381
        watershed services in Paraná Basin
            (Brazil), 378–379
        policy and institutional support, 381–383
        technology and knowledge support, 383–384
Sustainable yield index (SYI)
    rainfed production systems, 452–458, 456t–457t
SYI. *See* Sustainable yield index (SYI)
Systemic nature, as attribute of TK, 115
System of Rice Intensification (SRI), 144, 149,
    150, 160, 372

T

TAP (total agricultural production), 236, 237
Technical yield potential, 6
Technology and environmental decision making,
    178–179

Index

Terrace-based land management, of *Apatanis*, 324
Terrace/bun cultivation, 326, 327f
T factor, 485–488
TFP (total food production), 236, 237
TFP index. *See* Total factor productivity (TFP) index
Thaer's "humus theory," 9, 11
"The law of optimum," 11
"The living soil," 352
Thematic Strategy for Soil Protection, 341
Theology, 285
Three Kings Preparation, 260
Tillage systems
    agricultural soil degradation and, 342–343
    in Cerrado, 71, 75
    traditional plows used in India, 314, 314f
TK. *See* Traditional knowledge (TK)
Total agricultural production (TAP), 236, 237
Total factor productivity (TFP) index, 213
Total food production (TFP), 236, 237
Traditional knowledge (TK), 304–331
    defined, 304
    dynamism, 115
    to enhance resilience of soils in SSA, 110–133. *See also* Resilience
    fertility management, 305–308
    gender connectedness, 114
    indigenous soil conservation practices, 318–321, 318f, 319t–321t
    integrated natural resource management, 321
        flash flood control in Himalayas, 322
        iron toxicity, practices to overcome, 324
        jhum cultivation, 324–325, 325t
        location-specific runoff control practices, 322, 323f
        marshy saline soils management, 326–327
        slash-and-burn agriculture, 325–326
        terrace-based land management of *Apatanis*, 324
        terrace/bun cultivation, 326, 327f
        women and *jhum* cultivation, 325, 326f
    land classification in Central Himalayas, 308t
    land use systems
        physiography and crop management, 329, 330t
        soil depth and crop management, 327–328, 328t
        soil texture and crop choice, 328–329, 329t
    livelihood connectedness, 115
    overview, 304–305
    practicability, 116
    renaissance of, 129–132
    role in development of modern science, 330–331
    sensitivity to context, 115–116
    and SK, lack of integration of
        idealization of TK in past studies, 112–116
        institutional failures, 110–111
        justification of promotion of TK, 117–120
        lack of framework/mechanism for, 117–119
    soil classification, 305–308
        in Andhra Pradesh and Maharashtra, 307t
        in Arunachal Pradesh, 308t
        in Assam, 308t
        in Eastern India, 307t
        in Rajasthan (Northwestern India), 306t
    soil fertility, 309–317
        classes, 309–310
        crop residues recycling and incorporation of weeds, 313–314
        crop rotation, 312, 312f
        fallowing, 311–312
        green manuring, 315
        indigenous organic/liquid manures, 316
        management through tree-based system, 316–317
        manuring/farmyard manure, 310–311, 311f
        organic farming, 312–313, 313f
        sheep, goat, and cow penning, 310, 310f
        tillage practices, 314, 314f
    spatial–temporal attributes, 114
    systemic nature, 115
    union of SK and, 132–133
    western science *vs.*, 118t
Tree-based system
    fertility management through
        alder in Himalayas, 316–317
        Aquilaria and bamboo in Meghalaya, 317
        khejri in northwest India, 317

U

Ultisols, 67
UNCCD (United Nations Convention to Combat Desertification), 204, 209, 445
Uncertainty, in rainfall in rainfed agroecosystems, 426, 427f
UN Convention on Biological Diversity (CBD), 204
UNFCCC (UN Framework Convention on Climate Change), 204
UN Framework Convention on Climate Change (UNFCCC), 204
United Nations Convention to Combat Desertification (UNCCD), 204, 209, 445
United States Agency for International Development (USAID), 241
United States Department of Agriculture—National Resource Inventory (USDANRI) report, 484

Universal soil loss equation (USLE), 235, 485, 487–488
Untapped water–nutrient synergy, in rainfed production systems, 428
Urbanization
 environmental decision making and, 173–174
Urban sprawl, 173
USAID (United States Agency for International Development), 241
US classification system, 481–482
USDANRI (United States Department of Agriculture—National Resource Inventory) report, 484
US Environmental Protection Agency (EPA), 101, 235
USLE. *See* Universal soil loss equation (USLE)
Ustoxes, Cerrado, 65–66, 67
Ustults, 66, 74
Utisols, 19

V

Varsha (rain), 304
Vedas, 305
 philosophy of sustainable agriculture in, 312–313
Vegetation
 of Cerrado, 62, 64
Vertisol, 19
von Bingen, Abbess Hildegard, 260, 264
Vrikshayurveda, 316
Vulnerability
 defined, 120
 resilience and, 120–121, 123

W

Water availability, 30
 variation in, crop productivity and, 30–33, 31f
Water erosion
 agricultural soil degradation in Australia, 349–350
 agricultural soil degradation in China, 350–352
Water Erosion Prediction Project (WEPP), 484, 485

Waterlogging, 172
Water–nutrient synergy, untapped in rainfed production systems, 428
Water security
 SQ and, 529–531
Water use efficiency (WUE), 31, 35, 36, 153, 447, 528, 530
Weeds
 recycling of crop residues and incorporation of, 313–314
WEPP (Water Erosion Prediction Project), 484, 485
Wind erosion, 8
 agricultural soil degradation in Australia, 349–350
 CA effect on, 371
Winter sorghum-based rainfed production system, 459
Wischmeier, W.H., 488
Wolfe, David, 352
World religious texts, humans and soil in, 289, 290t
Worldviews, land management decision making and, 180
Worldwatch Institute, 97
WUE (water use efficiency), 31, 35, 36, 153, 447, 528, 530

Y

Yadav, 310
Yield gap(s), 6–8, 205
 defined, 6
 determinants, 6–7, 7f
 land degradation and, relationship of, 218–219, 218f
 in rainfed agriculture, 432–433, 432t–434t
 soil quality management, 7–8
Young, Harry, 346

Z

Zero net land degradation, 2
Zingg, Austin W., 486
Zoroastrianism, 290t